To Convert

from	to
Joules	calories
Calories	joules
Nanometers	angstro...
Angstroms	nanome...
Atmospheres	mmHg
mmHg (Torr)	atmosp...
Electron volts	kilojou...
Liter-atmospheres	joules

Physical Quantities

Atomic mass unit	amu
Avogadro number	N
Boltzmann constant	k
Electron charge	e
Electron mass	m
Electron volt	eV
Faraday constant	$\mathcal{F}$
Gas constant	R
Neutron mass	n
Planck's constant	h
Proton mass	p
Speed of light	c

Chemical
Principles and Properties

McGraw-Hill Book Company
New York   St. Louis
San Francisco   Düsseldorf
Johannesburg
Kuala Lumpur   London
Mexico   Montreal   New Delhi
Panama   Paris   São Paulo
Singapore   Sydney   Tokyo
Toronto

Michell J. Sienko
Professor of Chemistry
Cornell University

Robert A. Plane
Professor of Chemistry
Cornell University

Chemical
Principles and Properties

Second Edition

1 2 3 4 5 6 7 8 9 0   DODO   7 9 8 7 6 5 4

This book was set in Bodoni Book by York Graphic Services, Inc.
The editors were Thomas Adams and Carol First;
the designer was Ben Kann;
the production supervisor was Joe Campanella.
The orbital drawings were done by Jim Egleson,
and all other drawings were done by Vantage Art, Inc.
R. R. Donnelley & Sons Company was printer and binder.

Library of Congress Cataloging in Publication Data

Sienko, Michell J
    Chemical principles and properties.

    First ed. published in 1966 under title:
Chemistry: principles and properties.
    Bibliography: p.
    1.   Chemistry.   I.   Plane, Robert A., joint
author.   II.   Title.
QD31.2.S558   1974   540   73-18479
ISBN   0-07-057364-6

The illustration on the cover is a photomicrograph of a silicon-carbide
crystal developed by vapor-phase condensation. The spiral pattern of
contour lines show the spiral pattern of crystal growth. The "rainbow"
lines indicate crystal growth along lines of stress. The vari-colored
patches at left bottom and right mark impurities in the crystal. Silicon
carbide is used extensively in heavy-duty industrial abrasives. The
magnification is approximately 1400 diameters. The micrograph was
made by the late F. Hubbard Horn and is shown here through the
courtesy of the General Electric Research and Development Center,
Schenectady, New York.

V

Part II   *Properties of the Elements and Their Compounds* topes of Hydrogen. Proton Magnetic Resonance. Occurrence of Oxygen. Preparation of Oxygen. Properties and Uses of Oxygen. Compounds of Oxygen. Water. Water as a Solvent. Hydrates.

The foundations of this book appeared almost 20 years ago as a set of notes for a new freshman chemistry course at Cornell University. It was the time of Sputnik, and there was excitement in the air as the traditional descriptive approach of beginning chemistry gave way to new points of view based on principles. A rather complete shift has occurred. Some have said it has gone too far, for, as ever is the case, high school courses have followed the college model, and descriptive chemistry has all but vanished from the curriculum.

We now stand in the wake of another upheaval, this time social and political but characterized also by disillusionment with science. Part of the disenchantment stemmed from the fact that science was over-sold—it did not produce quick and easy answers as expected to the hard questions of society in turmoil. Also, however unjustified it may have been, it was easy to point to science as the initiator of material demands that led to pollution and degradation of the quality of the environment. A large segment of society turned its back on science, labeling it intrinsically bad because it was nonhumanitarian, dispassionate, and not involved. In the most simplistic view, it was said to be not relevant. Still, the problems did not go away. Populations increase; the tasks of feeding, housing, and giving health care become more pressing, more complicated, and more frustrating. The banning of DDT in one country leads to an increase in the incidence of malaria in another. How will these problems be solved, and by whom? Rather than rejecting science, which over the past 3 centuries has consistently demonstrated ability to solve problems, society is better served by turning to science to see how its methods can be applied to maximize the common good.

The first edition of "Chemical Principles and Properties" (called "Chemistry: Principles and Properties") was written to take advantage of increased sophistication in the prior chemical training of entering college students. This second edition recognizes that this sophistication is neither so deep nor so wide as previously supposed. Therefore, some of the subjects (e.g., atomic structure, chemical bonding, and thermodynamics) are given more gradual development than before. Others, such as electron descriptions, solid-state chemistry, and reaction mechanisms, have been extended closer to present practice. Descriptive chemistry has been reemphasized with expansion in some areas (e.g., silicates) and compression in others (e.g., noble gases). Completely new topics (e.g., wave functions, symmetry, molecular vibrations, and chain reactions)

IX

have been introduced. Much of the discussion is quantitative; it has been made more quantitative, both in breadth and depth.

The book is divided into three parts—Part I develops the principal concepts of chemistry; Part II applies these concepts to a systematic survey of the characteristic behavior of the chemical elements; Part III synthesizes the principles and properties into a critical overview of chemical limitations on the quality of life. Although the assumption has been made that the student begins with a rudimentary knowledge of chemistry, it is not assumed that he or she is headed for a career in chemistry. Rather, the principles established and the descriptive chemistry used as illustrations are broadly based so as to be appropriate as well for the engineer, the medical student, the agronomist, and the molecular biologist.

The general arrangement of Part I is that of increasing complexity— atoms, molecules, states of matter, solutions, reactions, equilibrium. However, should an instructor wish, he can start with Chapter 5 on states of matter and then go back to Chapter 1 or 2, depending on the sophistication level of his class. A few of the sections (i.e., 2.2, 3.10, and 7.2) have fine-print parts that are more difficult but can be omitted without loss of continuity. Some calculus is introduced, but only gradually and sparingly, and then only where needed. Numerous worked-out problems are included in the body of the text. Extended lists of exercises (over 800 in all, all new for this edition) are appended to each chapter. Each list of exercises is presented in order of increasing difficulty, culminating with a challenge for even the best of students. Numerical answers have been furnished for about half the problems; the remaining answers as well as worked-out solutions are included in an instructor's manual available from the publisher.

The second part of the book deals intensively with the elements of the periodic table. While all the elements are covered, emphasis is on the elements that are typical, those that constitute important resources, and those that are important in life and in materials science. Hydrogen, oxygen, and water are given special treatment; they are followed by group discussion from left to right across the table. The transition elements are considered in detail, first in general and then individually. Organic chemistry appears in depth but not in breadth, mainly as an interesting and unusual aspect of carbon chemistry. Equilibrium computations and thermodynamic applications are used repeatedly in the second half of the book, especially in the problem sets at the ends of chapters. Because qualitative analysis is still one of the best ways to learn solution chemistry and is often the vehicle for much laboratory work of the second semester, there has been included considerable information for detection of the common elements.

Part III is an entirely new part of the book which examines the special problems of man in his chemical environment. Air, water, food, and pollution are considered inputs for man; drugs and radiation are then treated as perturbing influences; finally, the problem of limits to growth is explored.

In the writing of this revision, the recommendations of the International Committee on Weights and Measures and the International Union of Pure and Applied Chemistry (IUPAC) have been largely followed, particularly as regards the consistent use of SI (Système International) units and the sign convention for electrode potentials. Complete use of SI units was not feasible for practical reasons, e.g., most chemists do not use all SI units and would not even recognize some of them. Exotic ones such as *pascal* and *siemens* have not been adopted yet, and commonly accepted alternatives are still widely preferred. In a few instances, where pro-SI and anti-SI forces are at a standoff, we have leaned in the direction of SI. Consequently, we have adopted joule rather than calorie for the energy unit and mole rather than gram-atom for amount of chemical material. Atmospheres have been retained for pressure units; Torr has been eliminated. A complete discussion of SI units, together with recommendations and a list of equivalents, is given in an extensive appendix. The sign of the reduction potential has been adopted throughout for electrode potentials.

In the previous edition we concluded with a chapter on nuclear chemistry with a final optimistic note on the probability of unlocking nuclear fusion for the good of mankind. We end here with a less optimistic view of the unsolved chemical problems of man in his environment. Teaching freshman chemistry has always been a challenging assignment; with the present generation it needs more than ever to be "like it is."

The invaluable guidance and assistance of Professor John P. Hunt, Florence Bernard, and Barbara Zabadal in the preparation of this edition is gratefully acknowledged. Professor R. E. Hughes is to be thanked for the X-ray photograph of $RbMmF_3$ used in Figure 7.2.

Michell J. Sienko

Robert A. Plane

Acquisition of knowledge is simplified by seeking basic principles. In chemistry, the principles include the generalizing statements that summarize observed phenomena and the theories that are proposed to account for the observations. Recognition of the principles of chemistry increases the depth of understanding and serves as a convenient framework for remembering a large body of information and also as a firm basis from which to make predictions about the unknown.

Principles of chemistry

Part I

Man has long speculated that matter in its ultimate state of subdivision is atomic—that it consists of discrete, indivisible particles. Early in the nineteenth century this idea was given quantitative expression by John Dalton. Working with the observed facts that in chemical reactions mass is conserved (law of conservation of matter) and that the combining ratio of elements is fixed by the compound formed, not by the available masses of starting elements (law of definite composition), Dalton was able to assign a self-consistent set of masses to the atoms of the various elements. Further, he was able to predict that if more than one compound is formed between two elements, the mass ratio of elements in one compound is a simple multiple of the mass ratio in the other compound (law of multiple proportions). We now know that none of the three aforementioned laws is precisely true; however, Dalton's idea of an indivisible, definite-mass atom persisted for almost a century and served as a most fruitful concept for the planning and execution of countless chemical experiments. The facts discovered, particularly that chemical properties varied in periodic fashion with increasing elemental atomic weight (the so-called "periodic law"), strongly hinted at an underlying structure for a divisible atom. The nature of this structure and how it has been deduced is the subject matter of this chapter.   3

1.1 Early atomic theory

The idea of an atom as an ultimate building block of matter appears to have originated with the early Greek philosophers. Two schools of thought were in lively contention: One believed that progressive subdivision of any sample of matter could be continued indefinitely; the other, that the process of subdivision had an ultimate limit. In the first view, even the smallest portion of a sample of matter is continuous, a sort of jelly that could be cut up into ever smaller and smaller regions, each being still jellylike in its consistency. The atomists, on the other hand, believed that any sample of matter is ultimately discontinuous. The process of chopping it up into bits would have to stop when it reached the ultimate, indivisible particles. These particles were called *atoms*, after the Greek word ατομοσ, *atomos*, meaning "indivisible."

Although very curious about the nature of their physical world, the Greek philosophers were not schooled to resort to controlled experimentation as a way to decide between conflicting theories. Also, their measuring instruments were too crude to deal with so subtle a concept as the atom. It remained for John Dalton to make the concept quantitative and apply it to an understanding of what matter is composed of and how it changes in the course of a chemical reaction. Although the idea of an atom as an ultimate building block has been very fruitful, it is not without ambiguity. In the modern view, the atom is believed to be composed of more "fundamental" units (e.g., protons, neutrons, and electrons), but the way these are held together or apart is not completely understood. Furthermore, the spatial extent of the assemblage of protons, neutrons, and electrons is not a clearly defined constant but depends on what else there is in the immediate environment. In particular, when two or more atoms "interact" with each other to form the more complicated unit we call a *molecule*, the distinct identity of the individual atoms is no longer retained; for many purposes, it is more useful to think in terms of the new unit formed. Still, we speak of the molecule as being composed of atoms, and the atom remains a useful, approximate concept that ties together a large body of observed behavior.

The Greek philosopher Empedocles taught in 440 B.C. that all matter is composed of four simple substances: earth, air, water, and fire. Aristotle, about 100 years later, replaced the concept of component elementary substances by "blendings" of abstract properties (e.g., coldness, hotness, dryness, and moistness). Robert Boyle, Irish chemist and physicist, writing in 1661 what was probably the first chemistry textbook, "The Sceptical Chymist," rejected the aristotelian view and gave the first modern definition of elements as "primitive and simple unmingled bodies which, not being made of any other bodies or of one another, are the ingredients of which mixt bodies are immediately compounded and into which they are ultimately resolved." By 1803, when Dalton proposed his atomic theory, approximately 10,000 compounds had been discovered; these were resolvable into about 35 elements, some of which, such as lime, subsequently proved to be com-

Fig. 1.1
Some elementary particles.

Name	Symbol	Mass, $\dot{m}_{e-}$	Charge, e
Electron	e^-	1	−1
Proton	p	1836	+1
Neutron	n	1839	0
Omega hyperon	Ω^-	3279	−1
Xi hyperon	Ξ	2586	−1
Sigma hyperon	Σ^-	2341	−1
Lambda hyperon	Λ°	2182	0
K meson	K°	974	0
Pi meson	π^+	275	+1
Muon	μ^-	208	−1
Neutrino	ν	<0.0005	0

pound in character. (The comparable modern figures are approximately 1 million compounds and slightly more than 100 elements.)

For his atomic theory of matter, Dalton made what amounts to the following assumptions: (1) All matter is ultimately composed of atoms. (2) Atoms can be neither subdivided nor changed one into another. (3) Atoms can be neither created nor destroyed. (4) Atoms of a particular element are identical with each other in size, shape, mass, and other properties; they differ from atoms of other elements in these properties. (5) Chemical change is the union or separation of undivided atoms.

Each of the above assumptions eventually proved to be wrong. For example, J. J. Thomson in 1897 showed that a common constituent of all atoms is a more elementary particle called the *electron*. It has a mass of 9.1096×10^{-28} g and an electric charge of -1.6022×10^{-19} coulomb. (Units and conversion factors are discussed in Appendix 1.) Subsequent work showed that atoms are composed of two other constituent particles: protons and neutrons. Although chemists in general now regard atoms as being composed only of electrons, neutrons, and protons, the actual situation is certainly more complicated. Figure 1.1 lists so-called "elementary particles" that have been obtained on the breakup of the atom. In the list, the masses of the particles are given as multiples of the electron mass; the charges are given in units based on the electron charge as −1.

Particle accelerators (e.g., the cyclotron and synchrotron) have enabled man to accelerate particles to such high energy that beams of energetic particles can be used to disrupt the fundamental forces holding the atom together. The daltonian atom has been chipped, split, smashed, and reassembled in anything but an indivisible manner. Fortunately for chemistry, most of the perturbation of atomic structure is of the chipping variety; so the great bulk of the atom remains recognizably unchanged in the course of a chemical transformation.

Dalton's key assumption that all the atoms of a given element are alike was shown to be incorrect by the discovery in 1907 of isotopes by H. N. McCoy and W. H. Ross. Isotopes, which are almost chemically inseparable from each other, have identical electronic makeups but differ from each other in mass because of subtle differences in the heart of the atom. The word *isotope* comes from the Greek ισος (same) and τοπος (place); it refers to the fact that isotopes occur in the same place when the chemical elements are arranged in a regular array according to their properties. For most chemical purposes, the existence of these isotopes can be ignored. Any collection of atoms such as encountered in chemical reactions can be characterized by an "average" atom which has a mass determined by the weighted mean of the various isotopes present. The principal effect of having identical atoms differing only in mass is to change the rate at which chemical transformations occur. For example, in the case of the element hydrogen, which is composed of three isotopes (protium, deuterium, and tritium), the lightest isotope generally undergoes chemical reaction most rapidly.

The final flaw of the Dalton theory was to assume that atoms remain unchanged in the course of a chemical reaction. This immediately ran into trouble because it begged the question of how atoms are able to stick together, either to their own kind or to atoms of a different kind. There followed an interesting period during which atoms were sometimes endowed with hooks and other mysterious "valence forces" to account for their combining capacity. At present, we know that rather minor electric modifications in the exterior reaches of an atom can explain why one atom binds to another; even so, we shall see that a simple particle picture will not suffice to account for chemical binding.

1.2 Atomic weight

The great concept that came out of the Dalton theory is that atoms can be characterized by their atomic weight. In this respect, we might note that we should be talking here about *mass*, not *weight*. What is the difference? Mass is an intrinsic inertial property, independent of the frame of reference. It measures the inertia of an object, i.e., its tendency to stay at rest or to continue in straight-line motion. Weight, on the other hand, is a force ($F = ma$; see Appendix 5.2) and tells how the object is being attracted in a given environment. An astronaut who has a weight of 70 kg on the face of the earth has a weight of zero in an orbiting satellite and a weight of 12 kg on the face of the moon. The gravitational pull is different in the three environments; his weight changes. However, his mass is the same in the three places. To push him away from another astronaut in the three environments would require the same effort for the same acceleration.

What about atoms? Strictly speaking, we would have to say that the atomic weight of a hydrogen atom, or any other atom for that matter, is zero in an orbiting satellite. Atomic-weight tables would then be meaningless. Actually they are not meaningless because we really are talking about atomic mass. So why not use "atomic mass" instead of "atomic weight"? There are two good reasons: For one thing, the term "atomic weight" and the related term "molecular weight" are so firmly ingrained in chemists' vocabularies that one needs to use them to communicate effectively. Secondly, the term "atomic mass" has taken on the connotation that it is the actual intrinsic mass of an individual atom that is meant. As chemists, we generally work with large numbers of atoms, and we need a relative number to describe an "average" atom in the collection. This relative number is based on reference to a given standard, currently the so-called "carbon-12" isotope of the element carbon. It is convenient to use atomic weight when referring to this relative number and to save atomic mass for referring to the absolute mass of an individual atom.

Dalton was the first to set up a scale of atomic weights for the elements. He chose as his reference the lightest element, hydrogen, and assigned it a value of unity on his arbitrary scale. Other elements were then given higher values depending on how much heavier their atoms were compared with hydrogen. To fix the relative weights, compounds were analyzed and their compositions as percent by weight used to apportion the distribution of mass. Unfortunately, Dalton made two mistakes: First, his analysis of water was faulty. On decomposing water into its component elements hydrogen and oxygen, he found 12.5% by weight hydrogen and 87.5% oxygen. (The true figures are 11.19% H and 88.81% O.)* He concluded that oxygen contributed seven times as much to the total weight as did the hydrogen. (He should have found eight times.) His second mistake was more serious. He assumed that a molecule of the compound water consisted of one atom each of hydrogen and oxygen. Since all the water molecules are alike, since all the hydrogen atoms are alike, and since all the oxygen atoms are alike, weight analysis of bulk water told him directly that the weight distribution in each water molecule is 1 part for the hydrogen and 7 parts for the oxygen. Hence the conclusion was that the atomic weight of oxygen is 7 units on a scale where hydrogen is 1. (In similar fashion he proceeded to examine other compounds where different elements were combined with oxygen. Having once decided that oxygen had an atomic weight of 7, the other elements could be assigned their appropriate values.) Actually, the water molecule contains *two* hydrogen atoms bound to a single oxygen atom. As a consequence, one-seventh of the mass of the water molecule has to be spread over two hydrogen atoms; each hydrogen atom contributes only one-fourteenth the total mass of the molecule. Had Dalton guessed the atomic makeup of water correctly, he would have had to assign a value of 14 for his atomic weight of oxygen. If we go further and correct his analysis result from one-seventh to one-eighth, we would get 16, a value which is not far from the presently accepted value, though the latter is based on quite a different method of assignment.

Shortly after the appearance of Dalton's list of atomic weights based

---

* In most cases the symbol for an element is just the capitalized first letter of the name, for example, H for hydrogen, O for oxygen, and P for phosphorus. If several elements have the same initial letter, a second small letter may be included in the symbol, for example, He for helium, Ho for holmium, and Ha for hahnium. In some cases, particularly for elements known since antiquity, the symbols are derived from Latin names, for example, K from *kalium* for potassium, Fe from *ferrum* for iron, and Au from *aurum* for gold. Symbols for all the elements are given on the inside back cover of the book.

on hydrogen as unity, the prestigious Swedish chemist Berzelius proposed that the standard reference be oxygen instead of hydrogen. Hydrogen was so light that analysis of it was very difficult; it made sense to pick a heavier standard. Unfortunately, Berzelius suggested a value of 100. The result was that many of the atomic weights came out to be such unwieldy, large numbers that Berzelius' scale never really gained acceptance.

Until 1960, atomic weights were generally expressed on a scale in which oxygen was assumed to have a value of 16 units. However, there were in practice two scales. Chemists used values referred to 16 for a "natural" oxygen atom—that is, a hypothetical "average" atom that gave proper weighting to various oxygen isotopes that are present in a "natural" mixture. The physicists, on the other hand, singled out the lightest of the isotopes, so-called "oxygen 16," and gave it an exact reference value of 16. Thus, on the chemists' scale, the unit for expressing atomic weights was one-sixteenth the mass of an "average" oxygen atom; on the physicists' scale the unit was one-sixteenth the mass of an oxygen-16 atom. Because the chemists' unit was slightly larger, values on the chemists' scale were slightly smaller than on the physicists' scale. As an example, the *chemical* atomic weight of natural oxygen was given as 16.0000; the *physical* atomic weight of natural oxygen was given as 16.0044.

As the determination of atomic weights became more precise, the small difference between chemical and physical values became more irksome. The result was a unified new scale, adopted by both disciplines in 1961, which gave up the oxygen standard and moved instead to the carbon-12 isotope. Natural carbon consists mainly of two isotopes, 98.89% by number of the carbon-12 variety and 1.11% of the carbon-13 variety. It is the atoms of the carbon-12 variety that are taken as standard; they are arbitrarily assigned a value of 12 atomic mass units. Accordingly, at the present time one *atomic mass unit* (amu) is defined as one-twelfth the mass of a carbon atom that is a carbon-12 isotope. As we shall see later, the carbon-12 isotope is believed to consist of six neutrons, six protons, and six electrons; the whole assembly has a mass of 12.0000 amu. The carbon-13 isotope, which is believed to consist of seven neutrons, six protons, and six electrons, has a total mass of 13.00335 amu. The natural mixture of carbon atoms, in which the relative abundance of carbon 12 and carbon 13 is 98.89% and 1.11%, respectively, is described as having an atomic weight of 12.011 amu. Actually, different samples of carbon may vary because of slight variations in isotopic composition.

Figure 1.2 lists the 105 elements presently known, together with their average natural atomic weights referred to the carbon-12 isotope. Also indicated, in the last column, are the isotopes that are present in the natural mixture and their relative abundances. The various isotopes are indicated by the symbol of the element with a superscript to give the numerical designation of that particular isotope. (The superscript corresponds to the sum total of neutrons and protons in the isotope.) The listing of elements is given in the order of increasing

Fig. 1.2

Element no.	Element name	Symbol	Atomic weight, amu	Isotope abundance, %
1	Hydrogen	H	1.0079	^{1}H (99.985), ^{2}H (0.015)
2	Helium	He	4.00260	^{3}He (0.00013), ^{4}He ($\approx$100)
3	Lithium	Li	6.941	^{6}Li (7.42), ^{7}Li (92.58)
4	Beryllium	Be	9.01218	^{9}Be (100)
5	Boron	B	10.81	^{10}B (19.78), ^{11}B (80.22)
6	Carbon	C	12.011	^{12}C (98.89), ^{13}C (1.11)
7	Nitrogen	N	14.0067	^{14}N (99.63), ^{15}N (0.37)
8	Oxygen	O	15.9994	^{16}O (99.759), ^{17}O (0.037), ^{18}O (0.204)
9	Fluorine	F	18.99840	^{19}F (100)
10	Neon	Ne	20.179	^{20}Ne (90.92), ^{21}Ne (0.257), ^{22}Ne (8.82)
11	Sodium	Na	22.98977	^{23}Na (100)
12	Magnesium	Mg	24.305	^{24}Mg (78.70), ^{25}Mg (10.13), ^{26}Mg (11.17)
13	Aluminum	Al	26.98154	^{27}Al (100)
14	Silicon	Si	28.086	^{28}Si (92.21), ^{29}Si (4.70), ^{30}Si (3.09)
15	Phosphorus	P	30.97376	^{31}P (100)
16	Sulfur	S	32.06	^{32}S (95.0), ^{33}S (0.76), ^{34}S (4.22), ^{36}S (0.014)
17	Chlorine	Cl	35.453	^{35}Cl (75.53), ^{37}Cl (24.47)
18	Argon	Ar	39.948	^{36}Ar (0.337), ^{38}Ar (0.063), ^{40}Ar (99.60)
19	Potassium	K	39.098	^{39}K (93.10), ^{40}K (0.0118), ^{41}K (6.88)
20	Calcium	Ca	40.08	^{40}Ca (96.97), ^{42}Ca (0.64), ^{43}Ca (0.145), ^{44}Ca (2.06), ^{46}Ca (0.0033)
21	Scandium	Sc	44.9559	^{45}Sc (100)
22	Titanium	Ti	47.90	^{46}Ti (7.93), ^{47}Ti (7.28), ^{48}Ti (73.94), ^{49}Ti (5.51), ^{50}Ti (5.34)
23	Vanadium	V	50.9414	^{50}V (0.24), ^{51}V (99.76)
24	Chromium	Cr	51.996	^{50}Cr (4.31), ^{52}Cr (83.76), ^{53}Cr (9.55), ^{54}Cr (2.38)
25	Manganese	Mn	54.9380	^{55}Mn (100)
26	Iron	Fe	55.847	^{54}Fe (5.82), ^{56}Fe (91.66), ^{57}Fe (2.19), ^{58}Fe (0.33)
27	Cobalt	Co	58.9332	^{59}Co (100)
28	Nickel	Ni	58.71	^{58}Ni (67.88), ^{60}Ni (26.23), ^{61}Ni (1.19), ^{62}Ni (3.66), ^{64}Ni (1.08)
29	Copper	Cu	63.546	^{63}Cu (69.09), ^{65}Cu (30.91)
30	Zinc	Zn	65.38	^{64}Zn (48.89), ^{66}Zn (27.81), ^{67}Zn (4.11), ^{68}Zn (18.57), ^{70}Zn (0.62)
31	Gallium	Ga	69.72	^{69}Ga (60.4), ^{71}Ga (39.6)
32	Germanium	Ge	72.59	^{70}Ge (20.52), ^{72}Ge (27.43), ^{73}Ge (7.76), ^{74}Ge (36.54), ^{76}Ge (7.76)
33	Arsenic	As	74.9216	^{75}As (100)
34	Selenium	Se	78.96	^{74}Se (0.87), ^{76}Se (9.02), ^{77}Se (7.58), ^{78}Se (23.52), ^{80}Se (49.82), ^{82}Se (9.19)
35	Bromine	Br	79.904	^{79}Br (50.54), ^{81}Br (49.46)
36	Krypton	Kr	83.80	^{78}Kr (0.35), ^{80}Kr (2.27), ^{82}Kr (11.56), ^{83}Kr (11.55), ^{84}Kr (56.90), ^{86}Kr (17.37)

Element no.	Element name	Symbol	Atomic weight, amu	Isotope abundance, %
37	Rubidium	Rb	85.4678	^{85}Rb *(72.15)*, ^{87}Rb *(27.85)*
38	Strontium	Sr	87.62	^{84}Sr *(0.56)*, ^{86}Sr *(9.86)*, ^{87}Sr *(7.02)*, ^{88}Sr *(82.56)*
39	Yttrium	Y	88.9059	^{89}Y *(100)*
40	Zirconium	Zr	91.22	^{90}Zr *(51.46)*, ^{91}Zr *(11.23)*, ^{92}Zr *(17.11)*, ^{94}Zr *(17.40)*, ^{96}Zr *(2.80)*
41	Niobium	Nb	92.9064	^{93}Nb *(100)*
42	Molybdenum	Mo	95.94	^{92}Mo *(15.84)*, ^{94}Mo *(9.04)*, ^{95}Mo *(15.72)*, ^{96}Mo *(16.53)*, ^{97}Mo *(9.46)*, ^{98}Mo *(23.78)*, ^{100}Mo *(9.63)*
43	Technetium	Tc	[98.9062]	*No stable isotope*
44	Ruthenium	Ru	101.07	^{96}Ru *(5.51)*, ^{98}Ru *(1.87)*, ^{99}Ru *(12.72)*, ^{100}Ru *(12.62)*, ^{101}Ru *(17.07)*, ^{102}Ru *(31.61)*, ^{104}Ru *(18.58)*
45	Rhodium	Rh	102.9055	^{103}Rh *(100)*
46	Palladium	Pd	106.4	^{102}Pd *(0.96)*, ^{104}Pd *(10.97)*, ^{105}Pd *(22.23)*, ^{106}Pd *(27.33)*, ^{108}Pd *(26.71)*, ^{110}Pd *(11.81)*
47	Silver	Ag	107.868	^{107}Ag *(51.82)*, ^{109}Ag *(48.18)*
48	Cadmium	Cd	112.40	^{106}Cd *(1.22)*, ^{108}Cd *(0.88)*, ^{110}Cd *(12.39)*, ^{111}Cd *(12.75)*, ^{112}Cd *(24.07)*, ^{113}Cd *(12.26)*, ^{114}Cd *(28.86)*, ^{116}Cd *(7.58)*
49	Indium	In	114.82	^{113}In *(4.28)*, ^{115}In *(95.72)*
50	Tin	Sn	118.69	^{112}Sn *(0.96)*, ^{114}Sn *(0.66)*, ^{115}Sn *(0.35)*, ^{116}Sn *(14.30)*, ^{117}Sn *(7.61)*, ^{118}Sn *(24.03)*, ^{119}Sn *(8.58)*, ^{120}Sn *(32.85)*, ^{122}Sn *(4.72)*, ^{124}Sn *(5.94)*
51	Antimony	Sb	121.75	^{121}Sb *(57.25)*, ^{123}Sb *(42.75)*
52	Tellurium	Te	127.60	^{120}Te *(0.089)*, ^{122}Te *(2.46)*, ^{123}Te *(0.87)*, ^{124}Te *(4.61)*, ^{125}Te *(6.99)*, ^{126}Te *(18.71)*, ^{128}Te *(31.79)*, ^{130}Te *(34.48)*
53	Iodine	I	126.9045	^{127}I *(100)*
54	Xenon	Xe	131.30	^{124}Xe *(0.096)*, ^{126}Xe *(0.090)*, ^{128}Xe *(1.92)*, ^{129}Xe *(26.44)*, ^{130}Xe *(4.08)*, ^{131}Xe *(21.18)*, ^{132}Xe *(26.89)*, ^{134}Xe *(10.44)*, ^{136}Xe *(8.87)*
55	Cesium	Cs	132.9054	^{133}Cs *(100)*
56	Barium	Ba	137.34	^{130}Ba *(0.101)*, ^{132}Ba *(0.097)*, ^{134}Ba *(2.42)*, ^{135}Ba *(6.59)*, ^{136}Ba *(7.81)*, ^{137}Ba *(11.32)*, ^{138}Ba *(71.66)*
57	Lanthanum	La	138.9055	^{138}La *(0.089)*, ^{139}La *(99.911)*
58	Cerium	Ce	140.12	^{136}Ce *(0.193)*, ^{138}Ce *(0.250)*, ^{140}Ce *(88.48)*, ^{142}Ce *(11.07)*
59	Praseodymium	Pr	140.9077	^{141}Pr *(100)*
60	Neodymium	Nd	144.24	^{142}Nd *(27.11)*, ^{143}Nd *(12.17)*, ^{144}Nd *(23.85)*, ^{145}Nd *(8.30)*, ^{146}Nd *(17.22)*, ^{148}Nd *(5.73)*, ^{150}Nd *(5.62)*
61	Promethium	Pm	[145]	*No stable isotope*

Element no.	Element name	Symbol	Atomic weight, amu	Isotope abundance, %
62	Samarium	Sm	150.4	^{144}Sm (3.09), ^{147}Sm (14.97), ^{148}Sm (11.24), ^{149}Sm (13.83), ^{150}Sm (7.44), ^{152}Sm (26.72), ^{154}Sm (22.71)
63	Europium	Eu	151.96	^{151}Eu (47.82), ^{153}Eu (52.18)
64	Gadolinium	Gd	157.25	^{152}Gd (0.20), ^{154}Gd (2.15), ^{155}Gd (14.73), ^{156}Gd (20.47), ^{157}Gd (15.68), ^{158}Gd (24.87), ^{160}Gd (21.90)
65	Terbium	Tb	158.9254	^{159}Tb (100)
66	Dysprosium	Dy	162.50	^{156}Dy (0.052), ^{158}Dy (0.090), ^{160}Dy (2.29), ^{161}Dy (18.88), ^{162}Dy (25.53), ^{163}Dy (24.97), ^{164}Dy (28.18)
67	Holmium	Ho	164.9304	^{165}Ho (100)
68	Erbium	Er	167.26	^{162}Er (0.136), ^{164}Er (1.56), ^{166}Er (33.41), ^{167}Er (22.94), ^{168}Er (27.07), ^{170}Er (14.88)
69	Thulium	Tm	168.9342	^{169}Tm (100)
70	Ytterbium	Yb	173.04	^{168}Yb (0.135), ^{170}Yb (3.03), ^{171}Yb (14.31), ^{172}Yb (21.82), ^{173}Yb (16.13), ^{174}Yb (31.84), ^{176}Yb (12.73)
71	Lutetium	Lu	174.97	^{175}Lu (97.41), ^{176}Lu (2.59)
72	Hafnium	Hf	178.49	^{174}Hf (0.18), ^{176}Hf (5.20), ^{177}Hf (18.50), ^{178}Hf (27.14), ^{179}Hf (13.75), ^{180}Hf (35.24)
73	Tantalum	Ta	180.9479	^{180}Ta (0.0123), ^{181}Ta (99.988)
74	Tungsten	W	183.85	^{180}W (0.14), ^{182}W (26.41), ^{183}W (14.40), ^{184}W (30.64), ^{186}W (28.41)
75	Rhenium	Re	186.2	^{185}Re (37.07), ^{187}Re (62.93)
76	Osmium	Os	190.2	^{184}Os (0.018), ^{186}Os (1.59), ^{187}Os (1.64), ^{188}Os (13.3), ^{189}Os (16.1), ^{190}Os (26.4), ^{192}Os (41.0)
77	Iridium	Ir	192.22	^{191}Ir (37.3), ^{193}Ir (62.7)
78	Platinum	Pt	195.09	^{190}Pt (0.0127), ^{192}Pt (0.78), ^{194}Pt (32.9), ^{195}Pt (33.8), ^{196}Pt (25.3), ^{198}Pt (7.21)
79	Gold	Au	196.9665	^{197}Au (100)
80	Mercury	Hg	200.59	^{196}Hg (0.146), ^{198}Hg (10.02), ^{199}Hg (16.84), ^{200}Hg (23.13), ^{201}Hg (13.22), ^{202}Hg (29.80), ^{204}Hg (6.85)
81	Thallium	Tl	204.37	^{203}Tl (29.50), ^{205}Tl (70.50)
82	Lead	Pb	207.2	^{204}Pb (1.48), ^{206}Pb (23.6), ^{207}Pb (22.6), ^{208}Pb (52.3)
83	Bismuth	Bi	208.9804	^{209}Bi (100)
84	Polonium	Po	[209.9829]	No stable isotope
85	Astatine	At	[210]	No stable isotope
86	Radon	Rn	[222.0175]	No stable isotope
87	Francium	Fr	[223.0198]	No stable isotope
88	Radium	Ra	[226.0254]	No stable isotope
89	Actinium	Ac	[227.0278]	No stable isotope

Element no.	Element name	Symbol	Atomic weight, amu	Isotope abundance, %
90	Thorium	Th	[232.0381]	No stable isotope
91	Protactinium	Pa	[231.0359]	No stable isotope
92	Uranium	U	[238.029]	No stable isotope
93	Neptunium	Np	[237.0482]	No stable isotope
94	Plutonium	Pu	[244]	No stable isotope
95	Americium	Am	[243.0614]	No stable isotope
96	Curium	Cm	[247]	No stable isotope
97	Berkelium	Bk	[247.0702]	No stable isotope
98	Californium	Cf	[251]	No stable isotope
99	Einsteinium	Es	[254.0881]	No stable isotope
100	Fermium	Fm	[257]	No stable isotope
101	Mendelevium	Md	[258]	No stable isotope
102	Nobelium	No	[255]	No stable isotope
103	Lawrencium	Lr	[256]	No stable isotope
104	Kurchatovium	Ku	[257]	No stable isotope
105	Hahnium	Ha	[260]	No stable isotope

Fig. 1.2

Atomic weights and natural abundances of isotopes.

number of electrons or protons in the neutral atoms. For some of the elements, the atomic weights are given in brackets. In these cases, there are no naturally occurring isotopes that are stable to spontaneous atomic disintegration; the bracketed number refers then to the most commonly available long-lived isotope. The listing of atomic weights is given again in alphabetic order on the inside back cover of the book.

How are atomic weights determined? In the old days, it was generally a question of analyzing a compound consisting of the unknown element combined with the reference element. If the reference element were oxygen, one could burn a given weight of the unknown element, say w_1 g, in an excess of oxygen so as to convert it to oxide (say w_2 g). From the gain in weight $w_2 - w_1$, one could calculate the weight of oxygen combined per w_1 g of metal, hence the weight of metal per standard weight of reference element. If the compound were a one-to-one compound, i.e., contained one atom of unknown element per atom of oxygen, then the atomic weight of the unknown would simply be equal to $16w_1/(w_2 - w_1)$. If, however, the compound were a two-to-one compound, i.e., contained two atoms of unknown element per atom of oxygen, the atomic weight of the unknown would be half as great. On the other hand, if there were two atoms of oxygen per atom of unknown, the atomic weight of the unknown would be $2(16)(w_1)/(w_2 - w_1)$.

Example 1

In the formation of scandium oxide, it is observed that 1.0000 g of scandium combines with oxygen to form 1.5338 g of oxide. What is the atomic weight of the scandium on a scale where the oxygen atomic weight is 15.9994 amu, assuming the oxide contains two atoms of scandium for every three atoms of oxygen?

Mass of Sc = 1.000 g

Mass of O = 1.5338 − 1.0000 = 0.5338 g

$$\text{Atomic weight of Sc} = \left(\frac{1.0000 \text{ g of Sc}}{0.5338 \text{ g of O}}\right)\left(\frac{15.9994 \text{ amu}}{1 \text{ atom of O}}\right)\left(\frac{3 \text{ atoms of O}}{2 \text{ atoms of Sc}}\right)$$

$$= 44.96 \text{ amu per atom of Sc}$$

Recall that a horizontal line is to be read as "per." Note also that the problem setup can be checked by an analysis of the dimensional units. Cancellation of the same units in the numerator and denominator gives

$$\left(\frac{\text{g}}{\text{g}}\right)\left(\frac{\text{amu}}{\text{atom}}\right)\left(\frac{\text{atom}}{\text{atom}}\right) = \frac{\text{amu}}{\text{atom}}$$

■ ■ ■

The critical feature for a calculation such as Example 1 is to decide how many atoms of unknown element have combined per atom of oxygen. How do we know that two atoms of scandium have combined with three atoms of oxygen? Since atoms are so small, it is impossible to count them directly. An indirect way to get the needed information is to use an observation embodied in the law of Dulong and Petit (1819). They noted that the *atomic heat* of a solid element at room temperature—that is, the amount of heat needed to raise the temperature of one atomic weight's worth of element by one degree Celsius —is generally about 26 J (or 6.2 cal).* The atomic heat is the product of atomic weight and heat capacity per gram. The heat capacity is the amount of heat required to raise the temperature of a substance by one degree Celsius. Since heat capacity is a directly measurable quantity, one can measure it for scandium and find it to be 0.556 J g^{-1} deg^{-1}. This immediately suggests that the atomic weight of scandium ought to be about 26/0.556, or 47, amu. If in Example 1 the ratio three atoms of oxygen per two atoms of scandium were not given, then it could have been figured out, since the fraction (1.000/0.5338)(15.9994) has to be multiplied by some simple atomic ratio so as to give an answer near 47. The closest such ratio is 3:2. If we use $\frac{3}{2}$ to multiply (1.0000/0.5338)(15.9994), we get 44.96 amu

---

* The International Union of Pure and Applied Chemistry (IUPAC) has recommended that the unit for expressing energy should no longer be the calorie (amount of heat it takes to warm one gram of water from 14.5°C to 15.5°C) but the joule (amount of energy it takes to move one kilogram of mass from rest through a distance of one meter in one second of time). The joule, which is best pronounced $j\overline{oo}l$ but is often heard as *jowl*, is named after the English physicist James P. Joule, who is famous for studying the interconversion of heat and work. Committees have done a great deal of work and generated much heat in disputation whether *joule* or *calorie* should be the preferred unit. In this present period of transition, one needs to know about both:

One calorie = 4.184 J
One joule = 0.2390 cal

See Appendix 1 for a description of systematic units.

as the atomic weight of scandium, not very far from the approximate value of 47 amu guessed from the law of Dulong and Petit.

The procedure outlined above for getting at the atomic weight is satisfactory as long as we can be absolutely certain that the ratio of atoms of oxygen to atoms of scandium is precisely 3:2. In other words, we have to assume that the atomic ratio is rational and simple. If we believe in undivided atoms, then the atomic ratio must be expressible as the ratio of whole numbers. Dalton argued further that these whole numbers must be small ones; so atomic ratios can only be very simple, for example, 1:1, 1:2, and 2:3. Such compounds, in which atomic ratios are fixed by ratios of small whole numbers, are called *stoichiometric*.* If the ratios are not fixed at simple values, the compounds are referred to as *nonstoichiometric;* they "deviate from stoichiometry." As an example, whereas ordinary stoichiometric scandium oxide has an oxygen-to-scandium ratio of 3:2, it is possible by going to very high temperatures to prepare a nonstoichiometric scandium oxide where the oxygen-to-scandium ratio is, for example, 1.496. Nonstoichiometric compounds generally have unusual and interesting electric properties; they have recently become the object of intensive study. It is amusing to note that John Dalton had a rival in his day, Claude Berthollet, who argued vehemently that compounds did not have to be stoichiometric. Berthollet lost the argument then, but the controversy is memorialized in modern nomenclature, where nonstoichiometric compounds are called *berthollides* and stoichiometric compounds are called *daltonides*.

Modern procedure for determining atomic weights does not depend on chemical combining weights and on guesses of atomic ratios but relies on direct determination of the absolute masses of isotopes and the counting of their relative abundances. The device used is called a *mass spectrometer*. Its operation will be understood better after consideration of Sec. 1.5, but the principle of operation is to whirl atoms in spiral tracks so that the more massive ones tend to collect on the outside because of their greater inertia to being deviated from straight-line motion. A beam of atoms composed of several isotopes can thus be split into several beams, each composed of only one isotope. From the positions of the beams and their relative intensities one can deduce the masses and the number of atoms of each type. The following example shows how mass spectrometric data can be used to calculate the atomic weight of an element.

Example 2
Natural chlorine consists of a mixture of two isotopes, designated, respectively, as chlorine 35 and chlorine 37. Mass spectrometric analysis indicates that 75.53% of the atoms are chlorine 35, having a mass of 34.968 amu , and 24.47% are chlorine 37, having a mass of 36.956 amu. Calculate the chemical atomic weight of natural chlorine.

* The first syllable *stoi* is pronounced *stoy* (rhymes with *boy*); it is often mispronounced *stow*. Stoichiometry comes from the Greek *stoicheion*, meaning "element" and *metron* meaning "measure."

Out of 10,000 atoms

$$7553 \text{ will have a mass of } 34.968 \text{ amu} = 264{,}110 \text{ amu}$$
$$2447 \text{ will have a mass of } 36.956 \text{ amu} = \underline{\ 90{,}430} \text{ amu}$$
$$\overline{10{,}000} \qquad\qquad\qquad\qquad\qquad\qquad\qquad 354{,}540 \text{ amu}$$

The average weight is 354,540/10,000, or 35.454, amu

■ ■ ■

In carrying out their experiments, chemists generally work with large numbers of atoms. To describe weight manipulations involving these large numbers, it is convenient to have a systematic unit that is related to the mass of the individual atoms. Such a unit is the *mole*. It is defined as the amount of material which has as many elementary entities of specified composition as there are atoms in 12 g of carbon-12 isotope. In other words, if we have 12 g of carbon 12, we have one mole of carbon-12 atoms. The number of atoms in a mole of atoms is referred to as the *Avogadro number*. Its numerical value is 6.0222×10^{23}, determined by methods which we shall discuss later. For the present, it is sufficient just to know that one mole of atoms of any element contains the Avogadro number of atoms and weighs as many grams as the atomic weight of that element.[*]

Once it is known that one mole of atoms of any element contains the Avogadro number of atoms, it is a simple matter to combine this information with an atomic weight as listed in the atomic-weight tables either to calculate masses of individual atoms or the number of atoms in any given weight of element. The following examples illustrate such calculations.

Example 3
Given that the atomic weight of carbon 12 is exactly 12 amu, what is the mass of an individual carbon-12 atom?

One mole of carbon-12 atoms weighs 12 g.

One mole of any element contains 6.0222×10^{23} atoms.

$$\frac{12 \text{ g}}{6.0222 \times 10^{23} \text{ atoms}} = 1.9926 \times 10^{-23} \text{ g/atom}$$

■ ■ ■

Example 4
Given that the atomic weight of iron is 55.847 amu, how many atoms are there in 6.02 g of iron?

[*] In the old days, some people preferred to use the term "gram-atom" for the amount of material that corresponded to the Avogadro number of atoms. The term "mole" was reserved for counting particles more complex than individual atoms. In the course of time, the term "gram-atom" lost favor, partly because the prefix "gram" suggested a weight instead of a number of particles. The *coup de grâce* was given in 1970 by the Commission on Symbols, Terminology, and Units of the Division of Physical Chemistry of the International Union for Pure and Applied Chemistry. In recommending the fundamental quantities for setting up SI (Système International) units, IUPAC specifically recommended that the mole be the unit for amount of material and that the gram-atom be phased out. *Requiescat in pace.*

One mole of iron atoms weighs 55.847 g.

One mole of any element contains 6.0222×10^{23} atoms.

$$\left(\frac{6.02\text{ g}}{55.847\text{ g/mol}}\right)(6.0222 \times 10^{23}\text{ atoms/mol}) = 6.49 \times 10^{22}\text{ atoms}$$

■ ■ ■

Example 5

You wish to carry out a chemical reaction in which two atoms of scandium (atomic weight 44.96 amu) combine with three atoms of oxygen (atomic weight 15.999 amu). If you have 1.00 g of scandium, what mass of oxygen must you take to satisfy all the scandium atoms?

One mole of scandium atoms weighs 44.96 g.

$$1.00\text{ g of scandium} = \frac{1.00\text{ g}}{44.96\text{ g/mol}} = 0.0222\text{ mol of scandium atoms}$$

The reaction requires two atoms of scandium per three atoms of oxygen, which is the same as two moles of scandium atoms per three moles of oxygen atoms.

$$(0.0222\text{ mol of scandium})\left(\frac{3\text{ mol of oxygen}}{2\text{ mol of scandium}}\right)$$
$$= 0.0333\text{ mol of oxygen atoms}$$

One mole of oxygen atoms weighs 15.999 g.

$$(0.0333\text{ mol})(15.999\text{ g/mol}) = 0.533\text{ g of oxygen}$$

■ ■ ■

1.3 Periodicity in chemical behavior

Not long after atomic weights started being assigned to the chemical elements, it was noted that certain regularities began to appear when the elements were arranged in order of increasing atomic weight. At first (early in the nineteenth century), these regularities appeared simply as mild curiosities, since relatively few elements were known and no grand design could be discerned. For example, of the three similar elements, calcium, strontium, and barium, it was remarked that the atomic weight of the middle one, strontium (atomic weight 88), is roughly the average of the other two, i.e., 40 for calcium and 137 for barium. Similarly, in the sequence chlorine (35.5), bromine (80), and iodine (127) three similar elements could be arranged so the middle one had an atomic weight that is the mean of the other two. Clearly there must be some regularity connecting atomic weight and similarity of properties.

As chemical research progressed and more elements were discovered and characterized, the regularities became more pronounced. By 1864, for example, Newlands was able to state that "if the elements are

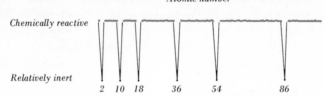

Fig. 1.3

Periodic occurrence of low reactivity in the elements.

arranged in the order of atomic weights, the eighth element, starting from a given one is a kind of repetition of the first, like the eighth note in an octave of music." It was only a few years later that, quite independently of each other, Lothar Meyer of Germany and Dmitri Mendeleev of Russia enunciated what is called the *periodic law*. As initially stated, the law described the observation that the elements if arranged according to their atomic weights showed a periodicity of properties. We now know that the original statement was not entirely exact; it is not really the atomic weight that decides the regular periodicity in properties but a more fundamental quantity called the *atomic number*. How this atomic number was discovered and what it represents we shall see in a later section. For the present we simply introduce it as a serial number, related to the atomic weight, which indicates the order in which the complexity of atoms is built up.

In its present form the periodic law states that the properties of elements recur periodically if the elements are arranged in order of increasing atomic number. To illustrate, we consider the property of chemical reactivity. Most elements react with other elements, but a few, such as argon, are relatively inert. Such elements react only under special conditions and then only with fluorine or oxygen. Figure 1.3 is a schematic representation of the distribution of relative inertness among the elements. Those showing low reactivity are helium, neon, argon, krypton, xenon, and radon, having atomic numbers 2, 10, 18, 36, 54, and 86, respectively. Since they are gases under usual conditions, they are sometimes called *inert gases*, reflecting the fact that prior to the discovery of their actual reactivity they were really thought to be completely inert. Although the name "inert gases" is still reasonably descriptive, most people now refer to them as *noble gases, rare gases,* or even *aerogens*.

The elements which directly follow the noble gases—lithium, sodium, potassium, rubidium, cesium, and francium, with atomic numbers 3, 11, 19, 37, 55, and 87, respectively—are metals; i.e., they have shiny luster and are good conductors of heat and electricity. As a group they are called the *alkali metals*. In their chemical properties the alkali metals bear strong resemblance to each other. For example, they all react vigorously with water to liberate hydrogen and form bitter-tasting, slippery, corrosive solutions that turn the dye litmus blue. If the solutions, which are described as being *basic*, are neutralized with hydrochloric acid and evaporated, a white solid is left behind in each case. These white solids are compounds of an alkali metal and chlorine, e.g., sodium chloride (NaCl) and potassium chloride (KCl).* They are called *salts* and are quite similar to each other; for instance, all dissolve readily in water to give an electrically conducting solution. The salts can also be made by direct reaction between the alkali metals and chlorine gas. All the alkali metals also form other series of similar compounds. The hydroxy compounds, such as sodium hydroxide (NaOH), are quite basic.

The elements which directly precede the noble gases—fluorine, chlorine, bromine, iodine, and astatine, with atomic numbers 9, 17, 35, 53, and 85, respectively—also resemble each other. As a group, they are called the *halogens*. (The element hydrogen, which directly precedes element number 2, helium, is not included in this group. As the first of all elements, hydrogen has unique properties.) Unlike the alkali elements, the halogens are nonmetals; i.e., they are poor conductors of heat and electricity. Under usual conditions, fluorine and chlorine are gases, bromine is a liquid, and iodine and astatine are solids.

The halogens resemble each other in that all react with hydrogen to form compounds, e.g., hydrogen fluoride (HF) or hydrogen chloride (HCl), which dissolve in water to give acid solutions (sour-tasting, turn litmus red). Neutralization of these acid solutions with sodium hydroxide, followed by evaporation of the water, leads to the formation of white sodium salts. These salts, e.g., sodium fluoride (NaF) and sodium iodide (NaI), can also be prepared by the direct reaction of the halogens with sodium. Like the alkali metals all the halogens form hydroxy compounds, but unlike the alkali hydroxides the hydroxy compounds of the halogens are all acidic. An example is ClOH, which is hypochlorous acid, more usually written HOCl.

The elements that fall between an alkali metal and the subsequent halogen show a progressive gradation of properties between two extremes. For example, the elements magnesium (atomic number 12), aluminum (13), silicon (14), phosphorus (15), and sulfur (16), which lie between sodium (11) and chlorine (17), represent such a gradation. In the sequence there is a decrease in metallic character. Magnesium and aluminum are metals; silicon is intermediate between a metal and a nonmetal; phosphorus and sulfur are nonmetals. Concurrently, there is a progressive change from basic to acidic character in the hydroxy compounds. The hydroxy compound of magnesium $Mg(OH)_2$, is basic; the hydroxy compound of aluminum, $Al(OH)_3$, is less basic; the hydroxy compound of silicon, $Si(OH)_4$, is even less basic and begins to show acidic properties; the hydroxy compounds of phosphorus and sulfur are distinctly acidic [for example, $PO(OH)_3$, or H_3PO_4, is phosphoric acid, and $SO_2(OH)_2$, or H_2SO_4, is sulfuric acid].

In order to emphasize the periodic reappearance of properties, it is customary to lay out the elements, not in a long straight line as

---

* Compounds are generally designated by *formulas*, which are combinations of symbols of elements. Parentheses indicate groupings of unlike atoms. Subscripts, for which the number 1 is understood when no subscript is given, indicate the relative number of atoms or groups of atoms. Thus, NaCl indicates the compound containing one atom of sodium (Na) per one atom of chlorine (Cl). Rules for the systematic naming of compounds are given in Appendix 2.

Group	I	II											III	IV	V	VI	VII	0
Period *1*	H 1																	He 2
2	Li 3	Be 4											B 5	C 6	N 7	O 8	F 9	Ne 10
3	Na 11	Mg 12					*Transition elements*						Al 13	Si 14	P 15	S 16	Cl 17	Ar 18
4	K 19	Ca 20	Sc 21	Ti 22	V 23	Cr 24	Mn 25	Fe 26	Co 27	Ni 28	Cu 29	Zn 30	Ga 31	Ge 32	As 33	Se 34	Br 35	Kr 36
5	Rb 37	Sr 38	Y 39	Zr 40	Nb 41	Mo 42	Tc 43	Ru 44	Rh 45	Pd 46	Ag 47	Cd 48	In 49	Sn 50	Sb 51	Te 52	I 53	Xe 54
6	Cs 55	Ba 56	* 57–71	Hf 72	Ta 73	W 74	Re 75	Os 76	Ir 77	Pt 78	Au 79	Hg 80	Tl 81	Pb 82	Bi 83	Po 84	At 85	Rn 86
7	Fr 87	Ra 88	† 89–103	Ku 104	Ha 105													

*	La 57	Ce 58	Pr 59	Nd 60	Pm 61	Sm 62	Eu 63	Gd 64	Tb 65	Dy 66	Ho 67	Er 68	Tm 69	Yb 70	Lu 71
†	Ac 89	Th 90	Pa 91	U 92	Np 93	Pu 94	Am 95	Cm 96	Bk 97	Cf 98	Es 99	Fm 100	Md 101	No 102	Lr 103

Fig. 1.4
Periodic table. (The number beneath the symbol of each element is the atomic number. The asterisk and the dagger represent the elements listed at the bottom of the table.)

in Fig. 1.3, but in what is known as a *periodic table.* The basic feature of a periodic table is the arrangement of the elements in order of increasing atomic number, with elements that are similar in properties placed under each other in a vertical column. There are many forms of the periodic table, one of which is shown in Fig. 1.4. The arrangement consists of vertical columns called *groups* and horizontal rows called *periods.* There are eight main groups, designated as across the top of Fig. 1.4 by I, II, III, IV, V, VI, VII, and 0. Group I includes H plus the alkali metals (Li, Na, K, Rb, Cs, and Fr); group VII, the halogens (F, Cl, Br, I, and At); and group 0, the noble gases (He, Ne, Ar, Kr, Xe, and Rn). The elements intervening between groups II and III are called collectively the *transition elements;* they include the elements designated by the asterisk and dagger at the bottom of the table. Each vertical column of transition elements is called a *subgroup* and is named after the head element. Thus, Zn, Cd, and Hg make up the *zinc subgroup.*

The periods are numbered from the top down. The first period contains but 2 elements (H and He); the second and third, 8 elements; the fourth and fifth, 18 elements. The elements denoted by the asterisk are part of the sixth period; those by the dagger, the seventh period.

The periodic table is a useful device for organizing the chemistry of the elements. Furthermore, the fact that the elements can be arranged systematically suggests a periodic recurrence of important features in the structure of individual atoms.

1.4 Experiments on the electrical nature of atoms

Michael Faraday, English physicist, reported in 1832 that when electric current is passed through a molten salt or a salt solution so as to decompose it into elementary substances, "the chemical action of the current of electricity is in direct proportion to the absolute quantity of electricity which passes." For example, a steady flow of electricity through molten sodium chloride decomposes it into sodium and chlorine and produces twice as much sodium and chlorine in 10 min as in 5 min. This is an example of what is called the *first law of electrolysis*. Faraday also discovered that the weight of element resulting from the decomposition is directly proportional to the atomic weight of the element divided by a small whole number. Thus, for example, if a suitable amount of electricity is used to decompose molten sodium chloride, one gets 23 g of sodium plus 35.5 g of chlorine; if the same amount of electricity is used to decompose molten calcium chloride, one gets 20 g of calcium plus 35.5 g of chlorine; molten aluminum chloride, 9 g of aluminum plus 35.5 g of chlorine. These numbers illustrate the *second law of electrolysis*; the masses of sodium, calcium, aluminum, and chlorine formed are the corresponding atomic weights (23, 40, 27, and 35.5) divided, respectively, by 1, 2, 3, and 1. The fact that chemical change is produced by electricity clearly indicates a relationship between atoms and electricity; the appearance of whole numbers as divisors suggests that the electric structure of atoms involves discrete particles of electricity.

The nature of these discrete particles was considerably elucidated during the latter half of the nineteenth century by the study of electric discharges through gases. Given a glass vessel containing metal electrodes sealed in at either side across which a high voltage can be imposed, one observes as one evacuates the vessel that the residual gas commences to glow when the gas pressure is sufficiently reduced. Subsequent study suggested that electric rays, called *cathode rays*, emanated from the negatively charged electrode (the *cathode*) and disappeared into the positive electrode (the *anode*). In traversing the space between cathode and anode, the cathode rays were believed to gain energy because of the accelerating effect of the high voltage. If the gas pressure in the tube were low enough, there would be so few gas molecules left in the vessel that the cathode rays could gain lots of energy before slamming into a gas molecule. When this energy gets big enough, a process called *ionization* occurs. "Negative electricity" is stripped off the neutral molecule, and the result is to form a positive fragment called a positive *ion*. The positive ions were then supposed to attract the "negative electricity," and the recombination process would emit energy in the form of the glowing light. Clever design,

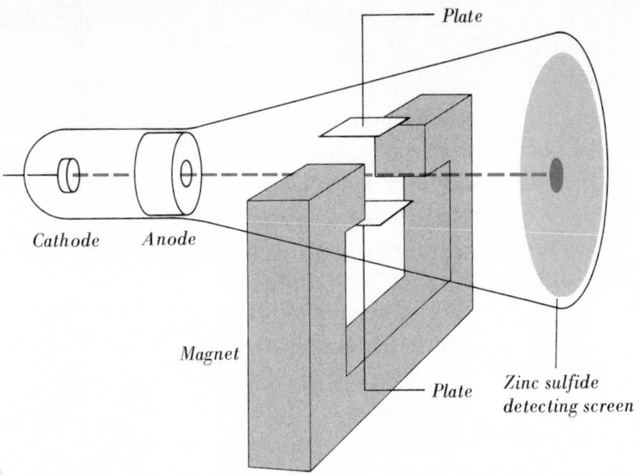

Plate

Cathode   Anode

Magnet

Plate   Zinc sulfide detecting screen

notably by Sir William Crookes of England and E. Goldstein of Germany, enabled one to select out a beam of charged fragments for study. Goldstein, for example, simply drilled a small hole through a cathode so that positive fragments which were attracted to the negatively charged cathode coasted right through the hole into the space on the other side of the cathode. By applying electric and magnetic forces to the beams of charged particles, it was possible to learn a great deal about the actual charges and masses of the particles that made up the beams. In this way the English physicist J. J. Thomson was led to the discovery of the electron in 1897. By precise quantitative study of beam behavior he was able to show that cathode rays consist of negative particles, electrons, which are all alike no matter what electrode material they come from and no matter what gas is in the tube. On the other hand, the beams of positive rays are different, depending on what gas is in the tube.

A typical cathode-ray tube, such as was used by Thomson for his quantitative studies, is shown in Fig. 1.5. The negatively charged cathode at the left emits electrons, which are accelerated to the right by attraction to the positively charged anode. The anode is a metal cylinder with a hole bored along its axis through which some electrons can coast. The result is to give a narrow beam continuing to the right and impinging on the zinc sulfide detecting screen covering the inner face of the tube. Zinc sulfide is a remarkable solid that has the property of converting the kinetic energy of incident electrons into visible light. Thus, a spot of light appears on the face of the tube where the electron beam intersects the surface. Electric plates in the tube, one located above the beam and the other below, can deflect the beam in a vertical direction, because the negative plate repels the negative beam and the positive plate attracts it. The amount of deflection, which can be monitored by noting the movement of the spot, is directly proportional to the voltage between the plates.

A magnet, which clamps around the tube from the outside, is set perpendicular to the tube so that there is a magnetic field perpendicular

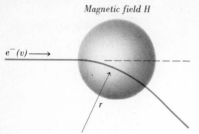

$e^-(v) \longrightarrow$

r

Fig. 1.6

Schematic representation showing how an electron with velocity v is deflected in a region of magnetic field H.

to the electron beam. The nature of electric-magnetic interactions is such that electric charges moving in a plane perpendicular to which there is a magnetic field tend to be bent into curved paths within that plane. The Thomson experiment consists of two parts: In one part a deflection is measured with plates uncharged; i.e., the deflection is due to the magnetic field alone. In the other part of the experiment the plates are charged in the proper direction and to the extent required to produce no net deflection; i.e., the electric field and the magnetic field exactly counteract each other.

Figure 1.6 shows the relation of the quantities involved. An electron with charge e moves to the right with velocity v. In the first part of the experiment it simply moves through the region of magnetic field H(i.e., there is still no voltage on the plates). H, being perpendicular to the plane of the diagram, bends the electron away from the dashed, straight-line path into a curved path having a radius of curvature r. The value of r can be determined by measuring how far the light spot on the face of the tube has been deviated from its undeflected (zero magnetic field) position. The force due to a magnetic field is Hev, which simply states that the force is directly proportional to the magnetic field, the charge on the electron, and the velocity of the electron. A basic law of physics states that always force = mass × acceleration. For a curving particle of velocity v moving along a radius of curvature r, the acceleration is v^2/r. Hence, we can write force = mass × acceleration = $m \times v^2/r$, where m is the mass of the particle. Setting this force equal to the magnetic force Hev, we get the equality

$$Hev = \frac{mv^2}{r}$$

from which

$$\frac{e}{m} = \frac{v}{Hr}$$

This relation tells us that the charge-to-mass ratio is equal to the velocity divided by the magnetic field and the radius of curvature that the magnetic field gives rise to. We can get a numerical value of e/m if we know the three related quantities in a particular experiment. The magnetic field H can be measured from the strength of the magnet; the radius of curvature r can be measured from the observed deflection of the light spot. How can we get the third quantity v? For this, we need to proceed to the second part of the experiment. We leave everything precisely as it was in the first part of the experiment, but now we crank up the voltage on the electric deflecting plates. A voltage on these plates, if in the right direction, produces an electric field E (equal to the voltage divided by the distance between the plates) that tends to force the electron back to its straight-line path. We simply adjust the voltage until we have brought the light spot back to its undeflected position. When there is no net deflection, the magnetic force Hev is just balanced by the electric force Ee. From this equality

$Hev = Ee$, we see that

$$v = \frac{E}{H}$$

In other words, knowing the values of E and H needed to balance electric and magnetic deflections of an electron beam tells us what is the velocity of the electrons in that beam. Substituting $v = E/H$ in the earlier equation $e/m = v/Hr$ gives

$$\frac{e}{m} = \frac{E}{H^2 r}$$

The numerical value of e/m so obtained for all electrons is equal to -1.7588×10^8 coulombs/g. (The coulomb, a unit for measuring electric charge, is the amount of charge passing a given point in one second when the electric current is one ampere. See Appendix 5.5.)

Once the charge-to-mass ratio has been measured, another experiment has to be performed to evaluate the charge separately from the mass. If, for example, the charge is measured independently, the mass can be calculated from the value of e/m.

The charge of the electron was first measured precisely in 1909 in a classic experiment by the American physicist R. A. Millikan. Figure 1.7 shows the essential features of the experiment. A cloud of oil

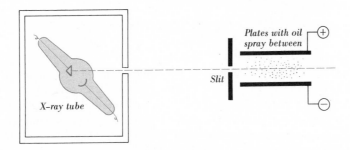

Fig. 1.7
Millikan's oil-drop
experiment.

droplets is sprayed between two charged plates. Because of gravity the droplets tend to settle. The gravitational force is equal to Mg, where M is the mass of a drop and g is the acceleration due to gravity. By irradiating the oil-drop chamber with X rays, the oil droplets can be given a negative charge. The X rays knock electrons off the molecules of air, and one or more of these electrons can be picked up by an oil droplet so as to make it negative. If the droplets carry a negative charge, they will experience an upward force because of the electric field between the charged plates. The magnitude of this electric force is Eq, where E is the electric field strength and q is the total electric charge on a particular drop. Different droplets may carry different charges, owing to the way the experiment is done.

The rate of motion of a particular oil droplet is observed in a transverse light beam. By adjusting the voltage on the plates, the particular droplet can be made stationary. Under this condition, the gravity force Mg (corrected for air buoyancy) just equals the electric force Eq. Since

E and g are known, q can be calculated once M is obtained. To evaluate M, the electric field is shut off, and the free fall of the droplet is observed. Because of air friction, there will be a drag on the droplet. The amount of drag increases with speed; so a limited rate of fall is reached. At this rate the net gravitational pull just equals the frictional drag. From a measurement of the rate, the radius of the drop can be calculated, since the limiting rate of free fall is proportional to the square of the radius. Once the droplet radius a is determined, the effective mass of the droplet can be calculated from $\frac{4}{3}\pi a^3$ times the density of the oil.

It is observed that, although the charges on different oil droplets vary, the values obtained are always small, whole-numbered multiples of -1.60×10^{-19} coulomb. This is the smallest possible charge that any one oil droplet can pick up and is assumed to be the charge of an individual electron. Combining the measured charge of the electron (-1.60×10^{-19} coulomb) with the measured charge-to-mass ratio (-1.76×10^8 coulombs/g) gives a measured mass for the electron of 9.1×10^{-28} g.

1.5 Atomic spectroscopy

At the same time that the above investigations into the electric nature of matter were occurring, parallel studies were going on into the nature of light given off when materials are heated. This kind of investigation, called *spectroscopy*, reached a peak late in the nineteenth century when Bunsen had perfected his burner for getting an almost colorless, very hot flame from the combustion of natural gas. Injection by spraying of salt solutions into such a flame produces various colors, which can be analyzed by passing the light through a prism or a diffraction grating. The prism separates component colors because light moves more slowly in glass than in air, depending on its color; the diffraction grating, which consists of a series of parallel grooves on a hard, flat surface, separates light because of selective reinforcement of light waves along particular directions.

Light is a form of electromagnetic radiation, composed of complicated periodic electric field and magnetic field pulsations transmitted through space. The pulsations, or waves, move through space at a speed of 2.998×10^8 m/sec and can be characterized in terms of a wavelength and a frequency. The *wavelength*, generally designated by λ, represents the distance between any pair of corresponding points, such as two successive wave crests or wave troughs. It is equal to the velocity at which the train of pulsations is propagated divided by the frequency of the wave. The *frequency*, generally designated by ν, can be pictured as the number of complete waves, or pulses, that passes a given point per second. As an illustration, if two wave trains differ in that wave 1 has twice the wavelength of wave 2, it follows that the frequency of the first wave train is half that of the second. The energy E of a light wave is proportional to its frequency; that is, $E = h\nu$, where

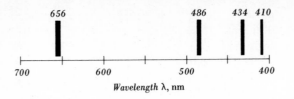

656   486   434 410

700   600   500   400

Wavelength λ, nm

Fig. 1.8
Hydrogen line spectrum.
(Numbers shown are
wavelengths in nanometers,
or tens of angstrom units.)

the proportionality constant h, a fundamental constant called the *Planck constant*, has the value of 6.6262×10^{-34} J-sec.

White light is a combination of waves of different wavelengths extending from about 400 nm (where 1 nm = 1 nanometer = 1×10^{-9} m) for violet light to about 700 nm for red light.* The range from 400 to 700 nm is called the *visible spectrum* since it is the part which the human eye can distinguish. Ultraviolet radiation lies on the short-wavelength side of 400 nm; infrared radiation, which we sense as heat waves, lies on the long-wavelength side of 700 nm. In terms of energy, ultraviolet radiation is on the high-energy side, and infrared radiation is on the low-energy side of the visible spectrum.

If white light as from a glowing solid is dispersed by being passed through a prism, it is observed that the resulting pattern consists of a *continuous spectrum* of colors, a gradual blending from one color to the next. If instead of white light the colored light from a heated gaseous sample is dispersed by a prism, the resulting pattern is not continuous but consists of a series of narrow lines of color. An example of such a *line spectrum* is shown in Fig. 1.8. The actual pattern obtained differs from element to element and is characteristic of the particular atoms present. As we shall see later, the absorption of energy by the atoms from the flame raises the energy of the atoms in a particular way; when the atoms return to lower energy states, they emit characteristic energies, and it is these characteristic energies that we see as the variously colored bands of light. Expressing the energy as a frequency (recall that energy E is proportional to frequency ν), we find that for every line spectrum of a particular element the observed frequencies can be described in a simple systematic way. For example, for hydrogen it is found that the frequencies seen in the visible part of the spectrum can be represented by the equation

$$\nu = 3.290 \times 10^{15} \left(\frac{1}{4} - \frac{1}{b^2} \right)$$

where b is simply any integer greater than 2, that is, 3, 4, 5, 6, etc. This is called the *Balmer series* and extends from a bright red line

* Up until recently, wavelengths have been generally expressed in units of 1×10^{-8} cm, or *angstrom units* (Å). IUPAC has recommended that angstrom be phased out. In its stead, the preferred unit is the nanometer, which is 1×10^{-9} m or 1×10^{-7} cm. Thus, we have

1 nanometer (nm) = 1×10^{-9} meter (m) = 1×10^{-7} centimeter (cm)
    = 10 angstrom units (Å)

The spectral range from 400 to 700 nm is also described as 4000 to 7000 × 10^{-8} cm, that is, 4000 to 7000 Å.

at a wavelength of 656.3 nm (corresponding to $b = 3$), through 486.1 nm ($b = 4$), 434.0 nm ($b = 5$), 410.1 nm ($b = 6$), and so forth, until it converges at 364.6 nm ($b =$ infinity) in the blue region of the spectrum.

In the ultraviolet region of the spectrum hydrogen shows another series of related lines fitting the equation

$$\nu = 3.290 \times 10^{15}\left(1 - \frac{1}{b^2}\right)$$

where b now can have values 2, 3, 4, 5, 6, etc. This is called the *Lyman series* and fits in the wavelength region between 121.5 nm and 91.2 nm. A third series, called the *Paschen series*, falls in the infrared region with a series of lines fitting the equation

$$\nu = 3.290 \times 10^{15}\left(\frac{1}{9} - \frac{1}{b^2}\right)$$

where $b = 4, 5, 6, 7$, etc.

It should be noted that the Lyman, Balmer, and Paschen series differ only in that the leading term inside the parentheses is, respectively, $\frac{1}{4}$, 1, and $\frac{1}{9}$, or expressed differently, $\frac{1}{2^2}$, $\frac{1}{1^2}$, and $\frac{1}{3^2}$. For a long time the integral character of these numbers and of the b values appeared to be just so much magic, but eventually they proved to be the main clue for unlocking the secret of the electron structure of atoms. Mysterious as it was, the observed regularities in the line spectra clearly suggested an underlying regularity in the energy levels of atoms.

1.6 Discovery of the nucleus

By the beginning of the twentieth century it was generally accepted that atoms can be fragmented into negatively charged electrons and positively charged residual fragments. Beam-deflection experiments such as those described in Sec. 1.4 indicated that electrons are light and identical, whereas the positive fragments are massive and dependent on the particular neutral atom they are derived from. By 1898, J. J. Thomson was able to propose a model for atoms in which the atom was considered to be a sphere of "positive electricity" in which negative electrons were embedded like so many jelly beans in a ball of cotton. Practically all the mass of the atom (better than 99.9 percent) was associated with the "positive electricity," but it was considered to be spread out more or less uniformly over the entire volume of the atom. The electrons contributed negligible mass to the total, and Thomson even went so far as to suggest that they were located in rings around the atom's center.

The English physicist Ernest Rutherford, while studying the effect of thin metal foils on the scattering of positive beams, discovered in 1911 that Thomson's model had to be corrected by pulling in all the positive charge and its associated mass into a tiny central core of the atom, the *nucleus*. The details of Rutherford's classic experiment are

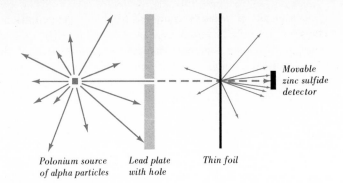

Movable zinc sulfide detector

Polonium source of alpha particles   *Lead plate with hole*   *Thin foil*

represented schematically in Fig. 1.9. A beam of *alpha particles*, which are the positively charged fragments formed when two electrons are subtracted each from neutral helium atoms, was obtained from a specimen containing the element polonium. This element, like all the very heavy elements, is radioactive; its atoms spontaneously disintegrate in random statistical fashion to give simpler fragments. One of the fragments is the alpha particle.

A beam of alpha particles can be obtained by placing the polonium in a lead box and letting the alpha particles come out through a pinhole in the lead. Lead is a good absorber for the particles resulting from radioactive decay; so it stops the alpha particles except for the ray that comes through the pinhole. Rutherford's experiment was to place thin sheets of metal in the path of the alpha ray in order to see how various metals would affect the alpha-particle trajectory. Alpha particles are very energetic; so they penetrate thin foils readily. What happens to them as they go through the foil, specifically how they are scattered from a straight-line path, can be monitored by placing a detector such as a zinc sulfide screen at various places on the exit side of the foil. On the basis of the Thomson model of the atom, it was thought that the metal foil consisted of atoms having their positive charge and mass distributed uniformly, so that alpha particles going through would have little reason to swerve aside from their original path. They should plow right through, and the detector would be expected to show at most only slight deflections. On the contrary, it was found that some of the alpha particles were deflected at astonishingly large angles. A few were actually reflected back toward the source. To Rutherford this was absolutely unbelievable. In his own words, "It was almost as incredible as if you fired a 15-inch shell at a piece of tissue paper and it came back and hit you."

The Thomson model could not account for such large deflections. If mass and positive charge were uniformly spread throughout the metal, a positively charged alpha particle would not encounter a large repulsion or major obstacle anywhere in its path. According to Rutherford, the only way to account for the large deflections is to say that the "positive electricity" and mass in the metal foil are concentrated in very small regions. Although most of the alpha particles go through without any deflection, occasionally one comes very close to the high

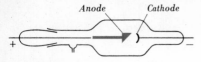

Fig. 1.10

An X-ray tube with interchangeable target.

concentration of positive charge. This high concentration of positive charge is essentially immovable because of its high mass. As the like charges get closer together, they repel each other, and the repulsion may be great enough to cause the relatively light alpha particle to swerve considerably from its original path. So, Rutherford suggested that an atom has a nucleus, or center, in which its positive charge and mass are concentrated.

The quantitative results of scattering experiments such as Rutherford's indicate that the nucleus of an atom has a radius of about 10^{-13} cm, which is about a hundred-thousandth the size ascribed to atoms in solids. Later, when neutrons and protons were discovered, it was noted that the radius of a particular nucleus could be expressed roughly as r (in centimeters) $\approx 1 \times 10^{-13} A^{1/3}$, where A is the total number of neutrons and protons in that nucleus.

About the same time that Rutherford was led to postulate the existence of the nucleus, H. G. J. Moseley was measuring the energies of X rays emitted by various elements. These measurements, which contributed greatly to further characterization of the nuclear atom, can be carried out with an X-ray tube such as the one shown in Fig. 1.10. Electrons from the curved cathode on the right are focused to impinge on the replaceable anode on the left, giving rise to the emission of X rays from the target material. It is found that wavelengths emitted are characteristic of the element from which the anode is made.

X rays, like light, are a form of electromagnetic radiation. They move through space with the speed of light $c = 2.998 \times 10^8$ m/sec, and they can be characterized in terms of a wavelength and a frequency. The wavelengths of X rays are much shorter than those of visible light, being generally about a thousandth as big. Hence, the frequencies, i.e., the energies, are some 1000 times as large. As a specific example, if an X-ray tube has an anode made of copper, the shortest-wavelength X ray that it emits has $\lambda = 0.1541$ nm, corresponding to a frequency $(\nu = c/\lambda)$ of 1.945×10^{18} sec^{-1} or an energy $(E = h\nu)$ of 1.289×10^{-15} J. On the other hand, if the anode is made of molybdenum, the shortest-wavelength X ray emitted has $\lambda = 0.0709$ nm, corresponding to $\nu = 4.228 \times 10^{18}$ sec^{-1} or $E = 2.802 \times 10^{-15}$ J. For comparison, blue light with $\lambda = 400$ nm corresponds to $\nu = 7.495 \times 10^{14}$ sec^{-1} or $E = 4.966 \times 10^{-19}$ J. Moseley's experiments, in which he successively substituted different anode materials, showed that there is a systematic relation between the wavelengths of the X rays emitted and the nature of the anode material. Specifically, he obtained a straight line when he plotted the square root of the X-ray frequency (for the short-wavelength limit) against the order in which the element appeared in the periodic table of the chemical elements.

As shown in Fig. 1.11, to get a smooth curve Moseley found it necessary to leave gaps in a few places (corresponding to then undiscovered elements) and to use an order based on trends of chemical properties rather than strict adherence to increasing atomic weight. For example, in the case of cobalt (58.93 amu) and nickel (58.71 amu), cobalt would come first on the basis of chemical trends, but nickel

would come first on the basis of its smaller atomic weight. From his
X-ray data Moseley was led to choose the order cobalt followed by
nickel, thus overturning the notion that atomic weight was most decisive
in fixing atomic properties. Instead, there is a characteristic number
called the *atomic number*, generally designated as Z, which is believed
to be equal to the positive charge of the nucleus and which determines
the ordering of the elements by chemical behavior.

Why is it in some cases that chemical properties of the elements
do not follow a strict order based on atomic weight but instead follow
an order based on atomic number? Put another way, why is there not
a direct parallel between increasing atomic number and increasing
atomic weight? The answer lies in the fact that different atoms of the
same element may differ in mass. In other words, as already pointed
out in Secs. 1.1 and 1.2, there are such things as isotopes, and the
actual atomic weight observed for an element depends on the isotope
distribution. The individual masses of the different isotopes and their
relative abundances can be measured by use of the *mass spectrometer*,
the essential features of which are represented schematically in Fig.
1.12. A positively charged filament on being heated by electric current
emits electrons which ionize the gas present in the vacuum chamber
near the filament. The positive ions are accelerated by attraction
through the negatively charged first slit and are bent into a circular
path by the magnetic field, which is perpendicular to the plane of the
diagram. Ions having different values of the charge-to-mass ratio follow
different paths, as shown by the curved lines in the figure. Separation
of ion paths occurs for the following reasons: For an accelerating
voltage V a particle with charge q is given an amount of energy equal
to qV. This energy is added to the particle as kinetic energy and so
equals $\frac{1}{2}mv^2$. From the equality

$$qV = \tfrac{1}{2}mv^2$$

the particle velocity v can be found to be

$$v = \sqrt{\frac{2qV}{m}} \ \text{ or } \ \left(\frac{2qV}{m}\right)^{1/2}$$

As shown on page 22, the charge-to-mass ratio of a charged particle
moving in a magnetic field H is related to the radius of curvature r
as follows:

$$\frac{q}{m} = \frac{v}{Hr}$$

Substituting $v = (2qV/m)^{1/2}$ in this equation, squaring both sides, and
canceling, we get

$$\frac{q}{m} = \frac{2V}{H^2 r^2}$$

This equation tells us that for fixed values of the voltage V and the
magnetic field H particles of larger charge-to-mass ratio follow paths

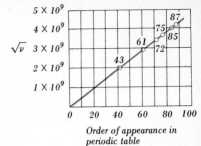

Fig. 1.11
Moseley plot of X-ray
frequencies for the elements.

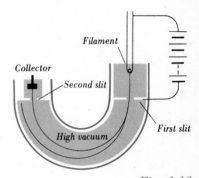

Fig. 1.12
Mass spectrometer (magnetic
field, not shown, is
perpendicular to the plane
of the diagram).

of smaller radius of curvature. In other words, for unit positive charge, light particles are bent more than heavy ones. Thus, particles of different charge-to-mass ratio can be separated in a mass spectrometer.

In practice, a mass spectrometer operates with a fixed collector to catch only ions of a particular radius of curvature. By varying either V or H, ions of a particular q/m can be focused into the collector. For a particular element it is found that as V is progressively increased, several beams are collected differing from each other in q/m value. For example, typical values for neon gas are 4.81×10^3, 9.62×10^3, and 4.33×10^3 coulombs/g. The second of these values, which is exactly twice the first, suggests that some of the neon atoms have lost two electrons instead of one; hence they have twice as large a charge-to-mass ratio. However, the third value, 4.33×10^3 coulombs/g, bears no simple relation to the other two and cannot be explained by assuming loss of an integral number of electrons. The fact that the charge-to-mass ratio differs by something other than a whole-number multiple indicates that particles of different mass must be present. In other words, the sample of the element neon contains nonidentical atoms, or isotopes.

For every element two or more isotopes are known, although all do not necessarily occur in nature. The relative abundances, which can vary widely from element to element, can be determined by measuring the relative intensity of the different isotope beams in the mass spectrometer. An illustration of isotope abundances is afforded by the element oxygen. Its natural mixture consists of three isotopes, which can be designated ^{16}O, ^{17}O, and ^{18}O. The superscripts indicate the *mass number*, which is the whole number closest to the isotopic mass. The masses of these isotopes are 15.9949 amu for ^{16}O, 16.9991 amu for ^{17}O, and 17.9991 amu for ^{18}O. The relative abundances are 99.759% ^{16}O, 0.037% ^{17}O, and 0.204% ^{18}O. The atomic weight used for calculations involving natural oxygen is a weighted average of the three, which turns out to be 15.9994 amu. Isotope abundances for the other elements are shown in Fig. 1.2.

Practically all the mass of an atom resides in the nucleus. However, there is not a one-to-one correspondence between increasing mass and increasing positive charge as determined by Moseley's X-ray method. For example, in the case of carbon in going from ^{12}C to ^{13}C the mass changes from 12 to 13.0034 amu, whereas the positive central charge stays constant at 6. Furthermore, in going from ^{12}C to ^{12}N the mass change is only from 12 to 12.0187 amu, whereas the central charge has changed from $+6$ to $+7$. The independence of nuclear charge from nuclear mass can be accounted for by assuming that the nucleus contains both protons and neutrons. Protons account for the positive charge, and protons and neutrons account for the mass. Counting the electron charge as -1, the proton has a charge of $+1$. The neutron, as the name implies, is electrically neutral.

The masses of protons, neutrons, and electrons are 1.00727, 1.00866, and 0.000549 amu, respectively. These masses apply strictly to isolated particles at rest and are not simply related to masses of atomic nuclei. In the formation of a nucleus from isolated protons and neutrons, some of the mass disappears as it is converted to energy. However, because the masses of the neutron and proton are so close to unity, the mass number (i.e., the whole number closest to the isotopic mass) gives directly the total number of neutrons plus protons in the nucleus. It follows, then, that isotopes of the same element have the same number of protons per nucleus but differ in the number of neutrons. Thus, for example, ^{12}C has six protons and six neutrons per nucleus, whereas ^{13}C has six protons and seven neutrons. Similarly, ^{12}N has seven protons and five neutrons, whereas ^{13}N has seven protons and six neutrons.

Frequently, nuclei are designated by symbols such as $^{18}_{8}O$. Here the subscript 8 indicates the atomic number Z, which is the nuclear charge, the number of protons in the nucleus, and the number of electrons in the neutral atom. The superscript 18 indicates the mass number A. A gives the total number of neutrons plus protons in the nucleus and is equal approximately to the atomic mass of the nucleus. $A - Z$ gives the number of neutrons, which in this case is 10.

1.7 The Bohr atom

Major credit for the first workable theory of atomic structure belongs to Niels Bohr, a Danish physicist who proposed what was then a revolutionary model in 1913. He was striving to explain three important sets of observations related to atomic behavior: periodic recurrence of properties when elements are ordered by atomic weight (or, more precisely, atomic number), systematic regularity of spectral-line frequencies, and the apparent contradiction between classical electrodynamics and its application to the problem of electrons in atoms. By this last point is meant the fact that electrically charged particles moving in curved paths are expected to radiate energy to the surroundings (e.g., the bluish glow emitted from a particle accelerator such as a cyclotron). Similarly, if an atom consists of negative electrons surrounding a positive nucleus and if these electrons are in motion so as to counteract the attraction into the nucleus, then the atom would be expected to radiate energy unceasingly. The loss of energy by such radiation would eventually result in the collapse of the atom. In fact, such a collapse is not observed, and an atom will not radiate unless it has previously been excited to a higher energy state.

In setting up Bohr's model of the atom, we go back to the fundamental principle of physics that force = mass × acceleration. The force is the force of electric attraction between a positively charged nucleus and a negatively charged electron. If a nucleus of positive charge Ze is distance r from an electron of negative charge e, then the force of electric interaction is $(Ze)(e)/r^2$. As seen on page 22, the acceleration experienced by an electron of mass m moving with velocity v along a path with radius of curvature r is v^2/r. Equating the appropriate terms goes as follows:

Force = mass × acceleration

$$\frac{Ze^2}{r^2} = m\frac{v^2}{r} \tag{1}$$

From this equation it would seem any value of r would be possible. However, Bohr then introduced the additional requirement, which was hard to accept at the time, that the angular momentum (see Appendix 5.3) of the electron can take on only certain permitted values. This requirement, referred to as the *quantum condition*, can be expressed as

$$mvr = \frac{nh}{2\pi} \tag{2}$$

Here mvr is the angular momentum, and n is a whole number called the *quantum number*, which can take on values 1, 2, 3, 4, 5, etc. h is the Planck constant. In words, Eq. (2) states that the angular momentum is restricted to taking on values that are whole multiples of $h/2\pi$. Each of these values would correspond to a permitted state of the atom. The corresponding permitted values of r can be obtained by combining Eqs. (1) and (2). To do this, Eq. (2) is solved for v, giving $v = nh/2\pi mr$, and the result is then substituted into Eq. (1):

$$\frac{Ze^2}{r^2} = \frac{m}{r}\left(\frac{n^2h^2}{4\pi^2m^2r^2}\right)$$

Canceling m from numerator and denominator and canceling r^2 from the denominator on each side gives on solving for r

$$r = n^2\left(\frac{h^2}{4\pi^2mZe^2}\right) = \frac{n^2}{Z}a_0 \tag{3}$$

Here all the constants have been lumped together and $h^2/4\pi^2me^2$ simply written as a_0. a_0, which is called the *Bohr radius*, has the numerical value 0.05292 nm. Accordingly, Eq. (3) states that the only permitted values for the radius of the electron path are those proportional to the square of a whole number n divided by the atomic number Z. For hydrogen, where $Z = 1$, allowed values of r are $1a_0$, $4a_0$, $9a_0$, $16a_0$, $25a_0$, ... In principle, nuclei with higher Z can also be treated by Eq. (3); however, since the derivation does not take into account repulsions between electrons in the same atom, Eq. (3) can strictly be used only for the combination of a nucleus and one electron.

Actually, the radius of an electron path in an atom, although interesting, is not an observable quantity. More accessible to experiment are the energy states of an atom, or rather differences between states. In the simple Bohr atom described above, the electron's energy in various states can be calculated. The total energy is the sum of kinetic and potential terms. The kinetic energy is $\frac{1}{2}mv^2$, the same as for any moving body; the potential energy is $-Ze^2/r$. (The reason for the minus sign is that when two opposite charges attract each other, the potential energy by convention becomes more negative as the charges come closer

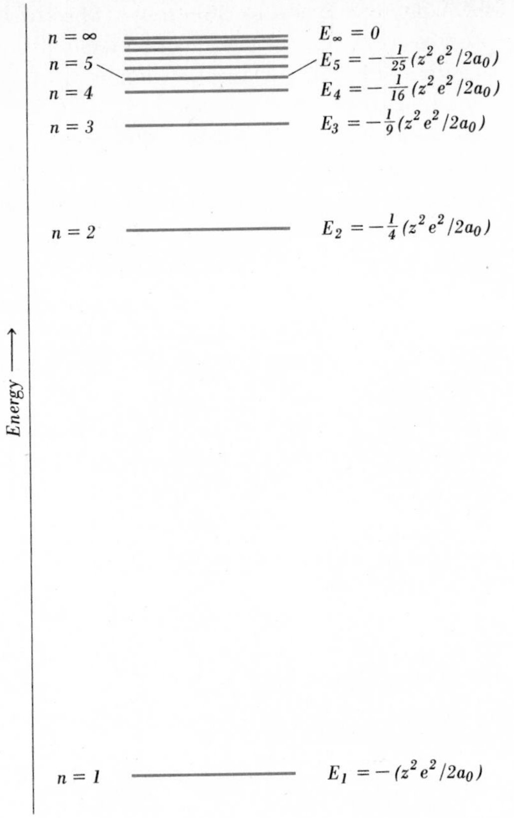

Fig. 1.13
Energy-level diagram
showing relative energies for
an atom consisting of a
nucleus with charge Z^+ plus
one electron with various
values of n.

$$n = \infty \quad\quad\quad\quad E_\infty = 0$$
$$n = 5 \quad\quad\quad\quad E_5 = -\tfrac{1}{25}(z^2 e^2/2a_0)$$
$$n = 4 \quad\quad\quad\quad E_4 = -\tfrac{1}{16}(z^2 e^2/2a_0)$$
$$n = 3 \quad\quad\quad\quad E_3 = -\tfrac{1}{9}(z^2 e^2/2a_0)$$

$$n = 2 \quad\quad\quad\quad E_2 = -\tfrac{1}{4}(z^2 e^2/2a_0)$$

Energy →

$$n = 1 \quad\quad\quad\quad E_1 = -(z^2 e^2/2a_0)$$

together.) Designating the total energy as E, we write

$$E = \text{kinetic energy} + \text{potential energy} = \frac{1}{2}mv^2 - \frac{Ze^2}{r} \tag{4}$$

From Eq. (1) we can see that mv^2 is Ze^2/r. Substituting $mv^2 = Ze^2/r$ into Eq. (4), we get

$$E = -\frac{Ze^2}{2r}$$

Finally, by putting in the expression for r from Eq. (3), we obtain

$$E = -\frac{Ze^2}{2n^2 a_0/Z} = -\frac{Z^2 e^2}{2n^2 a_0} \tag{5}$$

Equation (5), with different values 1, 2, 3, 4, etc., substituted for n, gives the values of the energy corresponding to the various permitted states of a one-electron atom. Figure 1.13 shows how these states are related to each other. The bottom line, marked $n = 1$, corresponds to the lowest permitted energy state. It is called the *ground state* and has an energy value of $E_1 = -Z^2 e^2/2a_0$. The next higher line, marked $n = 2$, corresponds to a less negative value of the energy $E_2 = -\frac{1}{4}Z^2 e^2/2a_0$. This represents an *excited state*; energy must be added to raise the atom from ground state E_1 to excited state E_2. As n

Section 1.7
The Bohr atom

33

subsequently increases, we move upward on the scale; E becomes progressively less negative and eventually approaches zero from the negative side. In the limit, when $n = \infty$, we reach the state corresponding to $E_\infty = 0$.

As can be seen from Eq. (3), the electron radius increases with increasing value of n. Electrons with $n = 1$ are closest to the nucleus; they are referred to as being in the K shell, in the K orbit, or in the innermost orbit. Electrons with $n = 2$ are next farther from the nucleus; they are referred to as being in the L shell, in the L orbit, or in a higher orbit. Higher orbits, or higher energy levels, are also referred to as outer energy levels. The letter designations K, L, M, N, . . . were originally introduced by workers in X-ray spectroscopy. Transitions of electrons from outer shells to the K shell give rise to the K series of X rays, which were the ones used by Moseley in working out his correlation of frequency with atomic number. As we shall see in the next section, the Bohr idea of orbits, or shells, for describing electron motions is no longer an acceptable one, but the terms "orbit" and "shell" are still often used to refer to energy levels.

Actually it is not possible to measure the energy of an individual atomic state. The lines observed in spectra correspond to transitions between such states—i.e., to energy differences between states. For calculating these differences we can make use of Eq. (5) to get the following:

$$E_b - E_a = -\frac{Z^2 e^2}{2 n_b^2 a_0} - \left(-\frac{Z^2 e^2}{2 n_a^2 a_0} \right)$$

$$E_b - E_a = \frac{Z^2 e^2}{2 a_0} \left(\frac{1}{n_a^2} - \frac{1}{n_b^2} \right) \tag{6}$$

Here E_b and E_a are the energies of two different states corresponding to quantum numbers n_b and n_a. Converting the energy difference to a frequency gives us

$$E_b - E_a = h\nu$$

where h is the Planck constant and ν is the frequency of the light wave emitted by the transition from state E_b to state E_a. Substituting for $E_b - E_a$ from Eq. (6) leads to

$$\nu = \frac{Z^2 e^2}{2 h a_0} \left(\frac{1}{n_a^2} - \frac{1}{n_b^2} \right)$$

If we choose $n_a = 1$, that is, select only those transitions that go down to the ground state, and substitute numerical values for the various constants, we finally get

$$\nu = 3.290 \times 10^{15} \left(\frac{1}{1} - \frac{1}{n_b^2} \right)$$

This is precisely the same as the expression given on page 26 for the observed series of ultraviolet lines in the hydrogen spectrum. In like manner, if E_a represents one of the higher states, it is possible to derive a similar equation for each of the other observed series of spectral lines. Thus, at least for the hydrogen atom the Bohr theory accurately describes observed atomic spectra. With some modification to approximate the effects due to electron-electron repulsion, the Bohr theory can be satisfactorily extended to a few other elements. However, as discussed in the next chapter, there are serious faults in the Bohr model, and they have led to its replacement by a more sophisticated theory of the atom.

1.8   Buildup of atoms on Bohr model

For a while, the Bohr model was extremely attractive; it not only explained away the puzzling regularities in line spectra, but it also made it possible to rationalize the periodic recurrence of chemical properties in the elements. The reasoning went as follows: Only specified energy levels (designated by $n = 1, 2, 3, 4, \ldots$) are permitted for electrons in atoms. Let us assume in addition that the maximum number of electrons allowed in any one energy level is also limited. Specifically, as Bohr discovered, let us assume that this maximum number is equal to $2n^2$. Then the lowest energy level ($n = 1$) will have a maximum population of $2(1)^2$; the second level, $2(2)^2$; the third level, $2(3)^2$; the fourth level, $2(4)^2$; and so on. If, as seems to be the case, chemical properties depend significantly on the number of electrons in the outermost energy level of an atom, then limitation of population to $2n^2$ automatically leads to periodic reappearance of properties. Each time a shell is completed, at 2, 8, 18, 32, . . . , respectively, addition of one more electron produces an atom with but one electron in its outermost occupied energy level.

Imagine the successive buildup of atoms (what in German is called *Aufbau*) by addition of electrons to a nucleus of the proper atomic number. Each electron enters the lowest energy level available. In the case of hydrogen ($Z = 1$) the lone electron goes into the K shell. The next electron, which would give an atom of helium ($Z = 2$), also enters the K shell. However, the third electron, which would give lithium ($Z = 3$), has to go into the L shell since the K shell is full when it has two electrons.

Figure 1.14 lists the first 18 elements in order of increasing atomic number and shows the number of electrons in the various energy levels. Since the K shell can accommodate only two electrons, it becomes completely populated in the chemically unreactive helium. Proceeding from helium the L-shell population increases from one to eight in the sequence lithium, beryllium, boron, carbon, nitrogen, oxygen, fluorine, and neon. In neon the situation is like that in helium. With two electrons in the K shell and eight electrons in the L shell, the shells which are occupied are completely filled, and the shells which are empty are completely empty. Neon is chemically unreactive. In other words, after a period, or a cycle of eight atoms, a repetition of the property of low reactivity appears.

In the next eight elements, electrons add to the third, or M, shell,

Fig. 1.14

Assignment of electrons to
various levels in atoms of
first eighteen elements.

Atomic no.	1	2	3	4	5	6	7	8
Element	H	He	Li	Be	B	C	N	O
Electron population								
K level	1	2	2	2	2	2	2	2
L level			1	2	3	4	5	6
M level								
		"inert"						

building it up gradually from one to eight electrons. The element argon, number 18, would not be expected to be unreactive because, according to the energy-level diagram, 10 more, or a total of 18, electrons can be put into the M shell. However, the observed fact is that argon is not very reactive. It must be that eight electrons in the third shell behave like a full shell. This point will be considered later.

That the properties of atoms are closely tied to the number of electrons in the outermost energy level can be seen further from the following examples. In the case of lithium there is one electron in the outermost level (the L shell); lithium is violently reactive. Sodium also has one electron in its outermost level (the M shell); it, too, is violently reactive. The properties of lithium and sodium are close to being identical. Similarly, beryllium ($Z = 4$) and magnesium ($Z = 12$) are similar to each other. Each has two electrons in its outermost level. In setting up the periodic table, elements with similar properties were placed under each other. This corresponds to grouping together the atoms which have the same number of electrons in their outermost levels. Thus, as can be seen on page 19, the element sodium, which has one electron in its outermost level, is placed under lithium in group I; magnesium is placed under beryllium in group II; aluminum under boron in group III; silicon under carbon in group IV; phosphorus under nitrogen in group V; sulfur under oxygen in group VI; chlorine under fluorine in group VII; and argon under neon in group 0. Because they have the same number of electrons in the outermost level, the atoms of each pair just mentioned have chemical similarity.

In the periodic table the first period contains but two elements (H and He). The second period contains eight elements (Li, Be, B, C, N, O, F, and Ne). The third period (Na, Mg, Al, Si, P, S, Cl, and Ar) also contains only 8 elements. How come there are only 8, whereas the simple energy-level picture suggests 18? The reason for this apparent discrepancy, as we shall see later, is associated with the fact that after eight electrons have been added to the third shell, the next two electrons go into the fourth shell, even though the third shell is not yet filled.

The fourth period, K through Kr, is more complicated. As can be seen from page 19, there are 18 elements in the fourth period, ranging from atomic number 19 to atomic number 36. Of these 18 elements,

9	10	11	12	13	14	15	16	17	18
F	Ne	Na	Mg	Al	Si	P	S	Cl	Ar
2	2	2	2	2	2	2	2	2	2
7	8	8	8	8	8	8	8	8	8
		1	2	3	4	5	6	7	8
	"inert"								*"inert"*

the first two (K and Ca) and the last six (Ga, Ge, As, Se, Br, and Kr) correspond to addition of electrons to the outermost (fourth) shell. The 10 intervening elements (Sc, Ti, V, Cr, Mn, Fe, Co, Ni, Cu, and Zn) involve belated filling of the next-to-outermost shell. There are two fourth-period elements, K ($Z = 19$) and Cu ($Z = 29$), both of which have one outermost electron. Similarly, Ca ($Z = 20$) and Zn ($Z = 30$) have two. K and Cu are similar, but only in some properties, because each has the same number of electrons in its outermost shell. However, they show differences in other properties, apparently because there are different numbers of electrons in the second-outermost shell. The second shell from the outside has an influence, sometimes quite large, on the chemical properties of an atom.

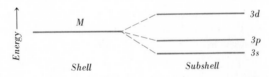

Fig. 1.15
Energy sublevels of the M shell.

Before going on, we need to clear up one question raised previously. In the third period we find 8 elements; from the simple energy-level diagram, we expect 18. Such an expectation rests on the assumption that all the electrons in a given shell (i.e., with the same value of n) are of the same energy. This is strictly true only for the hydrogen atom, where there are no electron-electron repulsions to worry about. However, in multielectron atoms electrons do have an effect on each other, and the result is to split some of the energy levels into sublevels, or subshells. Even in the case of hydrogen, the effect can be simulated by applying an external electric or magnetic field. If, for example, an electric field is applied to a specimen whose flame spectrum is being observed, what was initially a single spectral line may split into two or more lines, indicating the existence of energy sublevels arising from the same main level. This effect, called the *Stark effect*, is duplicated in certain atoms where the electrons themselves cause splitting into sublevels. Schematically, the separation of sublevels can be illustrated as in Fig. 1.15.

The number of subshells in any main shell turns out to be equal to the principal quantum number n. Thus, the K shell ($n = 1$) consists

Section 1.8
Buildup of atoms on
Bohr model

37

Fig. 1.16
*Energy-level diagram
showing component
subshells.*

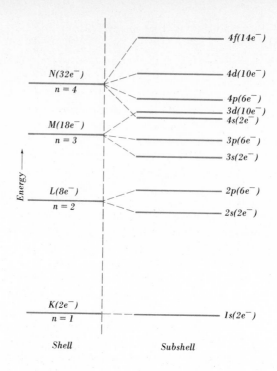

Shell   *Subshell*

of only one energy level. The L shell ($n = 2$) consists of two subshells, one group of electrons having an energy somewhat higher than that of the other group. In the M shell ($n = 3$) there are three possible energy levels; in the N shell, four energy levels; etc. The subshells are generally designated by a number followed by a letter, for example, $3s$. The number specifies the main shell to which the subshell belongs; the letter specifies the subshell. The lowest subshell is labeled s; the next higher p; the next higher d; the one above that f.* Superscripts are often used to indicate the electron population of a particular subshell. Thus, $3s^2$ means two electrons in the s subshell of the main shell having $n = 3$.

The energy-level diagram given on page 33 may now be redrawn. To the left of the dashed line in Fig. 1.16 are shown the main shells. To the right of the dashed line are shown component subshells. The relative spacing of the subshells is not the same for all elements but varies with Z. A distinctive feature of the diagram is the overlapping of the higher-energy subshells, an overlapping which gets more complicated as the fifth and sixth main shells are added to the picture.

Just as the number of electrons that can be put in any main shell is limited to $2n^2$, so the population of a subshell is similarly limited. An s subshell can hold 2 electrons; a p subshell, 6; a d subshell, 10;

---

* The letters s, p, d, and f were originally chosen on the basis of observations of line spectra. Certain lines were observed on being split to belong to a *sharp* series, and these were associated with energy transitions involving the s subshell; other spectrum lines belonged to what were called *principal, diffuse,* and *fundamental* series; hence the designations p, d, and f.

and an f subshell, 14. In Fig. 1.16 the numbers in parentheses indicate the maximum populations.

How does the existence of subshells affect the building of atoms from electrons and nuclei? As far as the first 18 elements are concerned, the number of electrons per main shell is as predicted before. Figure 1.17 shows, for example, that element 18, argon, has electron configuration 2, 8, 8, corresponding to two electrons in the $1s$ subshell, two in the $2s$, six in the $2p$, two in the $3s$, and six in the $3p$. We can also write this $1s^2 2s^2 2p^6 3s^2 3p^6$. The next subshell $3d$ is so much higher in energy (see Fig. 1.16) that argon behaves as an "inert" atom. There is stability associated with a grouping of eight electrons in any main shell everywhere in the periodic table.

Element	Atomic no.	Electron population									
		1s	2s	2p	3s	3p	3d	4s	4p	4d	4f
Argon	18	2	2	6	2	6					
Potassium	19	2	2	6	2	6	...	1			
Calcium	20	2	2	6	2	6	...	2			
Scandium	21	2	2	6	2	6	1	2			
Titanium	22	2	2	6	2	6	2	2			

Fig. 1.17

Electron configurations of atoms of some elements of the third period.

In the next element, potassium, number 19, the nineteenth electron goes into the $4s$ subshell, since $4s$ is lower in energy than $3d$ (Fig. 1.16), even though the third shell is not yet completely populated. In calcium, element 20, another electron is added to the $4s$ energy level. In element 21, scandium, the twenty-first electron goes into the next available state, the $3d$ level. With minor irregularities the buildup of the third subshell proceeds in this fashion in the next eight elements. The addition of electrons to the $3d$ subshell while the $4s$ subshell is occupied has the interesting effect on the chemistry of the elements from calcium to zinc that the chemical properties of these elements do not change as drastically as might be expected with increasing atomic number. In the sixth period there is an even better example of this. The elements 57 to 71, called the *lanthanides*, or *rare-earth elements*, are built up by the addition of electrons primarily to the third-outermost shell. Such changes deep within the atom do not affect chemical properties very much; all the lanthanides have nearly identical properties. Elements 89 to 103, called the *actinides*, also exhibit electron buildup in the third-outermost shell.

The electronic configurations of all the elements are shown in Fig. 1.18. These configurations apply to the atoms in their lowest energy states (in their ground states). The detailed assignment is based on observations of the spectra and of the magnetic properties of the individual elements. The question marks that occur in Fig. 1.18 denote cases where the assignment is in doubt. It should be noted, furthermore, that electronic configurations by themselves do not account for chemical properties of all the elements. Predictions made from these configurations are sometimes not borne out.

Section 1.8
Buildup of atoms on Bohr model

Fig. 1.18

Z	Element	1	2		3			4				5				6				7
		s	s	p	s	p	d	s	p	d	f	s	p	d	f	s	p	d	f	s
1	H	1																		
2	He	2																		
3	Li	2	1																	
4	Be	2	2																	
5	B	2	2	1																
6	C	2	2	2																
7	N	2	2	3																
8	O	2	2	4																
9	F	2	2	5																
10	Ne	2	2	6																
11	Na	2	2	6	1															
12	Mg	2	2	6	2															
13	Al	2	2	6	2	1														
14	Si	2	2	6	2	2														
15	P	2	2	6	2	3														
16	S	2	2	6	2	4														
17	Cl	2	2	6	2	5														
18	Ar	2	2	6	2	6														
19	K	2	2	6	2	6		1												
20	Ca	2	2	6	2	6		2												
21	Sc	2	2	6	2	6	1	2												
22	Ti	2	2	6	2	6	2	2												
23	V	2	2	6	2	6	3	2												
24	Cr	2	2	6	2	6	5	1												
25	Mn	2	2	6	2	6	5	2												
26	Fe	2	2	6	2	6	6	2												
27	Co	2	2	6	2	6	7	2												
28	Ni	2	2	6	2	6	8	2												
29	Cu	2	2	6	2	6	10	1												
30	Zn	2	2	6	2	6	10	2												
31	Ga	2	2	6	2	6	10	2	1											
32	Ge	2	2	6	2	6	10	2	2											
33	As	2	2	6	2	6	10	2	3											
34	Se	2	2	6	2	6	10	2	4											
35	Br	2	2	6	2	6	10	2	5											
36	Kr	2	2	6	2	6	10	2	6											
37	Rb	2	2	6	2	6	10	2	6			1								
38	Sr	2	2	6	2	6	10	2	6			2								
39	Y	2	2	6	2	6	10	2	6	1		2								
40	Zr	2	2	6	2	6	10	2	6	2		2								
41	Nb	2	2	6	2	6	10	2	6	4		1								
42	Mo	2	2	6	2	6	10	2	6	5		1								

Z	Element	1	2		3			4				5				6				7
		s	s	p	s	p	d	s	p	d	f	s	p	d	f	s	p	d	f	s
43	Tc	2	2	6	2	6	10	2	6	6		1?								
44	Ru	2	2	6	2	6	10	2	6	7		1								
45	Rh	2	2	6	2	6	10	2	6	8		1								
46	Pd	2	2	6	2	6	10	2	6	10										
47	Ag	2	2	6	2	6	10	2	6	10		1								
48	Cd	2	2	6	2	6	10	2	6	10		2								
49	In	2	2	6	2	6	10	2	6	10		2	1							
50	Sn	2	2	6	2	6	10	2	6	10		2	2							
51	Sb	2	2	6	2	6	10	2	6	10		2	3							
52	Te	2	2	6	2	6	10	2	6	10		2	4							
53	I	2	2	6	2	6	10	2	6	10		2	5							
54	Xe	2	2	6	2	6	10	2	6	10		2	6							
55	Cs	2	2	6	2	6	10	2	6	10		2	6			1				
56	Ba	2	2	6	2	6	10	2	6	10		2	6			2				
57	La	2	2	6	2	6	10	2	6	10		2	6	1		2				
58	Ce	2	2	6	2	6	10	2	6	10	2	2	6			2?				
59	Pr	2	2	6	2	6	10	2	6	10	3	2	6			2?				
60	Nd	2	2	6	2	6	10	2	6	10	4	2	6			2				
61	Pm	2	2	6	2	6	10	2	6	10	5	2	6			2?				
62	Sm	2	2	6	2	6	10	2	6	10	6	2	6			2				
63	Eu	2	2	6	2	6	10	2	6	10	7	2	6			2				
64	Gd	2	2	6	2	6	10	2	6	10	7	2	6	1		2				
65	Tb	2	2	6	2	6	10	2	6	10	9	2	6			2?				
66	Dy	2	2	6	2	6	10	2	6	10	10	2	6			2?				
67	Ho	2	2	6	2	6	10	2	6	10	11	2	6			2?				
68	Er	2	2	6	2	6	10	2	6	10	12	2	6			2?				
69	Tm	2	2	6	2	6	10	2	6	10	13	2	6			2				
70	Yb	2	2	6	2	6	10	2	6	10	14	2	6			2				
71	Lu	2	2	6	2	6	10	2	6	10	14	2	6	1		2				
72	Hf	2	2	6	2	6	10	2	6	10	14	2	6	2		2				
73	Ta	2	2	6	2	6	10	2	6	10	14	2	6	3		2				
74	W	2	2	6	2	6	10	2	6	10	14	2	6	4		2				
75	Re	2	2	6	2	6	10	2	6	10	14	2	6	5		2				
76	Os	2	2	6	2	6	10	2	6	10	14	2	6	6		2				
77	Ir	2	2	6	2	6	10	2	6	10	14	2	6	7		2				
78	Pt	2	2	6	2	6	10	2	6	10	14	2	6	9		1				
79	Au	2	2	6	2	6	10	2	6	10	14	2	6	10		1				
80	Hg	2	2	6	2	6	10	2	6	10	14	2	6	10		2				
81	Tl	2	2	6	2	6	10	2	6	10	14	2	6	10		2	1			
82	Pb	2	2	6	2	6	10	2	6	10	14	2	6	10		2	2			
83	Bi	2	2	6	2	6	10	2	6	10	14	2	6	10		2	3			
84	Po	2	2	6	2	6	10	2	6	10	14	2	6	10		2	4?			
85	At	2	2	6	2	6	10	2	6	10	14	2	6	10		2	5?			
86	Rn	2	2	6	2	6	10	2	6	10	14	2	6	10		2	6			

Z	Element	1	2		3			4				5				6				7
		s	s	p	s	p	d	s	p	d	f	s	p	d	f	s	p	d	f	s
87	Fr	2	2	6	2	6	10	2	6	10	14	2	6	10		2	6			1?
88	Ra	2	2	6	2	6	10	2	6	10	14	2	6	10		2	6			2
89	Ac	2	2	6	2	6	10	2	6	10	14	2	6	10		2	6	1		2?
90	Th	2	2	6	2	6	10	2	6	10	14	2	6	10		2	6	2		2
91	Pa	2	2	6	2	6	10	2	6	10	14	2	6	10	2	2	6	1		2?
92	U	2	2	6	2	6	10	2	6	10	14	2	6	10	3	2	6	1		2
93	Np	2	2	6	2	6	10	2	6	10	14	2	6	10	4	2	6	1		2?
94	Pu	2	2	6	2	6	10	2	6	10	14	2	6	10	6	2	6			2?
95	Am	2	2	6	2	6	10	2	6	10	14	2	6	10	7	2	6			2?
96	Cm	2	2	6	2	6	10	2	6	10	14	2	6	10	7	2	6	1		2?
97	Bk	2	2	6	2	6	10	2	6	10	14	2	6	10	8	2	6	1		2?
98	Cf	2	2	6	2	6	10	2	6	10	14	2	6	10	10	2	6			2?
99	Es	2	2	6	2	6	10	2	6	10	14	2	6	10	11	2	6			2?
100	Fm	2	2	6	2	6	10	2	6	10	14	2	6	10	12	2	6			2?
101	Md	2	2	6	2	6	10	2	6	10	14	2	6	10	13	2	6			2?
102	No	2	2	6	2	6	10	2	6	10	14	2	6	10	14	2	6			2?
103	Lr	2	2	6	2	6	10	2	6	10	14	2	6	10	14	2	6	1		2?
104	Ku	2	2	6	2	6	10	2	6	10	14	2	6	10	14	2	6	2		2?
105	Ha	2	2	6	2	6	10	2	6	10	14	2	6	10	14	2	6	3		2?

Fig. 1.18

Electron configurations of atoms of all the elements.

Exercises

Here and at the end of succeeding chapters are given questions and problems designed to test and amplify your mastery of the foregoing material. Each list is ordered in terms of increasing difficulty: *easy, **moderate, ***hard. Answers are provided for some but not all of the problems. Special care should be given to understanding each step toward the final numerical answer. Needed constants and conversion units are given in the chapter or in the appendix section at the end of the book.

*1.1 **Atomic structure.** What is the main aspect of atomic structure that justifies considering molecules as being composed of atoms?

*1.2 **Atomic weights.** Given that John Dalton found the percent analysis of water to be 12.5% hydrogen by weight, what should he have assigned to the atomic weight of oxygen if he assumed water contained two hydrogen atoms per every three oxygen atoms?

Ans. 4.67 amu

*1.3 **Heat capacity.** What value would you predict for the heat capacity of uranium?

*1.4 **Atomic weights.** Rubidium, which consists 72.15% of the 85 isotope and 27.85% of the 87 isotope, has an atomic weight of 85.4678 amu. If the mass of ^{85}Rb is 84.9117 amu, what must be the mass of ^{87}Rb?

Ans. 86.91 amu

*1.5   Isotope abundance. If an element X having an atomic weight of 210.197 amu consists of only the two isotopes ^{210}X and ^{212}X, having masses of 209.64 and 211.66 amu, respectively, what must be the relative abundance of the two isotopes?

*1.6   Isotopes. The isotope abundances given in Fig. 1.2 indicate that some of the elements consist of but a single isotope. What systematic regularity can you discover among such elements?

*1.7   Atom masses. Females of many insect species secrete chemical compounds called *pheromones* which attract the males for mating. One such compound has the formula $C_{19}H_{38}O$. Field tests indicate it takes but 10^{-12} g of this material to be effective. How many molecules does this correspond to?   *Ans. 2 × 10⁹*

*1.8   Atom ratios. How many grams of calcium must you combine with 1.00 g of phosphorus to make a compound in which the atom ratio is three atoms of calcium per two atoms of phosphorus?

*1.9   Periodic table. Indicate for each of the following the element that satisfies the given criteria:

a   Belongs to group I and second period
b   Belongs to manganese subgroup and is in the second period of transition elements
c   Is a halogen of the third period
d   Is an alkali element in the same period as bromine
e   Is a noble gas in the same period as sodium

*1.10   Oil-drop experiment. In a particular Millikan type experiment, the following charges were observed on the oil droplets: 1.44×10^{-18}, 2.56×10^{-18}, 4.80×10^{-19}, and 9.60×10^{-19} coulomb. What should you logically conclude for the charge of the electron? Justify.

*1.11   Frequency. What would be the frequency of a light wave that has the same wavelength as the radius of the first Bohr orbit?   *Ans. 5.66 × 10¹⁸*

*1.12   Light. It has been suggested that the time required for light to travel a distance of one centimeter be referred to as a *jiffy*. How many seconds is one jiffy?

*1.13   Light. If an Avogadro number of atoms each simultaneously emits a light wave of 400 nm wavelength, how much energy is emitted?   *Ans. 299 kJ*

*1.14   Spectroscopy. Calculate the wavelengths for the first three hydrogen spectral lines of the Paschen series. What will be the wavelength of the series limit?

*1.15   Atomic weight. If the mass of the electron is 9.109×10^{-28} g, what is its atomic weight in atomic mass units?

*1.16   Symbols. Indicate the number of neutrons, protons, and electrons in each of the following atoms: $^{7}_{3}$Li, $^{11}_{5}$B, $^{23}_{11}$Na, $^{31}_{15}$P, and $^{56}_{26}$Fe.

*1.17 *Bohr atom.* Draw a scale model showing how the radii of the first five Bohr orbits might compare for hydrogen and for oxygen.

*1.18 *Bohr atom.* According to the simple Bohr theory, at what value of Z would the innermost orbit be expected to be pulled inside a nucleus of radius 1×10^{-13} cm?   *Ans. 50,000*

*1.19 *Electron configurations.* Which elements of the periodic table have the following outermost electron configurations?
  a ns^2np^2
  b ns^2np^4
  c $(n-1)d^1ns^2$
  d $(n-1)d^5ns^1$

*1.20 *Electron configurations.* Which atoms have as many s electrons as p electrons?

*1.21 *Electron configuration.* What must be the values of x, y, and z for an atom having the electron configuration $1s^22s^22p^x3s^23p^6 3d^y4s^24p^65s^z5p^2$ in its ground state?

*1.22 *Atomic structure.* Give the nuclear makeup and the ground-state electron configuration for each of the following: ^7_3Li, $^{12}_6\text{C}$, $^{19}_9\text{F}$, and $^{40}_{20}\text{Ca}$.

**1.23 *Mass and weight.* What is the difference between mass and weight? How can one justify the fact that atomic *weights* are generally given in atomic mass units?

**1.24 *Atomic mass unit.* What is the mass in grams of one amu?

**1.25 *Atomic weights.* The element phosphorus, atomic weight 30.97 amu, combines with another element so that 1.0000 g of phosphorus needs 0.7764 g of the other element. If the atomic ratio in the compound is four atoms of phosphorus per three atoms of unknown element, what is the atomic weight of the unknown element?

**1.26 *Atomic-weight determination.* An unknown metal X, which has a heat capacity of 0.128 J/g, reacts with oxygen so that 1.3625 g of X forms 1.4158 g of oxide. What is the atomic weight of X?

Ans. 204

**1.27 *Mole.* How many moles of carbon atoms are there in 1.000 g of each of the following?
  a   Carbon-12 isotope
  b   Carbon-13 isotope
  c   Natural mixture of carbon-12 and carbon-13 isotopes

**1.28 *Avogadro number.* In principle, the Avogadro number could be determined experimentally by collecting a beam of alpha particles (doubly charged helium atoms). Suppose in such an experiment you observed that a beam equivalent to 6.40×10^{-5} coulomb/sec gave after 24 h of collection 1.14×10^{-4} g of neutral helium. What should you deduce for the value of the Avogadro number?

Ans. 6.06×10^{23}

**1.29   Cathode-ray tube. Given a cathode-ray tube in which deflection is being produced by a magnetic field alone, all other things being equal, what effect would there be on the observed deflection of increasing each of the following?

   a The mass of the particles
   b The velocity of the particles
   c The magnetic field
   d The charge on the particles

**1.30   Cathode-ray tube. Given a cathode-ray tube in which deflection is being produced by an electric field alone, all other things being equal, what effect would there be on the observed deflection of increasing each of the following?

   a The voltage on the plates
   b The distance between the plates
   c The velocity of the particles
   d The mass of the particles

**1.31   X rays. Predict the wavelength of the shortest X ray emitted by a chromium target, given that copper emits 0.154 nm and molybdenum 0.0709 nm.

**1.32   Mass spectrometer. Given the data of Fig. 1.2, what charge-to-mass ratios might one reasonably expect to see if a beam of chlorine atoms were analyzed in a mass spectrometer?

**1.33   Nuclei. Combining the data of Fig. 1.2 with the fact that the mass of the proton, neutron, and electron is 1.00728, 1.00867, and 0.000549 amu, respectively, show how the amount of missing mass (calculated per nuclear-contained particle) changes in the sequence $^{19}_{9}F$, $^{31}_{15}P$, and $^{59}_{27}Co$.   *Ans. 0.00836, 0.00911, 0.00942 amu per particle*

**1.34   Energy levels. Draw a scale model showing how the energy levels for neutral hydrogen compare with those for singly charged helium on a simple Bohr model. Refer the two sets of levels to the same zero, corresponding to infinite removal of the electron.

**1.35   Transitions. Compare the energy of the $n = 5$ to $n = 4$ transition for an electron plus nucleus with $Z = 3$ with the energy of the $n = 2$ to $n = 1$ transition for an electron plus nucleus with $Z = 2$.   *Ans. 6.8% as large*

**1.36   Energy units. If in Eq. (6) you want to be able to use coulombs for e and centimeters for a_0, what constant would you have to put before the parenthesis so that the energy difference would come out to be in joules?

**1.37   Spectroscopy. What value of Z would you have to pick for a one-electron atom so that excitation from the ground state to infinite ionization would correspond to a wavelength of 0.1 nm?   *Ans. 30*

**1.38   Buildup of atoms. Suppose that the maximum number of electrons permitted in a Bohr shell was set at n^3 instead of at $2n^2$. Which elements would then be equivalent to the noble gases?

***1.39   *Nucleus.* Suppose that in a Rutherford type of scattering experiment, an alpha particle must come within 10^{-12} cm of a nucleus to suffer appreciable deflection. What would be the rough probability that an alpha particle would be deflected in going through a foil 0.01 mm thick. Assume the atoms have a radius of 10^{-8} cm.

Ans. One chance in two or three thousand

***1.40   *Particle deflection.* In a given experiment such as is diagramed in Fig. 1.6, an electron beam enters the magnetic field 30 cm away from the detecting screen and travels 25 cm after leaving the magnetic field. If the beam deflection is 4.0 mm on the face of the cathode-ray tube and the magnetic field is 0.18 g coulomb^{-1} sec^{-1}, what is the electron velocity?   *Ans. 1.2×10^{10} cm/sec*

One of the great weaknesses of the Bohr theory of the atom (Bohr himself recognized and repeatedly called attention to it) is that it offered no real explanation of why the electrons were restricted to only certain orbits. Bohr *assumed* that the angular momentum was limited to certain values; he did not pretend to know why. He simply noted that if he made that assumption, then the mathematics worked out to give good agreement between predictions and experimental observations. In this chapter, we consider further developments in the aspects needed to describe the modern atom.

47

2.1 Wave nature of the electron

In 1924, the French physicist Louis de Broglie suggested a possible explanation for the quantized nature of electrons in atoms. He proposed that every moving particle has associated with it a wave nature like that of light. The wavelength of the de Broglie wave is given by

$$\lambda = \frac{h}{mv}$$

where h is the Planck constant and mv is the momentum of the particle. For describing the motion of massive particles the wave character is of little practical consequence since the associated wavelength is very small relative to particle dimensions. However, for describing the motion of low-mass particles, such as electrons, the wavelength is of a magnitude comparable to the dimensions of the atom in which the electron finds itself.

Although it represents a highly oversimplified argument, the following line of reasoning suggests why the existence of de Broglie waves leads to the quantum condition. Imagine an electron moving in a Bohr orbit. It has associated with it a wave of wavelength λ given by the de Broglie condition. If this wavelength did not divide into the path length a whole number of times, then the wave would destructively interfere with itself. A stable orbit would correspond only to one in which there is an integral multiple of wavelengths along the circumference of the orbit. Mathematically this corresponds to saying that

$$n\lambda = 2\pi r$$

where n is a whole number and r is the radius of the orbit. Substituting $\lambda = h/mv$, we get

$$n\left(\frac{h}{mv}\right) = 2\pi r$$

and rearranging we obtain

$$mvr = n\left(\frac{h}{2\pi}\right)$$

This is the same as Eq. (2) on page 32 for the quantum condition assumed by Bohr!

The concept of a de Broglie wave for a particle such as an electron means that the particle cannot be precisely localized. Instead, the electron must be thought of in the somewhat tenuous manner we use for imagining waves. The problem, a general one for particles of low mass, has been treated mathematically by Heisenberg. In a famous theorem, called the *uncertainty principle*, Heisenberg postulated (in 1927) that there is an inherent indeterminancy in knowing the combination of a particle's position and momentum. This indeterminacy can be expressed by saying that the uncertainty in the position times the uncertainty in the momentum is of the order of the Planck constant h.

What this means is that the more precisely we try to specify one of the variables, momentum or position, the more uncertain we are of the other. In the limit, if we precisely specify the position, we do not know the momentum; if we know the momentum; we do not know where the particle is. For large masses of large dimensions the uncertainty of position or momentum is a trivial fraction of the total, but for small masses of atomic dimension the uncertainty principle places a real restriction upon the extent of permitted knowledge.

To illustrate, suppose we allow an electron in an atom to have an uncertainty in momentum of the same order of magnitude as its momentum. It turns out then that the uncertainty of its position would be about equal to 0.1 nm, or about the size of an atom. To locate the electron more precisely than this would be meaningless; to locate it within the order of nuclear dimensions, 10^{-13} cm, is completely out of the question. Stated differently, any experiment, no matter how perfectly designed, to measure the location or the momentum of the electron must by the measuring process itself change either the momentum or the location by an amount comparable to the value sought.

Since tracks cannot be drawn for electrons in atoms, the best we can do is to speak of the probability or relative chance of finding an electron at a given location within the atom. The calculation of such a probability is quite involved mathematically. It uses equations which describe the motion of waves and applies them to the de Broglie wave associated with an electron. This procedure is a basic concern of the field called *wave mechanics.*

The probability of finding an electron can be specified by a probability distribution. There are three ways to represent the probability distribution. These are shown in Fig. 2.1 for the case of a 1s electron. At the top of the figure, the probability of finding the electron at a given location in space is plotted as a function of the distance of that location from the nucleus. The position of greatest probability is at the nucleus. As the distance from the nucleus increases, the probability of finding the electron at any location decreases. Nowhere, however, is the probability equal to zero. Even at points very far from the nucleus there is some chance, although it is small, of finding the electron. Figure 2.1b is another way of representing the same electronic distribution. Here the decreasing relative probability of finding the electron is represented by a decrease in the intensity of shading. Consistent with this picture, one visualizes the electron as forming a rather fuzzy charge cloud about a central nucleus. Sometimes it is convenient simply to indicate the shape of the charge cloud. This can be done by a boundary contour as in Fig. 2.1c. Remembering that atoms are three-dimensional, we should think of Fig. 2.1c as a sphere within which the chance of finding the electron is great.

The distributions shown in Fig. 2.1 are three different ways of representing the spatial distribution of an electron in a 1s energy level. Since these representations replace the Bohr idea of a simple orbit, they can properly be said to represent 1s orbits. To reduce any possible

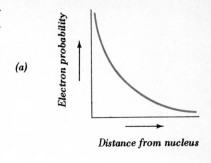

(a)

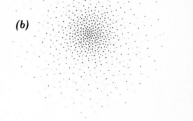

(b)

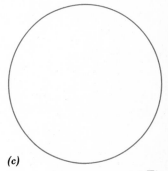

(c)

Fig. 2.1

Various ways of representing the spatial distribution of a 1s electron.

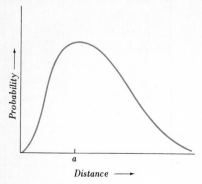

Fig. 2.2

Probability of finding a 1s electron in a spherical shell at distance r from the nucleus.

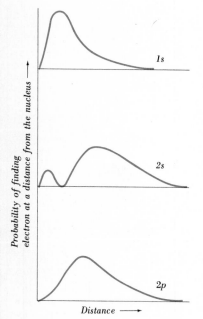

Fig. 2.3

Probability plots for various electron orbitals.

confusion between the old and new ideas, it has become customary to use the term *orbital* when referring to a given electronic probability distribution in space.

There is still another way of describing an electron in a 1s orbital, a way that serves to relate the old idea of electronic shells to probability concepts. First, we ask this question: Starting from the nucleus and working along a straight line from the nucleus to the outside of the atom, how does the chance of finding the 1s electron change? Obviously, the chance decreases, which is consistent with Fig. 2.1a. But suppose, as we move out of the atom, at each radial distance r from the nucleus we investigate all the possible locations in three-dimensional space at that distance r from the nucleus and determine the chance of finding the 1s electron. Then we move farther from the nucleus and investigate all the locations at a slightly bigger r. How does the chance of finding the 1s electron change? The answer is not immediately obvious, since we have to consider both of the following factors. The chance of finding the electron at a given location decreases as we move away from the nucleus, but the number of locations to be investigated increases as we move away from the nucleus. Mathematically, this is equivalent to considering the atom to be divided into concentric layers and multiplying the probability per unit volume in a given layer by the volume of that layer. The result for a 1s electron is the probability curve shown in Fig. 2.2. As can be seen, the probability of finding the electron goes up to a maximum and then decreases. The distance *a*, at which the probability reaches a maximum, can be thought of as corresponding to the radius of an electron shell.

Electrons that are in different energy levels differ from each other in having different probability distributions. For example, Fig. 2.3 shows the probability that a 1s, a 2s, and a 2p electron are at various distances from a given nucleus. It should be noted that the distances of maximum probability for the 2s and 2p electrons are roughly the same; they are larger than the distance of maximum probability for the 1s electron. This is consistent with the fact that the 2s and 2p electrons are of about the same energy and that this energy is greater than that of the 1s electron. The extra little bump at small distances in the 2s distribution indicates that the 2s electron spends on the average more of its time close to the nucleus than does the 2p electron. This can account for the fact that the 2s electron is bound more tightly to the nucleus (is of lower energy) than the 2p electron. Furthermore, it should be noted that all three of the distributions shown in Fig. 2.3 overlap, implying that electrons can "occupy" the region already "occupied" by other electrons. Specifically, electrons that are in outer orbitals can penetrate into the region occupied by electrons that are in inner orbitals.

Actually, there is another essential difference between s and p electrons which is not evident from Fig. 2.3. The spatial distribution of an s electron is spherically symmetric; i.e., its probability of being found is identical in all directions from the nucleus. On the other hand, p

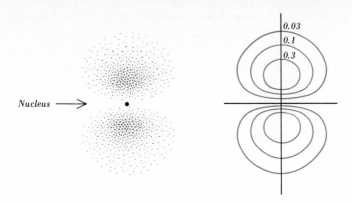

electrons are more probably found in some directions than in others. In fact, the probability distribution of a p electron can be thought of as forming two diffuse blobs, one on each side of the nucleus, as shown in Fig. 2.4. This is called a p orbital, and the electron in a p orbital has equal probability of being found in either half of it. A p subshell consists of three such orbitals all perpendicular to each other. The one symmetric along the x axis is called the p_x orbital; the one along the y axis, the p_y orbital; the one along the z axis, the p_z orbital. The respective orientations are shown in Fig. 2.5. Although it is not obvious, the combined distribution of three electrons (one each in p_x, p_y, p_z) or of six electrons (two each in p_x, p_y, and p_z) is a sphere.

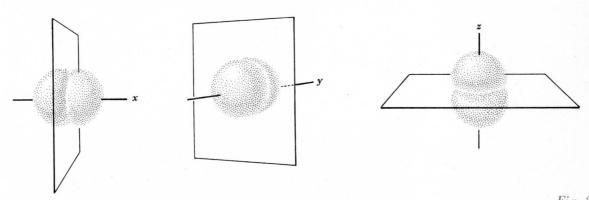

The d subshell, which can accommodate 10 electrons, can be resolved into five orbitals. These are somewhat more complicated than p or s orbitals. The 3d set is represented in Fig. 2.6. One of the 3d orbitals, the one labeled d_{z^2}, is symmetric about the z axis and can be visualized as a squashed doughnut around the pinched waist of an hourglass. The other four 3d orbitals look like inflated four-leaf clovers. The one designated $d_{x^2-y^2}$ has its maximum electron probability density along the x and y axes. The other three, d_{xy}, d_{yz}, and d_{zx}, have their probability maximums lying along lines that make $45°$ angles with the axes. As with the set of p orbitals, it turns out that having

Fig. 2.5
Relative orientations of p_x, p_y, and p_z orbitals.

Section 2.2
Wave functions

51

one electron in each (or two electrons in each) of the five *d* orbitals gives a spherical electronic distribution.

The *f* subshell consists of seven orbitals. The distributions are even more complex than those of the *d* orbitals. The geometry of the 4*f* orbitals is such that equal population of all seven orbitals again adds up to a sphere.

2.2   Wave functions

In the preceding section, electron probability distributions have been described in a qualitative way by using visual analogies as much as possible. The visualizations, although helpful in fitting observed facts together, should not be taken too literally. The Heisenberg uncertainty principle puts definite limitations on how much information can be specified about an atomic system; our pictures and diagrams frequently suggest more definite information than we are entitled to possess. For example, Fig. 2.6 suggests blobs of electron density where we can clearly distinguish between high-probability regions and low-probability

Fig. 2.6
Relative orientations in space of the five 3d orbitals.

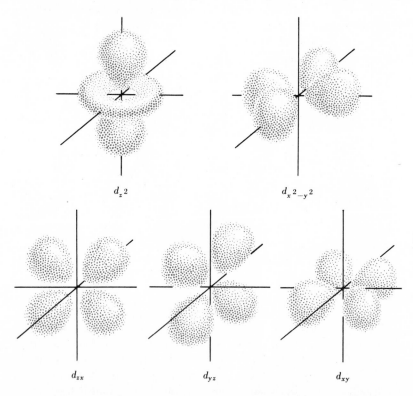

d_{z^2} $d_{x^2-y^2}$

d_{zx} d_{yz} d_{xy}

regions. In truth, the only thing we can be sure of is the location of the zero-probability features and the spatial symmetry of the electron distribution.

The most powerful way to give the requisite descriptions would be to make full use of mathematical representations. Unfortunately, the representations often get lost in the abstractions, and some chemists give up on them too soon (1) because they appear to be too difficult or (2) because they seem to be too far removed from reality. However, a disciplined effort to master the mathematical representations of electron distributions can prove very rewarding. The result can be deeper insight into nature than would otherwise be obtained and, as a minimum, acquisition of a very efficient way of representing quantitative information.

de Broglie's idea that a particle such as an electron behaves like a wave immediately suggested that the whole complex of mathematical formalism that had been developed to describe wave motion could be taken over and applied to the case of the electron. There are two important aspects to describing wave motion: How steeply does the height of the wave (amplitude) build up? How fast do the crests and troughs propagate through space? Because of the requirements that energy and mass be conserved, there is obviously a connection between the sizes of crests and troughs and how rapidly they can change in space and time. Stated another way, the buildup of a crest comes at the expense of a trough; so their repetitive annihilation and creation must be interconnected.

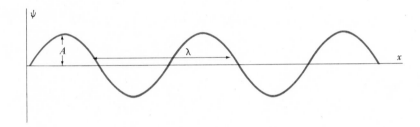

Fig. 2.7
Standing wave.

To get a better feeling for how waves are described, let us consider the wave shown in Fig. 2.7. The wiggly line that traces out the wave can be described mathematically by the equation

$$\psi = A \sin bx \tag{1}$$

ψ, which is a Greek letter that is correctly pronounced "psē" but is almost invariably mispronounced "si" or "psi," is called the *wave function;* it measures for each value of the distance x along the horizontal axis the corresponding excursion of the wavy line away from the horizontal axis. The letter A represents the amplitude of the wave and is the maximum value of ψ anywhere along the wave. The symbol "sin" stands for the sine function of trigonometry; b is a constant characteristic of the wave and is given by $2\pi/\lambda$, where λ is the wavelength of the wave. That $b = 2\pi/\lambda$ can be seen by recalling the periodic nature of trigonometric functions and noting that $\psi = A$ when $x = \lambda/4$, $\psi = 0$ when $x = \lambda/2$, and $\psi = -A$ when $x = 3\lambda/4$. (Recall that $\sin \theta$ oscillates in value 0, 1, 0, -1, 0 as $\theta = 0$, $\pi/2$,

Fig. 2.8

Running wave. (Position of arrow shows displacement of crest as time goes on.)

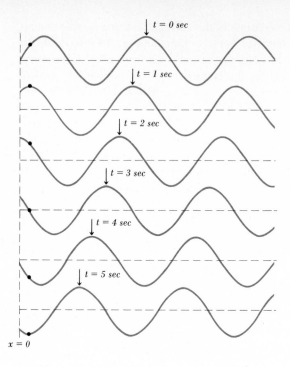

$t = 0$ sec

$t = 1$ sec

$t = 2$ sec

$t = 3$ sec

$t = 4$ sec

$t = 5$ sec

$x = 0$

π, $3\pi/4$, 2π, respectively.)* As shown in Fig. 2.7, the wave is a *stationary*, or *standing*, wave; it does not change with time.

Let us imagine, now, that the wave shown in Fig. 2.7 moves progressively to the left with time. We then have a moving, or *running*, wave, the progressive snapshots of which look like the sequence shown in Fig. 2.8. As noted by the marker arrow, the crests steadily advance to the left, even though the distance between successive crests (i.e., the wavelength) stays the same. Another feature to note is that the value of the wave function at any fixed value of x now oscillates with time. In Fig. 2.8, we have illustrated this by indicating with a large black dot what happens to the ψ value at a particular x. As time goes on, the point bobs up and down very much like a cork that is agitated by passing water waves in a pond. How can we express the mathematical description of the running wave? Each of the snapshots in Fig. 2.8 is like the picture of Fig. 2.7 except for the successive displacement of the waveform. To allow for the change with time, we need to modify Eq. (1) by adding a time-dependent term. Specifically, we can write

$$\psi = A \sin (bx + kt) \tag{2}$$

*The variable x for trigonometric functions such as $\sin x$ can be expressed either in degrees or in radians. In the wave functions given here, the angles are expressed in radians. One *radian* is the angle subtended at the origin by one radial length r along the perimeter of a circle $2\pi r$. Since there are 2π radians in a full circle of $360°$, one radian must be equal to $57.30°$. The following radian equivalents are useful to remember: $\pi/6 = 30°$; $\pi/4 = 45°$; $\pi/3 = 60°$; $\pi/2 = 90°$; $\pi = 180°$; and $2\pi = 360°$. After each cycle of 2π radians, trigonometric functions come back to their original value.

which is the equation for a running wave. Just as the constant b is related to the periodicity in space, the constant k is related to the periodicity in time. k, in fact, is related to the frequency of the wave, as can be seen from the following: Let us consider what happens at $x = 0$ as time goes on. At $t = 0$, $\psi = 0$. ψ builds up to its maximum value A when $kt = \pi/2$, then dies off again to zero at the time $kt = \pi$. Subsequently, as time progresses, ψ becomes negative, reaching a value of $-A$ at the time $kt = 3\pi/2$, and returning to zero at $kt = 2\pi$. If we represent with the Greek letter "tau" (τ) the length of time it takes ψ to go through the full cycle $\psi = 0$, $+A$, 0, $-A$, 0, then we have $k\tau = 2\pi$, or $k = 2\pi/\tau$. τ, which is called the *period* of the wave, measures the time needed for the wave to travel a distance λ and is just equal to the reciprocal of the wave frequency; that is, $\tau = 1/\nu$. The higher the frequency of the wave, the shorter the time it takes to go through a full cycle. Substituting $b = 2\pi/\lambda$ and $k = 2\pi/\tau = 2\pi\nu$ in Eq. (2) gives us the following expression for describing a running wave:

$$\psi = A \sin 2\pi \left(\frac{x}{\lambda} + \nu t \right) \tag{3}$$

When this equation is used to describe sound waves, the wave function ψ tells us about the pressure and how it alternates between compression and rarefaction; for electromagnetic waves, such as light waves or X rays, ψ tells us about the strength of the electric field or of the magnetic field and how it pulses in space and time. Figure 2.9 shows some

Kind of wave	Typical wavelength, cm	Typical frequency, cycles/sec
X rays	1×10^{-8}	3×10^{18}
Ultraviolet light	2000×10^{-8}	1.5×10^{15}
Visible light	5000×10^{-8}	0.6×10^{15}
Infrared	1×10^{-3}	3×10^{13}
Radar	1	3×10^{10}
Radio	3×10^{5}	1×10^{5}

of the characteristic constants that go into describing various waves of the electromagnetic spectrum.

Equation (3) can also be used to describe the motion of particles. All we need do is to put in the de Broglie relation $\lambda = h/mv$, relating the wavelength associated with a particle to its momentum mv. There is now a problem, however, in what we mean by the wave function ψ when we use Eq. (3) to describe particle waves. Actually, there is then no physical significance to ψ itself, but it turns out that ψ^2 can be related to the probability of locating the particle. As the associated wave moves through space and time, the square of the wave function tells us how the probability of finding the particle is transmitted from place to place as time goes on. In free space, there is no special problem in using Eq. (3), and, in fact, it can be used directly to describe, for example, how electrons are transferred through an infinitely large piece of metal. Complications arise when attractive forces come into the picture. In real metals, electrons that come to the surface of the specimen get attracted back into the metal; so a running electron wave that was coming out of the metal gets reflected on itself and is converted from a running wave into a standing wave.

The problem of describing an electron in a metal is illustrative of a famous kind of problem called the *particle-in-a-box* problem. Because it is related to the problem of describing an electron in an atom (which, after all, is a kind of box), it is instructive to consider more deeply the problem of a particle in a box. In its simplest form, the problem is posed as follows: Suppose we have a one-dimensional box, of length L, as suggested in Fig. 2.10. We assume we have trapped in this box a particle, such as an electron. To make sure the particle stays in the box, we simply say that the potential energy of the particle sharply rises to infinity to the left of $x = 0$ and to the right of $x = L$. Between $x = 0$ and $x = L$, the potential energy is assumed constant, say at a value of zero. What can we say about the kinetic energy of the particle in the box?

Fig. 2.10
Particle in a box.

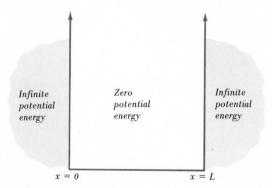

To describe the particle in the box, we use the standing-wave representation given by Eq. (1). (It may be that when we first put the particle in the box, a running-wave description might be more appropriate, but after the particle rattles around a bit, it should reach some steady state which is independent of time.) So, the wave function for the particle in the box can be written

$$\psi = A \sin 2\pi \left(\frac{x}{\lambda} \right) \tag{4}$$

where ψ as a function of x tells us about ψ^2, the probability of finding the particle. At the two edges of the box, at $x = 0$ and at $x = L$, the probability of finding the particle must vanish. In other words, ψ goes to zero at $x = 0$ and at $x = L$. Substituting the value $x = L$ in Eq. (4) gives us $\psi = A \sin 2\pi(L/\lambda) = 0$. The nature of trigonometric functions is such that this condition will be satisfied whenever $2\pi L/\lambda$ is an integral multiple of π. If we let n represent this integral multiple,

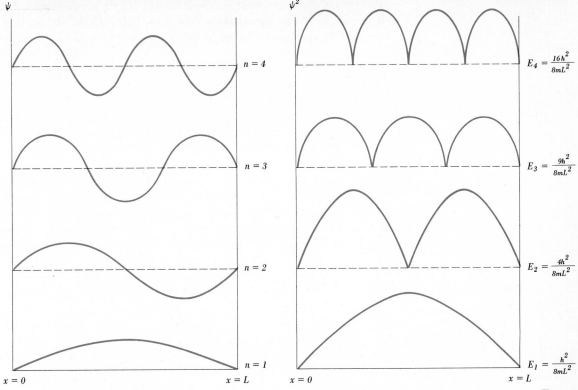

ψ $n = 4$ $n = 3$ $n = 2$ $n = 1$ $x = 0$ $x = L$

ψ^2 $E_4 = \dfrac{16h^2}{8mL^2}$ $E_3 = \dfrac{9h^2}{8mL^2}$ $E_2 = \dfrac{4h^2}{8mL^2}$ $E_1 = \dfrac{h^2}{8mL^2}$ $x = 0$ $x = L$

Fig. 2.11

Permitted wave functions ψ and probability distributions ψ^2 for particle in a box.

then we end up with the condition that $2\pi L/\lambda = n\pi$. Canceling the π gives $L/\frac{1}{2}\lambda = n$, which says that given the requirement that the wave function of the particle must go to zero at the boundaries of the box a whole number of half wavelengths must fit into the box. Stated differently, the only permitted values of the wavelength λ are those given by $2L/n$, n being able to take on values 1, 2, 3, 4, etc.

The restriction to only certain permitted values of λ is reminiscent of Bohr's quantum condition. The only difference is that, whereas Bohr introduced the quantum condition on an *ad hoc* basis to explain spectral lines, it comes out quite naturally here just from assuming a wave nature for the particle.

Figure 2.11 shows schematically what the wave functions ψ look like for the several lowest permitted states of a particle in the box. Shown also are the corresponding plots of ψ^2, where ψ^2 is related to the probability of finding the particle. As can be seen, for the lowest state $n = 1$ the particle is most probably found in the center of the box. For the second state $n = 2$, there are two regions of high probability, one in each half of the box; for this state the probability of finding the particle in the exact center of the box is zero.

What can we say about the energies of the particle in the box in its various states? The only permitted values of λ are those given by $2L/n$. The de Broglie condition says that $\lambda = h/mv$. Equating these two, we have the restriction that $h/mv = 2L/n$, or that permitted values of the momentum mv are given by $nh/2L$. Recalling that kinetic energy is $\frac{1}{2}mv^2$, i.e., the square of the momentum divided by $2m$, we can write for permitted values of the energy

$$E = \frac{n^2h^2}{8mL^2} \tag{5}$$

The energy values corresponding to the four lowest states are shown on the right side of Fig. 2.11. There are several points to note: (1) Since the potential energy inside the box is zero, the total energy E, which is the sum of kinetic plus potential, is just equal to the kinetic energy. (2) The total energy goes up as the square of the quantum number n. (3) The total energy decreases when the box is made bigger (that is, L is increased). This last point is well worth noting. It reflects an important principle of quantum mechanics: The smaller the space into which you try to squeeze a particle, the higher its kinetic energy gets to be. This has significance for electrons in molecules: Allowing an electron that was confined to a single atom to spread out its probability distribution over several atoms generally lowers the kinetic energy of the electron and contributes to stabilizing the system by lowering the total energy.

Actually the particle-in-a-box problem is a fairly straightforward calculation. The problem gets significantly harder when we tackle our central question: How can we describe an electron in an atom? Superficially, it is like the particle-in-a-box problem, except of course the box, i.e., the atom, is no longer a simple one-dimensional box but is spherical. There is a more serious complication. Whereas in the particle-in-the-box problem we had a uniform potential energy that could be set everywhere equal to zero, the potential energy of the electron in the atom decreases as one gets closer to the nucleus. In other words, there is a strong force pulling the electron into the nucleus. How can the wave description of an electron in an atom take this force into account?

The problem was resolved brilliantly in 1926 by Erwin Schrödinger. He took the equation of motion for waves, put in the de Broglie condition, and set the total energy as the sum of kinetic and potential terms, where the potential term was given the form $-Ze^2/r$ (Ze being the charge of the nucleus and e the charge of the electron at distance r from the nucleus). It would take us too far afield to trace through his mathematical arguments, some of which get to be rather abstruse, but we can look at his final statement of the problem described in mathematical terms. This is the famous *Schrödinger wave equation*, which can be written as follows:

$$\frac{d^2\psi}{dr^2} + \frac{8\pi^2 m}{h^2}\left(E + \frac{Ze^2}{r}\right)\psi = 0 \tag{6}$$

The notation here is that of differential calculus, where $d^2\psi/dr^2$ is the *second derivative of ψ with respect to r*. The letter d stands for an infinitesimal increment in a property, and $d\psi/dr$ stands for the infinitesimal increment in ψ that corresponds to an infinitesimal increment

Fig. 2.12

Wave functions ψ_{nlm} for one-electron atom.

1s $\qquad \psi_{100} = \dfrac{1}{\sqrt{\pi}} \left(\dfrac{Z}{a_0}\right)^{3/2} e^{-Zr/a_0}$

2s $\qquad \psi_{200} = \dfrac{1}{4\sqrt{2\pi}} \left(\dfrac{Z}{a_0}\right)^{3/2} \left(2 - \dfrac{Zr}{a_0}\right) e^{-Zr/2a_0}$

2p $\qquad \psi_{210} = \dfrac{1}{4\sqrt{2\pi}} \left(\dfrac{Z}{a_0}\right)^{5/2} e^{-Zr/2a_0} r \cos\theta$

$\qquad \psi_{21(\pm1)} = \dfrac{1}{8\sqrt{\pi}} \left(\dfrac{Z}{a_0}\right)^{5/2} e^{-Zr/2a_0} r \sin\theta\, e^{\pm i\phi}$

3s $\qquad \psi_{300} = \dfrac{1}{81\sqrt{3\pi}} \left(\dfrac{Z}{a_0}\right)^{3/2} \left(27 - \dfrac{18Zr}{a_0} + \dfrac{2Z^2 r^2}{a_0{}^2}\right) e^{-Zr/3a_0}$

3p $\qquad \psi_{310} = \dfrac{\sqrt{2}}{81\sqrt{\pi}} \left(\dfrac{Z}{a_0}\right)^{5/2} \left(6r - \dfrac{Zr^2}{a_0}\right) e^{-Zr/3a_0} \cos\theta$

$\qquad \psi_{31(\pm1)} = \dfrac{1}{81\sqrt{\pi}} \left(\dfrac{Z}{a_0}\right)^{5/2} \left(6r - \dfrac{Zr^2}{a_0}\right) e^{-Zr/3a_0} \sin\theta\, e^{\pm i\phi}$

3d $\qquad \psi_{320} = \dfrac{1}{81\sqrt{6\pi}} \left(\dfrac{Z}{a_0}\right)^{7/2} r^2 e^{-Zr/3a_0} (3\cos^2\theta - 1)$

$\qquad \psi_{32(\pm1)} = \dfrac{1}{81\sqrt{\pi}} \left(\dfrac{Z}{a_0}\right)^{7/2} r^2 e^{-Zr/3a_0} \sin\theta \cos\theta\, e^{\pm i\phi}$

$\qquad \psi_{32(\pm2)} = \dfrac{1}{162\sqrt{\pi}} \left(\dfrac{Z}{a_0}\right)^{7/2} r^2 e^{-Zr/3a_0} \sin^2\theta\, e^{\pm 2i\phi}$

in r. In other words, if we had a graph of ψ plotted as a function of r, the slope of the curve would be given by $d\psi/dr$. This is called the *first derivative of ψ with respect to r.* The second derivative simply tells how fast the first derivative is changing. Stated another way, the first derivative $d\psi/dr$ tells us the value of the slope at a given value of r; the second derivative $d^2\psi/dr^2$ tells us the curvature at that point. For wave motions, second derivatives essentially give the shapes of the waveforms.

The other symbols in Eq. (6) have been previously encountered. ψ is the wave function; E is the total energy; m is the mass of the electron; h is the Planck constant. Equation (6) tells how the wave function must behave but does not specify what ψ must be. ψ has a different form for each different value of E. Again, it turns out that only certain values of E are allowed, but now E is the sum of potential and kinetic contributions. What this means is that for an electron in a given energy state, as the electron moves in closer to the nucleus, its kinetic energy increases as its potential energy decreases. Stated otherwise, the electron speeds up as it approaches the nucleus and slows down as it moves farther away.

Figure 2.12 gives detailed mathematical expressions for some of the

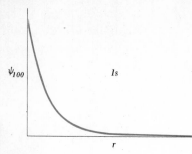

Fig. 2.13
Graph of wave function ψ_{100} for 1s electron as a function of distance r from nucleus.

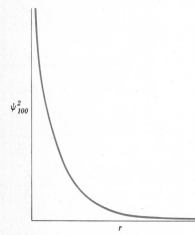

Fig. 2.14
Graph of square of the wave function ψ_{100}^2 as a function of r. Note that ψ^2 gives probability per unit volume.

wave functions allowed for a one-electron atom—i.e., an atom consisting of but one electron and a nucleus of charge Z. Each of the functions designates its corresponding orbital. Thus, the first function is the one appropriate to a 1s orbital. The subscript 100 in the designation ψ_{100} stands for the three important quantum numbers that are needed to describe this particular electron state. The first digit gives the value of the principal quantum number, which corresponds roughly to saying what shell the electron is in; the second digit gives information about the shape of the orbital; the third says something about how it is oriented in space. (Quantum numbers are discussed further in Sec. 2.4.)

■ The wave function ψ_{100}, for the 1s orbital, consists of three multiplicative terms: The first of these, $1/\sqrt{\pi}$, is a fractional constant which is basically figured out from the condition that if we put an electron in an orbital, then summing up over the whole orbital probability should lead to complete certainty of finding the electron; the second term $(Z/a_0)^{3/2}$, where Z is the atomic number and a_0 is the Bohr radius, 0.0529 nm, takes care of the fact that as Z increases, the electron is pulled closer to the nucleus so that the wave function must be built up at smaller r; the final term e^{-Zr/a_0}, the so-called "exponential" term, takes care of the fact that the chance of finding the electron drops off extremely rapidly as one goes to larger values of r. In this final term, e is the base for natural logarithms (see Appendix 4.2) and has a value of 2.718 . . . ■

Figure 2.13 shows what a graph of the wave function ψ_{100} looks like plotted as a function of r. Actually, it is not so much ψ as ψ^2 that we are interested in since it is ψ^2 that is proportional to the probability of finding the electron per unit volume. The course of ψ^2 as a function of r is shown in Fig. 2.14; it should be compared with Fig. 2.1.

Again, as was done on page 50, we need to make a distinction between the probability per unit volume and the total probability in a spherical shell of increasing radius. The probability per unit volume is just represented by ψ^2; so Fig. 2.14 tells how the probability per unit volume falls off as one goes out to bigger r. But now let us consider a spherical shell of thickness dr. (Recall that dr represents an infinitesimally small increment in r.) The surface area of a spherical shell is $4\pi r^2$. If we multiply this area by the thickness of the shell, we get $4\pi r^2\ dr$ as the volume of the shell. Each unit of volume in this shell corresponds to electron probability ψ^2; so the total probability of finding the electron in the shell is $(4\pi r^2\ dr)(\psi^2)$. Figure 2.15 shows how this quantity changes with increasing r. The dotted curves show, respectively, how $4\pi r^2\ dr$ increases and ψ^2 decreases as one goes to larger r. Multiplying the two contributions together gives rise to a maximum in $4\pi r^2\psi^2\ dr$ at an r value equal to a, which we likened before (page 50) to the Bohr radius. In fact, then, except for language there really is not much difference in describing the innermost shell of an atom by the Bohr picture, the qualitative probability picture, or the wave-function approach.

There is a real difference that shows up when we consider the first excited state, the one described by the second entry in Fig. 2.12. The wave function for the 2s orbital ψ_{200} contains a parenthetic term $2 - Zr/a_0$ which will go through zero as r increases. This means that ψ, and hence ψ^2, will vanish at some value of r. In other words, the probability will vanish. The place where the probability vanishes is called a *node;* in this case because of the spherical symmetry we have a *nodal surface.* Figure 2.16 shows respective plots of ψ, ψ^2, and $4\pi r^2 \psi^2\, dr$ for a 2s electron. The node occurs when ψ switches from positive to negative at a certain value of r. Since the probability is proportional to the *square* of the wave function, it does not matter whether ψ is positive or negative as far as describing the probability of finding the electron is concerned. (The sign of a wave function is important for telling the *phase* of the wave—i.e., whether it corresponds to trough or crest. This can be very important when combining one wave with another.) For a 2s electron, there are two regions of high probability, consisting of an inner sphere and an outer sphere, separated by a region of zero probability. This is one and the same electron. Because of its wave nature, its probability density is piled up in two regions of space, the inner of which is already "occupied" by the 1s electron.

The third entry in Fig. 2.12 shows that the wave functions for 2p electrons introduce still another feature—a variation with angular direction. For the wave function labeled ψ_{210} we have a dependence on θ, where θ is the polar angle, which measures the angular deviation away from the vertical, or z, axis. When θ is zero—that is, when looking along the z axis—$\cos\theta$ is a maximum; so the chance of finding the ψ_{210} electron is greatest when looking along the z axis. As θ increases, $\cos\theta$ decreases; ψ_{210} decreases, and the chance of finding the electron diminishes. When θ reaches $\pi/2$, that is, $90°$ away from the z direction, $\cos\theta$ goes to zero; we have a node. In fact, when $\theta = \pi/2$, we are looking in the xy plane, and we have *no* chance of finding this particular electron. It corresponds to the p_z electron, as was shown in Fig. 2.5.

When it comes to the p_x and p_y electrons, the wave-function representation gets more complicated. The fourth entry in Fig. 2.12 is labeled $\psi_{21(\pm1)}$ and contains, besides a $\sin\theta$ dependence, one that is $e^{\pm i\phi}$. What this means is that there are two possible wave functions, one depending on $e^{+i\phi}$ and the other on $e^{-i\phi}$. ϕ is an angle, which can be called the *longitude* angle; it measures the angular deviation out of the xz plane. i stands for the square root of -1 and is an imaginary mathematical quantity as opposed to real ones. The simplest way of visualizing the $e^{i\phi}$ and $e^{-i\phi}$ wave functions is that they correspond to electron running waves, one circulating clockwise around the z axis and the other counterclockwise. They are pretty much concentrated in the xy plane, since the $\sin\theta$ term ensures that as θ goes from $\pi/2$ to 0, the wave function will die off. Figure 2.17 shows crudely what ψ^2 looks like for the $\psi_{21(+1)}$ and $\psi_{21(-1)}$ cases; the pictures are identical except that in one electric charge is going around clockwise and in the other, counterclockwise.

For most chemical purposes, running waves are not very useful ways of representing electron probability distributions.* Mathematically equivalent representations can be generated by the simple expedient of combining the two

---

* When it comes to explaining magnetism, the running waves are necessary.

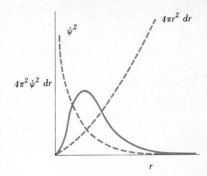

Fig. 2.15
Graph of total probability of finding 1s electron in a spherical shell of radius r and thickness dr.

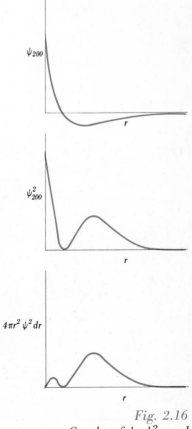

Fig. 2.16
Graphs of ψ, ψ^2, and $4\pi r^2\psi^2\, dr$ for a 2s electron.

Fig. 2.17

Probability distributions for 2p electrons corresponding to $\psi_{21(+1)}$ and $\psi_{21(-1)}$. On the left, the running wave goes clockwise; on the right, counterclockwise.

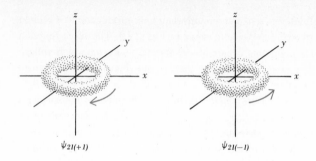

$\psi_{21(+1)}$ $\quad\quad\quad$ $\psi_{21(-1)}$

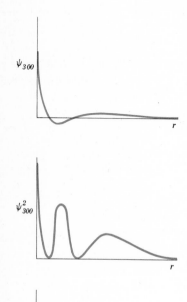

ψ_{300}

ψ_{300}^2

$4\pi r^2 \psi^2\, dr$

Fig. 2.18

Graphs of ψ, ψ^2, and $4\pi r^2\psi^2\, dr$ for a 3s electron.

running waves, either additively or subtractively. In other words, if we add the wave function $\psi_{21(+1)}$ to the wave function $\psi_{21(-1)}$, then we are combining the clockwise running wave with the counterclockwise running wave. The result is a standing wave, which piles up the charge along the x axis. We get what corresponds to a p_x electron (see Fig. 2.5). If we subtract $\psi_{21(+1)}$ from $\psi_{21(-1)}$, we get what corresponds to a p_y electron. The p_x, p_y, p_z representation given in Fig. 2.5 is fully equivalent to the ψ_{210}, $\psi_{21(+1)}$, $\psi_{21(-1)}$ representation given in Fig. 2.12. The p_x, p_y, p_z representation is particularly useful when talking about the establishment of chemical bonds in particular directions. No matter which representation is used, putting *one* electron into *each* (or two electrons into each) of the equivalent orbitals gives *in toto* a completely spherical charge distribution.

For principal quantum number $n = 3$, corresponding to what Bohr called the third shell, we have three different kinds of wave functions. The one labeled ψ_{300} describes the 3s electron. As shown in Fig. 2.18, the wave function starts out positive, goes negative, and then goes back to being positive again at very large values of r. This means there are two nodes. The effect is to produce concentric structure for the 3s electron, consisting of an innermost sphere, a middle spherical shell, and finally an outermost spherical shell. The 3p electrons, for which the wave functions ψ_{310}, $\psi_{31(+1)}$, and $\psi_{31(-1)}$ are given in Fig. 2.12, are much like 2p electrons in their angular variations, but in addition they have superposed on the angular variation a radial variation that has a node in it. In other words, the $3p_x$, $3p_y$, and $3p_z$ electrons are directed, respectively, along the p_x, p_y, and p_z axes; but as one goes out from the nucleus along any of these axes, the probability density first increases then decreases to zero, then increases to another maximum.

The 3d electrons are of special interest since they are mainly responsible for the sequence of change in the properties of the first row of transition elements. As noted in Fig. 2.12, there are basically three kinds of wave functions for 3d electrons—they are designated ψ_{320}, $\psi_{32(\pm1)}$, and $\psi_{32(\pm2)}$, respectively. They all have the same radial dependence, given by $r^2 e^{-Zr/3a_0}$, which corresponds to just one big bump in electron probability distribution as one goes out to larger r. They differ in angular dependence: ψ_{320} piles up charge mostly along the z axis, $\psi_{32(\pm1)}$ piles up charge halfway between the z axis and the xy plane, and $\psi_{32(\pm2)}$ piles up charge in the xy plane. Most striking of all, however, is the dependence on the longitude angle. ψ_{320} has no such dependence; it is already a *real* orbital, and there is no problem in visualizing it. $\psi_{32(+1)}$ and $\psi_{32(-1)}$ contain the terms $e^{i\phi}$ and $e^{-i\phi}$, respectively;

as before with p orbitals, they can be visualized as currents of electric charge circulating around the z axis in clockwise and counterclockwise directions. $\psi_{32(+2)}$ and $\psi_{32(-2)}$ contain the terms $e^{2i\phi}$ and $e^{-2i\phi}$, respectively; they also can be thought of as corresponding to clockwise and counterclockwise circulations of charge. Figure 2.19 shows how the various $3d$ orbitals can be visualized.

Again, as done before with the p wave functions, we can combine clockwise and counterclockwise circulations to end up with real standing waves, which are easier to visualize. If we add $\psi_{32(+1)}$ and $\psi_{32(-1)}$, we get what we have already designated as the d_{yz} orbital in Fig. 2.6; if we subtract these two wave functions, we get the orbital labeled d_{xz}. Combining the wave functions $\psi_{32(+2)}$ and $\psi_{32(-2)}$ additively or subtractively leads to $d_{x^2-y^2}$ and d_{xy}, respectively. The orbital labeled d_{z^2} in Fig. 2.6 is just the one that corresponds to ψ_{320}. No matter which way we choose to write their mathematical descriptions, there are five orbitals in the $3d$ set. If we put one electron into each of the orbitals described by ψ_{320}, $\psi_{32(+1)}$, $\psi_{32(-1)}$, $\psi_{32(+2)}$, and $\psi_{32(-2)}$, respectively, we end up with a spherical charge distribution. ∎

2.3 Electron spin

In Sec. 2.1 it was noted that there is one orbital in an s subshell, three orbitals in a p subshell, five in a d, and seven in an f. Since these subshells can accommodate 2, 6, 10, and 14 electrons, respectively, it follows that any orbital can hold 2 electrons. Actually the 2 electrons in the same orbital differ in one important respect: They have opposite *spin*. The reason for talking about electron spin comes from observations on the magnetic behavior of substances.

It is a familiar observation that certain solids, such as iron, are strongly attracted to magnets. Such materials are called *ferromagnetic*. Other substances, such as oxygen gas and copper sulfate, are weakly attracted to magnets. These are called *paramagnetic*. Still other substances, such as sodium chloride, are very feebly repelled by magnets and are called *diamagnetic*. Ferromagnetism is a special property of the solid state, but all three types of magnetic behavior just described are believed to arise from electrons in atoms.

Information about the magnetic behavior of individual atoms can be obtained from an experiment like the one first performed by the German physicists Otto Stern and Walter Gerlach in 1921. In their experiment, shown in Fig. 2.20, a beam of neutral silver atoms (from the vaporization of silver) was passed between the poles of a specially designed magnet. The beam was found to be split into two separate beams; i.e., half of the atoms were deflected in one direction, and the rest in the opposite direction.

In interpreting this experiment, the obvious inference is that the silver atoms are behaving like tiny magnets. Otherwise there would be no deflection of the beam. But the beam is split into two. This means that some of the magnets are pointing one way and the rest, in the opposite direction. How do these tiny magnets come about? There are two ways electrons could give rise to magnetism: They could circulate in a loop (as was shown in Fig. 2.17), or they could rotate on

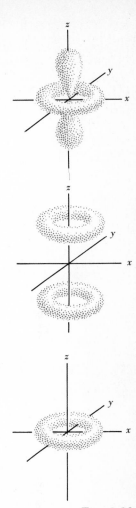

Fig. 2.19

Probability distributions for 3d electrons corresponding to ψ_{320}, $\psi_{32(\pm1)}$, *and* $\psi_{32(\pm2)}$. *For both* $\psi_{32(\pm1)}$ *and* $\psi_{32(\pm2)}$ *there are actually two orbitals apiece, corresponding to clockwise and counterclockwise circulation of charge about the z axis.*

their own axes. The outermost electron in silver is an *s*-type electron; as such, it has no circulation component in its motion, i.e., no orbital angular momentum. *s* electrons are spherically symmetric about the nucleus, and there is no net paramagnetism that arises from motion in the orbital. The only thing left is to think of the electron as spinning on its own axis. Since any spinning charge is magnetic, behavior as a tiny magnet suggests that the electron has a spin. Apparently, only two directions of spin are possible: clockwise or counterclockwise. In the clockwise case, the electron behaves as a magnet pointing in one direction; in the counterclockwise, as a magnet pointing in the opposite direction. If we have two electrons of opposite spin, we might expect them to attract each other, as two magnets would; but the electric repulsion due to like negative charges is very much greater than the magnetic attraction. When electrons are required to be together, as in a completely filled subshell of an atom, each electron will pair up with another electron of opposite spin. The electron pair in an orbital is nonmagnetic because the magnetism of one spin is canceled by the magnetism of the opposite spin.

In silver atoms, as shown by the electronic configuration given in Fig. 1.18, all the electrons are found in completed subshells except the one 5*s* electron. This electron obviously cannot be paired with another. Hence, its uncanceled spin gives magnetism to the silver atom. The two deflections observed in the Stern-Gerlach experiment presumably result from a separation of silver atoms of two types which differ in the direction of spin of the unpaired electron, one having clockwise spin and the other counterclockwise.

Any atom which, like the silver atom, contains an odd number of electrons must be paramagnetic. Furthermore, atoms which have an even number of electrons can also be paramagnetic provided that there is an unfilled subshell of electrons. These more complex cases will be considered in Sec. 17.5. When all the electrons in an atom are paired, there is no paramagnetism. There is only diamagnetism, which occurs in all matter, even though in paramagnetic substances it may be obscured. Diamagnetism arises not from the spin of electrons but from their electric charge. A detailed discussion of diamagnetism is beyond the scope of this book.

2.4 *Quantum numbers and orbital filling*

From the complete mathematical treatment of the wave nature of electrons in atoms there emerges a description of each electron in terms of four index numbers, the so-called "quantum numbers." The first of these, the *principal* quantum number n, gives, other things being equal, the order of increasing distance of the average electron distribution from the nucleus, and hence it is related to the order of the electronic energies. In most cases, however, the electronic energy depends significantly on the second quantum number also.

The second quantum number, usually designated by l, is called the *orbital* quantum number. It denotes the subshell which the electron occupies and indicates the angular shape of the electron distribution. Values permissible for l are 0, 1, 2, . . . , $n - 1$. $l = 0$ corresponds to an s subshell; $l = 1$, to a p subshell; $l = 2$, to a d subshell; $l = 3$, to an f subshell.

The third quantum number, usually designated by m_l, is called the *magnetic* quantum number. It tells something about the circulation of the electric charge, which gives rise to magnetism. This magnetism causes the orbitals within the same subshell to separate into discrete energy levels when a magnetic field is applied. The splitting, called the *Zeeman effect*, can be observed experimentally by measuring spectral lines in a magnetic field and noting how one line in the absence of a field may split into several on application of the field. For a given value of l, permitted values of m_l are the integers, including 0, from $+l$ to $-l$. Thus, for example, in the d subshell, where l is 2, allowed m_l values are $+2$, $+1$, 0, -1, and -2. As noted in Sec. 2.2, there is not a single unique set of orbitals within a subshell; i.e., there are many ways in which a spherical distribution can be broken down into component parts. Specifically, the set of $3d$ orbitals shown in Fig. 2.6 is but one such set and is not the same set that would be simply

Fig. 2.20
Stern-Gerlach experiment showing magnetic splitting of a beam of silver atoms.

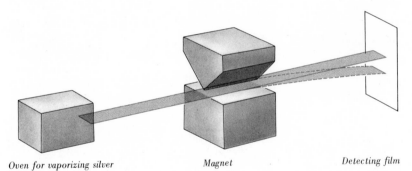

Oven for vaporizing silver   *Magnet*   *Detecting film*

designated by m_l values $+2$, $+1$, 0, -1, and -2. However, for our purposes what is important is the fact that there can be but five orbitals in any d subshell.

The fourth quantum number, usually designated m_s, is called the *spin* quantum number. It can have values of either $+\frac{1}{2}$ or $-\frac{1}{2}$, corresponding to the two possible orientations of electron spin. Instead of $+\frac{1}{2}$ and $-\frac{1}{2}$, the orientations are often designated by arrows pointing in opposite directions: ↑ and ↓.

An electron in an atom is completely described once its four quantum numbers have been specified. Furthermore, a fundamental principle, called the *Pauli exclusion principle*, states that no two electrons in the same atom can be completely identical, i.e., have the same values of all four quantum numbers. Because of this limitation and the limitation on permitted values of quantum numbers, the number of electrons in a shell is limited to $2n^2$, as mentioned on page 35. How this comes about is shown in Fig. 2.21. The number of entries in the last column indicates the number of electrons that can be accommodated in each shell. For the K shell, where we can have one electron with $m_s = +\frac{1}{2}$

Section 2.4
Quantum numbers and orbital filling

65

n	l	m	m_s
1 (K shell)	0 (s subshell)	0	$+\frac{1}{2}, -\frac{1}{2}$
2 (L shell)	0 (s subshell)	0	$+\frac{1}{2}, -\frac{1}{2}$
	1 (p subshell)	+1	$+\frac{1}{2}, -\frac{1}{2}$
		0	$+\frac{1}{2}, -\frac{1}{2}$
		-1	$+\frac{1}{2}, -\frac{1}{2}$
3 (M shell)	0 (s subshell)	0	$+\frac{1}{2}, -\frac{1}{2}$
	1 (p subshell)	+1	$+\frac{1}{2}, -\frac{1}{2}$
		0	$+\frac{1}{2}, -\frac{1}{2}$
		-1	$+\frac{1}{2}, -\frac{1}{2}$
	2 (d subshell)	+2	$+\frac{1}{2}, -\frac{1}{2}$
		+1	$+\frac{1}{2}, -\frac{1}{2}$
		0	$+\frac{1}{2}, -\frac{1}{2}$
		-1	$+\frac{1}{2}, -\frac{1}{2}$
		-2	$+\frac{1}{2}, -\frac{1}{2}$
4 (N shell)	0 (s subshell)	0	$+\frac{1}{2}, -\frac{1}{2}$
	1 (p subshell)	+1	$+\frac{1}{2}, -\frac{1}{2}$
		0	$+\frac{1}{2}, -\frac{1}{2}$
		-1	$+\frac{1}{2}, -\frac{1}{2}$
	2 (d subshell)	+2	$+\frac{1}{2}, -\frac{1}{2}$
		+1	$+\frac{1}{2}, -\frac{1}{2}$
		0	$+\frac{1}{2}, -\frac{1}{2}$
		-1	$+\frac{1}{2}, -\frac{1}{2}$
		-2	$+\frac{1}{2}, -\frac{1}{2}$
	3 (f subshell)	+3	$+\frac{1}{2}, -\frac{1}{2}$
		+2	$+\frac{1}{2}, -\frac{1}{2}$
		+1	$+\frac{1}{2}, -\frac{1}{2}$
		0	$+\frac{1}{2}, -\frac{1}{2}$
		-1	$+\frac{1}{2}, -\frac{1}{2}$
		-2	$+\frac{1}{2}, -\frac{1}{2}$
		-3	$+\frac{1}{2}, -\frac{1}{2}$

and another with $m_s = -\frac{1}{2}$, the total number of electrons is seen to be 2; for the L shell, where we can have $+\frac{1}{2}$ and $-\frac{1}{2}$ in the s orbital and $+\frac{1}{2}$ and $-\frac{1}{2}$ in each of the three p orbitals, the total number is 8; for the M, 18; and for the N, 32. In terms of the principal quantum number n, the maximum population is $2(1)^2$, $2(2)^2$, $2(3)^2$, $2(4)^2$, . . . , $2n^2$.

The assignment of electrons to orbitals for building up the periodic table is governed by several factors. These include the Pauli exclusion principle, the relative energies of subshells, and the repulsions between electrons in orbitals belonging to the same subshell. Figure 2.22 shows an orbital-filling diagram in which each circle represents schematically an orbital belonging to the subshell for which the designation is given in the left margin. There is a single circle for the s subshell since there is but one s orbital for each value of n; there are three circles for each of the p subshells, standing respectively for p_x, p_y, and p_z orbitals or their equivalent; there are five circles standing for the five orbitals of a d subshell. The orbitals are filled from the bottom of the diagram working upward. At any one horizontal level, i.e., within any subshell, the electrons are spread out one to an orbital before any pairing (two electrons ↑↓ in the same orbital) is allowed to occur. The reason for this is that electrons, being of the same negative electric charge, repel each other; hence, they try to spread over as much space as possible. This they can do by occupying separate orbitals, each of

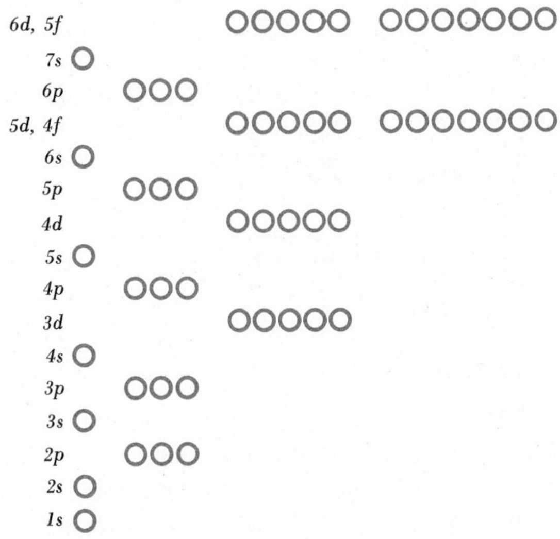

Fig. 2.22
Orbital-filling diagram.

which corresponds to a different distribution in space. After all the orbitals of a given subshell are populated by one electron, pairing generally begins to occur because repulsion due to two electrons in the same orbital is usually less than the energy jump to the next higher subshell.

It should be especially noted that Fig. 2.22 is a filling diagram and not an energy-level diagram. The reason for this distinction is that there is no single energy-level diagram applicable to all elements. The relative energies depend on the nuclear charge Z and the number of other electrons present. In building up the periodic table by adding an electron to a previously formed atom, the nuclear charge is also increased by one unit. When this is done, the energy levels shift, and in some cases the relative order changes.

The order given in Fig. 2.22 is that appropriate for the stepwise

Section 2.4
Quantum numbers
and orbital filling

buildup of the elements. It gives the relative order of adjacent subshells only when electron filling is occurring in that region. As an example, the chart shows $4s$ to lie lower than $3d$. $4s$ begins to fill at potassium ($Z = 19$); $3d$ does not begin to fill until scandium ($Z = 21$). This means that in potassium and scandium the $4s$ subshell is lower than the $3d$. However, it says nothing about the relative ordering of the same subshells in a later element such as zinc ($Z = 30$), in which, in fact, the $3d$ is the lower. A related point to note is that Fig. 2.22 is not an ionization diagram; i.e., it should not be used to decide which electrons are to be removed when an ion is formed from a particular atom (constant Z). Specifically, in forming Ti^{2+} from a neutral titanium atom ($Z = 22$, $1s^2 2s^2 2p^6 3s^2 3p^6 4s^2 3d^2$), it is not the two $3d$ electrons that are lost, but the two $4s$ electrons. The point is that in the Ti^{2+} ion, the $3d$ level is lower in energy that the $4s$, contrary to what might be expected from just looking at the order-of-filling diagram.

2.5   Many-electron atoms

The preceding discussion implicitly assumed that electrons in atoms could be described by a superposition of hydrogenlike descriptions. In other words, it was assumed that any particular orbital retained all its characteristic features no matter how many other electrons there were in the same atom. This cannot be quite true because, of course, electrons repel each other. Adding a second electron to a one-electron atom, for example, changes the problem quite drastically. It is no longer a simple question of considering the attraction of a light, negative electron to a heavy, positive nucleus (like a buzzing fly attracted to a fixed piece of sugar), but now one needs to consider that the light, negative electron is repelled by another light, negative electron, neither of which is even approximately fixed in space. The problem is evidently a complex one because the force acting on each electron is a combination of the attraction to the nucleus and repulsion from the other electron. As a matter of fact, the problem cannot be solved exactly, and we have to be satisfied with approximate solutions. The gist of these approximation methods is that each electron is considered separately to move in an average field generated by the nucleus and the other electrons. The other electrons are then adjusted one at a time to allow for what was found out in each step of the calculation. The process is continued until a self-consistent set of solutions is obtained, wherein small adjustments of one electron do not appreciably affect the others. This method of describing electron distributions is called the *Hartree-Fock method*. At present, with the advent of high-speed computers results can be obtained to any desired degree of accuracy, but of course the expense quickly mounts up as more and more electrons are added to the problem. The interesting point to note is that the elegant results are not much different from what one gets by assuming the electrons move independently of each other in hydrogenlike orbitals and then adding on repulsive interactions as perturbations.

The Pauli exclusion principle is actually a way of minimizing electron-electron repulsions. For example, in the case of the lithium atom, where the ground-state electron configuration is $1s^2 2s^1$, there is much lower electron-electron interaction than if all three electrons were tried to be squeezed into the $1s$ orbital. At first sight, it might even seem unfavorable to have two electrons in an orbital. The Pauli exclusion principle states that that is possible as long as the two electrons differ in spin. But how does the difference in spin compensate for the fact that just putting both electrons in the same orbital implies they are occupying the same space? The point is that our picture is still weak: We can talk about the time-averaged electron distributions for the two electrons in an orbital to be identical, but this does not allow for the fact that the electron motions are highly correlated with each other. Negative-negative repulsion assures that when one electron moves through one portion of space at a given time, the chance is great that the other electron is on the opposite side of the atom at that same time. The electrons do not maintain fixed separation in their motion, but on the average they can be regarded as staying as far apart as possible. Since the electrons differ in spin, one might be tempted to hypothesize that one portion of an orbital might be magnetically opposite to the other, but we need to recall that interchange of the electron spins is equally probable. In some cases, however, one does find magnetic spin separation for a single orbital. This happens, for example, with p orbitals in the solid state. If we have a magnetic ion near one lobe of the p orbital, we tend to attract the electron of one spin into spending more time in that lobe. The other electron in that same p orbital will on the average spend more of its time in the opposite lobe. This effect is very important in magnetic materials and is one of the reasons why magnetic elements in the memory core of high-speed computers behave as they do.

Besides the Pauli exclusion principle, there is another important guiding rule for getting proper electron distributions. It is called *Hund's rule* and simply states that the lowest energy state generally corresponds to that for which the spin magnetism is the greatest. Consider, for example, a case in which we have two p electrons in the same subshell. Using short lines to represent the three orbitals of a p subshell, we can envisage the following possibilities for a p^2 configuration: $\downarrow\uparrow$ __ __, $\uparrow$ $\downarrow$ __, or $\uparrow$ $\uparrow$ __. The last of these is the lowest energy state. It corresponds to a total spin of $(m_s = +\frac{1}{2})$ plus $(m_s = +\frac{1}{2}) = 1$, whereas the others correspond to a total spin of $(m_s = +\frac{1}{2})$ plus $(m_s = -\frac{1}{2}) = 0$. The high-spin state is favored because it is most effective at keeping the electrons apart. The state $\downarrow\uparrow$ __ __ suffers from the fact that both electrons are in the same orbital, so electron-electron repulsion is a maximum; the state $\uparrow$ $\downarrow$ __ suffers from the fact that since the spins are not the same, one electron can "leak" over into the other's orbital; the state $\uparrow$ $\uparrow$ __ does not permit such leak-over since the Pauli exclusion principle forbids two electrons of the same spin ever to occupy the same orbital.

Figure 2.23 shows schematically the magnetic spin situation for the ground states of the first 10 elements.

Fig. 2.23
Ground-state spin configurations for the first ten elements.

Element	Electron configuration	Spin arrangement				
H	$1s^1$	↑	—	—	—	—
He	$1s^2$	↑↓	—	—	—	—
Li	$1s^2 2s^1$	↑↓	↑	—	—	—
Be	$1s^2 2s^2$	↑↓	↑↓	—	—	—
B	$1s^2 2s^2 2p^1$	↑↓	↑↓	↑	—	—
C	$1s^2 2s^2 2p^2$	↑↓	↑↓	↑	↑	—
N	$1s^2 2s^2 2p^3$	↑↓	↑↓	↑	↑	↑
O	$1s^2 2s^2 2p^4$	↑↓	↑↓	↑↓	↑	↑
F	$1s^2 2s^2 2p^5$	↑↓	↑↓	↑↓	↑↓	↑
Ne	$1s^2 2s^2 2p^6$	↑↓	↑↓	↑↓	↑↓	↑↓

2.6 Atomic radii

The size of an atom is a difficult property to determine. For one thing, the electronic probability distribution never becomes exactly zero, even at great distances from the nucleus. Therefore, the distance designated as the boundary of the atom is an arbitrary choice. For another thing, the electronic probability distribution is affected by neighboring atoms; hence, the size of the atom may change in going from one condition to another, as, for example, in going from one compound to another. Therefore, in examining any table of atomic radii we must remember that the values listed may be meaningful only in providing a relative comparison of sizes. Figure 2.24 gives such a set of atomic radii. They have been deduced from observed distances between centers of adjacent atoms. These interatomic spacings can be determined from X-ray and spectral studies of bound atoms.

In general, the atomic radii decrease in going from left to right across the periodic table and increase in going from top to bottom. How can we explain these trends? Figure 2.25 shows the change of atomic radius within the second period. It also gives the nuclear charge and the electronic configuration. Within the period the nuclear charge increases from +3 to +9. What effect might this have on the K electrons? In each of these elements there are two K electrons. The two electrons are attracted to the nucleus by a force proportional to the nuclear charge. As the nuclear charge increases, the pull on the electrons is increased, and the maximum in the K probability distribution curve gets closer to the nucleus.

What about the L electrons? Here the problem is complicated by the fact that the L electrons are screened from the nucleus by the K electrons; so the attractive force of the nuclear positive charge is reduced by the intervening negative charges. In lithium, for example, the outermost electron is attracted not just by a charge of +3 but by a charge of +3 screened by two intervening negative electrons.

Periodic table of atomic radii:

1 H .037																	2 He —
3 Li .123	4 Be .089											5 B .080	6 C .077	7 N .074	8 O .074	9 F .072	10 Ne —
11 Na .157	12 Mg .136											13 Al .125	14 Si .117	15 P .110	16 S .104	17 Cl .099	18 Ar —
19 K .203	20 Ca .174	21 Sc .144	22 Ti .132	23 V .122	24 Cr .117	25 Mn .117	26 Fe .117	27 Co .116	28 Ni .115	29 Cu .117	30 Zn .125	31 Ga .125	32 Ge .122	33 As .121	34 Se .117	35 Br .114	36 Kr —
37 Rb .216	38 Sr .191	39 Y .162	40 Zr .145	41 Nb .134	42 Mo .129	43 Tc —	44 Ru .124	45 Rh .125	46 Pd .128	47 Ag .134	48 Cd .141	49 In .150	50 Sn .141	51 Sb .141	52 Te .137	53 I .133	54 Xe —
55 Cs .235	56 Ba .198	*	72 Hf .144	73 Ta .134	74 W .130	75 Re .128	76 Os .126	77 Ir .126	78 Pt .129	79 Au .134	80 Hg .144	81 Tl .155	82 Pb .154	83 Bi .152	84 Po .153	85 At —	86 Rn —
87 Fr —	88 Ra —	†	104 Ku —	105 Ha —													

*	57 La .169	58 Ce .165	59 Pr .165	60 Nd .164	61 Pm —	62 Sm .166	63 Eu .185	64 Gd .161	65 Tb .159	66 Dy .159	67 Ho .158	68 Er .157	69 Tm .156	70 Yb .170	71 Lu .156
†	89 Ac —	90 Th .165	91 Pa —	92 U .142	93 Np —	94 Pu —	95 Am —	96 Cm —	97 Bk —	98 Cf —	99 Es —	100 Fm —	101 Md —	102 No —	103 Lr —

*Fig. 2.24
Atomic radii of the elements. (Number above symbol is atomic number; that below is atomic radius in units of nanometers. 1 nm = 10 Å.)*

The net attractive charge is closer to a $+1$ charge than to a $+3$ charge. In the beryllium atom, the L electrons are attracted by a $+4$ nucleus screened by two negative charges, or effectively a $+2$ charge. Despite screening, in going from left to right across the period the L electrons have a higher and higher positive charge attracting them to the center of the atom. Just as the K shell becomes smaller because of this effect, the L shell gets smaller also.

How does the size of atoms change within a group? Figure 2.26 gives the data for the alkali elements. There is an increase of size from 0.123 nm to 0.235 nm in going from top to bottom. Going down the sequence, the number of shells populated is increasing stepwise. The more shells used, the bigger the atom. Because the nuclear charge progressively increases down the sequence, the individual shells get smaller, but adding a shell is apparently such a big effect that it dominates. Similar behavior is found for many of the other groups of the periodic table. There are, however, some places in the periodic table where the size does not change much within the same group. This is

	Li	Be	B	C	N	O	F
Atomic radius, nm	0.123	0.089	0.080	0.077	0.074	0.074	0.072
Nuclear charge	+3	+4	+5	+6	+7	+8	+9
K-level population	2e⁻	2e⁻	2e⁻	2e⁻	2e⁻	2e⁻	2e⁻
L-level population	1e⁻	2e⁻	3e⁻	4e⁻	5e⁻	6e⁻	7e⁻

Fig. 2.25
Change of atomic radius
within a period.

particularly true when elements 57 to 71 intervene between the two
atoms compared.

2.7 Ionization potential

When an electron is pulled off a neutral atom, the particle which
remains behind is a positively charged particle, or a positive ion. The
process, called *ionization*, can be described by writing

$$Na \longrightarrow Na^+ + e^-$$

In the process the positive sodium ion, shown on the right with a
superscript $^+$ to indicate a $+1$ charge, is formed. The electron is shown
separately as e^-. The *ionization potential* is the work that is required
to separate the negatively charged electron from the positively charged
sodium ion that is attracting it. In other words, the ionization potential

Fig. 2.26
Change of atomic radius
within a group.

Element	Atomic radius, nm	Nuclear charge	Electronic configuration
Li	0.123	+3	$1s^2 2s^1$
Na	0.157	+11	$1s^2 2s^2 2p^6 3s^1$
K	0.203	+19	$1s^2 2s^2 2p^6 3s^2 3p^6 4s^1$
Rb	0.216	+37	$1s^2 2s^2 2p^6 3s^2 3p^6 3d^{10} 4s^2 4p^6 5s^1$
Cs	0.235	+55	$1s^2 2s^2 2p^6 3s^2 3p^6 3d^{10} 4s^2 4p^6 4d^{10} 5s^2 5p^6 6s^1$

is the *energy required to pull an electron off an isolated atom*. It can
be measured experimentally by observing the spectrum of light emitted
by excited atoms. As has been noted in Sec. 1.5, such a spectrum
gives information on excited electronic states of atoms. These higher
states converge on the ultimate limit which corresponds to complete
removal of the electron from the atom, or ionization. Measurement of
the spectral line at the convergence limit gives the energy released
when an electron returns to the ground state. This energy must equal
that required to remove the electron completely from the atom. Usually,
the ionization potential is expressed in units of *electron volts* (eV), one
electron volt being the amount of energy an electron gains when accelerated through one volt. One electron volt is equal to 1.602×10^{-19} J.
One electron volt per atom is equivalent to 96.49 kJ (or 23.06 kcal) per
Avogadro number, 6.02×10^{23}, of atoms.

Figure 2.27 shows the values of the ionization potential for the first
60 elements. Within each period, for example, lithium to neon, there is,
with some exceptions, a fairly steady increase from left to right. Why
should it be harder to pull an electron off a neon atom, for example,

than off a lithium atom? At least two factors must be considered. First, the nuclear charge increases from left to right across a period. By itself this predicts that the ionization potential should increase from lithium to neon. Second, the atomic radius decreases from left to right. The size effect by itself would also predict that the ionization potential should increase, since the closer an electron is to the nucleus, the harder it is to pull off.

A shell of eight electrons, the so-called "octet," is a grouping particularly difficult to break up. It is especially hard to pull an electron off an atom having eight electrons in its outermost shell, and atoms such as neon have very high ionization potentials. Many of the apparent irregularities in the ionization potential can be correlated with the fact that completed subshells and half-completed subshells appear to have extra stability. Thus, the ionization potentials of beryllium ($2s$ subshell completed) and nitrogen ($2p$ half completed) appear to be higher than expected. In general, it is sufficient to remember that the elements of high ionization potential are on the right side of the periodic table and those of low ionization potential are on the left.

How about the trend within a group? Figure 2.27 shows the values of the ionization potential for the alkali elements and the noble gases. In both cases there is a progressive decrease of the ionization potential in going from the top to the bottom of the group. This is as predicted by the size change alone. The helium atom is quite small; the electron which is being pulled off is close to the nucleus. It is more firmly bound than in neon, in which the electron being pulled off is much farther from the nucleus. The increase in nuclear charge essentially cancels out because of the screening effect of the intervening electrons.

Values of the ionization potentials of the elements are given in Fig. 2.28. For each element the value given refers to the first ionization, i.e., the removal of but one electron from the neutral atom. In discussing the chemistry of the elements it is occasionally necessary to refer to second and higher ionizations, corresponding to further removal of

First ionization potentials table (values in electron volts):

1	2	3	4	5	6	7	8	9	10	11	12	13	14	15	16	17	18
1 H 13.6																	2 He 24.6
3 Li 5.4	4 Be 9.3											5 B 8.3	6 C 11.3	7 N 14.5	8 O 13.6	9 F 17.4	10 Ne 21.6
11 Na 5.1	12 Mg 7.6											13 Al 6.0	14 Si 8.2	15 P 11.0	16 S 10.4	17 Cl 13.0	18 Ar 15.8
19 K 4.3	20 Ca 6.1	21 Sc 6.5	22 Ti 6.8	23 V 6.7	24 Cr 6.8	25 Mn 7.4	26 Fe 7.9	27 Co 7.9	28 Ni 7.6	29 Cu 7.6	30 Zn 9.4	31 Ga 6.0	32 Ge 8.1	33 As 9.8	34 Se 9.8	35 Br 11.8	36 Kr 14.0
37 Rb 4.2	38 Sr 5.7	39 Y 6.4	40 Zr 6.8	41 Nb 6.9	42 Mo 7.1	43 Tc 7.3	44 Ru 7.4	45 Rh 7.5	46 Pd 8.3	47 Ag 7.6	48 Cd 9.0	49 In 5.8	50 Sn 7.3	51 Sb 8.6	52 Te 9.0	53 I 10.5	54 Xe 12.1
55 Cs 3.9	56 Ba 5.2	*	72 Hf 7	73 Ta 7.9	74 W 8.0	75 Re 7.9	76 Os 8.7	77 Ir 9	78 Pt 9.0	79 Au 9.2	80 Hg 10.4	81 Tl 6.1	82 Pb 7.4	83 Bi 7.3	84 Po 8.4	85 At —	86 Rn 10.7
87 Fr —	88 Ra 5.3	†	104 Ku —	105 Ha —													

*	57 La 5.6	58 Ce 6.9	59 Pr 5.8	60 Nd 6.3	61 Pm —	62 Sm 5.6	63 Eu 5.7	64 Gd 6.2	65 Tb 6.7	66 Dy 6.8	67 Ho —	68 Er —	69 Tm —	70 Yb 6.2	71 Lu 5.0
†	89 Ac 6.9	90 Th —	91 Pa —	92 U 4	93 Np —	94 Pu —	95 Am —	96 Cm —	97 Bk —	98 Cf —	99 Es —	100 Fm —	101 Md —	102 No —	103 Lr —

Fig. 2.28
First ionization potentials of the elements (in electron volts).

electrons. In every case, subsequent ionizations require increasingly large amounts of energy per electron. Furthermore, if the ionization requires breaking into a noble-gas configuration, an extra-large increase is observed. As an illustration, the successive ionization potentials for beryllium ($Z = 4$) are 9.32, 18.21, 153.85, and 217.66 eV, corresponding, respectively, to removal of the first, second, third, and fourth electrons.

2.8 Electron affinity

Also important for determining chemical properties is the tendency of an atom to pick up additional electrons. This property can be measured by the *electron affinity*, the *energy released when an electron adds to an isolated neutral atom*. When a neutral atom picks up an electron from some source, it forms a negative ion, as indicated by writing

$$X + e^- \longrightarrow X^-$$

The amount of energy released in this process is the electron affinity. Thus, the electron affinity measures the tightness of binding of an additional electron to an atom. The values for the halogen elements are given in Fig. 2.29.

Group VII elements are expected to have high electron affinity because addition of one electron leads to formation of a stable octet. The decrease of electron affinity observed from chlorine to iodine is not unexpected because the size increases in going down the group. In iodine the electron to be added goes into the fifth shell. Being farther from the nucleus, the added electron is not so tightly bound as one added to the other elements of the group. The unexpectedly low value for fluorine reflects the fact that for very small atoms there are significant mutual repulsions between the electrons in the outermost shell and these must be considered when adding another electron.

Element	Electron affinity, eV	Electronic configuration
F	3.45	$1s^2 2s^2 2p^5$
Cl	3.61	$1s^2 2s^2 2p^6 3s^2 3p^5$
Br	3.36	$1s^2 2s^2 2p^6 3s^2 3p^6 3d^{10} 4s^2 4p^5$
I	3.06	$1s^2 2s^2 2p^6 3s^2 3p^6 3d^{10} 4s^2 4p^6 4d^{10} 5s^2 5p^5$

Fig. 2.29
Electron affinities for group VII elements.

A knowledge of electron affinities can be combined with a knowledge of ionization potentials to predict which atoms can remove electrons from others. Unfortunately, the measurement of electron affinity is difficult and has been carried out for only a few elements. A method for describing the electron-attracting ability of other atoms will be discussed in the next chapter.

Exercises

*2.1 *Electron shell.* What feature of a $1s$ electron probability is most in support of the old Bohr concept of an electron shell?

*2.2 *d orbitals.* What is the difference between $d_{x^2-y^2}$ and d_{xy} orbitals? If you had an atom containing two such electrons inside an array of negative ions at the corners of an octahedron, what effect would you predict on the energies of the two orbitals of bringing in the negative ions closer to the central atom? Explain.

*2.3 *Electron spin.* Predict the total electron spin for each of the following electron configurations: (a) $1s^2$, (b) $1s^2 2s^1$, (c) $1s^2 2s^2 2p^3$, (d) $1s^2 2s^2 2p^5$, and (e) $1s^2 2s^2 2p^6 3s^2 3p^6 3d^5 4s^2$.   *Ans.* $0, \frac{1}{2}, \frac{3}{2}, \frac{1}{2}, \frac{5}{2}$

*2.4 *Magnetism.* Assuming all other orbitals to be vacant, rank the following configurations in order of increasing paramagnetism due to electron spin: p^1, p^3, p^5, d^1, d^3, and d^5.

*2.5 *Quantum numbers.* Tabulate the quantum numbers for each of the electrons in the ground-state oxygen atom.

*2.6 *Quantum numbers.* By analysis of permitted combinations of quantum numbers, show an f subshell can have but 14 electrons.

*2.7 *Orbital-filling diagram.* Draw orbital-filling diagrams for each of the following elements: C, F, Si, P, and Fe.

*2.8 *Electron configurations.* Give the proper ionic symbol for each of the following:

a $(Z = 8)$ $1s^2 2s^2 2p^6$
b $(Z = 11)$ $1s^2 2s^2 2p^6$
c $(Z = 25)$ $1s^2 2s^2 2p^6 3s^2 3p^6 3d^5$
d $(Z = 29)$ $1s^2 2s^2 2p^6 3s^2 3p^6 3d^{10}$

*2.9 *Atomic radii.* Make a graph showing atomic radius as a function of atomic number using the data of Fig. 2.24. What are the significant features of your graph, and how can they be accounted for?

*2.10 *Electron transfer.* How much energy is required to transfer one electron from an isolated sodium atom to an isolated chlorine atom?
Ans. 2.4 × 10⁻¹⁹ J

**2.11 *de Broglie wave.* Calculate the wavelength in centimeters associated with each of the following particles:

a   A 10,000-kg truck moving at 100 km/h
b   A 50-mg flea flying at 1 m/sec
c   A water molecule moving at 500 m/sec
d   A hydrogen atom moving at 5×10^7 cm/sec
e   An electron moving at 1×10^9 cm/sec

Recall that one joule is one kg-m²/sec².
Ans. 2.4 × 10⁻³⁷, 1.3 × 10⁻²⁷,
4.4 × 10⁻⁹, 7.9 × 10⁻¹¹, 7.2 × 10⁻⁹ cm

**2.12 *Electron probability.* State specifically what features of the electron probability distribution for a $1s$ electron are illustrated in each of the parts of Fig. 2.1. How would you modify each picture to make it appropriate for a $2s$ electron instead?

**2.13 *p orbitals.* Suppose you had an atom containing three equal-energy orbitals p_x, p_y, and p_z in free space. If you gradually brought up two negative ions one each on the positive and negative directions of the z axis, how would you affect the energy levels? Draw an appropriate energy-level diagram.

**2.14 *Wave functions.* We are given a wave function described by $\psi = A \sin 2\pi(x/\lambda + \nu t)$ for a wave traveling along a string. Let $A = 3.0$ cm, $\lambda = 12.0$ cm, and $\nu = 4.0$ sec⁻¹. What will be the value of the wave function under the following conditions?

a $t = 0$, $x = 1.0$ cm
b $t = 0$, $x = 2.0$ cm
c $t = 1.0$ sec, $x = 1.0$ cm
d $t = 2.0$ sec, $x = 2.0$ cm   *Ans. 1.5, 2.6, 1.5, 2.6 cm*

**2.15 *Particle in the box.* In a box of dimension 50 nm you have a particle with wavelength 20 nm. Sketch the cross section of the box, indicating the regions of maximum probability and the regions of minimum probability.

2.16   Particle in the box. Suppose you have an electron in a box of dimension 50 nm. What are the energies of its five lowest states? Recall that one joule = one kg-m^2/sec^2.

Ans. 2.4 × 10^{-21} J (× 1, 4, 9, °16, 25, respectively)

2.17   Schrödinger equation. Tell the meaning of each term in the Schrödinger wave equation. How would you modify Eq. (6) as shown on page 58 so that it would apply to the particle-in-the-box problem?

2.18   Electron shell. How do the electron probability distributions for 1s and for 2s electrons reconcile with the old Bohr picture of the shell?

2.19   Node. What is meant by the term "nodal surface" as applied to a 2s electron? Show that the node occurs at a distance that is inversely proportional to the charge on the nucleus.

2.20   Wave function. Show how the wave function for a $2p_z$ electron changes with increasing angle away from the z axis.

2.21   p orbitals. In the second shell, the p orbitals can be represented as p_x, p_y, and p_z or as $+1$, 0, and -1. What is the difference between these two representations, and how can they be made equivalent to each other?

2.22   Wave functions. In Bohr's third shell, there are three kinds of electrons. What is the essential difference between their respective wave functions and their respective electron probability distributions?

2.23   Stern-Gerlach experiment. How would you expect beams of sodium atoms and beams of magnesium atoms to differ in their behavior in a Stern-Gerlach type of experiment? Explain.

2.24   Electron spin. Suppose there were no such thing as electron spin. How would the structure of the periodic table presumably have to be modified, assuming all other quantum rules remain in effect?

2.25   Magnetism. Which of the following atoms in their ground states are likely to be paramagnetic: H, N, O, Ne, Ca, Al, and Zn? Arrange your choices in probable order of increasing paramagnetism.

2.26   Pauli exclusion principle. What is the Pauli exclusion principle? How does it account for the existence of groups and periods in the periodic table?

2.27   Atomic radii. Suggest a reason why the atomic radius for hydrogen as shown in Fig. 2.24 is so much smaller than the radius of the first Bohr orbit.

2.28   Ionization potential. What wavelength of light would it take to ionize a collection of sodium atoms?   *Ans. 243 nm*

2.29   de Broglie wave. Explain how the concept of the de Broglie wave allows a neutron or proton but not an electron to "exist" in a nucleus.

***2.30   *d orbitals*. What is the single essential feature that distinguishes *d* orbitals from *p* orbitals?

***2.31   *Wave functions*. You are given a vibrating violin string fixed at both ends vibrating so that two complete wavelengths fit into the string length, 90 cm. Plot the wave function at 10-cm intervals along the string, given that the amplitude is 1.5 mm.

***2.32   *Particle in the box*. Suppose you have an electron confined in a box of dimension 0.2 nm, i.e., about as big as an atom. How much energy would be emitted if the electron went from its first excited state to the ground state for this system? What wavelength of light would this energy correspond to?   *Ans. 44 nm*

***2.33   *Schrödinger equation*. Show that the wave function shown on page 56 for the particle-in-the-box problem satisfies the Schrödinger wave equation shown on page 58. Show that the only permitted values of E are given by $n^2h^2/8mL^2$. (*Hint*: Derivative of the sine is the cosine; derivative of the cosine is the negative sine.)

***2.34   *Wave function*. The atomic number enters twice in the expression for the wave function of a 1*s* electron. Indicate how each of these would affect the form of ψ, and tell how it affects the probability of finding the electron.

***2.35   *Wave functions*. At what values of r would you predict nodes in the wave function for a 3*s* electron bound to a nucleus of charge $+11$?   *Ans. 0.0092, 0.034 nm*

***2.36   *Wave function*. At what value of the polar angle θ would you need to look to have zero probability of finding the $3d_{z^2}$ electron?   *Ans. 54°45′*

***2.37   *Pauli exclusion principle*. Discuss critically the following statement: The Pauli exclusion principle is the main reason why atoms do not collapse to a point.

***2.38   *Many-electron atoms*. Suppose you had an atom consisting of a nucleus charged Z and one 2*s* electron. What would be the probable effect on the 2*s* orbital of adding a $2p_z$ electron to the same atom? Mention at least two effects.

***2.39   *Many-electron atoms*. Explain why adding the group of electrons $3p_x$, $3p_y$, and $3p_z$ affects a 4*s* orbital less than adding the group of electrons 3*s*, $3p_z$, and $3d_{z^2}$.

***2.40   *Electron affinity*. You are given the Avogadro number of X atoms. If half the X atoms transfer one electron to the other half of the X atoms, 409 kJ must be added; if all the resulting X$^-$ ions are subsequently converted to X$^+$ ions, an additional 733 kJ must be added. Figure out in electron volts the ionization potential and electron affinity of X.   *Ans. IP = 11.84 eV; EA = 3.36 eV*

The concept of a molecule is an old one, the term itself apparently having been introduced in about 1860 by Stanislao Cannizzaro in his description of gases as consisting of tiny aggregates of atoms. The idea of atomic clusters is, of course, much older; Dalton, for example, talked about "compound atoms" around 1800. At present, any electrically neutral aggregate of atoms held together strongly enough to be considered as a unit is called a *molecule*. The net attractive interaction between two atoms within a molecule is called a *chemical bond*.

Hydrogen gas is composed of aggregates of two hydrogen atoms. Water vapor is composed of aggregates of two hydrogen atoms and one oxygen atom. Solid sulfur consists of aggregates of eight sulfur atoms. In each of these cases the aggregate is usefully called a "molecule." On the other hand, in solid sodium chloride there are no simple aggregates consisting of a few atoms. All the sodium ions and all the chlorine ions in a given crystal are bound into one giant aggregate. The term "molecule" is not useful in such cases.

The fact of molecules is generally agreed on. The problem is what is the best way to give an adequate description of the electron distribution in them. Why do they stick together? Why do they have the shape and properties they do? Why is the hydrogen molecule H_2 and not H_3 or H_4? Why is H_2O, in its ground state, neither linear nor right-angled but bent at an angle of $104°31'$? The answers to these questions not only are important for a fundamental understanding of the behavior of matter but also lay the bases for considering applied problems. Why do many materials of construction generally show failure to tensile or shear stress at less than 1 percent of their theoretical strength? Why is the genetic-code-carrying DNA molecule folded into a double helix? Why does advancing the spark on an internal-combustion engine help alleviate the hydrocarbon emission problem but aggravate pollution by oxides of nitrogen? An intelligent route to solving these practical problems requires knowing something about electrons in molecules.   79

3.1 Electrons in molecules

In an isolated atom each electron is under the influence of only one nucleus and the other electrons. When two atoms come together, the electrons of one atom come under the influence of the electrons and nucleus of the other. The interaction might produce an attraction between the two atoms. If that is so, an electronic rearrangement must have occurred to give a more stable state. In other words, the formation of a chemical bond suggests that the molecule represents a state of lower energy than the isolated atoms represent.

A detailed description of electrons in molecules is a difficult problem. Two general approaches can be used: One is to consider the entire molecule as a unit with all the electrons moving under the influence of all the nuclei and all the other electrons. This approach recognizes that each electron belongs to the molecule as a whole and may move throughout the entire molecule. The spatial distributions of the electrons in the molecule are called *molecular orbitals* (MO's) and can be thought of in the same way as the orbitals of electrons in isolated atoms. The other approach to describing molecules is simpler but less correct. It assumes that the atoms in a molecule are very much like isolated atoms except that one or more electrons from the outer shell of one atom are accommodated in the outer shell of another atom. This method of describing molecules is called the *valence-bond* (*VB*) *method;* it utilizes directly the orbitals of isolated atoms.

In order to point up the difference in the two ways of viewing molecules, let us consider the hydrogen molecule. It is formed from two hydrogen atoms, each with one proton and one electron. In the molecular-orbital approach H_2 is visualized as consisting of two protons at some distance apart with the two electrons in an energy level that is spread out over the whole molecule. In the valence-bond approach, H_2 is visualized as consisting of two hydrogen *atoms* side by side with the electron shell of each atom the same as if the hydrogen atom were isolated except that now it may contain part of the time both electrons. No matter which picture is used, MO or VB, the molecule is held together partly because the electrons have been spread over more space and partly because the attraction of the two positive protons for the two negative electrons exceeds the repulsion between the two protons plus the repulsion between two electrons.

In the more complicated case of hydrogen chloride the molecule is formed from one hydrogen atom and one chlorine atom. The hydrogen atom contributes a $+1$ nucleus and 1 electron; the chlorine atom, a $+17$ nucleus and 17 electrons. In the molecular-orbital approach, the molecule is visualized as consisting of the two nuclei at some distance apart, with 18 electrons placed in various energy levels of the molecule as a whole. In the valence-bond approach, the molecule is visualized as consisting of one hydrogen *atom* and one chlorine *atom* side by side. The hydrogen atom is assumed to be the same as when it is alone except that part of the time it now may contain, besides its own electron, one of the electrons from the chlorine atom. As for the chlorine atom, the two inner shells are assumed to be unchanged. However, part of the time the outer shell contains, besides the original seven electrons, one additional electron from the hydrogen atom. The one electron from the hydrogen and one electron from the chlorine are considered as a pair of electrons shared between the atoms. The pair is regarded as holding the molecule together because it is attracted to both nuclei.

Certainly in a molecule the energy levels of many, if not all, of the electrons are changed from those of isolated atoms. Therefore, it would be desirable to discuss chemical bonding exclusively in terms of molecular orbitals. However, the valence-bond approach is so simple that it remains in great use among chemists. For either approach, it is convenient to consider only outer electrons as the ones involved in the *valence,* or chemical binding. These electrons are therefore referred to as the *valence electrons.* Sometimes they are shown as dots around the symbol for the element.

3.2   Ionic bonds

In discussing chemical bonds, we shall assume that they can be described as *ionic,* in which case electrons are completely transferred from one atom to another, or as *covalent,* in which case electrons are shared between atoms.

The formation of an ionic bond is favored in the reaction of an atom of low ionization potential with an atom of high electron affinity. An example of such a reaction is the one between atoms of cesium and chlorine. Cesium has low ionization potential; i.e., not much energy is required to pull off the outer electron. Chlorine has high electron affinity; i.e., considerable energy is released when an electron adds to the neutral atom to produce a noble-gas configuration (so-called "shell of eight," or octet). When a cesium atom and a chlorine atom combine in chemical reaction, an electron is transferred from the cesium to the chlorine. The cesium ion with its resulting positive charge attracts the chloride ion with its resulting negative charge. The attraction is called an *ionic bond* or, sometimes, an *electrovalent bond.**

The energy changes associated with formation of the ionic bond in cesium chloride can be thought of in three steps:

$$Cs(5s^2 5p^6 6s) \longrightarrow Cs^+ (5s^2 5p^6) + e^- \tag{1}$$

$$Cl(3s^2 3p^5) + e^- \longrightarrow Cl^-(3s^2 3p^6) \tag{2}$$

$$Cs^+ + Cl^- \longrightarrow [Cs^+][Cl^-] \tag{3}$$

* In the solid compound cesium chloride each Cs^+ is at the center of a cube surrounded by eight Cl^-'s, and each Cl^- in turn is at the center of a cube surrounded by eight Cs^+'s. The force of attraction is equal to each of the eight near neighbors. However, it is not generally useful to describe this situation as involving eight distinguishable bonds since the ionic bond is not confined to a particular orientation in space. Only in the case of gaseous CsCl, in which ion pairs Cs^+Cl^- may exist, does it make sense to speak of a single ionic bond between a particular positive ion and a particular negative ion.

The first step requires energy equal to the ionization potential of cesium, 3.89 eV; the second step liberates energy equal to the electron affinity of chlorine, 3.61 eV. It should be noted that the energy liberated in step (2) is not enough to compensate for the energy required in step (1); furthermore, this cesium-chlorine case is the most favorable one for electron transfer, since cesium has the lowest ionization potential measured and chlorine has the highest known electron affinity. Even in this case ionic-bond formation can occur only because positive and negative ions attract each other to liberate additional energy.

For step (3), energy is liberated when a positive ion of q_1 electronic charge units attracts a negative ion of q_2 electronic charge units to bring the nuclear centers to r nm apart; the magnitude of the liberated energy in electron volts is $1.44 \, q_1 q_2 / r$. In the case of Cs^+ and Cl^- the internuclear spacing will be about 0.35 nm, and thus the energy liberated by the ion pairing is $1.44(1)(1)/0.35$, or 4.1 eV. This more than makes up the deficit in the first two steps; so, clearly, it is step (3) that makes ionic-bond formation possible. Actually, for forming the solid compound step (3) is even more favorable than calculated here for simple ion pairing, since in the solid each positive ion has more than one negative-ion neighbor, and vice versa.

In forming compounds by this process of electron transfer, there must be a balance of electrons gained and lost. The reaction between cesium and chlorine requires one atom of cesium for every atom of chlorine. When barium reacts with chlorine, each barium atom loses two valence electrons $(6s^2)$ to form Ba^{2+}. Hence, two chlorine atoms, each picking up one electron, are required to balance this. The compound formed, barium chloride, contains one double-positive barium ion for every two singly negative chloride ions, as indicated by the formula $BaCl_2$. It should be noted that after reaction all these atoms are left with noble-gas octets in their outermost shells. It is a general rule that when ionic bonds are formed, enough electrons are transferred that the ions produced have octets of electrons.

Since, in general, the elements on the left of the periodic table have low ionization potential and the elements on the right have high electron affinity, ionic bonds are favored in reactions between these elements. Thus, any alkali metal (group I) reacts with any halogen (group VII) to form an ionic compound. Similarly, most of the group II elements react with the halogens or with group VI elements to form ionic bonds. In general, these ionic compounds resemble sodium chloride in that at room temperature they are white, brittle solids which dissolve in water to give conducting solutions. They melt at relatively high temperatures.

3.3 Covalent bonds

Most bonds cannot be adequately represented by assuming a complete transfer of electrons from one atom to another. For example, in the hydrogen molecule H_2 it would be unreasonable to say that one hydrogen atom pulls an electron from the other, identical hydrogen atom.

Fig. 3.1
Valence-bond view of covalent-bond formation in H_2.

Rather it is assumed that the electrons are shared between the atoms, and the bond is called *covalent*. The problem, of course, is what is meant by "sharing of the electrons." In the molecular-orbital description the shared electrons are attracted to both nuclei and are distributed in an orbital that encompasses the whole molecule. On the other hand, in the valence-bond description each of the shared electrons is regarded as being at any given time associated with one of the nuclei or the other but not with both simultaneously. Stated otherwise, in the VB description each electron is found in the atomic orbital of one atom or the other, and there is electron exchange in which the two electrons trade places. (In certain cases the valence-bond description also involves having both shared electrons on one atom; so there is a so-called "ionic" contribution to the binding.) There is a considerable difference in the concept of electron sharing as viewed from the two approaches, molecular-orbital and valence-bond. We shall examine each of these in turn, starting first with the simpler valence-bond approach and then returning to the molecular-orbital approach.

Figure 3.1 shows schematically the formation of a covalent bond in the H_2 molecule. Two H atoms which retain their identity come together and are bound into a diatomic unit. Figure 3.2 shows how the potential energy changes during bond formation. When the two atoms are far apart, their potential energies are independent of each other and are arbitrarily set at zero (off the far right of the diagram). As the two atoms approach, there is an attraction between them; the energy of the system decreases. This decrease of potential energy is shown by the solid line which drops slowly first, then steeply from right to left to a minimum value. To the left of the minimum, the potential energy shoots up steeply because of repulsion between like charges at every close distances of approach. The position of the minimum which occurs at 0.074 nm corresponds to the bond length, i.e., the average distance between the nuclei in the H_2 molecule. Because the potential energy is lowest at this distance, the two H atoms tend to favor this spacing.

Why should there be a minimum in the potential-energy curve? Why are the two atoms attracted? These basic questions are extremely difficult to answer, and no single method gives a completely satisfactory answer in all cases. The valence-bond method assumes that when the two H atoms come closer, there is an increasing chance that an electron from the $1s$ orbital of one atom transfers to the $1s$ orbital of the other atom. However, because of repulsion between electrons, it is not likely that both electrons should stay on the same atom at the same time. An exchange occurs; either electron may be found on either atom. This is tantamount to saying that both electrons occupy the same space and, according to the Pauli exclusion principle, is possible only if the two

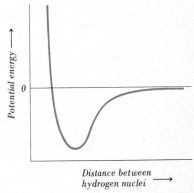

electrons exchanging have opposite spins. Consequently, the two electrons forming the covalent bond must have their spins opposite, or paired.

However, the question that still remains is why the energy decreases on bond formation. Magnetic attraction between opposite spins is much too feeble by several orders of magnitude to account for the energies involved. There are two major contributions to energy lowering: (1) A general result of wave mechanics is that confining an electron to a smaller volume raises its energy whereas spreading it out over a larger space lowers its energy. Thus, as the electron distribution changes from confinement on one atom to being spread over two atoms, the energy of the electron diminishes. (See the particle-in-the-box discussion on page 58.) (2) As the two H atoms come together, each electron feels the attraction of the other nucleus, an effect which more than compensates for the repulsion between like charges. At small internuclear separation, electron confinement and like-charge repulsion become dominant; hence, there is the steep rise in potential energy shown on the left side of Fig. 3.2.

The sharing of an electron pair as described above constitutes a *single bond*. It is customarily represented by a single dash or, in some cases, by a pair of dots. Since the sharing results from electron exchange between *two* atoms, the covalent bond is restricted as joining *two* atoms. In the case of a hydrogen atom, containing but one electron, only one bond can be formed to another atom. This is the reason why hydrogen does not normally form H_3 or H_4, since to do so at least one hydrogen atom would have to form more than one covalent bond.

The fluorine molecule F_2 is somewhat more complicated. It can be visualized as resulting from the sharing of an electron pair between the p orbitals of two fluorine atoms. A fluorine atom has in its ground state the electron configuration $1s^2 2s^2 2p^5$. All the subshells are filled except for the $2p$, which can accommodate six electrons. Of the three orbitals making up the $2p$ subshell, two are fully occupied by electron pairs and are not involved in the bonding. The third orbital has a single electron which can exchange with a corresponding electron of opposite spin on a second atom. Thus, the valence-bond description leads to a single covalent bond which, like the bond in H_2, involves a single pair of electrons but, unlike H_2, involves p orbitals rather than s orbitals.

When a hydrogen atom and a fluorine atom are brought together, the molecule HF is formed. This molecule may be visualized as resulting from the sharing of an electron pair between the $1s$ orbital of the hydrogen and a $2p$ orbital of the fluorine. The HF bond is also a single covalent bond, but it differs in a fundamental way from the bonds in H_2 and F_2. The reason for the difference is the unequal attraction of hydrogen and fluorine for a shared pair of electrons. In this case the electron pair spends more time on the fluorine than on the hydrogen. We shall return to this question in the following section.

In each of the covalent bonds so far described, the atoms involved can be visualized as having attained a stable, i.e., noble-gas, electron configuration. In the bonding of hydrogen the additional electron can be thought of as completing the $1s$ subshell, as occurs in the noble gas helium. In the bonding of fluorine the additional electron can be thought of as completing the $2p$ subshell, as occurs in the noble gas neon. In the majority of chemical compounds, covalent bonding occurs so as to produce noble-gas configurations. Except for bonds to hydrogen, this means completing the s and p subshells of the outermost shell so that they contain eight electrons. This is the basis of the so-called "octet rule," which states that *when atoms combine, the bonds formed are such that each atom is surrounded by an octet of electrons.** In F_2, for example, each fluorine has completed its octet and does not bind additional fluorine atoms.

Two parameters are generally used to describe covalent bonds: One is the length of the bond, i.e., the distance between the nuclei of the atoms being bonded together. The other is the strength of the bond, i.e., the energy that is required to break the bond apart so as to form the component atoms. Figure 3.3 tabulates bond lengths l and bond

Molecule	l, nm	D, kJ/mol
H_2	0.074	432
Li_2	0.267	105
Be_2 (*nonexistent*)	. . . . .	
B_2	0.159	289
C_2	0.131	628
N_2	0.109	941
O_2	0.121	494
F_2	0.143	151
Ne_2 (*nonexistent*)	. . . . .	

Fig. 3.3
Bond lengths l and bond strengths D for some diatomic molecules.

strengths D for some representative molecules. The bond strengths are given in kilojoules per mole of molecules. As can be seen, the diatomic molecules of the second period of the periodic table show a minimum in length and a maximum in strength at about the middle of the period. A question that might be raised is why such effects occur.

* An interesting modification of the octet rule has been suggested by J. W. Linnett, professor of chemistry at Cambridge University in England, who proposes that an octet can be considered as two quartets differing in electron spin. The four electrons of each quartet are disposed tetrahedrally about the atom, and the two quartets may be independent of each other. In forming bonds, electron sharing occurs to complete each quartet. In the case of F_2 there are seven electrons of each spin type in the whole molecule—three on each fluorine atom and one shared between the atoms. Each fluorine atom thus completes both quartets by sharing, and there is no net electron spin, since the two sets just cancel. The spin quartets are completed by having two tetrahedra joined through a common corner. The two separate spin systems can be pictured as follows:

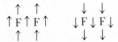

Besides single bonds, *double* and *triple* bonds may be formed in order that an atom can complete its octet. These correspond, respectively, to sharing of two pairs and three pairs of electrons between the bonded atoms. A triple bond, for example, would be typified by N_2. In this case each N atom (ground-state configuration $1s^2 2s^2 2p^3$) completes its octet by sharing three pairs of electrons between its p orbitals and the p orbitals of the other N. This is sometimes represented by the *dot formula* $:N:::N:$, where each N atom is regarded as being surrounded by three pairs of shared electrons and one pair of unshared $2s$ electrons. (In dot formulas, inner-shell electrons are not shown.)

In general, for the same pair of bonded atoms triple bonds are shorter than double bonds and double bonds are shorter than single bonds. Experiment shows, for example, that the carbon-carbon distance (center to center) is 0.120 nm in acetylene ($HC:::CH$), 0.133 nm in ethylene ($H_2C::CH_2$), and 0.154 nm in ethane ($H_3C:CH_3$).

In all the cases discussed above each shared pair of electrons involves one electron from each of the bonded atoms. There are also cases in which one atom in the bond contributes both of the electrons that are to be shared. Some examples are

$$
\begin{array}{c}
H \\
\cdot\cdot \\
H:N: \\
\cdot\cdot \\
H
\end{array}
+ H^+ \longrightarrow
\left[
\begin{array}{c}
H \\
\cdot\cdot \\
H:N:H \\
\cdot\cdot \\
H
\end{array}
\right]^+
$$

$$
\begin{array}{cc}
H & :F: \\
\cdot\cdot & \cdot\cdot\ \cdot\cdot \\
H:N: + & B:F: \\
\cdot\cdot & \cdot\cdot\ \cdot\cdot \\
H & :F:
\end{array}
\longrightarrow
\begin{array}{c}
H:F: \\
\cdot\cdot\ \cdot\cdot \\
H:N:B:F: \\
\cdot\cdot\ \cdot\cdot \\
H:F:
\end{array}
$$

$$
\left[
\begin{array}{c}
\cdot\cdot \\
:S: \\
\cdot\cdot
\end{array}
\right]^{2-}
+
\begin{array}{c}
\cdot\cdot \\
\cdot S: \\
\cdot\cdot
\end{array}
\longrightarrow
\left[
\begin{array}{c}
\cdot\cdot\ \ \cdot\cdot \\
:S:S: \\
\cdot\cdot\ \ \cdot\cdot
\end{array}
\right]^{2-}
$$

Such bonds are sometimes called *coordinate covalent,* or *donor-acceptor,* bonds. The use of such names is generally unnecessary, since the final bond is independent of the way it was formed. For example, in the first equation above the four bonds in the ammonium ion NH_4^+ are identical, although only one of them seems to be a so-called "coordinate" bond.

3.4   Polarity of bonds

Because electrons may be shared unequally between atoms, it is necessary to have some way of describing the electric-charge distribution in a bond. The usual way is to classify bonds as *polar* or *nonpolar.* As examples, the bonds in H_2 and F_2 are called nonpolar; the bond in HF, polar.

Why are the covalent bonds in H_2 and F_2 called nonpolar? The term "pole" generally refers to a center of charge distribution. In both H_2 and F_2 the "center of gravity" of the negative-charge distribution, which is at the center of the molecule since the electron pair is just as probably found with one nucleus as with the other, is the same as the center of the positive-charge distribution. The molecule is electrically neutral in two senses of the word. Not only does it contain an equal number of positive and negative charges (protons and electrons), but also the center of the positive charge coincides with the center of the negative charge. The molecule is a *nonpolar molecule*; it contains a *nonpolar bond* in which an electron pair is *shared equally* between two atomic kernels.

In the case of HF the bond is polar; the center of positive charge does not coincide with the center of negative charge. The molecule as a whole is electrically neutral because it contains equal numbers of positive and negative particles. However, owing to unequal sharing of the electron pair the F end of the molecule appears negative with respect to the H end. Polarity arises because the shared pair of electrons spends more time on the F than on the H, *not because* F *has more electrons than* H. (The charge of the unshared electrons is balanced by the greater positive charge of the F nucleus.)

As another example of a polar covalent bond, consider the bond between F and Cl in the molecule ClF. Cl, like F, has one vacancy in its p subshell. Exchange involving the p orbitals of the Cl and F can occur to produce a single covalent bond. However, the covalent bond between Cl and F has the pair of electrons shared unequally, spending more of its time on the F than on the Cl. The F end of the molecule therefore appears negative with respect to the Cl end. In Fig. 3.4 this polarity is indicated by a + and a − to show where the centers of positive and negative charge are and by a positive-tailed arrow in the direction of electron shift. Each molecule as a whole is electrically neutral—there are just as many positive charges as there are negative charges in the whole molecule—but there is a dissymmetry in the electric distribution. Molecules in which positive and negative centers do not coincide are called *polar molecules,* and any bond in which sharing between two atoms is *unequal* is a *polar bond.*

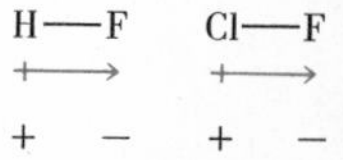

In molecules such as HF and ClF there are two centers of charge. Such molecules (or such bonds) are called dipoles. A *dipole* consists of a *positive and an equal negative charge separated* by some distance. Quantitatively, a dipole is described by giving its *dipole moment,* which is equal to the *charge times the distance between the positive and negative centers.* The unit for measuring dipole moments is called the *debye,* after Peter J. W. Debye, a Dutch chemist who first described polar molecules. One debye corresponds to the dipole moment which would be produced by a negative charge equal to 0.208 that of an electron separated by a distance of 1×10^{-8} cm from an equal but opposite charge. (The factor 0.208 comes from the fact that the electron charge used to be given as 4.80×10^{-10} esu, and one positive electron that was 1×10^{-8} cm away from one negative electron was assigned a dipole moment of 4.80×10^{-18} esu-cm, or 4.80 debyes. The reciprocal of 4.8 is 0.208.) The magnitude of the dipole moment measures the tendency of the dipole to turn when placed in an electric field.

H——F   Cl——F

⟶   ⟶

+   −   +   −

Fig. 3.4

Polar molecules.

As shown in Fig. 3.5 each dipole turns because its positive end is attracted to the negative plate and its negative end to the positive plate. Since the positive and negative centers are part of the same molecule, the molecules can only turn; there is no migration toward the plates.

The behavior of dipoles in an electric field gives an experimental method for distinguishing between polar and nonpolar molecules. The experiments involve the determination of a property called the *dielectric constant* (see Appendix 5.6), which can be measured as follows: It is observed that an electric capacitor (two parallel, metallic plates, like those shown in Fig. 3.5) has the ability to store electric charge. The *capacitance*, i.e., the amount of charge that can be put on the plates for a given voltage, depends upon the substance between the plates. The dielectric constant is defined as the ratio of the capacitance with the substance between the plates to the capacitance with a vacuum between them.

Fig. 3.5
Behavior of dipoles.

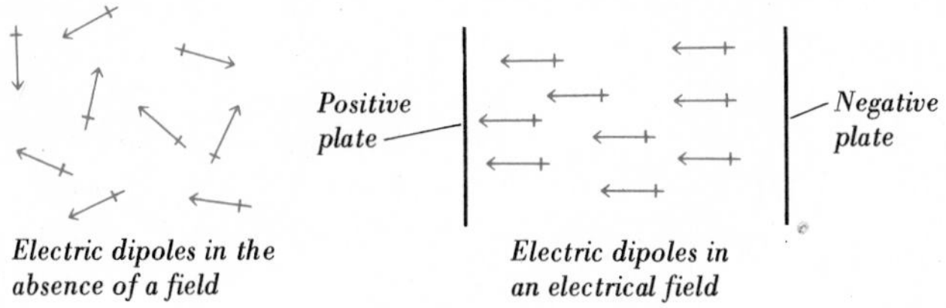

Electric dipoles in the absence of a field

Positive plate

Negative plate

Electric dipoles in an electrical field

In general, a substance which consists of polar molecules has a high dielectric constant; i.e., a capacitor can store much more charge when such a substance is between its plates. This high dielectric constant can be thought of as arising in the following way: As shown in Fig. 3.5 dipoles tend to turn in a charged capacitor so that negative ends are near the positive plate and positive ends are near the negative plate. This partially neutralizes the charge on the plates and permits more charge to be added. Thus, measurement of the dielectric constant gives information about the polarity of molecules. The fact that hydrogen gas has little effect on the capacity of a capacitor (dielectric constant 1.00026 as compared with 1.00000 for a perfect vacuum) confirms the idea that H_2 molecules are nonpolar.

The quantitative calculation of dipole moments of individual *bonds* from measured dielectric constants is complicated, because unshared electrons may contribute to the electric dissymmetry of the molecule and because the presence of charged plates can *polarize*, or temporarily distort, the charge distribution in molecules. Polarizability such as this can be described in terms of *induced* dipoles. The induced dipole can be distinguished from the permanent dipole by alternating the charges on the capacitor plates from positive to negative at such a high frequency that the molecules cannot turn their permanent dipole moments rapidly enough to keep up with the high-frequency field. In such case only the polarizability contributes to the dielectric constant. The permanent moment can then be found by taking the difference. The permanent dipole can also be found by measuring the *static dielectric constant* (i.e., the one at very low frequency) as a function of temperature. The contribution of the permanent dipole moment decreases as the temperature is raised because of increasing disorder at high temperatures. This hinders the lining up of dipoles with the field.

It is easy to predict whether a diatomic molecule will be polar or nonpolar. If the two atoms are alike, the *bond* between them must be nonpolar, and therefore the *molecule* is nonpolar. If the two atoms are different, the *bond* is polar, and the *molecule* is also polar. The degree of polarity of diatomic molecules increases as the atoms become more unlike in electron-pulling ability. It is not so easy to predict the polar nature of a molecule containing more than two atoms. Such a *molecule* can be nonpolar even though all the *bonds* in the molecule are individually polar. Carbon dioxide, CO_2, is an example. As shown in Fig. 3.6 the two oxygen atoms are bonded to the carbon atom. Since oxygen attracts the shared electrons more than carbon does, each carbon-oxygen bond is polar, with the shared electrons spending more time near the oxygen than near the carbon. The polarity of each bond is as shown in the figure. Because the molecule is linear, the effect of one dipole cancels the effect of the other. As a result, when carbon dioxide molecules are placed in an electric field, they do not line up, because any turning action of one bond is counteracted by the opposite turning action of the other bond. Carbon dioxide has a low dielectric constant.

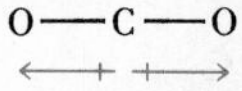

Water, H_2O, is a triatomic molecule in which two hydrogen atoms are bonded to the same oxygen atom. There are two different possibilities for its structure: The structure may be linear, with the three atoms arranged in a straight line, or the atoms may be arranged in the form of a bent chain. The two possibilities are shown in Fig. 3.7. The fact that water has a very high dielectric constant supports the bent structure. The linear structure would represent a nonpolar molecule in which the two polar bonds would be placed in line; so there would be no net dipole moment. In the actual molecule of water the two bond dipoles do not cancel out but instead, owing to the bent structure, give a resultant moment as shown on the right of the figure.

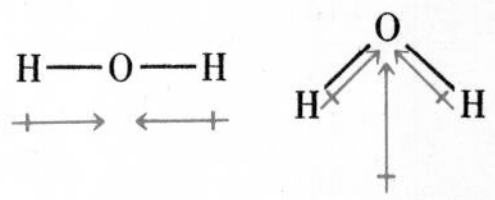

Ammonia, NH_3, is also a polar molecule. This rules out the possibility that the molecule might be planar with three H—N bonds pointing toward the middle of an equilateral triangle. Instead, the true structure has the three H atoms lying at the corners of the base of a triangular pyramid with the N atom at the apex. Each bond dipole H—N points upward toward the apex, and the three combine to give a resultant moment along the altitude of the pyramid.

In the light of the foregoing discussion of polar bonds, it is important to note that *there is no sharp distinction between ionic and covalent bonds.* In a chemical bond between atoms A and B, all gradations of polarity are possible, depending on the nature of A and B. If A and B have the same ability to attract electrons, the bond is nonpolar. If the electron-pulling ability of B is greater than that of A, the shared electrons spend more time on B, and the bond becomes more polar the greater the difference. If the electron-pulling ability of B greatly exceeds that of A, the electron pair will not be shared at all but will

O—C—O

Fig. 3.6

Nonpolar CO_2 *molecule in which individual C—O bond dipoles compensate each other.*

H—O—H

No net
moment

Resultant
moment

Fig. 3.7

Possible configurations of an H_2O *molecule.*

spend all its time on B. The result will be a negative ion B⁻ and a positive ion A⁺; the bond will be ionic.

3.5 Electronegativity

In the preceding section we referred to the electron-pulling ability of atoms in molecules. This *relative ability to attract shared electrons* is known as the *electronegativity*. R. S. Mulliken suggested that a quantitative measure of this property could be obtained by taking the average of the ionization potential and the electron affinity of the individual atoms. That both these quantities must be considered can be seen from the following argument: The bond in ClF consists of an electron pair shared unequally between F and Cl. The preference of the electron pair for one atom or the other depends on how much energy is required to pull an electron from one atom (the ionization potential) and how much energy is released when the electron is added to the other atom (the electron affinity). In ClF the electron pair spends more time on F than on Cl because the net energy required to transfer an electron from Cl to F is less than the net energy required to transfer an electron from F to Cl. In calculating the energy required for these transfers it is necessary to know both the ionization potential and the electron affinity of each atom. Unfortunately, electron affinities have been measured for only a very few elements; so the Mulliken method of evaluating the electron-pulling ability of atoms in molecules can be used in but a few cases.

The concept of electronegativity was actually first introduced by Linus Pauling in 1932. By using, as described in the next section, the various properties of molecules, such as dipole moments and energies required to break bonds, he was able to set up a useful scale of electronegativity comprising most of the elements. The numerical values, which have to be used with caution, are shown in Fig. 3.8. They describe roughly the relative tendency of an atom in forming a bond to go to a negative condition, i.e., to attract a shared electron pair. Fluorine (4.0) has been assigned the highest electronegativity of any element in the periodic table. The noble-gas elements have only recently been found to form chemical bonds, and their values have not yet been agreed upon. Otherwise, as we go from left to right across a period (increasing nuclear charge), there is a general increase in the electronegativity. The elements at the far left of the periodic table have low values. The elements at the right have high values. For the group VII elements, which are assigned the values F, 4.0; Cl, 3.0; Br, 2.8; and I, 2.5, the decreasing order is regular, unlike the order of electron affinities. In general, electronegativity decreases as we go down a periodic group (size increases).

Of what use are these values of electronegativity? For one thing, they can be used in predicting which bonds are ionic and which covalent. Since the electronegativity indicates the relative attraction for electrons, two elements of very different electronegativity, such as Na (0.9) and Cl (3.0), are expected to form ionic bonds. Thus, electro-

1 H 2.1																	2 He —
3 Li 1.0	4 Be 1.5											5 B 2.0	6 C 2.5	7 N 3.0	8 O 3.5	9 F 4.0	10 Ne —
11 Na 0.9	12 Mg 1.2											13 Al 1.5	14 Si 1.8	15 P 2.1	16 S 2.5	17 Cl 3.0	18 Ar —
19 K 0.8	20 Ca 1.0	21 Sc 1.3	22 Ti 1.5	23 V 1.6	24 Cr 1.6	25 Mn 1.5	26 Fe 1.8	27 Co 1.8	28 Ni 1.8	29 Cu 1.9	30 Zn 1.6	31 Ga 1.6	32 Ge 1.8	33 As 2.0	34 Se 2.4	35 Br 2.8	36 Kr —
37 Rb 0.8	38 Sr 1.0	39 Y 1.2	40 Zr 1.4	41 Nb 1.6	42 Mo 1.8	43 Tc 1.9	44 Ru 2.2	45 Rh 2.2	46 Pd 2.2	47 Ag 1.9	48 Cd 1.7	49 In 1.7	50 Sn 1.8	51 Sb 1.9	52 Te 2.1	53 I 2.5	54 Xe —
55 Cs 0.7	56 Ba 0.9	57–71 — 1.1–1.2	72 Hf 1.3	73 Ta 1.5	74 W 1.7	75 Re 1.9	76 Os 2.2	77 Ir 2.2	78 Pt 2.2	79 Au 2.4	80 Hg 1.9	81 Tl 1.8	82 Pb 1.8	83 Bi 1.9	84 Po 2.0	85 At 2.2	86 Rn —
87 Fr 0.7	88 Ra 0.9	89–103 Ac–Lr 1.1–1.3	104 Ku —	105 Ha —													

Fig. 3.8

Pauling scale of electronegativities for the various elements.

negativities support the expectation that the alkali elements and the group II elements form essentially ionic bonds with the elements of groups VI and VII. Two elements of about equal electronegativity, such as C (2.5) and H (2.1), are expected to form covalent bonds.

Furthermore, electronegativities can be used to predict polarity of covalent bonds. The further apart in electronegativity two elements are, the more polar the bond should be. Thus, the bond between H (2.1) and N (3.0) is more polar than that between H (2.1) and C (2.5). In both cases the H end should be positive, since H has the lower electronegativity.

3.6 Bond energies and the scale of electronegativity

One method for setting up the scale of electronegativities involves the use of bond energies. *Bond energy* is defined as the *energy required to break a bond so as to form neutral atoms*. It can be determined experimentally by measuring the heat involved in the decomposition reaction or by measuring spectroscopically the energy difference between the molecule in its lowest vibrational state and the completely dissociated state. The relation between bond energy and electronegativity can be seen from the following example: It is found that 431 kJ of heat is required to break the Avogadro number of H_2 molecules into individual atoms. Thus, the bond energy of H_2 is 431 kJ per Avogadro number of bonds, or 7.16×10^{-22} kJ per bond. Because the sharing of the electron pair is equal between the two H atoms, it would be reasonable

Section 3.6
Bond energies and the
scale of electronegativity

91

to assume that each bonded atom contributes half the bond energy, or 3.58×10^{-22} kJ. Furthermore, it would be reasonable to assume that in any bond in which H shares an electron pair *equally* with another atom the contribution by H to the bond energy should be 3.58×10^{-22} kJ. Similarly, from the bond energy found for Cl_2, 239 kJ per Avogadro number of bonds, we deduce that a Cl atom should contribute 1.99×10^{-22} kJ to any bond in which the sharing of an electron pair is equal.

Suppose we now consider the bond in HCl. This bond is polar, but for the moment let us imagine that the electron pair is shared equally. This amounts to picturing H in HCl to be the same as in H_2 and Cl the same as in Cl_2. If H contributes 3.58×10^{-22} kJ and if Cl contributes 1.99×10^{-22} kJ, the expected bond energy of HCl should be the sum of these contributions, or 5.57×10^{-22} kJ. Actually, the bond energy of HCl found by experiment is 427 kJ per Avogadro number of bonds, or 7.09×10^{-22} kJ per bond. The fact that the observed bond energy, 7.09×10^{-22} kJ, is significantly greater than the calculated value, 5.57×10^{-22} kJ, suggests that the electrons are *not* equally shared in HCl. The bond in question is actually more stable (requires more energy to break) than would be predicted by equal sharing.

The enhanced stability of HCl can be attributed to unequal sharing of the electron pair. If the electron pair spent more time on the Cl, that end of the molecule would become negative, and the H end positive. Since the positive and negative ends would attract each other, there would be additional binding energy. The amount of additional binding energy would depend on the relative electron-pulling ability of the bonded atoms since the greater the charge difference between the ends of the molecule, the greater the additional binding energy. Thus, it should be possible to estimate relative electronegativities from the difference between experimental bond energies and those calculated by assuming equal sharing.

In Fig. 3.9 experimental values of bond energies of the hydrogen halides (HX) are compared with values calculated by assuming equal sharing of electrons. It is evident that the discrepancy is greatest in HF and least in HI. This implies that the sharing of electrons between H and F is more unequal than the sharing between H and I. We could say that HF is more ionic than HI.

Numerical values of electronegativity have been selected by a complex procedure so as to account for the differences listed in Fig. 3.9. The most satisfactory procedure is to calculate first the difference between a bond dissociation energy $D(A—B)$ and the geometric mean of the bond dissociation energies $D(A—A)$ and $D(B—B)$:

$$\Delta' = D(A—B) - [D(A—A)\ D(B—B)]^{1/2}$$

This difference turns out to be proportional to the square of the difference in electronegativities x_A and x_B of the two elements:

$$\Delta' \sim (x_A - x_B)^2$$

Fig. 3.9
Bond energies.

| | Energy, kJ *per Avogadro number of bonds* | | | |
Bond	X = F	X = Cl	X = Br	X = I
H—H	431.8	431.8	431.8	431.8
X—X	151	239	190	149
H—X (calculated)	293	336	311	290
H—X (observed)	565	427	359	295
Difference	272	91	48	5

The proportionality constant is chosen in order to give a consistent set of values ranging up to 4.0 for F. As shown in Fig. 3.8, the electronegativity value assigned to H is 2.1. The values assigned to F (4.0), Cl (3.0), Br (2.8), and I (2.5) are consistent with the trend toward equal sharing in the sequence HF, HCl, HBr, HI.

Support for the assignment of electronegativity values comes from measurements of dipole moments. For the hydrogen halides the observed dipole moments are HF, 1.94; HCl, 1.08; HBr, 0.78; and HI, 0.38, expressed in debyes. The decreasing polarity from HF to HI also indicates a trend toward equal sharing of electrons, which is consistent with decreasing electronegativity from F to I.

3.7 Resonance

From the two preceding sections it should be evident that there is generally no simple way to describe the electron distribution in a molecule or in a bond so as to describe completely all its properties. Thus, we are led to qualifying descriptions such as "the bond in HCl is polar covalent," not just covalent. Alternatively, we can even attempt quantitative descriptions such as "the bond in HCl has a 17 percent ionic-bond character." What this means is that frequently no single picture of a molecule will be adequate but it will best be represented as a composite of several pictures. Such a problem is encountered in more obvious terms in the case of a molecule such as sulfur dioxide, SO_2. This molecule has a high dipole moment; hence, we conclude that it is nonlinear, with the atoms arranged in a bent chain. Sulfur has six outer-shell electrons, and oxygen also has six. There are thus a total of 18. These can be disposed in several ways:

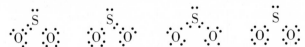

Neither formula (1) by itself nor formula (2) by itself is consistent with experimental fact because each formula indicates that the sulfur dioxide molecule has one double bond and one single bond. Experiments show the two bonds to be exactly the same length. Formula (3) is excluded because it contains unpaired electrons. Molecules containing unpaired electrons are paramagnetic; sulfur dioxide is not. Formula (4) is traditionally excluded because of the convenience of maintaining the sanctity of the octet rule.

A situation in which *no single electronic formula conforms both to observed properties and to the octet rule* is described as *resonance*. The SO_2 molecule can be described as a combination of formulas (1) and (2) in which the actual electronic distribution in the molecule is said to be a *resonance hybrid* of the contributing formulas. The choice of the word "resonance" for this situation is unfortunate because it encourages people to think that the molecule resonates from one structure to the other or that the extra electron pair jumps back and forth from one bond to the other. *Such is not the case.* The molecule has only one real electron structure. The problem is in describing it. The properties of a resonance hybrid do not oscillate from those of one contributing resonance structure to those of the other. The properties are fixed and are those of the actual hybrid structure.

3.8 Shapes of molecules and hybrid orbitals

Molecules which contain two atoms are necessarily linear, but those containing three or more atoms present complications. For example, why is the water molecule nonlinear? To answer this question, we must consider the nature of the orbitals involved in bonding the hydrogen to the oxygen and specifically the spatial distribution of the electronic charge clouds about each of the nuclei. Imagine assembling the molecule H_2O from two H atoms and one O atom. Each H atom has originally a single electron in a $1s$ orbital, which is spherically symmetric about the nucleus. The O atom has originally in its outer shell two $2s$ electrons (spherically symmetric) and four $2p$ electrons. Recalling the three p-type orbitals shown in Fig. 2.5, we find two electrons in one of the p orbitals and one electron in each of the other two. The O atom distribution thus looks as follows:

$$\underset{1s}{\underline{\uparrow\downarrow}} \qquad \underset{2s}{\underline{\uparrow\downarrow}} \qquad \underset{2p}{\underline{\uparrow\downarrow}\;\underline{\uparrow}\;\underline{\uparrow}}$$

In the valence-bond description the O—H bond arises from sharing of the $1s$ electron of hydrogen with one of the unpaired $2p$ electrons of the oxygen. Such sharing favors bonding along the direction of the $2p$ orbital used. To tie on two H atoms requires use of two $2p$ orbitals, which are at right angles to each other. Thus, on this simple picture we expect the two O—H bonds in H_2O to be perpendicular to each other. Actually, they form the somewhat greater angle of $104°31'$. We return to this discrepancy later.

Methane, CH_4, has a tetrahedral shape, as shown in Fig. 3.10, with the carbon at the center of a tetrahedron and the four hydrogens at the corners. The angles between the C—H bonds are $109°28'$. If we are to use p orbitals, why are the bond angles not right angles? Furthermore, why are there four equivalent C—H bonds formed, whereas we have but three p orbitals? To form four bonds to the central C atom, we need to use four orbitals of the central C, not just the three $2p$ orbitals but the $2s$ orbital as well.

The reason we get into the above problems is that the concept of

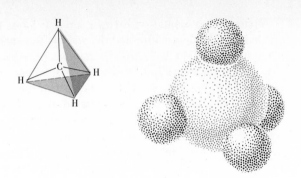

Fig. 3.10
Tetrahedral CH_4 *molecule.*

independent electron orbitals is an oversimplification when there is more than one electron in an atom. In other words, even though it is all right to talk about independent 2s and 2p orbitals in a hydrogen atom, where there is but one electron to worry about, such is not the case when both kinds of electrons are present simultaneously. The reason is that the presence of the 2s electron perturbs the motion of the 2p electron, and vice versa. Specifically, the presence of a 2s electron makes the 2p electron take on some s-like character—i.e., the 2p orbital becomes more spherically symmetric and less elongated. Similarly, the presence of a 2p electron makes the 2s electron take on some p-like character—i.e., the 2s orbital becomes less spherically symmetric and more elongated. The net result is that the original hydrogenlike 2s and 2p orbitals have to be replaced by *new orbitals that contain the combined characteristics of the original orbitals.* These new orbitals are called *hybrid orbitals.*

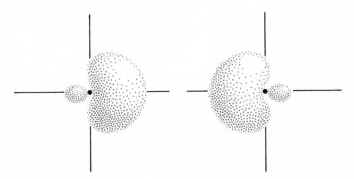

The process of hybridization is represented mathematically by the addition or subtraction of the wave functions for the individual orbitals. As a general rule, the number of hybrid orbitals resulting from a hybridization is equal to the number of orbitals that are being mixed together. As an example, if we mix an s orbital with a p orbital, we get as a result two hybrid orbitals, designated as *sp.* These are shown in Fig. 3.11. They correspond, respectively, to having most of the electron density on one side or the other of the nucleus and hence would be ideal for binding on two other atoms along a 180° line from the central atom.

Section 3.8
Shapes of molecules
and hybrid orbitals

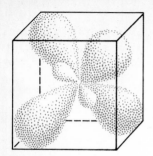

Fig. 3.12
sp^3-hybrid orbitals.

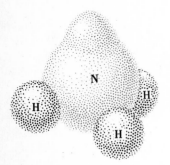

Fig. 3.13
Ammonia molecule.

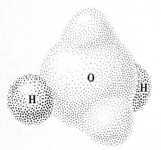

Fig. 3.14
Water molecule.

Mixing one *s* orbital with two *p* orbitals gives what are known as sp^2 hybrids. In designating hybrid orbitals the superscripts do not tell electron population but give the number of orbitals of a given type that go into the hybridization. They have the same shape as those shown in Fig. 3.11, but instead of there being two directed outward at 180° angles there are three directed outward at 120° angles. Stated otherwise, an atom at the center of a triangle has its three sp^2 hybrids directed outward toward the vertices of the triangle.

The most common situation for us will be the one where we have one *s* orbital and three *p* orbitals simultaneously occupied. In such case we have sp^3 hybrids. As shown in Fig. 3.12, there are four of them, and they are directed out to the corners of a tetrahedron. The formation of the CH_4 molecule can be pictured as involving the replacement of the one $2s$ and the three $2p$ orbitals on the carbon by this new set of four, equivalent hybrid orbitals directed toward the corners of a tetrahedron, and involving electron sharing between each of the sp^3 hybrids and the $1s$ orbital of a hydrogen to give the observed tetrahedral shape. Schematically, this corresponds to replacing

carbon (unhybridized) $\underline{\uparrow\downarrow}$ $\underline{\uparrow\downarrow}$ $\underline{\uparrow}$ $\underline{\uparrow}$ $\underline{}$

          $1s$ $2s$ $2p$

by

carbon (hybridized) $\underline{\uparrow\downarrow}$ $\underline{\uparrow}$ $\underline{\uparrow}$ $\underline{\uparrow}$ $\underline{\uparrow}$

         $1s$ sp^3

The use of hybrid tetrahedral orbitals can account for the observed shapes of molecules other than methane even when there are not four attached atoms. For example, the NH_3 molecule can be imagined as having been built from an N atom with the five outer-shell electrons $2s^2 2p^3$ distributed among four equivalent tetrahedral orbitals in such a way that two of the electrons are paired in one sp^3 orbital and the other three electrons are singly placed in the other three sp^3 hybrids. The three unpaired electrons are shared with the H atoms. The result, as shown in Fig. 3.13, is a pyramidal molecule in which the three hydrogens form the base and the lone pair of electrons the apex. The observed angles between N—H bonds in NH_3 are 108°, which is very nearly what is expected for a tetrahedron (109°28′).

Tetrahedral orbitals can also help explain the observed bond angle in H_2O. Following the reasoning of the preceding paragraph, we expect to find that H_2O is similar to NH_3 except that there are two lone pairs of electrons in the case of H_2O. Figure 3.14 attempts to show that the two lone pairs of electrons and the two bound hydrogens are directed approximately toward the corners of a tetrahedron.

The sp^3-hybrid orbitals are useful not only for describing simple molecules such as CH_4 but also for describing more complicated molecules such as C_2H_6, C_3H_8, and C_4H_{10}. In these chainlike hydrocarbons each C atom can be regarded as having four sp^3 hybrids directed toward the corners of a tetrahedron. Consequently, the preferred bonding directions are toward the corners of a tetrahedron. Electron sharing with a $1s$ orbital of an H atom forms a C—H bond in that direction,

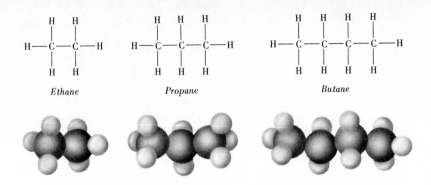

Fig. 3.15
Shapes of chain
hydrocarbons.

Ethane Propane Butane

whereas electron sharing with an sp^3 hybrid of another C atom forms a C—C bond at 109°28′ to the C—H bond. Structures that may result are shown in Fig. 3.15. The ethylene molecule, often represented as $H_2C{=}CH_2$, is planar, in that all six atoms lie in the same plane. The bond angles are all close to 120°; the C—C—H angles are actually 122°, and the H—C—H angles 116°. This shape can be accounted for reasonably well by assuming use of sp^2-hybrid orbitals. Mixing one s orbital and two p orbitals of a carbon atom produces the three

Fig. 3.16
A valence-bond formulation
of ethylene (unhybridized p
orbital lies above and below
the plane).

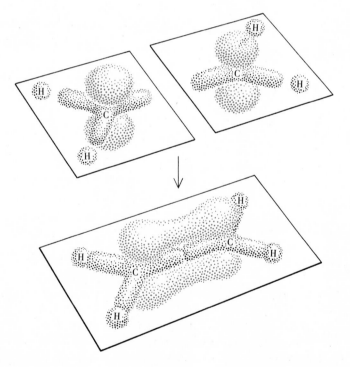

equivalent sp^2 orbitals all in a plane at angles of 120° to each other. The bonding directions of the sp^2-hybrid orbitals relative to the un-hybridized p orbital are shown in Fig. 3.16. Electron exchange between an sp^2 orbital of a carbon and an s orbital of a hydrogen gives a C—H bond. The four C—H bonds so obtained use four of the six available sp^2-hybrid orbitals. The other two sp^2-hybrid orbitals, one from each

Section 3.8
Shapes of molecules
and hybrid orbitals

97

atom, couple together so as to allow electron exchange between the carbon atoms to give a C—C bond.

Electron exchange can also occur between the unhybridized p orbital of one carbon and the unhybridized p orbital of the other carbon. This produces another bond, which has a peculiar shape since it results from side-to-side pairing of p orbitals rather than end-to-end pairing. Thus, the two carbon atoms are seen to be held together by two bonds of different shape, one of which is concentrated directly between the two nuclei and the other of which is split into two regions lying to the sides of the internuclear line. A bond of the first type is called a σ (*sigma*) *bond*; a bond of the second type, a π (*pi*) *bond*. It must be noted that a π bond, like a σ bond, is only one bond, in that it involves but a single pair of electrons shared between two atoms. The combination of one σ bond and one π bond constitutes what is called a *double bond*. In summary, for the molecule C_2H_4 the four hydrogen atoms are held to the carbons by four σ bonds, and the two carbons are held together by the combination of a σ plus a π bond.

Fig. 3.17
A valence-bond formulation of acetylene (unhybridized p orbitals p_y and p_z are perpendicular to each other and perpendicular to the bonding axis x going from upper left to lower right).

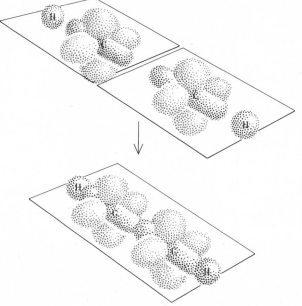

The acetylene molecule, which has the linear structure H—C≡C—H, can be described by a combination of sp hybridization and π bonding. For each carbon we imagine formation of two sp-hybrid orbitals from a carbon $2s$ and one of its $2p$ orbitals (let us call it p_x). One of the sp-hybrid orbitals is used to bind on a hydrogen atom by means of a σ bond; the other sp orbital forms a σ bond to the other carbon. As shown in Fig. 3.17, the unhybridized p orbitals (p_y and p_z) of the carbon are perpendicular to each other and in a plane perpendicular to the direction along which bonding occurs, i.e., the line through the four atoms. Electron exchange between the p_y orbital of one carbon and the p_y orbital of the other carbon gives one π bond; electron exchange between the p_z of one carbon and the p_z of the other carbon

Designation	Typical combination	Bond angle, °	Geometry
sp	$s + p_x$	180	Linear
sp^2	$s + p_x + p_y$	120	Trigonal
sp^3	$s + p_x + p_y + p_z$	109.47	Tetrahedral
dsp^2	$d_{x^2-y^2} + s + p_x + p_y$	90	Square-planar
d^2sp^3	$d_{x^2-y^2} + d_{z^2} + s + p_x + p_y + p_z$	90	Octahedral

Fig. 3.18
Hybrid orbitals.

gives another π bond. Thus, the carbon atoms in acetylene are held together by a triple bond composed of one σ bond and two π bonds.

In the examples discussed above, hybrid orbitals that consisted of various combinations of s and p orbitals were utilized. The geometry of these hybrids is summarized in Fig. 3.18. Included also are two commonly used sets of hybrids having an admixture of d orbitals. The dsp^2 hybrids are four orbitals directed toward the corners of a square, or, in other words, at $90°$ angles. These are best visualized as being concentrated along the x and y axes. An example of square-planar geometry is afforded by $PtCl_4^{2-}$ in K_2PtCl_4, where the anion consists of four chlorine atoms arranged in a square around the platinum. The d^2sp^3-hybrid set consists of six equivalent orbitals directed along the x, y, and z axes, as toward the corners of an octahedron. The set is constructed from the two d orbitals that are directed along the axes d_{z^2} and $d_{x^2-y^2}$ and a full set of s, p_x, p_y, and p_z orbitals. The molecule SF_6, which has an octahedral structure, can be described as resulting from formation of six σ bonds about the sulfur, each of these arising from electron exchange between a d^2sp^3-hybrid orbital on the sulfur and a p orbital on the fluorine.

3.9 Molecular orbitals

One of the surprising aspects of valence-bond descriptions of molecules is that such a simple picture adequately deals with such a large variety of cases. However, it is obvious from the nature of the assumptions that it must be incorrect. To assume that an electron orbital characteristic of an isolated atom is not changed by the presence of a second attracting center is at best only a crude approximation. In molecular-orbital theory the problem is met by placing the nuclei at the sites they occupy in the final molecule and then allowing the electrons to distribute themselves in the electric field arising from the nuclei and the other electrons in the molecule.

Such molecular orbitals are difficult to calculate exactly. However, they can be approximated by realizing that the part of a molecular orbital that lies close to one atomic nucleus will greatly resemble an atomic orbital centered on that nucleus. Likewise, the part of the molecular orbital that lies near a second nucleus will resemble an atomic orbital centered on the second nucleus. For the region where the electron is about equally far from both nuclei, the molecular orbital must take account of a mutual attraction. The most common way to approximate these conditions is to assume that the molecular orbital

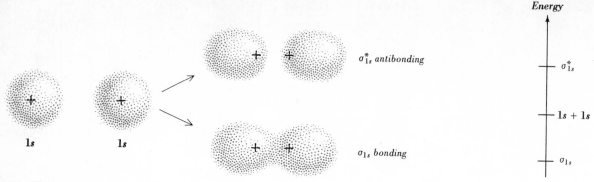

σ_{1s}^{*} antibonding

σ_{1s} bonding

Energy

σ_{1s}^{*}

$1s + 1s$

σ_{1s}

1s   1s

Fig. 3.19
Molecular-orbital formation in H_2 by combination of 1s atomic orbitals[2].

is a sum of atomic orbitals of the bonded atoms. The addition of orbitals is not simple, however, since *two* atomic orbitals when combined must produce *two* molecular orbitals. In other words, as with formation of hybrid orbitals, the number of resulting orbitals must equal the number of orbitals fed into the combination. This comes about because of the requirement in quantum mechanics that the total number of possible energy states must be conserved.

Let us construct the molecular orbitals for the H_2 molecule. We start by fixing the two protonic nuclei at the observed internuclear distance, 0.07415 nm. Molecular orbitals for electrons can be set up by combining the 1s orbital of one H atom with the 1s orbital of the other H atom. Figure 3.19 shows the two ways in which this can be done. The plus signs refer to the positions of the positively charged nuclei; the shaded area represents the region of high electronic probability. The lower molecular orbital results from simple addition of the two 1s distributions. In it the region between the two nuclei, where the two individual 1s orbitals would overlap, is correspondingly intensified. This indicates that the lower molecular orbital has appreciable probability density between the nuclei, producing a net bonding effect. For this reason it is called a *bonding orbital*. The upper molecular orbital represents the other possible way of combining two 1s orbitals. Recalling the wave nature of electrons, we can imagine this upper orbital as resulting from the addition of two waves of opposite phase, so that destructive interference occurs where they overlap.

In any case, the second way to combine orbitals produces a molecular orbital which differs from the first in that electron probability density in the region of overlap cancels rather than reinforces. The diminished probability for finding electron density in the region between the nuclei means that the two positive nuclei are not so well shielded from each other. Consequently, there is relatively a large repulsion between the nuclei, which tends to push them apart. For this reason, this type of molecular orbital is called an *antibonding orbital*.

Antibonding orbitals are generally marked with asterisks. The designation σ is a general one used for any molecular orbital in which electron density is symmetric all the way around the line drawn through the two nuclei. σ orbitals are said to have cylindrical symmetry about the bond axis. The subscript $1s$ in the designations σ_{1s} and σ_{1s}^* denotes the atomic orbitals from which the molecular orbitals were formed. Finally, it should be noted that the molecular orbital σ_{1s}, with enhanced electron density between the nuclei, is lower in energy than either the other molecular orbital σ_{1s}^*, with depleted electron density between the nuclei, or the two isolated atomic orbitals. This lowering of the total energy by transferring two electrons from two isolated $1s$ H orbitals to the σ_{1s} molecular orbital of H_2 corresponds to the bond energy of the H_2 molecule.

The molecular-orbital diagram of Fig. 3.19 can be used to describe what happens when two helium atoms come together. Each helium has two electrons; so there is a total of four electrons to be accommodated. Molecular orbitals, just like atomic orbitals, can hold only a single pair of electrons, i.e., two electrons of opposite spin. If there were such a thing as He_2, one pair of electrons would have to be in the σ_{1s}^* orbital since only one pair can be in the σ_{1s} orbital. The bonding effect due to the pair in the σ_{1s} orbital would be canceled by the antibonding effect (repulsion due to insufficiently shielded nuclei) due to the pair in the σ_{1s}^* orbital. Actually, it is generally true that antibonding orbitals are a bit more antibonding than bonding orbitals are bonding. In other words, for Fig. 3.19 the energy of σ_{1s}^* is a bit further above the energy line of the isolated orbitals than that of σ_{1s} is below it. As a result, He_2 is energetically unstable with respect to two separated helium atoms.

In the elements beyond helium the second quantum shell ($n = 2$) is involved, which means that both s and p orbitals are available. Just as with $1s$ orbitals, the $2s$ orbitals give rise to two molecular orbitals σ_{2s} and σ_{2s}^*. With lithium atoms (ground state $1s^2 2s^1$), we assume that the inner-shell electrons ($1s^2$) are not appreciably affected when two lithium atoms come together but instead remain in atomic orbitals. However, the outer electron of each atom must be accommodated in a molecular orbital. Both electrons (one from each atom) go into the σ_{2s}, which, being a bonding orbital, lowers the energy and allows formation of the molecule Li_2. This molecule has actually been detected in the vapor state and is estimated to make up some 10 percent of the molecules at the boiling point of lithium.

For the next element beryllium ($Z = 4$, $1s^2 2s^2$), formation of Be_2 would require one pair of electrons in the σ_{2s}^* molecular orbital as well as one pair in σ_{2s}. Just as in He_2, the antibonding effect is somewhat greater than the bonding effect; so no stable molecule is formed.

For the element boron ($Z = 5$, $1s^2 2s^2 2p^1$) and the subsequent elements, p orbitals must be considered. From the three kinds of p orbitals p_x, p_y, and p_z two kinds of molecular orbitals will result: σ orbitals that are cylindrically symmetric about the bond axis and π orbitals that come from side-to-side overlap. Furthermore, some of the orbitals will be bonding, while others will be antibonding. Figure 3.20 shows how σ and π orbitals can arise from combining the different p orbitals. Let us designate as the x axis the line joining the two nuclei; the y and z axes are then perpendicular to the internuclear line. As shown

ATOM 1   ATOM 2   MOLECULE

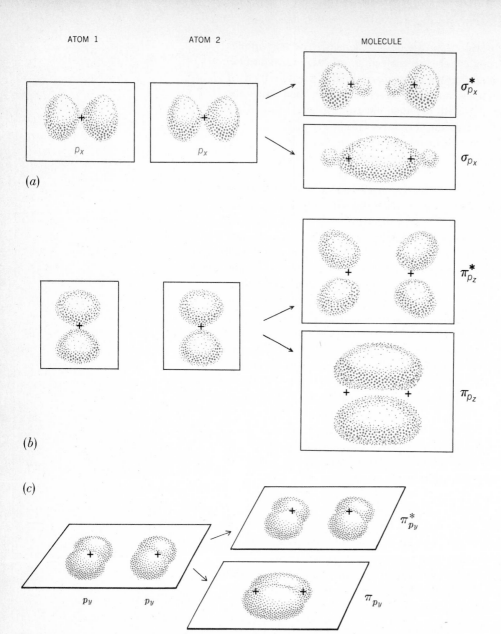

Fig. 3.20
Formation of molecular
orbitals by combination of
2p atomic orbitals (y axis is
perpendicular to plane of
paper).

in Fig. 3.20a combination of the p_x orbital of atom 1 with the p_x orbital of atom 2 produces two molecular orbitals σ_{p_x} and $\sigma^*_{p_x}$ corresponding, respectively, to bonding and antibonding possibilities. Both of these molecular orbitals are symmetric about the line connecting the two nuclei. For this reason they are both sigma and differ only in that σ_{p_x} has enhanced electron density between the nuclei whereas $\sigma^*_{p_x}$ has it depleted.

Figure 3.20b and c represents the cases in which p orbitals overlap, not end to end, but side to side. With side-to-side overlap the p_z orbital of atom 1 combines with the p_z orbital of atom 2 to produce the two molecular orbitals π_{p_z} and $\pi_{p_z}^*$. These are not symmetric around the bond line; instead, one of them, the bonding orbital π_{p_z}, looks like two sausage-shaped clouds lying on either side of the bond line; the other one, the antibonding orbital $\pi_{p_z}^*$, has four toplike lobes extending outward from the two nuclei. Side-to-side combination of the p_y orbitals forms π_{p_y} and $\pi_{p_y}^*$ in the same way except that the plane of the orbitals is rotated by $90°$ so as to be perpendicular to the plane of the paper. Except for the rotation, π_{p_y} and π_{p_z} orbitals are identical, as are $\pi_{p_y}^*$ and $\pi_{p_z}^*$.

Figure 3.21 shows the relative placing of the molecular orbitals on an energy diagram. (The exact order of the energy levels, specifically whether σ_{p_x} is higher or lower than π_{p_y} and π_{p_z}, has been in dispute. The arrangement shown in Fig. 3.21 is the one most useful for electron filling according to the most recent experiments.) In building up diatomic molecules by adding electrons to molecular orbitals the principles followed are the same as those for building up atoms by adding electrons to atomic orbitals: (1) No more than one pair of electrons may occupy a particular molecular orbital. (2) The lowest-energy molecular orbital that is available will fill first. (3) If there is more than one molecular orbital at the same level of energy, electrons spread out insofar as possible into separate orbitals. To ensure this, electron spins are as uncoupled as possible—i.e., the total spin is at its maximum.

Returning to the element boron ($Z = 5$, $1s^2 2s^2 2p^1$), two boron atoms will have two p electrons that need to be accommodated in the molecular orbitals of Fig. 3.21. There are two lowest-lying orbitals π_{p_y} and π_{p_z}; so the two electrons distribute themselves one to each. Consequently, the B_2 molecule is expected to be both stable (since the orbitals are bonding ones) and paramagnetic (since there are two unpaired electrons). In the case of two carbon atoms ($Z = 6$, $1s^2 2s^2 2p^2$), there are four p electrons to be accommodated. The two lowest-energy orbitals accommodate a pair each and are filled. Consequently, C_2 is diamagnetic. Furthermore, because C_2 has twice as many electrons in bonding orbitals as B_2 has, the molecule C_2 is expected to be considerably more stable than B_2 with respect to separated atoms. In point of fact, the bond energy for C_2 (628 kJ per Avogadro number of molecules) is about twice as great as that for B_2 (290 kJ).

The next three elements give the diatomic molecules N_2, O_2, and F_2. Their respective bond energies are 941, 494, and 151 kJ per Avogadro number of molecules. How can these bond energies be accounted for? The atoms N ($Z = 7$, $1s^2 2s^2 2p^3$), O($Z = 8$, $1s^2 2s^2 2p^4$), and F($Z = 9$, $1s^2 2s^2 2p^5$) have three, four, and five p electrons, respectively. The assignment of these electrons in the diatomic molecules is as shown in Fig. 3.22. In N_2 each N contributes three electrons; so there are six electrons, which fill the three lowest orbitals shown in the figure. All the electrons are paired, so N_2 is diamagnetic; all three of the occupied orbitals are bonding, so N_2 is more stable than C_2.

In O_2 there are two additional electrons, one from each atom, to be accommodated. Since all the bonding orbitals have been filled, the additional two electrons must be placed in antibonding orbitals. Of

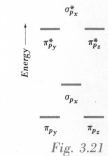

Fig. 3.21
Relative energies of some molecular orbitals.

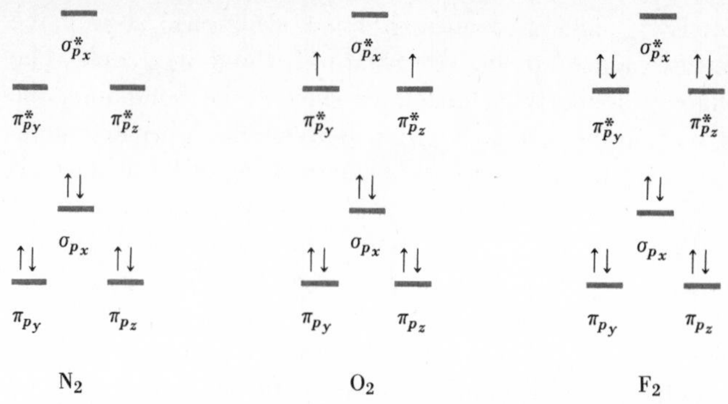

Fig. 3.22
Molecular-orbital occupancy in N_2, O_2, *and* F_2.

N_2 O_2 F_2

the antibonding orbitals the next lowest lying are $\pi^*_{p_y}$ and $\pi^*_{p_z}$. To minimize electric repulsion between like charges, one electron goes into each orbital. The electron spins are unpaired; so the O_2 molecule is paramagnetic. The presence of two antibonding electrons weakens the bonding between O atoms relative to N atoms; hence O_2 is more weakly bound than is N_2. Whereas N_2 can be considered to have a triple bond (three bonding pairs, or six bonding electrons), O_2 can be thought of as having a double bond (six bonding electrons minus two antibonding ones is about equivalent to four bonding electrons). The *bond order*, which is defined as half the number of bonding electrons less half the number of antibonding ones, drops from three for N_2 to two for O_2.

Finally, F_2 has two more electrons, which go to complete the $\pi^*_{p_y}$ and $\pi^*_{p_z}$ orbitals. The F_2 bond is thus even weaker and is equivalent to at most a single bond. In fact, the bond in F_2 is among the weakest of single covalent bonds, presumably because the effect of the two antibonding pairs is greater than that of two of the three bonding pairs. The dominance of antibonding orbitals over equal numbers of bonding orbitals also shows itself in the nonexistence of Ne_2. Neon ($Z = 10$, $1s^2 2s^2 2p^6$) has six p electrons; if there were such a thing as Ne_2, there would have to be three pairs of bonding electrons and three pairs of antibonding electrons, which would lead to no net attraction.

For the above diatomic molecules the molecular-orbital approach is a real improvement over the valence-bond approach. For example, even the simple molecular-orbital theory accounts for both the paramagnetism and the bond strength of O_2, whereas simple valence-bond theory does not.*

* Linnett's double-quartet theory (footnote, page 85) does, however, account for both properties. Each O atom contributes 6 electrons to give a total of 12 for O_2. These 12 can be thought of as being disposed so that 7 (3 unshared on each atom and 1 shared) belong to one spin system and 5 (1 unshared on each atom and 3 shared) belong to the opposite spin system. The "up-spin" and "down-spin" systems

$$\uparrow\uparrow \qquad \downarrow$$
$$\uparrow O \uparrow O \uparrow \quad \downarrow O \downarrow O \downarrow$$
$$\uparrow\uparrow \qquad \downarrow$$

can be pictured, respectively, as resulting from joining two tetrahedra *via* a common corner and joining two tetrahedra *via* a common face. With the two spin systems superimposed on each other, the final molecule has two uncompensated "up" spins and four shared electrons. Thus, the paramagnetism and the double bond order of O_2 are accounted for.

For more complex molecules, molecular-orbital descriptions are generally at least as satisfactory as valence-bond ones. They have the advantage of not requiring resonance (Sec. 3.7) and giving more reasonable pictures of electron sharing. However, there is a serious drawback to the molecular-orbital descriptions. Two electrons in a molecular orbital repel each other and hence tend to stay as far from each other as possible, yet the simple molecular-orbital description does not take account of this fact that the motions of the two electrons of a shared pair must be highly correlated. The valence-bond method, on the other hand, does this too well; it keeps the two electrons of a normal covalent bond apart by having them on different atoms and exchanging them. Hence, neither the molecular-orbital nor the valence-bond method fully describes the electric-charge distribution; each is, in a sense, an opposite extreme for describing chemical bonds.

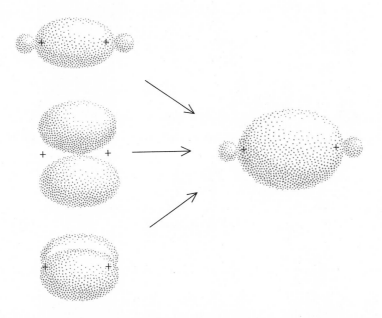

Fig. 3.23
Electron distribution of some
molecular orbitals in N_2.

Even if we ignore correlation effects, the final representation of electric-charge distribution in a molecule may be very difficult when electron density from one molecular orbital overlaps that from another. As an illustration, we can consider the N_2 molecule. If we disregard the s orbitals, the picture that emerges is something like that shown in Fig. 3.23. There are no discontinuities that separate one atom from the other or one orbital from another. As shown on the right, the total picture is that of two positive centers embedded in a diffuse, ellipsoidal charge cloud somewhat like a football. The cross section perpendicular to the bond axis is circular in symmetry. In other words, the two π orbitals overlap each other to produce uniform electron density all around the molecule axis. Similarly, there is a blending of the two π orbitals and the σ orbital.

For a more complex case, such as ethane, the molecular-orbital

Fig. 3.24
Electron distribution in the
C_2H_6 *molecule.*

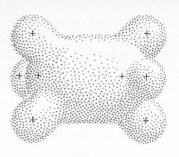

End view   *Side view*

picture leads to a charge distribution like that shown in Fig. 3.24 (which is to be contrasted with the valence-bond picture of Fig. 3.15). In contrast to the valence-bond description, in which atomic orbitals sit side by side, the molecular-orbital picture shows electron orbitals which may be between atoms and extend over several atoms. These molecular orbitals are constructed by overlapping atomic orbitals, which themselves are hybrid atomic orbitals. In the ethane case, sp^3 hybrids of the two carbons are combined to give the central orbital, which is mainly concentrated between the two carbons. Combination of another sp^3 hybrid of each carbon with the $1s$ orbital of a hydrogen gives an orbital connecting each carbon with each of its hydrogen atoms. The whole distribution fuzzes out and is even more symmetric than that shown in the figure, which has been simplified in order to show perspective. A better representation of the charge distribution in a C_2H_6 molecule would be a cylinder-shaped cloud slightly bulging at the two ends.

3.10 Wave functions of molecules

Because of its mathematical complexity, a full wave description of molecular binding is manifestly impossible in a text of this level. However, the wave-function language, despite its abstractness, gives much additional insight into the problem of why atoms stick together; so it is worth developing, even if only briefly. As a minimum, the wave-function description clarifies the essential difference between valence-bond and molecular-orbital descriptions.

First we need to recall that the wave function (refer to Sec. 2.2) represents a mathematical statement about electron probability distribution. Although ψ itself has no physical significance, the square of the wave function ψ^2 can be identified with the probability of finding the electron in a particular volume of space. The problem for molecules is to find the wave functions suitable for describing an electron's distribution in the molecule in the permitted energy states. As with atoms, the permitted energy states are determined from the Schrödinger equation, such as given on page 58, except that the potential-energy term is no longer Ze^2/r (corresponding to attraction to one nucleus) but some more complicated expression which describes simultaneous attraction to two nuclei plus repulsion by other electrons. The whole problem is actually a very complicated one, even for simple molecules. High-speed computer methods can in some cases generate molecular data that are more precise than attainable experimental accuracy, but, in general, we have to be satisfied with approximate solutions. The valence-bond and molecular-orbital methods represent two different approaches to getting such approximate solutions. The simplest molecule is the H_2 molecule; so in the ensuing discussion we consider it as the specific example. The test will be not how good a wave function we can get, since the wave function by itself is not an observable, but how well we can predict the energy of H_2 relative to two widely separated H atoms. We start first with the valence-bond approach.

The essence of the valence-bond treatment is that the H atoms retain their identity in the final molecule. This means that the atomic orbital used to describe a $1s$ electron in an isolated H atom can also be used to describe a $1s$ electron when it is in the molecule-bound state. The new principle that must be introduced is that if we have two completely independent systems separately described by the wave functions ψ_A and ψ_B, then the wave function for the two together is equal to the product $\psi_A\psi_B$. In other words, for two H atoms far apart the wave function is just the ψ_{1s} on one atom times the ψ_{1s} on the other atom. What happens as the two atoms are brought closer together? If we assume the individual wave functions ψ_{1s} do not change (which is the essence of strict valence-bond theory), there will be a slight attraction between the two H atoms as the $1s$ electron on one atom begins to feel the pull of the second nucleus and the $1s$ electron on the second atom begins to feel the pull of the first nucleus. The attraction is only very slight; if we calculated the energy by putting $\psi_A\psi_B$ into the Schrödinger equation and suitably modifying the potential-energy term, we would find that the H_2 molecule would be only about 4×10^{-20} J lower in energy than two isolated H atoms. This does not look too good because experimentally H_2 is found to be 72×10^{-20} J lower than 2H. So, even though we calculate that there is a feeble attraction, the simple procedure described above is not adequate for accounting for the observed bond strength of H_2.

What have we forgotten? This is where exchange comes in. When the two H atoms get close enough together, electron 1 on atom A and electron 2 on atom B may exchange places with each other. Mathematically, this can be expressed by writing the total wave function as the sum of two products:

$$\psi_A(1)\psi_B(2) \pm \psi_A(2)\psi_B(1)$$

The first of these products describes the situation when electron 1 is on atom A while electron 2 is on atom B; the second of the products describes the situation when electron 2 is on atom A and electron 1 is on atom B. Both of the terms must be used together to describe the molecule because electrons are indistinguishable from each other and we have no way of telling electron 1 from electron 2.

There are two possibilities for the total wave function—i.e., two possible states for the molecule—depending on whether we take the

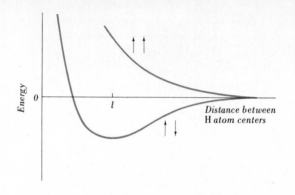

Fig. 3.25

Energy of H_2 *molecule as a function of distance between the* H *atom centers. (Upper curve holds when both electrons have the same spin; lower curve, when they have opposite spin.)*

plus or minus sign in forming the above sum. The plus sign holds when electrons 1 and 2 are opposite in spin; the minus sign holds when electrons 1 and 2 have the same spin. Figure 3.25 shows how the energy* of the two states differs as the distance between the two H atoms is changed. As can be seen, the state in which both electrons have the same spin is everywhere of greater energy than the far-apart H atoms; the molecule in that state tends to fly apart. On the other hand, the state in which the electrons have opposite spin shows a clear energy minimum at some finite spacing between the H atom centers. Calculations show that the minimum in the energy occurs at about 0.08 nm, which is not far from the experimentally observed H_2 bond length, 0.074 nm. More important, if we calculate the depth of the energy minimum [by putting $\psi_A(1)\psi_B(2) + \psi_A(2)\psi_B(1)$ into the Schrödinger equation], we find that the H_2 molecule is 50×10^{-20} J lower than two isolated H atoms. We are still far from the experimental value of 72×10^{-20} J, but the simple process of putting in exchange of the electrons has significantly improved the valence-bond calculation. Further refinements, such as slightly distorting the $1s$ functions that are used for ψ_A and ψ_B, allowing both electrons 1 and 2 to reside simultaneously for a short time on either atom A or atom B, and correcting for the fact that one electron screens the other from full attraction by the nucleus, can be used to improve further the valence-bond calculation. Even so, none of the refinements is as dramatically effective in stabilizing the H_2 molecule as is the exchange.

It is worth looking at this exchange stabilization more critically. As done above, it is but a mathematical device for allowing electron 1, which was originally on atom A, to be on atom B also. In other words, exchange is a mathematical device for spreading out the electron wave function from being localized on one atom to being spread over the whole molecule. As noted for the particle-in-the-box problem on page 58, spreading out the box for the wave function lowers the energy, and that is precisely what we are doing here. Still, it must not be forgotten that we would not have had to spread out the wave function

*This is the total energy, composed of the potential and kinetic energies of the electrons plus the potential energy arising from nuclear-nuclear repulsion.

(i.e., introduce the concept of exchange) if we had not insisted on starting out with the assumption that the wave function of each electron is confined to its own atom. Stated another way, we would not need to put in such a whopping big correction term (exchange) if we did not make such a whopping big mistake in our first assumption. As can be seen in the following discussion of molecular-orbital theory, exchange does not come into molecular-orbital theory; it is only an artifice of valence-bond theory.

How do we get a wave function for the H_2 molecule in molecular-orbital theory? The total wave function must satisfy the Schrödinger equation and must correspond to the permitted energy states of the H_2 molecule. Conceptually, the molecular-orbital procedure is simpler than that of the valence-bond method. For H_2, we set the two positive centers some distance apart (0.074 nm), allow the electrons to be attracted to the two nuclei, and calculate the wave function appropriate to this kind of double-attraction double-repulsion problem. Whereas the nuclei, because they are massive, can be considered to stay put, the two electrons are light. There is enormous complication in calculating how one of the electrons affects the wave motion of the other. In fact, the problem of two electrons plus two nuclei cannot be solved exactly, and we again have to be content with approximate solutions. However, a problem that can be solved exactly is that of the hydrogen-molecule ion $H_2{}^+$, which consists of but one electron plus two nuclei. We shall find it instructive to look at the problem of the hydrogen-molecule ion.

Unlike an atomic orbital, which has one attracting center, a molecular orbital is polycentric. For $H_2{}^+$, the molecular orbital encompasses the whole molecule and has two attracting centers. How is it best to set up a description that extends into the neighborhood of two nuclei? Clearly, when the electron is near one of the nuclei, the effect of the other nucleus becomes negligible. The simplest procedure is to assume that the electron near nucleus a is describable by an atomic orbital centered on a whereas the electron near nucleus b is describable by an atomic orbital centered on b.

■ Near nucleus a the wave function has the form (see Fig. 2.12)

$$\psi = \psi_{1s \, on \, a} = \frac{1}{\sqrt{\pi}} \left(\frac{1}{a_0} \right)^{3/2} e^{-r/a_0}$$

where a_0 is the Bohr radius, 0.052 nm, and r is the distance from nucleus a. Near nucleus b the wave function has the form

$$\psi = \psi_{1s \, on \, b} = \frac{1}{\sqrt{\pi}} \left(\frac{1}{a_0} \right)^{3/2} e^{-r/a_0}$$

where r now measures the distance from nucleus b. ■

The total description of the electron can be approximated by superposing the separately centered atomic orbitals. In other words, for the

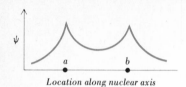

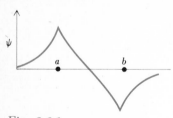

Location along nuclear axis

Fig. 3.26

Wave functions for $H_2{}^+$ by LCAO. For the top curve the 1s orbitals centered on a and b have been superposed in phase; for the bottom curve, out of phase.

molecular orbital we can write the total wave function as a sum of the individual wave functions:

$$\psi = \psi_{1s\,on\,a} \pm \psi_{1s\,on\,b}$$

Such a method of arriving at the total wave function is known as the *linear combination of atomic orbitals*; it is generally designated as LCAO. There are two possible ways of making the combination, indicated by the $\pm$, depending on whether we superpose the separate functions in phase or out of phase. Figure 3.26 illustrates what the resulting functions look like. In one case (top part of Fig. 3.26) the total wave function is everywhere positive, has two big maxima at a and b, and a finite value midway between the two nuclei. In the other case (bottom part of Fig. 3.26), the total wave function changes sign in going from the region around nucleus a to the region around nucleus b. Of necessity, for the latter case the total wave function has to go through a value of zero. Actually, as we have repeatedly noted, it is not the wave function itself that has physical significance; it is the square of the wave function, which is identified with the probability of finding the electron. Figure 3.27 shows what ψ^2 looks like for $H_2{}^+$. As can be seen, in the one case the probability density ψ^2 is great in the region between the two nuclei; in the other case it goes to zero. The contour diagrams of constant probability density, which are given in the lower portion of Fig. 3.27, emphasize the difference in the two cases. The situation at the left is characterized by finite electron probability density midway between the nuclei a and b; so it corresponds to a bonding state. It is what we have previously called σ_{1s} (see page 100). The situation at the right is characterized by zero electron probability density between the nuclei a and b; so it corresponds to an antibonding state. It is what we previously called σ_{1s}^*.

■ The difference between the two states lies in the fact that, when compared with just the superposed atomic orbitals, in the bonding case electron density is augmented between the nuclei whereas in the antibonding case electron density is withdrawn from this region; this can be seen by writing the following:

$$\psi_{\sigma_{1s}}{}^2 = (\psi_{1s\,on\,a} + \psi_{1s\,on\,b})^2$$
$$= \psi_{1s\,on\,a}^2 + \psi_{1s\,on\,b}^2 + 2\psi_{1s\,on\,a}\psi_{1s\,on\,b}$$

Because we need to square the sum of the components that go in to make the total wave function, we end up with the extra cross term $2\psi_{1s\,on\,a}\psi_{1s\,on\,b}$, which enhances the electron probability distribution over that obtained by just adding $\psi_{1s\,on\,a}^2$ and $\psi_{1s\,on\,b}^2$. The cross term has significant value only in the region between nuclei a and b. Once we get away from this region, either $\psi_{1s\,on\,a}$ or $\psi_{1s\,on\,b}$ becomes vanishingly small; so the product vanishes also. For the antibonding case we would write

$$\psi_{\sigma_{1s}^*}{}^2 = (\psi_{1s\,on\,a} - \psi_{1s\,on\,b})^2$$
$$= \psi_{1s\,on\,a}^2 + \psi_{1s\,on\,b}^2 - 2\psi_{1s\,on\,a}\psi_{1s\,on\,b}$$

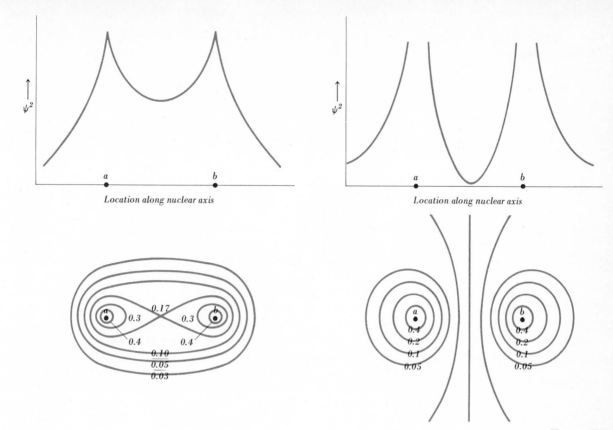

Fig. 3.27

Square of the wave functions for $H_2{}^+$ and corresponding contour diagrams showing lines of constant probability of finding the electron. a and b represent the two nuclei.

In this case the cross term depletes the electron density below what one would find by simple superposition of $\psi^2_{1s\,on\,a}$ and $\psi^2_{1s\,on\,b}$. ∎

Each of the total wave functions $\psi_{\sigma_{1s}}$ and $\psi_{\sigma^*_{1s}}$ corresponds to a different energy state (as found by solving the Schrödinger wave equation). The energy values, as is shown in Fig. 3.28, depend on how far apart the nuclei are. At infinitely large internuclear distance, the energies of the two states are identical. As the distance decreases, the two states separate—the bonding state decreases in energy, and the antibonding state increases. However, at very small internuclear spacings electron confinement and nuclear-nuclear repulsion begin to get important; so on the left side of Fig. 3.28 both σ_{1s} and σ^*_{1s} states rise very steeply. The whole situation is very much like that already discussed in Fig. 3.25 for the H_2 molecule; the main difference here is that the minimum in the energy curve is much shallower (28.2×10^{-20} J per $H_2{}^+$, as compared with 71.7×10^{-20} J per H_2) and occurs at a larger bond length (0.132 nm for $H_2{}^+$, as compared with 0.074 nm for H_2). The comparison illustrates a general fact: A one-electron bond (one electron shared between two nuclei) is weaker than a two-electron bond (an electron pair shared between two nuclei).

Once we have the wave functions for the two energy states of $H_2{}^+$,

Section 3.10
Wave functions of molecules

111

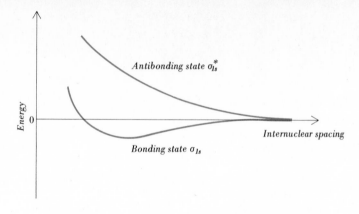

Antibonding state σ_{1s}^*

Energy

0

Internuclear spacing

Bonding state σ_{1s}

we can use them to discuss the behavior and buildup of diatomic molecules. For H_2^+ in its ground state, the electron is in the σ_{1s} molecular orbital. Excitation of the electron into the σ_{1s}^* orbital gives an excited state in which the species H_2^+ would be unstable with respect to the isolated components H^+ and H. If we added a second electron to H_2^+, we would get $H_2^+ + e^- \longrightarrow H_2$. Since a molecular orbital can accommodate two electrons provided they are of opposite spin, both electrons in H_2 would be described by the $\psi_{\sigma_{1s}}$ wave function. The assumption is implicit that putting the second electron into the orbital does not perturb the single electron that was there originally. As noted before, on page 105, this is the main weakness of simple molecular-orbital theory.

■ For diatomics more complicated than H_2, we proceed qualitatively in the same fashion. Quantitatively, we have to put in larger values of Z for the atomic wave functions, and if there are many electrons, we have to allow for screening. Still, we can easily get a crude picture of what goes on. For example, He_2 would have two electrons described by σ_{1s}-type wave functions and two electrons described by σ_{1s}^*-type wave functions. The actual molecule He_2 would be unstable (with respect to falling apart into two He atoms) because, as noted previously, antibonding raises the energy a bit more than bonding lowers it. In Li_2, the electron population would be $\sigma_{1s}^2\sigma_{1s}^{*2}\sigma_{2s}^2$, where the superscripts indicate there are pairs of electrons describable by the three kinds of wave functions σ_{1s}, σ_{1s}^*, and σ_{2s}. The σ_{1s} and σ_{1s}^* would be formed by plus-and-minus combination of the $1s$ orbitals of the Li atoms; the σ_{2s} would be formed by additive combination of the $2s$ orbitals.

What do we do in molecular-orbital theory to get the wave functions for a diatomic molecule such as AB, where A is not identical to B? The procedure is essentially the same except that we have to put in a weighting factor to take care of differences in electronegativity. This is expressed as follows:

$$\psi_{MO} = \psi_A + \lambda\psi_B$$

where ψ_{MO} is the total wave function for the molecular orbital, ψ_A and ψ_B are suitably matched atomic orbitals on atoms A and B, respectively, and λ is a constant that describes the polarity of the orbital. Whereas, as before, when A equals B, λ has values only of ± 1, it can now have a whole range of values. If, for example, $\lambda > 1$, then we say that ψ_B contributes more to the molecular orbital than does ψ_A. This corresponds to having the electron spend more time on A than on B. Values of λ are generally deduced from observations on the dipole moments of the bonds. ∎

3.11 Molecular vibrations

In Fig. 3.25 of the preceding section, a curve was presented showing how the energy of a diatomic molecule in its ground state varies as a function of the distance between the nuclei. The energy that is shown there is the kinetic energy of the electrons plus the potential energy of the electron-nuclei system. What is not shown is the kinetic energy associated with the motion of the nuclei. Even at absolute zero there is a certain oscillation of the atoms of the molecule. With this oscillation, or vibration, there is associated a kinetic energy; so somehow we need to show this in the diagram. In any case, we should reconcile Fig. 3.25 with the well-known quantization of molecular energy states. How can a molecule be in a fixed, constant-energy state if its energy is always changing with internuclear spacing? The answer, of course, is that it is the total energy (kinetic plus potential) which is constant. Figure 3.25 is not the whole story.

It is sometimes useful to visualize a diatomic molecule as being represented by two balls connected by a spring. The spring acts at the same time as the chemical bond holding the atoms together and the repulsive force between electron-electron and nucleus-nucleus. When the two balls are pulled apart, the spring acts as a restoring force to pull the balls back together: In analogous fashion, when two atoms are pulled apart, the chemical bond acts as a restoring force to bring them back together to the equilibrium separation. When two balls connected by a spring are compressed together, the spring tends to force them apart: Analogously, when two atoms of a molecule are pushed together, coulomb repulsion between electrons and especially between the positively charged nuclei forces the atoms apart. Such a system in which there is a restoring force approximately proportional to the amount of distortion is called a *harmonic oscillator*. Many of the classical ideas about a harmonic oscillator can be carried over directly to describing molecular vibrations.

What happens to a harmonic oscillator when an external force is applied to produce a distortion? In producing the distortion the force does work on the system and increases its potential energy. Similarly, when a molecule is stretched, its potential energy is raised with respect to that of the undistorted molecule. This is shown in Fig. 3.29. A is the undistorted molecule, and B shows where it is when it is stretched. The curve is a parabola, and the energy at B is greater than that of the undistorted molecule at A. If the external stretching force is now removed, the molecule tends to relax. The chemical bond pulls the atoms together. The potential energy drops as the molecule goes from B to A, but in the process the atoms have been set into motion toward each other. In other words, kinetic energy now appears in the

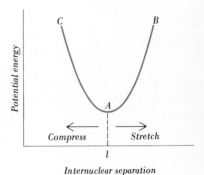

Fig. 3.29
Potential energy of a molecule as a function of distortion.

system in the guise of relative motion of the two nuclei. Since the total energy of the molecule stays constant, what is lost in the decrease of the potential energy shows up as an increase in kinetic energy. Thus, in Fig. 3.29 the molecule in going from B to A has converted the potential-energy difference between B and A into kinetic energy. By the time the molecule has reached A, its kinetic energy is a maximum. The kinetic energy at A then acts as the source of the driving force to overshoot the undistorted arrangement and compress the molecule; so it rides up the curve to C. At C all the kinetic energy has been used up and converted into potential-energy gain of the molecule. The situation is very much like that of a pendulum, for which the potential energy is greatest at the extremes of the swing, where the kinetic energy is zero, and minimum at the bottom of the swing, where the kinetic energy is a maximum. Like a pendulum or any other harmonic oscillator, molecules constantly interchange their energy between potential and kinetic, the total, however, staying constant.

Figure 3.30 shows how the total energy of a vibrating diatomic molecule is apportioned between potential and kinetic energy during the course of vibration about the internuclear distance that corresponds

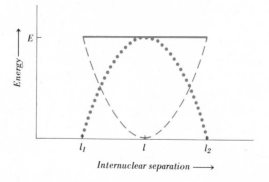

Fig. 3.30

Energy of a molecule during vibration. Solid line shows total energy; dashed curve shows potential energy; dotted curve shows kinetic energy. l_1 and l_2 represent, respectively, the most compressed and the most stretched lengths of the molecular bond.

to the average bond length l. The internuclear distance constantly changes between the two extremes l_1 and l_2, but the total energy remains constant at a value E. E, however, is the sum of potential energy and kinetic energy. So, as the potential energy slides back and forth along the dashed line through the trough, the kinetic energy moves back and forth along the dotted line over the hump.

For classical harmonic oscillators, any value of the total energy E is a permitted value. However, when quantum mechanics was applied to the problem, it was found that only certain values of E are allowed. In other words, the energy of a harmonic oscillator is quantized just as the electronic energy states of an electron in a hydrogen atom. The permitted values for vibrational energy states are given by

$$E = (v + \tfrac{1}{2})h\nu$$

where v is the so-called "vibrational quantum number." v can take on values of 0, 1, 2, 3, etc. (i.e., any positive integer). As before, h is the Planck constant, and ν is a frequency. The value of ν represents

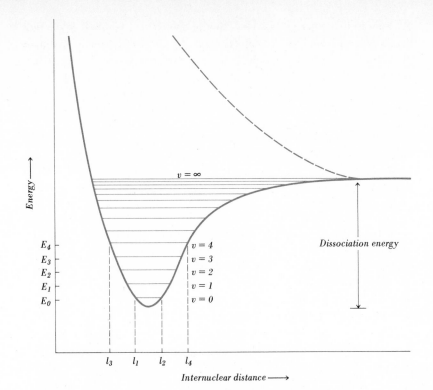

Fig. 3.31
Vibration states of a
diatomic molecule.

the vibration frequency that is characteristic of the particular bond. It depends on the "stiffness" of the bond and on the masses of the vibrating atoms. Specifically, $\nu = (\frac{1}{2\pi}) \sqrt{k/m_{\text{eff}}}$, where k is the force constant of the bond (i.e., a measure of the force required to deform a bond by a given amount) and m_{eff} is an effective mass for the vibrating atoms. For molecule AB where the atomic masses are unlike, the effective mass, also called the *reduced mass*, is given by $m_A m_B/(m_A + m_B)$.

Figure 3.31 shows some of the permitted values of vibrational energy states superposed on the potential-energy curve for a typical diatomic molecule. It might be noted that there are no vibrational states associated with the antibonding electronic state of the molecule (shown by the dashed curve). For the antibonding state, the molecular energy is everywhere greater than that of two separated atoms; so the antibonding state tends to fall apart into the constituent atoms. In the bonding state only a few of the vibrational states are shown. In these states, designated, for example, as $v = 0, 1, 2, 3,$ or 4, the total energy would be given, respectively, by E_0, E_1, E_2, E_3, or E_4, while the potential energy would more or less follow the trough through the minimum. It might be noted that in the lowest energy state, where $v = 0$, the molecule vibrates so that its bond length changes only between l_1 and l_2. In a higher state, such as the one where $v = 4$, the vibration is considerably more vigorous; the excursion of the bond length in the $v = 4$ state goes all the way from l_3 to l_4. As is evident, the more energetic the vibrational state, the wider the swings of the vibration.

As vibrational energy is added to the molecule, the molecule tends to stretch to ever greater elongations. In the limit, where v equals infinity, the elongation would be so great that the molecule would fly apart.

As is indicated in Fig. 3.31, the energy difference between the molecule in its lowest vibrational state and the $v = \infty$ state is called the *dissociation energy* of the molecule.

It is well to keep in mind the order of magnitude of the numbers involved in describing molecular vibrations and to have a qualitative idea of how they are likely to change in going from one molecule to another. A typical molecule, such as HCl, has a characteristic vibration frequency of about 10^{14} vibrations per second. At room temperature most of the molecules would be in the lowest vibration state ($v = 0$); only about one in a million would be in the first excited state ($v = 1$). Infrared light, which has a wavelength roughly in the range 0.01 to 0.0001 cm, corresponds in energy to the difference between molecular vibration states. Hence, infrared radiation (i.e., heat waves) can be absorbed by molecules in lifting them from a low vibration state to the next higher. In general, transitions change v by only one unit. Conversely, "hot" molecules (i.e., those in excited vibrational states) can give off infrared radiation as they make transitions to lower vibration states. Thus the study of absorption and emission of infrared radiation gives information about vibration states in the same way that the study of visible light gives information about electron states in atoms.

In general, the stronger a chemical bond, the greater will be its characteristic vibration frequency. Triple bonds generally vibrate faster than double bonds; double bonds, faster than single bonds. The heavier the atomic masses at the ends of the bond, the lower the characteristic frequency. Stated conversely, a low vibration frequency suggests that the bond is weak and/or that the bonded atoms are heavy.

Figure 3.32 gives pertinent data for some common bonds. Included are the bond length, the force constant, and the dissociation energy. The bond length corresponds to the distance at which the potential-energy curve reaches a minimum; it is equal to the average distance between the nuclei when the bond is vibrating in its lowest vibrational state. The force constant, which depends on the degree of curvature of the potential energy curve, measures the "stiffness" of the bond; it is given as the force in newtons needed to distort the bond per centimeter. (One newton is the force required to accelerate a mass of one kilogram by one meter per second per second. See Appendix 1 for units.) The dissociation energy measures the energy required to break the bond from its lowest vibration state to infinitely separated atoms; it is given as the number of kilojoules per Avogadro number of bonds broken.

As can be seen from Fig. 3.32, the force constant and the dissociation energy generally go the same way, increasing for stronger bonds and decreasing for weaker bonds. A clear exception is the case of F_2, where, for example, the force constant is bigger than for Cl_2 but the

Fig. 3.32
Vibration parameters for
some common bonds.

Bond	Length, nm	Force constant, newtons/cm	Dissociation energy, kJ/mol
H—H	0.074	5.1	432
H—F	0.092	8.8	561
H—Cl	0.128	4.8	428
H—Br	0.141	3.8	362
H—I	0.160	2.9	295
H—CH$_3$	0.109	5.0	423
H—NH$_2$	0.101	6.4	427
H—OH	0.096	7.7	492
N≡N	0.109	22.4	941
O=O	0.121	11.4	494
F—F	0.143	4.5	151
Cl—Cl	0.199	3.2	239
Br—Br	0.228	2.4	190
I—I	0.267	1.7	149

dissociation energy is less. The apparent anomaly probably comes from the fact that the F—F bond is so short that the repulsions between the unshared electrons on one F atom and on the other F atom are more important than usual in reducing the bond strength. The result is to give a potential-energy curve for which the trough is less deep than usual but is characterized by sharper curvature at its bottom.

Although it is not shown in Fig. 3.32, the vibration characteristics of a bond depend somewhat on what other atoms, if any, are joined to the pair in question. Thus, the C—C bond has force constants ranging from 4.5 to 5.6 newtons/cm depending on what other atoms are attached to the carbons; the C=C bond, from 9.5 to 9.9 newtons/cm. These are relatively slight perturbations; so the actual value of the force constant is fairly characteristic of the particular bond. Since the force constant decides the characteristic vibration frequency $\nu = (\frac{1}{2\pi}) \sqrt{k/m_{\text{eff}}}$, and in turn the energy spacing between the vibrational levels $E = (v + \frac{1}{2})h\nu$, the pattern of levels is characteristic of the bond. The result is that the pattern of infrared absorption can be used as a fingerprint for identifying complex molecules.*

Besides the vibrational motion discussed above, molecules can rotate in space. As with vibration, the energy of rotation is restricted to certain permitted values. The result is that there will be a characteristic pattern of allowed rotational energy levels for each molecule and this will affect the observed spectra. In general, rotations are of lower frequency than are vibrations. A typical rotation frequency, as exemplified by HCl, is 6.3×10^{11} sec^{-1}, about 100 times slower than the vibrations. Dur-

* Not every vibration necessarily shows up in the spectrum. One of the subsidiary requirements is that the bond vibration in question must result in a change of dipole moment. Since bonds between identical atoms are nonpolar, a vibration such as C—C may not show up directly in the infrared spectrum. In such cases, other phenomena such as sideways scattering of incident radiation (Raman effect) may have to be used to get information about the vibration.

ing the time it takes the molecule to complete one rotation, several hundred vibrations occur; hence, the vibration motion and the rotation motion are often treated as being approximately independent of each other. The energy spacing between rotation levels is considerably smaller than that between vibration levels. Transitions between rotation levels generally occur toward the low-energy end of the spectrum, in what is called the *microwave* region. Typical wavelengths are on the order of 0.1 to 10 cm, corresponding to radar waves. As contrast, wavelengths associated with vibrational transitions are on the order of 0.01 to 0.0001 cm. As with vibrational spectra, rotational spectra can give considerable information about molecules. Whereas vibrational spectra are most useful for getting at force constants and bond energies, rotational spectra are most informative about bond lengths and bond angles.

3.12   Symmetry

One of the most powerful ways to describe a molecule, a wave function, or anything else, for that matter, which has spatial extent is to give a description of its symmetry. The power of the method comes from the fact that if an object, such as a molecule, is endowed with symmetry, its component properties and associated descriptions have to be consistent with that same symmetry. For example, suppose a molecule is said to possess a *mirror plane*—i.e., one half of the molecule is a mirror image of the other half. Then the square of the wave function that describes an electron in one half of the molecule must have a mirror-image part in the other half of the molecule. Molecular vibrations and distortions of the electron cloud during interaction between molecules need to be consistent with symmetry requirements. Recently, great progress has been achieved in understanding mechanisms of chemical reactions by giving careful study to the symmetry changes during the reactions.

To describe the symmetry of a molecule, one needs to specify the symmetry elements that the molecule possesses. A *symmetry element* is a geometric entity (a point, a line, or a plane) with reference to which various superimposable parts of the molecule can be related. A *symmetry operation* is the act of bringing one superimposable part of the molecule into coincidence with another, either physically by manipulating a model, or mentally. Symmetry operations are always associated with symmetry elements.

There are three important kinds of symmetry elements: axis of rotation, plane of reflection, and inversion center. The meaning of these is as follows:

1   A molecule is said to have a *rotation axis* of symmetry if the molecule can be brought into self-coincidence by rotating it around an axis by $360°/n$. n, called the *order* of the axis, is 1, 2, 3, 4, 5, and 6, respectively, for rotations of 360, 180, 120, 90, 72, and 60°. The corresponding axes are designated C_1, C_2, C_3, C_4, C_5, and C_6.

In an alternative designation, a C_1 axis is called a onefold axis; C_2, a twofold axis; C_3, a threefold; etc. As a specific illustration, the H_2O molecule has a C_2 axis; it goes through the O atom and bisects the line between the two H atoms. Rotating an H_2O molecule about its C_2 axis brings one of the H atoms into coincidence with the other.

A onefold axis C_1 is rather special. It really does not imply much symmetry since, of course, rotating any object by 360° brings it back to self-coincidence. The rotated object is completely identical with the starting object, and this operation of rotation about C_1 is called the *identity* operation. As noted below, the identity operation is important when considering the net effect of several consecutive operations. Thus, for example, two consecutive rotations about C_2, each by 180°, bring a molecule back to its starting orientation and are equivalent to a single identity operation C_1.

2 A *mirror plane*, or a plane of reflection symmetry, implies that half the molecule is a reflection of the other half. The usual symbol for a mirror plane is σ or m. As illustration, the H_2O molecule has two mirror planes. One is trivial, and that is the plane containing the H and O atoms. The other is the plane passing through the O atom and perpendicularly bisecting the line between the two H atoms. The C_2 axis lies along the intersection of the two mirror planes.

3 An *inversion center* is the same as a center of symmetry. Its presence in a molecule implies that every point on the molecule can be reflected through the center of the molecule to match an identical point on the opposite side of the molecule. A center of inversion is usually symbolized by i. The H_2O molecule does not possess a center of inversion: A CO_2 molecule does.

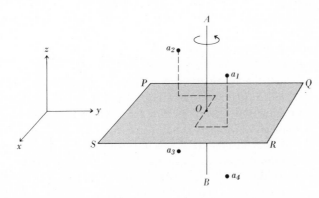

Fig. 3.33
Relation of symmetry elements relating four equivalent points on a molecule.

In general, a molecule has several symmetry elements. They have to be consistent with each other in the sense that an operation about one symmetry element cannot be in contradiction with another symmetry operation about some other symmetry element. How the self-consistency comes about can be seen in Fig. 3.33. Here we show a typical possible arrangement of four equivalent points on a molecule. The points could be four identical atoms as, for example, the four Cl atoms in the square planar complex $PtCl_4{}^{2-}$; we designate them as a_1, a_2, a_3, and a_4 so as to be able to refer to them individually.

Point O (where the Pt would be located) is an inversion center. Point a_1 (which has coordinates x, y, z) is related by inversion through the center O to a_3 (which has coordinates $-x$, $-y$, $-z$); a_2 (with coordinates $-x$, $-y$, z) is related by inversion to a_4 (with coordinates x, y, $-z$). As can be seen, inversion is equivalent to changing all positive coordinates to negative ones, and vice versa.

The line AB is a twofold axis of rotation. Rotation by 180° counterclockwise around the line AB sends point a_1 into a_2, and a_4 into a_3. Simultaneously, a_2 is sent into a_1, and a_3 into a_4.

The plane $PQRS$ is a reflection plane. Point a_1 is said to be related by reflection to point a_4—that is, a_1 and a_4 are on opposite sides of and equidistant from the reflection, or mirror, plane $PQRS$. Similarly, $PQRS$ is a reflection plane relating a_2 to a_3.

What happens if we carry out successive operations? Suppose, for example, we first rotate the molecule about C_2 and then reflect it in the mirror plane. The net result would be the same as if we had gone directly through the inversion center. Specifically, rotation of a_2 about the line AB sends a_2 into a_1; subsequent reflection of a_1 through $PQRS$ sends a_1 into a_4. The net result of $a_2 \longrightarrow a_1$ followed by $a_1 \longrightarrow a_4$ is the same as inverting a_2 directly through O into a_4. This illustrates a general principle: Two successive symmetry operations on a molecule are equivalent to some other symmetry operation characteristic of that molecule.

The complete set of symmetry operations that characterizes a molecule is said to constitute a *group*. The term "group" is used here in the special sense of mathematics, where there is a branch called *group theory* which concerns itself with how the elements of a group must be related to each other. The elements of a group may be numbers, mathematical functions, or physical operations. The main requirement is that the combination of two elements of the group (e.g., the product of two numbers or one symmetry operation followed by another) must be equal to some other element of the group. Usually, the interrelations between the elements of a group are represented by what is known as a *multiplication table*. This is simply a matrix where the columns represent one operation, the rows another operation, and the intersection of a row and column indicates the combined operation.

Let us consider as a specific example the group composed of the symmetry operations C_2, i, and σ. This group would be used to describe a molecule containing a twofold rotation axis, an inversion center, and a mirror plane. The multiplication table for the group would be set up as follows:

	C_2	i	σ
C_2	1	σ	i
i	σ	1	C_2
σ	i	C_2	1

The symbols C_2, i, and σ across the top of the table identify the columns; the symbols C_2, i, and σ along the left edge identify the rows. Where a column and a row cross, we have put in the entry which indicates the combined result of first carrying out the operation indicated by the column heading and then following it by the operation indicated by the row designation. The number 1 stands for the identity operation and indicates that the molecule has been returned to its original orientation. As can be seen (from the entry 1 where the column C_2 and the row C_2 cross), the result of carrying out a C_2 operation followed by a C_2 operation is to restore the molecule to its original orientation. This, of course, is consistent with the fact that a $180°$ rotation followed by a $180°$ rotation gives a $360°$ rotation. We can express this symbolically by writing

$$C_2 C_2 = 1$$

corresponding to the statement "the product of a C_2 operation and a C_2 operation is equal to the identity operation." "Product" here means the combined result of two consecutive operations, the convention being that the operation shown on the right of the product is the first one carried out followed by the one shown on the left. In similar fashion we have

$$iC_2 = \sigma$$

which means that a $180°$ rotation C_2 followed by inversion through the origin i is the same as reflection in the plane σ. In this case the order in which we carry out the operations is immaterial; so we have also

$$C_2 i = \sigma$$

In general, the order of carrying out the operations is important. The multiplication tables are useful in giving a quick overview of how the symmetry operations of a group are related to each other; they are also very useful when it comes to classifying properties, such as molecular vibrations, into symmetry groups.

As a further illustration of symmetry descriptions, let us consider the nitrate ion NO_3^-, shown in Fig. 3.34. The three oxygen atoms lie at the corners of an equilateral triangle, the center of which is occupied by nitrogen. The plane of the nitrate ion is a symmetry plane and can be designated as σ_h. There is a threefold rotation axis C_3 perpendicular to the plane; it is shown by the dashed line in Fig. 3.34. When there is a mirror plane perpendicular to a rotation axis, it is customary to designate the plane by a subscript h. (The letter h stands for "horizontal" and is to be distinguished from v standing for "vertical." By convention, the main rotation axis is aligned up and down—i.e., vertical—and mirror planes are described with respect to this up-and-down direction.)

Besides the threefold axis and associated horizontal mirror plane, NO_3^- is characterized by three other axes of symmetry. These are less

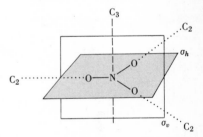

Fig. 3.34
Symmetry characteristics of nitrate ion.

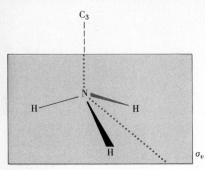

Fig. 3.35
Symmetry characteristics of ammonia molecule. The vertical plane σ_v passes through the left N—H bond and bisects the H—N—H angle at the right.

symmetric than the threefold axis and are called *subsidiary* axes, as distinguished from the main, or principal, symmetry axis. As indicated by the dotted lines in Fig. 3.34, they are C_2 axes and are disposed in the plane perpendicular to the C_3 axis. They pass through the O—N bonds; so they form angles of $120°$ with respect to each other. All three of the C_2 axes are equivalent to each other by symmetry, but they can be distinguished if necessary as C_2', C_2'', and C_2'''. For each of them, rotation about them by $180°$ interchanges two oxygen atoms.

Finally, to complete the description of NO_3^-, there are three more mirror planes defined, respectively, by the C_3 axis and each of the C_2 axes. These mirror planes, which can be designated σ_v', σ_v'', and σ_v''', are all vertical, in the sense that they are parallel to the main axis C_3. It might be noted that there is no center of inversion in NO_3^-.

As contrast to the NO_3^- case, Fig. 3.35 shows the ammonia molecule NH_3. The number of atoms is the same, but they are disposed quite differently. The H atoms lie at the vertices of an equilateral triangle, but the N atom does not lie in this plane. The immediate result is that there is no horizontal reflection plane. There is still a threefold axis, passing (as shown in Fig. 3.35) through the N and the middle of the triangular base, but the horizontal mirror plane σ_h and the three subsidiary C_2 axes are gone. Only the three vertical mirror planes (defined, respectively, by the C_3 axis and each of the N—H bonds) remain.

In summary, NO_3^- has the symmetry elements C_3, σ_h, C_2', C_2'', C_2''', σ_v', σ_v'', and σ_v'''; whereas NH_3, much less symmetric, has but C_3, σ_v', σ_v'', and σ_v'''.

Exercises

*3.1   *MO versus VB.* What is the essential difference between the molecular-orbital and valence-bond descriptions of, say, the diatomic chlorine molecule?

*3.2   *Covalent bond.* Show how in the valence-bond picture application of the Pauli exclusion principle ensures that the hydrogen molecule will be H_2 rather than H_3.

*3.3   *Octet rule.* What is the octet rule? How does it apply to each of the following molecules: H_2O, N_2, C_2H_4, and SO_3?

*3.4   *Covalent bonds.* Show with examples what is meant by single, double, and triple bonds.

*3.5   *Electronegativity.* Distinguish clearly, using specific examples, between electronegativity and electron affinity.

*3.6   *Resonance.* The ozone molecule consists of a bent chain of three oxygen atoms. The bond angle is $116.8°$, and the O—O bond length is 0.1277 nm. Give the electron-dot formulations of the resonance forms that account for this molecule.

*3.7   *Hybrid orbital.* Using hybrid orbitals describe the bonding in linear, covalent $BeCl_2$ and planar, covalent BCl_3. What difference is there in the hybrid orbitals used?

*3.8 *Molecular orbitals.* Show clearly, using examples, the essential difference between the members of each of the following pairs:
a   Sigma and pi orbitals
b   Bonding and antibonding orbitals
c σ_{1s} and σ_{1s}^*
d σ_{1s} and σ_{2s}
e σ_{2s} and σ_{2p_x}

**3.9 *Molecular vibrations.* Assuming all other things equal, what difference in characteristic vibration frequency would you expect for a diatomic molecule of mass 20 amu if (a) the mass is equally distributed between the two ends or (b) 1 amu is at one end and 19 amu are at the other?   *Ans. (a) is 44% of (b)*

**3.10 *Molecular vibrations.* Describe what influence each of the following would have on the characteristic vibration frequency of a diatomic molecule AB (where A = 2 amu and B = 10 amu):
a   Doubling the force constant
b   Doubling the mass at the light end
c   Doubling the mass at the heavy end
d   Doubling the mass at both ends   *Ans. 1.4, 0.76, 0.96, 0.71*

**3.11 *Ionic bond.* Given an ionic bond between positive ion A and negative ion B, how would you expect the strength of the bond to be affected by each of the following changes?
a   Doubling the charge on A
b   Doubling the charge on B
c   Simultaneously doubling the charge on A and on B
d   Doubling the radius of A
e   Doubling the radius of B
f   Simultaneously doubling the radius of A and of B

** 3.12 *Ionic bond.* Given that the radii of Na^+, K^+, F^-, and Cl^- are, respectively, 0.097, 0.133, 0.133 and 0.181 nm, how would you expect the ionic bond strength of NaCl to compare with that of KF?

**3.13 *Ionic bond.* If A has an ionization potential of 5.5 eV and B has an electron affinity of 1.5 eV, what is the maximum size of A^+ that would go with a 0.20 nm size of B^- to give energetically favorable bond formation?   *Ans. 0.16 nm*

**3.14 *Double quartets.* Suggest how the Linnett spin quartets might be disposed in the C_2H_4 and C_2H_2 molecules.

**3.15 *Polarity of molecules.* Account for the fact that the dipole moment decreases in the sequence HF, HCl, HBr, HI even though bond length and the number of electrons increase.

**3.16 *Dipole moment.* Given that the observed net dipole moment of H_2O is 1.84 debyes and the bond angle is 104.45°, what moment should be assigned to each O—H bond? *Ans. 1.50 debyes*

**3.17 *Electronegativity.* By using the data of Figs. 2.28 and 2.29, show that the Mulliken definition of electronegativity leads to the same sequence of relative values for F, Cl, Br, and I as the Pauling scale.

**3.18 *Molecular orbitals.* Explain how the molecular-orbital description of molecules accounts for the fact there is a maximum bond strength observed in the middle of period 2 (see data in Fig. 3.3).

**3.19 *Molecular orbitals.* From the molecular-orbital picture, how would you expect the bond strength of NO to compare with that of O_2? (The observed bond strengths are 678 and 494 kJ/mol, respectively.)

**3.20 *Bond order.* What is meant by bond order? How does it vary in the sequence NO, NO^+, NO^-?

**3.21 *Bond strength.* Explain why removal of an electron from O_2 strengthens the bond but removal of an electron from N_2 weakens it. What effect on bond strength would you predict for electron removal from NO?

**3.22 *MO versus VB.* Comment on the verity of the following statement: The VB model is better than the MO model in taking care of interelectronic repulsion.

**3.23 *Bond strength.* By reference to Fig. 3.25, indicate specifically what would correspond to bond dissociation energy.

***3.24 *Bond strength.* By referring to the data in Fig. 3.3, speculate on the reasons why single-bond strengths for Li_2 and F_2 are about the same even though bond lengths are quite different.

***3.25 *Dipole moment.* Calculate the net dipole moment (in debyes) for each of the following arrangements:
a One $+1$ and one -1 ion separated by 2×10^{-8} cm
b One $+2$ and one -2 ion separated by 2×10^{-8} cm
c Two $+1$ and one -2 ion located, respectively, at the vertices of an equilateral triangle that is 2×10^{-8} cm on edge
Ans. 9.6, 19.2, 16.6 debyes

***3.26 *Electronegativity and bond strength.* Given that the bond dissociation energies are 151, 239, and 253 kJ/mol for F_2, Cl_2, and FCl, respectively, calculate the bond dissociation energy expected for Br—F. The bond dissociation energy of Br_2 is 190 kJ/mol.
Ans. 260 kJ/mol

***3.27 *Hybrid orbitals.* In forming the d^2sp^3 set of hybrid orbitals for explaining the octahedral bonding in SF_6, why is it necessary to use d_{z^2} and $d_{x^2-y^2}$ instead of d_{xy}, d_{yz}, or d_{zx}?

***3.28 *Wave functions.* Describe how the valence-bond formulation of a chemical bond brings exchange into the picture.

***3.29   *Wave functions.* Draw graphs showing how the electronic wave functions ψ and ψ^2 change along an internuclear line for bonding and antibonding states of $H_2{}^+$.

***3.30   *Wave functions.* How would Fig. 3.26 have to be modified if the molecular orbital in question were based on $2p_x$ wave functions instead of $1s$?

***3.31   *Molecular vibrations.* How would you expect the characteristic vibration frequency of $H_2{}^+$ to compare with that of H_2? Explain your reasoning.

***3.32   *Molecular vibrations.* Using data from Fig. 3.32 calculate the characteristic vibration frequency and the energy of the lowest vibrational state for the H_2 molecule. What fraction of the bond dissociation energy is represented by the energy difference between successive vibrational energy states?

***3.33   *Symmetry.* List the symmetry elements that characterize an equilateral triangle. Which of these would be missing for an isosceles triangle?

***3.34   *Symmetry.* Find all the symmetry elements of a regular tetrahedron.

> *Ans. Three twofold axes, four threefold axes, six mirror planes, each of which includes a pair of threefold axes*

***3.35   *Symmetry group.* Work out the multiplication table for the group composed of the elements $C_4{}^1$, $C_4{}^3$, $C_2{}^1$, and 1. The notation is as follows: $C_4{}^1$ corresponds to a $90°$ rotation clockwise around the fourfold axis; $C_4{}^3$ corresponds to three consecutive $90°$ rotations clockwise (or, what amounts to the same thing, one $90°$ rotation counterclockwise) around the fourfold axis.

Stoichiometry is the aspect of chemistry that concerns itself with weight relations in chemical reactions. It rests on the concept of a chemical equation, which in turn is dependent on the idea of a chemical formula. The quantitative interpretation of chemical formulas is tied to the concept of atomic weight. Thus, stoichiometry is the manipulation of atomic weights to account for the amounts of material used up and produced in chemical changes.

4.1 Moles of atoms

Since atoms are extremely small, any laboratory experiment dealing with weighable amounts of elements must necessarily involve tremendous numbers of atoms. For example, in making hydrogen fluoride, it is not possible to weigh out one hydrogen and one fluorine atom. It is, however, possible to get equal numbers of hydrogen and fluorine atoms by using the appropriate relative masses of these atoms. From the atomic weights (1.0079 amu for hydrogen and 18.9984 amu for fluorine) we know that an average fluorine atom is 18.9984/1.0079 times as heavy as an average hydrogen atom. Suppose we take any definite number of fluorine atoms and an equal number of hydrogen atoms. The weight of the entire collection of fluorine atoms is 18.9984/1.0079 times as great as the weight of the entire collection of hydrogen atoms.

Conversely, any weight of fluorine that is 18.9984/1.0079 times as great as a weight of hydrogen must contain just as many fluorine atoms as there are hydrogen atoms. For example, 18.9984 g of fluorine contains the same number of atoms as does 1.0079 g of hydrogen. In general, when we take weights equal to the relative atomic weights of different elements, we always have the same number of atoms. We can take these weights in grams, pounds, or any other convenient unit. In other words, 18.9984 lb of fluorine contains the same number of atoms as 1.0079 lb of hydrogen.

The *mole* is defined as the amount of chemical material that has the same number of "elementary units" as 12 g of pure ^{12}C. The "elementary units" may be atoms,* molecules, ions, electrons, formula-units, etc. A mole of atoms is a collection of atoms whose total mass is the number of grams numerically equal to the atomic weight. Since sulfur has an atomic weight of 32.06 amu, a collection of sulfur atoms weighing 32.06 g is one mole of sulfur atoms. Since the atomic weight of iron is 55.847 amu, a collection of iron atoms weighing 55.847 g is one mole of iron atoms. The collections have different weights, but each has the same number of atoms. Suppose now we wish to make a compound in which there is one atom of iron for each atom of sulfur. If we take one mole of iron atoms and one mole of sulfur atoms, there are exactly enough iron atoms to match the sulfur atoms. Furthermore, the weights taken are of a size that can be handled with usual laboratory apparatus. Because equal numbers of moles of different elements contain equal numbers of atoms, it is convenient to refer to amounts of elements in terms of numbers of moles. For instance 3.2 g of sulfur is 3.2/32.06, or 0.10, mol.

---

* There used to be a special name, *gram-atom*, for the amount of material when the "elementary units" counted were individual atoms. The only caution that needs to be made in giving up the term "gram-atom" and using moles instead is that the identity of the particle being counted should be specifically stated. In other words, one should avoid simply saying "mole of hydrogen" but rather "mole of hydrogen atoms" or "mole of H_2 molecules." The term "mole" merely gives the number of particles; it is necessary to give their identity also.

Example 1

In a chemical reaction requiring three atoms of Mg for two atoms of N, how many grams of N are required by 4.86 g of Mg? (The atomic weight of Mg is 24.3 amu, and that of N is 14.0 amu.)

$$\frac{4.86 \text{ g of Mg}}{24.3 \text{ g per mole of Mg}} = 0.200 \text{ mol of Mg}$$

We need two-thirds as many moles of N as moles of Mg. 0.200 mol of Mg needs $\frac{2}{3}(0.200) = 0.133$ mol of N.

$$(0.133 \text{ mol of N})(14.0 \text{ g per mole of N}) = 1.86 \text{ g}$$

■ ■ ■

With modern techniques, it is possible to determine the *number of particles in one mole*. This number, referred to as the *Avogadro number*, is 6.0222×10^{23}. It should be remembered to at least three significant figures: 6.02×10^{23}.

Probably the most accurate determination of the Avogadro number is based on the study of solids. From the measured mass per unit volume (density) of the solid, the volume of one mole of atoms can be calculated. The spacing of the atoms in the solid can be found by using X rays. This enables a precise determination of the number of atoms in the volume which contains one mole. The method is illustrated in the following example, which was chosen for simplicity and does not give as precise a value as can be obtained from substances with more complicated structures.

Example 2

The density of solid AgCl is 5.56 g/cm³. The solid is made up of a cubic array of alternate Ag⁺ and Cl⁻ ions at a spacing of 2.773 × 10⁻⁸ cm between centers. From these data calculate the Avogadro number.

The mass of AgCl that contains a mole of each element is 107.868 g plus 35.453 g, or 143.321 g. From the density, 5.56 g/cm³, we find that the volume occupied is

$$\frac{143.321 \text{ g}}{5.56 \text{ g/cm}^3} = 25.78 \text{ cm}^3$$

This corresponds to a cube of edge length

$$\sqrt[3]{25.78 \text{ cm}^3} = 2.954 \text{ cm}$$

Along one edge there are

$$\frac{2.954 \text{ cm}}{2.773 \times 10^{-8} \text{ cm/ion}} = 1.065 \times 10^8 \text{ ions}$$

The total number of ions in the cube, which has 1.065×10^8 ions along each edge, is

$$(1.065 \times 10^8)^3 = 1.209 \times 10^{24}$$

Since the cube contains one Avogadro number of Ag^+ and one Avogadro number of Cl^-, the Avogadro number from these data is calculated to be 6.04×10^{23}.

■　■　■

A knowledge of the number of atoms in a mole of atoms, 6.02×10^{23}, can be used to calculate the mass of individual atoms as well as the number of atoms in any given mass of the element.

Example 3
The heaviest atom so far prepared has an atomic weight of 260 amu. What is its mass in grams?
One mole has a mass of 260 g and contains 6.02×10^{23} atoms. One atom has a mass of

$$\frac{260 \text{ g/mol}}{6.02 \times 10^{23} \text{ atoms/mol}} = 4.32 \times 10^{-22} \text{ g/atom}$$

■　■　■

Example 4
The dot at the end of this sentence has a mass of about one microgram (1×10^{-6} g). Assuming that the black stuff is carbon, calculate the approximate number of atoms of carbon needed to make such a dot.
The atomic weight of carbon is 12.0 amu.
One mole of carbon atoms weighs 12.0 g and contains 6.02×10^{23} atoms.

$$\left(\frac{1 \times 10^{-6} \text{ g}}{12.0 \text{ g/mol}}\right)(6.02 \times 10^{23} \text{ atoms/mol}) = 5 \times 10^{16} \text{ atoms}$$

■　■　■

4.2　Simplest formulas

Of the various types of chemical formulas, the *simplest formula*, also called the *empirical formula*, gives the bare minimum of information about a compound. It states only the *relative* number of atoms or moles of atoms in the compound. The convention used in writing the simplest formula is to write the symbols of the elements and affix subscripts to designate the relative numbers of moles of atoms of these elements. The formula A_xB_y represents a compound in which there is x mol of A atoms for every y mol of B atoms. Because of the relationship between moles of atoms and number of atoms, the simplest formula also gives information about the relative number of atoms in the compound. In A_xB_y there are x atoms of element A for every y atoms of type B. Nothing is to be inferred about the nature of this association—in particular, nothing about the size of makeup of the molecular aggregate—except the *relative* number of atoms in it.

The simplest formula is usually the direct result of an experiment. This procedure is illustrated in the following examples.

Example 5

When metal M *(atomic weight, 121.75 amu) is heated with excess sulfur, a chemical reaction occurs between the metal and sulfur. The excess sulfur is then driven off, leaving only the compound, consisting of combined metal and sulfur. From the weight of the metal and weight of compound, the weight of combined sulfur in the compound can be deduced. Here are some sample figures from an experiment of this type:*

Weight of metal	*2.435 g*
Weight of compound	*3.397 g*
Weight of combined sulfur	*0.962 g*

What is the simplest formula of the compound?

Since the simplest formula gives the relative numbers of moles of atoms in the compound, we must calculate how many moles of M atoms and S atoms have combined. If the atomic weight of M is 121.75 amu and that of S is 32.06 amu, then

$$\text{Number of moles of M atoms} = \frac{\text{weight of M}}{\text{weight of 1 mol of M atoms}}$$

$$= \frac{2.435 \text{ g}}{121.75 \text{ g/mol}}$$

$$= 0.0200 \text{ mol of M atoms}$$

$$\text{Number of moles of S atoms} = \frac{\text{weight of S}}{\text{weight of 1 mol of S atoms}}$$

$$= \frac{0.962 \text{ g}}{32.06 \text{ g/mol}}$$

$$= 0.0300 \text{ mol of S atoms}$$

In this compound there is 0.0200 mol of M atoms combined with 0.0300 mol of S atoms. Since the relative number of moles of M atoms to S atoms is 0.0200:0.0300, or 2:3, the simplest formula is M_2S_3.

■ ■ ■

Example 6

The analysis of a compound is often given in terms of percentage composition. What is the simplest formula of a compound which on analysis shows 50.05% S and 49.95% O by weight?

For the simplest formula we need the relative numbers of moles of atoms of sulfur and oxygen in the compound. Since only relative numbers are involved, we may consider any amount of compound, 1 g, 32.06 g, or any other weight. We shall work this problem in two ways for illustration:

(a) In *32.06 g of compound* there are 16.05 g of sulfur (50.05% of 32.06 g) and 16.01 g of oxygen (49.95% of 32.06 g).

$$\text{Number of moles of S atoms} = \frac{\text{weight of S}}{\text{weight of 1 mol of S atoms}}$$

$$= \frac{16.05 \text{ g}}{32.06 \text{ g/mol}}$$

$$= 0.5005 \text{ mol of S atoms}$$

$$\text{Number of moles of O atoms} = \frac{\text{weight of O}}{\text{weight of 1 mol of O atoms}}$$

$$= \frac{16.01 \text{ g}}{15.999 \text{ g/mol}}$$

$$= 1.001 \text{ mol of O atoms}$$

The simplest formula is $S_{0.5005}O_{1.001}$, or SO_2.

(b) In *100.0 g of compound* there are 50.05 g of sulfur and 49.95 g of oxygen.

$$\text{Number of moles of S atoms} = \frac{\text{weight of S}}{\text{weight of 1 mol of S atoms}}$$

$$= \frac{50.05 \text{ g}}{32.06 \text{ g/mol}}$$

$$= 1.561 \text{ mol of S atoms}$$

$$\text{Number of moles of O atoms} = \frac{\text{weight of O}}{\text{weight of 1 mol of O atoms}}$$

$$= \frac{49.95 \text{ g}}{15.999 \text{ g/mol}}$$

$$= 3.122 \text{ mol of O atoms}$$

The simplest formula is $S_{1.561}O_{3.122}$, or SO_2.

■ ■ ■

4.3 Molecular formulas

A second type of formula is the molecular formula. In the molecular formula the subscripts give the *actual* number of atoms of an element in one molecule of the compound. The molecule was defined previously as an aggregate of atoms bonded together tightly enough to be conveniently treated as a recognizable unit. In order to write the molecular formula, it is necessary to know how many atoms constitute the molecule. To find the actual number of atoms in a molecule, various experimental techniques can be used. For instance, X-ray determination of the positions of atoms in solids can give this information. Furthermore, as discussed in Chaps. 5 and 8, some of the properties of gases and solutions depend on the number of atoms in each molecular aggregate. A few molecular formulas thus determined are shown in Fig. 4.1, where they are compared with the corresponding simplest formulas. In some

Fig. 4.1
Molecular and simplest
formulas.

Substance	Molecular formula	Simplest formula
Benzene	C_6H_6	CH
Acetylene	C_2H_2	CH
Phosphorus	P_4	P
Gaseous water	H_2O	H_2O
Sucrose (a sugar)	$C_{12}H_{22}O_{11}$	$C_{12}H_{22}O_{11}$
Glucose (a sugar)	$C_6H_{12}O_6$	CH_2O
Nicotine	$C_{10}H_{14}N_2$	C_5H_7N

cases, e.g., gaseous water and sucrose, the molecular and simplest formulas are identical. In other cases they are not. We cannot tell from a formula whether it is molecular or simplest. However, if the subscripts given have a common divisor other than 1, the chance is good that it is a molecular formula. The molecular formula gives all the information the simplest formula gives, and more besides.

4.4   Formula weights and molecular weights

The *formula weight* is the *sum of all the atomic weights in the formula* under consideration, be it simplest or molecular. For NaCl, the formula weight is the atomic weight of sodium, 22.9898 amu, plus the atomic weight of chlorine, 35.453 amu, a total of 58.443 amu. For $C_{12}H_{22}O_{11}$, the formula weight is equal to 12 times the atomic weight of carbon plus 22 times the atomic weight of hydrogen plus 11 times the atomic weight of oxygen, or 342.299 amu. In all cases, the formula weight depends on which formula is written. If the molecular formula is used, the formula weight is called the *molecular weight*. For example, 342.299 amu is the molecular weight of sucrose.

As noted above, the formula weight is given in atomic mass units. The amount of substance whose mass in grams numerically equals the formula weight is called a *gram-mole*. In the case of $C_{12}H_{22}O_{11}$, where the formula weight is 342.299 amu, one gram-mole weighs 342.299 g. The formula weight can also be used in terms of tons, ounces, pounds, or other units of mass to give ton-moles, ounce-moles, etc. As example, 342.299 tons of sugar is one ton-mole of sugar. Since we are primarily concerned with gram-moles, it is convenient to omit the word "gram" and to understand that "mole" means "gram-mole."

There is an important relationship between number of gram-moles and number of particles. Let us consider sulfur chloride, the molecular formula of which is S_2Cl_2. One gram-mole of S_2Cl_2 weighs 135 g and contains 64 g of S and 71 g of Cl. In 64 g of S (atomic weight 32), there is 64/32, or 2, mol of S atoms; in 71 g of Cl (atomic weight 35.5), there is 71/35.5, or 2, mol of Cl atoms. Since one mole of atoms contains the Avogadro number of atoms, two moles of S atoms contain $2 \times 6.02 \times 10^{23}$ atoms of S, and two moles of Cl atoms contain $2 \times 6.02 \times 10^{23}$ atoms of Cl. The molecular formula S_2Cl_2 indicates that two atoms of S and two atoms of Cl make up one molecule of S_2Cl_2. Therefore, $2 \times 6.02 \times 10^{23}$ atoms of S and $2 \times 6.02 \times 10^{23}$ atoms of Cl make up 6.02×10^{23} molecules. For any substance whose molecular formula is known, *one gram-mole contains the Avogadro number of molecules.*

Let us consider the relationship between number of gram-moles and number of particles for a compound whose molecular formula is not known. Such a compound is phosphorus dichloride, whose simplest formula is PCl_2. One gram-mole of PCl_2 weighs 102 g and contains 31 g of P and 71 g of Cl. In 31 g of P (atomic weight 31) there is 31/31, or 1, mol of P atoms; in 71 g of Cl (atomic weight 35.5), there is 71/35.5, or 2, mol of Cl atoms. One mole of P atoms contains 6.02×10^{23} atoms of P; two moles of Cl atoms contains $2 \times 6.02 \times 10^{23}$ atoms of Cl. Since the molecular formula of PCl_2 is not known, the number of atoms in the molecule is not known. Therefore, we cannot state the number of molecules in one gram-mole of PCl_2. If we define a *formula-unit* as consisting of one P atom and two Cl atoms, there are then 6.02×10^{23} such formula-units in one gram-mole of PCl_2. For any substance, *one gram-mole contains the Avogadro number of formula-units.* Only if the formula is a molecular formula is the formula-unit the same as the molecule.

Even though formulas are based on experimentally determined percentage composition, it is sometimes necessary to calculate percentage composition from a formula.

Example 7
What is the percent composition of $Al_2(SO_4)_3$? Atomic weights are Al, 26.98; S, 32.06; and O, 16.00 amu.
One gram-mole of $Al_2(SO_4)_3$ contains

2 mol of Al atoms or 2×26.98 g Al, or	53.96 g Al
3 mol of S atoms or 3×32.06 g S, or	96.18 g S
12 mol of O atoms or 12×16.00 g O, or	192.00 g O
	342.14 g total mass

$$\text{Percent Al} = \frac{53.96 \text{ g}}{342.14 \text{ g}} \times 100 = 15.77\%$$

$$\text{Percent S} = \frac{96.18 \text{ g}}{342.14 \text{ g}} \times 100 = 28.11\%$$

$$\text{Percent O} = \frac{192.00 \text{ g}}{342.14 \text{ g}} \times 100 = 56.12\%$$

■ ■ ■

4.5 Chemical reactions

The other principal division of stoichiometry is concerned with weight changes in chemical reactions. Before considering these quantitative aspects, we need to examine ways of describing chemical change. It is possible to group chemical reactions into two broad, somewhat arbitrary, classes: (1) reactions in which there is no electron transfer and (2) reactions in which there is electron transfer from one atom to another atom.

Reactions in which no electrons are transferred usually involve the joining or separating of ions or molecules. An example of a reaction with no electron transfer occurs when an aqueous solution of sodium chloride is mixed with an aqueous solution of silver nitrate. The sodium chloride solution contains sodium ions and chloride ions; the silver nitrate solution contains silver ions and nitrate ions. When the two solutions are mixed, a chemical reaction occurs, as shown by the formation of a white precipitate. This white solid consists of silver and chloride ions clumped together in large aggregates. In the final solution, sodium ions and nitrate ions remain as they were initially. In the chemical reaction, the silver ions have combined with chloride ions to form solid silver chloride, which is insoluble in the water. In shorthand form, the reaction is indicated as

$$Ag^+(soln) + \overline{NO_3^-(soln)} + \overline{Na^+(soln)} + Cl^-(soln) \longrightarrow$$

$$AgCl(s) + \overline{Na^+(soln)} + \overline{NO_3^-(soln)}$$

where the abbreviation "(soln)" indicates that the ion is in solution and the notation (s) emphasizes the fact that silver chloride is formed as a solid. The strikeovers indicate cancellation of ions which do not change in the course of the reaction. The *net* reaction is

$$Ag^+(soln) + Cl^-(soln) \longrightarrow AgCl(s)$$

Reactions in which electrons are transferred from one atom to another are known as *oxidation-reduction reactions*. They are also sometimes known as *redox* reactions. Many of the most important chemical reactions fall into this class. For example, the combining of a sodium atom with a chlorine atom can be regarded as resulting from the transfer of an electron from the sodium to the chlorine, as shown schematically in Fig. 4.2. A neutral sodium atom gives up an electron to a neutral chlorine atom so as to form positively charged Na^+ and negatively charged Cl^-.

A less obvious example of an oxidation-reduction reaction is that in which hydrogen and oxygen combine to form water. In this case there is a change in the sharing of the electrons during the course of the reaction:

$$H:H + :\overset{\cdot\cdot}{\underset{\cdot}{O}}: \longrightarrow H:\overset{\cdot\cdot}{\underset{H}{O}}:$$

What happens to the hydrogen in the course of this reaction? In the initial state two hydrogen nuclei share a pair of electrons. Since the two hydrogen nuclei are identical, they share the electron pair equally; each hydrogen atom has a one-half time share of two electrons. In the final state the hydrogen shares a pair of electrons with oxygen. Since

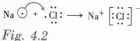

Fig. 4.2
Chemical reaction with electron transfer.

oxygen is more electronegative than hydrogen, the electron pair is not shared equally but belongs more to the oxygen than to the hydrogen. In the course of the reaction there is thus a change in the electron sharing, i.e., a partial transfer of electrons from hydrogen to oxygen.

4.6   Oxidation numbers

In order to keep track of electron shifts in oxidation-reduction reactions, it is convenient to introduce what is called the *oxidation number*. The *oxidation number*, which may also be referred to as *oxidation state*, is defined as the charge which an atom *appears* to have (with emphasis on the word "appears") when electrons are counted according to the following rules: (1) Electrons shared between two unlike atoms are counted with the more electronegative atom. (2) Electrons shared between two like atoms are divided equally between the sharing atoms.

What is the oxidation number of hydrogen in the H_2 molecule? The electron pair is shared by two identical atoms, and so, according to the second rule, half the electrons are counted with each atom. Since the hydrogen nucleus has a $+1$ charge and since one negative charge is counted with the nucleus, the apparent charge of each hydrogen atom is 0. The oxidation number of hydrogen in H_2 is therefore 0.

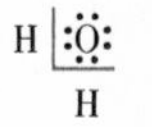

What are the oxidation numbers of hydrogen and oxygen in H_2O? Oxygen is the more electronegative, and so, according to the first rule, the shared electrons are counted with the oxygen, as shown by the line in Fig. 4.3. The hydrogen appears to have a charge of $+1$ and is assigned an oxidation number of $+1$. Since eight electrons are counted with the $+6$ oxygen kernel (nucleus plus inner-shell electrons), the apparent charge of oxygen is -2. Oxygen, therefore, has an oxidation number of -2 in H_2O.

In principle, electronic pictures can be drawn for all molecules and electrons counted in this way to deduce the oxidation numbers of the various atoms. This is laborious. It is more convenient to use the following operational rules, which are derived from rules (1) and (2):

1   In free elements, each atom has an oxidation number of 0, no matter how complicated the molecule is. Hydrogen in H_2, sodium in Na, sulfur in S_8, and phosphorus in P_4 all have oxidation numbers of 0.

2   In simple ions (i.e., those which contain but one atom) the oxidation *number is equal to the charge on the ion.* In these cases, the apparent charge of the atom is the real charge of the ion. In the tripositive aluminum ion, the oxidation number of the aluminum ion is $+3$. Iron, which can form a dipositive or a tripositive ion, sometimes has an oxidation number of $+2$ and sometimes $+3$. In the dinegative oxide ion, the oxidation number of oxygen is -2. It is useful to remember that elements of group I of the periodic table, lithium, sodium, potassium, rubidium, cesium, and francium form only $+1$ ions. Their oxidation number is $+1$ in all compounds. The group II elements, beryllium, magnesium, calcium, strontium, barium, and radium,

H $| \; \ddot{\text{:}}\ddot{\text{O}}\text{:}$ O *kernel 6 +*
H   H *kernel 1 +*

Fig. 4.3

Assignment of oxidation numbers in H_2O.

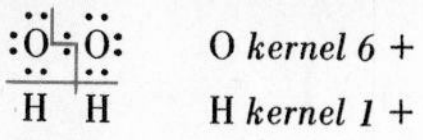

:Ö⁝Ö: O *kernel 6 +*
‾H‾‾H‾   H *kernel 1 +*

Fig. 4.4

Assignment of oxidation numbers in H_2O_2.

:Ö⁝F: O *kernel 6 +*
⁝F⁝   F *kernel 1 +*

Fig. 4.5

Assignment of oxidation numbers in OF_2.

form only $+2$ ions and hence always have oxidation numbers of $+2$ in all compounds.

3   In compounds containing oxygen, the oxidation number of each oxygen atom is generally -2. There are two kinds of exceptions: One arises in the case of the peroxides and superoxides, compounds of oxygen in which there is an oxygen-oxygen bond. In peroxides, e.g., hydrogen peroxide (H_2O_2), only seven electrons are counted with the $+6$ kernel of oxygen. Figure 4.4 shows how the electrons are assigned. In the hydrogen-oxygen bond the electrons are counted with oxygen, the more electronegative atom. In the oxygen-oxygen bond the electron pair is shared between two like atoms and is split equally between the sharing partners. The apparent charge of the oxygen is thus -1. Oxygen has an oxidation number of -1 in all peroxides. In superoxides, such as the rarely encountered KO_2, the oxidation number of oxygen is $-\frac{1}{2}$. The second exception is even less common. It occurs when oxygen is bonded to fluorine, the only atom which is more electronegative than oxygen. When oxygen is bonded to fluorine, shared electrons are counted with the fluorine. The assignment of oxidation numbers in the compound oxygen difluoride is shown in Fig. 4.5. The oxidation number of fluorine is -1, and the oxidation number of oxygen is $+2$.

4   In compounds containing hydrogen, the oxidation number of hydrogen is generally $+1$. This rule covers practically all the hydrogen compounds. It fails in the case of the hydrides, in which hydrogen is bonded to an atom less electronegative than hydrogen. For example, when hydrogen is bonded to sodium in the compound sodium hydride, NaH, the hydrogen is the more electronegative atom, and two electrons are counted with it. In hydrides, the oxidation number of hydrogen is -1.

5   All oxidation numbers must be consistent with the conservation of charge. Charge must be conserved in the sense that the sum of all the apparent charges in a particle must equal the net charge of that particle. This leads to the following conditions: (*a*) *For neutral molecules, the oxidation numbers of all the atoms must add up to zero;* (*b*) *for complex ions (charged particles which contain more than one atom), the oxidation numbers of all the atoms must add up to the charge on the ion.* As an example of a neutral molecule, we consider the case of H_2O. The oxidation number of hydrogen is $+1$. There are two hydrogen atoms. The total apparent charge contribution by hydrogen is $+2$. The oxidation number of oxygen is -2. The whole molecule appears to be neutral.

The neutrality rule enables us to assign oxidation numbers to any atom. For example, what is the oxidation number of sulfur in H_2SO_4? The oxidation number of hydrogen is $+1$; the oxidation number of oxygen is -2. The two hydrogens give an apparent charge of $+2$; the four oxygens give an apparent charge of -8. For neutrality the sulfur must contribute $+6$. Since there is but one sulfur atom, the oxidation number of sulfur is $+6$. In $H_2S_2O_3$ hydrogen contributes a total apparent charge of $+2$; oxygen contributes a total apparent charge of -6. For neutrality the sulfur contribution must be $+4$. Since there are two sulfur atoms, the oxidation number of each is $+2$.

Since oxidation numbers are quite arbitrary, they may have values which at first sight appear strange. For example, in cane sugar, $C_{12}H_{22}O_{11}$, the oxidation number of carbon is 0. The total apparent charge of 22 hydrogen atoms, $22(+1) = +22$, is canceled by that of 11 oxygen atoms, $11(-2) = -22$. According to the oxidation number, each carbon atom appears to contribute no charge to the molecule. Fractional oxidation numbers are also possible, as in $Na_2S_4O_6$, where the oxidation number of sulfur is $+\frac{10}{4}$.

In complex ions the apparent charges of all the atoms must add up to equal the charge on the ion. This is true in hydroxide ion OH^-, for example, where the superscript "minus" indicates that the ion has a net charge of -1. Since oxygen has an oxidation number of -2 and since hydrogen has an oxidation number of $+1$, the total apparent charge is $(-2) + (+1) = -1$, which is the same as the actual charge of the ion. In $Cr_2O_7^{2-}$, a dinegative ion, the seven oxygen atoms contribute -14. Chromium must contribute $+12$ in order to make the ion have a net charge of -2. Since there are two chromium atoms in the complex, each chromium has an oxidation number of $+6$.

In order to avoid confusion with the actual charge on an ion, which is written as a superscript, the oxidation number of an atom, when needed, is written beneath the atom to which it applies. For example, in

$$\underset{+5}{P_2} \quad \underset{-2}{O_7^{4-}}$$

the charge on the ion is -4; the oxidation numbers are $+5$ and -2. It should be emphasized strongly that oxidation numbers are not actual charges of atoms. In the specific case of $P_2O_7^{4-}$ it can be shown experimentally that the aggregate carries a -4 charge; however, it cannot be shown experimentally that the charge of P is $+5$ and that of the O is -2. The $+5$ and -2 are arbitrarily assigned numbers, and we must not conclude that $P_2O_7^{4-}$ contains P^{5+} ions and O^{2-} ions.

Although they are frequently confused with each other, oxidation number is not the same thing as valence. Valence, or *combining capacity*, can be interpreted in several ways. For example, it represents the number of hydrogen atoms which can be combined with a given atom. It also represents the number of single bonds which an atom can form. In any case, valence is a pure number and has no plus or minus associated with it. On the other hand, oxidation number is positive or negative. For example, in water the valence of oxygen is 2, but its oxidation number is -2. Furthermore, there may actually be a difference between the magnitude of the valence and the oxidation number. In hydrogen peroxide, for example, (Fig. 4.4) each oxygen atom has two single bonds, one that goes to oxygen and one to hydrogen. The valence of oxygen is therefore 2. However, as indicated before, the oxidation number of oxygen in H_2O_2 is -1.

4.7 Oxidation-reduction

The term *oxidation* refers to any chemical change in which there is an algebraic *increase in oxidation number*. For example, when hydrogen, H_2, reacts with oxygen to form water, H_2O, the hydrogen atoms change oxidation number from 0 to $+1$. The H_2 is said to undergo oxidation. When sucrose, $C_{12}H_{22}O_{11}$, is burned to give carbon dioxide, CO_2, carbon atoms increase in oxidation number from 0 to $+4$. The sucrose is oxidized. The term *reduction* applies to any algebraic *decrease in oxidation number*. For example, when oxygen, O_2, reacts with hydrogen to form H_2O, oxygen atoms change oxidation number from 0 to -2. This is a decrease in oxidation number; hence O_2 is said to undergo reduction.

In oxidation and reduction the increase and decrease of oxidation numbers result from a shift of electrons. The only way by which electrons can be moved away from one atom is for them to be moved toward another atom. In this process the oxidation number of the first atom increases, and the oxidation number of the second atom decreases. Oxidation and reduction must therefore always occur together and must just compensate each other.

Fig. 4.6

Terms used in describing oxidation-reduction.

Term	Oxidation-number change	Electron change
Oxidation	*Increase*	*Loss of electrons*
Reduction	*Decrease*	*Gain of electrons*
Oxidizing agent	*Decrease*	*Picks up electrons*
Reducing agent	*Increase*	*Supplies electrons*
Substance oxidized	*Increase*	*Loses electrons*
Substance reduced	*Decrease*	*Gains electrons*

The *oxidizing agent* is, by definition, the *substance that does the oxidizing; it is the substance containing the atom which shows a decrease in oxidation number*. For example, if in a reaction $KClO_3$ is converted to KCl, each chlorine atom decreases in oxidation number from $+5$ to -1. This amounts to getting six electrons (six negative charges) from other atoms. Thus, $KClO_3$ must cause oxidation and is acting as an oxidizing agent. Similarly, a *reducing agent is the substance that does the reducing; it is the substance containing the atom which shows an increase in oxidation number*. In the reaction of $C_{12}H_{22}O_{11}$ to give CO_2, $C_{12}H_{22}O_{11}$ is a reducing agent, because it contains carbon atoms which increase in oxidation number. It should be evident that when a substance acts as a reducing agent, it itself must be oxidized in the process. Figure 4.6 summarizes the terms used to describe oxidation-reduction.

Listed in Fig. 4.7 are some examples of oxidation-reduction processes. The numbers below the formulas indicate the oxidation numbers of interest. It must be emphasized that the terms "oxidizing agent"

Fig. 4.7

Examples of oxidation-reduction reactions.

Oxidizing agents	+	Reducing agents	$\longrightarrow$	Products
O_2		H_2		H_2O
0		0		$+1 -2$
Cl_2		Na		Na Cl
0		0		$+1 -1$
H_3O^+		Mg		$Mg^{2+} + H_2 + H_2O$
$+1$		0		$+2 \qquad 0$
$KClO_3$		$C_{12}H_{22}O_{11}$		$KCl + CO_2 + H_2O$
$+5$		0		$-1 \quad +4$
H_2O_2		H_2O_2		$H_2O + O_2$
-1		-1		$-2 \qquad 0$

and "reducing agent" refer to the entire substance and not to just one of the atoms contained therein. For example, in the next-to-last reaction of the table, the oxidizing agent is $KClO_3$, not $+5$ Cl. It can be shown that $KClO_3$ picks up electrons and therefore is an oxidizing agent; because the rules for assigning oxidation numbers are quite arbitrary, it *cannot* be shown that it is the chlorine atom of $KClO_3$ that picks up the electrons.

In the last reaction listed in Fig. 4.7, H_2O_2 is seen to act both as a reducing agent and an oxidizing agent. In oxidizing and reducing itself, it is said to undergo *autooxidation,* or *disproportionation.*

4.8   Balancing chemical equations

Chemical equations are shorthand designations which give information about a chemical reaction. We shall generally use *net equations, which specify only the substances used up and the substances formed in the chemical reaction.* Net equations omit anything which remains unchanged. The convention used in writing equations is to place what disappears (the *reactants*) on the left side and what appears (the *products*) on the right side. The reactants and products are separated by a single arrow $\rightarrow$, an equals sign $=$, or a double arrow $\rightleftharpoons$, depending on what aspect of the chemical reaction is being emphasized. An example of a net equation is

$$Cl_2(g) + 2H_2O + Ag^+ \longrightarrow AgCl(s) + HOCl + H_3O^+$$

The reactants and products are designated by symbols or formulas. The symbol can be thought of as representing either one atom or one mole of atoms. The formula represents either one formula-unit or one mole of formula-units. The notation (g) indicates the gas phase, and (s) the solid phase. *When no such phase notation appears, the aqueous phase is understood.*

To be complete, a chemical equation must satisfy three conditions: First, it must be consistent with the experimental facts; i.e., it must state what chemical species disappear and appear. Second, it must be consistent with the conservation of mass. (Since we cannot destroy mass, we must account for it. If an atom disappears from one substance, it must appear in another.) Third, the chemical equation must be consistent with the conservation of electric charge. (Since we cannot destroy electric charge, we must account for it.) The second and third conditions are expressed by saying that the equation must be *balanced*. A balanced equation contains the same number of atoms of the different kinds on the left and right sides; furthermore, the net charge is the same on both sides.

How do we go about writing balanced equations? One method, usually reserved for simple reactions, is to balance the equation by inspection. For example, in the reaction between a solution of silver nitrate and a solution of sodium chloride, silver ions and chloride ions disappear, and solid silver chloride appears. The equation for the reaction is

$$Ag^+ + Cl^- \longrightarrow AgCl(s)$$

Since there is one silver atom on the left and one on the right and since there is one chlorine atom on the left and one on the right, mass balance is satisfied. The net electric charge on the left is 0 ($+1$ for the silver ion plus -1 for the chloride ion totals 0), and the net charge on the right is 0. Therefore, the equation is also electrically balanced.

In the reaction between solid sodium and gaseous diatomic chlorine, solid sodium chloride is formed; so we write

$$Na(s) + Cl_2(g) \longrightarrow NaCl(s)$$

To balance this equation, we note that we have two chlorine atoms on the left; so we ought to have two chlorine atoms on the right. We cannot change the subscript of Cl in the formula NaCl because that would give the formula of a different compound. We can change only the coefficients; hence we put 2 in front of the NaCl. With two sodium atoms on the right we now need two sodium atoms on the left; therefore, we also place a 2 in front of the Na. The equation now reads

$$2Na(s) + Cl_2(g) \longrightarrow 2NaCl(s)$$

and has been balanced by inspection.

There are more complicated reactions involving electron transfer where balancing by inspection gets to be quite a chore. For example, suppose that in the reaction which occurs between potassium dichromate, sulfur, and water the products are taken to be sulfur dioxide, potassium hydroxide, and chromic oxide:

$$K_2Cr_2O_7(s) + H_2O + S(s) \longrightarrow SO_2(g) + KOH(s) + Cr_2O_3(s)$$

Although the equation can be balanced by inspection, it is easier to balance it by matching up the electron transfer, i.e., the oxidation and the reduction. As far as electron transfer is concerned, we have to worry only about those atoms which change oxidation number. On applying the rules for assigning oxidation numbers, we see that sulfur changes oxidation number from 0 to $+4$ and chromium from $+6$ to $+3$. As indicated below, each sulfur atom appears to lose four electrons, and each chromium atom appears to gain three electrons:

$$K_2Cr_2O_7(s) + H_2O + S(s) \longrightarrow SO_2(g) + KOH(s) + Cr_2O_3(s)$$

$+6$ 0 $+4$ $+3$

3e⁻ | *per atom*

6e⁻ | *per formula unit*

4e⁻

Since each formula-unit of $K_2Cr_2O_7$ contains two chromium atoms, a formula-unit will pick up 2×3, or 6 electrons. These electrons must be furnished by the sulfur. In order that the electron loss and the electron gain be equal, we take two $K_2Cr_2O_7$ formula-units for every three sulfur atoms. For every two $K_2Cr_2O_7$ formula-units that disappear, 12 electrons are picked up; for every three sulfur atoms used up, 12 electrons are furnished. Hence we write 2 in front of the $K_2Cr_2O_7$ and the Cr_2O_3 and 3 in front of the S and the SO_2 to give

$$2K_2Cr_2O_7(s) + H_2O + 3S(s) \longrightarrow 3SO_2(g) + KOH(s) + 2Cr_2O_3(s)$$

Although the tough part is over, the equation is not yet balanced. To complete the job, the other coefficients must be made consistent with those already selected. We can do this by inspection. From the above equation we see that there are four potassium atoms on the left; so we need four potassium atoms on the right; hence, we place a 4 in front of the KOH. The result

$$2K_2Cr_2O_7(s) + H_2O + 3S(s) \longrightarrow 3SO_2(g) +$$
$$4KOH(s) + 2Cr_2O_3(s)$$

is still not balanced. Balance can be achieved by counting up either the hydrogen atoms or the oxygen atoms on the right, where all the coefficients are already fixed. This shows that two molecules of H_2O are required on the left. The balanced equation is

$$2K_2Cr_2O_7(s) + 2H_2O + 3S(s) \longrightarrow 3SO_2(g) +$$
$$4KOH(s) + 2Cr_2O_3(s)$$

Here in summary for future reference are the steps followed:

1   Assign oxidation numbers for those atoms which change.
2   Decide on number of electrons to be shifted per atom.
3   Decide on number of electrons to be shifted per formula-unit.
4   Compensate electron gain and loss by writing appropriate co-efficients for the oxidizing agent and the reducing agent.
5   Insert other coefficients consistent with the conservation of matter.

In aqueous solution, balancing oxidation-reduction equations is different because we usually know in advance only the formulas of the oxidizing and the reducing agents as well as their products. The final balanced equation will involve these species, but it may also include

H_2O and H_3O^+ or OH^-, depending on whether the solution is acidic or basic. The insertion of these additional species is part of the process of balancing equations in aqueous solutions. As an illustration, we consider the problem of balancing the equation for the oxidation of H_2SO_3 by $Cr_2O_7^{2-}$ in *acidic solution* to form HSO_4^- and Cr^{3+}. The stepwise procedure is as follows:

1 Assign oxidation numbers:

$$Cr_2O_7^{2-} + H_2SO_3 \longrightarrow HSO_4^- + Cr^{3+}$$

$+6 \qquad\qquad +4 \qquad\qquad +6 \qquad +3$

2 Balance the electron transfer for those atoms that change oxidation number:

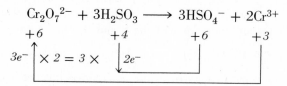

After the coefficient 3 has been placed in front of H_2SO_3 on the left side, the right side is made consistent with regard to the chromium and sulfur atoms.

3 Count up oxygen atoms and add H_2O to side that is deficient in oxygen. There are $7 + 3(3) = 16$ oxygens on the left and $3(4) = 12$ on the right; so we need $4H_2O$ on the right:

$$Cr_2O_7^{2-} + 3H_2SO_3 \longrightarrow 3HSO_4^- + 2Cr^{3+} + 4H_2O$$

4 Count up hydrogen atoms and add H^+ to side that is deficient in hydrogen. On the left there are 6 hydrogen atoms in the $3H_2SO_3$; on the right there are 3 hydrogen atoms in $3HSO_4^-$ plus 8 hydrogen atoms in $4H_2O$, making 11 hydrogen atoms in all. The left side, consequently, is deficient by 5 hydrogens; so we add $5H^+$:

$$Cr_2O_7^{2-} + 3H_2SO_3 + 5H^+ \longrightarrow 3HSO_4^- + 2Cr^{3+} + 4H_2O$$

5 Check by comparing the net charge on left and right sides of the equation. If the balancing has been done properly, net charge on both sides is the same. Left side has one dinegative plus three neutrals plus five monopositive species, which comes to $1(-2) + 3(0) + 5(+1) = +3$. Right side has three mononegative plus two tripositive plus four neutral species, which comes to $3(-1) + 2(+3) + 4(0) = +3$.

6 If the reaction is said to be in acidic solution, change each H^+ into H_3O^+, and add an equal number of H_2O molecules to the opposite side of the equation:

$$Cr_2O_7^{2-} + 3H_2SO_3 + 5H_3O^+ \longrightarrow 3HSO_4^- + 2Cr^{3+} + 9H_2O$$

On the other hand, if the reaction is said to be in basic solution, add enough OH^- to neutralize each H^+ to H_2O, and add an equal number of OH^- to the opposite side of the equation. Cancel any duplication of H_2O from left and right sides of the equation.

To see how the above method works for balancing an oxidation-reduction reaction in *basic solution*, we consider the same reaction carried out in basic medium. Instead of $Cr_2O_7^{2-}$, the oxidizing agent will be in the form of CrO_4^{2-}. The reducing agent, instead of being H_2SO_3, will be in the form of SO_3^{2-}. The corresponding products can be written CrO_2^- and SO_4^{2-}. The sequence of steps follows:

1 $\quad CrO_4^{2-} + SO_3^{2-} \longrightarrow SO_4^{2-} + CrO_2^-$

$\qquad +6 \qquad\quad +4 \qquad\qquad +6 \qquad\quad +3$

2 $\quad 2CrO_4^{2-} + 3SO_3^{2-} \longrightarrow 3SO_4^{2-} + 2CrO_2^-$

$\qquad +6 \qquad\qquad +4 \qquad\qquad\quad +6 \qquad\qquad +3$

$2 \times \left| 3e^- = 3 \times \right| 2e^-$

3 $\quad 2CrO_4^{2-} + 3SO_3^{2-} \longrightarrow 3SO_4^{2-} + 2CrO_2^-$

There are $2(4) + 3(3) = 17$ oxygen atoms on the left and $3(4) + 2(2) = 16$ oxygen atoms on the right. Place one H_2O on the right to balance oxygen atoms:

$2CrO_4^{2-} + 3SO_3^{2-} \longrightarrow 3SO_4^{2-} + 2CrO_2^- + H_2O$

4 Count up hydrogen atoms. There is none on the left, two on the right. We need two more on the left; so insert $2H^+$ on the left side:

$2CrO_4^{2-} + 3SO_3^{2-} + 2H^+ \longrightarrow 3SO_4^{2-} + 2CrO_2^- + H_2O$

This is a balanced equation, as can be checked by counting up net charges on left and right. However, it is wrong because it does not agree with the chemistry. It shows $2H^+$, which might be all right for acidic solution, but we were told this is a basic solution.

5 To get rid of the $2H^+$, we add $2OH^-$ on the left (giving $2H_2O$), but to avoid disturbing the balance we also add $2OH^-$ on the right. In sequence, this looks as follows:

$2CrO_4^{2-} + 3SO_3^{2-} + 2H^+ \longrightarrow 3SO_4^{2-} + 2CrO_2^- + H_2O$

$\qquad\qquad\qquad\qquad + 2OH^- \qquad\qquad\qquad\qquad\qquad\quad + 2OH^-$

$\overline{2CrO_4^{2-} + 3SO_3^{2-} + 2H_2O \longrightarrow 3SO_4^{2-} + 2CrO_2^- + H_2O + 2OH^-}$

$2CrO_4^{2-} + 3SO_3^{2-} + H_2O \longrightarrow 3SO_4^{2-} + 2CrO_2^- + 2OH^-$

To get the last line, we have simply subtracted one H_2O molecule from both sides of the equation in the preceding line.

There is an alternative method of balancing equations which avoids completely the problem of assigning oxidation numbers. This is the *method of half-reactions* (sometimes called the *ion-electron* method). The method is based on splitting the reaction into two parts—an oxidation half and a reduction half. The two half-reactions are balanced separately, showing electrons, and then combined in order to eliminate the electrons from the final balanced equation. To illustrate, we consider the same reaction as before, i.e., the oxidation of H_2SO_3 by $Cr_2O_7^{2-}$ in acidic solution to form HSO_4^- and Cr^{3+}. The detailed steps, *for acidic solution*, are as follows:

1   Separate the change into half-reactions.
2   Balance each half-reaction separately:
    a   Adjust coefficients so as to balance all atoms except H and O.
    b   Add H_2O to side deficient in O.
    c   Add H_3O^+ (one H_3O^+ for each needed H atom) to side deficient in H and an equal number of H_2O to the opposite side.
    d   Add e^- to side deficient in negative charge.
3   Multiply each half-reaction by an appropriate number so as to balance the electron gain and loss, and add.
4   Subtract any duplications on left and right.

Step 1

$$H_2SO_3 \longrightarrow HSO_4^- \qquad\qquad Cr_2O_7^{2-} \longrightarrow Cr^{3+}$$

Step 2a

$$H_2SO_3 \longrightarrow HSO_4^- \qquad\qquad Cr_2O_7^{2-} \longrightarrow 2Cr^{3+}$$

Step 2b

$$H_2SO_3 + H_2O \longrightarrow HSO_4^- \qquad Cr_2O_7^{2-} \longrightarrow 2Cr^{3+} + 7H_2O$$

Step 2c

$$H_2SO_3 + 4H_2O \longrightarrow \qquad\qquad Cr_2O_7^{2-} + 14H_3O^+ \longrightarrow$$
$$HSO_4^- + 3H_3O^+ \qquad\qquad\qquad 2Cr^{3+} + 21H_2O$$

Step 2d

$$H_2SO_3 + 4H_2O \longrightarrow \qquad\qquad Cr_2O_7^{2-} + 14H_3O^+ + 6e^- \longrightarrow$$
$$HSO_4^- + 3H_3O^+ + 2e^- \qquad\qquad 2Cr^{3+} + 21H_2O$$

[Two electrons have been added to the right side since, in step (2c), the left side has a net charge of 0 and the right side has a net charge of +2. The right side of (2c) is deficient in negative charge by two units.]

[Six electrons have been added to the left since, in step (2c), the left side has a net charge of +12 and the right side has +6.]

Step 3

$$3(H_2SO_3 + 4H_2O \longrightarrow HSO_4^- + 3H_3O^+ + 2e^-)$$

$$Cr_2O_7^{2-} + 14H_3O^+ + 6e^- \longrightarrow 2Cr^{3+} + 21H_2O$$

$$\overline{3H_2SO_3 + 12H_2O + Cr_2O_7^{2-} + 14H_3O^+ + 6e^- \longrightarrow}$$

$$3HSO_4^- + 9H_3O^+ + 6e^- + 2Cr^{3+} + 21H_2O$$

(The top half-reaction has been multiplied by 3 to get six electrons in each half of the reaction and then the two half-reactions have been added.)

Step 4

Since $12H_2O$, $9H_3O^+$, and $6e^-$ are duplicated on left and right sides, these can be subtracted to give

$$3H_2SO_3 + Cr_2O_7^{2-} + 5H_3O^+ \longrightarrow 3HSO_4^- + 2Cr^{3+} + 9H_2O$$

If the reaction occurs in *basic solution,* the equation should not contain H_3O^+. In order to add H atoms in step (2c), add H_2O molecules equal in number to the deficiency of H atoms and an equal number of OH^- ions to the opposite side. The rest of the method is the same. An example of reaction in basic solution is the change

$$Cr(OH)_3(s) + IO_3^- \longrightarrow I^- + CrO_4^{2-}$$

The half-reactions are

$$Cr(OH)_3(s) + 5OH^- \longrightarrow CrO_4^{2-} + 4H_2O + 3e^-$$

$$IO_3^- + 3H_2O + 6e^- \longrightarrow I^- + 6OH^-$$

and the net equation is

$$2Cr(OH)_3(s) + IO_3^- + 4OH^- \longrightarrow 2CrO_4^{2-} + I^- + 5H_2O$$

4.9 Calculations using chemical equations

A chemical equation is valuable from two standpoints: It gives information on an atomic scale and on a laboratory scale. For example,

$$8KClO_3(s) + C_{12}H_{22}O_{11}(s) \longrightarrow 8KCl(s) + 12CO_2(g) + 11H_2O(g)$$

Atomic scale

$$8 \begin{pmatrix} \text{formula-units} \\ \text{of } KClO_3 \end{pmatrix} + 1 \begin{pmatrix} \text{formula-unit} \\ \text{of } C_{12}H_{22}O_{11} \\ \textit{molecule} \end{pmatrix} \longrightarrow$$

$$8 \begin{pmatrix} \text{formula-units} \\ \text{of } KCl \end{pmatrix} + 12 \begin{pmatrix} \text{formula-units} \\ \text{of } CO_2 \\ \textit{molecule} \end{pmatrix} + 11 \begin{pmatrix} \text{formula-units} \\ \text{of } H_2O \\ \textit{molecule} \end{pmatrix}$$

Lab scale

$$8\left(\genfrac{}{}{0pt}{}{\text{mol}}{\text{of }KClO_3}\right) + 1\left(\genfrac{}{}{0pt}{}{\text{mol}}{\text{of }C_{12}H_{22}O_{11}}\right) \longrightarrow$$

$$8\left(\genfrac{}{}{0pt}{}{\text{mol}}{\text{of }KCl}\right) + 12\left(\genfrac{}{}{0pt}{}{\text{mol}}{\text{of }CO_2}\right) + 11\left(\genfrac{}{}{0pt}{}{\text{mol}}{\text{of }H_2O}\right)$$

$$980.394\text{ g} + 342.299\text{ g} \longrightarrow 596.408\text{ g} + 528.120\text{ g} + 198.165\text{ g}$$

On an atomic scale, the equation states that 8 formula-units of $KClO_3$ (each formula-unit containing a potassium atom, a chlorine atom, and three oxygen atoms) react with 1 formula-unit of $C_{12}H_{22}O_{11}$ to produce 8 formula-units of KCl, 12 formula-units of CO_2, and 11 formula-units of H_2O. Since the numbers are important only in a *relative* sense, the equation also indicates, for example, that 8 *dozen* formula-units of $KClO_3$ react with 1 *dozen* formula-units of $C_{12}H_{22}O_{11}$ to produce 8 *dozen* formula-units of KCl, 12 *dozen* formula-units of CO_2, and 11 *dozen* formula-units of H_2O. Multiplying the equation through by the Avogadro number converts it from the atomic scale to something which is useful in the laboratory. The Avogadro number of formula-units is one mole; so the equation signifies that 8 mol of $KClO_3$ reacts with 1 mol of $C_{12}H_{22}O_{11}$ to give 8 mol of KCl, 12 mol of CO_2, and 11 mol of H_2O. From the formula weights of the various compounds we can get further quantitative information from the equation. Eight moles of $KClO_3$ weighs eight times the formula weight, or 8×122.549 g, or 980.394 g; 1 mol of $C_{12}H_{22}O_{11}$ weighs 342.299 g; 8 mol of KCl weighs 8×74.551, or 596.408 g; 12 mol of CO_2 weighs 12×44.101, or 528.120 g; and 11 mol of H_2O weighs 11×18.015, or 198.165 g. The total mass on the left-hand side of the equation is 1322.693 g, and that on the right side, 1322.693 g. Mass is conserved, as it must be.

Once a balanced chemical equation is obtained, it can be used for solution of problems involving weight relationships in chemical reactions. This is illustrated by the following examples.

Example 8

How many grams of $KClO_3$ must be decomposed to give 0.96 g oxygen?

It is known that on heating, the white solid $KClO_3$ decomposes to form the white solid KCl and the gas oxygen O_2. To answer the question, we need the equation for the decomposition. In this equation $KClO_3$ is placed on the left, and KCl and O_2 on the right.

$$KClO_3(s) \longrightarrow KCl(s) + O_2(g)$$

$$\begin{array}{ccc} +5 -2 & -1 & 0 \end{array}$$

$6e^-$ $2e^- \times 3$

The chlorine atom changes oxidation number from $+5$ to -1. It appears to gain six electrons. Oxygen changes oxidation number from -2 to 0; each atom appears to lose two electrons. The formula-unit is such that there are three oxygen atoms for every chlorine atom; so the compound itself has taken care of the electron gain and the electron loss. One potassium and one chlorine on the left require one potassium and one chlorine on the right. Three oxygen atoms on the left require three oxygen atoms on the right. We can get these three oxygen atoms on the right by placing the coefficient $\frac{3}{2}$ before the formula O_2, giving

$$KClO_3(s) \longrightarrow KCl(s) + \tfrac{3}{2}O_2(g)$$

Multiplying through by 2 to get rid of the fraction gives

$$2KClO_3(s) \longrightarrow 2KCl(s) + 3O_2(g)$$

We now have the balanced equation and can proceed to solve the problem. Since a chemical equation may always be read directly in terms of moles, it is convenient to solve problems in terms of moles.

$$0.96 \text{ g of } O_2 = \frac{0.96 \text{ g}}{32 \text{ g/mol}} = 0.030 \text{ mol}$$

$$(0.030 \text{ mol of } O_2)\left(\frac{2 \text{ mol KClO}_3}{3 \text{ mol O}_2}\right) = 0.020 \text{ mol of KClO}_3$$

$$(0.020 \text{ mol of KClO}_3)(122.55 \text{ g/mol}) = 2.5 \text{ g}$$

■ ■ ■

Example 9
On heating, 4.90 g of $KClO_3$ shows a weight loss of 0.384 g. What percent of the original $KClO_3$ has decomposed?
The weight loss is due to the fact that a gas is driven off. The only gas formed in this reaction is oxygen, as seen from the equation obtained in Example 8.

$$2KClO_3(s) \longrightarrow 2KCl(s) + 3O_2(g)$$

$$0.384 \text{ g of } O_2 = \frac{0.384 \text{ g}}{32.0 \text{ g/mol}} = 0.0120 \text{ mol of } O_2$$

To get 0.0120 mol of O_2, we need to decompose

$$(0.0120 \text{ mol of } O_2)\left(\frac{2 \text{ mol KClO}_3}{3 \text{ mol O}_2}\right) = 0.00800 \text{ mol of KClO}_3$$

We originally had

$$\frac{4.90 \text{ g KClO}_3}{122.6 \text{ g/mol}} = 0.0400 \text{ mol of KClO}_3$$

$$\text{Percent decomposed} = \frac{\text{moles decomposed}}{\text{moles available}} \times 100$$

$$= \frac{0.00800}{0.0400} \times 100$$

$$= 20\%$$

■ ■ ■

Example 10

In the reaction of vanadium oxide, VO, with iron oxide, Fe_2O_3, the products are V_2O_5 and FeO. How many grams of V_2O_5 can be formed from 2.00 g of VO and 5.75 g of Fe_2O_3?

In solving this problem, we first write the balanced equation:

$$2VO(s) + 3Fe_2O_3(s) \longrightarrow 6FeO(s) + V_2O_5(s)$$

Next we decide which reactant limits the amounts of products and which reactant is present in excess. To do this, we convert the data into moles. The formula weight of VO is 66.94 amu; the formula weight of Fe_2O_3 is 159.69 amu.

$$2.00 \text{ g of VO} = \frac{2.00 \text{ g}}{66.94 \text{ g/mol}} = 0.0299 \text{ mol of VO}$$

$$5.75 \text{ g of } Fe_2O_3 = \frac{5.75 \text{ g}}{159.69 \text{ g/mol}} = 0.0360 \text{ mol of } Fe_2O_3$$

From the equation, 0.0299 mol of VO requires

$$(0.0299 \text{ mol VO})\left(\frac{3 \text{ mol } Fe_2O_3}{2 \text{ mol VO}}\right) = 0.0449 \text{ mol of } Fe_2O_3$$

Therefore, 0.0360 mol of Fe_2O_3 is limiting, and the 0.0299 mol of VO provides an excess. Calculation from the equation is based on Fe_2O_3.

$$(0.0360 \text{ mol of } Fe_2O_3)\left(\frac{1 \text{ mol } V_2O_5}{3 \text{ mol } Fe_2O_3}\right) = 0.0120 \text{ mol of } V_2O_5$$

$$(0.0120 \text{ mol of } V_2O_5)(181.9 \text{ g/mol}) = 2.18 \text{ g}$$

■ ■ ■

Example 11

The reaction $Cl_2(g) + S_2O_3^{2-} \longrightarrow SO_4^{2-} + Cl^-$ is to be carried out in basic solution. Starting with 0.15 mol Cl_2, 0.010 mol $S_2O_3^{2-}$, and 0.30 mol OH^-, how many moles of OH^- will be left in solution after the reaction is complete? Assume no other reactions take place.

First balance the equation to find the following:

$$4Cl_2(g) + S_2O_3^{2-} + 10OH^- \longrightarrow 2SO_4^{2-} + 8Cl^- + 5H_2O$$

Decide whether Cl_2 or $S_2O_3^{2-}$ is in excess. The equation requires that there be 4 mol Cl_2 per mol of $S_2O_3^{2-}$. The given 0.15 mol of Cl_2 per 0.010 mol of $S_2O_3^2$ is considerably in excess over the needed ratio; so Cl_2 is in excess, and, of the two reagents, $S_2O_3^{2-}$ is limiting. Next decide whether $S_2O_3^{2-}$ or OH^- is in excess. The equation requires that there be 10 mol OH^- per mole of $S_2O_3^{2-}$. 0.010 mol of $S_2O_3{}^{2-}$ requires

$$(0.010 \text{ mol } S_2O_3{}^{2-})\left(\frac{10 \text{ mol } OH^-}{1 \text{ mol } S_2O_3{}^{2-}}\right) = 0.10 \text{ mol of } \dot{O}H^-$$

We, therefore, have left $0.30 - 0.10 = 0.20$ mol of OH^-.

■ ■ ■

4.10 Equivalents

In solving the above problems it was necessary to use balanced chemical equations. In many cases balancing the equation can be bypassed by introducing a new quantity, the *equivalent* (equiv). *One equivalent of an oxidizing agent is defined as that mass of the substance that picks up the Avogadro number of electrons* in a particular reaction. *One equivalent of a reducing agent is defined as that mass of the substance that releases the Avogadro number of electrons* in a particular reaction. The equivalents are defined in this way so that one equiv of any oxidizing agent reacts exactly with one equiv of any reducing agent.

In the reaction of aluminum, Al, and oxygen, O_2, to produce Al_2O_3, Al changes oxidation number from 0 to $+3$ and O changes oxidation number from 0 to -2. Each atom of Al releases three electrons; so one mole of Al atoms (which is the Avogadro number of Al atoms) releases three times the Avogadro number of electrons. That mass of Al which releases the Avogadro number of electrons is one-third of a mole of atoms. So, for Al one equivalent is one-third of a mole of atoms, or $\frac{1}{3}(26.98)$, or 8.993 g. Each atom of O picks up two electrons. Each O_2 molecule picks up four electrons. One mole of O_2 picks up four times the Avogadro number of electrons. That mass of O_2 which picks up the Avogadro number of electrons is one-fourth of a mole. So, for O_2 one equivalent is one-fourth of a mole, or $\frac{1}{4}(32.00)$, 8.000 g. In the reaction of Al with O_2, 8.993 g of Al reacts exactly with 8.000 g of O.

Example 12
When magnesium burns in oxygen, it forms magnesium oxide. In a given experiment 1.2096 g of oxide is formed from 0.7296 g of magnesium. What is the mass of one equivalent of magnesium in this reaction?

Mass of oxygen combined $= 1.2096 - 0.7296 = 0.4800$ g

Since oxygen changes oxidation number from 0 to -2, each oxygen atom appears to gain two electrons. To gain the Avogadro number of electrons requires one-half of a mole of atoms, or 8.000 g of oxygen.

$$0.4800 \text{ g of } O = \frac{0.4800 \text{ g}}{8.000 \text{ g/equiv}} = 0.06000 \text{ equiv of } O$$

0.06000 equiv of O requires 0.06000 equiv of Mg.

0.7296 g of Mg used is 0.06000 equiv of Mg.

$$\frac{0.7296 \text{ g of Mg}}{0.06000 \text{ equiv}} = 12.16 \text{ g per equivalent of Mg}$$

■ ■ ■

For compounds, the weight of one equivalent can be calculated by dividing the weight of one mole by the electron gain or loss per formula-unit. This calculation requires knowledge of products. As an illustration, when HNO_3 (formula weight 63.013 amu) is reduced to NO, the change in oxidation number of N is from $+5$ to $+2$; therefore, the mass of one equivalent of HNO_3 is 63.013/3, or 21.004 g. However, when HNO_3 is reduced to NH_3, the N changes oxidation number from $+5$ to -3, and the mass of one equivalent of HNO_3 is 63.013/8, or 7.8766 g. Thus, the mass of one equivalent depends on what product is formed.

Example 13

How many grams of hydrogen sulfide, H_2S, react with 6.32 g of potassium permanganate, $KMnO_4$, to produce K_2SO_4 and MnO_2?

Mn changes oxidation number from $+7$ to $+4$ in this reaction.

One equivalent of $KMnO_4$ weighs 158.03/3, or 52.678 g.

$$6.32 \text{ g of } KMnO_4 = \frac{6.32 \text{ g}}{52.678 \text{ g/equiv}} = 0.120 \text{ equiv}$$

0.120 equiv of $KMnO_4$ requires 0.120 equiv of H_2S.

S changes oxidation number from -2 to $+6$ in this reaction.

One equivalent of H_2S weighs 34.076/8, or 4.259, g.

$$0.120 \text{ equiv of } H_2S = (0.120 \text{ equiv})(4.259 \text{ g/equiv}) = 0.511 \text{ g}$$

■ ■ ■

In the above examples, two things might be noticed: In the first place, it was not necessary to write balanced equations to solve the type of problem given. Secondly, although oxidation-number changes were used to find the mass of an equivalent, we could equally well have written the half-reaction and noted directly the number of electrons gained or lost per formula-unit. For example, when $H_2C_2O_4$ is oxidized to CO_2, its mass per equivalent is 45.018 g. This corresponds to the half-reaction $H_2C_2O_4 + 2H_2O \longrightarrow 2CO_2 + 2H_3O^+ + 2e^-$, which shows that the molecular weight of $H_2C_2O_4$, 90.036 g, needs to be divided by 2 to give the mass per equivalent.

Equivalents are also useful for acid-base reactions. For example, in the complete neutralization of $Ca(OH)_2$ by H_3PO_4, the nonnet equation is

$$3Ca(OH)_2 + 2H_3PO_4 \longrightarrow Ca_3(PO_4)_2 + 6H_2O$$

Since each mole of $Ca(OH)_2$ furnishes two moles of OH^- and each mole of H_3PO_4 furnishes three moles of H_3O^+, complete neutralization occurs if three moles of $Ca(OH)_2$ per two moles of H_3PO_4 are used.

From such an equation the usual stoichiometric calculations can be made. It is more convenient, however, to consider neutralization reactions by fixing attention only on the H_3O^+ ion and OH^- ion. For this purpose, equivalents are convenient. *One equivalent of an acid is the mass of acid required to furnish one mole of H_3O^+; one equivalent of a base is the mass of base required to furnish one mole of OH^- or to accept one mole of H_3O^+.* One equivalent of any acid exactly reacts with one equivalent of any base.

One of the simplest acids is HCl, hydrochloric acid, one mole of which weighs 36.5 g. Since one mole of HCl can furnish one mole of H_3O^+, 36.5 g of HCl is one equivalent. For HCl, and for all other monoprotic acids, one mole is the same as one equivalent. For a diprotic acid, such as H_2SO_4, one mole of acid can furnish on demand two moles of H_3O^+. By definition, two moles of H_3O^+ is the amount furnished by two equivalents of acid. Therefore, for complete neutralization one mole of H_2SO_4 is identical with two equivalents. Since one mole = 98 g = two equivalents, one equivalent of H_2SO_4 weighs 49 g. For complete reaction of a triprotic acid, such as H_3PO_4, one mole is equal to three equivalents. The situation is similar for bases. If the base is NaOH, one mole gives one mole of OH^-. Therefore, one mole of NaOH is one equivalent. If the base is $Ca(OH)_2$, one mole is two equivalents.

Example 14

1.00 g of the acid $C_6H_{10}O_4$ requires 0.768 g of KOH for complete neutralization. How many neutralizable protons are in this molecule?

$$0.768 \text{ g of KOH} = \frac{0.768 \text{ g}}{56.1 \text{ g/equiv KOH}} = 0.0137 \text{ equiv of KOH}$$

0.0137 equiv of base neutralizes 0.0137 equiv of acid.

$$\frac{1.00 \text{ g of } C_6H_{10}O_4}{0.0137 \text{ equiv acid}} = 73.0 \text{ g/equiv}$$

One mole of $C_6H_{10}O_4$ weighs 146.1 g.

$$\frac{73.0 \text{ g/equiv}}{146.1 \text{ g/mol}} = 0.500 \text{ mol/equiv}$$

Therefore, each mole furnishes two moles of H_3O^+—that is, $H^+(H_2O)$—or each molecule of $C_6H_{10}O_4$ contains two replaceable protons.

■   ■   ■

Exercises

4.1   Moles. How many moles are there in one atom?

Ans. 1.66 × 10⁻²⁴

4.2   Gram-atom. In what sense was it possible to claim that the term "gram-atom" is less ambiguous than the term "mole"?

*4.3   Moles. In a chemical reaction requiring two atoms of phosphorus for five atoms of oxygen, how many grams of oxygen are required by 3.10 g of phosphorus?

*4.4   Moles. How many mercury atoms would there be in a 100-g piece of swordfish said to contain 0.1 ppm (part per million by weight) of mercury?   *Ans. 3×10^{16}*

*4.5   Moles. Arrange the following samples in order of increasing number of atoms: (a) 6.50 g of N, (b) 2.50×10^{23} atoms of C, (c) 0.65 mol of O atoms, and (d) 6.50 g of O_2. What is the total mass of the entire collection?

*4.6   Simplest formula. When metal M is heated in halogen X_2, a compound MX_n is formed. In a given experiment 1.00 g of titanium reacts with chlorine to give 3.22 g of compound. What is the corresponding value of n?

*4.7   Simplest formula. What is the simplest formula of a compound that is composed of 72.4% iron and 27.6% oxygen by weight?   *Ans. Fe_3O_4*

*4.8   Simplest formula. A given sample of pure compound contains 9.81 g of zinc, 1.8×10^{23} atoms of chromium, and 0.60 mol of oxygen atoms. What is its simplest formula?

*4.9   Oxidation numbers. What is the oxidation number of nitrogen in each of the following: NH_3, NO_2, N_2O_3, HNO_3, $Ca(NO_3)_2$, N_2H_4, N_2O, NH_2OH, KNO_2, Na_3N, and KNH_2?

*4.10   Oxidation-reduction. In each of the following reactions, indicate the oxidizing agent, the reducing agent, and the substance oxidized:

a $N_2 + 3H_2 \longrightarrow 2NH_3$
b $2NO + O_2 \longrightarrow 2NO_2$
c $Cl_2 + 2OH^- \longrightarrow ClO^- + Cl^- + H_2O$
d $O_2 + 2F_2 \longrightarrow 2OF_2$
e $Cl_2 + 2NaI \longrightarrow 2NaCl + I_2$

*4.11   Equation calculations. Heating of $NaNO_3$ decomposes it to $NaNO_2$ and O_2. How much $NaNO_3$ would you have to decompose to produce 1.50 g of O_2?   *Ans. 7.97 g*

*4.12   Equation calculations. What is the maximum weight of SO_3 that could be made from 25.0 g of SO_2 and 6.00 g of O_2 by the reaction $2SO_2 + O_2 \longrightarrow 2SO_3$?

**4.13   Moles. How many moles are there in one amu of magnesium?   *Ans. 6.84×10^{-26}*

**4.14   Molecular formula. The active constituent of marihuana (hashish) is tetrahydrocannabinol, which analyzes by weight to 80.21% C, 9.62% H, and 10.28% O. If the molecular weight is 314.45 amu, what is the correspondingly appropriate molecular formula?

4.15 *Molecular formula.* The most common constituent of gasoline is iso-octane. It is a hydrocarbon, composed by weight of 84.12% carbon and 15.88% hydrogen. Given that it contains 5.27×10^{21} molecules per gram, what is its molecular formula?   *Ans.* C_8H_{18}

4.16 *Molecular weights.* What is the molecular weight of a substance, each molecule of which contains 9 carbon atoms and 13 hydrogen atoms, and 2.33×10^{-23} g of other components?

Ans. 135 amu

4.17 *Formula-units.* How many formula-units of NO_2 are there in (*a*) 1.00 g of NO_2 and (*b*) 1.00 g of N_2O_4?

4.18 *Percent composition.* What is the elemental percent composition (by weight) of a mixture that contains 20.0 g of $KAl(SO_4)_2$ and 60.0 g of K_2SO_4?   *Ans. 2.6% Al, 20.0% S, 39.9% O, 37.4% K*

4.19 *Oxidation numbers.* What is the oxidation number of X in each of the following: $NaXO_3$, Na_2XO_3, $Na_3(XO_3)_2$, Na_3XO_4, and $Na_3(XO_4)_2$?

4.20 *Balancing equations.* Balance each of the following by use of the oxidation-number method:

a $C_8H_{18} + O_2 \longrightarrow CO + H_2O$
b $C_8H_{18} + O_2 \longrightarrow CO_2 + H_2O$
c $C_8H_{18} + O_2 \longrightarrow C_8H_{14} + H_2O$

4.21 *Balancing equations.* Balance each of the following by use of the oxidation-number method:

a $H_2C_2O_4 + KMnO_4 \longrightarrow CO_2 + K_2O + MnO + H_2O$
b $H_2CO_2 + KMnO_4 \longrightarrow CO_2 + K_2O + MnO + H_2O$
c $H_2C_2O_4 + KMnO_4 \longrightarrow CO_2 + K_2O + Mn_2O_3 + H_2O$
d $H_2C_2O_4 + K_2MnO_4 \longrightarrow CO_2 + K_2O + Mn_2O_3 + H_2O$

4.22 *Balancing equations, aqueous solutions.* Write complete balanced equations for each of the following changes in acidic aqueous solution:

a $ClO_3^- + Fe^{2+} \longrightarrow Cl^- + Fe^{3+}$
b $NO_3^- + H_2S \longrightarrow HSO_4^- + NH_4^+$
c $N_2O_4 + BrO_3^- \longrightarrow NO_3^- + Br^-$
d $S_2O_3^{2-} + Sb_2O_5 \longrightarrow SbO^+ + H_2SO_3$

4.23 *Balancing equations, basic solutions.* Write complete balanced equations for each of the following changes in basic aqueous solution:

a $SO_3^{2-} + CrO_4^{2-} \longrightarrow SO_4^{2-} + Cr(OH)_3$
b $H_2 + ReO_4^- \longrightarrow ReO_2 + H_2O$
c $Fe + N_2H_4 \longrightarrow Fe(OH)_2 + NH_3$
d $S_2O_4^{2-} + Ag_2O \longrightarrow Ag + SO_3^{2-}$

4.24 *Half-reactions.* Write balanced half-reactions for each of the following changes under the conditions indicated:

a $MnO_4^- \longrightarrow Mn^{2+}$ (acidic)

b $ClO_3^- \longrightarrow Cl^-$ (basic)

c $I_2 \longrightarrow IO_3^-$ (acidic)

d $S_2O_3^{2-} \longrightarrow SO_2$ (basic)

**4.25 *Equation calculations.* How many pounds of air (which is 23.19% O_2 and 75.46% N_2 by weight) would be needed to burn a pound of gasoline by a reaction whereby C_8H_{18} reacts with O_2 to form CO_2 and H_2O? *Ans. 15.1 lb*

**4.26 *Equation calculations.* Suppose the change $HC_2O_4^- + Cl_2 \longrightarrow CO_3^{2-} + Cl^-$ is to be carried out in basic solution. Starting with 0.10 mol of OH^-, 0.10 mol of $HC_2O_4^-$, and 0.05 mol of Cl_2, how many moles of Cl^- would be expected to be in the final solution?

**4.27 *Equation calculations.* Some solid CaO in a test tube picks up water vapor from the surroundings to change completely to $Ca(OH)_2(s)$. An observed total initial weight (CaO + test tube) of 10.860 g goes eventually to 11.149 g. What is the weight of the test tube? *Ans. 9.959 g*

**4.28 *Equivalents, oxidation-reduction.* For the oxidation of VO by Fe_2O_3 to form V_2O_5 and FeO, what is the weight of one equivalent of VO and of Fe_2O_3?

**4.29 *Equivalents, oxidation-reduction.* When V_2O_5 is reduced, what would be its equivalent weight in going to each of the following products: VO_2, V_2O_3, VO, and V? *Ans. 90.94, 45.47, 30.31, 18.19 g*

**4.30 *Equivalents, oxidation-reduction.* In acting as a reducing agent a piece of metal M, weighing 16.00 g, gives up 2.25×10^{23} electrons. What is the weight of one equivalent of the metal?

**4.31 *Equivalents, acid-base.* How many equivalents of acid would there be in each of the following: (*a*) 1.50 g of HCl, (*b*) 1.50 g of H_2SO_4, (*c*) a mixture of 0.0030 mol of HNO_3 and 0.015 mol of H_3PO_4, (*d*) amount of acid needed to neutralize 1.50 g of NaOH, and (*e*) amount of acid needed to neutralize 0.010 mol of $Ca(OH)_2$ plus 1.50 g of KOH?

***4.32 *Avogadro number.* The density of KF is 2.48 g/cm³. The solid is made up of a cubic array of alternate K^+ and F^- ions at a spacing of 2.665×10^{-8} cm between centers. From these data calculate the apparent value of the Avogadro number. *Ans. 6.18×10^{23}*

***4.33 *Simplest formula.* One of the major atmospheric pollutants emitted by fuel-combustion power stations is a sulfur oxide mixture generally designated as SO_x. It consists mainly of SO_2 but may contain anywhere from 1 to 10% SO_3 computed as percent by weight of the total mass. What would be the appropriate range to assign to *x*?

***4.34 *Equation calculations.* On being heated in air, a mixture of FeO and Fe_3O_4 picks up oxygen so as to convert completely to

Fe_2O_3. If the observed weight gain is 5.00 percent of the initial weight, what must have been the composition of the initial mixture?

Ans. 79.9% Fe_3O_4 and 20.1% FeO by weight

***4.35   *Equivalents, acid-base.* A mixture consisting only of KOH and $Ca(OH)_2$ is neutralized by acid. If it takes exactly 0.100 equiv to neutralize 4.221 g of the mixture, what must have been its initial composition in percent by weight?

The properties of a substance are dependent not only on its chemical identity but also on its state of aggregation—i.e., whether solid, liquid, or gas. For example, in the gaseous state oxygen has oxidizing properties dependent on its chemical identity, but it also has the properties characteristic of any substance in the gaseous state, e.g., compressibility, rapid diffusion, and high thermal expansion. In this chapter we consider the general properties of all gases and try to account for them in terms of kinetic theory. Because of relatively small intermolecular attractions, the gaseous state has proved more amenable to theoretical attack than

the other states of matter.

5.1 Volume

The volume of a substance is the space occupied by that substance. If the substance is a gas, the volume of a sample is the same as the volume of the container in which the sample is held. Ordinarily, this volume is specified in units of liters, milliliters (ml), or cubic centimeters (cm^3). As the name implies, one *cubic centimeter* is the volume of a cube one centimeter on an edge. One *liter* is the volume of a cube that is one decimeter (1 dm = 0.1 m = 10 cm) on edge. Thus, a liter is exactly 1000 times as great as a cubic centimeter. It follows that one *milliliter*, which is a thousandth of a liter, is exactly equal to one cubic centimeter. This equality was not always true, since prior to 1964 the liter was defined as the volume occupied by one kilogram of water at the temperature of its maximum density. However, the difference is only 27 ppm.

The volume of liquids and solids does not change much with change of pressure or temperature. Consequently, to describe the amount of solid or liquid being handled, e.g., the number of moles, it is usually sufficient to specify only the volume of the sample. For gases, this is not enough. As an example, one liter of hydrogen at one atmosphere pressure and 0°C contains 0.0446 mol, whereas one liter at two atmospheres and 25°C contains 0.0817 mol. In order to fix the number of moles in a given sample of gas, it is necessary to know the pressure and temperature as well as the volume.

When solids or liquids are mixed together, the total volume is roughly equal to the sum of the original volumes. This is not necessarily true of gases, where the final volume after mixing depends strongly on the final pressure. If the final pressure is allowed to rise sufficiently, two or more gases can occupy the same volume as each gas does alone. Since all gases *can mix in any proportion*, they are said to be *miscible*.

5.2 Temperature

It is a familiar observation that a hot substance and a cold substance placed in contact with each other change so that the hot substance gets colder and the cold substance hotter. This is interpreted as resulting from a flow of heat energy from the hot body to the cold body. The hot body is said to have a higher temperature; the cold body, a lower temperature. Thus, *temperature is a property that determines the direction of heat flow*; heat always flows from a region of higher temperature to one of lower temperature.

The international scale for measuring temperature is an absolute scale; it starts with *absolute zero*. Absolute zero is the lower limit of temperature, and temperatures lower than it are unattainable. The scale is sometimes called the *Kelvin scale* after the English thermodynamicist Lord Kelvin, who proposed it in the year 1848. It is defined

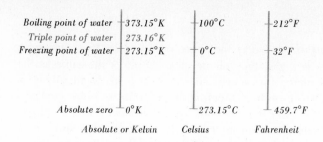

Fig. 5.1
Temperature scales.

	Absolute or Kelvin	Celsius	Fahrenheit
Boiling point of water	373.15°K	100°C	212°F
Triple point of water	273.16°K		
Freezing point of water	273.15°K	0°C	32°F
Absolute zero	0°K	273.15°C	459.7°F

by assigning the value 273.16°K* (degrees Kelvin) to the temperature at which H_2O coexists in the liquid, gaseous, and solid states. This point, called the *triple point* (Sec. 7.11), corresponds to the temperature at which liquid water, water vapor, and ice are all in equilibrium with each other. Thus, the size of the degree Kelvin is defined as 1/273.16 of the temperature difference between absolute zero and the triple point of H_2O. The triple point of H_2O turns out to be 0.01° higher than the normal† freezing point of H_2O; so on the Kelvin scale the normal freezing point of H_2O is 273.15°K.

The normal freezing point of H_2O (273.15°K) is set as the zero point of the Celsius scale. On the Celsius scale the size of the degree, designated by °C, is taken to be the same as the degree Kelvin. The normal boiling point of H_2O—i.e., the boiling point at an atmosphere of pressure—on the Celsius scale turns out to be 100°C. Because there are 100 degrees between the normal freezing and boiling points of H_2O, this temperature scale is sometimes called the *centigrade* scale. A comparison of the Kelvin scale with the Celsius and Fahrenheit scales is shown in Fig. 5.1.‡ While the size of one degree is the same on the Celsius and Kelvin scales, the Fahrenheit degree is only five-ninths as large. Temperature on the Celsius, or centigrade, scale is converted to temperature on the Kelvin scale by adding 273.15°:

$$°C + 273.15 = °K$$

To convert Fahrenheit temperature to Kelvin temperature, it is also necessary to correct for the difference in the size of the degree

$$(°F - 32) \times \tfrac{5}{9} + 273.15 = °K$$

* The number 273 was picked on the basis of observations that under ideal conditions all gases expand uniformly with rise in temperature. At 0°C the rate of expansion is always $\frac{1}{273}$ of the volume observed. It might also be noted that IUPAC has recommended that the unit on the Kelvin scale be called a "kelvin," not a "degree Kelvin." In line with this recommendation, we should write 273.16 K instead of 273.16°K. Physicists are generally moving to accept this recommendation, but chemists appear to be resisting it. For pedagogic reasons, we shall stick to °K in this book.
† The normal freezing point is the temperature at which H_2O freezes under one atmosphere of pressure; the triple point is the temperature at which H_2O freezes under its own vapor pressure, which is only 0.006 atm.
‡ The Fahrenheit scale, which is much used in English-speaking countries, was originally set up by the German physicist Gabriel Fahrenheit on the basis of zero degrees for a mixture of snow and salt.

5.3 Pressure

Just as temperature determines the direction of heat flow, so pressure is a property which determines the direction of mass flow. Unless otherwise constrained, matter tends to move from a place of higher pressure to a place of lower pressure. Quantitatively, *pressure* is defined as *force per unit area.* Force is defined as that which tends to change the state of rest or motion of an object. The fundamental unit of force is the *newton* (after Isaac Newton), which is the force required to accelerate one kilogram of matter by one meter per second in a time interval of one second. A related unit of force is the *dyne*, which is the force required to accelerate one gram of matter by one centimeter per second in one second. As can be seen

$$1 \text{ newton} = \frac{\text{kg-m}}{\text{sec}^2} = \frac{(10^3 \text{ g})(10^2 \text{ cm})}{\text{sec}^2} = 10^5 \text{ dyn}$$

Force can also be expressed in terms of pounds weight. Units for expressing pressure can be derived from any of the above. The fundamental unit would be newtons per square meter, for which the name *pascal* (after the French scientist Blaise Pascal) has recently been recommended. More traditionally, one would speak of dynes per square centimeter or pounds per square inch. The relative equivalence of the units is as follows:

$$1 \text{ pascal} = \frac{\text{newton}}{\text{m}^2} = \frac{10^5 \text{ dyn}}{(10^2 \text{ cm})^2} = 10 \text{ dyn/cm}^2$$

$$1 \text{ pound/in}^2 = \frac{(453.6 \text{ g})(980.6 \text{ cm/sec}^2)}{(2.54 \text{ cm})^2} = 68{,}947 \text{ dyn/cm}^2$$

In high-pressure work, a frequently encountered unit is the *bar*, which is 10^6 dyn/cm^2. A bar is very close to one atmosphere.

In *fluids,* a general term which includes *liquids and gases,* the pressure at a given point is the same in all directions. This can be visualized by considering a swimmer under water. At a given depth, no matter how he turns, the pressure exerted on all sides of him by the water is always the same. However, as he increases his depth, the pressure increases. This comes about because of the pull of gravity on the water above him. We can picture his body as being compressed by the weight of the column of water above him. In general, for all fluids, the greater the depth of immersion, the greater the pressure.

The earth is surrounded by a blanket of air approximately 500 mi thick. In effect, we live at the bottom of a fluid, the atmosphere, which exerts a pressure. The existence of this pressure can be shown by filling a long tube, closed at one end, with mercury and inverting it in a dish of mercury. (Any other liquid would do, but mercury has the advantage of not requiring too long a tube.) Some of the mercury runs out of the tube, but not all of it. This setup, called a *barometer,* is represented in Fig. 5.2. No matter how large the diameter of the tube

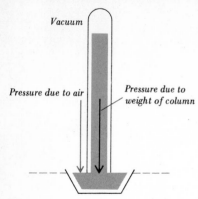

Vacuum

Pressure due to air

Pressure due to weight of column

Fig. 5.2
Barometer.

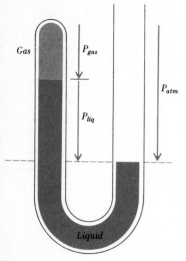

Gas

P_{gas}

P_{atm}

P_{liq}

Liquid

Fig. 5.3
Manometer.

and no matter how long the tube, the *difference* in height between the mercury level inside and outside the tube is the same. The fact that all the mercury does not run out shows that there must be a pressure exerted on the surface of the mercury in the dish sufficient to support the column of mercury.

To a good approximation, the space above the mercury level is a vacuum (contains only a negligible amount of mercury vapor) and exerts no pressure on the upper mercury level. The pressure at the bottom of the mercury column, therefore, is due only to the weight of the mercury column. As noted, it is a general property of fluids that at any given level in the fluid the pressure is constant. In Fig. 5.2 the dashed line represents the level of interest. At this level, outside the tube, the force per unit area is due to the atmosphere and can be labeled as P_{atm}. The pressure inside the tube is due to the pressure of the column of mercury and can be labeled P_{Hg}. The equality $P_{atm} = P_{Hg}$ provides us with a method for measuring the pressure exerted by the atmosphere.

Atmospheric pressure changes from day to day and from one altitude to another. A *standard atmosphere,* referred to as 1 atm, is defined as the pressure which supports a column of mercury that is 760 mm high at 0°C at sea level.*

Pressure can be expressed either in terms of number of atmospheres or number of millimeters of mercury (mmHg). The pressure unit mmHg is also frequently referred to as a *Torr,* in honor of Evangelista Torricelli, the inventor of the barometer. (However, IUPAC recommends that Torr be abandoned.) We can also express pressure by the height of a water column. Since water has a density of 1 g/ml, whereas mercury has a density of 13.6 g/ml, a given pressure supports a column of water that is 13.6 times as high as one of mercury. One atmosphere of pressure supports 76 cm of mercury or 76(13.6) cm of water, the latter being roughly 34 ft. In terms of pounds per square inch, one standard atmosphere is 14.7 psi. In terms of pascals, one standard atmosphere is 1.01×10^5 pascals.

The device shown in Fig. 5.3 is a *manometer,* used to measure the pressure of a trapped sample of gas. The manometer is constructed by placing a liquid in the bottom of a U tube with the gas sample in one arm of the U. If the right-hand tube is open to the atmosphere, the pressure which is exerted on the right-hand surface is atmospheric pressurre P_{atm}. At the same liquid level in both arms of the tube, the pressures must be equal; otherwise, there would be a flow of liquid from one arm to the other. At the level indicated by the dashed line in Fig. 5.3, the pressure in the left arm is equal to the pressure of

* If we think of pressure as weight per unit area, we can see why it is necessary that both 0°C and sea level be specified in defining the standard atmosphere. The density of liquid Hg changes with temperature, and therefore the weight of a 760-mm-high Hg column of fixed cross section changes with temperature. Hence the temperature must be specified. Similarly, the force of gravity changes slightly with altitude, and hence the weight of the Hg column changes when moved away from sea level.

the trapped gas P_{gas} plus the pressure of the column of liquid above the dashed line P_{liq}. We can therefore write

$$P_{atm} = P_{gas} + P_{liq}$$

or

$$P_{gas} = P_{atm} - P_{liq}$$

The atmospheric pressure can be measured by a barometer, and P_{liq} can be obtained by measuring the difference in height between the liquid level in the right and left arms and correcting for the known density of the liquid. P_{atm} and P_{liq} must be expressed in the same units. For example, if P_{atm} is in millimeters of mercury and the manometer liquid is water instead of mercury, the difference in water levels must be converted to its mercury equivalent by dividing by 13.6. If the bottom of the U tube consists of flexible rubber tubing, the right arm can be raised with respect to the left arm until the two liquid levels are at the same height, in which case $P_{liq} = 0$ and $P_{gas} = P_{atm}$.

5.4 P-V relation

A characteristic property of gases is their great compressibility. This behavior is summarized quantitatively in Boyle's law (1662). *Boyle's law* states that *at constant temperature a fixed mass of gas occupies a volume inversely proportional to the pressure exerted on it.* Boyle's law can be summarized by a pressure-volume, or *P-V*, plot like that shown in Fig. 5.4. The abscissa (horizontal axis) represents the pressure of a given sample of gas, and the ordinate (vertical axis), the volume occupied by it. The curve is a hyperbola, the equation for which is

$$PV = \text{constant}$$

or $V = \text{constant}/P$. The size of the constant is fixed once the mass of the sample and its temperature are specified.

The behavior specified by Boyle's law is not always observed. For most gases the law is best followed at low pressures and at high temperatures; as the pressure is increased or as the temperature is lowered, deviations may occur. This can be seen by considering the experimental data listed in Fig. 5.5. There are two series of experiments, one at 100°C and the other at −50°C. The pressure is measured when a fixed mass of gas, 39.95 g of argon, is contained in different volumes. The *PV* products in the last column, obtained by multiplying the observed values in the second and third columns, should, according to Boyle's law, be constant at each temperature. The data indicate that, although at the high temperature Boyle's law is closely obeyed, at the low temperature the *PV* product is not constant but drops off significantly as the pressure increases. In other words, Boyle's law is not obeyed. As the temperature of argon is decreased, its behavior deviates from that specified in Boyle's law.

That deviations from Boyle's law also increase at higher pressures can be seen from some experimental data for acetylene as given in

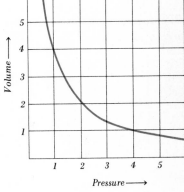

Fig. 5.4
P-V plot for a gas.

Fig. 5.5
Pressure-volume data for
39.95 g of argon gas.

Temperature, °C	V, liters	P, atm	P × V
100	2.000	15.28	30.560
	1.000	30.52	30.520
	0.500	60.99	30.500
	0.333	91.59	30.499
−50	2.000	8.99	17.980
	1.000	17.65	17.650
	0.500	34.10	17.050
	0.333	49.50	16.500

Fig. 5.6. When the pressure is doubled from 0.5 to 1.0 atm, the *PV* product is essentially unchanged; so in the low-pressure range acetylene follows Boyle's law reasonably well. However, when the pressure is doubled from 4.0 to 8.0 atm, the *PV* product does not remain constant but decreases by more than 3 percent; in this higher-pressure range, Boyle's law is not followed so well. For any gas, the lower the pressure, the closer the approach to Boyle's law behavior. When the law is obeyed, the gas is said to show *ideal* behavior.

Fig. 5.6
PV products for a sample of
acetylene at 0°C.

P, atm	0.5	1.0	2.0	4.0	8.0
PV	1.0057	1.0000	0.9891	0.9708	0.9360

A careful study of the deviations from ideal behavior led the Dutch scientist J. D. van der Waals to propose a modified version of the *P-V* relation. His modification is based on the observation that observed pressures generally are smaller than predicted by Boyle's law and observed volumes are generally greater than so predicted. The van der Waals relation is usually written

$$\left(P + \frac{n^2 a}{V^2}\right)(V - nb) = \text{constant}$$

and is related to the Boyle's law expression

$$P_{\text{ideal}} V_{\text{ideal}} = \text{constant}$$

by a correction term $n^2 a / V^2$ added to the observed pressure and a correction term nb subtracted from the observed volume. n is the number of moles in the sample; a and b are constants that are characteristic of the gas in question. As we shall see in Sec. 5.12, a can be related to molecular attractions, and b to molecular volumes. Representative values of the van der Waals constants are given in Fig. 5.7. Finally, it should be noted that at very high pressures and low temperatures even the van der Waals relation fails to represent the observed data. Other special relations have been proposed to handle such cases.

5.5 V-T relation

Another characteristic property of gases is their thermal expansion. Like most other substances, gases increase in volume when their temperature

Gas	a, liter^2 atm/mol^2	b, liter/mol
Helium	0.0341	0.0237
Argon	1.35	0.0322
Nitrogen	1.39	0.0391
Carbon dioxide	3.59	0.0427
Acetylene	4.39	0.0514
Carbon tetrachloride	20.39	0.1383

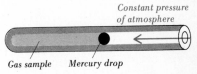

Fig. 5.7
van der Waals constants.

is raised. Experimentally, the dependence can be measured by confining a fixed mass of gas in a glass tube with a blob of liquid mercury, as shown in Fig. 5.8. The mercury blob, which is free to move, acts as a freely moving piston to confine the gas sample at constant pressure. As the gas is heated, the mercury moves out, and the volume of the gas sample increases.

Typical numerical data are plotted in Fig. 5.9. The points fall on a straight line, indicating that the volume varies linearly with temperature. If the temperature is lowered sufficiently, the gas liquefies, and no more experimental points can be obtained. However, if the straight line is extended, or extrapolated, to lower temperatures, as shown by the dashed line, it reaches a point of zero volume. The temperature at which the dashed line reaches zero volume is $-273.15°C$. It is significant that the value, $-273.15°C$, does not depend on the kind of gas used or on the pressure at which the experiment is performed. Designating $-273.15°C$ as absolute zero is thus reasonable, since temperatures below this would correspond to negative volume.

The volume-temperature data of Fig. 5.9 can be expressed by the relation

$$V = V_0 + \alpha t$$

where V is the observed volume at any temperature t (in degrees Celsius), V_0 is the volume at $0°C$, and α is a constant to be determined. Since V approaches zero when t approaches $-273.15°C$, we can write for that special point

$$0 = V_0 + \alpha(-273.15)$$

Solving for α, we get

$$\alpha = \frac{V_0}{273.15}$$

In other words, the constant α, which represents the slope of the graph in Fig. 5.9, has a value equal to $1/273.15$ of the volume at $0°C$. Substituting for α in the original equation, we get

$$V = V_0 + \frac{V_0}{273.15}t$$

which can be written

$$V = V_0\left(1 + \frac{t}{273.15}\right) = V_0\left(\frac{273.15 + t}{273.15}\right)$$

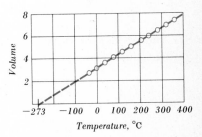

Constant pressure
of atmosphere

Gas sample   Mercury drop

Fig. 5.8
Simple device for measuring gas volume as a function of temperature.

Fig. 5.9
Plot of gas volume as a function of temperature.

Since absolute temperature, generally designated by T, is equal to $273.15 + t$, we can also write

$$V = \frac{V_0}{273.15} T = \text{constant} \times T$$

In words, this summarization of gas behavior, called *Charles' law* (1787), is as follows: *At constant pressure, the volume occupied by a fixed mass of gas is directly proportional to the absolute temperature.* The value of the proportionality constant depends on pressure and on the mass of gas.

Actually, Charles' law, like Boyle's law, represents the behavior of an *ideal*, or *perfect*, gas. For any *real* gas, especially at high pressures and at temperatures near the liquefaction point, deviations from Charles' law are observed. Near the liquefaction point the observed volume is less than that predicted by Charles' law.

Because of the Charles' law relation of volume to absolute temperature, calculations involving gases require that temperatures be expressed on the Kelvin scale. It is also convenient in working with gases to have a reference point. The customary reference point is at $273.15°K$ (0°C) and one standard atmosphere (760 mmHg) pressure. These conditions are called *standard temperature and pressure (STP)*.

5.6   Partial pressures

The behavior observed when two or more gases are placed in the same container is summarized in *Dalton's law of partial pressures* (1801). *Dalton's law* states that the *total pressure exerted by a mixture of gases is equal to the sum of the partial pressures* of the various gases. The *partial pressure* is defined as the *pressure the gas would exert if it were alone in the container.* As illustration, suppose a sample of hydrogen is pumped into a one-liter box and its pressure is found to be 0.065 atm. Suppose further that a sample of oxygen is pumped into a second one-liter box and its pressure is found to be 0.027 atm. If both samples are now transferred to a third one-liter box, the pressure is observed to be 0.092 atm. For the general case, Dalton's law can be written

$$P_{\text{total}} = P_1 + P_2 + P_3 + \cdots$$

where the subscripts denote the various gases occupying the same volume. Actually, Dalton's law is an idealization, but it is closely obeyed by most mixtures of nonreacting gases.

In laboratory experiments dealing with gases, the gases may be collected above water, in which case the water vapor contributes to the total pressure measured. Figure 5.10 illustrates an experiment in which oxygen gas is collected by water displacement. If the water level is the same inside and outside the bottle, then we may write

$$P_{\text{atm}} = P_{\text{oxygen}} + P_{\text{water vapor}}$$

$$P_{\text{oxygen}} = P_{\text{atm}} - P_{\text{water vapor}}$$

P_{atm} is obtained from a barometer. As we shall see later (Sec. 6.2), $P_{water\ vapor}$ is a function only of the temperature of the water and dissolved materials. This so-called "vapor pressure of water" has been measured at various water temperatures and is recorded in tables such as the one given in Appendix 3. Thus, the partial pressure of oxygen can be determined from a measured total pressure, a temperature, and reference to a table of vapor-pressure data. The following example shows how Dalton's law of partial pressures may enter into calculations involving gases.

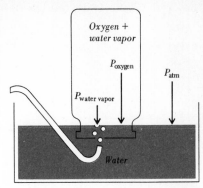

Fig. 5.10
Pressure contributions when oxygen is collected over water.

Example 1

If 40.0 liters of nitrogen are collected over water at 22°C when the atmospheric pressure is 0.957 atm, what is the volume of the dry nitrogen at standard temperature and pressure, assuming ideal behavior?

The initial volume of the nitrogen is 40.0 liters. The final volume is unknown. The initial pressure of the nitrogen gas is the atmospheric pressure, 0.957 atm, minus the vapor pressure of water. From Appendix 3 it is noted that at 22°C water has a vapor pressure of 0.026 atm; so this makes the initial pressure of nitrogen 0.957 − 0.026, or 0.931, atm. The initial temperature of the nitrogen is 22°C, or 273 + 22 = 295°K. Final conditions are standard; i.e., the final pressure is 1.000 atm, and the final temperature is 273°K. The problem is solved by considering separately how the volume is affected by a change in pressure and then by a change in temperature.

Pressure changes to 1.000/0.931 of its original value.
Volume changes inversely, or to 0.931/1.000 of its original value.
Temperature changes to 273/295 of its original value.
Volume changes proportionally, or to 273/295 of its original value.

$$V_{final} = V_{initial} \times \left(\begin{array}{c}\text{correction for}\\\text{pressure change}\end{array}\right) \times \left(\begin{array}{c}\text{correction for}\\\text{temperature change}\end{array}\right)$$

$$= 40.0 \text{ liters} \times \frac{0.931}{1.000} \times \frac{273}{295}$$

$$= 34.4 \text{ liters}$$

■ ■ ■

5.7 Avogadro principle

In the preceding section we assumed that when gases are mixed, they do not react with each other. However, sometimes they do react. For example, when a spark is passed through a mixture of hydrogen and oxygen gas, reaction occurs to form gaseous water. Similarly, when a mixture of hydrogen and chlorine gas is exposed to ultraviolet light, reaction occurs to form the gas hydrogen chloride. In any such reaction involving gases it is observed that with temperature and pressure fixed the volumes of the individual gases which actually react are simple multiples of each other.

As a specific example, in the reaction of hydrogen with oxygen to form water two liters of hydrogen are required for one liter of oxygen. In the reaction of hydrogen with chlorine, each liter of hydrogen requires one liter of chlorine; two liters of hydrogen chloride gas are formed. These observations are summarized in *Gay-Lussac's law of combining volumes* (1809), which states that *at a fixed pressure and temperature gases combine in simple proportions by volume and the volume of any gaseous product bears a whole-number ratio to that of any gaseous reactant.*

Just as in the law of multiple proportions the observation of simple ratios between combining weights of elements implied that matter is atomic, the occurrence of simple ratios between combining volumes of gases suggested that there is a simple relation between gas volume and number of molecules. Avogadro, in 1811, was the first to propose this relation: *Equal volumes of gases at the same temperature and pressure contain equal numbers of molecules.* That this principle accounts for Gay-Lussac's law can be seen from the following argument: When one volume of hydrogen combines with one volume of chlorine, the product, hydrogen chloride, contains equal numbers of hydrogen and chlorine atoms. These hydrogen and chlorine atoms come from the original molecules of hydrogen gas and chlorine gas. If we assume that both the hydrogen and the chlorine molecules are diatomic, then equal numbers of them are required for reaction. According to the Avogadro principle, these occupy equal volumes, consistent with the observation that the combining volumes of hydrogen and chlorine gas are equal.

The assumption that hydrogen and chlorine molecules are diatomic rather than monatomic can be justified as follows: If hydrogen were monatomic, i.e., consisted of individual hydrogen atoms, and if chlorine were also monatomic, then one liter of hydrogen (x atoms) would combine with one liter of chlorine (x atoms) to give one liter of HCl gas (x molecules). This is contrary to observation; the volume of HCl formed is *twice* as great as the volume of hydrogen or of chlorine used up. It must be that the hydrogen and chlorine molecules are more complex than monatomic, at least to the extent of containing an even number of atoms. If, as in fact turns out to be the case, hydrogen and chlorine are *diatomic*, then one liter of hydrogen (x molecules, or $2x$ atoms) will combine with one liter of chlorine (x molecules, or $2x$ atoms) to form two liters of hydrogen chloride ($2x$ molecules). This agrees with experiment.*

As first shown by Cannizzaro (1858), the Avogadro principle can be used as a basis for the determination of molecular weights. If two gases at the same temperature and pressure contain an equal number

---

* This reasoning can also be used to show that water is H_2O and not HO. It is observed that two volumes of hydrogen react with one volume of oxygen to form two volumes of gaseous water. Since one volume of oxygen gives two volumes of water, the oxygen molecule must contain an even number of oxygen atoms. If oxygen, like hydrogen, is diatomic, the fact that two volumes of hydrogen are needed per volume of oxygen implies that the water molecule contains twice as many hydrogen atoms as oxygen atoms.

of molecules in the same volume, then the relative weights of the volumes give directly the relative weights of the two kinds of molecules. For example, at STP 1 liter of acetylene is observed to weigh 1.17 g, whereas 1 liter of oxygen weighs 1.43 g. Since, according to the Avogadro principle, the number of molecules is the same in both samples, the acetylene molecule must be 1.17/1.43, or 0.818, times as heavy as the oxygen molecule. Since the diatomic oxygen molecule has a molecular weight of 32.0 amu, the molecular weight of acetylene must be 26.2 amu.

The volume occupied at STP by 32.00 g of oxygen (one mole) has been determined by experiment to be 22.4 liters. This is called the *molar volume* of oxygen at STP, and within the limits of ideal behavior it should be the volume occupied by one mole of any gas at STP. Figure 5.11 shows observed molar volumes for several gases. They are all approximately 22.4 liters. The value of 22.414 shown for the ideal gas is obtained from measurements made on gases at high temperatures and low pressures (where gas behavior is more nearly ideal) and then extrapolating to STP by using Boyle's and Charles' laws. For the first three gases, hydrogen, nitrogen, and oxygen, agreement with ideality is quite satisfactory. Even for the fourth, carbon dioxide, the agreement is better than 1 percent. Consequently, we usually assume that at STP the molar volume of any gas will be 22.4 liters. The following example shows how this molar volume can be used to determine molecular weight and molecular formulas.

Gas	Molar volume, liters
Hydrogen	22.432
Nitrogen	22.403
Oxygen	22.392
Carbon dioxide	22.263
Ideal gas	22.414

Fig. 5.11
Molar volumes at STP.

Example 2
Chemical analysis shows that ethylene has a simplest formula corresponding to one atom of carbon per two atoms of hydrogen. It has a density of 1.25 g/liter STP. What would be the corresponding molecular weight and molecular formula for ethylene?

At STP one mole of gas (if ideal) has a volume of 22.4 liters. Each liter of ethylene weighs 1.25 g; so one mole of ethylene would weigh 22.4×1.25 g, or 28.0 g. One mole is equal to the molecular weight in grams. Since the simplest formula is CH_2, the molecular formula must be some multiple of that, or $(CH_2)_x$. The formula weight of CH_2 is the atomic weight of carbon plus twice the atomic weight of hydrogen, or 14.0 amu. For $(CH_2)_x$, the formula weight is equal to x times 14.0. By experiment, this is equal to 28.0; so x must be equal to 2. The molecular formula of ethylene is therefore $(CH_2)_2$, or C_2H_4.

■   ■   ■

5.8   Equation of state

Boyle's law, Charles' law, and the Avogadro principle can be combined to give a general relation between the volume, pressure, temperature, and number of moles of a gas sample. Such a general relation is called an *equation of state*, because it tells how in going from one gaseous state to another the four variables V, P, T, and n change. Boyle's law,

Charles' law, and the Avogadro principle can be written, respectively,

$$V \sim \frac{1}{P} \qquad \text{at constant } T \text{ and } n$$

$$V \sim T \qquad \text{at constant } P \text{ and } n$$

$$V \sim n \qquad \text{at constant } T \text{ and } P$$

Combining these, we write

$$V \sim \frac{1}{P} Tn$$

(That this last relation embodies each of the other three can be seen by imagining any two of the variables, such as T and n, to be constant and noting the relation of the other two.) Written as a mathematical equation, the general relation becomes

$$V = R\frac{1}{P} Tn \qquad \text{or} \qquad PV = nRT$$

R is inserted as the constant of proportionality; it is called the *universal gas constant*. The equation $PV = nRT$ is called the *equation of state for an ideal gas*, or the *perfect-gas law*.

The numerical value of R can be found by substituting experimental quantities in the equation. At STP, $T = 273.15°K$, $P =$ one atm, and for one mole of gas $(n = 1)$ $V = 22.414$ liters. Consequently,

$$R = \frac{PV}{nT} = \frac{(1)(22.414)}{(1)(273.15)} = 0.082057$$

The units of R in this case are liter-atmospheres per degree Kelvin per mole. In order to use this value of R in the equation of state, P must be expressed in atmospheres, V in liters, n in moles, and T in degrees Kelvin.

Example 3
The density of an unknown gas at 98°C and 0.974 atm pressure is 2.50 g/liter. What is the molecular weight of this gas, assuming ideal behavior?

Temperature $= 98 + 273 = 371°K$

Pressure $= 0.974$ atm

From the equation of state $PV = nRT$, we can calculate the number of moles in one liter:

$$\frac{n}{V} = \frac{P}{RT} = \frac{0.974}{(0.0821)(371)} = 0.0320$$

Since 0.0320 mol weighs 2.50 g, 1 mol weighs 2.50/0.0320, or 78.1 g. Therefore, the molecular weight is 78.1 amu.

■ ■ ■

From the equation of state just described $PV = nRT$, it is seen that the constant for Boyle's law ($PV =$ constant) is just equal to nRT. As noted in Sec. 5.4, the work of van der Waals showed that it is not the simple product PV that remains constant at constant temperature but the more complex expression $(P + n^2a/V^2)(V - nb)$. Setting this product equal to nRT, we obtain the so-called "van der Waals equation of state":

$$\left(P + \frac{n^2a}{V^2}\right)(V - nb) = nRT$$

Example 4
By using the data in Fig. 5.7, calculate the pressure exerted by 0.250 mol of carbon dioxide in 0.275 liter at 100° C, and compare this value with the value expected for an ideal gas.
    $P = ?$; $n = 0.250$ mol; $a = 3.59$ liter^2-atm/mol^2; $V = 0.275$ liter; $b = 0.0427$ liter/mol; $R = 0.08206$ liter-atm deg^{-1} mol^{-1}; $T = 373°$K.

$$\left[P + \frac{(0.250)^2(3.59)}{(0.275)^2}\right][0.275 - (0.250)(0.0427)]$$

$$= (0.250)(0.08206)(373)$$

$P = 26.0$ atm

From $PV = nRT$, $P_{ideal} = 27.8$ atm. (The actual observed value is 26.1 atm.)

■ ■ ■

5.9 Graham's law of diffusion

A gas spreads to occupy any volume accessible to it. This spontaneous spreading of a substance throughout a phase is called *diffusion*. Diffusion can readily be observed by liberating some ammonia gas in a room. Its odor soon fills the room, indicating that the ammonia has become distributed throughout the entire volume of the room. Furthermore, it is found for a series of gases that the lightest gas (i.e., the one of lowest molecular weight) diffuses most rapidly. Quantitatively, under the same conditions the *rate of diffusion of a gas is observed to be inversely proportional to the square root of its molecular weight*. This is *Graham's law of diffusion* (1829), and in mathematical form it is written

$$R = \frac{constant}{\sqrt{m}} \qquad or \qquad \frac{R_1}{R_2} = \frac{\sqrt{m_2}}{\sqrt{m_1}}$$

R_1 and R_2 are the rates of diffusion of gases 1 and 2, and m_1 and m_2 are the respective molecular weights. In the case of oxygen gas and hydrogen gas

$$\frac{R_{H_2}}{R_{O_2}} = \frac{\sqrt{m_{O_2}}}{\sqrt{m_{H_2}}} = \sqrt{\frac{32}{2}} = \sqrt{16} = 4$$

The fact that heavier gases diffuse more slowly than light gases has been applied on a mammoth scale to effect the separation of uranium isotope ^{235}U from ^{238}U. Natural uranium consisting of 99.3% ^{238}U and 0.7% ^{235}U is converted to the gas UF_6, and the mixture of the gases is passed at low pressure through a porous solid. The heavier $^{238}UF_6$ diffuses less rapidly than $^{235}UF_6$; hence, the gas mixture which first emerges from the solid is richer in the ^{235}U isotope than was the starting mixture. Since the square root of the ratio of molecular weights is only 1.0043, the step must be repeated thousands of times, but eventually substantial enrichment of the desired ^{235}U isotope is obtained.

5.10   Kinetic theory

One aspect of observed gas behavior which originally gave the strongest clue to the nature of gases is the phenomenon known as *brownian motion*. This motion, first observed by the Scottish botanist Robert Brown (1827), is the *irregular zigzag movement of extremely minute particles when suspended in a liquid or gas*. Brownian motion can be observed by focusing a microscope on a particle of smoke illuminated from the side. The particle does not settle to the bottom of its container but moves continually to and fro and shows no sign of coming to rest. The smaller the suspended particle observed, the more violent is this permanent condition of irregular motion. The higher the temperature of the fluid, the more vigorous is the movement of the suspended particle.

The existence of brownian motion contradicts the idea of matter as a quiescent state and suggests that the molecules of matter though invisible are constantly moving. A particle of smoke appears to be jostled by its unseen neighboring molecules; thus, indirectly, the motion of the smoke particle reflects the motion of the submicroscopic, invisible molecules of matter. Here then is powerful support for the suggestion that matter consists of extremely small particles which are ever in motion. This "moving-molecule" theory is known as the *kinetic theory of matter*. Its two basic postulates are that molecules of matter are in motion and that heat is a manifestation of this motion.

Like any theory, the kinetic theory represents a model which is proposed to account for an observed set of facts. In order that the model be practical, certain simplifying assumptions must be made about its properties. The validity of each assumption and the reliability of the whole model can be checked by how well the facts are explained. For a perfect gas the following assumptions are made:

1   Gases consist of tiny molecules, which are so small and so far apart on the average that the actual volume of the molecules is negligible compared with the empty space between them.

2   In the perfect gas there are no attractive forces between molecules. The molecules are completely independent of each other.

3   The molecules of a gas are in rapid, random, straight-line motion, colliding with each other and with the walls of the container. In each collision it is assumed that there is no net loss of kinetic energy, although there may be a transfer of energy between the partners in the collision.

4 At a particular instant in any collection of gas molecules, different molecules have different speeds and, therefore, different kinetic energies. However, the average kinetic energy of all the molecules is assumed to be directly proportional to the absolute temperature.

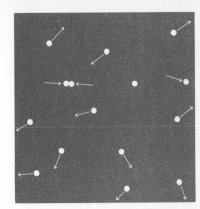

Fig. 5.12
Kinetic model of a gas.

Before discussing each of these assumptions, we might ask how the model is related to the observable quantities *V*, *P*, and *T*. The accepted model of a gas is that it is mainly empty space in which tiny points representing molecules are in violent motion, colliding with each other and with the walls of the container. Figure 5.12 shows an exaggerated version of this model. The *volume* of a gas is mostly empty space, but it is *occupied* in the sense that moving particles occupy the entire region in which they move. *Pressure*, defined as force per unit area, is exerted by gases because the molecules collide with the walls of the container. Each collision produces a tiny impulse, and the sum of all the impulses on one cm^2 of wall in one sec is the pressure. *Temperature* gives a quantitative measure of the average motion of the molecules.

That assumption (1) is reasonable can be seen from the fact that the compressibility of gases is so great. In oxygen gas, for example, 99.96 percent of the total volume is empty space at any instant. Since at STP there are 2.7×10^{19} molecules per cubic centimeter, the average spacing between the molecules is about 3.7 nm, which is about 13 times the oxygen molecular diameter. When oxygen or any other gas is compressed, the average spacing between molecules is reduced; i.e., the fraction of free space is diminished.

The validity of assumption (2) is supported by the observation that gases spontaneously expand to occupy all the volume accessible to them. This behavior occurs even with a highly compressed gas, where the molecules are fairly close together and any intermolecular forces would be expected to be greatest. It must be that there is no appreciable binding of one molecule of a gas to its neighbors.

As already indicated, the observation of brownian motion implies that molecules of a gas move, in agreement with assumption (3). Like any moving body, a molecule has an amount of kinetic energy equal to $\frac{1}{2}ms^2$, where *m* is the mass of the molecule and *s* is its speed. That molecules move in straight lines follows from the assumption of no attractive forces. Only if there were attractions between them could molecules be swerved from straight-line paths. Because there are so many molecules in a gas sample and because they are moving so rapidly (at $0°C$ the average speed of oxygen molecules is about 1700 km/h), there are frequent collisions between the molecules.

It is necessary to assume that molecular collisions are *elastic* (like collisions between billiard balls). Otherwise, kinetic energy would be lost by conversion to potential energy (as, for example, in distorting

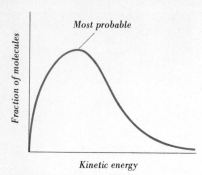

Fig. 5.13
Energy distribution in a gas.

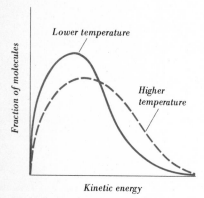

Fig. 5.14
Comparison of energy distribution in a gas sample at two different temperatures.

the molecules); motion of the molecules would eventually stop, and the molecules would settle to the bottom of the container. It might be noted that the distance a gas molecule has to travel before colliding elastically with another gas molecule is much greater than the average spacing between molecules because the molecules have many near misses. In oxygen at STP the average distance between successive collisions, called the *mean free path,* is approximately 300 times the molecular diameter; as already mentioned, the average spacing is only 13 times the molecular diameter.

Assumption (4) has two parts: (*a*) There is a distribution of kinetic energies (suggested by Maxwell), and (*b*) the average kinetic energy is proportional to the absolute temperature (suggested by Boltzmann). The distribution of energies comes about because molecular collisions continually change the speed of a particular molecule. One given molecule may move along with a certain speed, but it soon hits another molecule, to which it may lose some of its kinetic energy; perhaps later it may get hit by a third molecule from which it gains kinetic energy; and so on. The exchange of kinetic energy goes on constantly; so it is only the total kinetic energy of the gas sample that stays constant (provided, of course, no energy is lost to the surroundings or added to the gas sample from the outside, as by heating). The total kinetic energy consists of the contributions of all the molecules, each of which may be moving at a different speed. The situation is summarized in Fig. 5.13, which indicates the usual distribution of kinetic energies in a gas sample. Each point on the curve tells what fraction of the molecules have the specified value of the kinetic energy.

The temperature of a gas can be raised by the addition of heat. What happens to the molecules as the temperature is raised? The heat which is added is a form of energy, and so it can be used to increase the speed of the molecules and, therefore, the average kinetic energy. This is shown in Fig. 5.14, where the dashed curve describes the situation at higher temperature. The fraction of molecules having higher kinetic energies is larger; so the average kinetic energy is increased. Temperature serves to measure the average kinetic energy.

The assumption that average kinetic energy is directly proportional to the absolute temperature is supported by the fact that predictions based on the assumption agree so well with experiment. For example, it immediately follows that two different gases at the same temperature must have equal average kinetic energies. If two gases A and B have different molecular masses m_A and m_B, then the average speeds s_A and s_B must be related as follows:

Average kinetic energy of A = average kinetic energy of B

$$\tfrac{1}{2}m_A s_A{}^2 = \tfrac{1}{2}m_B s_B{}^2$$

$$\frac{s_A}{s_B} = \sqrt{\frac{m_B}{m_A}}$$

It is reasonable to assume that the rate of diffusion of a gas is directly proportional to the average speed of its molecules. Thus, we can write

$$\frac{\text{Rate of diffusion of A}}{\text{Rate of diffusion of B}} = \frac{s_A}{s_B} = \sqrt{\frac{m_B}{m_A}}$$

This inverse proportionality between diffusion rate and the square root of the molecular mass was found experimentally as Graham's law.

5.11 · Kinetic theory and equation of state

The equation of state of an ideal gas $PV = nRT$ can be derived from kinetic theory by considering in detail how the pressure of a gas arises from molecular impacts. Suppose we imagine a gas confined in a cubic box, as in Fig. 5.12. The pressure of the gas is proportional to the *number of molecular impacts* per square centimeter of wall area per second times the *impulse* (see Appendix 5.3) *of each impact.*

Let N be the number of molecules in the box; m, the mass of each molecule; s, the average speed of each molecule; and l, the length of one edge of the cubic box. The motion of any molecule in the box can be resolved into three components along the three edge directions of the box. We thus assume that the net effect on the walls is the same as if one-third of the molecules in the box were constrained to move normal to a pair of opposite faces. In other words, $\frac{1}{3}N$ molecules will be colliding with any chosen face of the box. How frequently will each of these molecules make a molecular impact on the wall of interest? Between successive impacts on the same wall the molecule has to travel the entire length of the box and back, i.e., the distance $2l$. Since it is traveling at a speed of s cm/sec, the total distance traveled per second is s cm. During one second, it will collide with the wall $s/2l$ times.

Impacts per face per second $= \dfrac{1}{3}N\dfrac{s}{2l}$

Area of face $= l^2$ square centimeters

Impacts per square centimeter per second $= \dfrac{1}{3}N\dfrac{s/2l}{l^2}$

Since the volume V of the box is l^3, the expression becomes

Impacts per square centimeter per second $= \dfrac{Ns}{6V}$

To get the impulse per impact, it is necessary to note that the impulse is equal to the change of momentum. In a collision in which a molecule of mass m and speed s bounces off the wall with the same speed but in an opposite direction (denoted by $-s$), the momentum changes from an initial value ms to a final value $-ms$, i.e., by an amount equal to $2ms$.

$$\text{Pressure} = \left[\frac{\text{impacts}}{(\text{area})(\text{second})}\right]\left[\frac{\text{impulse}}{\text{impact}}\right]$$

Section 5.11
Kinetic theory and
equation of state

173

$$P = \frac{Ns}{6V}(2ms) = \frac{N}{3V}ms^2$$

The proportionality between average kinetic energy $\frac{1}{2}ms^2$ and absolute temperature T can be written as

$$\tfrac{1}{2}ms^2 = \tfrac{3}{2}kT$$

where the constant k, called the Boltzmann constant, has the value 1.3806×10^{-23} J/deg, or 1.363×10^{-25} liter-atm/deg. By replacing ms^2 by $3kT$ in the pressure equation given above we find

$$P = \frac{N}{V}kT$$

The number of molecules N is just equal to n, the number of moles, times 6.0222×10^{23}. Since the Boltzmann constant k is the gas constant R divided by 6.0222×10^{23}, we have

$$P = \frac{nRT}{V}$$

Thus we have derived the equation of state $PV = nRT$ from first principles of molecular motion and molecular collisions.

5.12 Deviations from ideal behavior

The model of a gas as an ensemble of point masses moving independently of each other leads to the ideal equation of state $PV = nRT$. Yet, as we have seen, gases are better described by other equations of state such as the van der Waals equation:

$$\left(P + \frac{n^2a}{V^2}\right)(V - nb) = nRT$$

What is incorrect about the kinetic molecular model? In the first place, molecules are not point masses but have finite dimensions. This means that the molecules are not free to move in all of volume V but in something less than that. The van der Waals equation shows this in the subtractive term nb. The quantity $V - nb$ can be thought of as representing the *free volume* actually accessible to the molecules, and the quantity nb as an *excluded volume* which is not accessible to molecular motion because molecules are impenetrable in each other.

Figure 5.15 suggests how one can show that the magnitude of b is directly related to the size of molecules. Let r be the radius of a gas molecule, which we will assume to be representable as a hard sphere. The volume of the molecule is thus $4\pi r^3/3$, and obviously no other molecule can come into this space. But the situation is more complicated than this. Even in direct collision, another molecule cannot approach closer than shown by the dotted sphere. In other words, the centers of other molecules are excluded not just from the volume $4\pi r^3/3$ (as shown by the colored line) but from the volume $4\pi(2r)^3/3$, or $8(4\pi r^3/3)$ (as shown by the dashed line). The excluded volume is eight times the volume of an individual molecule. For n mol of molecules, the excluded volume would be as follows:

$$V_{excl} = 8\left(\frac{4\pi r^3}{3}\right)(6.02 \times 10^{23} \, n)\left(\frac{1}{2}\right)$$

The factor $\frac{1}{2}$ comes in because the excluded volume shown per molecule in Fig. 5.15 is actually generated by a pair of molecules; we have to be careful not to count the molecules twice. Since $nb = V_{excl}$, we can use the experimentally determined values of b to get an idea of molecular sizes. Figure 5.16 lists for several representative gases values of the van der Waals constant b and the molecular radius r derived therefrom.

A second reason for nonideal behavior is that, contrary to the assumption noted previously, there *are* attractive forces between molecules. The effect of such forces is to reduce the pressure exerted by a collection of gas molecules. The quantity $P + n^2a/V^2$ can be thought of as being equivalent to an ideal pressure as specified by the ideal equation of state. The actual pressure observed P is less than the ideal pressure by the amount n^2a/V^2. This correction term contains the quantity n/V squared, where n/V represents the number of moles, and hence the number of gas molecules, per unit volume. n/V, the *concentration* of gas molecules, when squared gives the probability of collisions between molecules (Sec. 10.5). a is a proportionality constant that indicates the magnitude of the cohesive force between molecules in collision. Thus, a, which gives a measure of deviation from ideality, is multiplied by the probability of collision to give a valuation of the drop from ideal pressure due to attractive forces.

The attractive forces described by the van der Waals constant a are special in that they must operate only at very short distances, i.e., during collision. For this reason, they are called *short-range forces*. In some cases it is easy to see the reason for attractive forces between molecules. For example, with polar molecules the positive end of one molecule may attract the negative end of another molecule. It is not surprising, therefore, that polar substances deviate markedly from ideal behavior. Water vapor, as an illustration, is so nonideal that even at room temperature it liquefies under slight pressure.

It is not so easy to see the reason for attractive forces between nonpolar molecules. Suppose we consider two neon atoms extremely close together, as shown in Fig. 5.17. We can imagine that *instantaneously* the electron distribution in atom 1 is unsymmetric, with a slight preponderance of electron charge density on one side. For a fraction of a picosecond, the atom would be in a state in which one end appears slightly negative with respect to the other end; i.e., the atom is momentarily a dipole. The neighboring atom, as a result, is distorted because the positive end of atom 1 displaces the electrons in atom 2. As shown in the figure, there is an instantaneous dipole in both of the atoms, with consequent attraction. This picture persists for only an extremely short time because the electrons are in rapid motion. As electrons in atom 1 move to the other side, electrons in

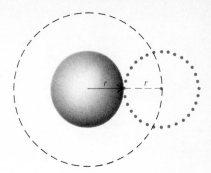

Fig. 5.15

Schematic diagram to show how b term of van der Waals equation is related to size of molecules. Dashed circle shows excluded volume.

	b, liters/mol	r, nm
Helium	0.0237	0.133
Hydrogen	0.0266	0.138
Nitrogen	0.0391	0.157
Water	0.0305	0.145
Benzene	0.115	0.225

Fig. 5.16

Values of van der Waals constant b and derived molecular radius r.

Atom 1   Atom 2

Instantaneous preponderance
of negative charge

Fig. 5.17
Model of van der Waals
attraction.

atom 2 follow. In fact, we can think of van der Waals forces as arising because electrons in adjacent molecules are beating in time, so as to produce synchronized fluctuating dipoles, which give rise to an instantaneous attraction. The attraction is strong when the particles are close together but rapidly weakens as they move apart. Also, the more electrons there are in a molecule and the less tightly bound these electrons are, the greater are the van der Waals forces.

The attractive forces become less significant as the temperature increases because a rise in temperature produces an effect that opposes the attractive forces. This effect is a disordering one because of molecular motion, which increases in speed as the temperature rises. Disordering arises because the molecules of a gas are in random motion. The attractive forces try to draw the molecules together, but the molecules, because of their motion, stay apart. As the temperature is raised, the molecules have more ability to overcome the attractive forces. The attractive forces are unchanged, but the motion of the molecules increases; hence, the attraction becomes relatively less important.

At sufficiently low temperature, attractive forces, no matter how weak, take over and draw the molecules together to form a liquid. This *temperature at which gas molecules coalesce to form a liquid* is called the *liquefaction temperature.* Liquefaction is easier at high pressures, where distances between molecules are smaller and hence intermolecular forces are greater. The higher the pressure, the easier the gas is to liquefy, and the less it needs to be cooled to accomplish liquefaction. Thus, the liquefaction temperature increases with increasing pressure.

5.13   Critical temperature

There is for each gas a temperature above which the attractive forces are not strong enough to produce liquefaction no matter how high a pressure is exerted on the system. This temperature is called the *critical temperature* and is designated by T_c. It is defined as the *temperature above which a substance can exist only as a gas.* Above the critical temperature molecular motion is so vigorous that, no matter how high the pressure, the molecules occupy the entire available volume as a gas. The critical temperature thus depends on the magnitude of the attractive forces between the molecules.

Figure 5.18 contains values of the critical temperature for some common substances. Listed also is the *critical pressure, which is the minimum pressure that must be exerted to produce liquefaction at the critical temperature.* At temperatures higher than the critical temperature, no amount of pressure can produce liquefaction. For example, above $647°K$ ($364°C$), H_2O exists only as a gas. Such a critical temperature is relatively high and indicates that the attractive forces between H_2O molecules are great enough that even at $647°K$ they can produce the liquid state. Attractive forces between SO_2 molecules are less than those between H_2O molecules; hence, the critical temperature of SO_2

Fig. 5.18
Critical constants.

Substance	Critical temperature, °K	Critical pressure, atm
Water, H_2O	647	217.7
Sulfur dioxide, SO_2	430	77.7
Hydrogen chloride, HCl	324	81.6
Carbon dioxide, CO_2	304	73.0
Oxygen, O_2	154	49.7
Nitrogen, N_2	126	33.5
Hydrogen, H_2	33	12.8
Helium, He	5.2	2.3

is lower than that of H_2O. Liquefaction of SO_2 cannot be achieved above 430°K (157°C).

In the extreme case of He the attractive forces are so weak that liquid He can exist only below 5.2°K (-267.9°C). At this very low temperature molecular motion is so slow that the van der Waals forces, weak as they are, can hold the atoms together in a liquid. In Fig. 5.18 the order of decreasing critical temperature is also the order of decreasing attractive forces, and we can think of the critical temperature as giving a measure of the attractive forces between molecules.

5.14 Cooling by expansion

Substances with high critical temperatures are easy to liquefy; substances with low critical temperatures must be cooled before they can be liquefied. For example, oxygen cannot be liquefied at room temperature, about 300°K. It must be cooled below 154°K (-119°C), the critical temperature, before liquefaction can occur. This cooling would be quite difficult were it not that gases sometimes cool themselves on expansion.* When a gas expands against a piston, the gas does work in pushing the piston. If the energy for this work comes from the kinetic energy of the gas molecules, a decrease in the kinetic energy of the molecules is observed as a lowering of the temperature. However, a temperature drop may be observed even for a gas which expands into a vacuum and therefore does no external work.

The cause of cooling by unrestrained expansion can be seen by considering the experiment shown in Fig. 5.19. The box shown is

* The word "sometimes" takes into consideration the fact that cooling does not occur at *any* temperature but only below what is called the *inversion temperature*. This for most gases is roughly six times the critical temperature. Above the inversion temperature, a gas warms when suddenly expanded; below the inversion temperature, it cools. Since most inversion temperatures turn out to be above room temperature, most gases at room temperature cool on expansion and heat on compression. However, the inversion temperatures of hydrogen and helium are well below room temperature (195 and 45°K, respectively); consequently, these gases heat up when allowed to expand (into a vacuum) at room temperature.

Fig. 5.19
Free expansion of a gas.

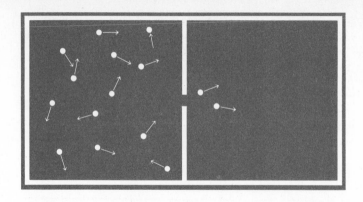

assumed perfectly insulated from its surroundings; so no heat can get in from the outside. It is divided into two compartments by a diaphragm. The left-hand compartment contains compressed gas; the right-hand compartment is originally empty. If a hole is now punched in the diaphragm, the gas streams into the vacuum. A thermometer in the path of the streaming gas would show a drop in temperature. As the gas streams into the empty space, molecules work against the attractive forces of their neighbors. This requires energy; and since no outside energy is available, the molecules must use up some of their kinetic energy. The average kinetic energy, as measured by the temperature, drops. If there were no attractive forces between molecules, there would be no cooling effect. Indeed, the fact that cooling is observed indicates that there *are* attractive forces between gas molecules.*

The commercial liquefaction of gases makes use of cooling by expansion. In order to liquefy air, for example, it is first compressed to high pressure, cooled with a refrigerant to remove the heat that accompanies compression, and then allowed to expand. Some of the air liquefies as a result of cooling on expansion; the rest is passed over the incoming pipes containing the compressed air to cool it further.

Exercises

*5.1   Pressure. When a chemist or physicist speaks of a high vacuum, he generally means a pressure that corresponds to about 1×10^{-6} mmHg. Express this in units of (*a*) atmospheres and (*b*) pascals.

*5.2   Volume. In what sense is there a difference in the meaning of the word "occupied" when we use it to describe the volume of a mole of liquid water vs. a mole of gaseous water?

* The opposite effect, warming on unrestrained expansion, may also be observed, above the inversion temperature. It indicates that there are also repulsive forces between gas molecules. Below the inversion temperature, repulsions are masked by the larger attractions; however, as the temperature is raised above the inversion temperature, the attractive forces become less important, and the repulsive forces, small as they are, dominate. The repulsions can be viewed as arising from the noninterpenetrability of molecules, and hence they are related to the *b* of the van der Waals equation.

*5.3 *Temperature.* In an ordinary mercury thermometer, the temperature is measured by noting the position of a fine mercury thread expanding into a capillary tube from a large sensing bulb. Suppose you had such a thermometer but it lacked any markings whatsoever. How would you proceed experimentally to put appropriate marks on it so that it would read degrees absolute directly?

*5.4 *Temperature.* Body temperature on the average is 98.6°F. What is this on (a) the Celsius scale and (b) the Kelvin scale?
Ans. (a) 37.0°C, (b) 310.2°K

*5.5 *Pressure.* A typical recommended tire gauge pressure is 24.0 psi. What would be the equivalent in each of the following units: (a) atmospheres, (b) Torr, (c) bars, and (d) pascals?

*5.6 *Manometer.* Given the setup of Fig. 5.3, what would be the pressure of the gas (in atmospheres) if P_{atm} is 745 Torr and P_{liq} is the equivalent of a mercury column 3.0 cm high? *Ans. 0.941 atm*

*5.7 *Boyle's law.* What would Boyle's law look like on a PV versus P plot?

*5.8 *Gas laws.* A gas-filled freely collapsible balloon is pushed from the surface level of a lake to a depth of 250 ft. Approximately what percent of its original volume will the balloon finally have? Assume ideal behavior.

*5.9 *Boyle's law.* Draw on the same graph PV versus P plots for (a) an ideal gas and (b) a nonideal gas such as acetylene at 0°C. (See Fig. 5.6 for data.)

*5.10 *Diffusion.* How would you expect the rate of diffusion of $N_2(g)$ to compare with that of $CO(g)$? Justify your answer.

*5.11 *Kinetic theory.* Given that the average speed of an O_2 molecule is 1700 km/h at 0°C, what would you expect would be the average speed of a CO_2 molecule at the same temperature?
Ans. 1450 km/h

*5.12 *Deviations from ideal behavior.* Which member of each of the following pairs is the more likely to deviate from ideal-gas behavior? Explain your choice.
a H_2 versus O_2
b N_2 versus CO
c   HF versus HCl
d CH_4 versus C_2H_6

*5.13 *Nonideal behavior.* When you let air out of an automobile tire, the issuing air feels cold to the touch. Why can you not use this as evidence for intermolecular attractions?

**5.14 *Temperature.* In engineering, a frequently used temperature scale is the Rankine scale (°R). Rankine temperature counts from zero as absolute zero, but the degree size is the same as the Fahrenheit degree. Deduce formulas for converting degrees Fahrenheit into degrees Rankine and degrees Kelvin into degrees Rankine.

**5.15   van der Waals equation. Using data from Figs. 5.6 and 5.7, show that the van der Waals product $(P + n^2a/V^2)(V - nb)$ is more nearly constant with rising pressure than is PV.

**5.16   Gas laws. How many full strokes of a bicycle pump (chamber 4.0 cm diameter and 40.0 cm long) would you need to make in order to pump up an automobile tire from a gauge pressure of zero to 24 psi? Assume temperature stays constant at $25°C$ and atmospheric pressure is one atmosphere. Note that gauge pressure measures only the excess over atmospheric pressure. A typical tire volume is about 25 liters.   *Ans. About 81 strokes*

**5.17   Gas laws. A 20-g chunk of Dry Ice (CO_2) is placed in an "empty" 0.75-liter wine bottle and tightly corked. What would be the final pressure in the bottle after all the CO_2 has evaporated and the temperature has reached $25°C$?

**5.18   Gas laws. Suppose the situation represented in Fig. 5.8 corresponds to $25°C$. To what temperature would you need to take the gadget so that the mercury drop would move all the way to the right end? Assume ideal behavior.   *Ans. About 300°C*

**5.19   Charles' law. Suppose you carried out volume measurements into the dashed-line regions as shown in Fig. 5.9. How would the data points probably appear?

**5.20   Gas laws. Generally, when a gas is collected by water displacement (Fig. 5.10), it is assumed that the water temperature is the same as that of the gas. Suppose in a given experiment you collected $535 cm^3$ of oxygen gas over water at a barometric pressure of 0.980 atm and a temperature of $26°C$. What percent error would you make in the STP volume of the sample if the water temperature were actually $11°C$ but all other conditions were as given?

**5.21   Gas laws. In collecting a sample of oxygen gas by water displacement (Fig. 5.10), what volume should you collect to get 0.355 g of O_2? Barometric pressure is 0.973 atm; temperature, $22°C$. Assume ideal behavior.   *Ans. 284 cm³*

**5.22   Avogadro principle. Show how the Avogadro principle can be applied to the observation "one volume of nitrogen gas reacts with three volumes of hydrogen gas to produce two volumes of ammonia gas" in order to deduce that nitrogen molecules must contain an even number of nitrogen atoms.

**5.23   Gas reactions. What volume of air (78% N_2, 21% O_2, 1% Ar) measured at $35°C$ and 0.965 atm would you need to supply in order to oxidize completely 1 gal of gasoline to $CO_2(g)$ and $H_2O(g)$? Density of gasoline is about 0.68 g/ml. Assume it is C_8H_{18} (1 gal = 3.79 liters).

5.24 *Molecular weight*. When a given sample of liquid weighing 0.382 g is completely evaporated and collected over water at barometric pressure 1.006 atm and temperature 28°C, the observed gas volume is 256 cm^3. What is the molecular weight of the material, assuming ideal behavior? *Ans. 38.0 amu*

5.25 *Molecular formula*. Analysis of a given hydrocarbon shows that it contains 83.63% C and 16.37% H by weight. The density of the pure vapor at 0.853 atm and 140°C is 0.00217 g/cm^3. What is the molecular formula of the hydrocarbon?

5.26 *Kinetic theory*. You are given samples of two different gases A and B. The mass of molecule A is twice the mass of molecule B; the average speed of the A molecules is twice the average speed of the B molecules. If both samples contain the same number of molecules per cubic centimeter, what must be the relative pressures in the two gas samples? *Ans. $P_A = 8P_B$*

5.27 *Kinetic theory*. Calculate in meters per second the average speed of a water molecule in water vapor at 25°C. Recall that one joule = 1 kg-m^2/sec^2.

5.28 *Kinetic theory*. Given that the average speed of an O$_2$ molecule is 1700 km/h at 0°C, what would you expect it to be at 100°C? *Ans. 2000 km/h*

5.29 *Molecular size*. What radius would you deduce for a carbon tetrachloride molecule CCl$_4$ from the fact the van der Waals b for the vapor is 0.138 liter/mol?

***5.30 *Partial pressure*.** What would be the final partial pressure of oxygen in the following experiment? A collapsed polyethylene bag of 30 liters capacity is partially blown up by the addition of 10 liters of nitrogen gas measured at 0.965 atm and 298°K. Subsequently, enough oxygen is pumped into the bag so that at 298°K and external pressure of 0.990 atm, the bag contains a full 30 liters. Assume ideal behavior. *Ans. 0.668 atm*

***5.31 *Equation of state*.** At STP, the observed molar volume of carbon dioxide is 22.263 liters. What fraction does this represent of the value calculated from (*a*) the ideal equation of state and (*b*) the van der Waals equation of state? (Refer to Fig. 5.7 for data.)

***5.32 *Diffusion*.** The relative rates of diffusion of NH$_3$(*g*) and HX(*g*) can be determined experimentally by simultaneously injecting NH$_3$(*g*) and HX(*g*), respectively, into the opposite ends of a glass tube and noting where a deposit of NH$_4$X(*s*) is formed. Given a 1-m tube, how far from the NH$_3$ injection end would you expect the NH$_4$X(*s*) to be formed when HX is (*a*) HF, (*b*) HCl, (*c*) HBr, or (*d*) HI? *Ans. (a) 52 cm, (b) 59 cm*

***5.33 *Kinetic theory*.** You are given 1 cm^3 of O$_2$ gas and 1 cm^3 of N$_2$ gas both at STP. Compare these two samples with respect to each of the following:

  a Number of molecules
  b Average speed of molecules
  c Number of impacts on container walls per second
  d Impulse of each molecular impact on container wall

***5.34 *Critical constants.* Why might one expect to find, as is generally the case, that the higher the critical temperature of a substance, the higher will be its critical pressure?

When a sample of gas is cooled or compressed, or both, it liquefies. In the process the molecules of the gas, originally far apart on the average, are slowed down and brought close enough together that attractive forces become significant. The individual molecules coalesce into a cluster, which settles to the bottom of the container as liquid. What are the general properties of liquids? How can they be accounted for by the kinetic molecular theory? In this chapter we consider the liquid state and compare it with the gaseous state. *183*

6.1 Properties of liquids

Liquids are practically incompressible. Unlike gases, there is little or no change in volume when the pressure on a liquid is changed. Theory accounts for this by the assumption that the amount of free space between the molecules of a liquid is almost a minimum. Any attempt to compress the liquid meets with resistance as the electron cloud of one molecule repels the electron cloud of an adjacent molecule.

Liquids maintain their volume no matter what the shape or size of the container. A 10-ml sample of liquid occupies a 10-ml volume whether it is placed in a small beaker or in a large flask, whereas a gas spreads out to fill the whole volume accessible to it. Gases do not maintain their volume because the molecules are essentially independent of each other and can move into any space available. In liquids the molecules are close together; so mutual attractions are strong. Consequently, the molecules are clustered together.

Liquids have no characteristic shape. In the absence of gravity, as in an orbiting space vehicle, the molecules of a liquid agglomerate into a spherical cluster. The cluster can readily be deformed, as by poking it, but left to itself it will gradually go back to a spherical shape. Not so when earth's gravity comes into the picture. On earth, a liquid sample assumes the shape of the bottom of its container. The implication of all this is that there are no fixed positions for the molecules of a liquid. They are free to slide over each other in order to occupy positions of the lowest possible potential energy with respect to each other and with respect to gravitational attraction by the earth.

Liquids diffuse, but slowly. When a drop of ink is released in water, there is at first a rather sharp boundary between the ink cloud and the water. Eventually the color diffuses throughout the rest of the liquid. In gases diffusion is much more rapid. Diffusion is able to occur because molecules have kinetic energy and move from one place to another. In a liquid, molecules do not move far before colliding with their neighbors. The mean free path, i.e., the average distance between collisions, is short and comparable to the average spacing between the molecules. Eventually each molecule of a liquid does migrate from one side of its container to the other, but it has to suffer many billions of collisions in doing so. In gases there is less obstruction to the migrating molecule. Because a gas is mostly empty space, the mean free path is much longer. Hence the molecules of one gas can mix rather quickly with those of another.

Liquids evaporate from open containers. Although there are attractive forces which hold molecules to each other, those molecules having kinetic energy great enough to overcome attractive forces can escape into the gas phase. In any collection, a given molecule does not have the same energy all the time. There is perpetually an exchange of energy between the colliding molecules. As an example, the collection might start out with all molecules of the same energy, but this situation would not persist long. Two or more molecules may simultaneously collide with a third one. Molecule 3 now has not only its initial energy but possibly some extra energy received from collision with its neighbors. Molecule 3 is now a molecule which is higher than average in kinetic energy. If it happens to be near the surface of the liquid, it may be able to overcome the attractive forces of neighbors and go off into the gas phase.

Figure 6.1 shows a typical energy distribution for the molecules of a sample of liquid at a given temperature. The curve is quite similar to the one given previously for a gas (Fig. 5.13). If the value marked E corresponds to the minimum kinetic energy required by a molecule to overcome attractive forces and escape from this liquid, then all the molecules in the shaded area of the curve have enough energy to overcome the attractive forces. These are the "hot" molecules, which have the possibility of escaping provided that they are close enough to the surface. If it is only highly energetic molecules that leave the liquid, then the average kinetic energy of those left behind must be lower. Each molecule that escapes carries with it more than an average amount of energy (part of which it uses in working against the attractive forces). Since the remaining molecules end up with lower average kinetic energy, their temperature must be lower. Thus, evaporation is generally accompanied by cooling.

When a liquid evaporates from a noninsulated container, such as a beaker, the temperature of the liquid does not fall very far before there is appreciable heat flow from the surroundings into the liquid. It the rate of evaporation is not too great, this flow of heat will be sufficient to match the energy required for evaporation. As a consequence, the temperature of the liquid remains close to that of the room, even though the liquid is evaporating. When evaporation proceeds from an insulated container, the heat flow from the surroundings is less; so the temperature of the liquid may drop significantly. An example of such an insulated container is the *vacuum bottle* (also called the *Dewar flask*, after the Scottish chemist who invented it, Sir James Dewar). The essential feature of the Dewar flask is double-walled construction with vacuum between the walls. The vacuum jacket acts as an insulator to inhibit heat flow into or out of the container.

When a liquid evaporates from a Dewar flask, the flow of heat from the surroundings is generally not fast enough to compensate for the evaporation. The temperature of the liquid will drop, the average kinetic energy of the molecules will decrease, and the rate of evaporation will diminish until the heat flow into the liquid just equals the heat required for evaporation. A lower temperature will be established at which the distribution of the kinetic energies will be shifted to the left, as shown by the dashed curve in Fig. 6.1. The fraction of energetic molecules will be less than before; so the rate of evaporation will be decreased. The temperature is constant but at a value which may be considerably lower than the original temperature. Liquid air, an extremely volatile liquid, remains in an open Dewar flask at approximately $-190°C$ for many hours.

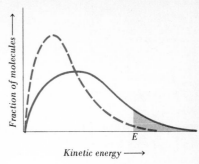

Fig. 6.1
Energy distribution in a liquid. Solid curve applies at one temperature; dashed curve, at a lower temperature.

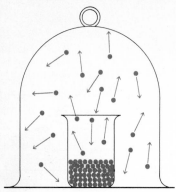

Fig. 6.2
Evaporation in a closed system.

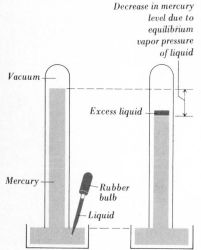

Decrease in mercury level due to equilibrium vapor pressure of liquid

Vacuum

Excess liquid

Mercury

Rubber bulb

Liquid

Fig. 6.3
Measurement of equilibrium vapor pressure.

6.2 Equilibrium vapor pressure

When a bell jar is placed over a beaker of evaporating liquid, as shown in Fig. 6.2, the liquid level drops for a while and then becomes constant. This can be explained as follows: Molecules escape from the liquid into the gas, or vapor, phase. After escaping, they are confined to a limited space. As the molecules accumulate in the space above the liquid, there is an increasing chance that in their random motion some of them will return to the liquid. Eventually, a situation is established in which molecules are returning to the liquid just as fast as other molecules are leaving it. At this point the liquid level no longer drops because the number of molecules evaporating per second is equal to the number of molecules condensing per second. A condition in which two changes exactly oppose each other is referred to as *dynamic equilibrium*. Although the system is not at a state of rest, it shows no net change. The amount of liquid in the beaker stays constant; the concentration of molecules in the vapor above the liquid is also constant. A particular molecule spends part of its time in the liquid and part in the vapor phase, but as molecules pass from liquid to gas, other molecules move from gas to liquid, keeping the number of molecules in each phase constant.

The molecules which are in the vapor exert a pressure. At equilibrium, this pressure is characteristic of the liquid. It is known as the *equilibrium vapor pressure*. As the term implies, it is the *pressure exerted by a vapor when the vapor is in equilibrium with its liquid*. The magnitude of the equilibrium vapor pressure depends (1) on the nature of the liquid and (2) on its temperature:

1 The nature of the liquid is involved since each liquid has characteristic attractive forces between its molecules. Molecules which have large mutual attractions have a small tendency to escape into the vapor phase. Such a liquid has a low equilibrium vapor pressure. Liquids composed of molecules with small mutual attractions have a high escaping tendency and therefore a high equilibrium vapor pressure.

2 As the temperature of a liquid is raised, the average kinetic energy of the molecules of the liquid increases. The number of high-energy molecules capable of escaping also becomes larger; so the equilibrium vapor pressure increases.

There are various devices for demonstrating equilibrium vapor pressure, one of which is shown in Fig. 6.3. It consists of a mercury barometer set up in the usual manner of inverting a tube full of mercury in a pool of mercury. The height difference between the upper mercury level and the lower mercury level represents the atmospheric pressure. Above the mercury there is initially a vacuum. By squeezing on the rubber bulb of the medicine dropper a drop of the liquid to be measured can be ejected into the mercury. Since practically any liquid is less dense than mercury, the drop will float to the top of the mercury, where enough of it will evaporate to establish an equilibrium between the liquid and vapor. The corresponding vapor pressure now pushes down the mercury column. (The excess liquid also acts to push down the mercury column, but this is a negligible effect, especially when there is little excess.)

The extent to which the mercury level is depressed gives a quantitative measure of the vapor pressure of the liquid. At 20°C, water depresses the column by 17.5 mm; hence, it has an equilibrium vapor pressure of 17.5/760, or 0.023, atm. Similarly, one would find for carbon tetrachloride, CCl_4, 0.120 atm and for chloroform, $CHCl_3$, 0.211 atm. These vapor-pressure values give an idea of the escaping tendencies of the molecules from the various liquids. In chloroform, for example, the attractive forces between molecules are smaller than in water, and chloroform evaporates to give a higher vapor pressure than water.

By repeating the above experiment at different temperatures, it is possible to determine the vapor pressure of a liquid as a function of temperature. Appendix 3 is a table showing the results for water in great detail. The general behavior of water, carbon tetrachloride, and chloroform is shown by the graph in Fig. 6.4. The vertical scale represents the vapor pressure, and the horizontal scale the temperature. As the temperature increases, the vapor pressure rises, first slowly and then more steeply, until at high temperatures it is rising almost vertically. (The curve continues but only up to the critical temperature. It does not go beyond the critical temperature, because above its critical temperature a liquid cannot exist.) A curve of this shape is frequently better presented by a plot in which the logarithm of the pressure (Appendix 4.2) is plotted against the reciprocal temperature, as shown in Fig. 6.5. It should be noted that, because of the reciprocal relation, high temperature is at the left of the figure and low temperature is at the right. (The temperatures in $T°K$ corresponding to various values of the abscissa $1000/T$ are shown across the top of the figure. The abscissa values at the bottom of the figure are shown as $1000/T$ instead of as $1/T$ in order to save writing a lot of zeros.) The analytic expression describing the lines in Fig. 6.5 is given by

$$\log p = \frac{-\Delta H'}{19.15T} + C$$

where p is the vapor pressure, T is the temperature, $\Delta H'$ is the heat required to transform one mole of liquid to the ideal-gas state,* and C is a constant dependent on the liquid and units used for expressing pressure. The number 19.15 comes from (2.303)(8.3143), where the number 2.303 arises from using base-10 logarithms instead of natural logarithms and the number 8.3143 is the value of the universal gas constant R in units of joules per mole per degree. Consistent with this,

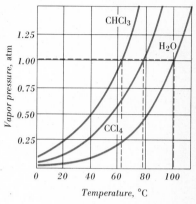

Fig. 6.4

Equilibrium vapor pressure as a function of temperature.

---

* The prime on ΔH emphasizes two assumptions: (1) The vapor is ideal, and (2) ΔH does not change with temperature. If both these assumptions are valid, straight-line plots are obtained as in Fig. 6.5. If either assumption is a poor one, then the ΔH will not generally agree with the directly measured heat of vaporization.

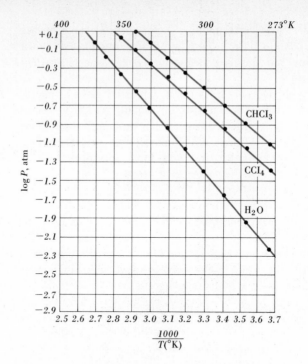

Fig. 6.5
Logarithmic plot of vapor
pressure as a function of
reciprocal temperature.

$\Delta H'$ is given in joules per mole. In a graph such as Fig. 6.5 the slope of the line is $-\Delta H'/19.15$. The slope will be constant—i.e., the line will be straight—as long as $\Delta H'$ has a constant value independent of change in temperature. This is generally true as long as the temperature range is not too large.

The constant C can be eliminated by subtracting the above equation for a liquid at a given temperature from that for the same liquid at another temperature. By denoting these conditions as T_1 and T_2 and the corresponding vapor pressures as p_1 and p_2, respectively, we get

$$\log p_1 - \log p_2 = \frac{-\Delta H'}{19.15T_1} + \frac{\Delta H'}{19.15T_2}$$

$$\log \frac{p_1}{p_2} = \frac{\Delta H'}{19.15}\left(\frac{1}{T_2} - \frac{1}{T_1}\right)$$

By use of this last equation, sometimes called the *Clausius-Clapeyron equation*, it is possible to evaluate $\Delta H'$ by substituting in the measured values of p_1 and p_2 at T_1 and T_2. Conversely, if $\Delta H'$ were known, the vapor pressure p_1 at one temperature T_1 could be calculated from a measured vapor pressure p_2 at some other temperature T_2.

Example 1
The vapor pressure of carbon tetrachloride is 0.132 atm at 23°C and 0.526 atm at 58°C. Calculate the $\Delta H'$ in this temperature range.

$p_1 = 0.132$ atm   at $T_1 = 296°$K

$p_2 = 0.526$ atm   at $T_2 = 331°$K

$$\log \frac{0.132}{0.526} = \frac{\Delta H'}{19.15} \left(\frac{1}{331} - \frac{1}{296} \right)$$

$$\Delta H' = 32,000 \text{ J/mol}$$

■ ■ ■

Example 2
Given that $\Delta H' = 32,000$ J/mol for carbon tetrachloride and that the vapor pressure is 0.132 atm at 23°C, calculate the vapor pressure expected at 38°C.

$$p_1 = 0.132 \text{ atm} \qquad \text{at } T_1 = 296°K$$

$$p_2 = ? \text{ atm} \qquad \text{at } T_2 = 311°K$$

$$\log \frac{0.132}{p_2} = \frac{32,000}{19.15} \left(\frac{1}{311} - \frac{1}{296} \right)$$

$$p_2 = 0.250 \text{ atm}$$

■ ■ ■

6.3   Boiling points

Boiling is a special case of vaporization; it is the rapid passage of a liquid into the vapor state by means of the formation of bubbles.* A liquid boils at its *boiling point, the temperature at which the vapor pressure of the liquid is equal to the prevailing atmospheric pressure.* At the boiling point (generally abbreviated bp) the vapor pressure of the liquid is high enough that the atmosphere can literally be pushed aside to make room for creation of bubbles of vapor in the interior of the liquid. The result is that vaporization can occur at any point in the liquid. In general, a molecule can evaporate only if two requirements are met. It must have enough kinetic energy, and it must be close enough to a liquid-vapor boundary. At the boiling point, bubbling enormously increases the liquid-vapor boundary, and therefore it is necessary only that molecules have enough kinetic energy in order to escape from the liquid. Any heat added to a liquid at its boiling point is used to give more molecules sufficient energy to escape; hence, the average kinetic energy of molecules remaining in the liquid does not increase. The temperature of a pure boiling liquid remains constant.

The boiling point of a liquid depends on the pressure to which the liquid is subjected. For instance, when the atmospheric pressure is 0.921 atm, water boils at 97.7°C; at 1.00 atm it boils at 100°C. To avoid ambiguity, it is necessary to define a *standard*, or *normal*, boiling point. The normal boiling point is the temperature at which the vapor pressure of a liquid is equal to one standard atmosphere. The normal

---

* When water is heated in an open container, it is usually observed that as the liquid is warmed, tiny bubbles gradually form at first, and then at a higher temperature, violent bubbling commences. The first bubbling should not be confused with boiling. It is due to the expulsion of the air usually dissolved in water.

boiling point is the one usually listed in tables of data. It can be determined from the vapor-pressure curve by finding the temperature which corresponds to 1 atm pressure. Figure 6.4 shows that the normal boiling point of water is 100°C, that of carbon tetrachloride is 76.8°C, and that of chloroform is 61°C. In general, the higher the normal boiling point, the greater must be the attractive forces between the molecules of a liquid.*

The change of boiling point with pressure can be computed from the Clausius-Clapeyron equation if we know either $\Delta H'$ and the normal boiling point or the vapor pressures at two temperatures near the boiling point. The following example illustrates how the boiling point of water can be calculated for an altitude of 2600 m, where the atmospheric pressure is about three-fourths of a standard atmosphere.

Example 3
Given that $\Delta H'$ of water at 100°C is 40,600 J/mol, calculate the boiling point of water at a pressure of 0.750 atm.

$$p_1 = 1.00 \text{ atm} \qquad \text{at } T_1 = 373°K$$

$$p_2 = 0.750 \text{ atm} \qquad \text{at } T_2 = ?$$

$$\log \frac{1.00}{0.750} = \frac{40,600}{19.15}\left(\frac{1}{T_2} - \frac{1}{373}\right)$$

$$T_2 = 365°K = 92°C$$

■ ■ ■

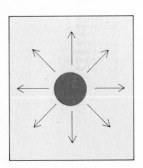

Molecule on surface (dashed line) of a liquid

6.4   Surface tension

The word *tension* comes from the Latin *tendere*, meaning "to stretch." Tension refers to the force that tends to stretch things out, and, in a sense, it can be thought of as the opposite of pressure, which tends to push things together. *Surface tension* refers specifically to the *force within a liquid* that acts parallel to the surface and *tends to stretch the surface out*. It arises from molecular interactions and, like the force that keeps a drumhead taut, has a major role in fixing the relative curvature of a liquid surface. As such it gives a direct measure of the strength of intermolecular attractions.

In the interior of a liquid, any particular molecule is subject to attractions to its neighbors. But the neighbors are distributed equally in all the directions of three-dimensional space; so there is no net force on the molecule. This is shown schematically in the bottom half of Fig. 6.6. Consider now a molecule on the surface of the liquid. The inter-

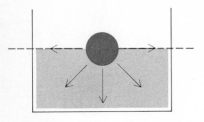

Molecule in interior of liquid

Fig. 6.6
Comparison of forces on molecule in interior and on surface of liquid.

---

* It is usually observed that in a series of similar compounds, such as CH_4, C_2H_6, and C_3H_8, the normal boiling point is highest for the compound of greatest molecular weight. Although it is tempting to explain a high boiling point in terms of large gravitational attraction between heavy molecules, gravitational attraction in reality is too small to explain the observations. A more reasonable explanation is that heavy molecules usually contain more electrons than light molecules and hence have greater van der Waals attractions.

molecular forces acting on it are unbalanced. There is first of all the net attraction into the bulk of the liquid (because all the attracting neighbors tend to be on the liquid side of the liquid-vapor interface). This attraction is, of course, why we have the liquid in the first place. Furthermore, there are forces parallel to the liquid-vapor interface (indicated by the dashed line in the top half of Fig. 6.6). These lie within the surface and tend to pull the given molecule to its neighbors within the surface. It is this latter force which is known as the surface tension. It is generally measured in dynes per centimeter and can be visualized as the surface-parallel force along a 1-cm line in the surface.

How can we measure the surface tension? There are two classical methods. In one, a wire ring, say of platinum, is lowered flat on the surface of the liquid and then gradually pulled out. Except for liquids like mercury which do not stick to platinum, the liquid generally adheres to the wire ring, and as the wire ring is pulled out of the surface, a column of liquid is dragged with it. The column sticks together because of molecule-molecule attractions, but eventually the column weight gets too great to be supported by the intermolecular attraction. At the instant that the column breaks away, one measures the force being used to pull it up, and this force gives a measure of the surface tension. Figure 6.7 gives a schematic representation of the ring-pulling method to measure surface tension.

The other classical method of measuring surface tension depends on the height of rise of a liquid in a glass capillary tube. Again, since mercury does not wet glass, it does not rise in a glass capillary. (It actually is depressed!) However, most liquids like water, alcohol, etc., tend to adhere to glass; so they form a film. This film curves upward; so the result is that there is a liquid film standing in a vertical direction. Figure 6.8 shows the geometry involved. The molecules in the liquid film attract each other, and the result is to pull liquid up into the capillary. Only a certain height can be pulled up because eventually the weight of the column gets so great that the molecule-molecule attractions cannot support it. Thus the height of the column, its size, and its weight tell us the surface tension, hence something about the molecular attractions.

Suppose we let γ (gamma) be the surface tension of a liquid in units of dynes per centimeter. If r is the radius of the capillary tube, then the dashed perimeter shown in Fig. 6.8, along which the molecule-molecule attractions are pulling upward, is $2\pi r$. Thus, the total upward force is $2\pi r\gamma$. The downward force is just the weight of the liquid column. If the column is h cm high and the liquid has a density ρ (rho) g/cm³, then the mass of the column is just $\pi r^2 h \rho$. Multiplying by g, the acceleration due to gravity, gives the total downward force. Equating the two forces, we get

$$2\pi r\gamma = \pi r^2 h \rho g$$

from which

$$\gamma = \tfrac{1}{2}rh \rho g$$

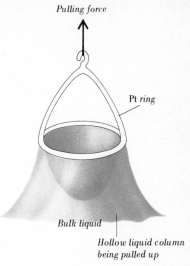

Fig. 6.7
Ring-pulling method of measuring surface tension.

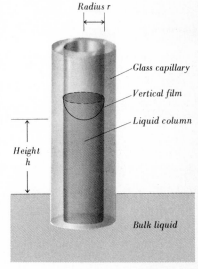

Fig. 6.8
Capillary-rise method of measuring surface tension.

Substance	Surface tension, dyn/cm
Benzene	28.89
Carbon tetrachloride	26.95
Ethyl alcohol	22.27
Methyl alcohol	22.61
Water	72.62

Fig. 6.9

Surface tension of various liquids at 20°C.

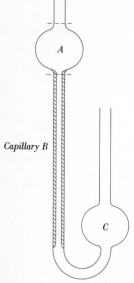

Fig. 6.10

Tube for measuring viscosity.

Thus measurement of r and h and knowledge of ρ and g enable us to determine the surface tension. Other things being equal, the higher a liquid rises in a given capillary, the bigger its surface tension. For a given liquid, the height of rise increases as the capillary is made smaller in diameter.

Values of the surface tension generally run from 20 to 40 dyn/cm at room temperature. Representative values for some liquids are listed in Fig. 6.9. As can be seen, the surface tension of water is unusually high. This, apparently, is a reflection of the large intermolecular attractions that exist between water molecules. When intermolecular attractions are extraordinarily small, surface-tension values are correspondingly low. Thus, for example, liquid helium has a surface tension of only 0.098 dyn/cm (4.3°K), and liquid nitrogen 6.2 dyn/cm (90.2°K). A rise in temperature generally decreases the surface tension.

Surface-tension effects can be very important. They explain why water soaks upward into a fine-grained material such as soil. Also, detergents and soaps owe their action mainly to their ability to lower the surface tension of water so that small droplets of oil can be stabilized.

6.5 Viscosity

Another characteristic property of liquids is *viscosity*, or *internal resistance to flow*. It is also characteristic of gases, but there the values are so low that it is hard to distinguish the small differences between the individual gases. Common liquids, such as water, are roughly 100 times as viscous as most gases.

Viscosity can be measured experimentally by a gadget such as that shown in Fig. 6.10. The liquid to be measured is sucked up into bulb A, and the time measured for it to pass through capillary B into collector C. The time required increases as the capillary length increases or the capillary diameter decreases, but, most important, it depends on the viscosity of the liquid. Water is a relatively nonviscous liquid, but benzene, chloroform, and ether are less so; the lubricating hydrocarbon oils are examples of relatively high viscosity liquids. There are no simple theories relating viscosity magnitudes to molecular structure, but in general small-molecule liquids move more rapidly than long molecules do.

Fluid flow is a complex phenomenon because the ideal kind of flow, *streamline flow*, where the sliding layers maintain their respective parallel geometries, can easily go over to *turbulent flow*, where, as the name implies, molecules get mixed up in their motion and do not stay in their respective concentric sheaths. Ideally, fluid flow in a tube would consist of a series of concentric cylindrical layers sliding past each other. The layer closest to the tube wall would be moving most slowly; the column in the middle of the tube would be moving most rapidly. The problem comes when molecules in one layer move to another, since the result is to equalize two velocities that should be different from each other. A possibly useful analogy is the jumping back and forth by passengers on two parallel trains; the interchange tends to equalize the speeds. This is the basic reason for viscosity; lateral interchange of moving molecules limits the downstream velocity.

Viscosity is generally measured in a unit called a *poise* (rhymes with "boys"), which is derived from the name of the French scientist Jean Louis Marie Poiseuille, who first studied quantitatively the flow of liquids in capillaries. One poise is one dyne-second per square centimeter, corresponding to the viscosity of a liquid which takes $8/\pi^*$ seconds for one cubic centimeter to pass through a tube one centimeter long and one centimeter in radius when there is a pressure difference of one dyne per square centimeter at the two ends of the tube. Actually the poise is too big to be a practical unit; so viscosity values are generally given in centipoises or millipoises. Figure 6.11 shows viscosity values for some typical liquids. A rise in temperature generally means a decrease in viscosity.

Substance	Viscosity, centipoises
Benzene	0.652
Carbon tetrachloride	0.969
Ethyl alcohol	1.200
Methyl alcohol	0.597
Water	1.002

Fig. 6.11

Viscosity of various liquids at 20°C.

Exercises

* 6.1 *Liquids vs. gases.* Show how kinetic molecular theory explains each of the following observations:

a Changing the pressure on a sample of liquid water from 1 to 100 atm reduces the water volume by less than 1 percent, but changing the pressure on a sample of gaseous nitrogen from 1 to 100 atm reduces the nitrogen volume by 99 percent.

b When 15 ml of liquid water is poured from a graduated cylinder to a beaker, the liquid volume remains at 15 ml, but when 15 cm^3 of hydrogen gas at STP is transferred from a 15-cm^3 bulb to a 100-cm^3 bulb, the gas volume expands to 100 cm^3.

c When a drop of red ink is carefully released at the bottom of a water sample without stirring, it takes several days for the red color to spread through the water, but a drop of brown bromine gas put into an evacuated container spreads throughout the container in a matter of seconds.

* 6.2 *Mean free path.* How do you account for the fact that the mean free path of an H_2O molecule in a sample of liquid water at 25°C is about 2×10^{-8} cm but the mean free path of an H_2O molecule in a sample of saturated water vapor at 25°C is about 1.5×10^{-6} cm?

* 6.3 *Cooling by liquids.* Explain why chloroform poured on your hands feels so much colder than water.

* 6.4 *Cooling.* Explain why 25°C liquid water poured into a Dewar

---

* The curious factor $8/\pi$ comes partly from the fact that the cross-sectional area of a capillary is πr^2 and partly from the fact that the main drag comes from an inner wall area of $2\pi r$ in circumference by one centimeter in height. Poiseuille first established on the basis of clinical studies of blood that the volume delivered is proportional to pressure, to the time, and to the fourth power of the radius, and is inversely proportional to the length of the capillary. Later theoretical work showed the proportionality constant is $\pi/8$ times the fluidity, where fluidity is just the reciprocal of viscosity. Besides inventing an apparatus for measuring viscosity Poiseuille invented the gadget for measuring blood pressure in the arteries.

Exercises

flask at 25°C becomes colder after a few hours whereas the same water poured into a beaker stays at 25°C.

*6.5 *Density.* The molecular diameter of an N_2 molecule, as deduced from the van der Waals b parameter, is 3.15×10^{-8} cm. The density of liquid nitrogen is 0.8081 g/cm³. On a hard-sphere model, what fraction of the liquid volume appears to be empty space?

Ans. 0.716

*6.6 *Surface tension.* What would be the surface tension of a liquid (density 0.876 g/cm³) which rises 2.0 cm in a capillary where water at 20°C rises 4.5 cm?

**6.7 *Evaporation.* Account for the following observation: When liquid nitrogen at 77°K is poured into a Dewar flask at 77°K, the liquid level drops faster immediately than, say, 2 or 3 h later.

**6.8 *Boiling point.* How would you account for the curious fact that when liquid nitrogen, bp 77°K, is used to provide a temperature reference point of 77°K, it helps to have a *heater* in the liquid?

**6.9 *Vapor pressure.* Suggest a molecular-structure reason why the vapor pressure at 20°C is greater for chloroform ($CHCl_3$) than for carbon tetrachloride (CCl_4). The respective values are 0.120 and 0.211 atm.

**6.10 *Water vapor pressure.* Using the data of Appendix 3, make a plot of log p versus $1/T$. What do you conclude about the heat of vaporization of water in this temperature range?

**6.11 *Vapor pressure.* If the vapor pressure of ethyl alcohol, C_2H_5OH, is 0.132 atm at 34.9°C and 0.526 atm at 63.5°C, what do you predict it will be at 19.0°C? *Ans. 0.0548 atm*

**6.12 *Vapor pressure.* If the vapor pressure of methyl alcohol, CH_3OH, is 0.0526 atm at 5.0°C and 0.132 atm at 21.2°C, what do you predict the normal boiling point will be? *Ans. 64.7°C*

**6.13 *Boiling points.* Two substances A and B have the same 0.132-atm vapor pressure at 15°C, but A has a heat of vaporization of 25,000 J/mol, whereas B, 35,000 J/mol. What will be the respective normal boiling points?

**6.14 *Boiling point vs. elevation.* At a 4000-m altitude the atmospheric pressure is about 0.605 atm. What boiling point would you expect for water under these conditions? *Ans. 86°C*

**6.15 *Surface tension.* Given that the surface tension of water is 72.62 dyn/cm at 20°C, how high should water rise in a capillary that is 1.0 mm in diameter?

**6.16 *Boiling point.* Liquid nitrogen is an excellent bath for keeping temperatures around 77°K, its normal boiling point. What pressure would you need to maintain over the liquid nitrogen if you wanted to set the bath temperature at 85°K? Heat of vaporization is about 5560 J/mol.   *Ans. 2.26 atm*

** 6.17   Surface tension. If you were assigned to tailor-make a liquid having surface tension higher than that of water, what structural features would you likely incorporate? What molecules might be likely candidates?

*** 6.18   Boiling. If a liquid is heated in a tightly sealed container, it never boils; but the same liquid when heated in an open container does boil. Explain the difference in behavior.

*** 6.19   Viscosity. In a steady-flow situation, how would each of the following factors affect the volume of liquid transferred through a capillary tube?
a   Double the time.
b   Double the radius of the capillary.
c   Double the viscosity of the liquid.
d   Double the length of the capillary.

*** 6.20   Viscosity. Why is it reasonable to expect that the viscosity of a gas would be considerably lower than that of a liquid?

Whereas the gaseous state is characterized by disorder, in that molecules of a gas are not constrained to occupy fixed positions in space, the solid state is characterized by order. Atoms in a solid are arranged in regular patterns, the existence of which simplifies the understanding of the solid state. The deciphering of the ordering pattern is a challenging problem, since the observed properties ultimately are determined by it. As a concept, ordered arrangement of atoms in the solid state is very old, being originally based on the observation that crystals show planar faces with characteristic angles between them. However, the working out of the detailed atomic arrangements was made possible only after the discovery of X rays and their application to the study of crystals. Actually there is no such thing as a perfect crystal. Imperfections are always present and, in fact, often are decisive in determining observed properties. Various solid-state devices, such as the transistor, owe their existence to defects in the solid state.

7.1 Properties of solids

Unlike a gas, which expands to occupy any accessible volume, a solid has a characteristic volume that does not change with change of container or with any but quite large changes of pressure or temperature. In other words, compared with gases solids are nearly incompressible and have low thermal coefficients of expansion. According to kinetic molecular theory, this results from the existence of strong attractive forces between closely spaced molecules. In addition, there are strong repulsive forces arising from the Pauli exclusion principle (Sec. 2.4) which keep electron clouds of molecules from penetrating each other significantly.

Another marked difference between the solid and gaseous states is in the diffusion rates. Whereas any gas can diffuse through another in a time short enough to allow easy observation, many solids diffuse so slowly as to make changes essentially imperceptible. For example, rock layers have been in contact with each other for millions of years and still retain sharp boundaries. In some solids, however, considerably more rapid diffusion has been demonstrated: e.g., some metals interpenetrate to a depth of 0.1 mm in a matter of hours at elevated temperatures that are still below the melting point. Based on the picture that solids consist of atoms in virtual contact, it is reasonable that solid-state diffusion should be a slow process. In fact, from this view one might ask why diffusion occurs at all. One principal reason is that the solids are imperfect, having, for example, vacancies where atoms or molecules should reside. Motion into the vacancies permits diffusion to occur at a rate proportional to the number of vacancies per unit volume. For many purposes, the existence of the vacancies can be ignored, but for certain properties, such as diffusion, conductivity, and mechanical strength, their presence determines the magnitude of the property.

Solids form *crystals*, definite geometric forms which are distinctive for the substance in question. The crystals show plane surfaces, called *faces*, which always intersect at an angle characteristic of the substance. As an example, sodium chloride generally crystallizes in the form of cubes with faces which intersect at an angle of 90°. When a crystal is broken, it splits, or shows *cleavage*, along certain preferred directions, so that the characteristic faces and angles result even when the material is ground to a fine powder. The same chemical substance can under different conditions form different kinds of crystals. For instance, although sodium chloride almost always crystallizes as cubes, it can also be made to crystallize in the shape of octahedra. The occurrence of different crystal forms of the same chemical is called *polymorphism*.

Not all solids are crystalline in form. Some substances, such as glass, have the solid-state properties of extremely slow diffusion and virtually complete maintenance of shape and volume but do not have the ordered crystalline state. These substances are sometimes called *amorphous solids*.

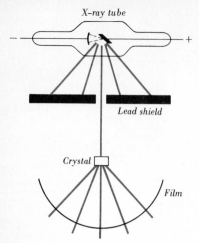

X-ray tube

Lead shield

Crystal

Film

Fig. 7.1
X-ray determination of
structure.

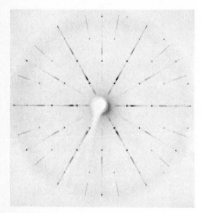

Fig. 7.2
X-ray photograph for a
typical crystal.

Chapter 7
The solid state and
changes of state
198

7.2   Determination of structure

Information about the arrangement of fundamental particles in solids can be gained from the external symmetry of crystals. However, much more information is obtained from X-ray diffraction. X rays are radiant energy much like light, but they are more energetic and have greater penetrating power (Sec. 1.6). They can be produced by bombarding a metal with energetic electrons. The energy of the electrons excites the atoms of the metal to a higher energy level. When the atoms fall from a high energy level to a low energy level, they emit energy in the form of X rays. The target metal is usually copper or molybdenum.

Figure 7.1 shows the setup for the study of crystalline materials. X rays are collimated into a beam by a lead shield with a hole in it. If sufficiently thick, the shield stops all X rays except the beam which comes through the hole. A small but well-formed crystal is mounted in the path of the X-ray beam. As the X rays penetrate the crystal, the atoms which make up the crystal scatter, or deflect, some of the X rays from their original path. The X rays are detected by a piece of photographic film at some distance from the crystal. The X-ray beam exposes the film, and when the film is developed, a spot appears at points where the beam struck it. The developed film shows not just one spot but a pattern which is uniquely characteristic of the crystal investigated.

In Fig. 7.2 is shown an X-ray photograph for a typical crystal. The big spot in the center corresponds to the main unscattered beam. (Generally a hole is cut in the film at this point so that the intense blackening that goes with the center spot is avoided.) The other spots represent a scattering of part of the original beam through various characteristic angles. The creation of many beams from one is like the effect observed when a light beam falls on a diffraction grating—a piece of glass on which are scratched thousands of parallel lines. The lines are opaque and leave, in effect, many narrow slits of unscratched surface for light to pass through. The spreading of the light wave from each slit results in interference between waves from different slits, giving rise to a spatial pattern of alternate regions of light and dark. The diffraction of X rays by crystals is similar to diffraction of light by a grating, and it suggests that crystalline materials consist of regular arrangements of atoms in space, in which the lines of atoms act like tiny slits for the X rays.

The diffraction of X rays is a complex phenomenon involving the interaction between incoming X rays and the electrons that make up atoms. We can consider X rays to consist of waves of electric (and associated magnetic) pulsations. Interaction of the incoming pulsations with the electrons in each atom produces corresponding electric (and magnetic) pulsations going out from that atom. Thus, each atom in the path of an X-ray beam receives and regenerates pulsations as waves in all directions. Consider (as shown in Fig. 7.3) two atoms side by side in an X-ray beam set into electric pulsation at the same frequency. Viewed head on (position A), the signals from the two atoms are both received in phase; so the signal from one reinforces that from the other. This need not be true when viewed at some other angle to the incident beam. For example, viewed from position B, the signals will be out of phase with each other because one signal has to travel farther than the other. If, however, as happens at position C, the path difference is equal to a whole wavelength, there will again be reinforcement of the signals.

For a real crystal consisting of many atoms regularly arranged, there will be certain angles (relative to the incident X-ray beam) at which there will be reinforcement in the emergent beam so as to produce spots on a photographic film. From the angles at which the spots appear it is possible to determine the distance between diffracting planes of atoms. The relation between the X-ray angles and the interplanar distances is given by what is called the *Bragg equation:*

$$n\lambda = 2d \sin \theta$$

Here n is a whole number (usually equal to 1), λ is the wavelength of the X-ray beam, and d is the distance between the planes of atoms that produce constructive interference along a direction that makes an angle 2θ with the direction of the incident beam.

■ The Bragg equation can be derived by using a reflection analogy, as shown in Fig. 7.4; however, it must be emphasized that X-ray diffraction is a much more complex process than simple reflection from planes of atoms. As shown in Fig. 7.4, the incident X rays make an angle θ_1 with the diffracting planes of atoms, and the reflected rays go off at the same angle θ_2 $(\theta_1 = \theta_2)$. For constructive reinforcement to occur ray 2, which arrives in phase with ray

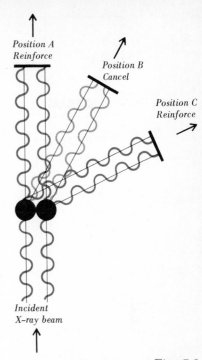

Fig. 7.3
Two-atom model for X-ray scattering.

Fig. 7.4
Reflection analogy for deriving Bragg's law.

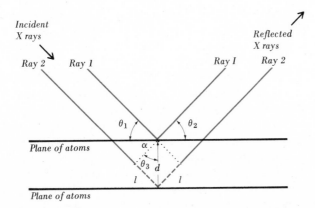

1, must also go off in phase with ray 1. Since ray 2 has a longer path length to travel, the only way in which phase matching can occur is if the extra path length (designated as $2l$ and shown dashed in Fig. 7.4) is equal to a whole number of X-ray wavelengths:

$$2l = n\lambda$$

Marking the wavefronts where the rays are in phase by dotted lines, we can see from Fig. 7.4 that the wavefronts are at right angles to the rays, just as

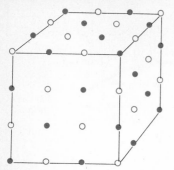

Fig. 7.5
Space lattice of NaCl.

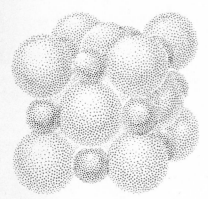

Fig. 7.6
Space-filling model of NaCl. Small spheres represent Na+ ions; large spheres, Cl− ions.

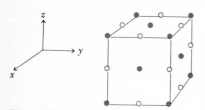

Fig. 7.7
Unit cell of NaCl.

d, the distance between planes of atoms, is at right angles to the planes of atoms. Since $\theta_1 + \alpha = 90° = \theta_3 + \alpha$, it must be true that $\theta_1 = \theta_3$. Noting that $\sin \theta_3 = l/d$, we can write $l = d \sin \theta$ and $2l = 2d \sin \theta = n\lambda$. ∎

7.3 Space lattice

A careful mathematical analysis of the spot patterns resulting from X-ray diffraction enables X-ray crystallographers to calculate the most probable positions of the particles that produce such a pattern. The process of calculation is an indirect one which involves guessing probable structures, calculating the X-ray patterns they would produce, and comparing these with experiment. The pattern of points which describes the arrangement of molecules or atoms in a crystal is known as a *space lattice*. In Fig. 7.5 is shown the space lattice of sodium chloride, NaCl.* Each of the points corresponds to the position of the center of an ion. The open circles locate the centers of the positive sodium ions, and the filled circles locate the positions of the negative chloride ions. It must be emphasized that the points do not represent the actual sodium ions and chloride ions; they represent only the positions occupied by the ion centers. In fact, in sodium chloride the ions are of different size and are practically touching each other, as shown in Fig. 7.6.

The space lattice has to be thought of as extending in all directions throughout the entire crystal. However, in discussing the space lattice it is sufficient to consider only enough of it to represent the order of arrangement. This small fraction of a space lattice, which sets the pattern for the whole lattice, is called the *unit cell*. It is defined as *the smallest portion of the space lattice which when moved repeatedly a distance equal to its own dimensions along the various directions generates the whole space lattice.* A unit cell of sodium chloride is the cube shown in Fig. 7.7. If this cube is moved through its edge length in the x direction, the y direction, and the z direction many times, the whole space lattice can be reproduced.

Several different kinds of unit-cell symmetry occur in crystalline substances. The simplest, known as *simple cubic*, is shown in Fig. 7.8a. Each point at the corner of the cube represents a position occupied by an atom or a molecule. The dashed lines marked a, b, and c represent the three characteristic directions of space, or *axes*, along which the structure must be extended to reproduce the entire space lattice. The c axis is generally oriented up and down; the b axis, left and right; the a axis, front and back. In the cubic system the directions of the a, b, and c axes are just the same as the x, y, and z directions of a cartesian coordinate system. Thus with simple cubic the points of the space lattice are arranged equally spaced along the three mutually perpendicular directions of space. Closely related to simple cubic symmetry is *body-centered cubic*, the unit cell for which is shown in Fig.

* In the strictest sense, a space lattice is concerned only with points, and all the points in a space lattice must be identical. In this sense, NaCl can be represented by two identical interpenetrating space lattices, one for the positions of the sodium ions and one for those of the chloride ions.

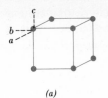
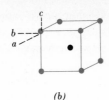
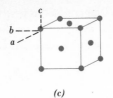

(a)   (b)   (c)

Fig. 7.8
Unit cells: (a) simple cubic,
(b) body-centered cubic, (c)
face-centered cubic.

7.8*b*. It is made up of points at the corners of a cube with an additional point in the center. In *face-centered cubic* symmetry there are points at the corners of the cube with additional points in the middle of each face, as shown in Fig. 7.8*c*.

Other kinds of symmetry are considerably more complicated. To produce *tetragonal* symmetry, the cube can be imagined to be elongated along the *c* direction. The lines of points still form right angles with each other, but the distance between points along one axis differs from that along the other two. Figure 7.9*a* shows a tetragonal unit cell. The separation of points along the *a* axis is the same as that along the *b* axis, but that along the *c* axis is different. Although it is not shown in Fig. 7.9, in *rhombic* (also called orthorhombic) symmetry, the unit cell retains mutually perpendicular edges, but the point separation is unequal in the *a*, *b*, and *c* directions. In *monoclinic* crystals the three axes *a*, *b*, and *c* are no longer perpendicular to each other. Monoclinic symmetry differs from rhombic symmetry in that the *c* axis does not make a right angle with the *ab* plane. An example is shown in Fig. 7.9*b*. In *triclinic* symmetry none of the three axes *a*, *b*, and *c* is perpendicular to any of the others. In the *hexagonal* type of symmetry, as shown in Fig. 7.9*c*, the points of the space lattice are arranged so as to form hexagons, which are stacked on top of each other. Each hexagon has an additional lattice point in its middle so as to create a rhombus-based prism, as indicated by the solid lines in Fig. 7.9*c*. For describing hexagonal symmetry, either the solid or dotted entity of Fig. 7.9*c* can be used as the unit cell.

7.4   *Packing of atoms*

The unit cells just discussed actually concern only points that locate atomic or molecular centers. Atoms are space-filling entities, and structures can be described as resulting from the packing together of representative spheres. The most efficient packing together of equal spheres, called *closest-packing*, can be achieved in two ways, each of which utilizes the same fraction (0.7406) of total space. One of these close-packing arrangements is called *hexagonal close-packed*. It can be envisioned as being built up as follows: Place a sphere on a flat surface. Surround it with six equal spheres as close as possible in the same plane. Looking down on the plane, the projection is as shown in Fig. 7.10*a*. Now form over the first layer a second layer of equally bunched spheres staggered as shown in Fig. 7.10*b* so that the second-layer spheres nestle into the depressions formed by the first-layer spheres. A third layer can now be added with each sphere directly above a

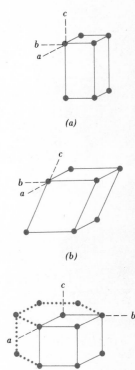

(a)

(b)

(c)

Fig. 7.9
Unit cells: (a) tetragonal,
(b) monoclinic, (c) hexagonal.

(a)

(b)

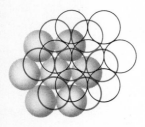

(c)

Fig. 7.10
Close-packing of spheres.

Chapter 7
The solid state and
changes of state

202

sphere of the first layer. Succeeding layers follow in alternating fashion *ababab* . . . until the hexagonal close-packed structure has been generated. Figure 7.11*a* shows how the hexagonal close-packed stacking layers are related to each other.

The other close-packed structure, called *cubic close-packed,* results if the buildup of layers *a* and *b* is the same as that described above but with the third layer added as shown in Fig. 7.10*c*. The spheres of the layer *c* (indicated by the black, open circles) are not directly above layer *a* (filled circles) or layer *b* (colored circles). For this cubic close-packed structure the sequence of layers is *abcabc* . . . Figure 7.11*b* shows how the cubic close-packed stacking layers are related to each other.

The two structures can also be described by unit cells. For hexagonal close-packing, the unit cell is like that of Fig. 7.9*c* except that a *b* layer has been inserted between the top and bottom faces of the unit cell. Thus, there is an additional lattice point at midheight in the solid-lined prism shown. For cubic close-packing the unit cell turns out to be the same as a face-centered cube. As can be seen from Fig. 7.11*b*, the body diagonal of a face-centered cube is perpendicular to the close-packed stacking layers.

Among the common materials that crystallize with hexagonal close-packing are many of the metals, such as magnesium, zinc, cadmium, and titanium. Also showing hexagonal close-packing is solid H_2, where apparently the tumbling H_2 molecules are equivalent to close-packing spheres. Cubic close-packing is shown by other metals, such as aluminum, copper, silver, and gold, as well as by simple rotating molecules such as CH_4 and HCl.

Close-packing of spheres can also be used to describe many ionic solids. For example, NaCl can be viewed as a cubic close-packed array of chloride ions with the sodium ions fitting into interstices between the chloride layers (Fig. 7.6). Interstices between layers of close-packed spheres are of two kinds, one of which (called a *tetrahedral hole*) has four spheres adjacent to it and the other of which (called an *octahedral hole*) has six spheres adjacent to it. The difference between the two kinds of holes is illustrated in Fig. 7.12, which is a projection view of two close-packed layers. The *a* layer is represented by filled balls, and the *b* layer above it by open balls. In the left part of the projection, marked by color, is shown a grouping of four adjacent balls (three from layer *a* and one from layer *b*) surrounding a tetrahedral hole. This is called a tetrahedral hole because a small atom inserted in the hole would have four neighboring atoms arranged at the corners of a regular tetrahedron. It should be recalled that a regular tetrahedron is a triangular-based pyramid in which each of the four faces is an equilateral triangle. The grouping of four spheres and its relation to a tetrahedron is shown again at the bottom left of the figure. In the right part of the layers of Fig. 7.12, also marked in color, is shown the grouping of six adjacent balls (three from layer *a* and three from layer *b*) that forms an octahedral hole. The octahedron, as indicated

Fig. 7.11
(a) Hexagonal close-packing;
(b) cubic close-packing.

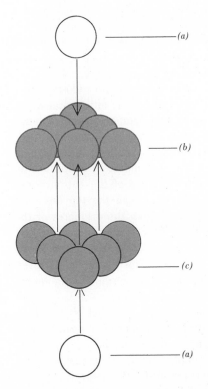

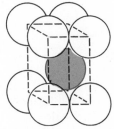

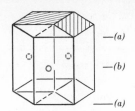

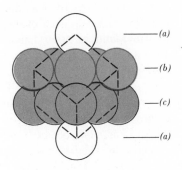

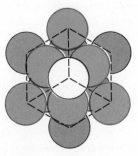

Expanded view of
stacking layers

(a)

Unit cell showing
hexagonal symmetry

Basic repeat unit of
hexagonal prism

Expanded view of
stacking layers

(b)

Edge view of stacking layers
showing face–centered
cubic unit cell

View perpendicular to
stacking layers showing
face-centered cubic
unit cell

Fig. 7.12
*Tetrahedral and octahedral
holes between layers of
close-packed spheres.*

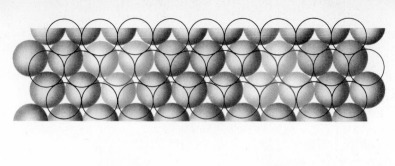

at the bottom right of the figure, is an eight-faced figure all of whose faces are equilateral triangles.

In the NaCl structure the Na^+ ions can be regarded as being in the octahedral holes created between the layers of Cl^- ions. Similarly, MgO, which has the same structure as NaCl (called the *rock-salt structure*), can be regarded as an array of close-packed oxide ions with a magnesium ion in each of the octahedral holes. The rock-salt structure is also found in other halides of Li, Na, K, and Rb and in oxides and sulfides of group II metals and of some transition metals (for example, TiO, MnO, and MnS).

The structure of Li_2O can be built up by taking a cubic close-packed array of oxide ions and inserting Li^+ into each of the tetrahedral holes. It turns out that, whereas in a close-packed arrangement of spheres there are as many octahedral holes as there are spheres, there are twice as many tetrahedral holes. This can be seen as follows: As shown at the bottom of Fig. 7.12, the tetrahedral hole indicated is formed by one sphere of the top layer *b* and three spheres of the bottom layer *a*. This gives a hole directly *under* each sphere of the top layer. In addition, there are tetrahedral holes formed from three spheres of the top layer and one from the bottom. Thus, there is also one tetrahedral hole *above* each sphere of the bottom layer. In the crystal as a whole there is one tetrahedral hole above and one below every close-packed atom, and, therefore, there are twice as many tetrahedral holes as close-packed atoms.

In the Li_2O structure, which is an example of what is called an *antifluorite* structure, every tetrahedral hole contains a positive ion. In other words, the positive ions are in holes created by the close-packing of negative ions. Other examples of this structure are the oxides, sulfides, selenides, and tellurides of Li, Na, and K. Closely related to the structure is what is called the *fluorite* structure, which is named after its most famous example, the mineral fluorite, CaF_2. This structure differs only in that the positive and negative ions have been interchanged; i.e., the negative ions (F^-) are in tetrahedral holes

*Chapter 7
The solid state and
changes of state*

204

made by the positive ions (Ca^{2+}). The fluorite structure is also found in other fluorides of group II elements and some oxides such as UO_2.

Another important solid-state structure is the *zinc blende structure*, named after the mineral ZnS. It can be visualized as consisting of a cubic close-packed array of sulfur atoms with zinc atoms disposed in one-half of the tetrahedral holes. If a hole above a particular sulfur atom is occupied, then the one below it is not. Examples of the zinc blende structure include CuCl, CuBr, CuI, AgI, and BeS, as well as the technologically important group III–V compounds. The latter are formed from an element of group III (for example, Al, Ga, and In) plus an element of group V (for example, P, As, and Sb), and they are frequently used in solid-state electric devices such as transistors and rectifiers.

In the compounds thus far discussed in this section, only the holes of one kind (i.e., tetrahedral *or* octahedral) have been utilized. More complicated structures can be built up by using both kinds of holes simultaneously. A most important example is the *spinel* structure, named after the mineral spinel, $MgAl_2O_4$. This consists of a cubic close-packed array of oxide ions with Mg^{2+} in tetrahedral holes and Al^{3+} in octahedral holes. Only one-eighth of the tetrahedral holes are occupied, and one-half of the octahedral. There are many spinels (for example, $NiAl_2O_4$, $MgCr_2O_4$, and $ZnFe_2O_4$) in which the dipositive ions (Ni^{2+}, Mg^{2+}, and Zn^{2+}) reside in the tetrahedral holes and the tripositive ions (Al^{3+}, Cr^{3+}, and Fe^{3+}) in the octahedral holes. Such compounds are referred to as *normal* spinels. There are other compounds in which the dipositive ions go to the octahedral holes and the tripositive ions are distributed half and half between the tetrahedral and the octahedral holes. Such compounds are called *inverse* spinels. Examples include $MgFe_2O_4$ and $MgIn_2O_4$, where the Mg^{2+} ions are in the octahedral holes and the Fe^{3+} or In^{3+} ions are split. Fe_3O_4, the mineral magnetite (also called lodestone), is an interesting example of an inverse spinel. It can be written $Fe(II)Fe(III)_2O_4$, corresponding to the presence of Fe^{2+} and Fe^{3+}. The Fe^{2+} ions occupy the octahedral holes, and the Fe^{3+} ions are divided between the tetrahedral and octahedral interstices. An electron can jump from Fe^{2+} to Fe^{3+} when each is in an octahedral hole. This gives rise to high electric conductivity, intense light absorption (responsible for the black color), and ferromagnetism arising from magnetic interaction between structurally equivalent ions. Information storage in the memory banks of electronic computers is largely based on manipulating the magnetic alignment of ions in spinel-type materials.

7.5 Solid-state defects

The preceding discussion implied ideal crystals. An ideal crystal is one which can be completely described by the unit cell; i.e., it contains no *lattice defects*. There are several important kinds of lattice defects. One, called *lattice vacancies*, arises if some of the lattice points are unoccupied. In other words, some of the atoms are missing. Another,

```
+ − + − + − +
− +   + − + −
+ − + − + − +
− +, −    − + −
+ − + − + − +
− + − + − + −
```
(a)

```
+ − + − + − +
− + − + − + −
+ −    − + − +
          +
−, + − + − + −
+ − + − + − +
− + − + − + −
```
(b)

Fig. 7.13
Lattice defects: (a) vacancies in NaCl, *(b) misplaced* Ag⁺ *in* AgBr.

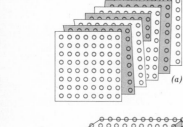

(a)

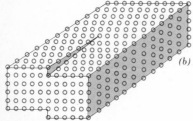

(b)

Fig. 7.14
Dislocations: (a) edge dislocation, (b) screw dislocation.

called *lattice interstitials*, arises if atoms are squeezed in so as to occupy positions between lattice points. All crystals are imperfect to a slight extent and contain defects. For example, in NaCl some of the sodium ions and chloride ions are always missing from the regular pattern (Fig. 7.13a). In silver bromide, AgBr, some of the silver ions are missing from their regular positions and are found at abnormal sites between other ions (Fig. 7.13b). Lattice vacancies occur to some extent in all crystals; their presence helps to explain how diffusion and ionic conductivity, small as they are, do occur in the solid state. Lattice interstitials are considerably less probable; they occur when small, positive ions can move into positions between normal planes of large, negative ions. The presence of interstitial Ag⁺ in AgBr and its enhanced ability to migrate are believed to be important for the formation of a photographic image when AgBr crystals are exposed to light (see Sec. 20.3).

Another kind of defect commonly found in solids, the importance of which is only now being recognized, is the *dislocation*. Dislocations are of two general types: *edge dislocations* and *screw dislocations*. These are shown in Fig. 7.14. A simple way to visualize an edge dislocation is as resulting from having a plane of atoms inserted only part way into a crystal. In a screw dislocation there is a line of atoms which represents an axis about which the crystal planes are warped, to give an effect similar to the threads of a screw. In the latter case, what would be a round trip about the screw axis results in a displacement to the crystal plane below. The points where edge or screw dislocation lines emerge to the surface of a crystal represent points of strain and enhanced chemical reactivity. Etching of metal crystals by acids occurs preferentially at such points. Current research aimed at producing materials more resistant to corrosive attack and of greater mechanical strength involves elimination of dislocations and other solid-state defects.

In addition to the defects arising from structural imperfections are defects of a more chemical nature. These are associated with the presence of chemical impurities, which may be there accidentally or might have been deliberately introduced. Such impurities can drastically change the properties of materials, and hence their controlled introduction is being exploited in producing new materials with desirable combinations of properties. As an example of how properties can be modified by impurities, it might be noted that the addition of less than 0.1% CaCl₂ to NaCl can raise the conductivity by 10,000 times. This comes about as follows: In the mixed crystal the Cl⁻ ion positions are unchanged, but the Ca²⁺ ions occupy positions that would normally be occupied by Na⁺. Because of the requirement for electric neutrality, each insertion of Ca²⁺ for Na⁺ means there will have to be creation of a lattice vacancy. The vacancy allows for ion mobility and hence increased conductivity.

A more practical utilization of impurity defects is made for Ge and Si crystals. The electric conductivity of these group IV elements in the pure state is extremely low. However, on addition of trace amounts of elements from either group III or group V the conductivity is greatly enhanced. Both Ge and Si have the diamond structure, as will be shown in Fig. 7.17. Each atom is bonded to four neighbors by four single covalent bonds. These require all four outer electrons of each group IV atom. In the pure elements Ge and Si there are no extra electrons, and therefore there is no conductivity (except at very high temperatures). When, however, a group V element—P, As, Sb, or Bi—is substituted for a Ge or Si atom, an extra electron above that required for forming four covalent bonds is introduced. This extra electron acts somewhat like a conduction electron in metals; hence, an As-doped Ge or Si crystal exhibits marked conductivity. If, on the other hand, a group III element—B, Al, Ga, or In—is substituted for a Ge or Si atom, an electron deficiency in the covalent bonding network is introduced. Such an electron vacancy is not confined to the impurity-atom site, but can move through the structure as another nearby electron moves in to fill it. The vacancy moves to the neighbor; and so on. In this way, the introduction of electron vacancies, or *holes* as they are sometimes called, allows electron motion to occur where it normally would not.

As will be discussed in Sec. 21.2, there is a class of materials called *semiconductors* for which the conductivity generally increases as the temperature is raised. This is opposite to the behavior of metals. Impurity-doped germanium and silicon act as semiconductors because there is weak binding of the excess electron or the hole to the impurity center responsible for it. Additional energy, as supplied by a temperature rise, is needed to free the electron or hole sufficiently for conductivity. Semiconductors in which electric transport is mainly by excess electrons are called *n* type (*n* for negative), and those in which electric transport is mainly by holes are called *p* type (*p* for positive). It must be emphasized that both *n*- and *p*-type semiconductors are electrically neutral. The electron excess or deficiency is with respect to that required for covalent bonding, not with respect to electric neutrality. To illustrate, substitution of As for Ge introduces not only one additional electron but also one additional unit of positive nuclear charge.

An interesting application of impurity semiconductors results from combination of *n*- and *p*-type materials to form a junction, the so-called "*n–p*" junction. This device can pass electric current more easily in one direction than in the reverse, and hence it can act as a *rectifier* for converting alternating current to direct current. Figure 7.15 shows schematically why there should be a difference in the ease of passing current in the two directions. The left side of the junction is *n* type, and the minuses represent the superfluous electrons arising from the group V impurity centers; the right side of the junction is *p* type, and the pluses represent the electron holes arising from the group III impurity centers. When an external voltage is applied (as shown in Fig. 7.15*a*) so as to favor motion of electrons from left to right and motion of holes from right to left, conductivity readily occurs. At the interface between the *n*- and *p*-type zones, electrons coming from the left annihilate holes coming from the right. Conductivity does not stop because the external voltage acts to supply more *n* carriers on the left

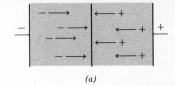

(a)

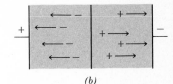

(b)

Fig. 7.15
Behavior of n-p junction when imposed voltage is reversed in sign.

and p carriers on the right. On the other hand, when the external voltage is reversed (as shown in Fig. 7.15b) so as to separate the n and p carriers, conductivity stops. There is no easy way in which new carriers can be created at the n-p interface.

Finally, we might note as a special kind of solid-state defect deviation from stoichiometry, or formation of nonstoichiometric compounds. Classic examples include $Cu_{1.87}S$, $MnO_{1.95}$, TiO_x (where x can range from 0.85 to 1.18), and Na_xWO_3 (where x can range from 0.3 to 0.98). In these nonstoichiometric materials there are generally vacancies that can be populated by an excess or deficiency of one element over what is required by simple rules of stoichiometry. For example, TiO is cubic with a rock-salt structure of Ti^{2+} and O^{2-} except that 15 percent of each kind of site is vacant. Addition of either element will change the Ti/O ratio by 15 percent in either direction. One striking result of deviations from stoichiometry is the change in color that occurs, for example, in sodium halide crystals heated in sodium vapor. Crystals of various colors result as the sodium-to-halide ratio increases to values such as $1.001:1$. It is believed that the sodium enters the structure of NaX to form an Na^+ ion that occupies a normal positive-ion site plus an electron that is trapped at a negative-ion site. Such trapped electrons are sometimes referred to as F centers (from the German *Farben,* or "color").

7.6   Bonding in solids

Often, instead of classifying solids by the symmetry of their arrangements or their mode of packing, it is more useful to classify them by the types of units that occupy the lattice points and their modes of binding. There are four such classifications of crystals: *molecular, ionic, covalent,* and *metallic.* Figure 7.16 lists for each classification the nature of the units that occupy the lattice points, the forces that bind these units together, the characteristic properties of the solids, and some typical examples.

Molecular solids are those in which the lattice points are occupied by molecules. In a molecular solid the bonding *within* the molecule is covalent and, in general, is much stronger than the bonding *between* the molecules. The bonding between the molecules can be of two types, dipole-dipole or van der Waals. Dipole-dipole attraction is encountered in those solids consisting of polar molecules. In the case of ice, for example, the negative end of one H_2O molecule attracts the positive end of a neighboring molecule. van der Waals attractions (Sec. 5.12) are present in all molecular solids. Because the total intermolecular attraction is generally small, molecular crystals usually have low melting temperatures. Furthermore, molecular substances are usually quite soft because the molecules can be easily displaced from one site to another. Finally, they are poor conductors of electricity because there is no easy way for an electron associated with one molecule to jump to another molecule. Most substances which exist as gases at room temperature form molecular solids.

Chapter 7
The solid state and
changes of state

208

	Molecular	Ionic	Covalent	Metallic
Units that occupy lattice points	Molecules	Positive ions Negative ions	Atoms	Positive ions in electron gas
Binding force	van der Waals Dipole-dipole	Electrostatic attraction	Shared electrons	Electric attraction between $+$ ions and $-$ electrons
Properties	Very soft Low melting point Volatile Good insulators	Quite hard and brittle Fairly high melting point Good insulators	Very hard Very high melting point Nonconductors	Hard or soft Moderate to very high melting point Good conductors
Examples	H_2 H_2O CO_2	NaCl KNO_3 Na_2SO_4 $MgAl_2O_4$	Diamond, C Carborundum, SiC Quartz, SiO_2	Na Cu Fe

Fig. 7.16
Classification of solids.

In an *ionic solid* the units that occupy the lattice points are positive and negative ions. For example, in Na_2SO_4 some of the lattice points are occupied by sodium ions, Na^+, and the others by sulfate ions, SO_4^{2-}. The forces of attraction are those between a positive and a negative charge and are high. Hence ionic solids usually have fairly high melting points, well above room temperature. Sodium sulfate, for example, melts at 884°C. Also, ionic solids tend to be brittle and fairly hard, with great tendency to fracture by cleavage. In the solid state, the ions are generally not free to move; therefore, these ionic substances are poor conductors of electricity. However, when melted, they become good conductors.

In a *covalent solid* the lattice points are occupied by atoms, which share electrons with their neighbors. The covalent bonds extend in fixed directions; so the result is a giant interlocking structure. The classic example of a covalent solid is diamond, in which each carbon atom is joined by pairs of shared electrons to four other atoms, as shown in Fig. 7.17. Each of these carbon atoms in turn is bound to four carbon atoms, etc., giving a giant three-dimensional molecule. In any solid of this type the bonds between the individual atoms are covalent and, usually, are quite strong. Substances with covalent structures generally have high melting points, are quite hard, and frequently are poor conductors of electricity.

In a *metallic solid* the points of the space lattice are occupied by positive ions. This array of positive ions is permeated by a cloud of highly mobile electrons derived from the outer atomic shells. In solid sodium, for example, the sodium ions are arranged in a body-centered cubic pattern with a cloud of electrons, or *electron gas,* as it is often called, arising from the contribution by each neutral sodium atom of its $3s$ outermost electron. The electron gas belongs to the whole crystal, for the electrons are waves extending over the whole crystal. A metal

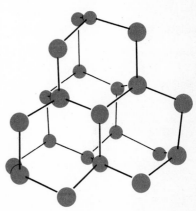

Fig. 7.17
Diamond structure.

crystal like that of sodium is held together by the attraction between the positive ions and the cloud of negative electrons. In other metals such as tungsten, there is also covalent binding between the positive ions superimposed on the ion-to-electron gas attraction. Because electrons can wander at will throughout the metal, a metallic solid is characterized by high electric and thermal conductivity. The other properties of metallic solids vary widely. Sodium, for example, has a low melting point; tungsten has a very high melting point. Sodium is soft and can be cut with a knife; tungsten is very hard.

What does the term "molecule" mean in regard to these various solids? In a molecular crystal, for example, solid CO_2, it is possible to distinguish discrete molecules. Each carbon atom has two relatively close oxygen atoms as neighbors, and all other atoms are at considerably greater distances. In an ionic substance, such as NaCl, this is not true. Each sodium ion is equally bound to its neighboring six chloride ions, as was shown in Fig. 7.6. The six neighboring chloride ions must all be considered as belonging equally to the same aggregate as the original sodium ion. But each chloride ion in turn is bonded to five other sodium ions besides the original one; so they also must be counted as part of the aggregate. Actually, all the ions in the whole crystal belong to the same aggregate, or giant molecule. A similar situation occurs in metallic and covalent crystals, in which all the ions or atoms are bound together as one giant aggregate. The term "molecule" is not particularly useful for ionic, metallic, or covalent solids.

7.7 Crystal energies

The magnitude of the attractive forces operative in crystals can be gauged by the *crystal energy* (also called *lattice energy*). This is the amount of energy required to convert one mole of material from the solid state to a gaseous state in which the gaseous species are the same as the units that occupy the lattice points. In the case of a molecular solid the crystal energy corresponds to the *energy of sublimation*, i.e., the energy needed to produce one mole of molecular gas. In the case of ionic solids the crystal energy is the energy needed to separate the positive and negative ions from one mole of solid into a gaseous mixture of positive and negative ions. The crystal energy of a covalent solid is usually considered to be that needed to form an atomic gas. In the case of a metal the problem is somewhat different in that positive ions occupy the original lattice points but the final gaseous state consists of neutral atoms.

Some typical values of the crystal energy are listed in Fig. 7.18. As can be seen, the values for the molecular substances, such as CO_2, are very low, indicating small intermolecular attractions. The largest values are generally found for the ionic solids.

■ These are especially interesting because they can be calculated in a rather simple way. Consider the case of NaCl. Each Na^+ in the crystal is attracted

Chapter 7
The solid state and
changes of state

210

Ne	2.5	NaCl	770	C	710
CO_2	23.4	MgO	3920	SiO_2	1720
Cl_2	31.0	CaF_2	2610	Na	105
H_2O	43.5	$MgAl_2O_4$	19,800	W	840

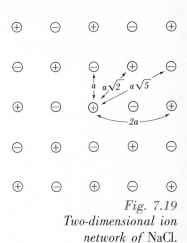

Fig. 7.18
Observed crystal energies in kilojoules per mole.

by every Cl⁻ and is repelled by every other Na⁺. All we have to do is to sum up over all the attractions and subtract out all the repulsions. This is easy to do because, given the fact that the ionic arrangement is regular and cubic, we can calculate precisely how many ions there are of a given kind at such and such a distance from the center of calculation. How this comes about can be seen by considering Fig. 7.19, which gives a two-dimensional representation of the NaCl arrangement. As can be seen, if a is the nearest distance between Na⁺ and Cl⁻, then for each Na⁺ there are in the two-dimensional plane four Cl⁻ at distance a, four Na⁺ at distance $a\sqrt{2}$, eight Cl⁻ at distance $a\sqrt{5}$, four Na⁺ at $2a$, four Na⁺ at $2a\sqrt{2}$, etc. In three dimensions, the number of ions in the successive layers is, of course, much greater. Given that the potential energy of an ion pair q_1 and q_2 separated by distance r is just q_1q_2/r, we can write the total potential energy of an NaCl crystal as the sum of a series of terms. In three dimensions we have around a given Na⁺ six Cl⁻ at distance a giving attractive potential energy of $-6q_1q_2/a$, twelve Na⁺ at distance $\sqrt{2}a$ giving repulsive potential of $+12q_1q_2/\sqrt{2}a$, eight Cl⁻ at distance $\sqrt{3}a$ giving attractive potential of $-8q_1q_2/\sqrt{3}a$, and so on. The series looks like this:

$$\frac{q_1q_2}{a}\left(-\frac{6}{\sqrt{1}}+\frac{12}{\sqrt{2}}-\frac{8}{\sqrt{3}}+\frac{6}{\sqrt{4}}-\frac{24}{\sqrt{5}}+\cdots\right)$$

There are trick ways to evaluate the complete series, but the numerical result is 1.748. In other words, the net effect of all the other ions on a particular ion in the NaCl structure can be represented by multiplying the attraction energy between an Na⁺ and Cl⁻ pair by 1.748. The number 1.748 is called the *Madelung constant,* after the German physicist who first evaluated it.

As indicated in Sec. 3.2, the attraction energy of an ion pair q_1 and q_2 separated by distance r is $1.44q_1q_2/r$ eV, where q_1 and q_2 are the unit charges and r is in nanometers. In solid NaCl the ionic charges are unity, and the internuclear spacing is 0.281 nm from which it follows the attraction energy is 1.44/0.281, or 5.12, eV. To convert electron volts per ion pair to kilojoules per mole requires multiplication by 96.49. This gives 494 kJ per mole of ion pairs. Putting these into the solid state increases the attraction energy to 494(1.748), or 864, kJ/mol. This calculation is valid only for incompressible spheres with the electric charge at the center. Realizing that we are dealing with atoms that can be compressed and electrically distorted to produce van der Waals attractions, we need to correct for these effects. There are several ways of estimating the correction terms resulting from these effects, but they all agree that the simple calculation above gives a result that is about 10 percent too great. When the 864 kJ is reduced by this 10 percent factor, the agreement with the experimental value of 770 kJ as given in Fig. 7.18 is not bad. ∎

Fig. 7.19
Two-dimensional ion network of NaCl.

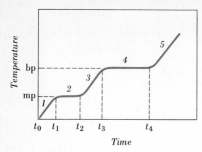

Fig. 7.20
Heating curve.

7.8 Heating curves

When heat is added to a solid, the solid warms up. However, when it reaches the melting point, even though heat may continue to be added, the solid is observed no longer to warm up but to stay at a fixed temperature while the melting process occurs. A similar phenomenon occurs at the boiling point of a liquid. The addition of heat to a boiling liquid does not raise the temperature but is used to convert liquid to gas. Why is it that at some temperatures addition of heat is accompanied by an increase in temperature and at other temperatures, not? To answer this question, we look at the general problem of adding heat to a sample of substance that starts at absolute zero as an ordered solid, melts to give the less ordered liquid, and finally boils to produce the completely disordered gaseous state. The temperature variations which accompany these changes of state are represented in Fig. 7.20. The curve shown is a *heating curve* corresponding to the uniform addition of heat to an initially solid substance. Since heat is added at a constant rate, distance on the time axis is also a measure of the amount of added heat.

At time t_0 the temperature is absolute zero. As heat is added, each particle vibrates back and forth about a lattice point, which thus represents only the center of this motion. As more heat is added, the vibration becomes more vigorous. Though no change is visible, because the amplitude of vibration is so small, the crystal progressively becomes slightly less ordered. The heat added increases the average kinetic motion of the particles. Since temperature measures average kinetic energy, temperature rises along portion 1 of Fig. 7.20. This continues until the melting point of the substance is reached.

At the melting point (abbreviated mp) the vibration of particles has become so energetic that any added heat serves to loosen the binding forces that keep neighboring particles in an ordered array. Consequently, from time t_1 to t_2, added heat goes not to increase the average kinetic energy but to increase the potential energy of the particles because of work done against the attractive forces. During this period, there is no change in average kinetic energy; so the substance stays at the same temperature. During the time interval t_1 to t_2 the amount of solid gradually decreases, and the amount of liquid increases. The *temperature at which solid and liquid coexist at equilibrium* is defined as the *melting point* of the substance.

The amount of heat necessary to melt one mole of solid is called the *molar heat of fusion*. It gives a measure of the difference in heat content between the solid and liquid states and is equal to the amount of heat energy that must be added to overcome the extra attraction that exists in the solid between the particles at the lattice sites. At the normal melting point, the heat of fusion of NaCl is 30.33 kJ/mol, and that of H_2O is 6.02 kJ/mol, values that reflect the greater attractive forces in NaCl.

Eventually (at time t_2), sufficient heat has been added to tear all the particles from the crystal structure. Along portion 3 of the curve,

Chapter 7
The solid state and
changes of state

212

added heat again increases the average kinetic energy; so the temperature rises. This continues until the boiling point (bp) is reached. At the boiling point added heat is used to overcome the attraction of one particle for its neighbors in the liquid. Along portion 4 of the curve there is an increase in the potential energy of the particles but no change in their average kinetic energy. During the time interval t_3 to t_4, the liquid sample is converted to gas. Finally, after all the liquid has been boiled off, added heat again raises the average kinetic energy of the particles; this is shown by the rising temperature along portion 5.

The amount of heat necessary to vaporize one mole of liquid is called the *molar heat of vaporization*. This quantity gives a measure of the attractive forces characteristic of the liquid. At the normal boiling point, the heat of vaporization of water is 40.7 kJ/mol, and that of chloroform is 29.5 kJ/mol. These values support the notion that attractive forces between water molecules are greater than those between chloroform molecules.

7.9 Cooling curves

The *cooling curve* results when heat is removed at a uniform rate from a substance. For a pure substance that is initially a gas, the temperature as a function of time looks like the curve shown in Fig. 7.21. As heat is removed from the gas, the temperature of the gas drops along the line marked g. During this time, the average kinetic energy of gas particles decreases in order to compensate for the removal of energy to the outside. This slowing down proceeds until the particles are so sluggish that the attractive forces become dominant.

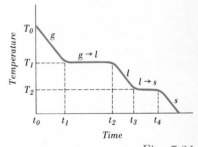

Fig. 7.21
Cooling curve.

At t_1 the particles coalesce to form a liquid. In the liquefaction process, particles leave the gas and enter the liquid state. Since energy is required to take a particle from the liquid to the gas state, the reverse process, in which a particle is taken from the gas to the liquid, releases energy. This decrease of potential energy on condensation supplies heat, which compensates for that being removed from the system. Thus, as liquefaction proceeds, the temperature does not fall; the particles on the average do not slow down in their motion. As a result, the gas and liquid are both at the same temperature, and the average kinetic energy of the particles in both phases is the same. During the time interval t_1 to t_2 the temperature remains constant at T_1, the *condensation*, or *liquefaction*, temperature.

At time t_2 all the gas particles have condensed into the liquid state. Continued removal of energy causes the particles to slow down further. The average kinetic energy decreases, and the temperature drops, as is shown along the line marked l. This drop continues until t_3, when the liquid begins to crystallize.

In crystallization the particles line up in a definite geometric pattern, and as they go from the liquid state to the solid state, their freedom of motion is diminished. As each succeeding particle is attracted into position to form the crystal structure, its potential energy drops. Thus, the removal of heat energy to the outside is compensated for by the energy available from a decrease in potential energy, and the average kinetic energy stays constant during the process. At the crystallization temperature the motion of the particles is not on the average slower in the solid phase than it is in the liquid phase, but it is a more restricted motion. From time t_3 to t_4 temperature remains constant as the liquid converts to solid. When all the particles have crystallized, further removal of heat drops the temperature, as shown, along the final part s of the curve.

The cooling curve is just the reverse of the warming curve. The temperature at which gas converts to liquid (liquefaction point) is the same as the temperature at which liquid converts to gas (boiling point). Similarly, the temperature at which liquid converts to solid (freezing point) is the same as the temperature at which solid converts to liquid (melting point).

7.10 Supercooling

Most cooling curves are not quite so simple as the one shown in Fig. 7.21. Complications usually occur on the portions of the cooling curve corresponding to transition from gas to liquid and from liquid to solid. Figure 7.22 shows the situation for the liquid-to-solid transition. Instead of following the dashed, flat portion, the temperature follows the dip. The liquid does not crystallize at the freezing point but instead *supercools*. Supercooling arises in the following way: The particles of a liquid have little recognizable pattern and move around in a disordered manner. At the freezing point they should line up in characteristic crystalline arrangement, but only by chance do they start crystallizing correctly. Often they do not snap into the correct pattern immediately, and when heat continues to be removed from the system without crystallizations occurring, the temperature falls below the freezing point. Particles continue moving through various patterns until enough of them hit the right one. Once the correct pattern has been built up to sufficient size, other particles very rapidly crystallize on it. When the multitude of particles crystallizes simultaneously, potential energy is converted to kinetic energy faster than energy is being removed to the outside; so the effect is to heat up the sample. The temperature *increases* until it coincides with the freezing-point temperature. From there on, the behavior is normal.

Supercooling may be reduced by two simple methods: One is to stir the liquid as vigorously as possible. This apparently increases the chance of forming the right crystal pattern. The other method involves introducing a seed crystal to provide a proper structure on which further crystallization can occur.

Some substances never crystallize in cooling experiments, but instead they remain permanently in the *undercooled*, or *supercooled*, state. Such substances are frequently called *glasses*, after their most famous example, but the term *amorphous material*, or *vitreous material*, is increasingly preferred. Amorphous materials owe their existence to the fact that supercooled atoms may be cooled so far that they are trapped in

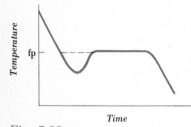

Fig. 7.22
Supercooling.

Chapter 7
The solid state and
changes of state
214

a disordered arrangement where they have relatively so little kinetic energy that they cannot move into the lower-energy, ordered array.

Amorphous materials are quite common. They include, besides glass, many plastics, such as polyethylene, vinyl polymers, and Teflon. They have many of the properties of solids, but their X-ray pictures are quite different from the X-ray pictures of crystalline solids. As discussed previously, an X-ray picture of a solid gives an orderly pattern of spots corresponding to diffraction from different planes of atoms. In an amorphous material there are no planes of atoms from which to get diffraction. Instead of a spot pattern, the X-ray pictures show concentric rings, like those for a liquid. The existence of these rings indicates that there is a certain amount of order, but it is far from perfect. Another indication that amorphous materials are not true solids is their behavior on being broken. Instead of showing cleavage with formation of flat faces and characteristic angles between faces, they generally break to give what are called *conchoidal* fractures. These are shell-shaped depressions such as are observed on the chips of a broken bottle.

7.11 Solid-gas equilibrium

As with a liquid, the particles of a solid can escape into the vapor phase to establish vapor pressure. The particles of a solid do not all have the same energy. There is a distribution in which most of the particles have energy near the average but some have less while others have more. Those particles which at any one time are of higher than average energy and are near the surface can overcome the attractive forces of their neighbors and escape into the vapor phase. If the solid is confined in a closed container, eventually there will be enough particles in the vapor phase that the rate of escape is equal to the rate of return. A dynamic equilibrium is set up; in it there is an equilibrium vapor pressure characteristic of the solid. Since the escaping tendency of the particles depends on the magnitude of the intermolecular forces in the particular solid considered, the equilibrium vapor pressure differs from one substance to another. If the attractive forces in the solid are small, as with a molecular crystal such as solid hydrogen, the escaping tendency is great, and vapor pressure is high. In an ionic crystal, on the other hand, the binding forces are usually large; so the vapor pressure is low.

The vapor pressure of solids also depends on temperature. The higher the temperature, the more energetic the particles, and the more easily they can escape. The more that escape, the higher the vapor pressure. Quantitative measurements of the vapor pressures of solids can be made in the same way as for liquids. Figure 7.23 shows how the vapor pressure of a typical substance changes with temperature. At absolute zero the particles have no escaping tendency; so the vapor pressure is zero. As the temperature is raised, the vapor pressure rises. It rarely gets to be very high before the solid melts. Above the melting point the vapor-pressure curve is just that of the liquid.

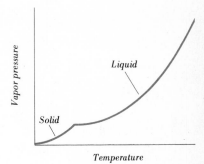

Fig. 7.23

Temperature variation of vapor pressure.

In Fig. 7.23 any point along the portion of the curve marked "Solid" corresponds to an equilibrium between vapor and solid. The number of particles leaving the solid is equal to the number returning. When the temperature is raised somewhat, more particles shake loose from the crystal per unit time than return. This causes a net increase in the concentration of particles in the vapor phase, which means a rise in vapor pressure. This in turn causes an increased rate of condensation to the solid. Eventually, equilibrium is reestablished with a higher vapor pressure at the new temperature.

The behavior of an equilibrium system when it is upset by an external action is the subject of a famous principle enunciated by Henry-Louis Le Châtelier in 1884. Le Châtelier's principle states that *if a stress is applied to a system at equilibrium, then the system readjusts, if possible, to reduce the stress.* Raising the temperature of a solid-vapor equilibrium system amounts to applying to the system a stress in the form of added heat. Since the conversion from solid to gas is endothermic (uses up heat),

$$\text{Solid} + \text{heat} \rightleftharpoons \text{gas}$$

the stress of added heat can be absorbed by converting some of the solid to gas. So, on addition of heat, the system moves to a state of higher vapor pressure.

At the point where the vapor-pressure curve of a liquid intersects the vapor-pressure curve of the solid (i.e., where the vapor pressure of the solid equals the vapor pressure of the liquid), there is simultaneously an equilibrium between solid and gas, between liquid and gas, and between solid and liquid. This point of intersection at which *solid, liquid, and gas coexist in equilibrium* with each other is called the *triple point.* Every substance has a characteristic triple point fixed by the nature of the attractive forces between its particles in the various phases. The triple-point temperature of H_2O is $273.16°K$ (or $0.01°C$), and the triple-point pressure is 0.00603 atm. As was noted in Sec. 5.2, the triple-point temperature of H_2O is not quite the same as its normal melting point, $273.15°K$. This difference comes about because the normal melting point is defined at one atmosphere pressure whereas at the triple point the only pressure exerted is the vapor pressure of the substance. As we shall see in the next section, the melting point of H_2O decreases under increasing pressure.

7.12 Phase diagrams

The relation between the solid, liquid, and gaseous states of a given substance as a function of the temperature and pressure can be summarized on a single graph known as a *phase diagram*. Each substance has its own particular phase diagram which can be worked out from experimental observations at various temperatures and pressures. Figure 7.24 gives the phase diagram of H_2O. On this diagram, which has coordinates of pressure and temperature, the substance H_2O is represented as being in the solid, liquid, or gaseous state, depending on the temper-

Chapter 7
The solid state and
changes of state

216

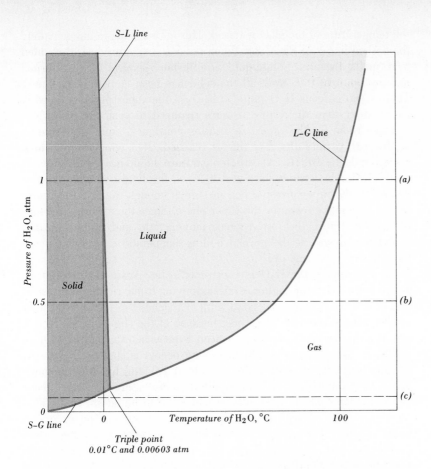

Fig. 7.24
Phase diagram of H_2O
(*scale of axes is somewhat distorted*).

S-L line

S-L line

L-G line

Pressure of H_2O, atm

1

0.5

0

Liquid

Solid

Gas

(a)

(b)

(c)

S-G line

0

Temperature of H_2O, °C

100

Triple point
0.01°C and 0.00603 atm

ature and pressure at which it is held. Each of the three differently labeled regions corresponds to a one-phase system. For all values of pressure and temperature falling inside such a single-phase region, the substance is in the state specified. For example, at a pressure of 0.5 atm H_2O at -10°C is in the solid state, at $+10$°C in the liquid state, and at $+100$°C in the gas state. The lines which separate one region from another are equilibrium lines, representing an equilibrium between two phases. In the diagram the *S-L* line represents equilibrium between the solid and liquid; the *L-G* line, equilibrium between liquid and gas; and the *S-G* line, equilibrium between solid and gas. The intersection of the three lines corresponds to the triple point, where all three phases are in equilibrium with each other.

The usefulness of a phase diagram can be illustrated by considering the behavior of H_2O when heat is added at a constant pressure. This corresponds to moving across the phase diagram from left to right. We distinguish three representative cases:

1 Pressure of the H_2O *is kept at 1.0 atm.* The experiment is this: A chunk of ice is placed in a cylinder so as to fill the cylinder completely. A piston resting on the ice carries a weight which corresponds to 1 atm of pressure. The H_2O starts as a solid. As heat is added, the temperature of the H_2O is raised. This corresponds to moving along the dashed line *a* in Fig. 7.24. When the *S-L* line is reached, the added heat melts the ice. Solid-liquid equilibrium persists at the normal melting point of 0°C until all the solid has been converted to liquid. There is no gaseous H_2O thus far because the vapor pressure of solid ice is lower than the 1 atm pressure required to push the piston out to make room for the vapor. As heating continues, liquid H_2O warms up from 0°C until the *L-G* line is reached, at a temperature which corresponds to 100°C. At this temperature liquid-gas equilibrium is established. The system stays at 100°C as the liquid converts to gas. At 100°C the vapor pressure becomes great enough to move the piston out so as to make room for the vapor phase. Since the external pressure is fixed at 1 atm, the piston keeps moving out, and liquid converts completely to gas at the normal boiling temperature. From then on, the gas simply warms up.

2   Pressure of the H_2O is kept at 0.5 atm. Again the H_2O starts as the solid. The temperature is raised, moving to the right along dashed line *b*. The H_2O stays as a solid until it reaches the temperature that corresponds to melting. This time, because of the slight tilt of the *S-L* line toward the left (the tilt has been somewhat exaggerated in Fig. 7.24), the temperature at which melting occurs is slightly higher. At 0.5 atm ice melts not at 0°C (as it would at 1.0 atm) but slightly above 0°C. The difference in melting-point temperature is only about 0.005°C. After all the ice has been converted to liquid at +0.005°C, further addition of heat warms the liquid up until boiling occurs at the *L-G* line. Boiling occurs when the vapor pressure of the water reaches 0.5 atm. The temperature at which this happens is 82°C, considerably lower than the normal boiling point of 100°C. Above 82°C at 0.5 atm only gaseous water exists.

3   Pressure of the H_2O is kept at 0.001 atm. If the pressure exerted by the H_2O is kept at 0.001 atm by suitable means, such as regulated pumping, the H_2O can exist only along the dashed line marked *c*. As the temperature is raised, solid H_2O warms up until it reaches the *S-G* line. Solid-gas equilibrium is established; solid converts to gas. For 0.001 atm, this would happen at −20°C. When all the solid has evaporated at this temperature, the temperature of the gas rises. There is no melting in this experiment and no passage through the liquid state.

An interesting aspect of the H_2O phase diagram is that the *S-L* line, representing equilibrium between solid and liquid, tilts to the left with increasing pressure. This is unusual; for most substances the *S-L* line tilts to the right. The direction of the tilt is important since it tells whether the melting point rises or falls with increased pressure. In the case of H_2O, as pressure is increased (this means moving up on the phase diagram), the temperature at which solid and liquid coexist decreases (toward the *left* on the phase diagram). The melting-point decrease is 0.01°C/atm.

The lowering of the ice melting point by increased pressure is predicted by Le Châtelier's principle. The argument goes as follows: The

Chapter 7
The solid state and
changes of state

218

density of ice is 0.9 g/cm^3; the density of water is 1.0 g/cm^3. In the solid state 1 g of H_2O occupies a volume of 1.1 cm^3; in the liquid state 1 g of H_2O occupies 1.0 cm^3. Thus, a given mass of H_2O occupies a larger volume as solid than as liquid. An equilibrium system consisting of water and ice at 1 atm will be at the normal melting point of $0°C$. If the pressure on the H_2O is now increased, there is a stress, which the system can relieve by shrinking in volume. It can shrink in volume by converting some of the ice to water; hence, melting is favored. But melting is an endothermic process; it requires heat. If the system is insulated, the only source of heat is the kinetic energy of the molecules. Consequently, the molecules must slow down, and the temperature must drop. The end result is that solid ice and liquid water under increased pressure coexist at lower temperature.

Another phase diagram of interest is that of carbon dioxide, CO_2. In general appearance, as shown in Fig. 7.25, it is similar to that of H_2O. However, as is more normal, the solid-liquid equilibrium line tilts to the right instead of to the left; i.e., the melting point of CO_2 rises with increased pressure. The triple-point pressure of CO_2 is 5.2 atm; the triple-point temperature is $-57°C$. Since the triple-point pressure is considerably above normal atmospheric pressure, the liquid phase of CO_2 is not generally observed. In order to liquefy CO_2, the pressure would have to be higher than 5.2 atm. At 1 atm, just the solid and gas exist. When solid CO_2 is used as a refrigerant in the form of Dry Ice, the conversion of solid to gas normally occurs at $-78°C$.

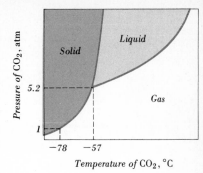

Fig. 7.25
Phase diagram of CO_2.

7.13 Entropy, free energy, and spontaneous change

It is generally taken for granted that any physical system tends toward a state of lower energy, yet when we look at a piece of ice at room temperature, we see it melt, i.e., spontaneously go to a state of higher energy. Of course, the reason it does this is that there is a flow of heat from the surroundings. But why does the chunk of ice use the added heat to melt rather than simply to warm up? Apparently, more than energy is involved in determining the direction of spontaneous change.

The additional factor which must be considered is the tendency of a system to assume the *most probable* state, which turns out to be the one with the most random molecular arrangement. The reason for this tendency is that there are many more ways by which a given arrangement can become more disordered than there are by which it can become more ordered. The disorder is described quantitatively in terms of a property called *entropy*, and we say that a disordered state has a higher entropy than an ordered state. As a specific example, H_2O in the form of liquid water (a random collection of various molecular arrangements) has a higher entropy than H_2O in the form of solid ice (a single ordered array). It is more probable that the higher-entropy state will be formed than the lower-entropy state.

Why is a state of high entropy (i.e., a disordered state) more probable than one of low entropy (i.e., an ordered state)? In ice there is but one recognizable repeat pattern that is unique for ice. In liquid water, there are millions of arrangements all of which are equally probable and all of which together constitute the state of the liquid. Any particular arrangement is just as probable as any other specified arrangement, and, in fact, each is as probable as the one arrangement we call "ice." The point is that there are many arrangements we call "liquid" compared with the one we call "ice." Hence, an arrangement characteristic of liquid water is more probable than the arrangement characteristic of ice.*

In the same sense that natural processes resulting in a *decrease of energy* are favored, those resulting in an *increase of entropy* are also favored. In certain cases, as in melting ice, the two tendencies oppose each other; so there is a question of which one wins out. Above the melting point the entropy increase is dominant, so spontaneous change is favored in the direction of melting; below the melting point the energy decrease is dominant, so spontaneous change is favored in the direction of freezing. It turns out that the temperature itself is the critical factor which governs how important the entropy increase is relative to the energy change. At the absolute zero of temperature $(T = 0°K)$, the entropy term contributes nothing; so the most stable state is the one of lowest energy. As the temperature rises, molecular motion increases, and the tendency to disorder becomes more important in determining the direction of spontaneous change. At sufficiently high temperatures (i.e., above the melting point of ice) the entropy factor becomes large enough to overcome even an unfavorable energy change.

Quantitatively, the interplay of energy and entropy can be described by using a concept called the *free energy*. The "free" emphasizes the fact that for conversion into useful work total energy needs to be distinguished into two parts: (1) a part that is freely available (i.e., the *free* energy) and (2) a part that is not (i.e., the *unavailable* energy). The unavailable energy is represented as the product of the entropy, generally designated by S, and a proportionality factor controlling the relative weighting to be given to the entropy term, which is just the absolute temperature T. Thus, for a mole of material the unavailable energy is TS, where S measures the probability of the state of the material and T gives the thermal disordering influence. The higher the temperature, the more the disorder, and the less available the energy becomes. The free energy per mole is generally symbolized by G;† so we can write the above relation

$$\text{Total energy} = \underset{\textit{Free energy}}{G} + \underset{\textit{Unavailable energy}}{TS}$$

---

* As an analogy, consider a pair of dice. Any specified configuration of the two dice is equally probable (considering die A to be distinguishable from die B), yet it is more likely that one rolls a 7 than a 2. The reason is that there are six ways to roll a 7 (viz., $6 + 1, 5 + 2, 4 + 3, 3 + 4, 2 + 5, 1 + 6$) but only a single way to roll a 2 (viz., $1 + 1$). We can say that the state called "seven" has a higher entropy than the state called "snake eyes." It is for this reason that rolling a 7 is favored.
† The symbol G is in honor of J. Willard Gibbs, a Yale professor of mathematics who founded the science of chemical thermodynamics. Sometimes the symbol F, associated with the word "free," is used in place of G.

Chapter 7
The solid state and
changes of state

220

In regard to the total energy, we must be careful to note that it is not just the *internal energy E* that we are talking about but also what might be called the *external energy*. This external energy, or, more precisely, external work, amounts to PV and comes about because substances generally have a volume V and are subject to an atmospheric pressure P. External work of the amount PV had to be done to push aside volume V of atmosphere, and any changes in PV have to be accounted for. This total energy, internal energy plus external work, is called *enthalpy* (or, sometimes, *total heat content*); it is generally designated by the letter H, again for one mole of substance. Writing

$$H \quad = \quad E \quad + \quad PV$$
Enthalpy   Internal energy   External work

we thus have for the free energy the following expression:

$$G \quad = \quad H \quad - \quad TS$$
Free energy   Enthalpy   Unavailable energy

As we shall see more fully in Chap. 13, it is not the absolute values of the thermodynamic quantities that are usually of prime importance, but the changes in them. To designate these changes, we introduce the symbol Δ before a property to designate a change, or, more exactly, an increase, in that property. Thus, for example, ΔS stands for the increase in molar entropy and represents the entropy of the final state less that of the initial state. If, by chance, the molar entropy in the final state is *less* than the molar entropy in the initial state, then ΔS would be a negative quantity. In other words, Δ defined as an increase can take on either positive or negative values.

For the increase in enthalpy ΔH, which is the molar enthalpy in the final state minus the molar enthalpy in the initial state, we can write $\Delta H = \Delta E + \Delta(PV)$, where ΔE is the increase in internal energy and $\Delta(PV)$ is the increase in the pressure-volume product going from the initial to the final state. The term $\Delta(PV)$ is generally small, especially at constant atmospheric pressure, where $\Delta(PV)$ becomes equal to $P\Delta V$, this being the work accompanying a change of volume ΔV at constant pressure P. For changes at constant pressure, therefore, we can write

$$\Delta H = \Delta E + P\Delta V$$

which states that the increase in enthalpy equals the increase in internal energy plus the work done by the system in expanding through ΔV against the pressure P imposed by the surroundings. In other words, if we wish to increase the energy E of a system by a certain amount ΔE, we must add enough energy ΔH to supply the energy needed to do the work $P\Delta V$ as well as increase the internal energy by the amount ΔE. Because $P\Delta V$ is generally small compared with ΔE, chemists frequently get away with substituting "energy changes" for "enthalpy changes," that is, ΔE for ΔH.

For an increase in free energy we can write

$$\Delta G = \Delta H - \Delta(TS)$$

which states that the increase in free energy from initial to final states of a substance ΔG is equal to the increase in enthalpy less the increase in the temperature-entropy product $\Delta(TS)$. For an isothermal change, i.e., one that occurs at a fixed temperature T, the free-energy change is given by

$$\Delta G = \Delta H - T \Delta S$$

The importance of this equation stems from the fact that *at constant temperature and pressure a chemical reaction or some physical change can occur spontaneously only if it is accompanied by a decrease in free energy*. In other words, ΔG should be negative, corresponding to a free energy lower in the final state than in the initial one. A negative ΔG can be produced by a decrease in enthalpy (i.e., a negative ΔH) or an increase in entropy (i.e., a positive ΔS). The former ($\Delta H < 0$) would correspond to an energetically favorable process, i.e., one in which energy decreases; the latter ($\Delta S > 0$) corresponds to one in which disorder increases and so is favored by random thermal motion.

For many processes ΔH and ΔS have the same sign; for example, for the melting of a solid both are positive. In such cases T will be all-important in deciding which term ΔH or $T \Delta S$ prevails and, therefore, whether ΔG is positive or negative. At low T, ΔH will predominate. Since for melting process ΔH is positive, ΔG will also have to be positive, and melting will not occur below the melting point. At very high temperature, however, the $T \Delta S$ term predominates. Since ΔS for melting is positive, $- T \Delta S$ will be negative, and ΔG will also be negative. Hence, melting should occur. Only at one temperature does $T \Delta S$ just match ΔH. ΔG at this temperature is equal to zero, and solid and liquid coexist in equilibrium. The temperature of this coexistence is, of course, the melting, or freezing, point. The quantitative application of the above ideas is illustrated in the following examples.

Example 1
For the melting of sodium chloride the heat required is 30.3 kJ/mol. The entropy increase is 28.2 J mol^{-1} deg^{-1}. Calculate the melting point from these data.
At the melting point

$$\Delta G = 0 = \Delta H - T_{mp} \Delta S$$

$$\Delta H = T_{mp} \Delta S$$

$$T_{mp} = \frac{\Delta H}{\Delta S} = \frac{30,300 \text{ J/mol}}{28.2 \text{ J mol}^{-1} \text{ deg}^{-1}} = 1070°\text{K}$$

■ ■ ■

Chapter 7
The solid state and
changes of state
222

Example 2
At $0°C$ ice has a density of 0.917 g/cm^3 and an entropy of $37.95 \text{ J mol}^{-1} \text{ deg}^{-1}$. At this temperature liquid water has a density of 0.9998 g/cm^3 and an entropy of $59.94 \text{ J mol}^{-1} \text{ deg}^{-1}$. Calculate

(a) the change of entropy ΔS, (b) the change of enthalpy ΔH, and (c) the change of energy ΔE for the conversion of one mole of ice to liquid water at the normal melting point.

(a) $\Delta S = S_{\text{liq}} - S_{\text{solid}} = 59.94 - 37.95 = 21.99 \text{ J mol}^{-1} \text{ deg}^{-1}.$

(b) $\Delta H = T_{\text{mp}}\Delta S = (273.15°)(21.99 \text{ J mol}^{-1} \text{ deg}^{-1})$
$= 6010 \text{ J/mol}.$

(c) $\Delta E = \Delta H - P\,\Delta V.$

$$\Delta V = V_{\text{liq}} - V_{\text{solid}} = \frac{18.015 \text{ g/mol}}{0.9998 \text{ g/cm}^3} - \frac{18.015 \text{ g/mol}}{0.917 \text{ g/cm}^3}$$

$\Delta V = -1.63 \text{ cm}^3/\text{mol} = -0.00163 \text{ liter/mol}$

$P\,\Delta V = (1.00 \text{ atm})(-0.00163 \text{ liter/mol})$
$= -0.00163 \text{ liter-atm/mol}$

One liter-atmosphere is 101.3 J.

$P\,\Delta V = (-0.00163 \text{ liter-atm/mol})(101.3 \text{ J liter}^{-1} \text{ atm}^{-1})$
$= -0.1651 \text{ J/mol}$

Therefore, $\Delta E = \Delta H - P\,\Delta V = 6010 + 0.1651 \approx$ 6010 J/mol.

■   ■   ■

Exercises

*7.1   Properties. How can crystallinity and cleavage be used to deduce ordering of atoms in the solid state?

*7.2   Diffraction. The diffraction of visible light by a grating was used to deduce the wave nature of light: Given the wave nature of X rays, what does X-ray diffraction by crystals indicate about the crystals?

*7.3   Space lattice. What is a space lattice? What is a unit cell? Even though the choice of a unit cell is somewhat arbitrary, why would it be incorrect to select as a unit cell a cube composed of four Na^+ ions and four Cl^- ions at alternate corners?

*7.4   Bonding in solids. Graphite is a form of carbon in which the carbon atoms are covalently bonded together in giant two-dimensional sheets which are stacked parallel to each other. Show how the bonding in such a structure can explain why graphite, unlike diamond, is very soft whereas, like diamond, it has a very high melting point.

*7.5   Crystal energies. How would you expect the crystal energy of a rock-salt-type compound MX to change with each of the following?
  a   Increase in size of M
  b   Increase in size of X
  c   Increase in charge on M and X

*7.6 *Heating curve.* Draw on the same graph two curves showing how the average potential energy and average kinetic energy per particle change on uniform addition of heat to a solid crystal starting at 0°K.

*7.7 *Cooling curve.* Explain why on the cooling of a liquid substance through its freezing point the temperature at which the temperature-time curve flattens out is *independent of the rate of cooling*.

**7.8 *Unit cell.* Draw a rectangular array of points for which $a \neq b$. Show that a rhombus-shaped unit cell is just as adequate for describing the lattice as is a conventional rectangular unit cell. Show also for both of the above nets that it is equally suitable to select a unit cell with one point in its interior or one point at each corner of the cell.

**7.9 *Unit cells.* In counting the number of points in cells other than hexagonal, a point on a face is counted $\frac{1}{2}$ inside the cell, a point on an edge is $\frac{1}{4}$ inside the cell, and a lattice point at a corner is $\frac{1}{8}$ inside the cell. Show that each of these fractions is justified. Tell also what is the net number of lattice points inside each of the following unit cells as presented in this chapter: (*a*) simple cubic, (*b*) face-centered cubic, (*c*) body-centered cubic, (*d*) rhombic, and (*e*) tetragonal.

**7.10 *Unit cells.* Suppose there were an atom located at each lattice point of the unit cells shown for the hexagonal system in Fig. 7.9*c*. If we allow for the different partition of atoms at faces, edges, or corners shared with adjacent unit cells, the equivalent of how many full atoms will there be in the (*a*) solid-lined and (*b*) dotted-lined choices given in Fig. 7.9*c*?   *Ans. (a) 1, (b) 3*

**7.11 *Rock-salt structure.* Assuming that Na^+ and Cl^- can be represented as hard spheres of radius 0.097 and 0.181 nm, respectively, decide whether any of the spheres would have to be in contact with each other to explain the observed fact that the edge length of the unit cell for NaCl is 0.5627 nm.

**7.12 *Rock-salt structure.* An early way to determine ionic sizes was to measure the X-ray spacings in a rock-salt-type material such as LiI and assume that the large, negative ions were in actual contact along a face diagonal of the face-centered cubic unit cell. Given that the unit-cell edge length of LiI is measured to be 0.6240 nm, what would you deduce for the ionic radius of I^-?   *Ans. 0.2206 nm*

**7.13 *Crystal energies.* Potassium fluoride, KF, has the same crystal structure as NaCl. The unit-cell edge length of KF is 0.533 nm. Estimate the crystal energy of KF, including a 10 percent correction for the repulsion. The experimental value is 803 kJ/mol.

Chapter 7
The solid state and
changes of state
224

**7.14 *Defects.* Silver chloride, AgCl, has the same structure as NaCl. A precise X-ray measurement of the lattice dimensions indicates the unit-cell edge length of AgCl is 0.55491 nm. Precision measurement of the density of a given crystal gives a value of 5.561 g/cm³. If we assume lattice vacancies are the only defects, what percent of the sites would appear to be empty? If lattice interstitials (Ag^+ ions) were also admitted, would the above percent be too high or too low?

Ans. 0.2%, too low

** **7.15** *Heating curve.* Suppose that 10 kJ is added to each of the following systems. What happens to the relative makeup of each system, assuming perfect heat insulation from the surroundings?

a   100 g $H_2O(s)$ + 100 g $H_2O(l)$ at 0°C
b   100 g $NaCl(s)$ + 100 g $NaCl(l)$ at 1070°K
c   100 g $H_2O(l)$ + 100 g $H_2O(g)$ at 100°C (1 atm)

** **7.16** *Supercooling.* When a supercooled liquid begins to crystallize, why does the subsequent temperature rise not overshoot the melting point?

** **7.17** *Phase diagram.* Draw a labeled phase diagram for a substance X which has the following properties: normal boiling point 220°C, normal freezing point 80°C, and triple point 60°C and 0.20 atm. What would you expect for the freezing point if the pressure were 0.80 atm? The boiling point?

Ans. 75, 189°C

** **7.18** *Entropy.* What entropy difference between solid and liquid states must there be for a substance melting at 100°C and having a heat of fusion of 10,000 J/mol.

** **7.19** *Enthalpy.* Under what circumstances would the ΔE for a melting process be identically equal to the ΔH for the same process?

** **7.20** *Free energy.* Draw curves showing how the entropy, enthalpy, and free energy change with time as heat is added uniformly at 1 atm to 18 g of H_2O starting at $-10°C$ and ending at $+10°C$.

** **7.21** *Free energy.* At 1 atm pressure, solid and gaseous CO_2 coexist at $-78°C$. What can you say about the relative free energies (per mole) at $-77°C$ and at $-79°C$?

*** **7.22** *Bragg's law.* How can you use Fig. 7.3 to deduce the Bragg's law dependence of diffraction angle on interplanar spacing?

*** **7.23** *Close-packing.* Calculate the distance between a plane passing through the centers of one close-packed layer of spheres and another plane passing through the centers of an adjacent close-packed layer. Assume all the spheres are identical with a radius of 0.200 nm.

Ans. 0.327 nm

*** **7.24** *Close-packing.* In a close-packed array of fluoride ions, assumed to be hard spheres of radius 0.133 nm, what is the radius of the largest positive ion that can fit into the tetrahedral hole? (*Hint:* Place F^- at alternate corners of a cube.)

*** **7.25** *Close-packing.* In a close-packed array of oxide ions, assumed to be hard spheres of radius 0.140 nm, what is the radius of the largest positive ion that can fit into the octahedral hole? (*Hint:* Place O^{2-} at face centers of a cube.)

*** **7.26** *Spinels.* $ZnFe_2O_4$ is a normal spinel; Fe_3O_4 is an inverse spinel. Given that it is also possible to prepare a continuous series of compounds of general formula $Zn_{1-x}Fe_{2+x}O_4$, what can you say about the charge and location of the Fe atoms in the structure as x is increased?

*** **7.27** *Crystal energies.* Unlike NaCl, CsCl crystallizes in a pattern in which the positive ions are at the centers of cubes and the negative ions are at the cube corners. Thus, the CsCl structure consists of two interpenetrating simple cubic arrays, one of Cs^+ ions and the other of Cl^- ions. If a represents the nearest distance between Cs^+ and Cl^-, what are the first five terms of the series to be used for computing the crystal energy?

Ans. $\dfrac{q_1 q_2}{a}\left(-8 + 3\sqrt{3} + 12\sqrt{3/8} - 24\sqrt{3/11} + 4 + \cdots\right)$

*** **7.28** *Le Châtelier.* According to Le Châtelier's principle, what would you expect to happen when an equilibrium system composed of 1 g of liquid H_2O and 1 g of gaseous H_2O in a constant-volume box at 100°C and 1 atm pressure is perturbed as follows?
a Add heat.
b Add ice.
c Raise pressure by reducing the volume of the box.

*** **7.29** *Triple point.* You are given a closed system containing ice, liquid water, and water vapor in equilibrium with each other at the triple point. In which phase, if any, will each of the following conditions be satisfied?
a Highest density
b Highest average potential energy of H_2O molecules
c Highest average kinetic energy of H_2O molecules
d Highest molar entropy
e Highest molar enthalpy

*** **7.30** *Phase diagram.* At what temperature will the solid-to-liquid transition occur when $H_2O(s)$ is heated as in Sec. 7.12 in a sealed container under the following conditions?
a Container was completely evacuated at start of experiment.
b Container had 1 atm inert-gas pressure at start of experiment.
c Container had 0.00603 atm inert-gas pressure at start of experiment.

*** **7.31** *Entropy.* Recalling that there are no attractive forces between the molecules of an ideal gas, predict qualitatively what must be the relation of the values of ΔE, ΔH, and ΔS for a process where an ideal gas expands into a vacuum at constant temperature. Justify your answers.

Chapter 7
The solid state and
changes of state

226

*** **7.32** *ΔS and ΔH.* Prove or disprove the following statement: Since the molar entropy of a substance always increases on melting, the molar enthalpy must also increase.

***7.33 *Symmetry.* It is a remarkable fact that molecules characterized by rotational symmetry axes C_n can be ordered in the solid state to give crystals of equal symmetry but only when $n = 1, 2, 3, 4,$ or 6, not when $n = 5$. By drawing two-dimensional arrays of the corresponding geometric figures (e.g., rectangles, triangles, squares, etc.), show why $n = 5$ is different.

The preceding discussion of the solid, liquid, and gaseous states was limited to pure substances. In practice, we continually deal with mixtures; hence the question that arises is the effect of mixing in a second component. A mixture is classified as *heterogeneous* or *homogeneous*. By its nature, a heterogeneous mixture consists of distinct phases, and the observed properties are largely the sum of those of the individual phases. However, a homogeneous mixture consists of a single phase which has properties that may differ in several important ways from those of the individual components. These homogeneous mixtures, or solutions, are of widespread importance in chemistry and deserve intensive study.

8.1 Types of solutions

Solutions, defined as *homogeneous mixtures of two or more components,* can be gaseous, liquid, or solid. Gaseous solutions are made by dissolving one gas in another. Since all gases mix in all proportions, any mixture of gases is homogeneous and is a solution. Air is a gaseous solution of nitrogen, oxygen, carbon dioxide, argon, and other molecules. The kinetic picture of a gaseous solution is like that of a pure gas, except that the molecules are of different kinds. Ideally, the molecules move independently of each other.

Liquid solutions are made by dissolving a gas, liquid, or solid in a liquid. If the liquid is water, the solution is called an *aqueous* solution. In the kinetic picture of a sugar-water solution, sugar molecules are distributed at random throughout the bulk of the solution. It is evident that on this molecular scale the term "homogeneous" has little significance. However, experiments cannot be performed with less than billions of molecules; so for practical purposes the solution is homogeneous.

Solid solutions are solids in which one component is randomly dispersed on an atomic or molecular scale throughout another component. As in a pure crystal, the packing of atoms is orderly, but there is no particular order in which the lattice points are occupied by one or other kind of atom. Solid solutions are of great practical importance since they make up a large fraction of the class of substances known as *alloys.* An alloy may be defined as a combination of two or more metallic elements which itself has metallic properties; an alloy may be a compound or a solution. Sterling silver, for example, is an alloy consisting of a solid solution of copper in silver. In brass, which is a general term for alloys of copper and zinc, it is possible to have a solid solution in which some of the copper atoms of the face-centered cubic structure of pure copper have been replaced by zinc atoms. Some kinds of steel are alloys of iron and carbon and can be considered as solid solutions in which carbon atoms are located in some of the spaces between iron atoms. The iron atoms are arranged in the regular structure of pure iron. Not all alloys are solid solutions. Some alloys, such as bismuth-cadmium, are heterogeneous mixtures containing tiny crystals of the constituent elements. Others, such as $MgCu_2$, are intermetallic compounds which contain atoms of different metals combined in definite proportions.

Two terms that are commonly used in the discussion of solutions are *solute* and *solvent.* The substance present in larger amount is generally referred to as the solvent, and the substance present in smaller amount, as the solute. However, the terms can be interchanged whenever it is convenient. For example, in speaking of solutions of sulfuric acid and water, sulfuric acid is referred to as the solute and water as the solvent even when the water molecules are in the minority. The term *solvated* is used to describe the species that result in solution from interaction of one or more molecules of solvent with the solute.

If the solvent is actually water, as will be true in most of the following discussion, the term *hydrated* is used instead.

8.2 Concentration

The properties of solutions, e.g., the color of a dye solution or the sweetness of a sugar solution, depend on the solution concentration. There are several common methods for describing concentration. They include mole fraction, molarity, molality, normality, percent solute, and formality:

1 The *mole fraction* is the ratio of the number of moles of one component to the total number of moles in the solution. For example, in a solution containing 1 mol of alcohol and 3 mol of water, the mole fraction of alcohol is $\frac{1}{4}$, and that of water $\frac{3}{4}$.

2 The *molarity* of a solute is the number of moles of solute per liter of solution and is usually designated by M. A 6.0-molar solution of HCl would be labeled 6.0 M HCl. Assuming the solvent is water, the label would be taken to mean that the solution had been made up in a ratio that corresponds to adding 6.0 mol of HCl to enough water to make a liter of solution.

3 The *molality* of a solute is the number of moles of solute per kilogram of solvent. It is usually designated by m. The label 6.0 m HCl is read "6.0-molal" and represents a solution made by adding to every 6.0 mol of HCl one kilogram of solvent.

4 The *normality* of a solute is the number of equivalents (Sec. 4.10) of solute per liter of solution. It is usually designated by N. The label 0.25 N KMnO$_4$ is read "0.25-normal" and represents a solution which contains 0.25 equiv of potassium permanganate per liter of solution. (As indicated previously, the size of an equivalent may be different for one reaction as compared with that for another.)

5 The *percent of solute* is a frequently used designation, but it is ambiguous since it may refer to percent by weight or percent by volume. If the former is meant, and it usually is, it is the percent of the total solution weight contributed by the weight of the solute. Thus, 3% H$_2$O$_2$ by weight means 3 g of H$_2$O$_2$ per 100 g of solution. Percent by volume is the percent of the *final* solution volume represented by the volume of the solute used to make the solution. For example, 12% alcohol by volume represents a solution made from 12 ml of alcohol plus enough solvent to bring the total volume up to 100 ml. Whereas percents by weight for a given solution add up to 100%, percents by volume generally do not. The reason for this is that when a mixture is formed, mass is conserved but volume generally is not.

6 Occasionally one finds use of still another concentration designation, the *formality*. This designation (abbreviated F) refers to the number of formula weights of solute per liter of solution and is used especially when one wishes to make a distinction between what is formally placed in a solution and what is actually there. Thus, for example, a 1-formal HCl solution, $1\,F$ HCl, made by dissolving 1 formula weight (or 36.5 g) of HCl in enough water to make a liter of solution, is actually $1\,M$ H_3O^+ and $1\,M$ Cl^-.

8.3 Properties of solutions

How are the properties of a solvent affected by the addition of a solute? Specifically, how are the properties of a typical solvent such as water affected by the addition of a typical solute such as sugar? Suppose we consider the following experiment: Two beakers, one containing pure water (beaker I) and the other containing a sugar-water solution (beaker II), are set under a bell jar, as shown in Fig. 8.1. As time goes on, it will be observed that the level of pure water in beaker I drops while the level of the solution in beaker II rises. Apparently there is a net transfer of water from pure solvent to solution, through the vapor phase. This transfer suggests that the escaping tendency of water molecules from pure water is higher than the escaping tendency of water molecules from the sugar-water solution. The escaping tendency is measured by the vapor pressure.

Another experimental observation which supports the idea that addition of a solute lowers the escaping tendency of solvent molecules is an observed lowering of the freezing point. For example, when sugar is added to water, it will be found necessary to cool below $0°C$ in order to freeze out ice. The implication is that the tendency of H_2O to escape from the liquid phase into the solid phase is also decreased by the presence of solute.

The *lowering of the freezing point* and the *reduction of the vapor pressure* are found, at least in dilute solutions, to be *directly proportional to the concentration of solute particles*. Why should there be such lowerings? The first impulse might be to ascribe the lowerings to greater attractive forces between the H_2O and the sugar than between H_2O and H_2O. The fact is, however, that the lowering of the vapor pressure and the reduction of the freezing point do not seem to depend on how strong are the interactions between the solute and the solvent. If one goes from one solute to another, say from sugar to sulfuric acid, one finds that it is not the strength of the intermolecular attractions that is most important (though there is an effect there) but the *relative number of solute particles*. Apparently, the most important effect of the solute is to reduce the concentration of H_2O molecules. In the solution less than 100% of the molecules are H_2O molecules, and therefore the escape of H_2O molecules from the solution is less probable than their escape from pure water. In terms of entropy we would say that the entropy of H_2O in the sugar-water solution is greater than in the pure water state. The addition of solute has increased the probability that H_2O molecules stay in the condensed phase.

Figure 8.2 shows in a phase diagram the typical effect on the solvent water of a particular concentration of a nonvolatile solute. The solid lines represent the phase diagram of pure H_2O; the dashed lines, that of the solution. The dashed line on the left corresponds to equilibrium

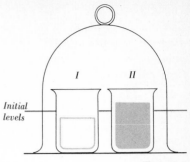

Fig. 8.1
Experimental proof that the escaping tendency of water from pure water (I) is greater than that from an aqueous sugar solution (II).

Fig. 8.2
Comparison of phase
diagram of water and an
aqueous solution. (Dashed
lines refer to solution.)

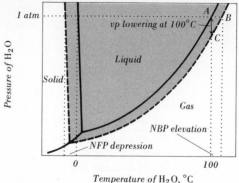

between solid H_2O (ice) and the liquid solution. It represents the temperatures at which pure solid H_2O freezes out when the particular solution is cooled at different pressures. The dashed line on the right corresponds to equilibrium between gaseous H_2O and the liquid solution. It represents the temperatures at which pure gaseous H_2O boils off when the solution is heated at various pressures.

The most striking feature shown on the phase diagram is the extension of the liquid range both to higher temperatures and to lower temperatures. The liquid phase of water has been made more probable by the dissolving of solute in it. Associated with this is the fact that the vapor pressure of the water has been reduced. For example, as seen from Fig. 8.2, at 100°C the vapor pressure of the solution is not 1 atm but less than that. Because the vapor pressure has been lowered, the solution does not boil at 100°C but at some higher temperature where the vapor pressure becomes equal to 1 atm. If the solute contributes nothing to the vapor pressure (is nonvolatile, as, for example, sugar), then the normal boiling point of the solution can be read directly from the phase diagram. The normal boiling point will be higher because of widening of the liquid-phase area. Similarly, there is a depression of the normal freezing point.

Quantitatively, the lowering of the vapor pressure of a solvent by addition of a solute is described by *Raoult's law* which states that *the fractional lowering of the vapor pressure of a solvent is equal to the mole fraction of solute present.* This can be written

$$\frac{p_1^{\,0} - p_1}{p_1^{\,0}} = x_2$$

where $p_1^{\,0}$ is the vapor pressure of the pure solvent and p_1 is the partial pressure of the solvent above a solution in which the mole fraction of solute is x_2. Since the mole fractions of solute and solvent x_1 and x_2 must add to give unity, we can write alternatively

$$\frac{p_1^{\,0} - p_1}{p_1^{\,0}} = 1 - x_1$$

$$1 - \frac{p_1}{p_1{}^0} = 1 - x_1$$

$$\frac{p_1}{p_1{}^0} = x_1$$

From the last relation it can be seen that Raoult's law amounts to a direct proportionality between the solvent vapor pressure and its mole fraction. In practice, Raoult's law is an idealization best realized in dilute solutions. As the concentration of solute is increased, the vapor pressure of the solvent component deviates from ideal behavior, usually in the positive sense (e.g., chloroform and ethyl alcohol) but for certain cases in the negative sense (e.g., chloroform and acetone). Typical behavior is shown in Fig. 8.3.

The Raoult's law proportionality between solvent vapor pressure and mole fraction of solvent explains why the freezing-point lowering and the boiling-point elevation are proportional to the concentration of solute molecules in solution. For either fp or bp the concentration of solute is important in determining the mole fraction of solvent. As the concentration of solute is increased, the mole fraction of the solvent is decreased, and its vapor pressure decreases proportionately. As shown in the phase diagram of Fig. 8.2, the extent to which the dashed curve falls below the solid curve is the amount of the vapor-pressure lowering; it controls both the boiling-point elevation and the freezing-point lowering. For the boiling point, because of the vapor-pressure lowering it becomes necessary to increase the temperature to get boiling to occur. Over the small range of values in question, the required temperature increase is directly proportional to the vapor-pressure decrease.* Since the vapor-pressure decrease in turn is proportional to the mole fraction of solute, it follows that the elevation of the boiling point should be directly proportional to the mole fraction of solute. These relations can be written as follows:

$$\Delta T_{bp} \sim \Delta p \sim x_2$$

where ΔT_{bp} is the elevation of the boiling point, Δp is the magnitude of the vapor-pressure lowering, and x_2 is the mole fraction of solute. By definition,

$$x_2 = \frac{n_{solute}}{n_{solute} + n_{solvent}}$$

where n is the number of moles. In dilute solution, n_{solute} is very small compared with $n_{solvent}$; so it can be neglected in the denominator. Given a solution of molality m, meaning m moles of solute per kilogram

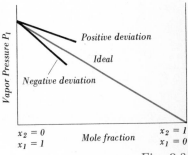

Positive deviation

Ideal

Negative deviation

$x_2 = 0$ $x_2 = 1$
$x_1 = 1$ *Mole fraction* $x_1 = 0$

Fig. 8.3
Raoult's law behavior for vapor pressure of a solvent as a function of mole fraction.

---

* To see how this comes about, look at triangle ABC in Fig. 8.2. The size of this triangle depends on how much solute is added to the solvent, but, no matter how big the triangle is, side AB will be proportional to side AC. In other words, boiling-point elevation is proportional to the drop in vapor pressure. The more dilute the solution, the better the argument holds. *De nihilo nihilum.*

of solvent, we can write

$$x_2 \approx \frac{n_{\text{solute}}}{n_{\text{solvent}}} = \frac{m}{1000/W}$$

where W is the molecular weight of the solvent. Since $1000/W$ is a constant for a particular solvent (being equal, for example, to $1000/18$, or 55.5, for H_2O), it follows that for dilute solutions in a given solvent, the boiling-point elevation will be proportional to the molality of the solute. This can be written

$$\Delta T_{\text{bp}} = K_{\text{bp}} m$$

where K_{bp} is a proportionality constant called the *molal boiling-point elevation constant*. The value of K_{bp} depends on the solvent, but not on the solute. For H_2O, K_{bp} equals $0.52°C$ per $1\,m$ solution. This value compares with $2.6°C/m$ for benzene, $1.24°C/m$ for ethyl alcohol, and $5.05°C/m$ for carbon tetrachloride.

The quantitative description of freezing-point lowering is worked out along lines similar to the above. Over small ranges of concentration the lowering of the triple-point temperature (see Fig. 8.2) can be taken to be proportional to the lowering of the triple-point vapor pressure. We assume further that the lowering of the triple-point temperature is directly reflected as the lowering of the normal freezing point. Thus, we can write the approximation

$$\Delta T_{\text{fp}} \sim \Delta P_{\text{triple point}} \sim x_2 = \frac{m}{1000/W}$$

or

$$\Delta T_{\text{fp}} = K_{\text{fp}} m$$

where K_{fp} is a proportionality constant called the *molal freezing-point lowering constant*. As with the boiling-point elevation constant K_{bp}, the freezing-point depression constant K_{fp} depends on the solvent, but not on the solute. For H_2O, K_{fp} equals $1.86°C/m$, which compares with $5.1°C/m$ for benzene and $6.9°C/m$ for naphthalene.

The above derivations of boiling-point and freezing-point formulas are approximate and rest on assumptions that are justified only for dilute solutions. In concentrated solutions K_{bp} and K_{fp} may vary somewhat. For the above derivations to be valid, two other conditions must be satisfied: The first is that the solute must be dissolved *only in the liquid phase*; in other words, the solute cannot be volatile, nor can it form a solid solution in the solid phase. The second point is that the number of solute particles in the solution should be the same as the number of molecules placed in the solution. If the dissolving process breaks up the molecules into smaller fragments, then it will be the *total concentration* of all kinds of particles that determines the freezing and boiling points of the solution. In the following example, where we show how freezing-point lowering can be used to determine molecular weight, we assume that there are no complications of the above kind.

Example 1

When 0.946 g of fructose (a sugar) is dissolved in 150 g of water, the resulting solution is observed to have a freezing point of $-0.0651°C$.

(a) *What must be the corresponding molecular weight of fructose?*

(b) *If the simplest formula of fructose is CH_2O, what must be its molecular formula?*

(a)   0.946 g of fructose per 150 g of H_2O is the same as

$$\frac{0.946 \text{ g of fructose}}{0.150 \text{ kg}} = 6.31 \text{ g of fructose per kilogram of } H_2O$$

The observed freezing point is $-0.0651°C$. For a $1\ m$ solution it would have been $-1.86°C$. Therefore, the concentration of the solution is

$$\frac{0.0651°C}{1.86°C/m} = 0.0350\ m$$

So the 6.31 g of fructose per kilogram of H_2O must represent $0.0350\ m$. Thus, 6.31 g of fructose must be 0.0350 mol of fructose.

$$\frac{6.31 \text{ g}}{0.0350 \text{ mol}} = 180 \text{ g per mole of fructose}$$

(b)   Molecular formula must be some multiple x of CH_2O, or $(CH_2O)_x$. The simplest-formula weight is $12 + 2 + 16 = 30$ amu. The molecular weight is 180 amu, or x times 30; so $x = 6$. The molecular formula is $(CH_2O)_6$, or $C_6H_{12}O_6$.

■ ■ ■

8.4   Electrolytes

As first shown by the Swedish chemist Svante Arrhenius in 1887, there are many cases in which the dissolving process is accompanied by dissociation, or breaking apart, of molecules. The dissociated fragments are usually electrically charged; so electrical measurements can show whether dissociation has occurred. Charged particles, or ions, moving in solution constitute an electric current; so a measurement of the electric conductivity of the solution is all that is required. Figure 8.4 is a schematic diagram of an apparatus for determining whether a solute is dissociated into ions. A pair of electrodes is connected in series with an ammeter to a source of electricity. As long as the two electrodes are kept separated, no electric current flows through the circuit, and the meter reads zero. When the two electrodes are joined by an electric conductor, the circuit is complete, and the meter deflects. When the electrodes are dipped into a beaker of water, the meter stays near zero, indicating that water does not conduct electricity appreciably. When sugar is dissolved in the water, the solution does not conduct; but when sodium chloride is dissolved in the water, the solution does conduct.

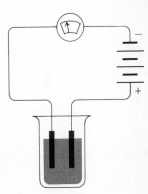

Fig. 8.4

Experiment to determine conductivity of a solution.

Fig. 8.5
Classification of solutes as electrolytes or nonelectrolytes.

Electrolytes		Nonelectrolytes	
HCl	*Hydrochloric acid*	$C_{12}H_{22}O_{11}$	*Sucrose*
H_2SO_4	*Sulfuric acid*	C_2H_5OH	*Ethyl alcohol*
$HC_2H_3O_2$	*Acetic acid*	N_2	*Nitrogen*
$NaOH$	*Sodium hydroxide*	O_2	*Oxygen*
$Ca(OH)_2$	*Calcium hydroxide*	CH_4	*Methane*
$NaCl$	*Sodium chloride*	CO	*Carbon monoxide*
Na_2SO_4	*Sodium sulfate*	CH_3COCH_3	*Acetone*

By such experiments it is possible to classify substances into two groups: (1) those which produce conducting solutions and (2) those which produce essentially nonconducting solutions. Solutes of the first class are called *electrolytes;* those of the second, *nonelectrolytes*. Figure 8.5 gives the names and formulas of several examples.

Electric conductivity requires the existence of charged particles. The greater the number of charges available for carrying electricity, the greater the conductivity observed. By quantitatively measuring the conductivity, it is possible to get information about the relative concentration of charges in the solution. When the conductivity of a solution labeled 1 *m* HCl is compared with the conductivity of 1 *m* $HC_2H_3O_2$ (acetic acid), it is found that 1 *m* HCl conducts to a greater extent than does 1 *m* $HC_2H_3O_2$. Since both solutions correspond to the dissolving of 1 mol of solute per kilogram of water, the inference is that HCl breaks up more fully into ions than does $HC_2H_3O_2$. From experimental observations of this kind, electrolytes may be subdivided into two more or less distinct groups: *strong* electrolytes, which give solutions that are good conductors of electricity, and *weak* electrolytes, which give solutions that are only slightly conducting. Figure 8.6 lists typical compounds of both classes. Strong electrolytes are essentially 100% dissociated into ions, whereas weak electrolytes may be dissociated only a few percent.

Fig. 8.6
Classification of solutes as strong electrolytes or weak electrolytes.

Strong electrolytes		Weak electrolytes	
HCl	*Hydrochloric acid*	$HC_2H_3O_2$	*Acetic acid*
$NaOH$	*Sodium hydroxide*	$TlOH$	*Thallous hydroxide*
$NaCl$	*Sodium chloride*	$HgCl_2$	*Mercuric chloride*
KCN	*Potassium cyanide*	HCN	*Hydrogen cyanide*
$BaSO_4$	*Barium sulfate*	$CdSO_4$	*Cadmium sulfate*

In a solution of a nonelectrolyte, where there is no dissociation, the molecules of solute retain their unbroken identity. For example, when sugar dissolves in water, as shown in Fig. 8.7, the sugar molecules persist in the solution as solvated, or hydrated, species consisting of sugar molecules surrounded by clusters of water molecules. This is an uncharged, or neutral, species; so when positive and negative electrodes are inserted in the solution, there is no reason for the particles to move one way or the other. Hence there is no electric conductivity.

Electrolytes before being dissolved may be ionic or molecular substances. In the case of *ionic substances* it is not surprising that there are charged particles in solution, because the undissolved solid is already made up of charged particles. The solvent simply rips the lattice apart into its constituent pieces. Figure 8.8 shows schematically the dissolving of the ionic solid sodium chloride in water. Positive ends (hydrogen atoms) of water molecules are attracted to the negative chloride ion. The chloride ion, weakened in its attraction to the crystal lattice, moves off into the solution with its associated cluster of water molecules. It is now a hydrated chloride ion. The entire species is negatively charged because the chloride ion itself is negatively charged. At the same time, the sodium ion undergoes similar hydration except that the negative, or oxygen, end of the water molecule faces the ion. Since the solution as a whole must be electrically neutral, the hydrated sodium ions and hydrated chloride ions must be formed in equal numbers. When electrodes are inserted into sodium chloride solution, the positively charged hydrated sodium ions are attracted to the negative electrode, and the negatively charged hydrated chloride ions are attracted to the positive electrode. There is net transport of electric charge as positive charge moves in one direction and negative charge moves in the opposite direction. (Electric conductivity in solutions is discussed in greater detail in Sec. 14.1.)

Ions may also be formed when certain *molecular substances* are dissolved in the proper solvent. For example, HCl in the pure solid, liquid, or gaseous state does not conduct electricity because it consists of neutral, distinct molecules; no ions are present. However, when HCl is placed in water, the resulting solution does conduct, indicating the formation of charged particles. The electrically neutral HCl molecules are believed to interact with the solvent as follows:

$$H\!:\!\overset{\cdot\cdot}{\underset{H}{O}}\!: \; + \; H\!:\!\overset{\cdot\cdot}{\underset{\cdot\cdot}{Cl}}\!: \; \longrightarrow \; \left[H\!:\!\overset{\cdot\cdot}{\underset{H}{O}}\!:\!H\right]^{+} + \left[:\!\overset{\cdot\cdot}{\underset{\cdot\cdot}{Cl}}\!:\right]^{-}$$

A proton has been transferred from HCl to H_2O to form the new species H_3O^+ and Cl^-. Thus positive and negative ions are formed, even though none is present in pure HCl. The positively charged H_3O^+ is referred to as a *hydronium*, or *oxonium*, ion. The negative ion is the chloride ion. Both H_3O^+ and Cl^- are hydrated, in that there are water molecules stuck on them just as on an ionic solute. The above reaction would be written

$$H_2O + HCl \longrightarrow H_3O^+ + Cl^-$$

thereby emphasizing that the dissociation of a solute by water is a chemical reaction between two substances. However, since H_3O^+ is nothing but a hydrated proton $H^+(H_2O)$ and since water of hydration is usually omitted from chemical equations, the argument is often made that the above equation can be more conveniently written as

$$HCl \longrightarrow H^+ + Cl^-$$

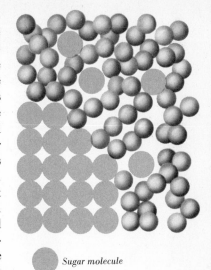

Sugar molecule

Water molecule

Fig. 8.7
Schematic representation of sugar dissolving.

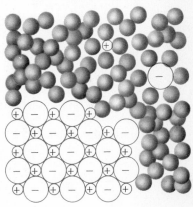

Fig. 8.8
Schematic representation of NaCl *dissolving.*

There is the tacit understanding that all species are hydrated, but the practice of omitting water of hydration from chemical equations often leads one to forget that in aqueous solution water is associated with all the dissolved species and may affect their properties. The picture is most distorted in the case of the hydrogen ion, where H^+ is nothing but a proton (nucleus of H atom) whereas H_3O^+ has a volume which is about 10^{15} times as great and is comparable in size with other ions.*

8.5 Percent dissociation

The extent of dissociation of solutes varies widely. Some substances, such as. HCl, are almost completely dissociated into ions in aqueous solution. Other substances, such as $HC_2H_3O_2$, are only slightly dissociated. The difference depends on relative bond strengths and entropy changes on dissociation. Percent dissociation can be determined by measuring any property that depends on the concentration of ions. Conductivity measurements can be used, as well as freezing-point lowering. The freezing-point lowering is somewhat easier in most cases and can be used because the freezing-point depression of a solvent is proportional to the molal concentration of particles dissolved in the solvent. As an example, consider the solution of an electrolyte AB in water. If some of the AB molecules are dissociated into A^+ and B^- ions, then the solution contains three kinds of dissolved particles: undissociated AB molecules, A^+ ions, and B^- ions. Each particle contributes to the freezing-point depression, and if the solution is dilute enough, it does not make any difference whether the particle is charged or not. By determining the freezing-point lowering of a specific solution of AB, the total concentration of particles and thus the percent dissociation of AB can be calculated.

Example 2
The freezing point of 0.0100 m AB *solution is* $-0.0193°C$. *What percent of the* AB *molecules have been dissociated by the water?*
     Freezing-point lowering constant of water $= 1.86°C/m$.

$$\text{Total concentration of particles} = \frac{0.0193°C}{1.86°C/m} = 0.0104 \ m$$

Let $x =$ concentration of AB particles that dissociate. This gives x molal A^+ and x molal B^-, leaving $(0.0100 - x)$ molal undissociated AB. Total concentration of particles is $x + x + (0.0100 - x) = 0.0100 + x$ molal. Equating, we get

$$0.0104 \ m = (0.0100 + x) \ m$$

$$x = 0.0004 \ m$$

The fraction of AB molecules dissociated is equal to the number

* The problem is even more complicated than this. There is substantial evidence that H_3O^+ in aqueous solution is bound to three other water molecules to form a fairly definable species $H_9O_4^+$.

of AB molecules dissociated divided by the number of AB molecules initially available:

$$\text{Fraction of AB dissociated} = \frac{0.0004}{0.0100} = 0.04 = 4\%$$

∎ ∎ ∎

Measurements on various electrolytes under different conditions indicate that the percent dissociation of an electrolyte depends on the nature of the solute, the nature of the solvent, the concentration of the solute, and the temperature:

1 *Nature of the solute.* When a molecule AB is dissociated into ions A^+ and B^-, the bond AB must be broken. The extent to which this process takes place depends on the nature of A and B. For example, at the same temperature and total concentration, HCl in water might be 100% dissociated, whereas HF in water would be 1% dissociated.

2 *Nature of the solvent.* This point is easily overlooked because we usually think only of water as the solvent. In solvents other than water the behavior of solutes may be different. As an example, under comparable conditions HCl in water is 100% dissociated, but HCl in benzene is less than 1% dissociated. Also, it is quite possible that an electrolyte which in water is less dissociated than HCl may in another solvent be more dissociated than HCl.

3 *Temperature.* There is no simple rule for the effect of temperature on percent dissociation. The percent dissociation of some substances increases as the temperature is raised; the percent dissociation of other substances decreases or is unchanged. Some substances show a combination of all effects. For example, acetic acid is 1% dissociated in a given solution at room temperature; at higher temperatures and at lower temperatures in that same solution it would be less than 1% dissociated.

4 *Concentration.* The percent dissociation of an electrolyte increases as the concentration of electrolyte decreases. The more dilute a solution, the higher the percent dissociation. Figure 8.9 gives numerical values of the percent dissociation of $HC_2H_3O_2$ into H_3O^+ and $C_2H_3O_2^-$ (acetate ion) in solutions of various total concentration. In the more dilute solutions a higher percentage of the electrolyte is dissociated into positive and negative ions. In infinitely dilute solutions, the percent dissociation approaches 100%. An extreme case of near-infinite dilution can be imagined as one made by dissolving a single molecule of AB in an ocean of solvent. If the molecule breaks up to form ions, the chance that the ions will come together again is vanishingly small. Since the one molecule is dissociated, the percent dissociation must be 100%. At infinite dilution, all electrolytes approach 100% dissociation.*

Concentration	Percent dissociation
1 M $HC_2H_3O_2$	0.4
0.1 M $HC_2H_3O_2$	1.3
0.01 M $HC_2H_3O_2$	4.3
0.001 M $HC_2H_3O_2$	15
0.00001 M $HC_2H_3O_2$	75

Fig. 8.9

Concentration dependence of percent dissociation of acetic acid.

* Strictly speaking, 100% dissociation will be approached only if the solvent has no dissociation fragments in common with the solute. For example, in water there is a trace of H_3O^+ and OH^- which comes from the self-dissociation of H_2O. The presence of this H_3O^+ prevents complete dissociation of solutes such as HX. For the example of Fig. 8.9 the maximum attainable percent dissociation cannot exceed 99.4%, a number which is fixed by the dissociation constant of acetic acid (see Sec. 12.9).

Definition of a strong electrolyte as one that is highly dissociated and a weak electrolyte as one that is slightly dissociated has to be qualified since in very dilute solutions even weak electrolytes tend toward complete dissociation. The ambiguity in the definition can be reduced by adopting the convention that a 1 M solution be the reference state. If a substance in 1 M solution is highly dissociated, it is generally called a strong electrolyte; if in 1 M solution it is slightly dissociated, it is generally called a weak electrolyte. It is a curious observation that if we work with this criterion, practically all substances fall into one or the other class—there are very few that are in-between.

8.6 Interionic attractions

Suppose we prepare a series of aqueous solutions of NaCl varying in concentration and carefully measure their freezing points. The results, when calculated as freezing-point lowering *per mole* of NaCl, will look like those shown in Fig. 8.10. The more dilute the solution, the more closely we approach the value 2(1.86)°C, or 3.72°C. Apparently, only in very dilute solution would one mole of NaCl be totally dissociated into two moles of particles (Na$^+$ and Cl$^-$). The more concentrated the solution, the smaller the molar freezing-point lowering. It almost looks as if the NaCl were not completely dissociated in the more concentrated solutions. In fact, from about 1890, when Arrhenius introduced his electrolytic dissociation theory, until 1923 it was generally believed that NaCl in water is a strong but not completely dissociated electrolyte.

Fig. 8.10

Concentration dependence of freezing-point lowering in NaCl *solutions.*

Concentration	Observed freezing point, °C	Freezing-point lowering, °C per molal NaCl
0.1 m NaCl	*−0.347*	*3.47*
0.01 m NaCl	*−0.0361*	*3.61*
0.001 m NaCl	*−0.00366*	*3.66*
0.0001 m NaCl	*−0.000372*	*3.72*

In 1923, Peter Debye and Ernst Hückel proposed an explanation, now known as the *Debye-Hückel theory*, which accounted for the NaCl results and laid the foundations for modern electrolyte theory. The argument was simply this: X rays show that the salt NaCl is 100 percent composed of ions in the solid state. It would be highly unlikely that NaCl were not completely composed of ions in the solution state as well. Therefore, any reduced effect of the ions Na$^+$ and Cl$^-$ in solution must be due, not to association into neutral NaCl molecules, but to some other effect which reduces the activity of the ions. What more logical conclusion than to suggest that the reduced activity comes from interionic attractions between Na$^+$ and Cl$^-$ in the solution state.

The electric force between two charges, call them q_1 and q_2, separated by a distance r is proportional to the product of the charges and inversely proportional to the square of the distance between them. This is a statement of *Coulomb's law* and can be expressed symbolically as

$$F = \frac{q_1 q_2}{\varepsilon r^2}$$

where F is the force and ε is the dielectric constant of the medium in which the charges are immersed. (For a vacuum, $\varepsilon = 1$, and for liquid water $\varepsilon \approx 80$.)

In a dilute solution of NaCl, the ions are relatively far apart. On the average, r is big; so F is small and can reasonably be neglected. What happens as the concentration of solute is increased? The average interionic distance decreases, and the attractive force between positive and negative ions increases. Simultaneously, the repulsive force between like charges also increases. The effect is, as concentration of NaCl increases, Na^+ and Cl^- ions draw together. However, Debye and Hückel did not propose that Na^+ and Cl^- pair up in concentrated solution; they recognized that kinetic motion would keep things moving in a constant state of flux; so the best they could do was to speak of an average distribution of ions. The point is that in an aqueous solution in the vicinity of, say, a *positive ion* there is considerably more chance of finding *negative ions* than other ions of the same charge. In other words, each Na^+ probably has more Cl^- ions than Na^+ ions in the near vicinity and can be considered to be immersed in an atmosphere that is relatively negatively charged. Similarly, the Cl^- ion has in its own vicinity an environment that is relatively rich in Na^+ ions. The ions, of course, being in solution, are not fixed but move in and out of any defined region; so it is a statistical preponderance of positive or negative ions that we are talking about. Each ion is in an atmosphere that is relatively rich in its *counterion* (i.e., the ion of opposite charge).

What would be the effect on an ion of having several counterions in the vicinity? The most obvious influence would be on the electric conductivity. If an external electric field were applied (e.g., put in some electrodes), the given ion would not move so freely as it would if the counterions were far away. There would be a drag effect, which would make the ions less mobile than expected. This, indeed, is what is found; the more concentrated a salt solution is, the smaller the solute conductance *per mole*. Thus, for example, if we double the concentration of an NaCl solution from 0.01 to 0.02 m, we do not double the conductivity but only raise it by 97%. If we double the concentration from 0.05 to 0.1 m, we only raise it by 96%. The more concentrated the solution, the less each mole of NaCl is worth in raising the conductivity. Figure 8.11 gives some data showing how the conductivity per mole increases as the concentration of NaCl decreases. The value at infinite dilution is obtained by extrapolating measurements at very low concentrations.

Similar considerations apply to freezing-point lowering. Although NaCl is 100% dissociated in solution so as to give two moles of particles

Concentration	Conductivity per mole (arbitrary units)
0.1 m NaCl	106.74
0.01 m NaCl	118.51
0.001 m NaCl	123.74
Infinite dilution	126.45

Fig. 8.11

Concentration dependence of conductivity per mole of NaCl *in* NaCl *solutions.*

(Na$^+$ and Cl$^-$) per mole of solute dissolved, the charged particles restrict each other's activity; so their full contribution to freezing-point lowering cannot be exerted. As noted in Sec. 8.3, freezing-point lowering comes about because the solute lowers the escaping tendency of the water molecules. In concentrated NaCl solutions, the ions are somewhat restricted in their freedom to lower the escaping tendency of water.

The Debye-Hückel interionic-attraction theory of electrolyte solutions accounts quantitatively for the attraction between completely dissociated ions and justifies the view that NaCl and similar ionic solids should be considered to be completely dissociated in dilute aqueous solution. Further, this theory can be applied to calculate the interionic attraction when the concentration of ions is rather low, as in a solution of a weak electrolyte. This calculation must be made if one wants to make a very accurate determination of percent dissociation from freezing-point data.

Even the Debye-Hückel theory does not account for the observed freezing-point lowering by electrolytes in concentrated solutions. One important effect is that there may not be sufficient water molecules to hydrate all the ions. Concentrated solutions are not well understood at present.

8.7   Solubility

The term "solubility" is used in several senses. It describes the qualitative idea of the dissolving process. It also is used quantitatively to describe the composition of the resulting solutions. The solutions considered up to now represent *unsaturated solutions*, to which solute can be added successively to produce a whole series of solutions which differ slightly in concentration. With any particular solute and any particular solvent a large number of unsaturated solutions are possible. However, in most cases, the process of adding solute cannot go on indefinitely. Eventually, a stage is reached beyond which addition of solute to a specified amount of solvent does not produce another solution of higher concentration. Instead, the excess solute remains undissolved. In these cases there is a limit to the amount of solute which can be dissolved in a given amount of solvent. The solution which represents this limit is called a *saturated solution*, and the concentration of the saturated solution is called the *solubility* of the given solute in the particular solvent used.

The best way to ensure having a saturated solution is to maintain an excess of solute in contact with the solution. If the solution is unsaturated, solute disappears until saturation is established. If the solution is indeed saturated, the amount of excess solute remains unchanged, as does the concentration of the solution. The system is in a state of equilibrium. Apparently, it is a state of dynamic equilibrium since, for example, an irregularly shaped crystal of solute dropped into the solution changes its shape although remaining constant in mass. In the equilibrium state, dissolving of solute is still occurring, but the dissolving is compensated for by precipitation of solute out of solution. The number of solute particles going into solution per unit time equals the number of solute particles leaving. The concentration of solute in solution remains constant; the amount of solute in excess remains constant. Needless to say, the amount of excess solute present in contact with the saturated solution does not affect the concentration of the saturated solution. In fact, it is possible to filter or separate the excess solute completely and still have a saturated solution. *A saturated solution may be defined as one which is or would be in equilibrium with excess solute.*

The concentration of the saturated solution, i.e., the solubility, depends on (1) the nature of the solvent, (2) the nature of the solute, (3) the temperature, and (4) the pressure. In considering these we should keep in mind that three important interactions operate in the dissolving process: Solute particles are separated one from the other in a process that takes energy; solvent particles are pushed apart to make a hole to accommodate the solute, and this also takes energy; finally, solvent particles attract the solute particles in the process that provides energy. This set of interactions—solute-solute, solvent-solvent, and solute-solvent—is useful in discussing the magnitude of the solubility. However, as mentioned in Sec. 7.13, the direction for spontaneous change is decided not only by energy balance but also by entropy changes. In the case of the solution process this means that due consideration must be given to the fact that the solution represents a more disordered and hence more probable state than that of the unmixed components. Thus, dissolving will frequently occur—because of this entropy increase—even when the solute-solvent interactions are not energetically able to compensate for the energy required in the sum of the solute-solute and solvent-solvent interactions.

1   Nature of the solvent. A useful generalization much quoted in chemistry is "like dissolves like." What this means is that high solubility occurs when molecules of solute are similar in structure and electric properties to molecules of solvent. When there is such similarity, e.g., both solute and solvent have high dipole moment, then solute-solvent attractions are particularly strong. When solute and solvent are dissimilar, solute-solvent attractions are likely to be weak. For this reason polar substances such as H_2O are usually good solvents for polar substances such as alcohol but poor solvents for nonpolar substances such as gasoline.

In general, an ionic solid has higher solubility in a polar solvent than in a nonpolar solvent. For example, at room temperature the solubility of NaCl in water is 311 g per liter of solution, whereas the solubility of NaCl in gasoline is essentially zero. Also, the more polar the solvent, the greater the solubility. For example, at room temperature, the solubility of NaCl in ethyl alcohol is 0.51 g per liter of solution, compared with 311 g per liter of solution in water. The difference is ascribed to the lower polarity (lower dipole moment) of the ethyl alcohol molecule, with resulting lower attractions for the ions.

As seen in the last section, high dielectric constant for a medium means small attraction between ions. Water, being more polar than ethyl alcohol, has a higher dielectric constant than ethyl alcohol. This means Na^+ and Cl^- attract each other less in water than in ethyl alcohol.

2 *Nature of the solute.* Changing the solute means changing the solute-solute and solute-solvent interactions. At room temperature the amount of sucrose that can be dissolved in water is 1311 g per liter of solution. This is more than four times as great as the solubility of NaCl in water, 311 g/liter. However, these numbers are rather misleading. The number of particles involved can better be seen by comparing the molar solubilities. A saturated solution of NaCl is 5.3 M, whereas a saturated solution of sucrose is 3.8 M. On a molar basis NaCl has a higher solubility in water than sucrose has. Since the attractions in solid NaCl (ion-to-ion) are greater than those in sucrose (molecule-to-molecule), the reason for the higher solubility of NaCl apparently lies in the fact that the interactions between Na^+ and Cl^- and water molecules are greater than the interactions between sucrose molecules and water molecules.

What effect does the presence of one solute in a solution have on the solubility of another solute in that same solution? As a crude generalization, unless the concentration of solute is high, there is little effect. For example, approximately the same concentration of NaCl can be dissolved in 0.1 M sucrose solution as in pure water. However, the solubility of NaCl is appreciably lowered by another solute *with an ion in common,* such as KCl or $NaNO_3$.

3 *Temperature.* The solubility of *gases in water* usually decreases as the temperature of the solution increases. The tiny bubbles which form when water is first heated are due to the fact that dissolved air becomes less soluble as the temperature is raised. The flat taste characteristic of boiled water is largely due to the fact that dissolved air has been expelled.* However, in the case of *gases in other liquid solvents* (and, in fact, even in water for some gases such as helium at higher temperatures) the solubility of gases may actually increase with rising temperature. Similarly, there is no general rule for the temperature change of solubility of *liquids* and *solids.* For example, with increasing temperature, lithium carbonate in water decreases in solubility, potassium chloride increases, and sodium chloride shows practically no change. Illustrative data are given in Fig. 8.12.

The change of solubility with temperature is closely related to the heat of solution of the substance. The *heat of solution* is usually defined as the heat evolved when a mole of solute dissolves to give the saturated solution; it can be written as the heat that accompanies the following process:

Solute + solvent $\longrightarrow$ saturated solution + heat of solution

* One of the problems in thermal pollution of our water resources is that a rise in temperature means less dissolved oxygen. For example, raising a lake's temperature from 20 to 40°C reduces dissolved oxygen from 9 to 6 ppm (by weight). This can drastically reduce the plant and fish life that can survive in the water.

Substance	0°C	20°C	40°C	60°C	80°C	100°C
$AgNO_3$	1220	2220	3760	5250	6690	9520
$Ca(OH)_2$	1.85	1.65	1.41	1.16	0.94	0.77
KCl	276	340	400	455	511	567
NaCl	357	360	366	373	384	398
Li_2CO_3	15.4	13.3	11.7	10.1	8.5	7.2
$O_2(g)$ at 1 atm	0.069	0.043	0.031	0.027	0.014	0.000
$CO_2(g)$ at 1 atm	3.34	1.69	0.97	0.058	....	
$He(g)$ at 1 atm	0.00167	0.00153	0.00152	0.00162	0.00177	

Fig. 8.12

Change of solubility with temperature (solubility in grams of solute per kilogram of H_2O).

The heat of solution as experimentally determined can be a positive quantity, in which case heat is evolved to the surroundings, or it can be a negative quantity, in which case heat is absorbed from the surroundings. For example, whereas the heat of solution of lithium carbonate in water is positive, the heat of solution of potassium chloride is negative. Whereas the dissolving of lithium carbonate evolves heat to the surroundings, the dissolving of potassium chloride absorbs heat from the surroundings. For the latter process we can write

$$KCl(s) + H_2O \longrightarrow \text{solution} - \text{heat}$$

or

$$\text{Heat} + KCl(s) + H_2O \longrightarrow \text{solution}$$

Heat must be supplied; i.e., the process is endothermic. When a substance with negative heat of solution is dissolved, there is usually a drop in the temperature of the solution. If the heat of solution is positive, the temperature rises.

Whether the heat of solution is positive or negative depends on the nature of the solute and the solvent. More precisely, the heat of solution depends on the relative magnitude of two energies: (1) the energy required to break up the solid lattice and (2) the energy liberated when the resulting particles are solvated. In the case of potassium chloride in water, the overall process can be imagined to occur in two consecutive steps:

$$KCl(s) \longrightarrow K^+(g) + Cl^-(g)$$

$$K^+(g) + Cl^-(g) \xrightarrow[water]{} K^+(aq) + Cl^-(aq)$$

The first step, vaporizing the solid, requires energy. Work must be done to separate positive and negative ions from each other. The amount of energy required per mole is call the *lattice energy*. The second step liberates energy. As water molecules are separated from each other and attracted to the ions, energy is liberated to the surroundings. This energy is called the *hydration energy*.* When the hydration energy

---

* Note that the hydration energy actually takes into account both the solvent-solvent interaction (the energy required to make a hole in the water) and the solvent-solute interaction (the energy released by the ion-water attraction). These are lumped together because experimentally they cannot be separated. In other words, we cannot hydrate an ion without first making room for it, any more than we can make a hole in the water without putting something in it.

Process	ΔH, kJ/mol	ΔS, kJ mol^{-1} °K^{-1}	ΔG, kJ/mol
NaCl(s) $\longrightarrow$ Na$^+$(g) + Cl$^-$(g)	+770	+0.227	+702
Na$^+$(g) + Cl$^-$(g) $\longrightarrow$ Na$^+$(aq) + Cl$^-$(aq)	−766	−0.184	−711
NaCl(s) $\longrightarrow$ Na$^+$(aq) + Cl$^-$(aq)	+4	+0.043	−9

Fig. 8.13

Thermodynamic parameters at 298° K. (The numbers apply only under specific ideal conditions: p = 1 atm; concentration = 1 m.)

is greater than the lattice energy, the overall dissolving process liberates energy to the surroundings; i.e., the net process is exothermic. When less energy is furnished by the hydration step than is required to break up the lattice, the overall solution process is endothermic. In a few cases the lattice energy is approximately equal to the hydration energy. For example, the lattice energy of NaCl is 770 kJ/mol and the hydration energy is 766 kJ/mol. These two energies just about balance each other; so the heat of solution, which is the difference between the two, is nearly zero.

We might ask why it is that NaCl dissolves in water at all if it takes 770 kJ of heat to break up the lattice whereas only 766 kJ is made available on hydrating the ions. The net process appears to be energetically unfavorable. The answer is that the dissolving occurs because there is a favorable entropy change. Figure 8.13 shows the quantitative values leading to this conclusion. The overall dissolving process, shown in the last line, is the sum of the two preceding steps. For each step, values are listed for the ΔH, the ΔS, and the ΔG, all at 298°K. The ΔH represents the increase in enthalpy, or heat content, of the system (Sec. 7.13); it is equal in magnitude but opposite in sign to what we have called the "heat of reaction." There is often confusion in the sign convention of ΔH and the heat of reaction because the former describes the system and the latter refers to the surroundings. We illustrate specifically with the second line of Fig. 8.13. The equation

Na$^+$(g) + Cl$^-$(g) $\longrightarrow$ Na$^+$(aq) + Cl$^-$(aq)

has associated with it $\Delta H = -766$ kJ/mol. This means that the enthalpy of the Na$^+$ and Cl$^-$ system has *decreased* by 766 kJ in going from the gaseous to the hydrated state. The *decrease* in H came about because of liberation of 766 kJ to the surroundings. So, the heat of reaction, i.e., the heat evolved to the surroundings, is +766 kJ, but the ΔH of the system is −766 kJ. In similar fashion, Fig. 8.13 shows that ΔS for the second step equals −0.184 kJ mol^{-1} deg.$^{-1}$. This means that the ion system decreases in entropy by 0.184 kJ in going from the gas to hydrated states.

Once we know ΔH and ΔS, we can calculate ΔG, the free-energy increase of the system, from $\Delta G = \Delta H - T \Delta S$ (Sec. 7.13). Since the temperature is specified as 298°K, we have for the hydration process $\Delta G = \Delta H - T \Delta S = -766 - (298)(-0.184)$, or −711, kJ/mol. In other words, when Na$^+$(g) and Cl$^-$(g) become hydrated, there is a decrease in the free energy of the system by 711 kJ. The important point to note in Fig. 8.13 is that for the final overall process, although ΔH is positive (and therefore energetically unfavorable), ΔG still turns out to be negative; hence, the overall dissolving reaction is favored. The favorable ΔG comes about because the $T \Delta S$ term, which is equal to $(298°)(+0.043 \text{ kJ mol}^{-1} \text{deg}^{-1}) = 13 \text{ kJ/mol}$, subtracts from the unfavorable ΔH of $+4 \text{ kJ/mol}$ to give -9 kJ/mol. The positive ΔS term arises from the fact that ions in the solution, although more restricted than in the gaseous state, have considerably more freedom than they do in solid NaCl. The overall dissolving change is one of increasing disorder and hence of increasing entropy.

Once the appropriate thermodynamic parameters* are known, the behavior of a saturated solution with respect to changing temperature can be deduced. Consider, for example, a solute such as NaCl whose solubility is limited by an unfavorable ΔH for the process of dissolving into the saturated solution. In the equilibrium state, $\Delta G = 0$; i.e., there would be no change in free energy per mole if an infinitesimal amount of solute were transferred from excess solute into the saturated solution, or vice versa. The condition $\Delta G = 0$ comes about because the unfavorable ΔH is just balanced out by a favorable $T \Delta S$. What happens if T is increased slightly? If we assume ΔH and ΔS stay constant, only T changes. Hence, $T \Delta S$ wins out in making ΔG negative, and more solute tends to dissolve. Stated differently, in the case where both ΔH and ΔS are positive, an increase of T at equilibrium makes $\Delta H - T \Delta S = \Delta G$ less than zero and so leads to spontaneous dissolving of additional solute.

In other cases, such as Li_2CO_3, solubility is limited not by an unfavorable ΔH but by an unfavorable $T \Delta S$. This happens if ΔH is negative (reaction is exothermic) and if ΔS is also negative (reaction produces order, as by making an orderly structure of solvent molecules about the solute ions). In such a case an increase of T (again if we assume ΔH and ΔS are constant) increases the unfavorable $T \Delta S$ term; so $\Delta H - T \Delta S = \Delta G$ becomes greater than zero. As a result, the reverse reaction should occur as the temperature rises.

To summarize, if the saturated solution is one for which ΔH for the process

Solute + solvent $\longrightarrow$ solution

is positive (endothermic reaction), the solubility increases with rising temperature; if it is one in which ΔH is negative (exothermic reaction), the solubility decreases with rising temperature. A convenient way of remembering the direction of the effects is in terms of Le Châtelier's

---

* As we shall see later (Sec. 13.7), the thermodynamic parameters, such as those shown in Fig. 8.13, vary somewhat with changes in temperature and concentration. The numerical values quoted are generally for standard reference states such as 1 atm pressure, 298°K, and dissolved species at 1 m concentration. The point that needs to be made here is that the numbers given in Fig. 8.13 are for such reference states, not for the saturated solution. Because of changes to saturation concentrations brought about by the dissolving process, the $\Delta G = -9\text{kJ}$ of Fig. 8.13 goes to $\Delta G = 0$ at saturation.

principle (Sec. 7.11). If the dissolving process absorbs heat, the stress of increased temperature can be relieved by favoring the dissolving reaction. If the dissolving process liberates heat, the stress of increased temperature can be relieved by the reverse process, namely, precipitation.

4 *Pressure.* The solubility of all gases is increased as the partial pressure of the gas above the solution is increased. For example, the concentration of CO_2 which is dissolved in a carbonated beverage (e.g., champagne) is dependent directly on the partial pressure of CO_2 in the gas phase. When a bottle is opened, the pressure of CO_2 drops, its solubility is diminished, and bubbles of CO_2 form and escape from the beverage. Quantitatively, this is expressed in *Henry's law*, which states that at constant temperature *the mole fraction of gas dissolved in a solvent is directly proportional to the partial pressure of the gas.* Henry's law is usually written

$$K = \frac{p}{x}$$

where K is the Henry's law constant, p is the partial pressure of the solute in the gas phase, and x is the mole fraction of gas in the solution. K is generally given in millimeters of mercury and changes somewhat with temperature. A typical value of K is 2.95×10^7, which is the value for oxygen in water at $20°C$.

As far as liquid and solid solutes are concerned, there is essentially no change of solubility with pressure. If there is a change, it can be predicted by Le Châtelier's principle, since it depends on the relative volume of the solution and the component substances. In general the volume change on solution is so small that pressures on the order of thousands of atmospheres are needed in order to change the solubility appreciably.

In closing this section on solubility we need to note that it is sometimes possible to prepare solutions which have a higher concentration of solute than would correspond to a saturated solution. Such solutions are referred to as being *supersaturated;* they are unstable with respect to separation of excess solute. A supersaturated solution of sodium acetate, $NaC_2H_3O_2$, for example, can be made as follows: A saturated solution of $NaC_2H_3O_2$ and H_2O in contact with excess solute is heated until the increase of solubility with temperature is sufficient to dissolve all the excess solute. At sufficiently high temperatures an unsaturated solution results. This unsaturated solution is then cooled very carefully. The system ought to return to its original equilibrium state with excess solute crystallizing out. This, in fact, does happen with most solids. However, with some, such as sodium acetate, cooling, if done carefully, can be accomplished without crystallization. The resulting solution has a concentration of solute higher than would correspond to the saturated solution at the lower temperature. It is supersaturated. The situation is reminiscent of that observed in the supercooling of a liquid below its freezing point. Supersaturation can usually be destroyed in the same manner, i.e., by seeding. When a tiny seed crystal of sodium acetate is placed in a supersaturated solution of sodium acetate, excess solute crystallizes on it until the remaining solution is just saturated. Occasionally, a mechanical disturbance such as a sudden shock may suffice to break the supersaturation. Dust particles or even scratches on the inner surface of the container may act as centers on which crystallization can start.

8.8 Colloids

In introducing the topic of solutions it was more or less implied that it is easily possible to distinguish between a homogeneous mixture and a heterogeneous one. However, this distinction is not a sharp one. There are systems which are neither obviously homogeneous nor obviously heterogeneous. They are classed as intermediate and are known as *colloids*. Examples are cigarette smoke, fog, emulsions, foam, and albumin. In order to get an idea of what a colloid is, imagine a process in which a sample of solid is placed in a liquid and progressively subdivided. As long as distinct particles of solid are visible to the naked eye, there is no question that the system is heterogeneous. On standing, the visible particles generally separate out. Depending on the relative density of solid and liquid, the particles float to the top or settle to the bottom. They can be separated easily by filtration. However, as progressive subdivision is continued, a state is finally reached in which the dispersed particles have been broken down to individual molecules or atoms. In this limit, two phases can no longer be distinguished; a solution has been produced. No matter how powerful a microscope is used, a solution appears uniform throughout, and individual molecules cannot be seen. Dissolved particles do not separate out on standing, nor can they be separated by filtration.

Between coarse suspensions and true solutions there is a region of change from heterogeneity to homogeneity. In this region dispersed particles are so small that they do not form an obviously separate phase, but they are not so small that they can be said to be in true solution. This state of subdivision is called the *colloidal state*. On standing, particles of a colloid do not separate out at an appreciable rate; they cannot be seen under a microscope; nor can they be separated by filtration. The dividing lines between colloids and solutions on one hand and between colloids and discrete phases on the other are not rigorously fixed since a continuous gradation of particle size is possible. Usually, the definition of colloid is based on size. When particle size lies between about 10^{-7} and 10^{-4} cm, a dispersion can be called a *colloid*, a *colloidal suspension*, or a *colloidal solution*.

The size of a dispersed particle does not tell anything about the constitution of the particle. It may consist of atoms, of small molecules, or of one giant molecule. For example, colloidal gold consists of various-sized particles each containing a million or more gold atoms. Colloidal sulfur can be made with particles containing a thousand or so S_8 molecules. An example of a giant molecule (also called *macro-*

molecule) is hemoglobin, the protein responsible for the red color of the blood. The molecular weight of this molecule is 66,800 amu, and its diameter is approximately 6×10^{-7} cm.

Colloids are frequently classified on the basis of the states of aggregation of the component phases, even though the separate phases are no longer visibly distinguishable once the colloid is formed. The more important classifications are *sols, emulsions, gels,* and *aerosols.* In *sols* a solid is dispersed through a liquid, so that the liquid forms the continuous phase and bits of solid form the discontinuous phase. Milk of magnesia is a sol consisting of solid particles of magnesium hydroxide dispersed through water. Sols can be made by *dispersion* (breaking down larger particles) or *condensation* (building up small particles to colloidal dimensions). Colloidal gold is a sol which can be made by striking an electric arc between two gold electrodes under water. It can also be made by chemical reduction of chloroauric acid, $HAuCl_4$, by a slow reducing agent such as hydrazine, N_2H_4. Investigation of gold sol by X rays has shown that the particles of gold which are dispersed throughout the water are crystalline in nature.

Emulsions are colloids in which a liquid is dispersed through a liquid. A common example is ordinary milk, which consists of butterfat globules dispersed through an aqueous solution. A *gel* is an unusual type of colloid in which a liquid contains a solid arranged in a fine network extending throughout the system. Both the solid and the liquid phases are continuous. Examples of gels are jellies, gelatin, agar, and slimy precipitates such as aluminum hydroxide. An *aerosol* is a colloid made by dispersing either a solid or a liquid in a gas. The former is called a *smoke,* and the latter a *fog. Smog,* in at least one of its types, the so-called "London smog," is essentially a combination of coal smoke and sulfuric acid fog.

8.9   *Light scattering and Brownian motion*

When a beam of light is passed through a solution or a pure liquid, the path of the beam is invisible when viewed from the side. However, when a beam of light is turned on a colloid, an observer to one side can see the path of the beam. The situation is shown in Fig. 8.14. The effect, called the *Tyndall effect,* can be produced readily by turning a column of light on an aqueous solution of sodium thiosulfate, $Na_2S_2O_3$, and adding a few drops of dilute acid. The ensuing chemical reaction produces elemental sulfur. The light beam is invisible until the sulfur particles aggregate to colloidal dimensions. In an ordinary solution the particles of solute are much smaller than the wavelength of visible light. Since visible light has a wavelength ranging from 400 to 700 nm, solute particles that are 0.5 nm or so in diameter are too small to reflect or scatter the wave to the side. However, when the solute particles grow to a size that is on the order of a wavelength, the light beam is scattered, or diffracted, and becomes visible from the side. Careful studies of the intensity of scattering as a function of angle of scattering can be used to determine not only the size but also the shape of macromolecules. In this way one can deduce that albumin (having molecular weight 69,000 amu) is ellipsoidal in shape whereas fibrinogen (a blood-clotting protein of molecular weight 400,000 amu) has an elongated, fibrillar shape. It must be emphasized that the shape of the molecule is not seen directly; it must be deduced from the light-scattering data.

When a microscope is focused on a Tyndall beam, the colloidal particle itself is too small to be seen, but its position may be fixed by noting where the point of light appears. When observed in this way, colloidal particles are seen to undergo brownian motion, the rapid, random, zigzag motion previously mentioned in discussing gases (Sec. 5.10). The smaller the colloidal particle, the more violent its brownian motion.

Under ordinary circumstances, a colloid in an uninsulated container shows no tendency to settle out. However, when the colloid is kept in a well-insulated container, after a time there will develop a gradation in the concentration of colloidal particles from the top to the bottom of the sample. This gradation in concentration develops because there are two opposing effects: (1) the attraction due to gravity, which tends to pull heavier particles down, and (2) the dispersing effect due to brownian motion. The more massive the particles, the more important is effect (1), and the more pronounced is the concentration gradient.* The main reason no appreciable concentration gradient is observed for colloids in uninsulated containers is that there are convection currents due to nonuniform temperature. These currents keep the colloidal suspension constantly stirred up.

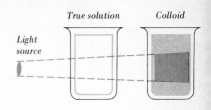

Fig. 8.14
Light scattering.

8.10 Adsorption

In some colloids the particles adsorb ions and thereby acquire electric charge. For example, ferric oxide sol (which is chemically closely related to the material we know as rust) consists of positively charged aggregates of ferric oxide units due to adsorption of H^+. The resulting positive charge enhances the stability of the colloid. Normally, when one neutral particle in its brownian motion hits another, the two coagulate to form a larger particle. The larger particle in turn grows by collision until it becomes so large that brownian motion cannot keep it in suspension. With ferric oxide, where the H^+ ions are believed to be stuck on oxygen atoms which protrude from the particles, the colloidal particles have a net positive charge and hence repel similarly charged particles. The charged ferric oxide particles try to stay as far apart from each other on the average as possible. Consequently, there is less chance that they will come together to form a large mass that settles out. In a similar way, arsenious sulfide As_2S_3, forms a beautiful

* Two extreme cases can be imagined. With rocks in water the settling is so pronounced that all the rocks are at the bottom. With a true solution, such as sugar in water, the gradation in concentration is so slight that only the most careful experiments involving very tall columns and precise temperature control could show any difference in concentration between the top and bottom of the sample.

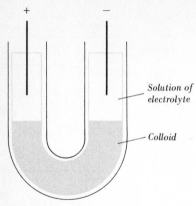

Solution of
electrolyte

Colloid

Fig. 8.15
Electrophoresis.

yellow negative sol by adsorbing SH⁻ or OH⁻ ions. As one might guess, mixing a positively charged ferric oxide sol with a negatively charged arsenious sulfide sol coagulates them both.

That some colloidal particles are electrically charged can be shown by studying *electrophoresis, the migration of colloidal particles in an electric field*. Figure 8.15 shows the experimental setup. A U tube is partly filled with the colloidal solution to be examined. Very carefully, so as not to disturb the colloid, the remainder of the U tube is filled with a solution of a suitable electrolyte. The electrolytic solution needs to be of lower density (so that it will stay on top), different in appearance from the colloid (so that the boundary can be clearly seen), and appreciably electrically conducting. The choice of electrolyte is limited by the fact that many electrolytes coagulate colloids, but usually a salt, such as sodium bromide, which does not affect the colloid can be found. Electrodes are inserted into the solution, with one electrode positively charged and the other negatively charged. After a time that may range from 30 min to 48 h, the boundaries between the colloid and the electrolyte solution will be observed to have shifted toward one or the other electrode because of net migration of the colloid through the solvent.

From the direction of migration of the colloid the sign of the charge of the colloidal particles can be determined. For example, when the electrophoresis of ferric oxide sol is observed in this cell, the boundary moves toward the negative electrode and away from the positive electrode, suggesting that the colloid is positively charged. By observing the rate at which migration occurs, it is also possible to get information about the size and the shape of the colloidal particles. The study of electrophoresis has been applied with great success to protein molecules. In acid solution protein molecules pick up hydrogen ions to become positively charged. From the migration in an electric field it has been possible to draw inferences about their size and shape.

Some colloidal particles adsorb films of molecules, which shield them from other particles. An example of this adsorption is found in gelatin. Gelatin is a high-molecular-weight protein (made by boiling skin, tendons, ligaments, bones, etc., in water) which has the property of tying to itself a sheath of water. This film of very tightly bound water protects the gelatin particle from coagulating with another gelatin particle. If two gelatin particles collide, they do not coagulate, because the shielded gelatin parts have not been able to get in contact with each other. This property of gelatin is used in stabilizing colloids of silver bromide in preparing photographic film. When finely divided silver bromide is stirred up with water, it settles out. However, if the silver bromide is first mixed with gelatin, the gelatin forms a film on the outside of each silver bromide particle. The gelatin in turn adsorbs a layer of water; so essentially two protective films have formed on the silver bromide to keep it in suspension.

There are two reasons for the high adsorptive properties shown by colloids: One is the extremely large surface area. For example, whereas one cubic centimeter of sulfur in the form of a single cube 1 cm on edge has a surface area of only 6 cm², when ground up into cubes which are 10^{-5} cm on the edge, it has a surface area of 6×10^5 cm². The second reason for great adsorption is that surface atoms in general have special properties. Valence forces of surface atoms are usually not satisfied since the atom is designed to bond in three dimensions, which it cannot do at the surface. The greater the state of subdivision, the greater the fraction of atoms there are on the surface with unsatisfied valence forces.

Charcoal is a substance in which the surface atoms are an appreciable fraction of the total number. It consists of porous carbon having a network of fine tunnels extending through the specimen. The surface area on the walls of these tunnels has been determined to be on the order of 100 m²/g. On all this large surface area are carbon atoms which have unsaturated valence. They can attract molecules, especially polar molecules, thus accounting for the high adsorption that is characteristic of charcoal. When a mixture of hydrogen sulfide, H_2S, and oxygen is passed over a charcoal surface, the H_2S is selectively adsorbed. Because H_2S is a polar molecule with the sulfur end more negative than the hydrogen end, it is more strongly adsorbed than the oxygen molecule, which is symmetric and nonpolar. The charcoal gas mask makes use of this principle of selective adsorption. The charcoal selectively adsorbs poisonous gases, which are usually complicated polar molecules, and lets the oxygen through for respiration.

At higher temperatures molecular motion makes adsorption more difficult. Thus, spent charcoal with its surface completely covered may be reactivated by heating it up to drive off the adsorbed gases. At low temperatures, where molecular motion is slight, adsorption increases. In fact, at very low temperatures selectivity is less pronounced, and even nonpolar molecules may be adsorbed, presumably because of van der Waals attraction. At liquid-nitrogen temperatures ($-196°C$) even oxygen gas is strongly adsorbed.

Exercises

* **8.1** *Terms.* What, if any, is the difference between the following terms?
  a Solute vs. solvent
  b Solution vs. alloy
  c Solvated vs. hydrated
  d Homogeneous vs. heterogeneous

* **8.2** *Mole fraction.* What will be the mole fraction of oxygen in an air sample that contains by volume 78.084% N_2, 20.946% O_2, 0.033% CO_2, and 0.934% Ar?

* **8.3** *Normality.* For a reaction in which $Cr_2O_7^{2-}$ goes to Cr^{3+}, how many grams of $Na_2Cr_2O_7$ would you need to make 500 ml of 3.00 N $Na_2Cr_2O_7$ (density 1.075 g/ml)? *Ans. 65.5 g*

* **8.4** *Concentrations.* A solution is 40.0% by weight C_2H_5OH and 60% by weight H_2O. Its density is 0.935 g/ml. Calculate for this

solution the concentration of C_2H_5OH in terms of each of the following: (a) molality, (b) molarity, and (c) mole fraction.

* 8.5 *Concentration percent.* When 10.00 ml of ethyl alcohol (C_2H_5OH, density 0.7893 g/ml) is mixed with 20.0 ml of water, the final solution has a density of 0.957 g/ml. What will be the concentration of the ethyl alcohol in solution expressed as (a) percent by weight and (b) percent by volume?

Ans. (a) 28.3% by weight, (b) 34.3% by volume

* 8.6 *Electrolytes.* Criticize the following definitions: An electrolyte is a substance that conducts electric current; a nonelectrolyte is a substance that does not.

* 8.7 *Electrolytes.* In what sense might it be more appropriate to use the term "ionization" than "dissociation" for the process that occurs when HCl is placed in water?

* 8.8 *Solubility.* You are given three unlabeled beakers containing aqueous solutions of sodium acetate that are, respectively, unsaturated, saturated, and supersaturated. You are also given but two tiny crystals of sodium acetate. How would you proceed to determine which solution is which?

* 8.9 *Solubility.* The solubilities of NaCl and $C_{12}H_{22}O_{11}$ in water are 311 and 1311 g per liter of solution, respectively. The corresponding densities of the saturated solutions are 1.197 and 1.473 g/ml. Calculate the molality and the weight percent of solute in each of these solutions.   *Ans. 6.00 m, 26.0% NaCl; 23.7 m, 89.0% $C_{12}H_{22}O_{11}$*

* 8.10 *Colloids.* Account for each of the following observations:
a   In a tall, well-insulated column the concentration of colloidal particles is greater at the bottom than at the top.
b   A ferric oxide sol that settles out in a matter of hours can be stabilized for days by a touch of hydrochloric acid.
c   Mixing a red ferric oxide sol with a yellow arsenious sulfide sol produces a colorless solution.
d   It takes half as many moles of barium chloride as it does of sodium chloride to precipitate an arsenious sulfide sol.

** 8.11 *Concentration.* You are given a solution made by dissolving 10.0 g of potassium dichromate, $K_2Cr_2O_7$, in 90.0 g of water. The density of the resulting solution is 1.070 g/ml. What are (a) the molarity and (b) the molality of the solution?

** 8.12 *Molarity.* How many grams of water should one add to 0.150 liter of 6.02 *M* HCl solution (density 1.10 g/ml) to make 2.42 *M* HCl (density 1.04 g/ml)?   *Ans. 223 g*

** 8.13 *Molality.* What will be the molality of the final solution made by mixing equal volumes of an HCl solution that is 10.00% HCl by weight (density 1.048 g/ml) with an HCl solution that is 20% HCl by weight (density 1.098 g/ml)? How would your answer be changed if you mixed equal weights of the two solutions instead of equal volumes?

** 8.14   Freezing-point lowering. Explain qualitatively why the vapor-pressure lowering produced by a solute dissolved in water necessarily leads to a freezing-point lowering.

** 8.15   Raoult's law. When 0.50 mol of NH_4Cl is dissolved in 1 kg of water, the observed vapor-pressure lowering of water is 0.0158 atm at a temperature of 100°C. Assuming NH_4Cl is 100% dissociated into NH_4^+ and Cl^-, what vapor-pressure lowering would you have expected? How might you explain the discrepancy?

Ans. 0.0177 atm

** 8.16   Raoult's law. Cholesterol is very soluble in ether. If 0.869 g of cholesterol dissolved in 4.44 g of diethyl ether ($C_4H_{10}O$) lowers the vapor pressure of ether from 0.526 to 0.507 atm at 17.9°C, what must be the molecular weight of cholesterol?

** 8.17   Boiling-point elevation. In making maple syrup by evaporating the sap of a maple tree, it is recommended that evaporation be continued until the boiling point becomes 4.0°C higher than that of the starting sap (essentially pure water). Assuming that the sugar in the final syrup is $C_{12}H_{22}O_{11}$, what would be the percent-by-weight composition of the final syrup?   *Ans. 72%*

** 8.18   Molecular weight. Skatole, one of the putrid-smelling constituents of human waste, analyzes to 82.41% by weight carbon, 6.92% hydrogen, and 10.67% nitrogen. When it dissolves in ethyl alcohol to the extent of 10.0 g per 100 g of C_2H_5OH, it raises its boiling point from 78.51 to 79.46°C. What is the molecular formula of skatole? The boiling-point elevation constant of C_2H_5OH is 1.24°C/m.

** 8.19   Percent dissociation. Acetic acid, $HC_2H_3O_2$, is dissociated slightly in water to give H_3O^+ and $C_2H_3O_2^-$. An aqueous solution labeled 0.2600 m $HC_2H_3O_2$ shows an observed freezing point of −0.488°C. Calculate the apparent percent dissociation of the $HC_2H_3O_2$.

** 8.20   Strong electrolytes. Taking the data of Fig. 8.10 at face value, what would appear to be the percent dissociation of NaCl in 0.1 and 0.01 m solutions? How does the Debye-Hückel theory reconcile these results with the assumption of 100% dissociation?

Ans. 87 and 94%

** 8.21   Solubility. Account for the fact that pouring some ethyl alcohol into a saturated aqueous solution of sodium chloride produces a precipitate but doing the same thing to a saturated aqueous solution of sucrose does not.

** 8.22   Temperature. If a kilogram of saturated aqueous NaCl solution at 100°C is cooled to 20°C, how many grams of NaCl should precipitate? Data are given in Fig. 8.12.

** 8.23   Gas solubility. In a pollution-free lake that is in equilibrium with air (containing 20.95% O_2 by volume) the concentration of dissolved oxygen at 20°C is 9 ppm by weight. What value of the Henry's law constant for oxygen-gas solubility would you deduce from this?   *Ans. 3 × 10⁷*

** 8.24   Gas solubility. The Henry's law constant for dissolving CO_2 in water at 20°C is 1.08×10^6 mm Hg per mole fraction. You are given 1.00 liter of gaseous CO_2 at 1.00 atm pressure and 20°C in contact with 1.00 liter of pure water. How many atmospheres of pressure would you have to exert to dissolve the CO_2 in the water?

** 8.25   Gas solubility. In discussions of water pollution, the content of dissolved oxygen is generally described in parts per million in terms of weight. Using data from Fig. 8.12, calculate the parts per million to be expected for dissolved oxygen in a freshwater lake at 0, 20, and 40°C. Note the fact that oxygen pressure in the atmosphere is normally about 0.21 atm.   *Ans. 14.5, 9.0, 6.5 ppm*

** 8.26 ΔH and ΔS. Given two experimental setups each at 25°C, one consisting of NaCl(*s*) in contact with pure water and the other consisting of NaCl(*s*) in contact with a saturated aqueous solution of NaCl, discuss what relations must exist between ΔH and ΔS for the two systems for the process NaCl(*s*) $\longrightarrow$ NaCl(*aq*).

** 8.27   Colloids. If one assumes that the hemoglobin molecule, molecular weight 66,800 amu, has a roughly spherical shape with a diameter of about 6.0 nm, what would be the average density of material in the molecule?

** 8.28   Light scattering. If one looks at smog by reflected light, it appears bluish; but if one looks at it by transmitted light, it appears reddish brown. Suggest a reason for this. (*Hint:* The scattering of electromagnetic radiation is inversely proportional to the fourth power of the wavelength.)

** 8.29   Colloids. Suppose you have a protein that consists of a distribution of macromolecules varying in molecular weight about some mean such as 600,000 amu. How might you set about to separate the molecules into different molecular-weight ranges?

** 8.30   Adsorption. The total surface area of finely divided materials can be determined by measuring the amount of simple molecules adsorbed on the surface. Assuming that a nitrogen molecule needs a cross-sectional area of 9×10^{-16} cm² to be adsorbed, what would be the total surface area of a 1.00-g sample of charcoal that is observed to adsorb 207 cm³ (STP) of N_2 under appropriate conditions?   *Ans. 500 m²*

*** 8.31   Freezing-point lowering. A 15-liter automobile cooling system is to be conditioned against freezing down to −30°C. Assuming ideal behavior and no volume change on mixing, what volume of glycerol ($C_3H_8O_3$, density 1.26 g/ml) would you need to provide?

***8.32 *Strong electrolytes.* Predict the effect on the conductivity per mole and the freezing-point lowering per mole of NaCl in water of reducing the dielectric constant of the water. Assume all other factors stay the same.

***8.33 *Henry's law.* For nitrogen gas dissolved in water, the Henry's law constant changes from 5.75×10^7 mm Hg per mole fraction at $20°C$ to 9.20×10^7 mm Hg per mole fraction at $80°C$. What volume of nitrogen gas measured at $80°C$ and 1 atm would you expect to see expelled on heating in air (78.08% nitrogen by volume) one liter of nitrogen-saturated water?   *Ans. 6.1 cm³*

***8.34 *ΔH and ΔS.* The ΔH for the process $KCl(s) \longrightarrow K^+(g) + Cl^-(g)$ is 699 kJ/mol; the ΔH for $K^+(g) + Cl^-(g) \longrightarrow K^+(aq) + Cl^-(aq)$ is -690 kJ/mol. From its resemblance to NaCl and the data in Fig. 8.13, estimate values of ΔS for each of the above processes, and use them to predict a value for ΔG for the process $KCl(s) \longrightarrow K^+(aq) + Cl^-(aq)$. By comparing the temperature-solubility data for NaCl and KCl given in Fig. 8.12, decide whether your estimates of ΔS are too large or too small. Discuss which of the steps, lattice breakup or ion hydration, is probably more at fault.

***8.35 *ΔH and ΔS.* From the following data predict whether $TlCl(s)$ is soluble or insoluble in water:

$TlCl(s) \longrightarrow Tl^+(g) + Cl^-(g)$ $\Delta H = 736$ kJ/mol
$\Delta S = +0.249$ kJ mol^{-1} deg^{-1}

$Tl^+(g) \longrightarrow Tl^+(aq)$ $\Delta H = -322$ kJ/mol
$\Delta S = -0.044$ kJ mol^{-1} deg^{-1}

$Cl^-(g) \longrightarrow Cl^-(aq)$ $\Delta H = -377$ kJ/mol
$\Delta S = -0.097$ kJ mol^{-1} deg^{-1}

Calculate ΔG for $TlCl(s) \longrightarrow Tl^+(aq) + Cl^-(aq)$ at $25°C$.

***8.36 *Colloids.* Suppose you have a gold sol in which the colloidal particles are approximately spherical and contain roughly a million gold atoms each. If one assumes the gold atoms are in cubic close-packing (fcc) with unit-cell edge length of 0.407 nm, approximately what percent of the gold atoms are surface atoms, as distinguished from interior atoms?   *Ans. About 4%*

One reason for the heavy emphasis on solutions is that a large fraction of all chemical reactions are carried out in solutions. In this chapter we consider in some detail the various kinds of solution reactions. Special attention is given to what species are present and what changes they undergo.

9.1 Acids and bases

Several definitions of acids and bases are in common use, and there is no general agreement on which is the most useful. It is therefore necessary to be familiar with each of them. We consider first the simplest definition, that suggested by Arrhenius, the discoverer of electrolytic dissociation. He accounted for the traditional acidic properties (viz., sour taste, red coloration of litmus, and reaction with metals to liberate hydrogen) by postulating that all acids have the formula HX and can dissociate to give H^+ and X^-:

$$HX \longrightarrow H^+ + X^-$$

It might be noted that X^- itself may contain a dissociable proton and itself be an acid. Thus, for Arrhenius, acids include HCl, HNO_3, H_2SO_4, HSO_4^-, and $HC_2H_3O_2$. Along with this definition of acid, Arrhenius proposed that all bases can be written MOH and can dissociate to give M^+ and OH^-:

$$MOH \longrightarrow M^+ + OH^-$$

Examples of Arrhenius bases are NaOH, $Ca(OH)_2$, and $Al(OH)_3$. Since an acid produces H^+ and a base produces OH^-, a base can neutralize an acid by forming water and the salt MX (in solution as M^+ and X^-).

The Arrhenius system has the attractive feature of simplicity, and for that reason it is widely used. However, it suffers from lack of generality. For example, it does not cover obviously acidic solutions such as CO_2 in H_2O or basic solutions such as those of NH_3 in H_2O. To handle such cases, the system requires that these substances form acids or bases on reaction with H_2O. Thus, for example, it was postulated that CO_2 reacts with H_2O to form H_2CO_3 (carbonic acid) and NH_3 reacts with H_2O to form NH_4OH (ammonium hydroxide). In neither case, however, is the situation so simple. In aqueous solutions of CO_2 less than 1% of the CO_2 exists as H_2CO_3; in aqueous ammonia no discrete NH_4OH species has ever been identified.

By slight modification, the Arrhenius system can be generalized to cover all acidic and basic solutions in water as well as in other solvents. This definition is sometimes referred to as the *general-solvent system of acids and bases*. In this system it is recognized that the solvent itself may be somewhat dissociated into positive and negative fragments. In such cases, the term "acid" is applied to any substance that raises the concentration of the positive fragments; the term "base" is applied to any substance that raises the concentration of the negative fragments.

Let us consider, for example, the situation in water. Even when very highly purified, water conducts electric current to a slight extent and is assumed to be slightly dissociated into positive and negative ions. This can be written as

$$H_2O \longrightarrow H^+ + OH^-$$

but because the hydrogen ion is so strongly hydrated (as discussed in Sec. 8.4), it is better to write the dissociation as

$$H_2O + H_2O \longrightarrow H_3O^+ + OH^-$$

The degree of dissociation is very small; in pure water the concentration of hydronium ion is 1.0×10^{-7} M. The concentration of hydroxide ion is, of course, the same, since each time a water molecule is split, one hydronium ion and one hydroxide ion are formed.

In a liter of H_2O at room temperature there is approximately 1000 g, or 55 mol, of H_2O. Thus, the fraction of H_2O dissociated is $1.0 \times 10^{-7}/55$, or 0.0000002%. On the average, only 2 out of 1 billion molecules of H_2O are dissociated.

In pure water, the hydronium-ion concentration and the hydroxide-ion concentration are equal, but the balance can be upset by the addition of other substances. Those *substances which increase the hydronium-ion concentration are called acids;* those substances which increase the *hydroxide-ion concentration* are called *bases*. HCl is obviously an acid because when placed in water it is converted to H_3O^+ and Cl^-, thus raising the concentration of H_3O^+. It is a *strong acid* because it is *completely converted* to H_3O^+. $HC_2H_3O_2$, on the other hand, is a *weak acid* because it is only *slightly converted* to H_3O^+. What about CO_2? It is also an acid because it raises the H_3O^+ concentration over what it would be in pure water. The reaction can be represented as

$$CO_2 + 2H_2O \longrightarrow H_3O^+ + HCO_3^-$$

where HCO_3^- represents the bicarbonate ion. Similarly, NH_3 is considered a base because its addition to water increases the hydroxide-ion concentration. The reaction can be represented as

$$NH_3 + H_2O \longrightarrow NH_4^+ + OH^-$$

Other more complex reactions (sometimes called *hydrolysis reactions*) can occur between salts and water so as to change the H_3O^+ and OH^- concentrations of the water. For example, when certain aluminum compounds are added to water, the solutions are acidic; when certain sulfides are added to water, the solutions are basic. For aluminum salts the increase of hydronium-ion concentration can be attributed to the net reaction

$$Al^{3+} + 2H_2O \longrightarrow H_3O^+ + AlOH^{2+}$$

and for sulfides the increase of hydroxide-ion concentration can be attributed to

$$S^{2-} + H_2O \longrightarrow OH^- + SH^-$$

Solvents other than H_2O also show self-dissociation. For example, in liquid ammonia there is slight conductivity because of the reaction

$$NH_3 + NH_3 \longrightarrow NH_4^+ + NH_2^-$$

In the ammonia system, any substance that raises the NH_4^+ concentration is an acid; any substance that raises the NH_2^- concentration, a base. Thus, NH_4Cl in liquid ammonia is an acid; KNH_2 is a base. As expected, "neutralization" occurs when liquid-ammonia solutions of NH_4Cl and KNH_2 are mixed in equivalent amounts.

A third commonly used system for defining acids and bases is that proposed in 1923 independently by J. N. Brønsted of Denmark and by T. M. Lowry of England. In the *Brønsted-Lowry system* an acid is defined as any species that acts as a *proton donor* and a base as any species that acts as a *proton acceptor*. As far as acids are concerned, the Brønsted-Lowry definition matches the Arrhenius definition discussed above. The new feature is its more general definition of base. For example, not only is OH^- a base in this system, but so is H_2O. When HCl is placed in H_2O and transfers a proton to H_2O

$$HCl + H_2O \longrightarrow H_3O^+ + Cl^-$$

the H_2O acts as a proton acceptor and is therefore a base. Similarly, when H_2O undergoes self-dissociation

$$H_2O + H_2O \longrightarrow H_3O^+ + OH^-$$

one of the H_2O molecules acts as the proton donor, and the other H_2O molecule acts as the proton acceptor.

As a general rule, any species that can donate a proton (and hence can be written HX) gives a species X^- that in turn can accept a proton to form HX. Thus, every Brønsted-Lowry acid is paired with a Brønsted-Lowry base. The two are said to constitute a *conjugate acid-base pair*. To illustrate, HCl is the conjugate acid of Cl^-; Cl^- is the conjugate base of HCl. Other pairs of conjugate acids and bases include the following: $HC_2H_3O_2$ and $C_2H_3O_2^-$, H_2SO_4 and HSO_4^-, HSO_4^- and SO_4^{2-}, NH_4^+ and NH_3, H_3O^+ and H_2O, and H_2O and OH^-. In the Brønsted-Lowry system, as contrasted to the simple Arrhenius view, "dissociation" of an acid is not a simple breakup of HX to give H^+ and X^- but rather a proton transfer from an acid HX to a molecule of solvent that functions as a base:

$$HX + H_2O \longrightarrow H_3O^+ + X^-$$

For the forward reaction, HX is the acid, and H_2O is the base; for the reverse reaction H_3O^+ is the acid, and X^- the base.

Different acids vary in their tendency to give up a proton. The relative acid strength depends on this tendency and, to a smaller degree, on the solvent in which the proton transfer occurs. Picking H_2O as the common solvent, we can write an ordered list of acids, as is done in Fig. 9.1. In the left column acids are arranged by decreasing acid strength; i.e., the strongest acid is at the top. In the right column is shown the conjugate base of each of the acids. Since a strong acid HX necessarily implies a weak conjugate base X^-, the order in the right column is inverted; i.e., the weakest base is at the top, and the strongest is at the bottom. From the ordered list it is possible to predict the direction in which proton transfer is favored. Any acid in the list has great tendency to transfer a proton to any base *below* it in the list; conversely any acid has small tendency to transfer a proton to

Fig. 9.1
Brønsted-Lowry acids and
bases.

		Conjugate acid	Conjugate base	
Strongest		$HClO_4$	ClO_4^-	Weakest
		H_2SO_4	HSO_4^-	
		HCl	Cl^-	
		H_3O^+	H_2O	
		HSO_4^-	SO_4^{2-}	
		HF	F^-	
		$HC_2H_3O_2$	$C_2H_3O_2^-$	
		H_2S	HS^-	
		NH_4^+	NH_3	
		HCO_3^-	CO_3^{2-}	
		H_2O	OH^-	
		HS^-	S^{2-}	
Weakest		OH^-	O^{2-}	Strongest

any base *above* it. For example, the order in the list shows that when HF is placed in water, there is but little tendency for the HF to transfer its proton to H_2O. (HF on the acid side of the list is lower than H_2O on the base side of the list.) This only confirms the fact that HF in H_2O is a weak acid. Similarly, the order of the list indicates that when HF in H_2O is mixed with Cl^- and NH_3, the proton will be preferentially transferred to NH_3 and not to Cl^-. Of all the acids listed in Fig. 9.1, only the so-called "strong" acids $HClO_4$, H_2SO_4, and HCl have greater tendency to give up a proton than does H_3O^+. They, therefore, transfer protons to the base H_2O to form H_3O^+. It might be noted that HSO_4^- in the acid column lies below H_3O^+; hence, HSO_4^- has but little tendency to transfer a proton to H_2O and is to be regarded as a weak acid.

Finally, we should note that even though tendencies may be small, reaction can occur to some extent in all cases. Thus, HSO_4^- in H_2O does undergo limited proton transfer to form some H_3O^+ and SO_4^{2-}. Similarly, for NH_3 in H_2O there is some proton transfer from H_2O to NH_3, leading to limited formation of OH^- and NH_4^+:

$$H_2O + NH_3 \xrightarrow[\text{slight}]{} NH_4^+ + OH^-$$

Likewise, solutions of carbonate are slightly basic because of the limited reaction

$$H_2O + CO_3^{2-} \longrightarrow HCO_3^- + OH^-$$

These cases of limited proton transfer to give slightly basic solutions are to be contrasted with the last two bases in the table. If either S^{2-} or O^{2-} is placed in water, there is marked production of OH^-:

$$H_2O + S^{2-} \longrightarrow HS^- + OH^-$$

$$H_2O + O^{2-} \longrightarrow OH^- + OH^-$$

In the last reaction conversion is 100 percent complete. In the other cases, where extent of conversion is limited, the problem can be treated quantitatively by use of equilibrium constants, as will be done in Sec. 12.10.

There is yet another commonly used definition of acid and base; it is called the *Lewis definition*. It applies the term "acid" to any species that acts as an *electron-pair acceptor* and the term "base" to an *electron-pair donor*. For instance, in the reaction

$$H^+ + \begin{array}{c} H \\ \cdot\cdot \\ :N:H \\ \cdot\cdot \\ H \end{array} \longrightarrow \left[\begin{array}{c} H \\ \cdot\cdot \\ H:N:H \\ \cdot\cdot \\ H \end{array} \right]^+$$

the Lewis acid H^+ accepts a share in the pair of electrons donated by the Lewis base NH_3. The great generality of the Lewis definition can be seen from the fact that it covers the reaction

$$\begin{array}{c} :\ddot{F}: \quad H \\ \cdot\cdot \quad \cdot\cdot \\ :\ddot{F}:B + :N:H \\ \cdot\cdot \quad \cdot\cdot \\ :\ddot{F}: \quad H \end{array} \longrightarrow \begin{array}{c} :\ddot{F}:H \\ \cdot\cdot \quad \cdot\cdot \\ :\ddot{F}:\ddot{B}:\ddot{N}:H \\ \cdot\cdot \quad \cdot\cdot \\ :\ddot{F}:H \end{array}$$

as well as

$$\begin{array}{c} :\ddot{Cl}: \\ \cdot\cdot \\ :\ddot{Cl}:Al + :\ddot{Cl}:^- \\ \cdot\cdot \\ :\ddot{Cl}: \end{array} \longrightarrow \left[\begin{array}{c} :\ddot{Cl}: \\ \cdot\cdot \quad \cdot\cdot \\ :\ddot{Cl}:Al:\ddot{Cl}: \\ \cdot\cdot \quad \cdot\cdot \\ :\ddot{Cl}: \end{array} \right]^-$$

and even

$$:\ddot{Br}:Hg:\ddot{Br}: + 2\left[:\ddot{Br}: \right]^- \longrightarrow \left[\begin{array}{c} :\ddot{Br}: \\ \cdot\cdot \\ :\ddot{Br}:Hg:\ddot{Br}: \\ \cdot\cdot \\ :\ddot{Br}: \end{array} \right]^{2-}$$

Since Lewis acids look for electrons, they are sometimes described as electron-seeking, or *electrophilic* (from the Greek *philos*, meaning "loving"). Similarly, Lewis bases, which look for appropriate atoms to bind their electrons to, are described as being *nucleophilic*.

The choice of definition of acids and bases can be defended equally in each of the above four systems. The Arrhenius system, because of its simplicity, has the widest usage to recommend it. Its minor modification into the general-solvent system produces a workable definition which is sufficiently general to cover most cases. The Brønsted-Lowry system has great merit, owing to its emphasis on the role of the solvent. However, it must be recognized that in half a century it has not yet displaced the older definitions from the chemical literature. The Lewis system also frequently appears in the literature, owing, no doubt, to its great generality. However, this generality in turn is a weakness. For aqueous systems, at least, it is doubtful that the Lewis system will ever replace other definitions of acids and bases.

9.2 Neutralization

No matter which definition of acids and bases is used, it is generally agreed that acids and bases react with one another. In the Arrhenius system this reaction is called *neutralization,* because as the H^+ (or, better yet, H_3O^+) of the acid reacts with the OH^- of the base, the acidic and basic properties of the acid and the base disappear. However, even here the term "neutralization" has to be taken with caution since the acid and base need to be of comparable strength to produce a strictly neutral solution, i.e., one that like water has a precise equality in H_3O^+ and OH^- concentrations. For example, when one mole of acetic acid reacts with one mole of sodium hydroxide, the result is not a neutral solution but a slightly basic one. The reason for this, as will be discussed in Sec. 12.10, is that acetate ion disturbs the water-dissociation equilibrium. As for the more general definitions, e.g., the Brønsted-Lowry one, the wide range of acid-base reactions makes the term "neutralization" more questionable. Thus, reaction of the acid HCl with the base H_2O can hardly be called a neutralization. Strictly speaking, then, one should define neutralization as the reaction of acids and bases of comparable strength. The following paragraphs consider four possible situations involving reaction between an acid and a base in aqueous solution: strong acid–strong base, weak acid–strong base, strong acid–weak base, and weak acid–weak base. As noted above, only the first and the last of these should properly be called neutralization. However, common usage often refers to all four as neutralization reactions.

1 Strong acid–strong base. The neutralization of HCl with NaOH is sometimes written as

$$HCl + NaOH \longrightarrow H_2O + NaCl$$

but since HCl, NaOH, and NaCl are all strong electrolytes, the species present in solution are ions. The equation is better written as

$$H^+ + Cl^- + Na^+ + OH^- \longrightarrow H_2O + Na^+ + Cl^-$$

or, still better, since we recognize that H^+ is more properly represented by H_3O^+, as

$$H_3O^+ + Cl^- + Na^+ + OH^- \longrightarrow 2H_2O + Na^+ + Cl^-$$

Since Na^+ and Cl^- appear on both sides of the equation, they can be dropped out to give the net equation

$$H_3O^+ + OH^- \longrightarrow 2H_2O$$

Neither the negative ion (Cl^-) nor the positive ion (Na^+) appears in the net equation; so it is general and applies to the neutralization of any strong acid by any strong base.

2 Weak acid–strong base. For the reaction of a weak acid with a strong base the net equation can be represented as

$$HA + OH^- \longrightarrow H_2O + A^-$$

where HA stands for a weak acid, such as acetic acid, $HC_2H_3O_2$. Since weak acids are only slightly dissociated into H_3O^+ and A^- ions, the original solution of the weak acid contains predominantly HA molecules. In the reaction it is the HA molecules that ultimately disappear, and this must be shown in the net equation. It may well be that the actual mechanism of the reaction involves, first, dissociation of HA into H_3O^+ and A^-, with subsequent union of H_3O^+ and OH^- to give H_2O. The net equation represents only the overall reaction.

   3   *Strong acid–weak base.* For the reaction of strong acid by a weak base, the net reaction can be written

$$H_3O^+ + MOH \longrightarrow M^+ + 2H_2O$$

where MOH represents a weak base such as thallium hydroxide, TlOH. There are not very many weak bases of the type MOH. Most of the weak bases are like NH_3. In such cases, the net reaction for neutralization by a strong acid is written

$$NH_3 + H_3O^+ \longrightarrow NH_4^+ + H_2O$$

   4   *Weak acid–weak base.* For the reaction between a weak acid and a weak base, the net equation is

$$HA + MOH \longrightarrow M^+ + A^- + H_2O$$

If the weak base is NH_3

$$HA + NH_3 \longrightarrow NH_4^+ + A^-$$

9.3   *Polyprotic acids*

The term *polyprotic acid* (also referred to as *polybasic acid*) is used to describe those acids which can furnish more than one proton per molecule. Two examples of polyprotic acids are the diprotic acid H_2SO_4, sulfuric acid, and the triprotic acid H_3PO_4, phosphoric acid. In reaction, polyprotic acids usually transfer only one proton at a time. For example, when placed in water, H_2SO_4 transfers a proton to H_2O:

$$H_2SO_4 + H_2O \longrightarrow H_3O^+ + HSO_4^-$$

This reaction is essentially 100 percent complete; so, in this sense, H_2SO_4 is called a strong electrolyte. When a solution containing one mole of NaOH is mixed with a solution containing one mole of H_2SO_4, one mole of H_3O^+ neutralizes one mole of OH^-. Evaporation of the resulting solution gives one mole of the salt $NaHSO_4$, sodium hydrogen sulfate. (This salt is also called sodium bisulfate and, less and less frequently, primary sodium sulfate.) The ion HSO_4^- is an acid in its own right. Although fairly weak, it can dissociate to give SO_4^{2-}. This reaction can be written

$$HSO_4^- + H_2O \xrightarrow[slight]{} H_3O^+ + SO_4^{2-}$$

It can be considered as the second step in the dissociation of the diprotic acid H_2SO_4 and occurs significantly only when there is a large demand for protons. For example, when one mole of H_2SO_4 is mixed in solution with two moles of NaOH, the two moles of OH^- neutralize two moles of H^+. Evaporation of the solution would produce the salt Na_2SO_4, sodium sulfate.

The triprotic acid H_3PO_4 undergoes dissociation in three steps:

$$H_3PO_4 + H_2O \xrightarrow{slight} H_3O^+ + H_2PO_4^-$$

$$H_2PO_4^- + H_2O \xrightarrow{slight} H_3O^+ + HPO_4^{2-}$$

$$HPO_4^{2-} + H_2O \xrightarrow{slight} H_3O^+ + PO_4^{3-}$$

The extent of dissociation is again governed by the demand for protons. It is possible to get three salts from this acid. The following are the net equations for reactions between one mole of H_3PO_4 and one, two, and three moles, respectively, of NaOH in solution:

$$H_3PO_4 + OH^- \longrightarrow H_2O + H_2PO_4^-$$

$$H_3PO_4 + 2OH^- \longrightarrow 2H_2O + HPO_4^{2-}$$

$$H_3PO_4 + 3OH^- \longrightarrow 3H_2O + PO_4^{3-}$$

Evaporation of the corresponding solutions would give the salts NaH_2PO_4, monosodium dihydrogen phosphate; Na_2HPO_4, disodium monohydrogen phosphate; and Na_3PO_4, trisodium phosphate. These are sometimes referred to as the *primary*, *secondary*, and *tertiary* sodium phosphates, respectively.

9.4   Hydrolysis

When the salt NaCl is placed in water, the resulting solution is observed to be neutral; i.e., the concentrations of H_3O^+ and OH^- are equal, 1×10^{-7} M, just as in pure water. However, when the salt $NaC_2H_3O_2$ is dissolved in water, the resulting solution is observed to be slightly basic. Other salts such as ammonium chloride, NH_4Cl, or aluminum chloride, $AlCl_3$, give slightly acid solutions. These interactions between salts and water are called *hydrolysis*. The term comes from the Greek *lysis*, which means "loosening;" so the literal meaning of hydrolysis is "loosening, or breaking up, by water."

Hydrolysis is not fundamentally different from any acid-base reaction as viewed in the Brønsted-Lowry system. In the case of sodium acetate, for example, the basic nature of the resulting aqueous solution can be understood from the following equation:

$$C_2H_3O_2^- + H_2O \xrightarrow{slight} HC_2H_3O_2 + OH^-$$

Here the Brønsted-Lowry base $C_2H_3O_2^-$ accepts a proton from the Brønsted-Lowry acid H_2O to form the conjugate acid $HC_2H_3O_2$ and the conjugate base OH^-. The basic nature of the solution is due to the formation of OH^-. This reaction goes only slightly from left to right because $C_2H_3O_2^-$ is too weak a base to compete significantly with OH^- for a proton (see Fig. 9.1).

In principle, any negative ion can act as a base. In practice, however, only a few negative ions compete effectively with OH^-. These include only the very strong bases (for example, O^{2-} and S^{2-}); so only solutions of these ions are likely to be as basic as solutions prepared from hydroxides. Most negative ions compete only slightly with OH^- for protons, and in aqueous solutions of these ions the extent of hydrolysis is small. Thus, solutions of Na_2CO_3, $NaC_2H_3O_2$, NaF, and Na_2SO_4, for example, are only slightly basic. Although the extent of hydrolysis increases with dilution, in solutions of the above salts at moderate concentration (order of one molar) the extent of hydrolysis rarely exceeds a percent or so. The extent of hydrolysis of a few negative ions (for example, ClO_4^-, Cl^-, and NO_3^-) is so small as to be undetectable. These anions are the weakest of bases, and their conjugate acids are the strongest of acids.

In a few cases the negative ion of a salt can act as an acid. For example, HSO_4^- can transfer a proton to H_2O to give an acidic solution containing H_3O^+ and SO_4^{2-}. Thus, aqueous solutions of $NaHSO_4$ are definitely acidic. In similar fashion, HCO_3^- (bicarbonate ion) could transfer a proton to H_2O to form H_3O^+ except that HCO_3^- can also accept a proton (to form ultimately H_2O and CO_2). The two effects occur about equally and so tend to cancel. Solutions of $NaHCO_3$ are very slightly basic.

Positive ions that hydrolyze generally produce acidic solutions. In some cases, for example, NH_4^+, the source of the proton that leads to H_3O^+ is obvious:

$$NH_4^+ + H_2O \xrightarrow[slight]{} NH_3 + H_3O^+$$

In other cases, e.g., solutions of Al^{3+} salts, the source of the acidity is more complicated. It is probably best attributed to splitting off a proton from the water molecules directly attached to the Al^{3+} of the hydrated aluminum ion. Although the configuration of hydrated aluminum ion is not completely known, it is usually assumed to be octahedral; i.e., the aluminum ion is thought to be attached to six water molecules at the corners of an octahedron. Figure 9.2 represents an octahedron oriented in the same way as hydrated aluminum ion. The oxygen atoms (large spheres) are centered on the corners of an octahedron, the center of which is occupied by the aluminum ion. The transfer of a proton from the hydrated aluminum ion to solvent can be written

$$Al(H_2O)_6^{3+} + H_2O \longrightarrow Al(H_2O)_5OH^{2+} + H_3O^+$$

The reaction, which leaves a doubly charged complex ion, is represented by Fig. 9.3.

The extent of loss of protons from bound water molecules in hydrated ions depends on how tightly the protons are held to the oxygen atoms. The effect of the aluminum ion is to pull electrons toward itself away from the water molecule, as shown in Fig. 9.4. The bonds between the hydrogen atoms and the oxygen of the water molecule are thereby weakened, and the protons may dissociate. The higher the charge of

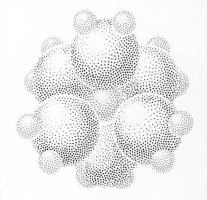

Fig. 9.2
Octahedron and $Al(H_2O)_6^{3+}$ ion. (The colored sphere buried in the middle is the Al; the small outer spheres are H atoms attached to the O atoms.)

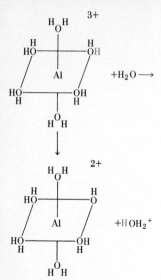

Fig. 9.3
Acid reaction of hydrated aluminum ion.

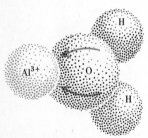

Fig. 9.4
Weakening of O—H bond by electron attraction to Al³⁺ ion.

the cation, the more pull exerted on the electrons, and the more easily the protons dissociate. Thus, most $+1$ ions (for example, Na^+, K^+, and Ag^+) show no detectable hydrolysis; most $+2$ ions (for example, Mg^{2+}, Zn^{2+}, and Cu^{2+}) are slightly hydrolyzed; $+3$ ions (for example, Al^{3+}, Fe^{3+}, and Cr^{3+}) are moderately hydrolyzed and are about as acidic as the moderately weak acetic acid.

In the case of very strong interaction between a central positive ion and bound water of hydration, all the protons from the attached water molecules may dissociate. The existence of oxyions, such as SO_4^{2-}, can be explained in this way. The argument is as follows: In SO_4^{2-} the sulfur is assigned an oxidation state of $+6$. Except under extreme conditions of high temperature, S^{6+} is not encountered as a chemical species. However, let us imagine having an S^{6+} ion, which is then placed in water. Molecules of water cluster about it in the usual process of hydration, but S^{6+} has such a high positive charge that electrons are strongly attracted to it. The hydrogen-oxygen bonds in the water are so weakened that the protons split off to leave the oxygen associated with the sulfur. Thus, the formation of SO_4^{2-} can be described by the hypothetical reaction

$$S^{6+} + 4H_2O \longrightarrow SO_4^{2-} + 8H^+$$

Of the eight dissociating protons, the last one is weakly held, thus accounting for hydrogen sulfate ion, HSO_4^-. The formation of nitrate ion, NO_3^-, can be similarly explained as resulting from the 100 percent hydrolysis of hypothetical N^{5+}.

9.5 Amphoterism

If a solution of sodium hydroxide is added dropwise to a solution of aluminum nitrate, it is observed that a white precipitate, aluminum hydroxide, is first formed, but on further addition of base or on addition of acid the precipitate dissolves. The net equations for the dissolving processes can be written

$$Al(OH)_3(s) + OH^- \longrightarrow Al(OH)_4^-$$

$$Al(OH)_3(s) + 3H_3O^+ \longrightarrow Al^{3+} + 6H_2O$$

These two equations indicate that $Al(OH)_3$ is *amphoteric*, i.e., *able to neutralize bases and acids*. In other words, aluminum hydroxide is able to act as an acid and as a base. Zinc hydroxide, $Zn(OH)_2$, lead hydroxide, $Pb(OH)_2$, and chromium hydroxide, $Cr(OH)_3$, are examples of other common amphoteric hydroxides. Oxides may also be classified as amphoteric if they react with water to form amphoteric hydroxides.

How can amphoterism be explained? The answer can be seen by writing the above equations in a way that emphasizes the probable species involved:

create

$$Al(OH)_3(H_2O)_3(s) + OH^- \longrightarrow Al(OH)_4(H_2O)_2^- + H_2O$$

$$Al(OH)_3(H_2O)_3(s) + 3H_3O^+ \longrightarrow Al(H_2O)_6^{3+} + 3H_2O$$

Each of these formulas represents a species containing Al^{3+} at the center of an octahedron with six oxygen-containing species, either OH^- or H_2O, at the octahedral corners. It should be noted that for all the complexes the number of near-neighbor oxygens is six. Figure 9.5 shows the relative arrangements of atoms in the various species. In the first of the reactions, a proton is transferred from $Al(OH)_3(H_2O)_3$ to OH^-; in the second, proton transfer is from H_3O^+ to $Al(OH)_3(H_2O)_3$. Figure 9.5 also shows (in color) the two possible proton transfers. In case A a proton from an H_2O bound to the central Al transfers to an external OH^-; in case B a proton from an external H_3O^+ transfers to an OH bound to the Al. For case A, the aluminum hydroxide species acts as a proton donor; for case B, a proton acceptor. In both cases, the chemical bond of interest is an

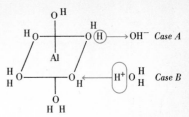

Fig. 9.5
Two possible modes of reaction leading to amphoteric behavior of aluminum hydroxide.

$$\overset{\displaystyle H}{\underset{}{|}}$$
Al—O-to-H

bond which is broken in case A and formed in case B.

For amphoterism to occur, this bond must be intermediate in strength between an H-to-OH bond and an H-to-OH_2 bond. The strength of the Al(OH)-to-H bond is determined by the electron-pulling ability of Al^{3+}, as shown in Fig. 9.4. If the electron pull is insufficient, the O-to-H bond of H_2O is not sufficiently weakened to enhance proton transfer from a bound water of hydration and to make it more favorable than proton transfer from a solvent H_2O. This is the reason why Na^+, Ca^{2+}, and even La^{3+} do not give amphoteric hydroxides. These ions do not hydrolyze very much, an indication that the O-to-H bond in their attached water of hydration has not been weakened sufficiently to show acidic properties. In general, positive ions that do not hydrolyze very much are unlikely to give amphoteric hydroxides. On the other hand, positive ions that hydrolyze too much (for example, S^{6+}) likewise do not give amphoteric hydroxides. This arises because a hydrated ion in which the O-to-H bond has been weakened too much (i.e., below the strength of the H-to-OH_2 bond in H_3O^+) will not accept a proton from H_3O^+. A good example is ClOH. It has so little affinity for a proton that it does not act as a base but only as an acid. The Cl end of the molecule pulls electrons away from the O so strongly that the O has little tendency to add an additional proton.

9.6 Stoichiometry of solutions

Labels on reagent bottles generally specify what the solution was made from, but usually not what the solution contains. For example, the label $0.5\ M$ HCl may appear on a solution made from 0.5 mol of HCl and sufficient water to give one liter of solution. Despite the label, *there are no HCl molecules in the solution.* HCl is a strong electrolyte and is completely converted to H_3O^+ and Cl^-. One might argue that the label might better read $0.5\ M\ H_3O^+$ and $0.5\ M\ Cl^-$. For most quantitative considerations, however, it is not necessary to know what species are actually in the solution. It is necessary to know only what is ultimately available. The label 0.5 M $HC_2H_3O_2$ also tells what the solution was made from, but in this case the solution actually contains $HC_2H_3O_2$ molecules; it is a weak electrolyte and is very slightly dissociated. There is only a trace of H_3O^+ and $C_2H_3O_2^-$ in the solution. However, if the solution were used for a neutralization reaction, not only the trace of H_3O^+ but also the $HC_2H_3O_2$ would be neutralized.

The use of solutions for keeping track of stoichiometry changes in chemical reactions requires a clear distinction between the *number of moles* of solute in a solution and its *concentration*. To illustrate, let us suppose 15.8 g of $KMnO_4$, potassium permanganate, is to be dissolved to make 1 liter of 0.100 M $KMnO_4$ solution. The formula weight of $KMnO_4$ is 158 amu; hence 15.8 g is equal to 0.100 mol. To make up the solution, we place the solute in a graduated container and add water to it. Not precisely 1 liter of water is added but only enough to bring the total volume to a liter. (Usually, the volume of solute plus the volume of the solvent is not exactly equal to the volume of the final solution.) The solution can now be labeled 0.100 M $KMnO_4$, since it contains 0.100 mol of $KMnO_4$ in 1 liter of solution. The concentration does not depend on how much of this solution is taken. Whether one drop or 200 ml is considered, the solution is still 0.100 M $KMnO_4$. However, the number of moles of $KMnO_4$ taken does depend on the volume of solution. If the volume and the concentration of a sample are known, the number of moles of solute in the sample is the number of moles per liter multiplied by the volume of the sample in liters. In 200 ml of 0.100 M $KMnO_4$, there is (0.200 liter)(0.100 mol/liter), or 0.0200 mol, of $KMnO_4$.

Solutions are convenient because they permit measuring amounts of solute, not by weighing the solute, which is a tedious procedure, but by measuring a volume of solution. For example, suppose a given chemical reaction requires 0.0100 mol of $KMnO_4$. This amount of $KMnO_4$ can be provided by weighing out 1.58 g of $KMnO_4$ or, more easily, by measuring out 100 ml of 0.100 M $KMnO_4$ solution.

In summary, we can write

Liters $\times$ molarity of solution = moles of solute in sample

Liters $\times$ normality of solution = equivalents of solute in sample

The following examples show how these quantitative relations are applied.

Example 1
To what volume must 50.0 ml of 3.50 M H_2SO_4 be diluted in order to make 2.00 M H_2SO_4?

50.0 ml of 3.50 M H_2SO_4 contains (0.0500 liter)(3.50 mol/liter), or 0.175 mol, of H_2SO_4.

We wish the final solution to be 2.00 M.
Therefore, the final volume must be

$$\frac{0.175 \text{ mol}}{2.00 \text{ mol/liter}} = 0.0875 \text{ liter} = 87.5 \text{ ml}$$

■ ■ ■

Example 2
If 50.0 ml of 0.50 M H_2SO_4 is added to 75.0 ml of 0.25 M H_2SO_4, what will be the concentration of the final solution, assuming its volume is 125 ml?

50.0 ml of 0.50 M H_2SO_4 contains (0.0500 liter)(0.50 mol/liter).
75.0 ml of 0.25 M H_2SO_4 contains (0.0750 liter)(0.25 mol/liter).

Total moles = (0.0500 liter)(0.50 mol/liter)
$$+ \ (0.0750 \text{ liter})(0.25 \text{ mol/liter}) = 0.044$$

Final concentration $= \dfrac{0.044 \text{ mol}}{0.125 \text{ liter}} = 0.35 \ M \ H_2SO_4$

■ ■ ■

Example 3
How many milliliters of 0.025 M H_3PO_4 are required to neutralize 25 ml of 0.030 M $Ca(OH)_2$? Assume complete neutralization.

Method 1   Moles

25 ml of 0.030 M $Ca(OH)_2$ contains

(0.025 liter)(0.030 mol/liter) = 0.00075 mol of $Ca(OH)_2$

The balanced equation is

$$3Ca(OH)_2 + 2H_3PO_4 \longrightarrow Ca_3(PO_4)_2 + 6H_2O$$

0.00075 mol of $Ca(OH)_2$ requires

$$(0.00075 \text{ mol}) \left(\frac{2 \text{ mol of } H_3PO_4}{3 \text{ mol of } Ca(OH)_2} \right) = 0.00050 \text{ mol of } H_3PO_4$$

To get this from 0.025 M H_3PO_4, we need

$$\frac{0.00050 \text{ mol } H_3PO_4}{0.025 \text{ mol/liter}} = 0.020 \text{ liter} = 20 \text{ ml}$$

Method 2   Equivalents

0.030 M $Ca(OH)_2$ means 0.060 N $Ca(OH)_2$.
25 ml of 0.060 N $Ca(OH)_2$ contains

(0.025 liter)(0.060 equiv/liter) = 0.0015 equiv of $Ca(OH)_2$

We need 0.0015 equiv of acid to neutralize 0.0015 equiv of base.
0.025 M H_3PO_4 means 0.075 N H_3PO_4.

To get 0.0015 equiv of acid, take

$$\frac{0.0015 \text{ equiv}}{0.075 \text{ equiv/liter}} = 0.020 \text{ liter} = 20 \text{ ml}$$

■　■　■

Example 4

A solution is made by mixing 10.0 ml of 0.0200 M Ca(NO₃)₂ and 15.0 ml of 0.0300 M NaNO₃. Assuming the final volume is 25.0 ml and that dissociation is complete, calculate the concentrations of Ca^{2+}, Na^+, *and* NO_3^- *in the final solution.*

$$Ca^{2+} = (0.0100 \text{ liter})(0.0200 \text{ mol/liter}) = 0.000200 \text{ mol}$$

$$= \frac{0.000200 \text{ mol}}{0.0250 \text{ liter}} = 0.00800 \ M$$

$$Na^+ = (0.0150 \text{ liter})(0.0300 \text{ mol/liter}) = 0.000450 \text{ mol}$$

$$= \frac{0.000450 \text{ mol}}{0.0250 \text{ liter}} = 0.0180 \ M$$

$$NO_3^- = (0.0100 \text{ liter})(2 \times 0.0200 \text{ mol/liter})$$
$$+ (0.0150 \text{ liter})(0.0300 \text{ mol/liter})$$

$$= 0.000400 + 0.000450 = 0.000850 \text{ mol}$$

$$= \frac{0.000850 \text{ mol}}{0.0250 \text{ liter}} = 0.0340 \ M$$

■　■　■

In the quantitative treatment of oxidation-reduction reactions in aqueous solution, electrolytic dissociation actually simplifies the calculations. Only the net reaction need be considered; other ions present in the solution can be ignored. As a specific case we consider the reaction of an acidified solution of $KMnO_4$ with a solution of ferrous sulfate, $FeSO_4$. Before reaction occurs, the mixture contains K^+, MnO_4^-, H_3O^+, HSO_4^-, Fe^{2+}, and SO_4^{2-}. After the reaction is complete, the mixture contains K^+, HSO_4^-, Mn^{2+}, Fe^{3+}, and SO_4^{2-}. The K^+, HSO_4^-, and SO_4^{2-} do not change in the course of the reaction and can be ignored. The net reaction shows the disappearance of MnO_4^-, Fe^{2+}, and H_3O^+ and the appearance of Mn^{2+}, Fe^{3+}, and H_2O. The equation can be written

$$5Fe^{2+} + MnO_4^- + 8H_3O^+ \longrightarrow 5Fe^{3+} + Mn^{2+} + 12H_2O$$

Net equations show both the species involved in the reactions and the stoichiometry. The number of moles of reactants can be calculated from the equation. The volumes of solutions necessary for complete reaction are thus specified. In general, principal interest is focused on the oxidizing agent and the reducing agent, since the acidic or basic nature of the solution is usually provided by an excess of an acid or base.

Example 5
How many milliliters of 0.20 M KMnO$_4$ *are required to oxidize 25.0 ml of 0.40 M* FeSO$_4$ *in acidic solution? The reaction which occurs is the oxidation of* Fe^{2+} *by* MnO$_4^-$ *to give* Fe^{3+} *and* Mn^{2+}.

Method 1   Moles

The balanced net equation is

$$5Fe^{2+} + MnO_4^- + 8H_3O^+ \longrightarrow 5Fe^{3+} + Mn^{2+} + 12H_2O$$

25.0 ml of 0.40 M FeSO$_4$ supplies

$$(0.0250 \text{ liter})(0.40 \text{ mol/liter}) = 0.010 \text{ mol of Fe}^{2+}$$

The equation states that one mole of MnO$_4^-$ is needed per five moles of Fe^{2+}.

0.010 mol of Fe^{2+} requires

$$(0.010 \text{ mol of Fe}^{2+})\left(\frac{1 \text{ mol of MnO}_4^-}{5 \text{ mol of Fe}^{2+}}\right) = 0.0020 \text{ mol of MnO}_4^-$$

To get this from 0.20 M KMnO$_4$ solution, we need to take

$$\frac{0.0020 \text{ mol of MnO}_4^-}{0.20 \text{ mol/liter}} = 0.010 \text{ liter} = 10 \text{ ml}$$

The problem can also be solved by using equivalents as defined in Sec. 4.10. In this reaction the reducing agent Fe^{2+} changes to Fe^{3+}. Each Fe^{2+} loses one electron to the oxidizing agent. The Avogadro number of Fe^{2+} ions, or one mole of Fe^{2+}, can furnish the Avogadro number of electrons. For this reaction, one mole of FeSO$_4$ is equal to one equivalent of FeSO$_4$, and 0.40 M FeSO$_4$ is 0.40 N. The oxidizing agent, MnO$_4^-$, changes to Mn^{2+} in the course of the reaction. As the manganese changes oxidation state from $+7$ to $+2$, each MnO$_4^-$ appears to gain five electrons. One mole of MnO$_4^-$ requires five times the Avogadro number of electrons and so is equal to five equivalents. Therefore, 0.20 M KMnO$_4$ is 1.0 N. In general, the *normality of a solution of an oxidizing or reducing agent is equal to the molarity times the electron change per formula-unit.*

Example 5 can be rephrased using normality instead of molarity: *How many milliliters of 1.0 N* KMnO$_4$ *are required to oxidize 25.0 ml of 0.40 N* FeSO$_4$ *in acidic solution?*

Method 2   Equivalents

25.0 ml of 0.40 N FeSO$_4$ supplies

$$(0.0250 \text{ liter})(0.40 \text{ equiv/liter}) = 0.010 \text{ equiv of reducing agent}$$

One equivalent of any reducing agent requires one equivalent of any oxidizing agent.

0.010 equiv of reducing agent requires 0.010 equiv of oxidizing agent

$$\frac{0.010 \text{ equiv}}{1.0 \text{ equiv/liter}} = 0.010 \text{ liter} = 10 \text{ ml}$$

■ ■ ■

In some cases a given oxidizing agent can be reduced to different products depending on conditions. For example, MnO_4^- can be reduced to Mn^{2+}, MnO_2, or MnO_4^{2-} if the medium is acid, neutral, or basic, respectively. In these reactions the corresponding number of electrons transferred is five, three, and one. A solution that is $1\ M$ $KMnO_4$ is $5\ N$ $KMnO_4$, $3\ N$ $KMnO_4$, or $1\ N$ $KMnO_4$ depending on which product is formed. Thus, the normality given for a particular solution may be ambiguous unless the reaction for which the solution is to be used is specified.

Example 6
A solution is made by mixing 200 ml of 0.100 M $K_2Cr_2O_7$, 250 ml of 0.200 M H_2SO_3 (a weak electrolyte), and 350 ml of 1.00 M $HClO_4$ (a strong electrolyte). A reaction occurs in which H_2SO_3 and $Cr_2O_7^{2-}$ convert in acidic solution to HSO_4^- and Cr^{3+}. Assuming the final volume is 0.800 liter, calculate the final concentration of each species that was present in the initial solution.

The best way to solve such a problem is to make a table listing in the first column all the species of interest, in the second column how many moles of each species are present at the start of the reaction, in the third column how many moles change, in the fourth column how many moles are left, and in the last column what the final concentrations are.

The balanced equation for the conversion of $H_2SO_3 + Cr_2O_7^{2-}$ in acidic solution to $HSO_4^- + Cr^{3+}$ (see Sec. 4.8) is as follows:

$$Cr_2O_7^{2-} + 3H_2SO_3 + 5H_3O^+ \longrightarrow 2Cr^{3+} + 3HSO_4^- + 9H_2O$$

To calculate the moles of species present initially, we proceed as follows:

Moles of K^+ = (0.200 liter)(0.100 mol of $K_2Cr_2O_7$ per liter)

$$\left(\frac{2 \text{ mol of } K^+}{1 \text{ mol of } K_2Cr_2O_7} \right)$$

$$= 0.0400 \text{ mol}$$

Moles of $Cr_2O_7^{2-}$ = (0.200 liter)(0.100 mol of $K_2Cr_2O_7$ per liter)

$$= 0.0200 \text{ mol}$$

Moles of H_2SO_3 = (0.250 liter)(0.200 mol H_2SO_3 per liter)

$$= 0.0500 \text{ mol}$$

Moles of H_3O^+ = (0.350 liter)(1.00 mol $HClO_4$ per liter)

$$= 0.350 \text{ mol}$$

Moles of $ClO_4^- = (0.350 \text{ liter})(1.00 \text{ mol } HClO_4 \text{ per liter})$

$$= 0.350 \text{ mol}$$

The equation states that we need three moles of H_2SO_3 for each mole of $Cr_2O_7^{2-}$. We have 0.0500 mol of H_2SO_3 for 0.0200 mol of $Cr_2O_7^{2-}$. Therefore, $Cr_2O_7^{2-}$ is present in excess, and the extent of reaction is limited by the moles of H_2SO_3. From the equation, we see that per mole of H_2SO_3 we use up $\frac{1}{3}$ mol of $Cr_2O_7^{2-}$ and $\frac{5}{3}$ mol of H_3O^+, or per 0.0500 mol of H_2SO_3 we use up $\frac{1}{3}(0.0500)$ mol of $Cr_2O_7^{2-}$ and $\frac{5}{3}(0.0500)$ mol of H_3O^+. To get the values in the last column, we divide the moles present at the end of the reaction (next-to-last column) by the final volume of solution, i.e., 0.800 liter.

Species	Moles at start	Moles used up	Moles at end	Final concentration, M
K^+	0.0400	0	0.0400	0.0500
$Cr_2O_7^{2-}$	0.0200	$\frac{1}{3}(0.0500)$	0.0033	0.0041
H_2SO_3	0.0500	0.0500	0	0
H_3O^+	0.350	$\frac{5}{3}(0.0500)$	0.267	0.334
ClO_4^-	0.350	0	0.350	0.438

■ ■ ■

Exercises

*9.1   *Acids and bases.* Define the terms "acid" and "base" in each of the following systems: Arrhenius, general-solvent, Brønsted-Lowry, and Lewis. Give a specific example in each case.

*9.2   *Arrhenius acids.* Both sulfur dioxide (SO_2) and carbon dioxide (CO_2) dissolve in water to give aqueous solutions. Contrast the way in which the Arrhenius and the general-solvent systems account for this acidity. Write balanced net equations where feasible.

*9.3   *Neutralization.* Write a balanced net equation for each of the following neutralizations:
  a HNO_3 and NaOH (aqueous)
  b HNO_3 and NH_3 (aqueous)
  c   NaOH and CO_2 (aqueous)
  d $NaNH_2$ and NH_4NO_3 (in liquid ammonia)

*9.4   *Brønsted-Lowry.* Nitrous acid, HNO_2, is a weak electrolyte in aqueous solution. Write an equation for its dissociation, and tell how the Brønsted-Lowry terms "acid," "base," "conjugate acid," "conjugate base," and "conjugate acid-base pair" apply to this system.

*9.5   *Nucleophilic, electrophilic.* In the reaction HF + H_2O $\longrightarrow$ H_3O^+ + F^-, which species is nucleophilic, and which is electrophilic? How would these terms apply for the reverse reaction?

*9.6   *Neutralization.* How would you write the neutralization reaction that occurs when aqueous solutions of Al^{3+} and S^{2-} are mixed with each other?

*9.7 *Polyprotic acids.* Pyrophosphoric acid, $H_4P_2O_7$, is tetra-protic. Write balanced net equations for the stepwise dissociation of this acid, and pick out all the conjugate acid-base pairs.

*9.8 *Hydrolysis.* Account for the fact that an aqueous solution of Na_2CO_3 is basic whereas a solution of NaCl is neutral and one of $NaHSO_4$ is acidic. First do this in the Brønsted-Lowry system, then do it in the Arrhenius system.

*9.9 *General solvent.* In pure liquid ammonia, the concentration of NH_4^+ and of NH_2^- is $1.0 \times 10^{-16} M$. If liquid ammonia has a density of 0.77 g/ml, what percent of the NH_3 is dissociated?

Ans. 2.2 × 10⁻¹⁶%

*9.10 *Neutralization.* What would be the molarity of an H_2SO_4 solution, 15.0 ml of which is completely neutralized by 26.5 ml of $0.100\ M$ NaOH?

*9.11 *Neutralization.* How many grams of NaOH would it take for complete neutralization of 35.0 ml of $0.165\ M$ H_3PO_4?

Ans. 0.694 g

*9.12 *Stoichiometry.* What would be the final concentration of H_3O^+ and ClO_4^- in a solution made by mixing 20.0 ml of $0.30\ M$ $HClO_4$ and 40.0 ml of $0.15\ M$ $HClO_4$? Assume volumes are additive.

*9.13 *Stoichiometry.* Given 0.500 liter of $0.200\ M$ $HClO_4$ and all the water you need, how much $0.125\ M$ $HClO_4$ can you make? Assume volumes are additive.

Ans. 0.800 liter

*9.14 *Stoichiometry.* Assuming volumes are additive, how much water should you add to 25.0 ml of $3.00\ M$ H_2SO_4 to make (*a*) $0.100\ M$ H_2SO_4 and (*b*) $0.100\ N$ H_2SO_4?

*9.15 *Stoichiometry.* For a reaction in which $Cr_2O_7^{2-}$ is oxidized by H_2SO_3 in acidic solution to give Cr^{3+} and HSO_4^-, how many milliliters of $0.20\ M$ H_2SO_3 would you need to make 0.050 liter of $0.10\ M$ $K_2Cr_2O_7$?

Ans. 75 ml

*9.16 *Stoichiometry.* How many grams of $Ca(NO_3)_2$ must you add to 25.0 ml of $0.25\ M$ $NaNO_3$ to double the nitrate-ion concentration? Assume no increase in volume.

**9.17 *General-solvent system.* Suppose we have a solvent of general formula AB which is slightly dissociated into A^+ and B^-. State specifically what would constitute an acid, a base, and a salt in the solvent AB.

**9.18 *Brønsted-Lowry.* By reference to Fig. 9.1, predict what would happen if a solution of $NaHSO_4$ were added to a mixed solution of NaHS and NaCl. What would happen if a solution of $NaHSO_4$ were mixed with a solution of NH_3?

9.19 *Lewis acids.* Silver ion, Ag^+, combines with two ammonia molecules to form a complex of formula $Ag(NH_3)_2^+$. Draw a possible structural formula for the complex, and show how the terms "Lewis acid" and "Lewis base" apply to its formation. Distinguish the components as nucleophilic or electrophilic.

9.20 *Polyprotic acids.* In the dissociation sequence for H_3PO_4 there appear three Brønsted-Lowry acids and three Brønsted-Lowry bases. Rank them in order of increasing strengths, and explain the basis of your ordering.

9.21 *Hydrolysis.* Write a balanced net equation for the hydrolysis of a positive ion M^{n+}. How would the extent of hydrolysis probably vary with (a) increasing n and (b) increasing radius of M^{n+}? Explain.

9.22 *Hydrolysis.* You are given each of the following ions at the same relative concentration in aqueous solution: CO_3^{2-}, Cl^-, Ca^{2+}, Na^+, Ga^{3+}, O^{2-}, OH^-, Al^{3+}, and F^-. Arrange the solutions in probable order of decreasing acidity.

9.23 *Amphoterism.* Chromic ion, Cr^{3+}, forms an amphoteric hydroxide. Assuming each species contains a central chromium bound to six near-neighbor oxygen atoms, write two equations that illustrate the amphoteric nature of the hydroxide.

9.24 *Amphoterism.* Account for the fact that in the series M^{2+}, M^{3+}, M^{4+} it is observed for a given element that the hydroxides are, respectively, basic, amphoteric, and acidic.

9.25 *Stoichiometry.* You are given 0.250 liter of 0.750 M NaOH and 0.750 liter of water. Using no other starting material what is the maximum amount of 0.150 M NaOH that you can make? Assume volumes are additive.

9.26 *Stoichiometry.* Given a change in which Sn^{2+} is oxidized by BrO_3^- in acidic solution to give Sn^{4+} and Br^-, what would be the final concentration of H_3O^+ in a solution made by mixing 25.0 ml of 0.500 M HNO$_3$ (strong electrolyte), 10.0 ml of 0.030 M Sn(NO$_3$)$_2$, and 20.0 ml of 0.02 M NaBrO$_3$? Assume volumes are additive.

Ans. 0.216 M

9.27 *General-solvent system.* Consider a solvent of general formula AB which is slightly dissociated into A^+ and B^-. A dynamic equilibrium exists in the solvent wherein the dissociation process $AB \longrightarrow A^+ + B^-$ is opposed by an association process $A^+ + B^- \longrightarrow AB$. Use the principle of Le Châtelier to explain what happens to the A^+ concentration and to the B^- concentration when an acid is dissolved in the solvent AB.

9.28 *Hydrolysis.* When the hypothetical ions Cl^{7+}, Cl^{5+}, and Cl^{3+} are placed in aqueous solution, the acids formed are HClO$_4$, HClO$_3$, and HClO$_2$. Predict the relative strengths of these acids on the basis of a hydrolysis model.

***9.29 *Amphoterism.* Zinc hydroxide is an amphoteric hydroxide for which the formula can be written $Zn(OH)_2(H_2O)_2$, corresponding to four near-neighbor oxygen atoms disposed tetrahedrally about the zinc. Draw pictures showing possible acidic and basic behavior by the hydroxide. What condition needs to be satisfied for the relative bond strengths ZnOH-to-H, H-to-OH, and H-to-H_2O? Explain your reasoning.

***9.30 *Stoichiometry.* Given the change $Fe^{2+} + MnO_4^- \longrightarrow Fe^{3+} + Mn^{2+}$ in acidic solution, calculate the concentration of Fe^{2+}, MnO_4^-, Fe^{3+}, Mn^{2+}, and H_3O^+ in a solution made by mixing 0.152 g of $FeSO_4$, 10.0 ml of 0.0100 M $KMnO_4$, and 90.0 ml of 0.100 M $HClO_4$. Assume final volume is 0.100 liter.

***9.31 *Stoichiometry.* A solution is made by mixing 4.00 g of NaOH, 1.58 g of $Na_2S_2O_3$, and 4.47 g of NaClO in enough water to make 100 ml of solution. A reaction occurs in which $S_2O_3^{2-} + ClO^- \longrightarrow SO_4^{2-} + Cl^-$. What is the concentration of each species in the final solution?

Ans. 1.80 M Na+, 0.20 M ClO−, 0.20 M SO₄²−,
0.40 M Cl−, 0.80 M OH−

Chemical kinetics is the branch of chemistry concerned with the velocity of chemical reactions and the mechanism by which chemical reactions occur. The term *reaction velocity* is used to describe the rate at which chemical change occurs. The term *reaction mechanism* is used to describe the sequence of steps by which the overall change is accomplished. In many reactions it is only the disappearance of starting materials and the appearance of final products that can be detected; i.e., only the net reaction is observable. In general, however, the net reaction is not the whole story but simply represents a summation of all the changes that occur. The net change may actually consist of several consecutive reactions, each of which constitutes a step in the formation of final products. In discussing chemical reactions it is important to keep clear the distinction between a net reaction and one step in that reaction.

When a reaction occurs in steps, intermediate species are probably formed, and they may not be detectable because they may be promptly used up in a subsequent step. However, by investigating the influence that various factors have on the rate at which the net change occurs, it is sometimes possible to elucidate what the intermediates are and how they are involved in the mechanism of the reaction.

What factors influence the rate of chemical reaction? The question is of great practical importance because, as we shall see in this and succeeding chapters, a determination that a reaction *should* go does not guarantee that it *will* go. The question of how fast a particular reaction will go has to be left completely open. Experiments show that four important factors that influence reaction rates are (1) the nature of reactants, (2) the concentration of reactants, (3) the temperature, and (4) catalysis.

10.1 Nature of reactants

In a chemical reaction, bonds are formed and bonds are broken. The rate should therefore depend on the specific bonds involved. Experimentally, the reaction velocity depends on the specific substances brought together in reaction. For example, the reduction of permanganate ion in acidic solution by ferrous ion is practically instantaneous. MnO_4^- disappears as fast as ferrous sulfate solution is added; the limiting factor is the rate of mixing the solutions. On the other hand, the reduction of permanganate ion in acidic solution by oxalic acid, $H_2C_2O_4$, is not instantaneous. The violet color characteristic of MnO_4^- persists long after the solutions are mixed. In these two reactions everything is identical except the nature of the reducing agent, but still the reaction rates are quite different.

The rates observed for different reactants vary widely. There are reactions, such as occur in acid-base neutralization, which may be over in a nanosecond; so the rate is difficult to measure. There are also very slow reactions, such as those occurring in geologic processes, which may not reach completion in a million years. The changes in a lifetime may be too small to be detected. Most information has been accumulated about reactions that occur at rates intermediate between these extremes.

10.2 Concentration of reactants

It is found by experiment that the rate of a *homogeneous* reaction, i.e., one which occurs in only one phase, depends on the concentration of reactants in that phase. *Heterogeneous* reactions, on the other hand, involve more than one phase, in which case it is found that the rate is proportional to the area of contact between the phases. An example is the rusting of iron. Rusting, a heterogeneous reaction involving a solid phase, iron, and a gas phase, oxygen, is slow when the surface of contact is small, as with an iron nail, but is rapid when there is greater area of contact as with steel (iron) wool.

The rate of a homogeneous reaction depends in general on the concentration (amount per unit volume) of reactants in solution. For gaseous solutions the concentrations can be changed by altering the pressure. For liquid solutions the concentration of an individual reactant can be changed either by its addition or removal or by changing the volume of the system, as by addition or removal of solvent. The specific effect on reaction rate has to be determined by experiment. Thus, in the reaction of substance A with substance B, the addition of A may cause an increase, a decrease, or no change in rate, depending on the particular reaction. Quantitatively, the rate may double, triple, become half as great, etc. A priori, it is not possible to look at the net equation for the chemical reaction and tell how the rate will be affected by a change in the concentration of the reactants. The quantitative influence of concentration on rate can be found only by experiment.

The determination of how the rate of reaction depends on the concentration of reactants is an experimental problem beset by many

Experiment	Initial molar concentration		Initial rate, atm/min
	NO	H_2	
I	0.006	0.001	0.025
II	0.006	0.002	0.050
III	0.006	0.003	0.075
IV	0.001	0.009	0.0063
V	0.002	0.009	0.025
VI	0.003	0.009	0.056

Fig. 10.1
Rate of reaction of NO and
H_2 at 800° C.

difficulties. The usual procedure is to do a series of experiments in which everything is kept constant except the concentration of one reactant. As the concentration of the one reactant is systematically changed, the reaction rate is measured. This may be done by noting the rate of disappearance of a reactant or the rate of formation of a product. Experimental difficulties usually arise in determining the instantaneous concentration of a component as it changes with time.

The reaction between hydrogen and nitric oxide

$$2H_2(g) + 2NO(g) \longrightarrow 2H_2O(g) + N_2(g)$$

is a homogeneous reaction which can be investigated kinetically by following the change in pressure of the gaseous mixture as the reaction proceeds. The pressure drops because four moles of gas reactants are converted to three moles of gas products. Typical data for several experiments at a constant temperature of 800°C are given in Fig. 10.1. Since reactants are being used up during the course of the reaction, their concentrations and their rate of reaction are constantly changing. The concentrations and rates listed are those at the very beginning of the reaction, when little change has occurred. The first three experiments have the same initial concentration of NO but differ in initial concentration of H_2. The last three experiments have the same initial concentration of H_2 but differ in initial concentration of NO.

The data for experiments I and II show that when the initial concentration of NO is kept constant at 0.006 M, doubling the concentration of H_2 from 0.001 M to 0.002 M doubles the rate from 0.025 to 0.050 atm/min. I and III show that tripling the concentration of H_2 triples the rate. The rate of reaction is therefore found to be proportional to the first power of the concentration of H_2. The data for experiments IV and V show that when the initial concentration of H_2 is kept constant, doubling the concentration of NO quadruples the rate; IV and VI show that tripling the concentration of NO triply triples the rate. The rate of reaction is therefore found to be proportional to the square, or second power, of the concentration of NO. Quantitatively, the data can be summarized by stating that the reaction rate is proportional to (concentration of H_2) × (concentration of NO)^2. This can be written mathematically as

$$\text{Rate} = k[H_2][NO]^2$$

The equation, known as the *rate law* for the reaction, states that the rate is equal to a proportionality constant k times the concentration of H_2 to the first power, $[H_2]$, times the concentration of NO to the second power, $[NO]^2$. The brackets around a formula are a conventional way to indicate concentration in moles per liter. The proportionality constant k, called the *specific rate constant*, is characteristic of a given reaction but varies with temperature.

The general form of rate law for the reaction $nA + mB \longrightarrow pC + qD$ can be written

$$\text{Rate} = k[A]^\alpha [B]^\beta [X]^\gamma$$

where α is the appropriate power to which the concentration of A must be raised and β is the appropriate power to which the concentration of B must be raised in order to summarize the observed data. X represents a species which does not appear in the net reaction but may be involved in the rate law. The exponents α, β, and γ are generally integers; but they may be fractions, they may be zero, or they may be negative. A negative exponent would express inverse proportionality, where an *increase* in concentration produces a *decrease* in rate. The values of α, β, and γ indicate what is called the *order of the reaction* in A, B, and X, respectively. If $\alpha = 1$, the reaction is said to be first order in A; if $\alpha = 2$, the reaction is said to be second order in A, etc. The total reaction order is $\alpha + \beta + \gamma$. Probably the most important thing to note is that the rate law is determined by experiment. A common misconception is that the coefficients in the balanced net equation, for example, n and m, are equal to the exponents in the rate law, that is, α and β. *This, in general, is not true.* For example, in the reaction between H_2 and NO the exponents in the rate law are 1 and 2, whereas the coefficients in the balanced equation are 2 and 2. The only way to determine the exponents in the rate law is to do the experiment.

In our discussion of rate laws we have so far been limited to the *initial* rate, i.e., the reaction rate at the very beginning of reaction before any significant changes of concentration have occurred. What happens as time goes on? The concentrations change, and the rate of reaction will change in accord with the rate law. To describe such a system, it is most convenient to use the formalism of calculus. The treatment goes as follows: Suppose, as shown in Fig. 10.2, we have a particular reaction in which a molecule of A converts to a molecule of B. Since A and B are stoichiometrically related, the rate of conversion can be expressed either as the rate of disappearance of A or as the rate of formation of B. Referring to the rate of change in the concentration of A, we can see from the two triangles drawn adjacent to the curve for A that the rate is not constant with time. For equal time intervals Δt, the concentration of A near the start of the reaction changes by an amount ΔC_A but later in the reaction by a smaller amount $\Delta C'_A$. This can also be expressed by saying the slope (i.e., the tangent) of the curve decreases as time goes on. Mathematically the

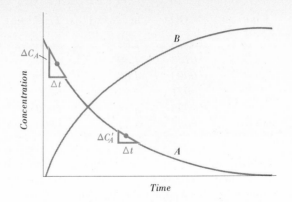

slope is expressed by the derivative of C_A with respect to t. This derivative is defined as follows:

$$\frac{dC_A}{dt} = \lim_{\Delta t \to 0} \frac{\Delta C_A}{\Delta t}$$

ΔC_A is the difference between C_A at some time $t + \Delta t$ and C_A at time t. The equation merely states that if Δt's are picked shorter and shorter ($\Delta t \longrightarrow 0$), then as Δt approaches zero, the limiting ratio of $\Delta C_A / \Delta t$ is dC_A/dt. With reference to Fig. 10.2, we can imagine the colored triangle as shrinking to zero so that the tangent line (hypotenuse of the triangle) shrinks to the point shown. The term dC/dt, called the derivative of C with respect to t, expresses the instantaneous rate at which C changes with t at any particular point. The symbolism dC/dt can be read as the increment in C divided by the increment in t. In other words, dC/dt is the slope of a curve of C plotted as a function of t. For a straight line, dC/dt is a constant; for a curve that gradually levels out dC/dt gradually approaches zero. For a reaction A $\longrightarrow$ B that is dependent on the first power of the A concentration (i.e., is a first-order reaction), the rate law can be written

$$-\frac{d[A]}{dt} = k[A] \qquad \text{or} \qquad \frac{d[B]}{dt} = k[A]$$

The minus sign comes from the fact that the A concentration is decreasing with time—in other words, $d[A]/dt$ is a negative number. Thus, the reaction rate can be expressed in terms of either the disappearance of a reactant or the formation of a product. The two derivatives are of opposite sign, and both change with time as long as the concentration of reactant is changing.

In order to get at the specific rate constant k, the above equation for $-d[A]/dt$ can be rearranged to give

$$\frac{d[A]}{[A]} = -k\,dt$$

Fig. 10.3
Rate data for decomposition
of nitrogen pentoxide at
45°C.

Time, sec	$[N_2O_5]$	$-\ln [N_2O_5]$	Time, sec	$[N_2O_5]$	$-\ln [N_2O_5]$
0	0.0176	4.04	3600	0.0029	5.84
600	0.0124	4.39	4200	0.0022	6.12
1200	0.0093	4.68	4800	0.0017	6.38
1800	0.0071	4.95	5400	0.0012	6.73
2400	0.0053	5.24	6000	0.0009	7.01
3000	0.0039	5.55	7200	0.0005	7.60

It turns out that $d[A]/[A]$, which represents the fractional change of $[A]$, can be represented as $d \ln [A]$, where ln stands for the natural logarithm (see Appendix 4.2). Therefore, we can write

$$d \ln [A] = -k \, dt$$

or, again rearranging,

$$\frac{d \ln [A]}{dt} = -k$$

This last equation says that for a first-order reaction the plot of $\ln [A]$ versus t is one of constant slope; in other words, it is a straight line. The slope of the line is equal to $-k$. Figure 10.3 shows some data for the reaction

$$N_2O_5(g) \longrightarrow 2NO_2(g) + \tfrac{1}{2}O_2(g)$$

which shows typical first-order behavior. As shown in Fig. 10.4, the plot of concentration of N_2O_5 versus t is not a straight line (i.e., the rate of change is not constant), but that of the negative natural logarithm is. The specific rate constant k that can be deduced from the slope of the line has a value of 4.9×10^{-4} sec^{-1}.

In general, for any first-order reaction the plot of logarithm of concentration vs. time will be a straight line, indicating that it is the fractional decrease of reactant concentration that is constant with time. Thus, no matter when during a reaction we choose to examine it, the time required to use up half of the remaining reactant will stay the same. Of course, as time progresses, there will be less and less reactant left, but the time needed to use up half of it stays the same. In other words, the reaction goes slower and slower. Consequently, for first-order reactions the rate can be specified either by giving a value for the specific rate constant k (for example, 4.9×10^{-4} sec^{-1} for the reaction above) or by specifying the half-life of the reaction $t_{1/2}$. The relation between these two quantities is

$$t_{1/2} = \frac{0.693}{k}$$

where the factor 0.693 is just the natural logarithm of 2. For the reaction given above the half-life is 1400 sec.

A *second-order reaction* is one in which the rate is proportional to the square of a reactant concentration or to the product of two reactant concentrations each taken to the first power. An example of a second-order reaction is

$$2HI \longrightarrow H_2 + I_2$$

For this reaction a plot of the logarithm of concentration vs. time will not give a straight line. Instead, to get a straight line, it is necessary to plot the reciprocal of the concentration, $[HI]^{-1}$, against time. That such a plot will give a straight line can be seen as follows: Since the reaction is second order, the rate law is

$$-\frac{d[HI]}{dt} = k[HI]^2$$

which can be rewritten

$$-\frac{d[HI]}{[HI]^2} = k\,dt$$

From differential calculus, $dx/x^2 = -d(1/x)$; so we can write

$$d\,\frac{1}{[HI]} = k\,dt$$

or

$$\frac{d(1/[HI])}{dt} = k$$

The last equation states that for this second-order reaction, a plot of $1/[HI]$ versus t is a straight line of slope k. Thus, k is readily obtained from the experimental data.

Another example of a second-order reaction is the reverse of the above change, namely,

$$H_2 + I_2 \longrightarrow 2HI$$

In this case a straight line will be obtained by plotting against time the reciprocal concentration of either H_2 or I_2. However, this works *only* if the two concentrations are equal. If $[H_2]$ is equal to $[I_2]$, either one can be used to measure the rate, both can be set equal to x, and the calculus is the same as above. However, if $[H_2]$ is not equal to $[I_2]$, this cannot be done. To appreciate the difference, we can consider the situation where H_2, for example, is present in such overwhelming excess that all the I_2 can react without appreciably changing the concentration of H_2. As a consequence, $[H_2]$ would be considered to be a constant. The reaction would be what is called *pseudo-first order*; i.e., a plot of $\ln[I_2]$ versus t would give a straight line. To see how this comes about, we can let x represent the concentration of I_2 and $x + a$, the concentration of H_2. As time goes on, x decreases, but a remains constant because the excess of H_2 over I_2 stays fixed. Letting the rate be dx/dt, we can write for the chemical reaction

$$H_2 + I_2 \longrightarrow 2HI$$

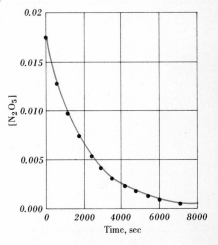

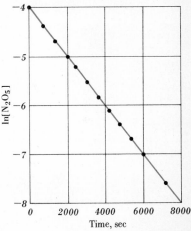

Fig. 10.4

Plots of rate data for decomposition of N_2O_5.

the rate law

$$-\frac{d[I_2]}{dt} = k[H_2][I_2]$$

or, substituting,

$$-\frac{dx}{dt} = k(x + a)(x)$$

If a is zero, this is a simple second-order case, and we have $-dx/dt = kx^2$, or $-dx/x^2 = k\,dt = d(1/x)$, as indicated above. If, on the other hand, a is much greater than x, then we can approximate $x + a$ by a and get

$$-\frac{dx}{dt} \approx kax$$

from which we get

$$-\frac{d \ln x}{dt} \approx ka$$

This last equation indicates that when H_2 is way in excess over I_2, a plot of $\ln [I_2]$ versus t gives a straight line of slope $-ka$, where k is the second-order rate constant and a is the excess H_2 concentration. In general, for rate analysis involving more than one reactant, it is convenient to use excess concentrations for all but one reactant and thus determine the order of the reaction with respect to that reactant. Successive experiments with other reagents in excess can tell about the other reactants.

In summary, the dependence of reaction rate on concentration can be obtained from graphic analysis of the experimental data. The data for concentration of a limiting reactant are successively plotted against time to see which plot gives the best straight line. If $\ln C$ versus t is linear, the reaction is first order in that reactant; if C^{-1} versus t is linear, the reaction is second order in that reactant. For higher orders the time plot of $1/C^n$ would have to be linear to indicate the reaction to be of order $n + 1$ for the reactant in question.

10.3 Temperature

How does the temperature of a reaction affect the reaction rate? Observations indicate that a rise in temperature almost invariably increases the rate of any reaction. Furthermore, despite a common misconception, a decrease in temperature almost invariably decreases the rate no matter whether the reaction is exothermic or endothermic.* The change of

---

* There are a few extraordinary cases in which a reaction rate apparently decreases with an increase of temperature. Such odd behavior can arise when there is a sequence of forward and backward steps in which the rate of a backward step increases more rapidly with a rise in temperature than does the rate of the subsequent forward step. By allowing for enough steps in a reaction sequence, any observation, no matter how strange, can be explained.

rate with temperature is expressed by a change in the specific rate constant k. For every reaction, k increases with increasing temperature. As to the magnitude of the effect, no generalization can be made. The magnitude varies from one reaction to another and from one temperature range to another. A rule, which must be used with great caution, is that a $10°C$ rise in temperature approximately doubles or triples the reaction rate. For each specific reaction it is necessary to determine from experiment the quantitative effect of a rise in temperature.

The relation between the specific rate constant k and the temperature T (in degrees Kelvin) is usually expressed by the so-called "Arrhenius equation"

$$k = Ae^{-E/RT}$$

Here A is a constant characteristic of the reaction; e is the base of natural logarithms (Appendix 4.2); E is an energy called the *activation energy* of the reaction; R is the gas constant equal to $8.314\,J\,mol^{-1}\,deg^{-1}$. The dependence of k on T is not linear; it is exponential. Consequently, small changes in T may produce relatively large changes in k. However, the connection is not a simple one. It can perhaps best be seen by taking the natural logarithm of both sides of the Arrhenius equation. This gives

$$\ln k = \ln A - \frac{E}{RT}$$

Since $\ln x = 2.303 \log x$ (where log stands for logarithm to the base 10), the last equation can be rewritten

$$\log k = \log A - \frac{E}{2.303RT}$$

For a typical first-order reaction, A might be $1.0 \times 10^{14}\,sec^{-1}$, and E might be $80\,kJ/mol$. At room temperature ($T = 300°K$)

$$\log k = \log (1.0 \times 10^{14}) - \frac{80,000}{(2.303)(8.314)(300)} = 0.07$$

which means that $\log k = 0.07$ or that $k = 1.2\,sec^{-1}$. But if the temperature is $10°K$ higher ($T = 310°K$), it turns out that $\log k = 0.52$ or that $k = 3.3\,sec^{-1}$. In other words, in this case a $10°K$ rise near room temperature roughly triples the rate.

What if we have a reaction the same as above except that E is 160 instead of $80\,kJ/mol$? In that case k will be $1.4 \times 10^{-14}\,sec^{-1}$ at $300°K$ and $1.1 \times 10^{-13}\,sec^{-1}$ at $310°K$. In other words, a $10°K$ rise near room temperature has increased k roughly eight times. Finally, it might be noted that the effect of a $10°K$ rise will be greater at low temperature than at high.

The value of the activation energy E for a particular reaction can be determined from experimental data by plotting the logarithm of the observed values of k against $1/T$ for several temperatures studied. From the equation

$$\log k = \log A - \frac{E}{2.303R}\left(\frac{1}{T}\right)$$

we can see that a plot of $\log k$ versus $1/T$ gives a straight line for which the slope is $-E/2.303R$. Figure 10.5 shows some typical data for the decomposition of N_2O_5 to NO_2 and O_2. The slope of the line turns out to be -5400, which corresponds to an activation energy of 100 kJ/mol.

10.4   Catalysis

It is found by experiment that some reactions can be speeded up by the presence of substances which themselves remain unchanged after the reaction has ended. Such substances are known as *catalysts*, and their effect is known as *catalysis*. Often only a trace of catalyst is sufficient to accelerate the reaction. However, there are many reactions in which the rate of reaction is proportional to some power of the concentration of catalyst. The actual dependence of rate on catalyst concentration must be determined by experiment. If the experiments show there is such a dependence, then the catalyst concentration to the appropriate power becomes part of the rate law in the same way as the reactants become a part of it.

There are numerous examples of catalysis. For instance, when $KClO_3$ is heated so that it decomposes into KCl and oxygen, it is observed that manganese dioxide, MnO_2, considerably accelerates the reaction. At the end of the reaction the $KClO_3$ is gone, but all the MnO_2 remains. It appears that the catalyst is not involved in the reaction because the starting amount can be recovered. However, the catalyst must take some part in the reaction, or else it could not change the rate.

When hydrogen gas escapes from a cylinder into the air, no change is visible. If, however, the escaping hydrogen is directed at finely divided platinum, the platinum glows, and eventually the hydrogen ignites. In the absence of platinum, the rate of reaction is too small to observe. In contact with platinum, H_2 reacts with O_2 from the air to form H_2O. As they react, energy is liberated and heats the platinum. As the platinum gets hotter, the hydrogen and oxygen heat up; so their rate of reaction increases until ignition eventually occurs, and the reaction of hydrogen with oxygen becomes self-sustaining.

Enzymes are complex substances in biologic systems which act as catalysts for biochemical processes. Pepsin in the gastric juice and ptyalin in the saliva are examples. Ptyalin is the catalyst which accelerates the conversion of starch to sugar. Although starch will react with water to form sugar, it takes weeks for the conversion to occur. A trace of ptyalin is enough to make the reaction proceed at a biologically useful rate.

A special type of catalysis occasionally encountered is *autocatalysis*, or *self-catalysis*. As the name implies, this is catalysis in which one of the products of the reaction is a catalyst for the reaction. For

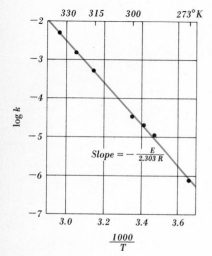

Fig. 10.5

Plot of log k versus 1/T to show how energy of activation can be obtained.

example, in the reaction of permanganate ion with oxalic acid

$$2MnO_4^- + 5H_2C_2O_4 + 6H_3O^+ \longrightarrow 2Mn^{2+} + 10CO_2 + 14H_2O$$
Violet   *Colorless*

the product Mn^{2+} catalyzes the reaction. This can readily be observed by mixing solutions of potassium permanganate, sulfuric acid, and oxalic acid. No appreciable decolorization occurs until a tiny crystal of manganous sulfate, $MnSO_4$, is dropped into the reaction mixture, whereupon fast decolorization occurs. The uncatalyzed reaction is quite slow. Until some Mn^{2+} is formed, the reaction does not proceed at an appreciable speed. Addition of a trace of Mn^{2+} enables the reaction to start. More Mn^{2+} is produced and speeds the reaction further.

Interesting catalysis is observed in the case of hydrogen peroxide. Hydrogen peroxide decomposes to water and oxygen. The reaction is rapid enough that solutions of hydrogen peroxide are ordinarily difficult to keep without decomposition. It has been observed that certain substances, such as phosphates, can be added in trace amounts to slow down the rate of decomposition. This looks like the reverse of catalysis, and, in fact, substances which slow down rates used to be called *negative catalysts*. The name is misleading, since the function of the phosphate in hydrogen peroxide is probably to destroy the action of catalysts already present in the hydrogen peroxide. For example, it is found by experiment that the decomposition of hydrogen peroxide is catalyzed by traces of Fe^{3+} ion. When phosphate is added, it combines with the Fe^{3+} and prevents the Fe^{3+} from functioning as a catalyst.

10.5   Collision theory

The observed facts of chemical kinetics can be interpreted in terms of *collision theory*. This theory makes the basic assumption that for a chemical reaction to occur, particles must collide. This may seem like an obvious assumption to make, but it was not many years ago that people believed in "reaction at a distance." It was then believed that reaction was probably initiated by passage of radiation from one reactant to another. At present we believe quite generally that for substance A to react with substance B, it is necessary that the particles A, be they molecules, ions, or atoms, collide with particles B. In the collision, atoms and electrons are rearranged. There is a reshuffling of chemical bonds that leads to production of other species.

According to collision theory, the rate of any step in a reaction is directly proportional to (1) the *number of collisions per unit time* between the reacting particles involved in that step and (2) the *fraction of these collisions that are effective*. That the rate should depend on the number of collisions per unit time seems obvious. For instance, in a box that contains A molecules and B molecules there is a certain frequency of collision between A and B molecules. If more A molecules are placed in the box, the collision frequency between A molecules and B molecules is increased. With more collisions between reacting molecules, the reaction between A and B should go faster. However, this cannot be the full story. Calculation of the number of collisions between particles indicates that the collision frequency is very high. In a mixture containing 1 mol of A molecules and 1 mol of B molecules as gases at STP, the number of collisions between A and B is about 10^{30} sec^{-1}. If every one of these collisions led to reaction, the reaction would be over like a shot; all reactions would be exceedingly fast. By observation, this is not true. It must be that only a rather small fraction of the collisions lead to reaction.

Why are some collisions effective and others not? Collisions between A molecules and B molecules when they are slow-moving may be so gentle that there is no change in the identity of the molecules during the collision. The colliding particles separate to resume their original identity. The electron cloud associated with A and the electron cloud associated with B repel each other because they are similarly charged. In a gentle collision the repulsion between the electron clouds may simply cause the molecules to bounce off each other. However, if A or B or both A and B have much kinetic energy before collision, they can easily use their kinetic energy to do work against the repulsive forces. If the kinetic energy available is big enough, the repulsive forces can be overcome, and the molecules penetrate into each other far enough that significant, large-scale electron and atom rearrangement ensues. One or more new species may be formed.

The extra amount of energy above the average level required in a collision to produce chemical reaction is called the *energy of activation*. Its magnitude depends on the nature of the reactants. Some reactions have a large energy of activation. Such reactions are slow, since only a relatively small fraction of the reactant particles will have enough kinetic energy to furnish the required energy of activation. Other reactions have a small energy of activation. Such reactions are fast, since a greater fraction of the collisions are effective. More of the particles have sufficient kinetic energy to furnish the required energy of activation.

Qualitatively, the collision theory quite satisfactorily accounts for the four factors listed at the beginning of this chapter as being observed to influence reaction rates:

1 The rate of chemical reaction depends on the *nature of the chemical reactants*, because the energy of activation differs from one reaction to another.

2 The rate of reaction depends on the *concentration of reactants*, because the number of collisions increases as the concentration is increased.

3 The rate of reaction depends on the *temperature*, because an increase of temperature makes molecules move faster. They collide more frequently, and, what is more important, the collisions are more violent and more likely to result in reaction. Any collection of molecules has a distribution of energies (Fig. 5.14). According to collision theory only the highly energetic molecules have enough energy to react. As the temperature is raised, the whole distribution curve shifts to higher energies, and a larger fraction of the molecules become highly energetic. More of the collisions are therefore effective at high temperature than at low temperature. In the Arrhenius formulation $k = Ae^{-E/RT}$, A, which is sometimes called the *frequency factor*, can be thought of as the number of collisions, and $e^{-E/RT}$ as the fraction of collisions that are effective.

4 The rate of reaction would be accelerated by the presence of *catalysts*, if somehow, a catalyst made collisions more effective. This could be done, for example, by a preliminary step in which one or more of the reactants reacted with the catalyst to give a species having lower activation energy. New reactants could be produced which would react more rapidly than the original reactants.

10.6 Stepwise reactions

One of the trickiest aspects of chemical kinetics is to account properly for the observed power dependence of rate on concentration. Why in some cases does rate depend on the first power of concentration and in other cases on the square of concentration? For simplicity, let us consider one step of a reaction. Suppose that in this step one molecule of A reacts with one molecule of B to form a molecule AB. The balanced equation for *this step* is

$$A + B \longrightarrow AB$$

According to collision theory, the rate of formation of AB is proportional to the rate at which A and B collide. Let us imagine that we have a box that contains some B molecules and a single A molecule. The rate at which the A molecule collides with B molecules is directly proportional to the number of B molecules in the box. (If we should double the number of B molecules in the box, we would then have twice as many A–B collisions per unit time.) Suppose now we place a second A molecule in the box. We now have twice as many A molecules in the box; so the total number of A–B collisions per unit time is doubled. In other words, the rate at which A and B molecules collide is directly proportional to the concentration of A and to the concentration of B. The rate of formation of AB should therefore be directly proportional to the concentration of A and to that of B. Thus, the rate law for this step is

$$\text{Rate} = \frac{d[\text{AB}]}{dt} = k[\text{A}][\text{B}]$$

It should be noted that the exponents of [A] and of [B] in the rate law are unity, just as the two coefficients are in the balanced equation *for the step*.

What is the situation if the balanced equation for a step involves coefficients larger than 1? Consider the reaction

$$2A \longrightarrow A_2$$

In this step an A molecule must collide with another A molecule to form A_2. The rate at which A_2 forms is thus proportional to the rate at which two A molecules collide. Again we imagine a box, this time containing only molecules of type A. The rate at which *any one* A molecule collides with any other A molecule is proportional to the number of other A molecules in the box. If we should double the number of other A molecules in the box, we would then double the rate at which collisions occur with the one molecule under observation. Now suppose we extend our observation to all the molecules in the box. The total number of collisions per second is proportional to the number of collisions per second made by one A molecule times the total number of A molecules in the box. Another way of saying the same thing is that the rate of collision is proportional to the number of molecules hitting multiplied by the number of molecules being hit. In any event, we can say that the rate at which two A molecules collide is proportional to the concentration of A times the concentration of A, i.e., to the square of the concentration of A. Consequently, for the step

$$2A \longrightarrow A_2$$

we can write the rate law

$$\text{Rate} = \frac{d[A_2]}{dt} = k[A]^2$$

It might be noted that the exponent of the concentration of A in the rate law is 2, just as the coefficient of A in the balanced equation for the step is 2.

For the general case of a single step for which the balanced chemical equation shows disappearance of n molecules of A and m molecules of B to form product P, we can write the rate law

$$\text{Rate} = \frac{d[P]}{dt} = k[A]^n[B]^m$$

This indicates that the rate of the step is proportional to the concentration of species A taken to the nth power times the concentration of species B taken to the mth power. We should note again that the overall chemical change generally consists of several consecutive steps. A knowledge of only the overall balanced chemical equation *does not* permit us to predict what the experimentally observed rate law will be. For example, in the reaction between NO and H_2, discussed in Sec. 10.2, the rate law obtained experimentally states that the reaction rate is proportional to the first power of the H_2 concentration times the second power of the NO concentration. Yet the balanced equation for the reaction is

$$2H_2(g) + 2NO(g) \longrightarrow N_2(g) + 2H_2O(g)$$

In a one-step collision theory, it would appear that reaction could occur by collision between two H_2 molecules and two NO molecules. The number of such collisions per second would be proportional to the molar concentration of H_2 squared times the molar concentration of NO squared. This would mean that doubling the H_2 concentration ought to quadruple the number of collisions per second and therefore quadruple the rate. This does not agree with experiment, which tells us the rate is proportional to the first power of H_2, not the second. It must be that the collisions that determine the rate are not between two H_2 molecules and two NO molecules, but are something simpler. In other words, a one-step explanation does not suffice.

To account for the actual observed rate law, we have to assume that this reaction, like many others, occurs in steps. In such stepwise reactions, it is always the slow step that determines the rate. It is the bottleneck. The following example shows a two-step reaction:

$$A(g) + B(g) \longrightarrow \text{intermediate} \tag{1}$$

$$\text{Intermediate} + B(g) \longrightarrow C(g) \tag{2}$$

Net reaction

$$A(g) + 2B(g) \longrightarrow C(g)$$

In step (1) a molecule of A collides with a molecule of B to form a short-lived intermediate. In step (2) the intermediate reacts with another molecule of B to form the final product C. If the first step is slow and if the second step is fast, the rate at which C forms will depend only on the rate at which the intermediate forms. As soon as the intermediate appears, it is used up in the second reaction. The rate at which the intermediate is produced is determined by the collision of A and B. Thus, the rate of the slow step is proportional to the concentration of A times the concentration of B. Since the slow step is the rate-determining step, the rate law for the overall change is

$$\text{Rate} = \frac{d[C]}{dt} = k[A][B]$$

Still the chemical equation for the overall change is determined by adding steps (1) and (2):

$$A(g) + B(g) + B(g) \longrightarrow C(g)$$

The intermediate cancels out because it occurs on both sides of the equation. The coefficients which appear in the net equation are thus different from the exponents in the rate law.

For the specific reaction

$$2H_2(g) + 2NO(g) \longrightarrow N_2(g) + 2H_2O(g)$$

the rate law is determined by experiment to be

$$\text{Rate} = \frac{d[N_2]}{dt} = k[H_2][NO]^2$$

The reaction must occur in steps. One set of steps, in which the first step would be slower and therefore rate-determining, might be as follows:

$$H_2(g) + NO(g) + NO(g) \xrightarrow[slow]{} N_2O(g) + H_2O(g)$$

$$H_2(g) + N_2O(g) \xrightarrow[fast]{} N_2(g) + H_2O(g)$$

However, many kinetics people object to such a mechanism because it requires a simultaneous collision of three molecules. As any billiard player knows, three-body collisions are quite improbable. An alternative set of steps without three-body collision would be the following:

$$2NO(g) \xrightarrow[fast]{} N_2O_2(g)$$

$$N_2O_2(g) + H_2(g) \xrightarrow[slow]{} N_2O(g) + H_2O(g)$$

$$N_2O(g) + H_2(g) \longrightarrow N_2(g) + H_2O(g)$$

The middle step would be rate-determining. Since the N_2O_2 would be rapidly formed by collision of two NO molecules, the concentration of N_2O_2 would be proportional to the square of the concentration of NO. The rate of the second step would thus depend on the square of the concentration of NO times the concentration of H_2, in agreement with the observed rate law. On the basis of the observed rate law, it is not possible to distinguish between the two mechanisms. In fact, it may be that neither is right and the actual mechanism may be much more complicated.

Frequently it happens that several alternative paths do exist by which the overall change is accomplished. In such cases all the paths must be considered in the rate law. As an example, let us consider a situation where A reacts to give products by three alternative paths:

$$A \xrightarrow{k_1} products \tag{3}$$

$$A + B \xrightarrow{k_2} products \tag{4}$$

$$A + A \xrightarrow{k_3} products \tag{5}$$

The products need not be formed directly as shown but may go through intermediates that react by fast following steps. As indicated, k_1, k_2, and k_3 are the respective rate constants for the slow step in each of the three paths. For Eq. (3), the rate of disappearance of A is just proportional to the concentration of A; for Eq. (4), to the product of A concentration times B concentration (B might be a catalyst); for Eq. (5), where the rate of disappearance of A is doubled because of the two molecules of A involved, the rate is proportional to the square of the concentration of A. For the total rate of disappearance of A by all three paths we can write

$$\frac{-d[A]}{dt} = k_1[A] + k_2[A][B] + 2k_3[A]^2$$

where each of the terms tells how fast A disappears by that particular path. Unraveling such a rate law can be quite a chore, but often judicious choice of experimental conditions can help. If the concentration of A were made very small, for example, as by diluting the system, the term in $[A]^2$ could be neglected with respect to the others. Alternatively, the concentration of B could be made much larger than that of A; so only the middle term would need to be considered. Finally, it is rare that all three of the rate constants k_1, k_2, and k_3 would be exactly equal. Generally, one would be considerably greater than the others (i.e., reaction along one of the paths would be much faster); so if, for example, k_1 were greater than k_2 or k_3, we might be able to neglect the last two terms with respect to the first term.

Help in unraveling mechanisms often comes from careful attention to how the concentrations of substances change with time. This is particularly true when there are consecutive reactions. For example, let us consider the sequence of elementary steps

$$A \xrightarrow{k_1} B \xrightarrow{k_2} C$$

where A converts to B with rate constant k_1 and then B converts to C with rate constant k_2. The rate at which A *disappears* can be written

$$-\frac{d[A]}{dt} = k_1[A]$$

The rate at which C *appears* can be written

$$\frac{d[C]}{dt} = k_2[B]$$

What can we say about the concentration of B? It *increases* with time because of being formed from A but *decreases* with time because of being converted to C. The net increase in B will be equal to the difference between these two processes; so we can write

$$\frac{d[B]}{dt} = k_1[A] - k_2[B]$$

Depending on whether the first or second term is the greater, $d[B]/dt$ will be positive or negative. If $d[B]/dt$ is positive, as when $k_1[A] > k_2[B]$, the concentration of B will increase with time; if $d[B]/dt$ is negative, as when $k_1[A] < k_2[B]$, the concentration of B will fall off with time. Both situations will occur, consecutively, because $[A]$ and $[B]$ are also changing with time. If we start with pure A at time $t = 0$, $d[B]/dt$ is certainly positive, but after long enough time for A to be substantially used up, $d[B]/dt$ will become negative. This means that the concentration of B will peak sometime along in the reaction. At that juncture, the condition will be $d[B]/dt = 0 = k_1[A] - k_2[B]$, or $[A]/[B] = k_2/k_1$. In other words, the behavior of the system will be decided by the relative values of k_1 and k_2. Figure 10.6 shows what the behavior looks like in a typical case. As the concentration of starting material A falls off and the concentration of product C rises, the concentration of the intermediate B rises to a maximum and then gradually falls off. The situation shown is for the specific situation $k_1 = \frac{1}{2}k_2$. The smaller k_1 becomes relative to k_2,

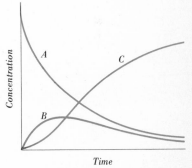

Fig. 10.6

Change of concentrations with time for system with consecutive reactions
$$A \rightarrow B \rightarrow C.$$

the less pronounced is the buildup in the concentration of B. In fact if $k_1 \ll k_2$, the intermediate B may never be detected.

There are but few chemical reactions that follow the simple sequence A $\longrightarrow$ B $\longrightarrow$ C, with each step being first order. A specific example is the decomposition of gaseous dimethyl ether, CH_3OCH_3, to give CH_4, H_2, and CO. The first step is believed to be $CH_3OCH_3 \longrightarrow CH_4 + HCHO$, which produces methane and formaldehyde, and then the formaldehyde goes on to decompose according to the reaction HCHO $\longrightarrow H_2 + CO$.

A more typical example of consecutive reactions is found in the radioactive decay of nuclear isotopes. In general, unless we stop at a stable nucleus, the isotope we get from the radioactive decay of one species goes on to decay further to give another species. For example, the uranium isotope ^{238}U spontaneously disintegrates by α-particle emission to give the thorium isotope ^{234}Th, and this in turn spontaneously converts by β-particle emission to give the protactinium isotope ^{234}Pa. (See Sec. 18.3.)

10.7   Chain reactions

One of the most interesting types of complication that may arise in chemical kinetics is the occurrence of a *chain reaction*. This comes about when an intermediate species that is consumed in one step is regenerated in a later step. The result is to set up a sequence of steps that endlessly repeat themselves, like the links of a chain, until the chain is terminated or the starting materials are exhausted. Examples of chain reactions are found in combustion flames, gas explosions, and, currently of avid interest, the generation of Los Angeles–type smog.

One of the first reactions which was recognized as involving a chain mechanism was the deceptively simple combination of hydrogen and bromine to form hydrogen bromide:

$$H_2(g) + Br_2(g) \longrightarrow 2HBr(g)$$

The rate law for HBr formation by this reaction turns out to be fiendishly complicated, viz.,

$$\frac{d[HBr]}{dt} = \frac{k[H_2][Br_2]^{1/2}}{1 + k'[HBr]/[H_2]}$$

A mechanism that fits this rate law goes as follows: First there is a *chain-starting step* which involves dissociation of diatomic bromine to give individual atoms

$$Br_2 \longrightarrow 2Br \tag{6}$$

The bromine atom then reacts with an H_2 molecule to form HBr and liberate a hydrogen atom. The hydrogen atom, being very reactive, proceeds to combine with a bromine molecule to abstract one bromine atom for forming HBr and liberating another bromine atom which can repeat the cycle. Alternatively, some of the hydrogen atoms combine with product HBr to form H_2 and liberate Br for carrying out the chain. In any event there are three subsequent reactions following Eq. (6) above, which are referred to as *chain-propagating steps*. They can be written

$$Br + H_2 \longrightarrow HBr + H \tag{7}$$

$$H + Br_2 \longrightarrow HBr + Br \tag{8}$$

$$H + HBr \longrightarrow H_2 + Br \tag{9}$$

As can be seen there is re-formed [in Eqs. (8) and (9)] the very reactive intermediate Br that is needed to start off Eq. (7). Thus, one Br atom can carry out its job of making HBr but then be regenerated in the process to do the whole thing over and over again. Sometimes this happens thousands of times before a Br gets sopped up by some extraneous reaction, as with some scavenger or perhaps with an impurity on the wall of the vessel. In any case the chain eventually is terminated, the number of cycles achieved before that being called the *chain length*. Even if there are no scavengers or impurities, propagation of the chain may come to a halt by a reaction such as

$$2Br \longrightarrow Br_2 \tag{10}$$

which is called the *chain-terminating step*. Usually the chain-terminating step also involves some third species, which may be some other molecule of the mixture, to carry off some of the energy when the two Br atoms come together.

Chain reactions can lead to spectacular explosions if the chain-propagating steps produce heat faster than it can be conducted away by the surroundings. In such cases there will be a constantly rising temperature, which drives the reactions to ever-increasing speed, resulting in a runaway situation. Another possibility is for *chain branching* to occur. This comes about when reactive intermediates are produced by the chain-propagating steps in greater number than go into these steps. As a consequence, additional pathways open up very rapidly. An example of a system that reacts by chain branching is the H_2-O_2 system. The reaction

$$2H_2(g) + O_2(g) \longrightarrow 2H_2O(g)$$

is tremendously complicated, as can be seen from the following facts: Below $400°C$, the reaction is too slow to be detected. Above $600°C$, it goes with explosive violence (except at very low pressures). At some temperature in between, say $500°C$, the behavior is most bizarre. At low pressure (0.001 atm), there is no explosion; at slightly higher pressure (0.01 atm), explosion; at still higher pressure (0.1 atm), again no explosion; at even higher pressure (10 atm), explosion again. The situation is represented schematically in Fig. 10.7, which shows the curve separating explosive and nonexplosive conditions for a stoichiometric mixture of H_2 and O_2 in a given container. The container, especially its wall area and the kind of material it is made of, is very important in determining the actual shape and position of the curve.

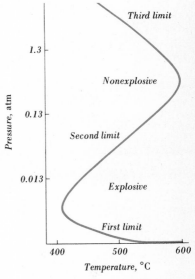

Fig. 10.7

Curve showing explosive and nonexplosive regions of pressure and temperature for a given hydrogen-oxygen mixture in a given container.

No one knows for sure just what goes on in the $H_2 + O_2$ reaction, but the following steps are believed to be included in the overall process:

Chain-starting reaction

$$H_2 + O_2 \longrightarrow 2OH$$

Chain-propagating reaction

$$OH + H_2 \longrightarrow H_2O + H$$

Chain-branching reaction

$$H + O_2 \longrightarrow OH + O$$

$$O + H_2 \longrightarrow OH + H$$

Chain breaking near second explosive limit

$$H + O_2 \longrightarrow HO_2$$

$$HO_2 \longrightarrow \text{destruction at the container wall}$$

Chain propagation near third explosive limit

$$H + O_2 \longrightarrow HO_2$$

$$HO_2 + H_2 \longrightarrow H_2O + OH$$

Chain breaking below first explosive limit

$$H \longrightarrow \text{destruction at surface}$$

$$OH \longrightarrow \text{destruction at surface}$$

$$O \longrightarrow \text{destruction at surface}$$

As can be seen, this mechanism invokes some exotic-looking species (for example, OH) which would not be encountered in normal chemistry work, but then explosions are rather exotic phenomena, very complex, very poorly understood. Unraveling them is of great practical interest because of possible application to resolving major technological dilemmas. For example, the powering of automobiles depends largely on a flame combustion reaction in which gasoline hydrocarbons are oxidized by chain-reaction mechanisms. Too often at the beginning of a power stroke in an internal-combustion engine the reaction gets away and explodes prematurely. The result is engine knock, an audible staccato in the engine. With it comes air pollution. To reduce engine knock, we add an amazing compound, tetraethyl lead, to the gasoline, but this spews lead into the environment, bad for man and beast. We can replace the lead by changing the formulation of the gasoline, but this costs money. There is a cheap way to do it, but that also spews out some special compounds that are the raw material for smog formation. The dilemma is still unresolved; its resolution calls for preliminary understanding of some complicated chain reactions.

Another major, unsolved chain-reaction problem is the generation of Los Angeles–type smog. Unlike London smog, which depends on

SO$_2$, fog, and smoke, the Los Angeles smog starts with sunshine, oxygen, and the products of combustion. Details of smog formation and its relation to the internal-combustion engine are considered further in Sec. 26.2, but here it will suffice to note that high-temperature combustion generates an oxide of nitrogen NO (nitric oxide) which on exposure to oxygen of the air converts to another oxide of nitrogen NO$_2$ (nitrogen dioxide). Unlike NO, NO$_2$ has the unfortunate property of being broken up by sunlight into NO and monatomic oxygen:

$$NO_2 \xrightarrow{hv} NO + O$$

The monatomic oxygen is highly reactive and promptly combines with diatomic O$_2$ from the air to form ozone, O$_3$, an irritating air pollutant:

$$O + O_2 \longrightarrow O_3$$

Normally not much O$_3$ accumulates because it is scavenged by the NO, which picks it up and reconverts it to O$_2$:

$$O_3 + NO \longrightarrow O_2 + NO_2$$

NO$_2$ is regenerated, and a sort of cycle is set up by which the O$_3$ is kept down to tolerable concentrations. But now comes the automobile. Among its exhaust products are unburned hydrocarbons, which react with the O$_3$ in a poorly understood series of chain reactions to produce a whole mess of air pollutants (aldehydes, ketones, organic peroxides, etc.—see Sec. 26.3).

More important, the balanced NO$_2$ cycle is broken. Among the products of the chain reactions are very reactive hydrocarbon fragments known as *free radicals*. These are unusual molecules which do not have a normal complement of covalent bonds but an unpaired, nonbonding electron available for quickly forming another bond. An example of a free radical is CH$_3\cdot$, the methyl radical, where three hydrogens are bound to a central carbon by regular electron-pair bonds and there is besides an additional unpaired, unsatisfied, lone electron (indicated by the dot). It is believed that this free radical and others like it react in complex chains to reduce the NO concentration in the atmosphere below what would correspond to a desirable steady state. With the NO depleted, there would be insufficient scavenger to mop up the O$_3$; so its concentration in the atmosphere would increase. Ozone in the stratosphere is a good thing because it absorbs the very short ultraviolet radiation of the sun, but ozone in the lower atmosphere is harmful and irritating to plants and animals (nose and throat irritation start at atmospheric concentrations of 0.2 ppm).

In an attempt to cut down on smog formation, strict limitations have been placed on the allowable exhaust emission of hydrocarbons. As discussed in Sec. 26.4, this has been achieved by regulating automobile engines to operate at higher air-fuel ratios. Unfortunately, the associated effect is to increase the emission of nitrogen oxides; so, even if only temporarily, we have simply exchanged one form of air pollution for another. The permanent solution awaits the solving of some chain-reaction problems.

10.8 Energy diagrams

A very useful way to visualize the changes that occur during a reaction step is to focus on the change in potential energy that occurs when reactants pass over to being products. To illustrate, we consider the one-step reaction in which one atom of H collides with one molecule of HBr to form one molecule of H_2 and one atom of Br:

$$H + H—Br \longrightarrow H—H + Br$$

For simplicity, the collision is assumed to occur along a straight line through all the nuclei. The H atom approaches the H—Br from the left and joins up with it to form some kind of transient complex particle H—H—Br (called the *activated complex*), and then the Br atom breaks away to move off toward the right. We can describe the geometry of the situation by specifying two distances—that between H and H, which we designate $r_{H\cdots HBr}$, and that between H and Br, which we designate $r_{HH\cdots Br}$. Figure 10.8 shows schematically how the potential energy of the system changes as these two distances are varied. Figure 10.8 can be thought of as a surface in which there are two valleys connected by a pass, or saddle point. The colored lines represent contour lines of constant potential energy. The arrows indicate the probable path followed by the system when it changes from being H + HBr to HH + Br. On the energy surface the system moves along the floor of one valley up a hill to the saddle point and then down a hill to another valley. The arrowed path can be described as the *reaction coordinate*, and distances along this path tell us how far we have gone from the reactants H + HBr to the products HH + Br. Figure 10.9 shows how the potential energy of the system changes along the reaction coordinate. In the initial state, H and HBr particles are far enough apart not to affect each other. The potential energy is the sum of the potential energy of H by itself plus that of HBr by itself. It makes no difference what the actual value is since potential energy is relative. As H and HBr come together, the forces of repulsion between the electron clouds become appreciable. Work must be done on the system to squash the particles together. This means the potential energy must increase. It increases until it reaches a maximum that corresponds to the activated complex. The activated complex then splits into HH and Br particles, and the potential energy drops as HH and Br go apart.

The difference (shown by the double-headed arrow in Fig. 10.9) between the potential energy of the initial state H plus HBr and the potential energy of the activated complex is a measure of the energy which must be added to the particles in order to get them to react. This is the activation energy of the reaction. It usually is supplied by converting some of the kinetic energy of the particles into potential energy. If H and HBr particles do not have much kinetic energy, on collision they are able to go only part way up the side of the hump. All the kinetic energy may be converted into potential energy without getting the pair pushed up to the activated complex. In such cases H and HBr slide back down the hump and fly apart unchanged. The

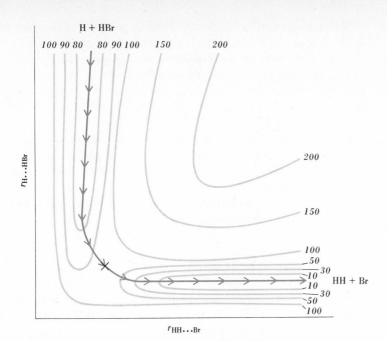

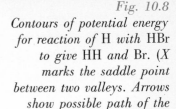

Fig. 10.8

Contours of potential energy
for reaction of H with HBr
to give HH and Br. (X
marks the saddle point
between two valleys. Arrows
show possible path of the
reaction.)

situation is similar to that of a ball rolled up the side of a hill. If
the ball is rolled slowly, it goes part way up, stops, and rolls back.
If the ball is rolled rapidly, it goes completely to the top of the hill
and down the other side. Similarly, if H and HBr particles have high
enough kinetic energy, they can attain the activated complex and get
over the hump from H and HBr to HH and Br. In a reaction at higher
temperatures more molecules get over the potential-energy hump per
unit time; so the reaction occurs faster.

Two other aspects of Fig. 10.9 are of interest. For the case represented, the final state HH and Br has lower potential energy than the
initial state H and HBr. There is a net decrease of 67 kJ in the potential
energy as the reaction proceeds. This energy usually shows up as heat;
so the particular change

$$H + HBr \longrightarrow HH + Br \qquad \Delta H = -67 \text{ kJ}$$

is exothermic. The amount of energy required to raise H and HBr
to the activated complex is more than made up for when the activated
complex springs apart to form HH and Br. However, since the activation energy is appreciable, the reaction takes time, even though the
system eventually goes to a lower potential-energy state.

Figure 10.9 was presented as a potential-energy diagram to be read
from left to right for the reaction of H and HBr to produce HH and
Br. It can also be read from right to left as a diagram for the reaction
of HH and Br to produce H and HBr; that is, the reaction is reversible.
As can be seen from the diagram, the leftward reaction to produce
H and HBr is endothermic. Also, this reaction between HH and Br
has a higher activation energy than does the reaction between H and
HBr.

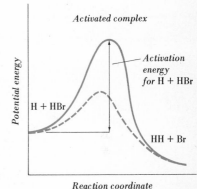

Fig. 10.9

Potential-energy change
during a reaction. (Broken
line refers to catalyzed
reaction.)

When a reaction is catalyzed, there is generally a change of path or mechanism. Since the rate is now faster, the activation energy for the new path must be lower than for the old path. The broken curve in Fig. 10.9 shows schematically what the potential-energy curve might look like for the new path. Since the barrier is lower, at any given temperature and concentration more particles per unit time would get over the hump; hence the reaction goes faster. As an example of catalytic change of path, the reaction between hydrogen and oxygen to form water is catalyzed by the presence of platinum. It has been suggested that the effect of the platinum is to react with H_2 molecules to produce H atoms. The oxygen molecules then collide with H atoms instead of with H_2 molecules. The new path has a lower activation energy in the rate-determining step.

10.9 Reactions in aqueous solutions

Most of the above discussions concerned the kinetics of chemical reactions in the gas phase. How does the situation change when one goes into the aqueous phase? In general, the reaction kinetics are more complicated and less well understood, but most of the above principles carry over directly. The biggest difference is that in the condensed phase mixing is a much slower process. It takes time to get reacting species to make contact with each other. Solvent neighbors jostle the reacting species and frequently cage them in so that they have more trouble making contact with another reactant. As a result, reactions in the aqueous phase are generally slower than equivalent reactions in the gas phase. Usually this turns out to be a help because slower rates mean more time to unravel the function of intermediates.

Still, it must be admitted that reactions in aqueous solutions are not very well understood. The prime reason for this is that, given the huge perturbing influence of the solvent, very rarely do we have a clear idea of what the reacting species is. It may have one, two, three, four, six, etc., H_2O molecules bound to itself, some so tightly that simple identity as H_2O may be lost. In cases such as that, *ad hoc* postulations of reaction mechanisms frequently founder on lack of supporting proof for the real existence of species that have been dreamed up.

Nevertheless, a great deal of progress has been made in unraveling the murky details of solution reaction mechanisms. As an indication of what can be accomplished we consider a truly classic problem, the mechanism of oxidation-reduction in aqueous solution. This is a ubiquitous problem, extending to and probably including the basic metabolic reactions that control life. What can we say about the mechanism of electron transfer? It would seem that this would be a perfectly straightforward process: An electron hops from the species that acts as the reducing agent to the species that acts as the oxidizing agent. What more is there to worry about? Simply this! Electron hopping is a surprisingly rare event. It can occur only when the symmetry of the environment around the receptor atom matches the symmetry of the environment around the donor atom. Electron descriptions (i.e., wave functions) fit certain symmetry requirements, and if the receiving atom is not ready with the appropriate environment, then the electron jump just will not occur. Knowledge of this comes as a shock to many people. They reason that the electron being lighter in mass has a greater probability of jumping to another atom than does an atom, but this is not true. In many cases, atom transfer occurs more rapidly than does electron transfer.

What is the experimental evidence for believing the ideas put forth in the preceding paragraph? To reduce complications to a minimum, we consider the oxidation of iron ion from the $+2$ state (ferrous) to the $+3$ state (ferric). Interested as we are in electron transfer from a reducing agent to an oxidizing agent we take up the simplest of all cases: electron transfer from Fe^{2+} to Fe^{3+}. The Fe^{2+} acts as the reducing agent, i.e., the electron donor; Fe^{3+} acts as the oxidizing agent, i.e., the electron acceptor. But of course when Fe^{3+} picks up an electron, it gets converted to Fe^{2+}; and when Fe^{2+} gives up an electron, it forms Fe^{3+}. So, the products, Fe^{2+} and Fe^{3+}, are the same as the reactants, Fe^{2+} and Fe^{3+}. To keep the bookkeeping straight, we do what is frequently done in such cases—we label one of the atoms by making it a radioactive species. Suppose we make the starting Fe^{3+} radioactive (we will designate this by an asterisk). We ask then how fast Fe^{2+} transfers an electron to radioactive Fe^{3+} producing radioactive Fe^{2+} and nonradioactive Fe^{3+}. Symbolically, the reaction would look as follows:

$$Fe^{2+} + Fe^{*3+} \longrightarrow Fe^{3+} + Fe^{*2+}$$

By following the radioactivity and measuring how fast it shifts from $+3$ iron to $+2$ iron we can get a direct handle on how fast electron transfers occur. It is not an easy experimental problem, but it has been done with enough precision that we can with confidence say that the rate law for the above electron transfer is given as follows:

$$\text{Rate} = k_1[Fe^{2+}][Fe^{3+}] + k_2\frac{[Fe^{2+}][Fe^{3+}]}{[H_3O^+]} + k_3[Fe^{2+}][Fe^{3+}][A^-]$$

where A^- stands for the negative ion, for example, Cl^-, present in the solution. The fact that there are three terms in the rate law indicates three alternative pathways for the reaction. The preference of one over the other depends on the relative values of k_1, k_2, and k_3. The first surprise is that k_1 is relatively small. Whereas $k_2 = 1.0 \times 10^3$ and $k_3 = 9.7$ (when A^- is chloride ion), k_1 is equal only to 0.87. In other words, the k_1 path is the least preferred. This one, incidentally, is the one that involves direct electron transfer from Fe^{2+} to Fe^{3+}. As can be seen, the rate of reaction along the k_1 path is equal to $k_1[Fe^{2+}][Fe^{3+}]$, suggesting that collision between one Fe^{2+} and one Fe^{3+} is the mechanistic event. Presumably, by this path Fe^{2+} collides directly with Fe^{3+}, an electron jumps from Fe^{2+} to Fe^{3+}, and the ions separate. The fact that reaction by this pathway is relatively slow indicates that direct electron transfer from one ion to another is not the most favorable process. (The reason for this is that Fe^{2+} probably clusters solvent molecules around itself quite differently from Fe^{3+}. If the environment symmetry seen by the electron is not fairly similar, the electron just will not jump. It waits until fluctuations adjust the receptor environment properly, and then it promptly flops over. But this may be a relatively slow process, which will have to play second fiddle to some other more probable event.)

The second term in the rate law is particularly interesting because of the inverse dependence on concentration of hydronium ion. As we shall see in Chap. 12, hydronium concentrations are inversely related to hydroxide-ion concentrations, since $[H_3O^+][OH^-]$ must be equal to a constant. So the second term, instead of being written $k_2[Fe^{2+}][Fe^{3+}]/[H_3O^+]$, is equal to $k_2'[Fe^{2+}][Fe^{3+}][OH^-]$. This is most revealing because it is known that Fe^{3+} and OH^- have a great tendency to combine to form the complex $FeOH^{2+}$, i.e., an Fe^{3+} ion with an OH^- attached. Hence, the second term of the rate law can be written $k_2''[Fe^{2+}][FeOH^{2+}]$, which implies collision between Fe^{2+} and a complex species $FeOH^{2+}$. The large value of k_2, namely, 1.0×10^3, suggests that this is *the* important pathway by which electron transfer is achieved.

Why should electron transfer by the k_2 path go faster than by the k_1 path? Let us look at probable mechanisms. Without doubt, Fe^{2+} is hydrated; so we should consider it as $Fe(H_2O)^{2+}$ or some more complicated hydrate species. What happens when such an $Fe(H_2O)^{2+}$ species is next to an $FeOH^{2+}$ species? The following scheme shows what might happen:

$$
\begin{bmatrix} Fe^{2+} \\ \mid \\ O \\ H \quad H \end{bmatrix}^{2+} + \begin{bmatrix} Fe^{*3+} \\ \mid \\ O^- \\ H \end{bmatrix}^{2+} \longrightarrow \begin{bmatrix} Fe^{3+} \\ \mid \\ O^- \\ H \end{bmatrix}^{2+} + \begin{bmatrix} Fe^{*2+} \\ \mid \\ O \\ H \quad H \end{bmatrix}^{2+}
$$

A hydrogen atom (H) might jump from the H_2O attached to Fe^{2+} to the OH^- that is attached to Fe^{*3+}. The result will be to form a species $Fe^*(H_2O)^{2+}$. Effectively the radioactive Fe would have been reduced, not by the jump of an electron, which is a relatively hindered process, but by the jump of an H atom. Unlike the electron transfer, which is hedged with symmetry restrictions, the H-atom transfer is relatively uncomplicated. This apparently is the explanation why k_2 is greater than k_1. Electron transfer *via* the indirect route of riding an H-atom transfer is a more probable route than direct electron jump from Fe^{2+} to Fe^{*3+}.

The third term in the rate law has a similar connotation. Dependence on $k_3[Fe^{2+}][Fe^{3+}][A^-]$ can be written $k_3'[Fe^{2+}][FeA^{2+}]$, suggesting complex-ion formation between Fe^{3+} and A^-. If the negative ion A^- is chloride ion, Cl^-, as it well might be, our rate law shows $k_3'[Fe^{2+}][FeCl^{2+}]$. This suggests a collision between Fe^{2+} and the complex species $FeCl^{2+}$ (presumably consisting of Fe^{3+} and Cl^-). If the activated complex $Fe^{2+} \cdots FeCl^{2+}$ proceeds to break apart by transfer of a neutral chlorine atom from the radioactive iron to the nonradioactive iron, then we have a process that looks like

$$Fe^{2+} + [Cl^- Fe^{*3+}]^{2+} \longrightarrow Fe^{2+} Cl^- Fe^{*3+} \longrightarrow$$
$$[FeCl]^{2+} + Fe^{*2+}$$

The neutral chlorine atom, as it jumps over from the radioactive-labeled Fe^{3+} to Fe^{2+}, carries no electron with it; given the way we count oxidation numbers, it looks like a -1 chlorine (Cl^-) is now attached to iron in an $FeCl^{2+}$ species; hence the iron must have been oxidized, not by an electron transfer away from it, but by neutral-atom transfer toward itself.

Direct proof for atom-transfer-type mechanisms has been obtained in the case of chromium ion. Unlike most positive ions that are placed in water, chromium only very slowly changes the neighbors in its immediate environment. This property enables us to observe at leisure what atoms are bound to the central chromium atoms. Thus, when it is found that mixing Cr^{*2+} with Cr^{3+} that is bound to Cl^- (as $CrCl^{2+}$) produces Cr^*Cl^{2+} and nonradioactive Cr^{3+}, then we know that the oxidation of the Cr^{*2+} to Cr^{*3+} has occurred not by electron transfer away from it but by neutral-chlorine-atom transfer toward itself.

A large number of oxidation-reduction reactions correspond to a two-unit change in oxidation state. Thus, for example, the reaction

$$SO_3^{2-} + ClO_3^- \longrightarrow SO_4^{2-} + ClO_2^-$$

shows an increase in sulfur oxidation state from $+4$ to $+6$ and a reduction in chlorine oxidation state from $+5$ to $+3$. Atom transfer is the most likely explanation since two-electron jumps are not very probable. This has been confirmed by using labeled isotopes as tracers. For example, in the above reaction, it is found that if the oxygen atoms in ClO_3^- are made ^{18}O instead of the conventional ^{16}O, then the ^{18}O isotope appears in the SO_4^{2-} product. The following scheme shows what probably is happening:

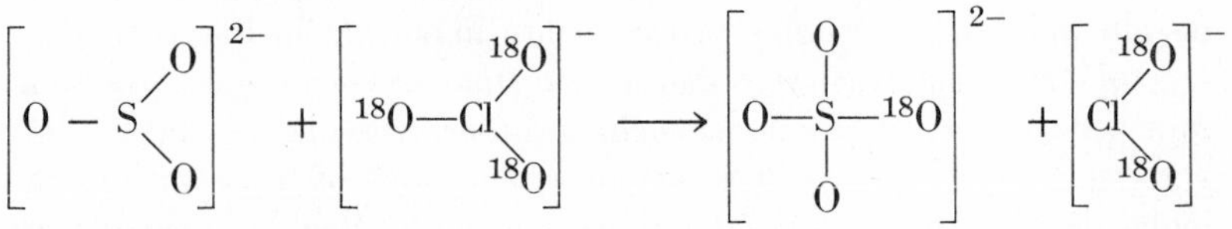

10.10 Very fast reactions

Many of the most important chemical reactions in solution, e.g., those in biologic systems, take place in times shorter than a microsecond (10^{-6} sec). This presents a real problem to the experimental investigator because it is not possible to mix solutions together in times much less than a millisecond (10^{-3} sec). The fastest mixing is produced by shooting together very fine streams of one solution containing one reactant and another containing the other reactant. Even so, the time required for the reactants to diffuse together is many times longer than the time required for the very fast reactions.

An example of a very fast reaction is the one between hydronium ion, H_3O^+, and hydroxide ion, OH^-. In a mix that would be 0.1 M H_3O^+ and 0.1 M OH^-, the reaction would be 99.9 percent complete in 10^{-8} sec. Because this time is so much shorter than that required to mix the solutions together, special techniques have to be used to study the reaction rate. One type of special technique which has been successfully employed, called *relaxation spectroscopy*, uses a sudden perturbation to bring about additional reaction in an already-reacted solution. For example, sudden compression, rapid temperature jump, or abrupt application of an electric field can disturb a system so as to bring about additional reaction. Measurements are then made while the system adjusts to the perturbation. The time connected with the adjustment process is called the *relaxation time*. As a specific illustration, if there is a large instantaneous increase in the electric field on a solution containing a partly dissociated electrolyte, there will be an increase of conductivity with time. If the field is abruptly switched off, the conductivity drops with a finite relaxation time. By comparing the time lag in the conductivity decay with that found for a strong electrolyte, it is possible to assess the rates at which association and dissociation occur.

From relaxation experiments it has been possible to determine rates for a variety of very fast reactions. As might be expected, reactions between oppositely charged simple ions go very fast. In fact, they generally occur at rates comparable to the rates at which collisions take place. In other words, the activation energy for the reaction is low compared with the average kinetic energy of the colliding particles. In the case of H_3O^+ plus OH^-, the activation energy is sufficiently low that virtually every encounter of H_3O^+ with OH^- leads to formation of H_2O.

Relaxation spectroscopy has become a major tool in unraveling chemical reactions in biologic processes. By comparing measured rates of biologic reactions with those for simple processes, it is possible to decide which of the simple processes are important in living systems. One of the major findings is that proton transfer is extremely prevalent in a wide variety of reactions, such as those involving enzymes. Along with such transfer steps there are rearrangements of the giant enzyme molecule itself. Relaxation studies are proving invaluable in elucidating the nature of these protein rearrangements.

Exercises

*10.1 *Concentration dependence.* For the reaction $2NO(g) + H_2(g) \longrightarrow N_2O(g) + H_2O(g)$ at $1100°K$, data as shown on the top of the next page were obtained. Find the rate law and the numerical value of the specific rate constant.

Ans. Rate = $k[NO]^2[H_2]$, $k = 2.2$ atm^{-2} min^{-1}

Initial Pressure of NO, atm	Initial Pressure of H$_2$, atm	Initial rate of pressure decrease, atm/min
0.150	0.400	0.020
0.075	0.400	0.005
0.150	0.200	0.010

*10.2 *Rate law.* In studying the kinetics of the reaction $X(g) + Y(g) \longrightarrow Z(g)$ at 800°K, the following data were observed:

Initial concentration of X, M	Initial concentration of Y, M	Initial rate of formation of Z, M/min
0.10	0.10	0.030
0.20	0.20	0.240
0.20	0.10	0.120

(a) Write the rate law for the reaction. (b) What is the numerical value of the specific rate constant? (c) What would be the initial rate of Z formation starting with 0.15 M X and 0.15 M Y? (d) How would the rate in (c) be changed if after the reaction had just begun the volume of the container were abruptly doubled?

*10.3 *Reaction rates.* When mixed at 700°K, $H_2(g)$ and $I_2(g)$ react to produce $HI(g)$. The reaction is first order in H_2 and first order in I_2. Suppose at time $t = 0$, one mole of H_2 and one mole of I_2 are simultaneously injected into a 1-liter box. One second later, before reaction is complete, the contents of the box are examined for the number of moles of HI. What would be the probable effect on this number if each of the following changes were made in the initial conditions?

a   Use two moles of H_2 instead of one.
b   Use two moles of I_2 instead of one.
c   Use a 2-liter box.
d   Raise the temperature to 750°K.
e   Add a platinum catalyst.
f   Add enough neon gas to double the initial pressure.

*10.4 *General.* Suppose you are given the overall reaction $A(g) + 2B(g) \longrightarrow C(g) + D(g)$. The reaction shows first-order kinetics for the concentration of A and first-order kinetics for the concentration of B.

a   Explain what experimental data must have been obtained to show the reaction is first order in A and in B.
b   Give a possible mechanism which would be consistent with the observed rate law. Indicate which is the slow step.
c   Explain how you would determine the activation energy for this reaction.

*10.5   General. There are very few examples of third-order reactions. When examples are encountered, third-order reactions are generally slower than second- or first-order reactions. Explain these statements in terms of collision theory.

*10.6   Temperature. Explain why the rate of a chemical reaction increases with temperature.

*10.7   Arrhenius equation. Give the Arrhenius equation for the specific rate constant. Tell what each symbol represents.

*10.8   Concentration. Give an appropriately labeled graph of concentration and time which would be valid for a second-order reaction only.

*10.9   Catalyst. Comment critically on the following statement: The presence of a catalyst affects the rate of a chemical reaction, but since the catalyst does not participate in the reaction, it is not used up.

*10.10   General. Assume that an A molecule reacts with two B molecules in a one-step process to give AB_2: (a) Write a rate law for this reaction. (b) If the initial rate of formation of AB_2 is 2.0×10^{-5} M/sec and the initial concentrations of A and B are 0.30 M, what is the value of the specific rate constant? (c) Suggest a more likely mechanism for this reaction.

Ans. (a) Rate $= k[A][B]^2$, (b) $k = 7.4 \times 10^{-4}$ M^{-2} sec^{-1}

*10.11   Reaction rate. A reacts with B in a one-step reaction to give C. The rate constant for the reaction is 2.0×10^{-3} M^{-1} sec^{-1}. If 0.50 mol of A and 0.30 mol of B are placed in a 0.50-liter box, what is the initial rate of the reaction?

*10.12   Catalysis. What is the essential difference between catalysis and autocatalysis? Give an example of each.

*10.13   Collision theory. Explain why when the rate-determining step involves collision of two A molecules, the rate will be proportional to the square of the A concentration.

**10.14   Temperature. Show by computation of a typical case that the effect of a 10°C rise in temperature will have a greater effect on the rate constant k at low temperature than it does at high.

**10.15   Half-life. When heated to 600°C, acetone (CH_3COCH_3) decomposes to give CO and various hydrocarbons. The reaction is found to be first order in acetone concentration with a half-life of 81 sec. Given at 600°C a 1-liter container into which acetone is injected at 0.48 atm, approximately how long would it take for the acetone pressure to drop to 0.45 atm?

Ans. 8 sec

**10.16   Rate constant. Explain why when giving values of the specific rate constant you do not have to be precise about the concentration units when you are talking about first-order reactions but you do have to be precise for higher-order reactions.

** *10.17* *Activation energy.* What activation energy should a reaction have so that raising the temperature by $10°C$ at $0°C$ would triple the reaction rate?

** *10.18* *Energy of activation.* For the gas-phase decomposition of acetaldehyde, $CH_3CHO \longrightarrow CH_4 + CO$, the second-order rate constant changes from $0.105\ M^{-1}\ sec^{-1}$ at $759°K$ to $0.343\ M^{-1}\ sec^{-1}$ at $791°K$. Calculate the activation energy that this corresponds to. What rate constant would you predict for $836°K$?

Ans. $185\ kJ$, $1.56\ M^{-1}\ sec^{-1}$

** *10.19* *Potential energy.* Consider a 100-kJ exothermic reaction between one H atom and one X_2 molecule to give an X atom and an HX molecule. Plot the potential energy of the system as a function of the course of the reaction. Indicate clearly on your diagram the magnitude of (*a*) the activation energy, (*b*) the heat of reaction, (*c*) the potential energy of the reactants and products, and (*d*) the potential energy of the activated complex. If the activation energy for the reaction $H + X_2 \longrightarrow HX + X$ is $50\ kJ/mol$, what would be the activation energy for the reaction of X with HX to give H and X_2?

** *10.20* *Consecutive reactions.* Suppose you have the change

$$A \xrightarrow{k_1} B \xrightarrow{k_2} C$$

corresponding to two consecutive reactions. Draw a graph analogous to Fig. 10.6 for the time variation of the concentrations of A, B, and C, assuming that $k_1 = k_2$.

** *10.21* *Chain reactions.* On the basis of the scheme proposed in Sec. 10.7 why should there be three explosive limits for the H_2–O_2 reaction?

** *10.22* *Aqueous reactions.* Show how atom transfer might accelerate an oxidation-reduction reaction over what would be expected for direct electron transfer. How does a determination of the rate law lead to such a picture?

** *10.23* *Very fast reactions.* If it takes at least 10^{-3} sec to mix solutions of two very fast reacting reagents, how can one determine rates of reactions that are completed in less than 10^{-6} sec?

*** *10.24* *Second-order reaction.* At $518°C$ and relatively low pressures the thermal decomposition of acetaldehyde, $CH_3CHO(g) \longrightarrow CH_4(g) + CO(g)$, is found to be second order in acetaldehyde. From the following data deduce a numerical value for the specific rate constant in units of $atm^{-1}\ sec^{-1}$.

Ans. $5.1 \times 10^{-3}\ atm^{-1}\ sec^{-1}$

Time, sec	0	42	105	242	480
Total pressure, atm	0.478	0.522	0.575	0.654	0.733

***10.25   *Catalysis.* At high pressures the decomposition of acet-
aldehyde, $CH_3CHO \longrightarrow CH_4 + CO$, goes by a first-order reaction,
but if molecular I_2 is present, it may also decompose by an I_2-catalyzed
second-order reaction involving CH_3CHO and I_2. Write a rate law for
the net decomposition, and tell how you might unravel the situation
experimentally.

***10.26   *Mechanisms.* The reaction of hydrogen peroxide with
iodide, $H_2O_2 + 2H_3O^+ + 2I^- \longrightarrow 4H_2O + I_2$, is found to be first
order in H_2O_2 and first order in I^-. Suggest a mechanism that would
correspond to such a rate law.

***10.27   *Complex rate laws.* Suppose you are given the reaction
$P + Q \longrightarrow R + S$ which is catalyzed by species T. The rate for
R formation is observed to follow the following rate law:

$$\frac{d[R]}{dt} = k_1[P] + k_2[P][Q] + k_3[Q][T]$$

What can you say about the mechanism of the reaction? What would
you infer about the relative value of k_3 as compared with k_1 and k_2?

***10.28   *Chain reaction.* What is meant by the term "chain
reaction"? With a specific illustration of your own devising (i.e., not
an example taken from the text), indicate the main distinction between
a chain-starting, a chain-propagating, and a chain-terminating step.
Discuss the importance of energy absorption and emission in the first
and last of these.

***10.29   *Chain reactions.* See if you can derive the observed
fiendishly complicated rate law for the $H_2 + Br_2$ reaction. To do this,
first write an equation for $d[HBr]/dt$ as governed by Eqs. (7) to (9).
Then remove the concentrations of Br and H by using the *steady-state*
approximation, which is that concentrations of the reactive intermediates that are regenerated do not change with time.

***10.30   *Smog.* On a typical day in Los Angeles, the hydrocarbon
content in the atmosphere starts to rise about 6 A.M., peaks about 8
A.M., and then dies off at around 10 A.M. A similar humped curve occurs
around 6 P.M. The ozone concentration rises in a hump about 2 h
following the hydrocarbon hump. How might you explain all this?

***10.31   *Energy diagrams.* Given a reaction $X + YZ \longrightarrow$
$XY + Z$, draw a probable contour diagram for the change of potential
energy with relevant separation distances. Trace a probable reaction
path on the potential-energy diagram, and speculate on the significance
of a wiggly path rather than a straight path over the saddle point.

***10.32   *Radioactive.* Using the kinetics of consecutive reactions
make a sketch of the probable time variation in the number of nuclei
of ^{238}U (half-life = 4.5×10^9 yr), ^{234}Pa (half-life = 6.7 h), and ^{234}U
(half-life = 2.5×10^5 yr) in a sample that starts out to be pure ^{238}U
at time $t = 0$.

It is found by experiment that when reacting species undergo chemical reaction, conversion of reactants to products is often incomplete, no matter how long the reaction is allowed to continue. In the initial state the reactants are present at some definite concentration. As the reaction proceeds, the concentrations decrease. Sooner or later, however, they level off and become constant. A state in which the concentrations no longer change with time is established. This state which persists as long as the system is free of external perturbations is known as the state of *chemical equilibrium*.

11.1 The equilibrium state

As an example of the attainment of equilibrium, we consider the reaction

$$A(g) + B(g) \longrightarrow C(g) + D(g)$$

in which one molecule of A reacts with one molecule of B to form one molecule each of C and D. For simplicity, all the substances are taken to be gases. At the start of the experiment, A and B (not necessarily in equal amounts) are mixed in a box. The concentration of A and the concentration of B are measured as time passes. Results of the measurements are plotted in Fig. 11.1, where concentration is the vertical axis and time is the horizontal axis. The initial concentration of A is some definite number, depending on the number of moles of A and the volume of the box. As time goes on, the concentration of A diminishes, at first quite rapidly but then less rapidly. Eventually, the concentration of A levels off and becomes constant. The concentration of B changes in similar fashion, though it may not start off at the same value as A. The initial concentrations of C and D are zero. As time goes on, C and D are produced. Their concentrations, which, of course, must be equal to each other, increase quite rapidly at first but then level off. At time t_e each of the concentrations [A], [B], [C], and [D] becomes essentially constant. Once this equilibrium state has been established, it persists indefinitely, and if no other slow reaction occurs, will last forever.

The constant state that characterized equilibrium vapor pressure (Sec. 6.2) was attributed to equality of opposing reactions involving evaporation and condensation of molecules. Similarly, the constant state that characterizes chemical equilibrium is due to equality of opposing reactions A + B $\longrightarrow$ C + D and C + D $\longrightarrow$ A + B. A and B molecules react to form C and D as long as A and B are present. The reaction does not stop at time t_e. As soon as an appreciable number of C and D molecules have formed, they start to react with each other to produce A and B. After time t_e, the forward and reverse changes occur at the same rate. The equality of opposing reactions is indicated by writing

$$A(g) + B(g) \rightleftharpoons C(g) + D(g)$$

or

$$A(g) + B(g) = C(g) + D(g)$$

As can be seen, reversibility and chemical equilibrium are inextricably tied up with each other.

11.2 Mass action

It is found by experiment that every particular reaction has its own specific equilibrium state, characterized by a definite relation between the concentrations of the materials. To illustrate this relation, we consider the equilibrium involving $PCl_3(g)$, $Cl_2(g)$, and $PCl_5(g)$. PCl_5 is

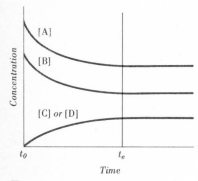

Fig. 11.1
Approach to equilibrium for the reaction
$A(g) + B(g) \rightarrow C(g) + D(g)$

thermally unstable; so it partially decomposes to PCl_3 and Cl_2. When PCl_3 and Cl_2 are mixed, they react partly to form PCl_5.

1 Suppose we inject one mole of PCl_5 into a 1-liter box at $546°K$ and wait for equilibrium. At equilibrium, we will find $0.764\ M\ PCl_5$, $0.236\ M\ PCl_3$, and $0.236\ M\ Cl_2$. Some of the PCl_5 has decomposed into PCl_3 and Cl_2.

2 We proceed next to inject one mole of PCl_3 plus one mole of Cl_2 into another 1-liter box at $546°K$ and again wait for equilibrium. We find $0.764\ M\ PCl_5$, $0.236\ M\ PCl_3$, and $0.236\ M\ Cl_2$. Some of the PCl_3 and Cl_2 have combined to form PCl_5. We note that the final state we obtain here is exactly the same as that observed in (1). It makes no difference whether we start with one mole of PCl_5 and let it decompose or with one mole of PCl_3 and one mole of Cl_2 and let them react. Stoichiometrically, these two starting systems are equivalent; they lead to the same final equilibrium state.

3 In our third experiment, we start at $t = 0$ with a random mixture of all three components, say one mole of PCl_5, two moles of PCl_3, and three moles of Cl_2. Again we put these in a 1-liter box at $546°K$ and wait for equilibrium to establish itself. On examining the contents, we find $2.83\ M\ PCl_5$, $1.175\ M\ Cl_2$, and $0.175\ M\ PCl_3$.

What do all three of these experiments have in common? Simply this: If we form the ratio $[PCl_5]/[PCl_3][Cl_2]$, where we recall brackets stand for concentrations in moles per liter, then in each case we get the numerical value 13.7. In other words, at $546°K$

$$\frac{[PCl_5]}{[PCl_3][Cl_2]} = \frac{0.764}{(0.236)(0.236)} = \frac{2.83}{(0.175)(1.175)} = 13.7$$

On the other hand, if we had formed the ratio $[PCl_3][Cl_2]/[PCl_5]$, then we would have had $1/13.7$, or 0.0730. Either of these numbers, 13.7 or 0.0730, characterizes the system composed of PCl_5, PCl_3, and Cl_2 and is called the *equilibrium constant* of the system. Whenever PCl_5, PCl_3, and Cl_2 are present together in an equilibrium situation at $546°K$ no matter what the individual values are, the expression

$$\frac{[PCl_5]}{[PCl_3][Cl_2]} = 13.7$$

must be satisfied.

For the more general case we can write the balanced equation

$$nA(g) + mB(g) + \cdots \rightleftharpoons pC(g) + qD(g) + \cdots$$

for any reaction where n molecules of A plus m molecules of B react to form p molecules of C and q molecules of D. The three dots represent other possible reactants or products. The letters n, m, p, and q represent numbers that are the coefficients of the balanced chemical equation. The letters A, B, C, and D represent the formulas of the various reactants and products. If the balanced equation is written in this way, the relationship that is constant at equilibrium is

$$\frac{[C]^p[D]^q \cdots}{[A]^n[B]^m \cdots}$$

In this ratio, which is called the *mass-action expression*,* the brackets designate concentrations in moles per liter, and the exponents are the powers to which the concentrations must be raised. In writing the mass-action expression, we follow the usual convention of putting the materials from the right side of the chemical equation in the numerator and the materials from the left side in the denominator. At equilibrium the expression is numerically equal to the equilibrium constant K for the particular reaction:

$$\frac{[C]^p[D]^q \cdots}{[A]^n[B]^m \cdots} = K$$

This equilibrium condition is called the *law of chemical equilibrium.* The law states that in a system at chemical equilibrium the concentrations of the materials which participate in the reaction must satisfy the condition expressed by the constancy of the mass-action expression. There is no other restriction on the individual concentrations.

The law of chemical equilibrium is an experimental fact. It can, however, be justified by using the principles of chemical kinetics and requiring at equilibrium the equality of the rates of forward and reverse reactions. For example, in the equilibrium

$$A(g) + B(g) \rightleftharpoons C(g) + D(g)$$

reaction may proceed in a single step or in a series of steps. The condition at equilibrium $K = [C][D]/[A][B]$ is the same in either case and can be written without any knowledge of the kinetics of the reaction. That this is justified can be shown as follows:

1 Suppose the reaction proceeds through a single reversible step:

$$A(g) + B(g) \rightleftharpoons C(g) + D(g)$$

If k is the specific rate constant for the forward reaction, we can write

Rate of forward reaction $= k[A][B]$

Similarly, if k' is the specific rate constant of the reverse reaction,

---

* The term "mass action" derives from the original work of Cato Maximilian Guldberg and Peter Waage, Norwegian chemists, who in 1864 proposed that the reaction $A + B = C + D$ could be treated as follows: The *action force* between A and B is proportional to the *active mass* of A times the active mass of B. This is called the *law of mass action.* Similarly, the action force between C and D is proportional to the active masses of C and D. At equilibrium the action force between A and B equals the action force between C and D. Although Guldberg and Waage were not clear in what was meant by "action force" and "active mass," their work was very influential in the development of a suitable description of chemical equilibrium.

we can write

Rate of reverse reaction $= k'[\text{C}][\text{D}]$

At equilibrium, the rate of the forward reaction must be equal to the rate of the reverse reaction. Consequently,

$$k[\text{A}][\text{B}] = k'[\text{C}][\text{D}]$$

or

$$\frac{[\text{C}][\text{D}]}{[\text{A}][\text{B}]} = \frac{k}{k'} = \text{a constant}$$

This proves that the mass-action expression is equal to a constant.

2 Suppose the reaction proceeds through more than one reversible step. For example, it might go through an intermediate Q by the steps

$$\text{A}(g) + \text{A}(g) \rightleftharpoons \text{C}(g) + \text{Q}(g) \tag{1}$$

$$\text{Q}(g) + \text{B}(g) \rightleftharpoons \text{A}(g) + \text{D}(g) \tag{2}$$

which add up to give

$$\text{A}(g) + \text{B}(g) \rightleftharpoons \text{C}(g) + \text{D}(g)$$

the same as above. If we let k_1 and k'_1 be the rate constants for the forward and reverse directions of step (1) and k_2 and k'_2 be the rate constants for the forward and reverse directions of step (2), then at equilibrium we can set forward and reverse rates to be equal as follows:

$$k_1[\text{A}][\text{A}] = k'_1[\text{C}][\text{Q}] \qquad \text{for the first step}$$

$$k_2[\text{Q}][\text{B}] = k'_2[\text{A}][\text{D}] \qquad \text{for the second step}$$

These simultaneous equations can be combined by solving for and eliminating the chemical intermediate Q. The result is

$$\frac{k_1[\text{A}][\text{A}]}{k'_1[\text{C}]} = \frac{k'_2[\text{A}][\text{D}]}{k_2[\text{B}]}$$

which, on rearranging and simplifying, gives

$$\frac{[\text{C}][\text{D}]}{[\text{A}][\text{B}]} = \frac{k_1 k_2}{k'_1 k'_2}$$

Again, the mass-action expression is shown to be equal to a constant.

11.3 Equilibrium constant

The numbers observed for equilibrium constants vary from very large to extremely small, depending on the specific reaction. If the equilibrium constant is small $(K < 1)$, the numerator of the mass-action expression is smaller than the denominator. This means that in the equilibrium state, the concentration of at least one of the products, materials on the right side of the chemical equation, is small. Therefore, a small equilibrium constant implies that the reaction does not proceed far from left to right. For example, $K = 1.0 \times 10^{-5}$ for

$$A(g) + B(g) \rightleftharpoons C(g) + D(g)$$

would mean that the mixing of A and B does not result in production of much C and D at equilibrium. If the equilibrium constant is large ($K > 1$), the denominator of the mass-action expression is smaller than the numerator. This means that in the equilibrium state, the concentration of at least one of the reactants, materials on the left of the chemical equation, is small. Therefore, a large equilibrium constant implies that the reaction proceeds from left to right essentially to completion. For example, $K = 1.0 \times 10^5$ for

$$E(g) + F(g) \rightleftharpoons G(g) + H(g)$$

would mean that the mixing of E and F results in practically complete conversion to G and H.

The numerical value of the equilibrium constant is determined by experiment. For example, the equilibrium involving hydrogen, iodine, and hydrogen iodide can be described by

$$H_2(g) + I_2(g) \rightleftharpoons 2HI(g)$$

for which the equilibrium condition is

$$K = \frac{[HI]^2}{[H_2][I_2]}$$

Measurement of all three concentrations in a particular equilibrium state at 490°C gave the following results:

Concentration of H_2 = 0.000862 mol/liter

Concentration of I_2 = 0.00263 mol/liter

Concentration of HI = 0.0102 mol/liter

Since these concentrations are equilibrium concentrations, they must satisfy the equilibrium condition

$$K = \frac{[HI]^2}{[H_2][I_2]} = \frac{(0.0102)^2}{(0.000862)(0.00263)} = 45.9$$

Once the number 45.9 has been evaluated, it can be used to describe any equilibrium system at 490°C containing H_2, I_2, and HI. If the mass-action expression is not equal to 45.9, then the mixture is not at equilibrium, and changes will occur until equilibrium is established.

In the above example it happens that the number of moles of gas on the two sides of the equation is equal. The concentration units in the numerator and denominator cancel; so the final equilibrium constant is dimensionless. In some cases, for example, $PCl_5(g) \rightleftharpoons PCl_3(g) + Cl_2(g)$, there are different numbers of moles on the two sides of the chemical equation; so K would appear to have units. At $546°K$

$$K = \frac{[PCl_3][Cl_2]}{[PCl_5]} = 0.073 \text{ mol/liter}$$

As we shall see in Sec. 13.7, concentration is often replaced by a dimensionless quantity called *chemical activity*, or simply *activity*. Activity can be evaluated by dividing the actual concentration of a species by the concentration of the species in a standard reference state; hence, the units cancel out. Activity can be considered to specify how many times as effective a species is as it would be in its standard state. In some cases activity is not exactly proportional to concentration because there are correction factors for nonideal behavior. We shall assume behavior is ideal and neglect such correction factors. We shall also use concentrations instead of activities, but we will not put down any units for K since these can obviously be figured out from the mass-action expression. If there may be ambiguity in what is meant by K, equilibrium constants using concentrations expressed in moles per liter are sometimes designated as K_C.

11.4 Equilibrium calculations

Given the equilibrium constant for a reaction, it can be used to describe any system containing the chemical components of that reaction at equilibrium. Thus, the value 45.9 can be used to describe any system containing H_2, I_2, and HI provided it is in chemical equilibrium at $490°C$.

Example 1
One mole of H_2 *and one mole of* I_2 *are introduced into a 1-liter box at a temperature of 490°C. What will be the final concentrations in the box when equilibrium has been established?*

Initially, there is no HI in the box. The system is not at equilibrium since the mass-action expression is zero instead of 45.9. In order to establish equilibrium, changes must occur to produce HI. HI can come only from the reaction

$$H_2(g) + I_2(g) \longrightarrow 2HI(g)$$

This reaction proceeds to produce enough HI to satisfy the equilibrium condition.

Let n equal the number of moles of hydrogen that disappear in the process of establishing equilibrium. Every time one mole of hydrogen disappears, one mole of iodine also disappears. So, n also represents the number of moles of iodine that disappear. According to the balanced equation, if one mole of hydrogen or iodine disappears, two moles of HI must be formed. If n moles of hydrogen disappear, $2n$ moles of HI must appear. Therefore, $2n$ is equal to the number of moles of HI formed in order to establish equilibrium. Since the volume of the

Initial concentration, mol/liter	Equilibrium concentration, mol/liter
$[H_2] = 1.000$	$[H_2] = 1.000 - n$
$[I_2] = 1.000$	$[I_2] = 1.000 - n$
$[HI] = 0$	$[HI] = 2n$

box is 1 liter, the concentration of each component is the same as the number of moles of that component. The situation is summarized in the table shown at the left.

The equilibrium concentrations must satisfy the condition

$$\frac{[HI]^2}{[H_2][I_2]} = 45.9$$

Substitution gives

$$\frac{(2n)^2}{(1.000 - n)(1.000 - n)} = 45.9$$

for which

$$n = 0.772*$$

Therefore, at equilibrium

$[H_2] = 1.000 - n = 0.228$ mol/liter

$[I_2] = 1.000 - n = 0.228$ mol/liter

$[HI] = 2n = 1.544$ mol/liter

That these values represent equilibrium concentrations can be checked by calculating the value of the mass-action expression

$$\frac{[HI]^2}{[H_2][I_2]} = \frac{(1.544)^2}{(0.228)(0.228)}$$

It must be 45.9 at 490°C.

■ ■ ■

To emphasize the fact that it makes no difference from which side of the stoichiometric chemical equation equilibrium is approached, we consider what happens when only HI is placed in the box at 490°C. Since initially there is no hydrogen or iodine in the system, decomposition of HI must occur in order to establish equilibrium.

Example 2

Two moles HI are injected into a box of 1-liter volume at 490°C. What will be the concentration of each species in the box at equilibrium?

The equilibrium can again be written

$$H_2(g) + I_2(g) \rightleftharpoons 2HI(g)$$

$$K = \frac{[HI]^2}{[H_2][I_2]} = 45.9$$

---

*The equation can be solved by taking the square root of both sides. For a more general case we can use the ordinary algebraic methods for solving quadratic equations (see Appendix 4.3). Of the two roots necessarily obtained for the quadratic equation, one can be discarded as physically impossible. In this case the root $n = 1.42$ would be thrown away as corresponding to more than 100 percent reaction.

Let x equal the number of moles of HI that must decompose in order to establish equilibrium. The above chemical equation shows for the back reaction that for each two moles of HI that disappear, one mole of hydrogen and one mole of iodine must be formed. If x moles of HI disappear, $x/2$ moles of hydrogen and $x/2$ moles of iodine appear. The initial and final concentrations are summarized in the table at the right. At equilibrium

Initial concentration, mol/liter	Equilibrium concentration, mol/liter
$[HI] = 2.000$	$[HI] = 2.000 - x$
$[H_2] = 0$	$[H_2] = x/2$
$[I_2] = 0$	$[I_2] = x/2$

$$\frac{[HI]^2}{[H_2][I_2]} = 45.9 = \frac{(2.000 - x)^2}{(x/2)(x/2)}$$

for which

$$x = 0.456$$

Therefore, at equilibrium

$$[H_2] = \frac{x}{2} = 0.228 \text{ mol/liter}$$

$$[I_2] = \frac{x}{2} = 0.228 \text{ mol/liter}$$

$$[HI] = 2.000 - x = 1.544 \text{ mol/liter}$$

■ ■ ■

The above two examples show that it makes no difference whether the equilibrium state is produced from the material on the left side of the chemical equation or from the material on the right side. Change occurs so as to produce the material that is missing in sufficient concentration to establish equilibrium. Sometimes the initial nonequilibrium system contains all the components, in which case the change necessary to establish equilibrium may not be obvious. In the following example it is not immediately clear whether the concentration of HI must increase or decrease in order to establish equilibrium.

Example 3
One mole of H_2, *two moles of* I_2, *and three moles of* HI *are injected into a 1-liter box. What will be the concentration of each species at equilibrium at 490° C?*
The equilibrium is

$$H_2(g) + I_2(g) \rightleftharpoons 2HI(g)$$

Let x be the number of moles of H_2 that must be used up in order to establish equilibrium. (If it turns out that we have guessed wrong and *more* H_2 must be formed, x would turn out to be a negative number.) According to the stoichiometry of the reaction, x is also the number of moles of I_2 that must be used up, and $2x$ is the number of moles of HI that must be formed. To reach equilibrium, the initial concentration of H_2 is reduced by the amount x, the concentration of I_2 is reduced by the amount x, and the concentration of HI is increased by $2x$. (See table on next page.)

Initial concentration, mol/liter	Equilibrium concentration, mol/liter
$[H_2] = 1.000$	$[H_2] = 1.000 - x$
$[I_2] = 2.000$	$[I_2] = 2.000 - x$
$[HI] = 3.000$	$[HI] = 3.000 + 2x$

At equilibrium

$$\frac{[HI]^2}{[H_2][I_2]} = 45.9 = \frac{(3.000 + 2x)^2}{(1.000 - x)(2.000 - x)}$$

for which

$$x = 0.684$$

Therefore, at equilibrium

$$[H_2] = 1.000 - x = 0.316 \text{ mol/liter}$$

$$[I_2] = 2.000 - x = 1.316 \text{ mol/liter}$$

$$[HI] = 3.000 + 2x = 4.368 \text{ mol/liter}$$

■ ■ ■

11.5 Heterogeneous equilibrium

Heterogeneous equilibria are those equilibria which involve two or more phases. For example, the equilibrium

$$2C(s) + O_2(g) \rightleftharpoons 2CO(g)$$

involves both gaseous and solid phases. The solid phase consists of pure carbon, and the gas phase consists of a mixture of oxygen and carbon monoxide. In mass-action expressions, the concentrations that are shown must correspond to the particular phase specified by the chemical equation. For example, for the above equilibrium the equilibrium condition is

$$K = \frac{[CO(g)]^2}{[C(s)]^2[O_2(g)]}$$

where $[CO(g)]$ refers to the concentration of CO *in the gas phase*, $[C(s)]$ refers to the concentration of C *in the solid phase*, and $[O_2(g)]$ refers to the concentration of O_2 *in the gas phase*.

A simplification of the above equilibrium condition is possible because the concentration of carbon in the solid phase of carbon is not a variable. The concentration of carbon monoxide in the gas phase can be changed, e.g., by the addition of carbon monoxide. If the volume remains constant and more carbon monoxide has been added, the concentration of carbon monoxide will be increased. Similarly, the concentration of oxygen could be changed, but for solid carbon this is not possible. If more solid carbon is added, the concentration is not changed because as the number of moles of carbon increases, the volume of carbon also increases. The number of *moles per unit volume of solid carbon* is the same number, no matter how many moles of carbon is present.

In the general case, at a given temperature the concentration of a substance that is a pure solid or a pure liquid cannot be changed and is a constant. This constant can be combined with the original equilibrium constant to give a new equilibrium constant for which the mass-action expression does not include the pure condensed phase. Thus, for the equilibrium

$$2C(s) + O_2(g) \rightleftharpoons 2CO(g)$$

$$K[C(s)]^2 = \frac{[CO]^2}{[O_2]}$$

where $[C(s)]$ is a constant. Therefore,

$$K[C(s)]^2 = K'$$

$$K' = \frac{[CO]^2}{[O_2]}$$

The last equation expresses the requirement that a system containing $CO(g)$, $O_2(g)$, and $C(s)$ is in equilibrium no matter how much $C(s)$ is present provided that $[CO]^2/[O_2]$ has the proper value. The simple rule is that for heterogeneous equilibria pure solids and pure liquids are omitted from the mass-action expression. Further examples are given below:

At 1000°C $\qquad H_2(g) + S(g) \rightleftharpoons H_2S(g) \qquad K_1 = \dfrac{[H_2S]}{[H_2][S]}$

At 200°C $\qquad H_2(g) + S(l) \rightleftharpoons H_2S(g) \qquad K_2 = \dfrac{[H_2S]}{[H_2]}$

At -100°C $\qquad H_2(g) + S(s) \rightleftharpoons H_2S(s) \qquad K_3 = \dfrac{1}{[H_2]}$

11.6 Equilibrium changes

When a system at equilibrium is disturbed, chemical reaction occurs so as to reestablish equilibrium. As an example, we consider the equilibrium system consisting of H_2, I_2, and HI in a sealed box:

$$H_2(g) + I_2(g) \rightleftharpoons 2HI(g)$$

$$K = \frac{[HI]^2}{[H_2][I_2]}$$

At 490°C, K is 45.9. The concentrations of HI, H_2, and I_2 do not change until conditions are changed. Several kinds of changes are possible: (1) H_2, I_2, or HI can be injected into the box. (2) H_2, I_2, or HI can be removed. (3) The volume of the box can be changed. (4) The temperature of the system can be changed. (5) A catalyst can be added. How is the equilibrium state affected by each of these changes?

  1 Suppose the concentration of one of the components is changed by addition of that component. For example, suppose that more H_2

is added to a box which already contains H_2, I_2, and HI in equilibrium at 490°C. What effect would such a concentration increase have on the other components? The problem can be explored three ways:

(a) *The equilibrium constant.* The equilibrium condition is of the form

$$\frac{[HI]^2}{[H_2][I_2]} = 45.9$$

By increasing the concentration of hydrogen, the denominator is made bigger. If everything else stayed the same, the fraction would become less than 45.9; therefore, the system would no longer be at equilibrium. To reestablish equilibrium, two things could happen: There could be a decrease in the concentration of I_2; so the denominator would be restored to its original value. Or there could be an increase in the concentration of HI; so the numerator would increase to compensate for the increased denominator. Since iodine atoms must exist either as I_2 or HI molecules, decrease of I_2 and increase of HI occur simultaneously.

(b) *Le Châtelier's principle.* According to the principle of Le Châtelier (Sec. 7.11), any equilibrium system subjected to a stress tends to change so as to relieve the stress. For a system in chemical equilibrium, changing the concentration of one of the components constitutes a stress. If in the present case hydrogen is added to the box, the equilibrium system

$$H_2(g) + I_2(g) \rightleftharpoons 2HI(g)$$

adjusts itself to absorb the effect of the added hydrogen. The stress can be absorbed if some hydrogen molecules combine with iodine molecules to form HI. This means that the concentration of HI would increase and the concentration of I_2 would decrease. Le Châtelier's principle thus leads to the same prediction as did use of the equilibrium constant in part (a) above.

(c) *Kinetics.* The effect of added H_2 can be predicted from a consideration of reaction rates. The argument here is relatively simple if the reaction proceeds by one step. In the equilibrium state, collisions between H_2 and I_2 molecules form HI, and simultaneously collisions between HI molecules form H_2 and I_2. These two rates are equal. By adding H_2 to the box, the chance for collision between H_2 and I_2 is increased. The more collisions there are, the faster HI is formed. The instantaneous effect of adding hydrogen therefore is to increase the rate of HI formation. There is, however, no instantaneous effect on the decomposition rate of HI. For a time, HI will be forming faster than it is decomposing, and so its concentration will increase. Eventually, the concentration of HI will increase to the point where there are more collisions between HI molecules; so the reverse reaction, the decomposition of HI, will begin to speed up. It will continue to speed up until it equals the increased rate of HI formation. Equilibrium will be reestablished with a net increase in HI and H_2 and a decrease in I_2.

2   *Suppose the concentration of one of the components is changed by removal of that component.* For example, suppose that some H_2 is removed from the box.

(a)   *The equilibrium condition*

$$\frac{[HI]^2}{[H_2][I_2]} = 45.9$$

predicts that a decrease in $[H_2]$ would be compensated for by an increase in $[I_2]$ and a decrease in $[HI]$.

(b)   *Le Châtelier's principle* predicts that the system would adjust to relieve the stress caused by the removal of H_2. Some HI would decompose to form H_2 to replace some of that removed. The effect would be to reduce the concentration of HI and increase the concentration of I_2.

(c)   *Kinetics* predicts that the removal of hydrogen from the container would reduce the rate at which H_2 and I_2 combine to form HI. This would mean that, instantaneously, HI would be forming from H_2 and I_2 more slowly than it would be decomposing to H_2 and I_2. The result would be a net decrease in HI concentration and a net increase in I_2 concentration.

3   *Suppose the volume of the box is decreased.* In cases (1) and (2) the volume of the box was kept constant; so the change in concentration and the change in number of moles was parallel. If the volume of the box were decreased, the *concentration* of all species would be increased. However, to determine how the number of moles changes, a detailed analysis of the specific reaction would be required.

(a)   *The equilibrium condition*

$$\frac{[HI]^2}{[H_2][I_2]} = 45.9$$

is expressed in terms of concentration. For each component, the concentration is equal to the number of moles n of that component divided by the volume of the box V. By substitution of n/V for concentration, the equilibrium condition can be rewritten

$$\frac{[HI]^2}{[H_2][I_2]} = \frac{(n_{HI}/V)^2}{(n_{H_2}/V)(n_{I_2}/V)} = \frac{n_{HI}^2}{(n_{H_2})(n_{I_2})} = 45.9$$

As can be seen, V cancels out. No matter what the volume of the box, the number of moles of HI squared divided by the number of moles of H_2 times the number of moles of I_2 must equal 45.9. Changing the volume of the box in this case does not change the number of moles of each species. The reason is that for the particular case

$$H_2(g) + I_2(g) \rightleftharpoons 2HI(g)$$

the number of gas molecules on the left side of the equation is the same as the number of gas molecules on the right side of the equation. If this condition were not true, the volume would not cancel out of the mass-action expression. An example of a case that is volume dependent is the equilibrium between nitrogen, hydrogen, and ammonia:

$$N_2(g) + 3H_2(g) \rightleftharpoons 2NH_3(g)$$

The equilibrium expression is

$$K = \frac{[NH_3]^2}{[N_2][H_2]^3}$$

Substituting n/V for concentration gives

$$K = \frac{(n_{NH_3}/V)^2}{(n_{N_2}/V)(n_{H_2}/V)^3} = \frac{n_{NH_3}{}^2}{(n_{N_2})(n_{H_2})^3} V^2$$

In this case the volume does not cancel out, and a change in V must be compensated for by a change in the ratio of the number of moles. Specifically, when volume is decreased, the fraction $n_{NH_3}{}^2/n_{N_2}n_{H_2}{}^3$ must increase in order to maintain K constant. The number of moles of ammonia would have to increase at the expense of the moles of nitrogen and hydrogen. However, the *concentrations* of ammonia, nitrogen, and hydrogen (as distinct from *number of moles*) all increase. In the case of nitrogen and hydrogen this comes about only because the volume decreases more than the number of moles decreases. In commercial production of ammonia from nitrogen and hydrogen, the reaction is carried out in as small a volume as possible in order to maximize the extent of conversion to ammonia.

(b) *Le Châtelier's principle* predicts the effect of reduced volume on the system H_2, I_2, and HI at equilibrium as follows: When the volume of the box is reduced, a stress is applied in that the molecules are crowded closer together. The stress can be relieved if the molecules could be reduced in number. In the case of

$$H_2(g) + I_2(g) \rightleftharpoons 2HI(g)$$

there is no device by which this can be accomplished since the disappearance of one molecule of hydrogen and one molecule of iodine is accompanied by the appearance of two molecules of HI. Neither the forward nor reverse reaction can absorb the stress. There is no net change; the number of moles of H_2, I_2, and HI stays constant. Of course, since the volume is diminished, the concentration of each component is increased. In the NH_3 equilibrium the situation would be different. When one molecule of N_2 reacts with three molecules of H_2, two molecules of NH_3 are formed. A decrease of the volume of the box could be compensated for by forming fewer molecules, i.e., by favoring the formation of NH_3. It is a general principle that for reactions in which there is a change in the number of gas molecules, a decrease in the volume favors the reaction direction that produces fewer molecules.

(c) *Kinetics* predicts the effect of decrease in volume by considering the effects on the rates of the forward and reverse reactions. For example, in the equilibrium system containing H_2, I_2, and HI a decrease of volume forces H_2 and I_2 molecules closer together; so they collide more frequently. There is consequently an increase in the rate of forward reaction. At the same time, HI molecules are also brought closer together; so they too collide more frequently. The back reaction, therefore, is also increased. If the number of gas molecules is the same on the left and the right of the equation, the rate of forward reaction is increased just as much as that of back reaction. Hence, there would be no net change in the number of molecules of any type. If there is a change in the number of gas molecules, the situation is more complicated. For example, in the case

$$N_2(g) + 3H_2(g) \rightleftharpoons 2NH_3(g)$$

it turns out that a decrease of volume increases the forward rate to a greater extent than the back rate. The net effect is to increase the number of NH_3 molecules present at equilibrium.

4 *Suppose the temperature of the system is changed.*

(a) The *equilibrium constant* has a specific value at a given temperature. If the temperature is changed, K generally changes value. For reactions which are endothermic, a rise in temperature causes K to increase; for those reactions which are exothermic, a rise in temperature causes K to decrease (how this comes about is discussed in Sec. 14.8). The reaction

$$H_2(g) + I_2(g) \rightleftharpoons 2HI(g) + 13 \text{ kJ}$$

is exothermic as written, and K decreases when the temperature increases. With increase of the temperature, the concentration of HI at equilibrium diminishes, while the concentrations of H_2 and I_2 increase. This is another way of saying that HI is less stable at higher temperatures.

(b) *Le Châtelier's principle* predicts that a rise in temperature favors the change that uses up heat. When one mole of H_2 and one mole of I_2 disappear, two moles of HI and 13 kJ of heat are liberated. The reverse process absorbs heat. At equilibrium the liberation of heat by the forward reaction is compensated for by the absorption of heat by the back reaction. If the temperature is increased, the system tries to relieve the stress by absorbing the added heat. Since the back reaction uses heat, it is the one that is favored. Favoring the back reaction causes a net decrease in the concentration of HI and a net increase in the concentration of I_2 and H_2.

(c) *Kinetics.* It is a general principle of chemical kinetics that the rate of a reaction is increased by an increase in temperature (Sec. 10.3). Furthermore, for a given equilibrium it is always found that the rate of the endothermic reaction (which necessarily is the one with larger activation energy) is increased relatively more than is the rate of the exothermic reaction. For the HI equilibrium, the rate of HI decomposition (which is endothermic) is increased more than is the rate of HI formation (which is exothermic). The result is a net decrease in the concentration of HI.

5 *What effect does a catalyst have on an equilibrium system?*

(a) *The equilibrium constant* is concerned only with the materials shown in the net equation. Intermediates may be involved, but the net equation ignores them. A catalyst may itself be an intermediate or at least affect intermediates, but it does not appear in the net equation or in the equilibrium-constant expression. Hence, insertion of a catalyst into an equilibrium system has no effect on equilibrium concentrations.

(b) *Le Châtelier's principle* has nothing to say about the presence of a catalyst.

(c) *Kinetics* gives the best argument for the fact that a catalyst cannot affect the composition of an equilibrium system. According to reaction-rate theory, the rate of chemical reaction depends on how fast particles can get over the potential-energy barrier between the initial and final states. For example, Fig. 11.2 shows the potential-energy barrier for the reaction

$$H_2(g) + I_2(g) \rightleftharpoons 2HI(g)$$

The dashed line represents the path in the presence of a catalyst. The rate of forward reaction depends on the height of the barrier between the initial and final states. Since the catalyst reduces the height of this barrier, it speeds up the rate. However, if the potential-energy barrier is lowered

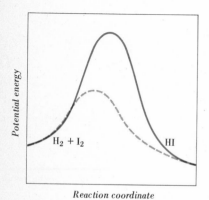

Fig. 11.2

Potential-energy diagram showing effect of catalysis (dashed line applies to catalyzed path).

for the change in the forward direction, it is likewise lowered for the change in the reverse direction. Thus, the
reverse change is also accelerated. The increase in the rates
of the forward and reverse reactions is the same; hence,
the equilibrium concentrations are unchanged.

Quantitatively, we can see this last point by considering
the Arrhenius equation (Sec. 10.3) for the specific rate
constant

$$k = Ae^{-E/RT}$$

where E is the activation energy. For the case shown in
Fig. 11.2, the activation energy of the forward process is
167 kJ per mole of H_2 or I_2 used, whereas the activation
energy of the reverse process is 180 kJ per mole of H_2
or I_2 formed. Suppose the catalyst acts to lower the activation barrier by 80 kJ/mol. For the forward process we can
write

$$k_f = A_f e^{-167/RT} \qquad \text{uncatalyzed}$$

$$k'_f = A_f e^{-(167-80)/RT} \qquad \text{catalyzed}$$
$$= A_f [e^{-167/RT}][e^{80/RT}]$$

For the reverse process we can similarly write

$$k_r = A_r e^{-180/RT} \qquad \text{uncatalyzed}$$

$$k'_r = A_r e^{-(180-80)/RT} \qquad \text{catalyzed}$$
$$= A_r [e^{-180/RT}][e^{80/RT}]$$

It should be noted that the catalyst effect of dropping the
activation energy by 80 kJ/mol is to multiply each rate
by the same factor $e^{80/RT}$. Clearly, the equilibrium constant, which is the ratio of k_f to k_r, is left unchanged.

Exercises

*11.1 *Approach to equilibrium.* Sketch a set of curves similar to
those in Fig. 11.1 for the concentration changes with time in the system

$$2A(g) + B(g) \longrightarrow 2C(g) + D(g)$$

Assume A and B start off at the same concentration.

*11.2 *Equilibrium concentrations.* Explain why for the equilib-
rium $PCl_5(g) \rightleftharpoons PCl_3(g) + Cl_2(g)$ halving the equilibrium concentrations would destroy the equilibrium situation, whereas for the equilibrium $H_2(g) + I_2(g) \rightleftharpoons 2HI(g)$, halving the equilibrium concentrations leaves an equilibrium situation intact. For the latter to
be true, what condition would be needed in the general case $nA(g) +
mB(g) \rightleftharpoons pC(g) + qD(g)$?

*11.3 *Equilibrium concentrations.* If $K = 1.60$ at 986°C for

$H_2(g) + CO_2(g) \rightleftharpoons H_2O(g) + CO(g)$, which of the following situations represents an equilibrium situation: (a) 0.442 M H_2, 0.442 M CO_2, 0.558 M H_2O, 0.558 M CO; (b) 0.884 M H_2, 0.884 M CO_2, 1.117 M H_2O, 1.117 M CO; or (c) 1.666 M H_2, 0.666 M CO_2, 1.334 M H_2O, 1.334 M CO?

*11.4   Terms. Indicate clearly, using examples, what is meant by each of the following terms: (a) mass-action expression, (b) equilibrium constant, and (c) equilibrium condition.

*11.5   Equilibrium constant. Why in giving the numerical value for an equilibrium constant is it also necessary to write the balanced chemical equation? Give examples to prove your point.

*11.6   Units. Discuss the problem of units for K with specific reference to $K = 13.7$ for $PCl_3(g) + Cl_2(g) \rightleftharpoons PCl_5(g)$ and $K = 0.0218$ for $2HI(g) \rightleftharpoons H_2(g) + I_2(g)$.

*11.7   Equilibrium constant. Given that $K = 45.9$ for $H_2(g) + I_2(g) \rightleftharpoons 2HI(g)$, calculate K for each of the following:
  a $2HI(g) \rightleftharpoons H_2(g) + I_2(g)$
  b $HI(g) \rightleftharpoons \frac{1}{2}H_2(g) + \frac{1}{2}I_2(g)$
  c $\frac{1}{2}H_2(g) + \frac{1}{2}I_2(g) \rightleftharpoons HI(g)$

*11.8   Equilibrium condition. Write an equation for the equilibrium condition in each of the following cases:
  a $COCl_2(g) \rightleftharpoons CO(g) + Cl_2(g)$
  b $2NO(g) + O_2(g) \rightleftharpoons 2NO_2(g)$
  c $KClO_3(s) \rightleftharpoons KCl(s) + \frac{3}{2}O_2(g)$
  d $CaCO_3(s) \rightleftharpoons CaO(s) + CO_2(g)$
  e $H_2(g) + 3Fe_2O_3(s) \rightleftharpoons H_2O(g) + 2Fe_3O_4(s)$
  f $2CaC_2(s) + 5O_2(g) \rightleftharpoons 2CaCO_3(s) + 2CO_2(g)$

*11.9   Equilibrium changes. $PCl_5(g)$, $PCl_3(g)$, and $Cl_2(g)$ are in equilibrium with each other in a box at 546°K. The decomposition of PCl_5 is endothermic. What effect would there be on the number of moles of PCl_5 and on the concentration of PCl_5 in the box at equilibrium if each of the following changes were made?
  a   Add some Cl_2 to the box.
  b   Add some inert gas to the box.
  c   Reduce the volume of the box.
  d   Raise the temperature of the system.
  e   Remove some PCl_3 from the box.

*11.10   Application. In the commercial production of sulfuric acid one of the common paths goes through the oxidation of sulfur dioxide to sulfur trioxide. Given $SO_2(g) + \frac{1}{2}O_2(g) \rightleftharpoons SO_3 + 98$ kJ, what are three specific things you might do to maximize the yield of sulfur trioxide?

*11.11   Equilibrium calculations. $K = 1.60$ at 986°C for $H_2(g) + CO_2(g) \rightleftharpoons H_2O(g) + CO(g)$. If at time $t = 0$, one mole of H_2 and one mole of CO_2 are injected into a 20-liter box and allowed to equilibrate at 986°C, what will be the final concentrations of H_2 and CO_2? *Ans. 0.022 M*

11.12 Equilibrium calculations. We are given the reaction $FeO(s) + CO(g) \rightleftharpoons Fe(l) + CO_2(g)$.

a   In an experiment at 1000°C, the equilibrium concentration of CO_2 is found to be $6.05 \times 10^{-4} M$, and that of CO is $1.45 \times 10^{-4} M$. What is the value of K at 1000°C?

b   What concentration of CO_2 would be in equilibrium with $2.50 \times 10^{-4} M$ CO at 1000°C?

c   What would be the equilibrium partial pressures of CO_2 and CO in (b)?

d   In an experiment in which the initial concentration of CO is $1.00 \times 10^{-3} M$ and CO_2 is zero, what would be the final equilibrium concentrations?

e   What weight of Fe would be formed at equilibrium in a furnace that is 30 m high and has an average horizontal area of 10 m^2 from 600 mol of CO and 600 mol of FeO?

11.13 Equilibrium calculation. When 1.000 mol of PCl_5 is injected into a 1-liter box at 500°K, it is found that 13.9% of the PCl_5 has decomposed into PCl_3 and Cl_2. Calculate K_C for $PCl_5(g) \rightleftharpoons PCl_3(g) + Cl_2(g)$ at 500°K. *Ans. 0.0224*

11.14 Equilibrium calculation. Given that $K_C = 1.03 \times 10^{-3}$ for $H_2S(g) \rightleftharpoons H_2(g) + \frac{1}{2}S_2(g)$ at 750°C, how many moles of S_2 must there be in equilibrium in a 3.68-liter box if the box contains 1.63 mol of H_2S and 0.864 mol of H_2?

** *11.15 Approach to equilibrium.* Draw a graph showing how the *rate* of forward change $k_f[A][B]$ varies with time in the interval t_0 to t_e for the reaction $A(g) + B(g) \longrightarrow C(g) + D(g)$ as presented in Fig. 11.1. On the same graph show how the net rate of C formation changes in the same time interval.

** *11.16 Approach to equilibrium.* Explain why it does not make any difference so far as the final equilibrium state is concerned from which side of a balanced chemical equation the starting materials are chosen. (*Hint:* See Fig. 11.2.)

** *11.17 Equilibrium and rates.* When $PCl_5(g)$ is placed in a box at 546°K, it decomposes at the following rate:

$$-\frac{d[PCl_5]}{dt} = k_f[PCl_5] - k_r[PCl_3][Cl_2]$$

Use this rate law to derive the equilibrium condition for $PCl_5(g) \rightleftharpoons PCl_3(g) + Cl_2(g)$.

** *11.18 Equilibrium and rates.* Nitrous oxide, N_2O, decomposes as follows: $2N_2O(g) \longrightarrow 2N_2(g) + O_2(g)$. The reaction is first order in N_2O. Take as a mechanism a slow step $N_2O \longrightarrow N_2 + O$ followed by a fast step $O + O \longrightarrow O_2$, and derive from it the equilibrium condition for $2N_2O(g) \rightleftharpoons 2N_2(g) + O_2(g)$.

** *11.19* **Equilibrium calculations.** Given that $K = 45.9$ at 490°C for $H_2(g) + I_2(g) \rightleftharpoons 2HI(g)$, calculate the equilibrium concentrations that will finally be established after 0.500 mol of H_2 and 0.500 mol of I_2 have been introduced into a 2-liter box at 490°C.
Ans. *0.057 M* H_2, *0.057 M* I_2, *0.386 M* HI

** *11.20* **Equilibrium calculations.** Given that $K = 45.9$ at 490°C for $H_2(g) + I_2(g) \rightleftharpoons 2HI(g)$, calculate the equilibrium concentrations that will finally be established after 0.500 mol of HI has been introduced into a 2-liter box at 490°C.

** *11.21* **Equilibrium calculations.** Given that $K = 45.9$ at 490°C for $H_2(g) + I_2(g) \rightleftharpoons 2HI(g)$, calculate the equilibrium concentrations that will finally be established after one mole of H_2, two moles of I_2, and three moles of HI are injected into a 2-liter box at 490°C.
Ans. *0.158 M* H_2, *0.658 M* I_2, *2.184 M* HI

** *11.22* **Units for K.** Expressed in moles per liter the equilibrium constant K_C has a value of 0.073 at 546°K for $PCl_5(g) \rightleftharpoons PCl_3(g) + Cl_2(g)$. Suppose you wanted to express the equilibrium condition not by using concentrations in moles per liter but partial pressures in atmospheres. What would be the appropriate numerical value to quote for K_p? Assume ideal behavior.

** *11.23* **Equilibrium calculations.** You are given that $K = 1.60$ at 986°C for $H_2(g) + CO_2(g) \rightleftharpoons H_2O(g) + CO(g)$. (*a*) If at time $t = 0$, one mole each of H_2, CO_2, H_2O, and CO are simultaneously injected into a 20-liter box and allowed to equilibrate at 986°C, what will be the final concentrations of all four species? (*b*) What would happen to these concentrations if subsequently another mole of H_2 were injected and a new equilibrium were established?
Ans. (*a*) *0.0442 M* H_2, *0.0442 M* CO_2, *0.0558 M* H_2O, *0.0558 M* CO

** *11.24* **Equilibrium calculation.** You are told that $K_C = 3.76 \times 10^{-5}$ for $I_2(g) \rightleftharpoons 2I(g)$ at 1000°K. You start an experiment by injecting 1.00 mol of I_2 into a 2.00-liter box at 1000°K. What will be the final equilibrium concentrations of I_2 and of I?

*** *11.25* **Percent conversion.** $K_C = 2.37 \times 10^{-3}$ at 1000°K for $N_2(g) + 3H_2(g) \rightleftharpoons 2NH_3(g)$. If we start with one mole of N_2 and one mole of H_2 in a 1-liter box at 1000°K, what percent of the hydrogen will be converted to NH_3 at equilibrium? Ans. *6.5%*

*** *11.26* **Equilibrium pressures.** In solid-state chemistry it is often necessary to adjust the oxygen pressure at which crystals grow to very low values. This can be done conveniently by using a gaseous equilibrium system where the oxygen pressure is regulated indirectly by controlling other components. Given that $K_C = 5.31 \times 10^{-10}$ mol/liter at 2000°K for the equilibrium $2H_2O(g) \rightleftharpoons 2H_2(g) + O_2(g)$, what pressures of H_2 and H_2O should you provide to set the oxygen pressure at 1.00×10^{-7} atm at $2000°K$? The total pressure is to be fixed at 1.00 atm.

*** **11.27** *Pressure calculations.* We are given that $K_C = 33.3$ at $760°K$ for $PCl_5(g) \rightleftharpoons PCl_3(g) + Cl_2(g)$. If 1.00 g of PCl_5 is injected into an evacuated 500-ml flask at $760°K$ and allowed to come to equilibrium, what percent of the PCl will decompose, and what will be the final total pressure in the flask?   *Ans. 99.97%, 1.20 atm*

*** **11.28** *Equilibrium calculations.* At high temperatures, phosgene, $COCl_2$, decomposes to give CO and Cl_2. In a typical experiment 0.631 g of $COCl_2$ is injected into an evacuated flask of volume 472.0 cm^3 at $1000°K$. When equilibrium has been established, it is found that the total pressure in the flask is 2.175 atm. Calculate K_C for $COCl_2(g) \rightleftharpoons CO(g) + Cl_2(g)$ at $1000°K$.

Perhaps the most common application of chemical-equilibrium principles is to reactions in aqueous solutions. The solute species in such solutions are susceptible to concentration variation just as gases are. Therefore, the principles developed in the preceding chapter for defining and describing an equilibrium state are directly applicable to solutions. Since reactions are often reversible and tend toward the equilibrium state, the principles of equilibrium give the key to understanding the many reactions that may occur in solution. In this chapter we consider equilibria in aqueous solutions and how they are related to the reactions that are observed. The important equilibria to be considered are of two fundamental types: *dissociation*, an equilibrium between a dissolved, undissociated species and its component parts, e.g., acetic acid molecules in equilibrium with hydronium ions and acetate ions; and *solubility*, an equilibrium between a pure phase, usually a solid, and its characteristic species in solution, e.g., solid barium sulfate in equilibrium with barium ions and sulfate ions. In addition, we shall consider the simultaneous establishment of two or more equilibria in the same solution.

12.1 Dissociation

When the weak acid HX is placed in water, some of it interacts with the H_2O to give H_3O^+ and X^-. When equilibrium is established, the back reaction of H_3O^+ with X^- to form HX and H_2O occurs at a rate just sufficient to balance the forward reaction. The equilibrium can be represented by the reversible equation

$$HX + H_2O \rightleftharpoons H_3O^+ + X^-$$

for which the equilibrium condition is

$$\frac{[H_3O^+][X^-]}{[HX][H_2O]} = K'$$

This equation can be simplified by taking note of the fact that usually in aqueous solutions by far the most abundant species is H_2O. Any reaction such as this so-called "dissociation of HX" uses up only a trifle of the H_2O; so the concentration of H_2O is essentially constant. It is therefore common practice to assume $[H_2O]$ is a constant number which can be absorbed in K'. Hence, we can write

$$\frac{[H_3O^+][X^-]}{[HX]} = K'[H_2O] = \text{a constant} = K$$

K, which is called the *dissociation constant* of HX, can also be designated as K_{diss}, K_{HX}, or K_a (where a stands for acid). Actually, the constancy of the above mass-action expression holds only for dilute solutions.* In concentrated solutions, interionic attractions must be considered (Sec. 8.6) as well as the change in the concentration of H_2O. Since we will confine ourselves to dilute solutions, we will assume the above equilibrium condition remains valid, independent of the total ionic concentrations in the solution. Figure 12.1 shows representative values of the dissociation constants of weak acids. Other values may be found in Appendix 6. Values are quoted for 25°C. The smaller the value of K_{diss}, the weaker the acid. Thus, HCN is a weaker acid than HF, and it is much less dissociated for a given concentration. When K_{diss} is 1 or greater, the acid is extensively dissociated, even in $1\ M$ solution, and is classified as moderately strong. When K_{diss} is 10 or greater, the acid is essentially 100% dissociated in all except very concentrated solutions. For example, perchloric acid, $HClO_4$, is one of our strongest acids and has K_{diss} greater than 10. Similarly, HNO_3, HCl, and H_2SO_4 are common acids with high dissociation constants.

The dissociation constant can also be applied to ions which dissociate as acids. For example, the hydrogen sulfate ion, HSO_4^-, can dissociate

---

* Furthermore, the mass-action expression is strictly constant only if we use the chemical activity of the various species in the expression. Just as for gases (page 317), the chemical activity of a species in the mass-action expression is frequently approximated by its concentration.

Fig. 12.1

Dissociation equilibria of
weak acids. (Values of K_{diss}
do not include H_2O in
mass-action expression.)

Acid	Reaction	K_{diss} (25°C)
Acetic	$HC_2H_3O_2 + H_2O \rightleftharpoons H_3O^+ + C_2H_3O_2^-$	1.8×10^{-5}
Nitrous	$HNO_2 + H_2O \rightleftharpoons H_3O^+ + NO_2^-$	4.5×10^{-4}
Hydrofluoric	$HF + H_2O \rightleftharpoons H_3O^+ + F^-$	6.7×10^{-4}
Hydrocyanic	$HCN + H_2O \rightleftharpoons H_3O^+ + CN^-$	4.0×10^{-10}
Sulfurous	$H_2SO_3 + H_2O \rightleftharpoons H_3O^+ + HSO_3^-$	1.3×10^{-2}

into H_3O^+ and SO_4^{2-} and must be in equilibrium with these ions:

$$HSO_4^- + H_2O \rightleftharpoons H_3O^+ + SO_4^{2-}$$

$$K_{HSO_4^-} = \frac{[H_3O^+][SO_4^{2-}]}{[HSO_4^-]} = 1.26 \times 10^{-2}$$

This dissociation constant of HSO_4^- is actually the second dissociation constant of H_2SO_4, since it applies to the second step of its dissociation.

$$H_2SO_4 + H_2O \rightleftharpoons H_3O^+ + HSO_4^- \qquad K_I > 10$$

$$HSO_4^- + H_2O \rightleftharpoons H_3O^+ + SO_4^{2-} \qquad K_{II} = 1.26 \times 10^{-2}$$

K_I and K_{II} are, respectively, the first and second dissociation constants of sulfuric acid. The large value of K_I means that H_2SO_4 is essentially completely dissociated into H_3O^+ and HSO_4^-. The moderate value of K_{II} means that a modest amount of the HSO_4^- (about 10% in 0.1 M H_2SO_4) is in turn dissociated into H_3O^+ and SO_4^{2-}. In a given solution of H_2SO_4 both equilibria exist simultaneously, and both constants must be satisfied by whatever is in solution, that is, H_2SO_4, H_3O^+, HSO_4^-, and SO_4^{2-}.

For weak bases the dissociation equilibria are treated in much the same way as above. If the base can be written MOH, then its dissociation can be represented as

$$MOH \rightleftharpoons M^+ + OH^-$$

The equilibrium condition for this would be

$$\frac{[M^+][OH^-]}{[MOH]} = K_{diss}$$

However, most weak bases are more complicated than this. For example, aqueous ammonia is best described by the equilibrium

$$NH_3 + H_2O \rightleftharpoons NH_4^+ + OH^-$$

for which the equilibrium condition is

$$\frac{[NH_4^+][OH^-]}{[NH_3]} = K = 1.8 \times 10^{-5}$$

In the mass-action expression, the concentration of H_2O does not appear because it is essentially invariant. The constant K is referred to as the dissociation constant for aqueous ammonia. The value given refers to 25°C, but its change with temperature is small.

Besides acids and bases, there are in chemistry a few simple salts that are weak electrolytes; i.e., they are only slightly dissociated in solution. An example is mercuric chloride, $HgCl_2$, which dissociates

$$HgCl_2 \rightleftharpoons HgCl^+ + Cl^-$$

for which

$$\frac{[HgCl^+][Cl^-]}{[HgCl_2]} = K_{diss} = 3.3 \times 10^{-7}$$

The second dissociation is

$$HgCl^+ \rightleftharpoons Hg^{2+} + Cl^-$$

for which

$$\frac{[Hg^{2+}][Cl^-]}{[HgCl^+]} = K_{II} = 1.8 \times 10^{-7}$$

Mercuric chloride is an exception to the usual rule that salts are 100 percent dissociated in solution. However, $HgCl_2$ is not unique. For instance, cadmium sulfate, $CdSO_4$, is a weak electrolyte, with a dissociation constant of 5×10^{-3}.

As pointed out in Sec. 11.4, it makes no difference whether equilibrium is approached by starting with material on the left side of the chemical equation or material on the right side. Change occurs to form any missing material in sufficient concentration to establish equilibrium. For the case of weak electrolytes this means that the same equilibrium state is produced by having the electrolyte dissociate as is produced by having the component ions associate. Specifically, the same final solution would result whether we placed one mole of acetic acid or one mole of H_3O^+ plus one mole of $C_2H_3O_2^-$ in a liter of water. In either case, the condition for equilibrium would be the same:

$$\frac{[H_3O^+][C_2H_3O_2^-]}{[HC_2H_3O_2]} = 1.8 \times 10^{-5}$$

Since $HC_2H_3O_2$ is a weak acid, the concentration of H_3O^+ and of $C_2H_3O_2^-$ in the final solutions must be small.

When ions are mixed and association occurs, a chemical equation can be written to stress the direction of the net reaction. For example, when solutions of HCl and $NaC_2H_3O_2$ are mixed, the equation can be written

$$H_3O^+ + C_2H_3O_2^- \rightleftharpoons HC_2H_3O_2 + H_2O$$

for which

$$\frac{[HC_2H_3O_2]}{[H_3O^+][C_2H_3O_2^-]} = K_{assoc}$$

The numerical value of K_{assoc} is 5.6×10^4, which is the reciprocal of K_{diss} for acetic acid.

Association occurs whenever the constituent parts of a weak electrolyte are mixed. Thus, when solutions of NH_4Cl and $NaOH$ are mixed, NH_4^+ ions associate with OH^- ions to form NH_3 and H_2O. Likewise, when solutions of $Hg(NO_3)_2$ and $NaCl$ are mixed, Hg^{2+} ions associate with Cl^- ions to form $HgCl_2$.

12.2 Calculations using K_{diss}

The methods of equilibrium calculation described in Sec. 11.4 apply to dissociation equilibria in aqueous solution. Like any equilibrium constant, K_{diss} must be experimentally determined. Once its value is known at a given temperature, it can be used for all calculations involving that equilibrium at the given temperature.

Example 1
What is the concentration of all solute species in a solution labeled 1.00 M $HC_2H_3O_2$? What percent of the acid is dissociated?

$$HC_2H_3O_2 + H_2O \rightleftharpoons H_3O^+ + C_2H_3O_2^- \qquad K_{diss} = 1.8 \times 10^{-5}$$

Let x equal the moles per liter of $HC_2H_3O_2$ that dissociate to establish equilibrium. According to the dissociation equation, each mole of $HC_2H_3O_2$ that dissociates produces one mole of H_3O^+ and one mole of $C_2H_3O_2^-$. If x moles of $HC_2H_3O_2$ dissociate, then x moles of H_3O^+ and x moles of $C_2H_3O_2^-$ must be formed. The initial and final equilibrium concentrations are summarized as follows:

Initial concentration, mol/liter	Equilibrium concentration, mol/liter
$[HC_2H_3O_2] = 1.00$	$[HC_2H_3O_2] = 1.00 - x$
$[H_3O^+] = 0$	$[H_3O^+] = x$
$[C_2H_3O_2^-] = 0$	$[C_2H_3O_2^-] = x$

At equilibrium

$$\frac{[H_3O^+][C_2H_3O_2^-]}{[HC_2H_3O_2]} = 1.8 \times 10^{-5} = \frac{xx}{1.00 - x}$$

Solving this equation by use of the quadratic formula (Appendix 4.3) gives $x = 0.0042$. Therefore, at equilibrium (with due regard for significant figures)

$$[HC_2H_3O_2] = 1.00 - x = 1.00\ M$$

$$[H_3O^+] = x = 0.0042\ M$$

$$[C_2H_3O_2^-] = x = 0.0042\ M$$

The percent dissociation is defined as 100 times the number of moles of $HC_2H_3O_2$ dissociated divided by the number of moles of $HC_2H_3O_2$ originally available:

$$\text{Percent dissociation} = \frac{100 \times 0.0042}{1.00} = 0.42\%$$

■ ■ ■

It might be noted that much of the algebraic work involved in solving equilibrium problems can be avoided by judicious attention to chemical facts which may suggest laborsaving approximations. Thus, in Example 1, since $HC_2H_3O_2$ is a weak acid, it cannot be much dissociated. In other words, x must be small compared with 1.00 and may be neglected when added to or subtracted from 1.00. Thus, instead of solving the exact equation

$$1.8 \times 10^{-5} = \frac{xx}{1.00 - x}$$

we can solve the approximate equation

$$1.8 \times 10^{-5} \approx \frac{xx}{1.00}$$

obtained by assuming that $1.00 - x \approx 1.00$. From

$$1.8 \times 10^{-5} \approx x^2$$

we quickly get

$$x \approx \sqrt{1.8 \times 10^{-5}} = \sqrt{18 \times 10^{-6}} = 4.2 \times 10^{-3}$$

Checking the approximation and paying due attention to significant figures, we find that $1.00 - x = 1.00 - 4.2 \times 10^{-3} = 1.00$, as assumed.

Example 2

Suppose that 1.00 mol of HCl and 1.00 mol of $NaC_2H_3O_2$ are mixed in enough water to make a liter of solution. What will be the concentration of the various species in the final solution?

Since HCl and $NaC_2H_3O_2$ are strong electrolytes, they are 100% dissociated in solution. The Na^+ and Cl^- do not associate and so can be ignored. The problem is thus one of associating H_3O^+ and $C_2H_3O_2^-$ to form $HC_2H_3O_2$ in sufficient concentration to satisfy the equilibrium

$$HC_2H_3O_2 + H_2O \rightleftharpoons H_3O^+ + C_2H_3O_2^-$$

which is described by $K_{diss} = 1.8 \times 10^{-5}$. The initial and equilibrium concentrations can be summarized as follows

Initial concentration, M	*Equilibrium concentration, M*
$[H_3O^+] = 1.00$	$[H_3O^+] = 1.00 - y$
$[C_2H_3O_2^-] = 1.00$	$[C_2H_3O_2^-] = 1.00 - y$
$[HC_2H_3O_2] = 0$	$[HC_2H_3O_2] = y$

where y is the moles of H_3O^+ and of $C_2H_3O_2^-$ that associate per liter to form y moles of $HC_2H_3O_2$.

At equilibrium

$$\frac{[H_3O^+][C_2H_3O_2^-]}{[HC_2H_3O_2]} = 1.8 \times 10^{-5} = \frac{(1.00 - y)(1.00 - y)}{y}$$

This equation can be solved by applying the quadratic formula (see Appendix 4.3), giving $y = 0.996$. Since y is not small compared with 1.00, the simplifying approximation made in solving Example 1 that $1.00 - y \approx 1.00$ cannot be used.

A better way to solve this problem is to make use of the fact that the final equilibrium state does not depend on what route you take to get to it. In other words, we simply note that when 1.00 mol of H_3O^+ and 1.00 mol of $C_2H_3O_2^-$ are mixed, the resulting system is exactly the same as if H_3O^+ and $C_2H_3O_2^-$ first completely react to form 1.00 mol of $HC_2H_3O_2$, which then dissociates to establish equilibrium. If x is defined as the moles per liter that dissociate of this hypothetical 1.00 mol of $HC_2H_3O_2$, the problem becomes identical with Example 1. We can therefore write down directly the equilibrium concentrations as determined by the simple calculation on page 336:

$[H_3O^+] = 0.0042 \ M$

$[C_2H_3O_2^-] = 0.0042 \ M$

$[HC_2H_3O_2] = 1.00 \ M$

$[Na^+] = 1.00 \ M$

$[Cl^-] = 1.00 \ M$

■ ■ ■

Example 3
What are the concentrations of species and the percent dissociation in 0.10 M $HC_2H_3O_2$?

Let x = moles of $HC_2H_3O_2$ that dissociate per liter.
Then x = final concentrations of H_3O^+ and $C_2H_3O_2^-$ formed.
$0.10 - x$ = final concentration of $HC_2H_3O_2$ left undissociated.
At equilibrium

$$\frac{[H_3O^+][C_2H_3O_2^-]}{[HC_2H_3O_2]} = 1.8 \times 10^{-5} = \frac{xx}{0.10 - x}$$

Assuming that x is small compared with 0.10,

$$\frac{x^2}{0.10} \approx 1.8 \times 10^{-5}$$

$x^2 \approx 1.8 \times 10^{-6}$

$x \approx 1.3 \times 10^{-3}$

Therefore, at equilibrium

$[H_3O^+] = x = 0.0013\ M$

$[C_2H_3O_2^-] = x = 0.0013\ M$

$[HC_2H_3O_2] = 0.10 - x = 0.10\ M$

$$\text{Percent dissociation} = \frac{100 \times 0.0013}{0.10} = 1.3\%$$

■ ■ ■

Comparison of Examples 1 and 3 illustrates the general fact that when a solution of a weak electrolyte is diluted, the percent dissociation *increases* although the concentration of each species *decreases*. It may be noted that there is a tenfold dilution in bulk acid in going from $1.00\ M\ HC_2H_3O_2$ to $0.10\ M\ HC_2H_3O_2$, but the concentration of H_3O^+ does not decrease tenfold, but decreases only from 0.0042 to 0.0013 M. This, of course, is consistent with the fact that in the more dilute solution a greater percentage of the acid is dissociated.

12.3   *Dissociation of water;* pH

In the preceding section we have ignored the fact that water is a weak electrolyte and is dissociated according to the equation

$$H_2O + H_2O \rightleftharpoons H_3O^+ + OH^-$$

In pure water and in all aqueous solutions, this equilibrium exists and must satisfy the condition

$$\frac{[H_3O^+][OH^-]}{[H_2O]^2} = K$$

In all dilute solutions the concentration of H_2O can be considered constant and combined with the constant K to give K_w as follows:

$$K[H_2O]^2 = K_w = [H_3O^+][OH^-]$$

K_w is usually called the dissociation constant, or *ion product*, of water. It has the value of 1.0×10^{-14} at 25°C.

In pure water all the H_3O^+ and the OH^- must come from the dissociation of water molecules. If x moles of H_3O^+ are produced per liter, x moles of OH^- must be simultaneously produced.

$[H_3O^+][OH^-] = 1.0 \times 10^{-14}$

$xx = 1.0 \times 10^{-14}$

$x^2 = 1.0 \times 10^{-14}$

$x = 1.0 \times 10^{-7}$

Thus, in pure water the concentrations of H_3O^+ and OH^- are each $1.0 \times 10^{-7}\ M$. This very small concentration is to be compared with the water concentration of approximately 55.4 mol/liter. (A liter of water at $25°C$ weighs $997\,g$, and a mole of water weighs $18.0\,g$; therefore, 1 liter contains $997/18.0$, or 55.4, mol.) This means that on the average there is one H_3O^+ ion and one OH^- ion for every 554 million water molecules.

If an acid is added to water, the hydronium-ion concentration increases above $1.0 \times 10^{-7}\,M$. The ion product must remain equal to 1.0×10^{-14}; consequently, the hydroxide-ion concentration decreases below $1.0 \times 10^{-7}\,M$. Similarly, when a base is added to water, the concentration of OH^- increases above $1.0 \times 10^{-7}\,M$, and the concentration of H_3O^+ decreases below $1.0 \times 10^{-7}\,M$. As a convenience for working with small concentrations, the *pH scale* has been devised to express the concentration of H_3O^+. By definition,

$$pH = -\log [H_3O^+] \qquad \text{or} \qquad [H_3O^+] = 10^{-pH}$$

For example, in pure water, where the concentration of H_3O^+ is $1.0 \times 10^{-7}\,M$, the pH is 7. All neutral solutions have a pH of 7. Acid solutions have pH less than 7; basic solutions have pH greater than 7. (For a review of logarithms as applied to pH, see Appendix 4.2.)

Example 4
What is the pH *of 0.20 M* HCl?
In 0.20 M HCl, practically all the H_3O^+ comes from the 100 percent dissociation of the strong electrolyte HCl. H_2O is such a weak electrolyte in comparison that it contributes a negligible amount of H_3O^+.

$$[H_3O^+] = 0.20\,M = 2.0 \times 10^{-1}\,M$$

$$pH = -\log (2.0 \times 10^{-1}) = 1 - 0.30 = 0.70$$

■ ■ ■

Example 5
What is the pH *of 0.10 M* NaOH?
NaOH is a strong electrolyte and accounts for essentially all the OH^- in the solution.

$$[OH^-] = 0.10\,M$$

$$[H_3O^+] = \frac{K_w}{[OH^-]} = \frac{1.0 \times 10^{-14}}{0.10} = 1.0 \times 10^{-13}\,M$$

$$pH = -\log (1.0 \times 10^{-13}) = 13.00$$

■ ■ ■

In Examples 4 and 5 the contributions from H_2O dissociation to $[H_3O^+]$ in the acidic solution and to $[OH^-]$ in the basic solution are negligible. This is true because acids and bases repress the dissociation of H_2O. To illustrate, in Example 5, the added OH^- represses the dissociation of H_2O so that only 1.0×10^{-13} mol of H_3O^+ per liter is produced. This means that only 1.0×10^{-13} mol of OH^- per liter comes from the H_2O dissociation, an amount that is indeed negligible compared with the 0.10 mol that comes from 0.10 M NaOH.

12.4   Titration and indicators

So far, we have emphasized the dissociation of H_2O to give H_3O^+ and OH^- ions. However, since equilibrium may be approached from the left or the right side of an equation, the same equilibrium constant that describes the dissociation of H_2O also describes the association of H_3O^+ and OH^- to form H_2O. Such association occurs in neutralization reactions, as discussed in Sec. 9.2, and is the basis of the process of *titration*, the progressive addition of an acid to a base, or vice versa. At each step in the titration the expression $[H_3O^+][OH^-] = 1.0 \times 10^{-14}$ must be satisfied in the solution. Figure 12.2 represents what happens to the concentration of H_3O^+ and OH^- as solid NaOH is added stepwise to 0.010 mol of HCl in a liter of H_2O. (We add solid NaOH instead of a solution of NaOH, which would be the more usual practice, in order to avoid the complication of dilution due to mixing two solutions.) As NaOH is progressively added, the original solution changes from acidic (pH less than 7) to basic (pH greater than 7). The titration can be represented graphically by plotting the concentration of H_3O^+ against the moles of added NaOH. However, since the H_3O^+ concentration changes by a factor of 10 billion during

Moles of NaOH added	$[H_3O^+]$	$[OH^-]$	pH
0.000	0.010	1.0×10^{-12}	2.00
0.001	0.009	1.1×10^{-12}	2.04
0.002	0.008	1.3×10^{-12}	2.10
0.003	0.007	1.4×10^{-12}	2.15
0.004	0.006	1.7×10^{-12}	2.23
0.005	0.005	2.0×10^{-12}	2.30
0.006	0.004	2.5×10^{-12}	2.40
0.007	0.003	3.3×10^{-12}	2.52
0.008	0.002	5.0×10^{-12}	2.70
0.009	0.001	1.0×10^{-11}	3.00
0.010	1.0×10^{-7}	1.0×10^{-7}	7.00
0.011	1.0×10^{-11}	0.001	11.00
0.012	5.0×10^{-12}	0.002	11.30
0.013	3.3×10^{-12}	0.003	11.48
0.014	2.5×10^{-12}	0.004	11.60
0.015	2.0×10^{-12}	0.005	11.70
0.016	1.7×10^{-12}	0.006	11.77
0.017	1.4×10^{-12}	0.007	11.85
0.018	1.3×10^{-12}	0.008	11.90
0.019	1.1×10^{-12}	0.009	11.96
0.020	1.0×10^{-12}	0.010	12.00

Fig. 12.2

Progressive addition of solid NaOH to one liter of 0.010 M HCl.

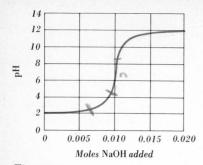

Fig. 12.3
*Titration curve for addition
of solid* NaOH *to one liter
of 0.010 M* HCl.

the experiment, it is hard to get all the values on the same scale. Not so with the pH. It changes only by a factor of 6 and is a convenient representation of what happens to the solution during the titration.

Figure 12.3 represents the change of pH as solid NaOH is added to a liter of 0.010 M HCl. The pH first rises very slowly, then rapidly through the neutral point, and finally very slowly as the solution gets more basic. Such a pH curve is typical of the titration of any strong acid with any strong base. The important thing to note is that as the neutral point is approached, there is a sharp rise in pH. At this point even a trace of NaOH adds enough moles of base to increase the pH greatly. Thus, any method which locates the point at which the pH changes rapidly can be used to detect the *equivalence point* of a titration, i.e., the point at which equivalent amounts of base and acid have been mixed.

One method for determining the equivalence point makes use of the fact that many dyes have colors that are sensitive to hydronium-ion concentration. Such dyes can be used as *indicators* to give information about the pH of a solution. Indicators can be considered to be weak acids, HIn, which dissociate to give H_3O^+ and In^-, where In^- is some complicated grouping of carbon and other atoms. As weak acids, HIn must satisfy the condition

$$\frac{[H_3O^+][In^-]}{[HIn]} = K \qquad \text{or} \qquad \frac{[In^-]}{[HIn]} = \frac{K}{[H_3O^+]}$$

from which it is evident that the ratio $[In^-]/[HIn]$ is inversely proportional to the hydronium-ion concentration of the solution. If the species In^- and HIn have different colors, the color of the solution depends on which species is predominant. For phenolphthalein, HIn is colorless, but In^- is red. In solutions of high hydronium-ion concentration, the ratio $[In^-]/[HIn]$ is small, and the colorless species HIn is dominant. Conversely, when $[H_3O^+]$ is small, the red species In^- is dominant. Figure 12.4 lists the characteristic colors of some common indicators.

In Fig. 12.3 the pH rises so sharply at the equivalence point that any one of the indicators of Fig. 12.4 except methyl violet and possibly

Fig. 12.4
*Typical indicators and their
color ranges.*

Indicator	pH at which color changes	Color at lower pH	Color at higher pH
Methyl violet	1	Yellow	Blue
Erythrosin	3	Orange	Red
Bromophenol blue	4	Yellow	Blue
Methyl orange	4	Red	Yellow
Methyl red	5	Red	Yellow
p-Nitrophenol	6	Colorless	Yellow
Bromothymol blue	7	Yellow	Blue
Phenolphthalein	9	Colorless	Red
Thymolphthalein	10	Colorless	Blue
Alizarin yellow	11	Yellow	Red

alizarin yellow could be used to tell when enough NaOH had been added to neutralize 1 liter of 0.010 M HCl.

The titration curve of Fig. 12.3 is general for strong acids and strong bases but does not apply when a strong acid is titrated with a weak base, when a strong base is titrated with a weak acid, or when a weak acid is titrated with a weak base. In the latter cases the shapes of the titration curves are quite different and require individual consideration before an indicator is chosen. It turns out, for example, that in titrating acetic acid with sodium hydroxide, phenolphthalein is satisfactory but methyl orange is not.

12.5 *Buffer solutions*

In practically all biologic processes, as well as in many other chemical processes, it is important that the pH not deviate very much from a fixed value. For example, the proper functioning of human blood in carrying oxygen to the cells from the lungs is dependent on maintaining a pH very near to 7.4. In fact, for a particular individual, there is a difference of but 0.02 pH unit between venous and arterial blood in spite of numerous acid- and base-producing reactions in the cells.

The near constancy of pH in a system to which acid or base is added is due to what is called *buffering* action of an acid-base equilibrium. The buffer contains both acid and base and can respond to addition of either one or the other. Let us consider, for example, a solution that contains acetic acid molecules and acetate ions (plus other ions, of course). The principal equilibrium in this solution is written

$$HC_2H_3O_2 + H_2O \rightleftharpoons H_3O^+ + C_2H_3O_2^-$$

for which

$$\frac{[H_3O^+][C_2H_3O_2^-]}{[HC_2H_3O_2]} = K$$

Solving this expression for $[H_3O^+]$, we get

$$[H_3O^+] = K\frac{[HC_2H_3O_2]}{[C_2H_3O_2^-]}$$

which indicates that the hydronium-ion concentration depends on K and on the ratio of the concentrations of undissociated acetic acid to acetate ion. Taking the negative logarithm of both sides, we get

$$pH = -\log K - \log\frac{[HC_2H_3O_2]}{[C_2H_3O_2^-]}$$

Sometimes the symbol pK is introduced to represent $-\log K$ of the acid. Hence, we can write

$$pH = pK - \log\frac{[HC_2H_3O_2]}{[C_2H_3O_2^-]}$$

In a particular solution made by dissolving equal numbers of moles of $HC_2H_3O_2$ and $NaC_2H_3O_2$ the ratio $[HC_2H_3O_2]/[C_2H_3O_2^-]$ is equal to unity; hence pH is just equal to pK, since log 1 equals zero. In any solution where the ratio $[HC_2H_3O_2]/[C_2H_3O_2^-]$ is not far from unity, the pH will not differ much from pK. Thus, a mixture of acetic acid and acetate ion is said to be a *buffer* for pH of $-\log(1.8 \times 10^{-5})$, or 4.74. If a small amount of strong acid is added to such a solution, some of the acetate ion is converted to acetic acid; if base is added, some of the acetic acid is converted to acetate ion. In either case the ratio $[HC_2H_3O_2]/[C_2H_3O_2^-]$ changes slightly from unity, and the pH changes even less—not nearly so much as in the absence of the buffer.

Example 6
Calculate the pH of a solution made by adding 0.0010 mol of NaOH *to 100 ml of a solution that is 0.50 M* $HC_2H_3O_2$ *and 0.50 M* $NaC_2H_3O_2$.
In the 100 ml there are originally 0.050 mol of $HC_2H_3O_2$ and 0.050 mol of $C_2H_3O_2^-$. We can assume that all the added 0.0010 mol of OH^- reacts to convert an equivalent amount of $HC_2H_3O_2$ into $C_2H_3O_2^-$. This would give 0.049 mol $HC_2H_3O_2$ and 0.051 mol $C_2H_3O_2^-$ in the final solution. Since the volume of the solution stays at 0.100 liter, the respective concentrations would be 0.49 M and 0.51 M. Hence the pH is

$$\text{pH} = -\log(1.8 \times 10^{-5}) - \log \frac{0.49}{0.51} = +4.74 + 0.017 = +4.76$$

In contrast, when 0.0010 mol of NaOH is added to 100 ml of *water*, the pH goes from 7.0 to 12.0.

■　■　■

In general, any solution of a weak acid which also contains a salt of that acid can function as a buffer. The buffer region—i.e., the region in which the pH changes most slowly—is located around the pK of the acid. In a similar way, a solution of a weak base plus a salt of that base can function as a buffer to keep the OH^- concentration equal to $K[MOH]/[M^+]$:

$$MOH \rightleftharpoons M^+ + OH^-$$

$$\frac{[M^+][OH^-]}{[MOH]} = K$$

$$[OH^-] = \frac{K[MOH]}{[M^+]}$$

$$[H_3O^+] = \frac{K_w}{[OH^-]} = \frac{K_w}{K}\frac{[M^+]}{[MOH]}$$

$$\text{pH} = -\log\frac{K_w}{K} - \log\frac{[M^+]}{[MOH]}$$

As is evident, there are as many possible buffers as there are weak acids and weak bases. In human blood there are a number of buffers acting simultaneously. These include (1) dissolved CO_2 and HCO_3^-, (2) $H_2PO_4^-$ and HPO_4^{2-}, and (3) the various proteins which can accept hydrogen ions.

12.6   Complex ions

The term *complex ion* refers to a charged particle which contains more than one atom. Certain complex ions, e.g., sulfate, SO_4^{2-}, are little different from simple ions in that for all practical purposes they do not dissociate into smaller fragments. Others, however, may dissociate to establish an equilibrium between the complex ion and its component pieces. Thus, for example, in a solution containing the silver-ammonia complex ion, $Ag(NH_3)_2^+$, there is an equilibrium between the complex ion, silver ion, and ammonia molecules. Although the dissociation probably occurs in steps, we can write the overall equilibrium as

$$Ag(NH_3)_2^+ \rightleftharpoons Ag^+ + 2NH_3$$

for which the equilibrium condition is

$$\frac{[Ag^+][NH_3]^2}{[Ag(NH_3)_2^+]} = 6 \times 10^{-8}$$

When silver nitrate, $AgNO_3$, and aqueous ammonia are mixed, enough silver-ammonia complex ion is formed to satisfy the above equilibrium condition. Furthermore, if the concentration of ammonia in the solution were increased, as by addition of more ammonia, the concentration of silver ion would have to decrease, as required by constancy of the mass-action expression.

Figure 12.5 lists some common complex ions and their overall equi-

Complex ion	Reaction	K
Copper-ammonia	$Cu(NH_3)_4^{2+} \rightleftharpoons Cu^{2+} + 4NH_3$	1.0×10^{-12}
Cobaltous-ammonia	$Co(NH_3)_6^{2+} \rightleftharpoons Co^{2+} + 6NH_3$	4.0×10^{-5}
Cobaltic-ammonia	$Co(NH_3)_6^{3+} \rightleftharpoons Co^{3+} + 6NH_3$	6.3×10^{-36}
Silver-ammonia	$Ag(NH_3)_2^+ \rightleftharpoons Ag^+ + 2NH_3$	6×10^{-8}
Silver-thiosulfate	$Ag(S_2O_3)_2^{3-} \rightleftharpoons Ag^+ + 2S_2O_3^{2-}$	6×10^{-14}
Silver-cyanide	$Ag(CN)_2^- \rightleftharpoons Ag^+ + 2CN^-$	1.8×10^{-19}
Ferric-thiocyanate	$FeNSC^{2+} \rightleftharpoons Fe^{3+} + NSC^-$	1×10^{-3}
Mercuric-cyanide	$Hg(CN)_4^{2-} \rightleftharpoons Hg^{2+} + 4CN^-$	4×10^{-42}

Fig. 12.5
Overall dissociation constants for some complex ions.

librium constants. These numbers give a measure of the stability of the respective complex ions with respect to dissociation. Of the three complex ions shown for silver, the silver-cyanide complex is least dissociated and is said to be the most stable. For example, in a solution containing silver ion, cyanide ion, thiosulfate ion, and ammonia, the silver-cyanide complex is preferentially formed.

12.7 Solubility of ionic solids

When an ionic solid is placed in water, an equilibrium is established between the ions in the saturated solution and the excess solid phase. For example, with excess solid silver chloride in contact with a saturated solution of silver chloride, the equilibrium is

$$AgCl(s) \rightleftharpoons Ag^+ + Cl^-$$

for which

$$\frac{[Ag^+][Cl^-]}{[AgCl(s)]} = K$$

The concentration of a pure solid is a constant number. Thus, the concentration of silver chloride *in the solid phase* is fixed and cannot change, no matter how much solid there is in contact with the solution. It follows that

$$[Ag^+][Cl^-] = K[AgCl(s)] = K_{sp}$$

The constant K_{sp} is called the *solubility product,* and the expression $[Ag^+][Cl^-]$ the *ion product*. The equation states that the ion product must equal K_{sp} when the saturated solution is in equilibrium with excess solid. It should be noted that there is no separate restriction on what the concentrations of Ag^+ and Cl^- must be. The concentration of Ag^+ can have any value as long as the concentration of Cl^- is such that the product of Ag^+ concentration and Cl^- concentration is equal to K_{sp}.

The numerical value of K_{sp}, as of any equilibrium constant, must be determined by experiment. Once determined, it can be tabulated for future use. (Appendix 6 contains some typical values.) The kind of experiment that could be done, although it would be impractical except perhaps with radioactive tracers, is illustrated as follows for the case of barium sulfate: Solid $BaSO_4$ is ground up and thoroughly agitated with a liter of H_2O at $25°C$ until the saturated solution is formed. The saturated solution is then filtered and evaporated to get rid of the solvent, and residual solid $BaSO_4$ is dried and analyzed. The solubility of $BaSO_4$ thus determined is 3.9×10^{-5} mol per liter of H_2O at $25°C$.

Like practically all salts, $BaSO_4$ is a strong electrolyte and so is 100% dissociated into ions. Therefore, when 3.9×10^{-5} mol of $BaSO_4$ dissolves, it forms 3.9×10^{-5} mol of Ba^{2+} and 3.9×10^{-5} mol of SO_4^{2-}. In the saturated solution the concentration of Ba^{2+} is $3.9 \times 10^{-5} M$, and the concentration of SO_4^{2-} is $3.9 \times 10^{-5} M$. Therefore, for the equilibrium

$$BaSO_4(s) \rightleftharpoons Ba^{2+} + SO_4^{2-}$$

we have the condition

$$K_{sp} = [Ba^{2+}][SO_4^{2-}] = (3.9 \times 10^{-5})(3.9 \times 10^{-5}) = 1.5 \times 10^{-9}$$

This means that in any solution containing Ba^{2+} and SO_4^{2-} in equilibrium with solid $BaSO_4$, the product of the concentrations of Ba^{2+} and

SO_4^{2-} should be equal to 1.5×10^{-9}. Since K_{sp} is a very small number, $BaSO_4$ may be called an insoluble salt. If $[Ba^{2+}]$ multiplied by $[SO_4^{2-}]$ is less than 1.5×10^{-9}, the solution is unsaturated, and $BaSO_4$ should dissolve to increase the concentrations of Ba^{2+} and SO_4^{2-}. If the product of $[Ba^{2+}]$ and $[SO_4^{2-}]$ is greater than 1.5×10^{-9}, the system is not at equilibrium. $BaSO_4$ should precipitate in order to decrease the concentrations of Ba^{2+} and SO_4^{2-}.

When $BaSO_4$ is placed in pure water, the concentrations of Ba^{2+} and SO_4^{2-} must be equal. On the other hand, it is possible to prepare a solution in which unequal concentrations of Ba^{2+} and SO_4^{2-} are in equilibrium with solid $BaSO_4$. As an illustration, suppose that unequal amounts of barium chloride and sodium sulfate are added to water. A precipitate of $BaSO_4$ forms if K_{sp} of $BaSO_4$ is exceeded. However, there is no requirement that $[Ba^{2+}] = [SO_4^{2-}]$, since the two ions come from different sources. Alternatively, barium sulfate solid might be shaken up with an Na_2SO_4 solution. Some barium sulfate would dissolve, but in the final solution the concentration of SO_4^{2-} would be considerably greater than the concentration of Ba^{2+}.

Example 7

Given that K_{sp} of radium sulfate, $RaSO_4$, is 4×10^{-11}, calculate its solubility in (a) pure water and (b) 0.10 M Na_2SO_4.

(a)   Let $x =$ moles of $RaSO_4$ that dissolve per liter of water. Then, in the saturated solution

$[Ra^{2+}] = x$ mol/liter

$[SO_4^{2-}] = x$ mol/liter

$RaSO_4(s) \rightleftharpoons Ra^{2+} + SO_4^{2-}$

$[Ra^{2+}][SO_4^{2-}] = K_{sp} = 4 \times 10^{-11}$

$xx = 4 \times 10^{-11}$

$x = \sqrt{40 \times 10^{-12}} = 6 \times 10^{-6}$ mol/liter

Thus, the solubility of $RaSO_4$ is 6×10^{-6} mol per liter of water, giving a solution containing 6×10^{-6} M Ra^{2+} and 6×10^{-6} M SO_4^{2-}.

(b)   Let $y =$ moles of $RaSO_4$ that dissolve per liter of 0.10 M Na_2SO_4. This dissolving produces y moles of Ra^{2+} and y moles of SO_4^{2-}. The solution already contains 0.10 M SO_4^{2-}. Thus, in the final saturated solution

$[Ra^{2+}] = y$ mol/liter

$[SO_4^{2-}] = y + 0.10$ mol/liter

where

$[Ra^{2+}][SO_4^{2-}] = y(y + 0.10) = K_{sp} = 4 \times 10^{-11}$

Since K_{sp} is very small, not much $RaSO_4$ dissolves, and y is so small that it is negligible compared with 0.10.

$$y + 0.10 \approx 0.10$$

$$[Ra^{2+}][SO_4^{2-}] \approx (y)(0.10) \approx 4 \times 10^{-11}$$

$$y \approx \frac{4 \times 10^{-11}}{0.10} = 4 \times 10^{-10} \text{ mol/liter}$$

Thus, the solubility of $RaSO_4$ in 0.10 M Na_2SO_4 is 4×10^{-10} mol/liter, giving a solution in which the concentration of Ra^{2+} is 4×10^{-10} M and that of SO_4^{2-} is 0.10 M.

■　■　■

It is interesting to note that $RaSO_4$ is less soluble in an Na_2SO_4 solution than it is in pure water. This is an example of the *common-ion effect*, whereby the solubility of an ionic salt is decreased by the presence of another solute that furnishes an ion in common. Thus, radium sulfate is less soluble than it is in water in any solution containing either radium ion or sulfate ion. The greater the concentration of the common ion, the less radium sulfate can dissolve. Of course, if the common ion is present in negligible concentration, it has no appreciable effect on the solubility. This is illustrated in the following example.

Example 8

Given that magnesium hydroxide, $Mg(OH)_2$, is a strong electrolyte and has a solubility product of 8.9×10^{-12}, calculate the solubility of $Mg(OH)_2$ in water.

Let x = moles of $Mg(OH)_2$ that dissolve per liter. According to the equation

$$Mg(OH)_2(s) \rightleftharpoons Mg^{2+} + 2OH^-$$

x moles of $Mg(OH)_2$ dissolve to give x moles of Mg^{2+} and $2x$ moles of OH^-. Some hydroxide ion is also furnished by the dissociation of H_2O. Since H_2O is a very weak electrolyte, we can assume, as we did in Example 5, that H_2O dissociation contributes only a negligible amount of OH^- compared with that furnished by the dissolving of $Mg(OH)_2$. Thus at equilibrium

$$[Mg^{2+}] = x \text{ mol/liter}$$

$$[OH^-] \approx 2x \text{ mol/liter}$$

For the saturated solution the equilibrium is

$$Mg(OH)_2(s) \rightleftharpoons Mg^{2+} + 2OH^-$$

and $K_{sp} = 8.9 \times 10^{-12} = [Mg^{2+}][OH^-]^2$.

Substituting, we get

$$(x)(2x)^2 = 8.9 \times 10^{-12}$$

$$4x^3 = 8.9 \times 10^{-12}$$

$$x = \sqrt[3]{2.2 \times 10^{-12}} = 1.3 \times 10^{-4} \text{ mol/liter}$$

Thus, 1.3×10^{-4} mol of $Mg(OH)_2$ dissolves per liter of H_2O. The saturated solution contains $1.3 \times 10^{-4}\ M$ Mg^{2+} and $2.6 \times 10^{-4}\ M$ OH^-.

■ ■ ■

As noted in Sec. 11.2, the mass-action expression for a given reaction contains the concentrations raised to powers that correspond to coefficients in the chemical equation. Since the ion product is a mass-action expression, it must be formed by raising the concentrations to powers that correspond to coefficients in the equation for the dissolving process. Thus

$$K_{sp} = [A^{2+}][B^{2-}] \qquad \text{for} \qquad AB(s) \rightleftharpoons A^{2+} + B^{2-}$$

but

$$K_{sp} = [A^{2+}][B^-]^2 \qquad \text{for} \qquad AB_2(s) \rightleftharpoons A^{2+} + 2B^-$$

An exponent applies to the concentration of the ion specified inside the brackets, no matter where that ion comes from. For example, in the following example practically all the OH^- comes from NaOH, but its concentration still must be squared.

Example 9
Calculate the solubility of $Mg(OH)_2$ *in 0.050 M NaOH.*
   Let $x =$ moles of $Mg(OH)_2$ that dissolve per liter. This forms x moles of Mg^{2+} and $2x$ moles of OH^-. Since the solution already contains 0.050 mol of OH^-, equilibrium concentrations would be as follows:

$[Mg^{2+}] = x$ mol/liter

$[OH^-] = 2x + 0.050$ mol/liter

$[Mg^{2+}][OH^-]^2 = (x)(2x + 0.050)^2 = K_{sp}$

$(x)(2x + 0.050)^2 = 8.9 \times 10^{-12}$

On the basis of the small value of K_{sp} we can guess that x is probably a very small number; so $2x$ can be neglected when added to 0.050. We would have then the approximate relation

$(x)(0.050)^2 \approx 8.9 \times 10^{-12}$

$x = 3.6 \times 10^{-9}$ mol/liter

Since x turns out to be small compared with 0.050, the assumption made above was indeed valid. The calculation indicates that 3.6×10^{-9} mol of $Mg(OH)_2$ can dissolve in 1 liter of 0.050 M NaOH to give a saturated solution containing $3.6 \times 10^{-9}\ M$ Mg^{2+} and 0.050 M OH^-.

■ ■ ■

12.8   Precipitation

One of the most useful applications of the solubility product is to predict whether or not precipitation should occur when two solutions are mixed.

In a saturated solution of a salt, the ion product should be equal to K_{sp}. If two solutions containing the ions of a salt are mixed and if the ion product then exceeds K_{sp}, then precipitation should occur.

Example 10
Should precipitation occur when 50 ml of 5.0×10^{-4} M $Ca(NO_3)_2$ is mixed with 50 ml of 2.0×10^{-4} M NaF to give 100 ml of solution? The K_{sp} of CaF_2 is 1.7×10^{-10}.

In order to solve such a problem, we first calculate the concentration of the ions in the mixture, assuming that no precipitation occurs. Thus, the Ca^{2+} from the 5.0×10^{-4} M $Ca(NO_3)_2$ solution is 2.5×10^{-4} M in the final mixture because of twofold dilution when the solutions are mixed. Likewise F^- is diluted to 1.0×10^{-4} M. Therefore, if no precipitation occurred, the final solution would have

$$[Ca^{2+}] = 2.5 \times 10^{-4} M \qquad \text{and} \qquad [F^-] = 1.0 \times 10^{-4} M$$

To determine whether precipitation should occur, it is necessary to see whether the appropriate ion product would exceed the solubility product. For a saturated solution of CaF_2 the equilibrium would be

$$CaF_2(s) \rightleftharpoons Ca^{2+} + 2F^-$$

for which the ion product is $[Ca^{2+}][F^-]^2$. In the present mixture the ion product has the numerical value

$$[Ca^{2+}][F^-]^2 = (2.5 \times 10^{-4})(1.0 \times 10^{-4})^2 = 2.5 \times 10^{-12}$$

Since this number does not exceed 1.7×10^{-10}, which is the K_{sp} of CaF_2, precipitation should not occur. The solution obtained as the final mixture is unsaturated with respect to precipitation of CaF_2.

■ ■ ■

In order to precipitate a salt, the ion product must be made to exceed the K_{sp} of that salt. This gives a method for driving ions out of solution. For example, if we are given a solution of $RaCl_2$, the Ra^{2+} can be made to precipitate as $RaSO_4$ by addition of Na_2SO_4. The more the concentration of SO_4^{2-} is increased in the solution, the lower the concentration of Ra^{2+} becomes. Practically all the valuable Ra^{2+} can be recovered from solution this way by adding a large excess of SO_4^{2-} ions.

12.9 Simultaneous equilibria

In the preceding discussions only one equilibrium was considered at a time. This is an idealized situation since usually aqueous solutions have two or more equilibria which must be satisfied simultaneously. For example, in a solution containing the weak acid $HC_2H_3O_2$ there are two dissociation equilibria:

$$HC_2H_3O_2 + H_2O \rightleftharpoons H_3O^+ + C_2H_3O_2^-$$

$$\frac{[H_3O^+][C_2H_3O_2^-]}{[HC_2H_3O_2]} = K_{diss}$$

and

$$H_2O + H_2O \rightleftharpoons H_3O^+ + OH^- \qquad [H_3O^+][OH^-] = K_w$$

The solution of acetic acid has a characteristic concentration of H_3O^+ which simultaneously satisfies K_{diss} and K_w. Strictly speaking, this H_3O^+ comes partly from the dissociation of $HC_2H_3O_2$ and partly from the dissociation of H_2O. However, H_2O is so slightly dissociated compared with $HC_2H_3O_2$ that we can consider the H_3O^+ as coming entirely from the $HC_2H_3O_2$. This assumption was implicitly made in the calculations of Sec. 12.2. The H_3O^+ concentration of $1\,M\ HC_2H_3O_2$ was calculated by assuming that negligible H_3O^+ is contributed by dissociation of H_2O. To see how this comes about quantitatively, we can do the following calculation for a related problem.

Example 11
Given that $K_{diss} = 6.7 \times 10^{-4}$, *calculate the* H_3O^+ *and* OH^- *concentration in* 0.10 M HF *solution.*
First we set up the two equilibria that are involved:

$$HF + H_2O \rightleftharpoons H_3O^+ + F^-$$

$$H_2O + H_2O \rightleftharpoons H_3O^+ + OH^-$$

Then we define two unknowns x and y for the concentration of H_3O^+ that is contributed by each of these equilibria. Of course, there is no way of distinguishing whether the H_3O^+ comes from the HF or from the H_2O; there is only one H_3O^+ concentration in the solution. Still it is instructive to see the relative contribution by each of these processes. If x is the moles per liter of H_3O^+ produced by the first equation, then x is also the moles per liter of F^- produced. Similarly, if y is the concentration of H_3O^+ produced by the second equation, then y must also be equal to the concentration of OH^- produced. For the equilibrium state, therefore, we can write

$[H_3O^+] = x + y$ mol/liter

$[F^-] = x$ mol/liter

$[OH^-] = y$ mol/liter

$[HF] = 0.10 - x$ mol/liter

These concentrations if they are equilibrium values must satisfy the two equilibrium conditions

$$K_{diss} = \frac{[H_3O^+][F^-]}{[HF]} = 6.7 \times 10^{-4} = \frac{(x + y)(x)}{0.10 - x}$$

$$K_w = [H_3O^+][OH^-] = 1.0 \times 10^{-14} = (x + y)(y)$$

The problem is to solve these two simultaneous equations for the two unknowns x and y. The exact solution is not easy, and the problem is especially difficult when x and y are about equal in value. Fortunately, one of the contributions usually is the dominant one; so the other can be neglected to give a quite good approximate solution. What creates trouble for the neophyte is the recognition of which contribution is dominant.

In this particular case we can find out whether x or y is dominant by noting what each equilibrium would produce if not affected by the presence of the other one. Specifically, $0.10\,M$ HF would lead to $[H_3O^+] = 7.7 \times 10^{-3}\,M$; pure water would lead to $[H_3O^+] = 1.0 \times 10^{-7}\,M$. Evidently, the H_3O^+ from the HF will be dominant. A fairly good general rule is that the larger the dissociation constant, the more it dominates the final equilibrium state. (However, one needs to be cautious if exponents of the mass-action expression are not all unity, in which case a smaller K may be the more important one.)

Once it has been decided that the H_3O^+ from the HF is more important than the H_3O^+ from H_2O, the problem falls apart. We need only calculate the HF dissociation problem, ignoring the H_2O dissociation, and then go back to calculate the other equilibrium. Specifically, for $0.10\,M$ HF we have

$$HF + H_2O \rightleftharpoons H_3O^+ + F^-$$

$0.10-x$ x x

$$K_{diss} = \frac{[H_3O^+][F^-]}{[HF]} = 6.7 \times 10^{-4} = \frac{(x + y)(x)}{0.10 - x}$$

Neglecting y with respect to x and solving for x gives us

$$x = 7.7 \times 10^{-3}\,M = [H_3O^+]$$

Then we consider the water equilibrium, which is a second-order effect:

$$H_2O + H_2O \rightleftharpoons H_3O^+ + OH^-$$

$$K_w = [H_3O^+][OH^-] = 1.0 \times 10^{-14} = (x + y)(y)$$

As we have seen, y is negligible compared with x; so we can write approximately

$$xy \approx 1.0 \times 10^{-14}$$

Since we have already found $x = 7.7 \times 10^{-3}\,M$, we can substitute for x and get

$$(7.7 \times 10^{-3})(y) = 1.0 \times 10^{-14}$$

from which it follows that

$$y = \frac{1.0 \times 10^{-14}}{7.7 \times 10^{-3}} = 1.3 \times 10^{-12}\, M = [OH^-]$$

As can be seen, the value for y is truly negligible compared with the value for x; so we were quite justified in approximating $x + y \approx x$.

■ ■ ■

Example 12
Calculate the concentrations of H_3O^+ and OH^- in a solution made by mixing 0.50 mol of $HC_2H_3O_2$ and 0.50 mol of HCN with enough water to make a liter of solution.
There are three simultaneous equilibria in the final solution:

$$HC_2H_3O_2 + H_2O \rightleftharpoons H_3O^+ + C_2H_3O_2^-$$
$$K_{HC_2H_3O_2} = 1.8 \times 10^{-5} \quad (1)$$

$$HCN + H_2O \rightleftharpoons H_3O^+ + CN^- \quad K_{HCN} = 4 \times 10^{-10} \quad (2)$$

$$H_2O + H_2O \rightleftharpoons H_3O^+ + OH^- \quad K_w = 1.0 \times 10^{-14} \quad (3)$$

Only acetic acid contributes an appreciable concentration of H_3O^+, because it has much the largest dissociation constant. Ignoring the other dissociations, let $x =$ moles of $HC_2H_3O_2$ that dissociate per liter. Then, at equilibrium

$[HC_2H_3O_2] = 0.50 - x$ mol/liter

$[H_3O^+] = x$ mol/liter

$[C_2H_3O_2^-] = x$ mol/liter

$$\frac{[H_3O^+][C_2H_3O_2^-]}{[HC_2H_3O_2]} = \frac{xx}{0.50 - x} = 1.8 \times 10^{-5}$$

$x = 3.0 \times 10^{-3}\, M$

Thus, the final solution has a hydronium-ion concentration of $3.0 \times 10^{-3}\, M$. Substituting this value in the equilibrium condition for Eq. (3) gives

$K_w = [H_3O^+][OH^-] = 1.0 \times 10^{-14}$

$(3.0 \times 10^{-3})[OH^-] = 1.0 \times 10^{-14}$

$[OH^-] = 3.3 \times 10^{-12}\, M$

■ ■ ■

Another common example of simultaneous equilibria occurs in solutions of polyprotic acids (Sec. 9.3). For example, suppose we are asked to calculate the pH of 0.10 M H_2SO_4. The temptation is very great to say that because H_2SO_4 is a strong acid, it is completely dissociated into 0.20 M H_3O^+, there being two protons per molecule of H_2SO_4. Consequently, the pH should be $-\log [H_3O^+] = -\log 0.20 = 0.70$.

However, this is wrong! The reaction is not $H_2SO_4 + 2H_2O \rightleftharpoons 2H_3O^+ + SO_4^{2-}$ but two successive steps

$$H_2SO_4 + H_2O \rightleftharpoons H_3O^+ + HSO_4^-$$

$$HSO_4^- + H_2O \rightleftharpoons H_3O^+ + SO_4^{2-}$$

whereby the H_3O^+ produced in the first step impedes the formation of H_3O^+ by the second step.

Quantitatively, the calculation goes like this: In 0.10 M H_2SO_4, the first reaction

$$H_2SO_4 + H_2O \longrightarrow H_3O^+ + HSO_4^-$$

goes 100 percent to the right with complete conversion of H_2SO_4 to HSO_4^-. This is not precisely correct because $K_{diss} \neq \infty$, but, even so, when $K \gg 1$, a reaction is essentially complete as written. Next we consider the second step of the dissociation

$$HSO_4^- + H_2O \rightleftharpoons H_3O^+ + SO_4^{2-}$$

for which

$$K_{II} = \frac{[H_3O^+][SO_4^{2-}]}{[HSO_4^-]} = 1.26 \times 10^{-2}$$

What are the equilibrium concentrations? The first step converts 0.10 M H_2SO_4 into 0.10 M H_3O^+ plus 0.10 M HSO_4^-. Let x = moles per liter of H_3O^+ generated by the second step. Then at equilibrium, we will have

$$[H_3O^+] = 0.10 + x \text{ mol/liter}$$

$$[HSO_4^-] = 0.10 - x \text{ mol/liter}$$

$$[SO_4^{2-}] = x \text{ mol/liter}$$

Substituting these concentrations into the equilibrium condition for K_{II} gives

$$K_{II} = \frac{[H_3O^+][SO_4^{2-}]}{[HSO_4^-]} = \frac{(0.10 + x)(x)}{0.10 - x} = 1.26 \times 10^{-2}$$

This is a relatively simple equation to solve even though x is not negligible compared with 0.10. Either the quadratic formula or the method of successive approximations [where x is first assumed negligible with respect to 0.10 to give an approximate value of x, which is fed back successively to get better and better guesses for $(0.10 + x)/(0.10 - x)$; see Appendix 4.4] leads to $x = 0.011$. Consequently, $[H_3O^+] = 0.10 + x = 0.11$ M, from which pH $= -\log 0.11 = 0.96$.

One of the most useful applications of simultaneous equilibria occurs in solutions of hydrogen sulfide, H_2S, where, besides the H_2O equilibrium, there are two equilibria that correspond to stepwise dissociation of H_2S:

$$H_2S + H_2O \rightleftharpoons H_3O^+ + HS^- \qquad K_I = 1.1 \times 10^{-7}$$

$$HS^- + H_2O \rightleftharpoons H_3O^+ + S^{2-} \qquad K_{II} = 1 \times 10^{-14}$$

Since H_2S is a weak acid, a solution of H_2S is slightly acidic. In order to calculate the acidity of the solution, is it necessary to consider both steps of the dissociation? The case is quite analogous to that of acetic acid or hydrofluoric acid in water, where the H_2O dissociation contributes a negligible concentration of H_3O^+. In H_2S the dissociation of HS^- contributes a negligible concentration of H_3O^+. Using only K_I, we can calculate that the concentration of H_3O^+ in $0.10\,M$ H_2S is approximately $1 \times 10^{-4}\,M$ and the HS^- concentration is $1 \times 10^{-4}\,M$.

Because of the second step of the dissociation of H_2S, there is a small trace of sulfide ion, S^{2-}, in the solution. Its numerical magnitude can be calculated by using K_{II}:

$$K_{II} = \frac{[H_3O^+][S^{2-}]}{[HS^-]} = 1 \times 10^{-14}$$

If, as is the case in $0.10\,M$ H_2S, the concentrations of H_3O^+ and HS^- are $1 \times 10^{-4}\,M$, they cancel each other out of the expression, and $[S^{2-}] = 1 \times 10^{-14}\,M$.

In any solution of H_2S, both K_I and K_{II} must be simultaneously satisfied, and this gives rise to the two simultaneous equations

$$\frac{[H_3O^+][HS^-]}{[H_2S]} = 1.1 \times 10^{-7} \qquad (4)$$

$$\frac{[H_3O^+][S^{2-}]}{[HS^-]} = 1 \times 10^{-14} \qquad (5)$$

Solving Eqs. (4) and (5) for $[HS^-]$, we get

$$[HS^-] = 1.1 \times 10^{-7}\, \frac{[H_2S]}{[H_3O^+]} \qquad (6)$$

$$[HS^-] = \frac{[H_3O^+][S^{2-}]}{1 \times 10^{-14}} \qquad (7)$$

and equating Eqs. (6) and (7) gives

$$\frac{1.1 \times 10^{-7}[H_2S]}{[H_3O^+]} = \frac{[H_3O^+][S^{2-}]}{1 \times 10^{-14}} \qquad (8)$$

Rearranging the terms in Eq. (8), we get the condition for any H_2S solution that

$$(1.1 \times 10^{-7})(1 \times 10^{-14}) = \frac{[H_3O^+]^2[S^{2-}]}{[H_2S]}$$

For a saturated solution of H_2S at atmospheric pressure and room temperature, the concentration of H_2S in solution is constant at 0.10 M. This means that for a saturated solution of H_2S

$$[H_3O^+]^2[S^{2-}] = (1.1 \times 10^{-7})(1 \times 10^{-14})(0.10)$$

$$[H_3O^+]^2[S^{2-}] = 1 \times 10^{-22} \tag{9}$$

Equation (9) is useful because it states that the sulfide-ion concentration of a saturated H_2S solution can be changed by changing the concentration of H_3O^+. For example, if enough HCl is added to a saturated H_2S solution to make the H_3O^+ concentration 1 M, the S^{2-} concentration becomes 1×10^{-22} M. This possibility of changing the S^{2-} concentration by juggling the concentration of H_3O^+ is the basis of the classic method of ion separation in qualitative analysis by sulfide precipitation.

Example 13
A solution contains Zn^{2+} and Cu^{2+}, each at 0.02 M. The K_{sp} of ZnS is 1 × 10⁻²²; that of CuS, 8 × 10⁻³⁷. If the solution is made 1 M in H_3O^+ and H_2S gas is bubbled in until the solution is saturated, should a precipitate form?

In a saturated H_2S solution $[H_3O^+]^2[S^{2-}] = 1 \times 10^{-22}$.
If $[H_3O^+] = 1$ M, $[S^{2-}] = 1 \times 10^{-22}$ M.
For ZnS, the ion product is

$$[Zn^{2+}][S^{2-}] = (0.02)(1 \times 10^{-22}) = 2 \times 10^{-24}$$

For CuS, the ion product is

$$[Cu^{2+}][S^{2-}] = (0.02)(1 \times 10^{-22}) = 2 \times 10^{-24}$$

Since the ion product of ZnS does not exceed 1×10^{-22}, the K_{sp} of ZnS, ZnS does not precipitate. Since the ion product of CuS does exceed 8×10^{-37}, the K_{sp} of CuS, CuS does precipitate.

■ ■ ■

The principles of simultaneous equilibrium can also be applied to dissolving solids by introducing appropriate secondary equilibria. For example, although ZnS is essentially insoluble in water, it can be made to dissolve by the addition of acid. The qualitative argument is as follows: If solid ZnS is added to pure water, the equilibrium is

$$ZnS(s) \rightleftharpoons Zn^{2+} + S^{2-} \tag{10}$$

When acid is added, the additional equilibria

$$H_3O^+ + S^{2-} \rightleftharpoons HS^- + H_2O \tag{11}$$

$$H_3O^+ + HS^- \rightleftharpoons H_2S + H_2O \tag{12}$$

become important. The added H_3O^+ reacts with S^{2-} to form HS^- and H_2S. As the concentration of S^{2-} is reduced, more ZnS can dissolve. The net reaction for the dissolving is the sum of Eqs. (10) to (12), or

$$ZnS(s) + 2H_3O^+ \rightleftharpoons Zn^{2+} + H_2S + 2H_2O$$

Similarly, although AgCl is insoluble in water, it can be dissolved by addition of sodium thiosulfate, $Na_2S_2O_3$. In water the equilibrium is

$$AgCl(s) \rightleftharpoons Ag^+ + Cl^-$$

Added thiosulfate ion, $S_2O_3{}^{2-}$, combines with Ag^+ to form the complex ion $Ag(S_2O_3)_2{}^{3-}$ by the equation

$$Ag^+ + 2S_2O_3{}^{2-} \rightleftharpoons Ag(S_2O_3)_2{}^{3-}$$

Since the concentration of Ag^+ is thereby reduced, more AgCl can dissolve. The net reaction for the dissolving is

$$AgCl(s) + 2S_2O_3{}^{2-} \rightleftharpoons Ag(S_2O_3)_2{}^{3-} + Cl^-$$

The last reaction is of practical value in photographic developing. An insoluble silver salt such as AgCl is the active ingredient in photographic emulsions and must be removed in order to *fix* the picture. Dissolution is accomplished through the use of *hypo*, a solution of $Na_2S_2O_3$, which converts the insoluble, light-sensitive AgCl into a soluble complex that can be washed away.

12.10 *Hydrolysis*

One of the most important applications of simultaneous equilibria is the quantitative description of hydrolysis. The subject was discussed qualitatively in Sec. 9.4, where it was pointed out that what is important is the relative proton affinity of a solute species compared with that of water. Specifically, for the case of sodium acetate, $NaC_2H_3O_2$, the solution is slightly basic owing to a reaction between $C_2H_3O_2{}^-$ and water:

$$C_2H_3O_2{}^- + H_2O \rightleftharpoons HC_2H_3O_2 + OH^-$$

In the forward reaction a proton is transferred from a water molecule to an acetate ion; in the reverse reaction a proton is transferred from an acetic acid molecule to a hydroxide ion. Clearly, what is involved is the relative proton affinity of $C_2H_3O_2{}^-$ and OH^-. The former can be described by K_{diss} of $HC_2H_3O_2$; the latter, by K_w. This can be seen explicitly by writing the equilibrium condition for the net hydrolysis reaction as given above:

$$\frac{[HC_2H_3O_2][OH^-]}{[C_2H_3O_2{}^-]} = K_{hyd}$$

Multiplying the numerator and denominator by $[H_3O^+]$ gives

$$\frac{[HC_2H_3O_2][OH^-][H_3O^+]}{[C_2H_3O_2{}^-][H_3O^+]} = K_{hyd}$$

or

$$\frac{[OH^-][H_3O^+]}{[C_2H_3O_2{}^-][H_3O^+]/[HC_2H_3O_2]} = K_{hyd}$$

In the last step the terms have been rearranged to emphasize that the numerator is K_w and the denominator is K_{diss}:

$$\frac{K_w}{K_{diss}} = K_{hyd}$$

In other words, the hydrolysis constant K_{hyd} is just the ratio of the water dissociation constant to the weak-acid dissociation constant.

The quantitative treatment of hydrolysis can also be approached as follows: For the hydrolysis of X^- ion the equilibria are

$$2H_2O \rightleftharpoons H_3O^+ + OH^- \qquad [H_3O^+][OH^-] = K_w$$

$$X^- + H_3O^+ \rightleftharpoons HX + H_2O \qquad \frac{[HX]}{[H_3O^+][X^-]} = \frac{1}{K_{diss}}$$

These two equilibrium conditions must be satisfied simultaneously and so can be combined into one. This is done by solving for $[H_3O^+]$ and equating to eliminate $[H_3O^+]$:

$$[H_3O^+] = \frac{K_w}{[OH^-]} = \frac{[HX]}{[X^-]} K_{diss}$$

$$\frac{[HX][OH^-]}{[X^-]} = \frac{K_w}{K_{diss}}$$

This final expression represents a condition that must be satisfied by the hydrolysis. The net reaction for hydrolysis can be written

$$X^- + H_2O \rightleftharpoons HX + OH^-$$

The equilibrium condition is

$$\frac{[HX][OH^-]}{[X^-]} = K_{hyd}$$

where H_2O is omitted because it is constant.

Once the numerical value of K_{hyd} has been obtained, it can be used for equilibrium calculations in the usual way. The following examples illustrate specific cases.

Example 14
Calculate the pH *of 0.10 M* $NaC_2H_3O_2$ *and the percent hydrolysis.*
The net hydrolysis reaction is

$$C_2H_3O_2^- + H_2O \rightleftharpoons HC_2H_3O_2 + OH^-$$

for which

$$\frac{[HC_2H_3O_2][OH^-]}{[C_2H_3O_2^-]} = \frac{K_w}{K_{diss}} = \frac{1.0 \times 10^{-14}}{1.8 \times 10^{-5}} = 5.6 \times 10^{-10}$$

Let x = moles of $C_2H_3O_2^-$ that hydrolyze per liter. This forms x moles of $HC_2H_3O_2$ and x moles of OH^- and leaves $0.10 - x$ moles of

$C_2H_3O_2^-$. At equilibrium

$[HC_2H_3O_2] = x$ mol/liter

$[OH^-] = x$ mol/liter

$[C_2H_3O_2^-] = 0.10 - x$ mol/liter

Substituting in the mass-action expression gives

$$\frac{xx}{0.10 - x} = 5.6 \times 10^{-10}$$

Assuming that x is small compared with 0.10 gives

$$\frac{x^2}{0.10} \approx 5.6 \times 10^{-10}$$

$$x = 7.5 \times 10^{-6}\, M$$

Since x represents the concentration of OH^-,

$$[H_3O^+] = \frac{K_w}{[OH^-]} = \frac{1.0 \times 10^{-14}}{x} = \frac{1.0 \times 10^{-14}}{7.5 \times 10^{-6}} = 1.3 \times 10^{-9}$$

$$pH = -\log [H_3O^+] = -\log (1.3 \times 10^{-9}) = 8.89$$

The percent hydrolysis of acetate ion in this solution is given as follows:

$$\frac{\text{Moles } C_2H_3O_2^- \text{ hydrolyzed} \times 100}{\text{Moles } C_2H_3O_2^- \text{ available}} = \frac{7.5 \times 10^{-6} \times 100}{0.10}$$

$$= 0.0075\%$$

■ ■ ■

Example 15
What is the concentration of H_3O^+ in 0.10 M $AlCl_3$? The hydrolysis constant of Al^{3+} is 1.4×10^{-5}.

$$Al^{3+} + 2H_2O \rightleftharpoons AlOH^{2+} + H_3O^+$$

$$\frac{[AlOH^{2+}][H_3O^+]}{[Al^{3+}]} = 1.4 \times 10^{-5}$$

Let x = moles of Al^{3+} that hydrolyze. At equilibrium

$[Al^{3+}] = 0.10 - x$ mol/liter

$[AlOH^{2+}] = x$ mol/liter

$[H_3O^+] = x$ mol/liter

$$\frac{xx}{0.10 - x} = 1.4 \times 10^{-5}$$

$$x = 1.2 \times 10^{-3}\, M$$

The concentration of H_3O^+ in 0.10 M $AlCl_3$ should be 1.2×10^{-3} M, according to this calculation. (For comparison, the concentration of H_3O^+ in 0.10 M $HC_2H_3O_2$ is 1.3×10^{-3} M.)

■ ■ ■

Exercises

*12.1 *Dissociation.* The weak acid HF is 7.9% dissociated in 0.100 M HF. Write the equation for the dissociation, set up the equilibrium constant, and calculate K_{diss}. Show that, within the significant figures allowed in the problem, the H_2O concentration does not change because of the above dissociation.   *Ans. 6.8 × 10⁻⁴*

*12.2 *Percent dissociation.* If the hydronium-ion concentration is fixed at 0.1 M, what percent of acid is in the dissociated form for each of the following cases: (a) $K_{diss} = 10$, (b) $K_{diss} = 1$, and (c) $K_{diss} = 0.1$?

*12.3 *Percent dissociation.* What K_{diss} must you pick for an acid HX so that it will be 1.00% dissociated in a solution where the hydronium-ion concentration is fixed at 0.100 M?   *Ans. 1.01 × 10⁻³*

*12.4 *Percent dissociation.* Given a solution that is 0.10 M $NaHSO_4$, to what H_3O^+ concentration should you adjust the solution so that 50.0% of the HSO_4^- is in the dissociated form?

*12.5 *Weak acid.* What is the concentration of H_3O^+, $C_2H_3O_2^-$, and $HC_2H_3O_2$ in a solution that is labeled 0.250 M $HC_2H_3O_2$?
          Ans. 2.1 × 10⁻³ M H₃O⁺, 0.248 M HC₂H₃O₂

*12.6 *Percent dissociation.* Given that K_{diss} for $HC_2H_3O_2$ is 1.8×10^{-5}, what concentration of x M $HC_2H_3O_2$ must you take so that the acid will be 1.0% dissociated?

*12.7 *Water.* The dissociation constant of water for $2H_2O \rightleftharpoons$ $H_3O^+ + OH^-$ changes from 1.00×10^{-14} at 25°C to 9.62×10^{-14} at 60°C. What happens to the pH of water when the temperature is raised from 25 to 60°C? What happens to the neutrality?
          Ans. pH goes from 7.00 to 6.51, stays neutral

*12.8 pH. What is the pH of each of the following solutions: (a) 0.10 M HCl, (b) 0.050 M HCl, (c) 3.5×10^{-5} M HCl, (d) 3.50×10^{-5} M NaOH, and (e) 0.050 M NaOH?

*12.9 *Buffer solutions.* What are the two minimal ingredients of a buffer solution? What factor decides at which pH a buffer solution will act as a buffer? Given a fixed pH at which buffer action is to occur, how can you increase the capacity of the buffer to resist pH change?

*12.10 *Solubility product.* Calculate K_{sp} for $Hg_2Cl_2(s) \rightleftharpoons$ $Hg_2^{2+} + 2Cl^-$ given that the solubility of Hg_2Cl_2 is equal to 3.1×10^{-4} g/liter.   *Ans. 1.1 × 10⁻¹⁸*

*12.11 *Solubility.* Given that the K_{sp} of AgCl is equal to 1.7×10^{-10}, calculate the number of grams of AgCl that can dissolve in 1000 liters of (a) water and (b) $0.25\ M$ AgNO$_3$.

*12.12 *Hydrolysis.* Calculate the pH of $0.50\ M$ NaC$_2$H$_3$O$_2$ and the percent hydrolysis.   *Ans. 9.22, 0.0033%*

**12.13 *Association.* Given that K_{diss} for CdSO$_4$ is 5×10^{-3}, what is the value of K for Cd^{2+} + SO$_4^{2-}$ $\rightleftharpoons$ CdSO$_4$? If 0.100 mol of Cd^{2+} is mixed with 0.100 mol of SO$_4^{2-}$ in a liter of solution, approximately what percent of the ions will associate?

**12.14 *Association.* Given that K_{diss} of HF is equal to 6.71×10^{-4}, calculate the concentration of the various species in a solution made by dissolving 0.500 mol of HCl and 0.500 mol of NaF in enough water to make 0.250 liter of solution.

Ans. 0.0362 M H$_3$O$^+$, 0.0362 M F$^-$,
1.96 M HF, 2.00 M Cl$^-$, 2.00 M Na$^+$

**12.15 *Percent dissociation.* Suppose you have the weak diprotic acid H$_2$X, where both K_I and K_{II} are considerably less than unity. Compare qualitatively what happens to the percent dissociation of the two acids in this system, H$_2$X and HX$^-$, with increasing dilution. Explain how this comes about.

**12.16 pH. What is the pH of each of the following solutions: (a) $0.150\ M$ H$_2$SO$_4$ and (b) $0.075\ M$ H$_2$SO$_4$?   *Ans. 0.793, 1.07*

**12.17 pH. What is the hydronium-ion concentration that corresponds to each of the following pH values: (a) 0.00, (b) 1.50, (c) -1.50, (d) 7.77, and (e) 15.00?

**12.18 pH. Suppose you are given some concentrated hydrochloric acid which is labeled $11.6\ M$ HCl. You wish to make up 50.0 ml of pH = 1.25 solution. What is the recipe?

Ans. 0.24 ml (about 5 drops) of 11.6 M HCl
plus enough water to make 50.0 ml

**12.19 *Self-dissociation.* Liquid ammonia is a waterlike solvent which dissociates according to the equilibrium 2NH$_3$(l) $\rightleftharpoons$ NH$_4^+$ + NH$_2^-$, for which $K = 1.0 \times 10^{-33}$ at $-33.4°$C. If the density of the liquid is 0.68 g/ml, what will be the equilibrium concentration of NH$_4^+$ and NH$_2^-$?

**12.20 *Indicators.* Usually, a particular indicator color will be dominant when the concentration of the corresponding species is 10 times as great as the concentration of the other species. For the indicator phenolphthalein HIn is colorless, and In$^-$ is red; K_{diss} is 10^{-9}. How many milliliters of $0.10\ M$ NaOH would you have to add to 50.0 ml of a just colorless phenolphthalein solution to make it just red?   *Ans. 0.05 ml (1 drop)*

**12.21 *Indicators.* The indicator thymol blue changes from red at pH 1.2 to yellow at pH 2.8 and then from yellow at pH 8.0 to blue at pH 9.6. How might you account for this?

** 12.22 Buffer solutions. What would be the pH of each of the following buffer solutions?

a   0.225 mol of $HC_2H_3O_2$ and 0.225 mol of $NaC_2H_3O_2$ in enough water to make 0.600 liter of solution

b   0.225 mol of $HC_2H_3O_2$ and 0.225 mol of $NaC_2H_3O_2$ in enough water to make 0.300 liter of solution

c   0.300 mol of $HC_2H_3O_2$ and 0.225 mol of $NaC_2H_3O_2$ in enough water to make 0.300 liter of solution

d   0.225 mol of $HC_2H_3O_2$ and 0.300 mol of $NaC_2H_3O_2$ in enough water to make 0.300 liter of solution

Ans. 4.74, 4.74, 4.62, 4.87

** 12.23 Complex ions. Suppose you have a $1\,M$ solution of each of the first four complex ions shown in Fig. 12.5. Arrange them in order of increasing NH_3 concentration that would be in equilibrium with each.

** 12.24 Complex ions. For the plating of silver mirrors on glass it is important to control the silver-ion concentration at low, steady values in order to get a suitable deposit. Suppose you wish to adjust the Ag^+ concentration to $1.0 \times 10^{-6}\,M$ in a solution that contains $0.15\,M\ Ag(NH_3)_2{}^+$. If the final solution also contains $0.30\,M\ NH_4{}^+$, to what pH must the solution have been brought to achieve the desired result?

Ans. 8.76

** 12.25 Solubility product. Normally the concentration of the solid in the solid phase is not included in writing K_{sp}. Under what circumstance would it be necessary to include it?

** 12.26 Solubility product. Saturated solutions of $Ca(OH)_2$, $Sr(OH)_2$, and $Ba(OH)_2$ show, respectively, pH values of 12.14, 12.94, and 13.33. What are the K_{sp} values for the corresponding hydroxides?

** 12.27 Precipitation. Given that the K_{sp} of $BaSO_4$ is 1.5×10^{-9}, calculate the concentration of Ba^{2+} and of $SO_4{}^{2-}$ in a solution made by mixing 10.0 ml of $0.0100\,M\ BaCl_2$ with 15.0 ml of $0.0025\,M$ Na_2SO_4. Assume volumes are additive.

Ans. $2.5 \times 10^{-3}\,M\ Ba^{2+}$, $6.0 \times 10^{-7}\,M\ SO_4{}^{2-}$

** 12.28 Dissociation. We are given that K_{diss} for HF is 6.71×10^{-4} and that of $HC_2H_3O_2$ is 1.8×10^{-5}. What is the relative percent dissociation of HF and $HC_2H_3O_2$ in a solution made by dissolving 0.10 mol of HF and 0.10 mol of $HC_2H_3O_2$ in 250 ml of solution?

** 12.29 Dissociation. What concentration of H_2SO_4 should you take to have a solution with a pH value of 1.000?   *Ans. 0.0899 M*

** 12.30 Simultaneous equilibria. Given that $K_I = 4.3 \times 10^{-7}$ for $CO_2 + 2H_2O \rightleftharpoons H_3O^+ + HCO_3{}^-$ and $K_{II} = 4.7 \times 10^{-11}$ for

$HCO_3^- + H_2O \rightleftharpoons H_3O^+ + CO_3^{2-}$, show that in an aqueous solution saturated with CO_2 at one atm pressure, where $[CO_2] = 0.034\,M$, $[H_3O^+]$ and $[CO_3^{2-}]$ satisfy the relation

$$[H_3O^+]^2[CO_3^{2-}] = 6.9 \times 10^{-19}$$

You are given a solution that is $0.020\,M$ Mg^{2+} and $0.020\,M$ Sr^{2+}. If $K_{sp} = 1 \times 10^{-5}$ for $MgCO_3$ and 7×10^{-10} for $SrCO_3$, to what pH should the given solution be adjusted so that CO_2 addition precipitates $SrCO_3$ but not $MgCO_3$?

** **12.31** *Hydrolysis.* For CO_2 in water, $K_I = 4.3 \times 10^{-7}$, and $K_{II} = 4.7 \times 10^{-11}$. Calculate the pH of $0.500\,M$ Na_2CO_3 and the percent hydrolysis.   *Ans. 12.01, 2.0%*

*** **12.32** *Dilution.* Given a solution of $0.500\,M$ HF for which K_{diss} is 6.71×10^{-4}, how far would you need to dilute this solution in order to double the percent dissociation of HF?

*** **12.33** *pH.* Calculate the pH of a solution made by mixing 10.0 ml of $0.150\,M$ H_2SO_4 with 15.0 ml of $0.150\,M$ NaOH. Assume volumes are additive.   *Ans. 2.12*

*** **12.34** *Titration.* Suppose that 50.0 ml of $0.200\,M$ HCl is added gradually to 25.0 ml of $0.200\,M$ NaOH. Calculate the pH of the initial $0.200\,M$ NaOH solution and after each successive addition of 5.00 ml of the acid. Assume volumes are additive. Make a plot of pH versus volume of acid added.

*** **12.35** *Titration.* Suppose that 50.0 ml of $0.200\,M$ $HC_2H_3O_2$ is added gradually to 25.0 ml of $0.200\,M$ NaOH. Calculate the pH of the initial $0.200\,M$ NaOH solution and after each successive addition of 5.00 ml of the acid. Assume volumes are additive. Make a plot of pH versus volume of acid added.

Ans. 13.301, 13.125, 12.933, 12.699,
12.347, 8.87, 5.44, 5.14, 4.97, 4.84, 4.74

*** **12.36** *Buffer solutions.* Calculate the pH of each of the following solutions before and after addition of 1.00 ml of $0.500\,M$ HCl:
  a   50.0 ml of H_2O
  b   50.0 ml of $0.100\,M$ HCl
  c   50.0 ml of $0.100\,M$ $HC_2H_3O_2$ and $0.100\,M$ $NaC_2H_3O_2$
  d   50.0 ml of $0.500\,M$ $HC_2H_3O_2$ and $0.500\,M$ $NaC_2H_3O_2$
  e   100.0 ml of $0.500\,M$ $HC_2H_3O_2$ and $0.500\,M$ $NaC_2H_3O_2$

*** **12.37** *Solubility.* Given that the K_{sp} for $Ca(OH)_2$ is 1.3×10^{-6}, calculate the number of moles of $Ca(OH)_2$ that can dissolve in one liter of each of the following: (*a*) H_2O, (*b*) $0.10\,M$ NaOH, (*c*) $0.10\,M$ $Ca(NO_3)_2$, and (*d*) a solution that is simultaneously $0.10\,M$ NH_3 and $0.10\,M$ NH_4NO_3.

Ans. 6.9×10^{-3}, 1.3×10^{-4}, 1.8×10^{-3}, 0.052

*** **12.38** *Precipitation.* Given that the K_{sp} of $Sr(OH)_2$ is

3.2×10^{-4}, what would be the final concentration of Sr^{2+} and of OH^- in a solution made by mixing 10.0 ml of 0.500 M $Sr(NO_3)_2$ and 20.0 ml of 0.75 M NaOH? Assume volumes are additive.

*** *12.39* *Simultaneous equilibria.* Given that for CO_2 in water $K_I = 4.3 \times 10^{-7}$ and $K_{II} = 4.7 \times 10^{-11}$, calculate the pH of 0.25 M NaHCO$_3$ solution. *Ans. 8.35*

*** *12.40* *Hydrolysis.* For H$_2$S in water, $K_I = 1.1 \times 10^{-7}$, and $K_{II} = 1.1 \times 10^{-14}$. Calculate the pH and the percent hydrolysis in 0.313 M Na$_2$S solution.

At several points in preceding chapters we have found it necessary to introduce concepts such as energy, enthalpy, and entropy in order to gain some insight into the fundamental reasons for observed phenomena. Such discussions of energy, entropy, and related functions make up the broad field of *thermodynamics*, which by its broad generalizations unifies all the sciences into a single magnificent structure. Unfortunately, the field of thermodynamics is necessarily abstract, and a reasonable appreciation of it can come only after the accumulation of considerable background information. In this chapter, we can at best only begin to lay a foundation for further study in this field.   *365*

13.1 Systems and functions

In thermodynamics the term *system* is used to designate that region of the physical world that is being considered. This might be, for example, one mole of CO_2 gas in a sealed container, a liter of 0.10 M $CuSO_4$ solution, or a particular crystal of NaCl containing $CaCl_2$ impurity. In general, we shall need to be concerned with processes that occur in the system and the relationship of the system to its surroundings. For example, there might be an endothermic reaction occurring in the system with enough heat flow from the surroundings to keep the system at a fixed temperature. Such a system would be referred to as an *isothermal* system. An alternative possibility would be to have the system completely insulated from its surroundings; such an isolated system would be called *adiabatic*.

To describe a system completely, we must specify a number of variables. The most frequently used variables in chemistry are temperature, pressure, volume, and chemical composition. The major concern of thermodynamics is with those properties that depend only on the present state of the system and not on its past history. This means that a thermodynamic property, such as the internal energy, is completely determined once the state of the system (i.e., temperature, pressure, volume, and composition) is specified. The internal energy does not depend on what path the system followed to reach its present condition. (An example of a nonthermodynamic property, one that does depend on past history, is the mechanical strength of crystals, where strains may be introduced and persist to different degrees depending on the heat treatment, etc.)

Because a thermodynamic property depends only on the state of the system, *changes* in thermodynamic properties are independent of how the system is taken from state 1 to state 2. For example, suppose one mole of O_2 gas is taken from the state where it occupies 100 liters at $273°K$ and 0.224 atm to the state where it occupies 100 liters at $546°K$ and 0.448 atm. The thermodynamic properties, such as energy, entropy, enthalpy, and free energy, change from the initial state to the final state by amounts that do not depend on whether the gas is first heated and then compressed or first compressed and then heated. Thus, if the energy of the initial state is designated E_1 and the energy of the final state E_2, the increase in energy $E_2 - E_1$, designated by ΔE, is independent of the path. It might be noted that for any thermodynamic property X, ΔX means the X of the final state minus the X of the initial state—in other words ΔX is always the increase in X in going from initial to final state. If in this process X should happen to decrease, ΔX will, of course, have a negative value.

13.2 First law

Two laws form the basis of thermodynamics. The first of these, and by far the simpler, is equivalent to the law of conservation of energy. It states that for any system the increase in energy ΔE is equal to the heat *absorbed by* the system q minus the work *performed by* the system w:

$$\Delta E = q - w$$

Since ΔE is an energy change, both q (heat) and w (work) must be in energy units. The major difficulty students have in understanding the first law is in keeping straight the sign convention used for q and w. When q is positive, heat is absorbed by the system, and the energy of the system increases; when w is positive, the system is doing work on its surroundings, and the energy of the system decreases. In other words, q and w act in opposition as far as the ΔE of the system is concerned. We might also note that $\Delta E = 0$ for any system completely isolated from its surroundings (i.e., heat cannot flow into or out of the system, no work can be done by the system on its surroundings, and no work can be done by the surroundings on the system). $\Delta E = 0$ means the energy of the system is constant.

Examples of heat energy q are discussed in the following sections. Before considering them we might mention some examples of w. Suppose a gas sample trapped in a cylinder with a sliding piston expands from initial volume V_1 to final volume V_2 against a constant atmospheric pressure P. The work the gas sample does against the surrounding atmosphere is $P(V_2 - V_1) = P\,\Delta V$. As a second example, suppose a lead storage battery is connected to drive an electric motor. The work done by the battery is equal to its voltage times the total charge conducted, where the total charge is simply current multiplied by the elapsed time.

13.3   Enthalpy and heat capacity

In chemistry most experiments are conducted at constant pressure (e.g., at atmospheric pressure). If there is a volume change ΔV of the system, then associated with this ΔV there will be work done by the system equal to $P\,\Delta V$. To take account simultaneously of the energy of a system and the work done by the system on the surroundings, there is a thermodynamic property called the *enthalpy*, designated by H. In general, $H = E + PV$, and $\Delta H = \Delta E + \Delta(PV)$. The latter equation states that the increase in enthalpy of a system is equal to its increase in energy plus its increase in the PV product. At constant pressure, to which we shall henceforth restrict our attention, $\Delta(PV)$ is just $P\,\Delta V$, which is the amount of work done on constant-pressure surroundings by a volume change ΔV. Thus, at constant pressure

$$\Delta H = \Delta E + P\,\Delta V$$

or

$$\Delta E = \Delta H - P\,\Delta V$$

Comparison of this last relation with the first law $\Delta E = q - w$ shows that if the only work done by the system is $P\,\Delta V$ work, then *at constant pressure $\Delta H = q$*. In other words, the enthalpy increase ΔH is equal to the heat absorbed by the system at constant pressure. For this reason H is sometimes referred to as the *heat content* of the system, although the term "enthalpy" is preferred. Another way to summarize the above relations is to say that if heat is added to a system at constant pressure, part of the heat is used to increase the internal energy of the system, and the rest is used to do work on the surroundings.

Example 1

For the decomposition $CaCO_3(s) \longrightarrow CaO(s) + CO_2(g)$ *at* $950°C$ *and* CO_2 *pressure of 1 atm, the* ΔH *is 176 kJ/mol. Assuming that the volume of the solid phase changes by little compared with the volume of the gas generated, calculate the* ΔE *for the decomposition at 1 atm.*

$$\Delta E = \Delta H - P\,\Delta V$$

$$\Delta V = V_{products} - V_{reactants} \approx V_{gas} = \frac{n_{gas}RT}{P}$$

where we have introduced the gas law $PV = nRT$ to calculate V_{gas}.

$$P\,\Delta V = n_{gas}RT = (1.00 \text{ mol})(8.31 \text{ J mol}^{-1} \text{ deg}^{-1})(1223 \text{ deg})$$
$$= 10.2 \text{ kJ}$$

$$\Delta E = 176 - 10.2 = 166 \text{ kJ}$$

■ ■ ■

In words, the preceding example states that if to a mole of $CaCO_3(s)$ at $950°C$ and 1 atm of pressure 176 kJ of heat is added, then 166 kJ goes into the chemical energy of decomposition, and 10.2 kJ goes into doing expansion work against the atmosphere. In those reactions in which gases are not involved or in which the number of moles of gas does not change, the $P\,\Delta V$ work will be so small as to be negligible; consequently, in such reactions practically all the heat added goes to chemical energy. This last statement is equivalent to saying $\Delta E \approx \Delta H$, as was shown on page 223 (Example 2 of Sec. 7.13).

Thus far we have considered what happens to the enthalpy of a system if the system is kept at constant temperature and pressure while a chemical reaction occurs. What happens if the temperature is allowed to change? Obviously, added heat may then be used to heat up the system to a higher temperature. To describe this heating-up process quantitatively, we use the concept of *heat capacity*, which is defined as the amount of heat required to raise the temperature of one mole of material by one degree. Throughout our discussion we shall restrict ourselves to the heat capacity at constant pressure, usually designated by C_p. The values of C_p are found to be dependent on the chemical identity of the material, its state (whether gaseous, liquid, or solid), and its temperature. Typical values are listed in Fig. 13.1. For most substances, the liquid state has the highest heat capacity, then the solid, then the gas. Finally, we should note that the heat capacity can also

$H_2O(s)$ at $239°K$	33.30	$SO_2(s)$ at $198°K$	69.04
$H_2O(s)$ at $271°K$	37.78	$SO_2(l)$ at $270°K$	86.61
$H_2O(l)$ at $273°K$	75.86	$Zn(s)$ at $693°K$	29.7
$H_2O(l)$ at $298°K$	75.23	$Zn(l)$ at $1000°K$	31.4
$H_2O(l)$ at $373°K$	75.90	$Hg(l)$ at $500°K$	27.6
$H_2O(g)$ at $383°K$	36.28	$Xe(g)$ at $165°K$	20.9

be given per gram instead of per mole, in which case it is sometimes called *specific heat.**

For simplicity, we consider one mole of a single substance. What happens if its temperature increases from T_1 to T_2, where ΔT is small enough that the heat capacity C_p is the same at T_2 as at T_1? The amount of heat absorbed by the substance is equal to $C_p \Delta T$. This heat goes to raise the enthalpy H of the material by an amount ΔH. Hence, we can write

$$\Delta H = C_p \Delta T$$

For example, to take one mole of H_2O from 20 to $30°C$ requires that $(75.23 \text{ J mol}^{-1} \text{ deg}^{-1})(10 \text{ deg})$, or 752.3 J, be added to the enthalpy of H_2O; that is $\Delta H = 752.3$ J. If C_p changes with temperature, then it is necessary to take this into account by the methods of integral calculus

$$\Delta H = \int_{T_1}^{T_2} C_p \, dT$$

where one sums up all the contributions of $C_p \, dT$ over the small increments of temperature T where C_p is constant. For mixtures of substances the total ΔH is calculated as the sum of the individual components.

Finally, if a phase change occurs somewhere in the interval between T_1 and T_2, then it is necessary to take into account the change in enthalpy associated with the phase change as well as the difference in heat capacity between the two phases.

---

* There is a mild problem here for those who like to worry about such things. Some handbooks give specific heat in units such as calories per gram; others give no units, arguing that specific heat was originally defined as the ratio of the heat capacity per gram of substance divided by the heat capacity per gram of H_2O. Because the heat capacity per gram of H_2O is very close to 1 cal/deg, dividing by this quantity does not change the numerical value but does get rid of the units. The problem is the same as that encountered in the distinction between density and specific gravity. Density is in grams per cubic centimeter; specific gravity has no units, being a ratio of the density of the material to the density of H_2O. Since the density of H_2O is very close to unity, dividing by 1 g/cm^3 only gets rid of the units. In spite of all this, most practicing chemists use the term "specific heat" as if it had the units of energy per gram. *Facilis descensus Averno.*

Example 2

By using data from Fig. 13.1, calculate the increase of enthalpy for 100 g of H_2O *going from ice at* $-10°C$ *to liquid water at* $+15°C$. *The molar heat of fusion of ice is 6.02 kJ.*

$$\frac{100 \text{ g}}{18.0 \text{ g/mol}} = 5.55 \text{ mol of } H_2O$$

To heat ice, one needs

$$(5.55 \text{ mol})(37.78 \text{ J mol}^{-1} \text{ deg}^{-1})(10 \text{ deg}) = 2100 \text{ J}$$

To melt ice, one needs

$$(5.55 \text{ mol})(6.02 \text{ kJ/mol}) = 33{,}400 \text{ J}$$

To heat liquid, one needs

$$(5.55 \text{ mol})(75.86 \text{ J mol}^{-1} \text{ deg}^{-1})(15 \text{ deg}) = 6300 \text{ J}$$

The total increase in enthalpy equals

$$\Delta H = 2100 + 33{,}400 + 6300 = 41{,}800 \text{ J}$$

■ ■ ■

13.4   Enthalpy changes in chemical reactions

In general, when a chemical reaction occurs, heat is either evolved to or absorbed from the surroundings; so the enthalpy of the system changes. As a specific example we consider the reaction between hydrogen and oxygen to form water. The initial state consists of one mole of $H_2(g)$ and $\frac{1}{2}$ mol of $O_2(g)$ each at 1 atm pressure and 25°C; the final state, one mole of $H_2O(l)$ at one atmosphere pressure and 25°C. The change can be written

$$H_2(g) + \tfrac{1}{2}O_2(g) \longrightarrow H_2O(l)$$

If the pressure is to remain constant at one atmosphere, then clearly the volume of the system must be allowed to shrink. This can be accomplished experimentally through use of a reaction chamber consisting of a cylinder fitted with a sliding piston which moves in as the reaction progresses. If the temperature is to remain constant, despite the fact that heat is liberated in the process, then a thermostated constant-temperature bath in which the reaction chamber is immersed must be provided. Both the pressure and the temperature need to be fixed, since the enthalpy of a material depends on its pressure and temperature. By convention, most data are quoted for one atmosphere and 25°C, these conditions being used to define what is called the *standard state* at 25°C. When the above reaction occurs at 25°C so that H_2 and O_2 in their standard states change to H_2O in its standard state, 286 kJ of heat is liberated to the surrounding bath per mole of H_2O formed. Clearly, the chemical system (initially H_2 and O_2, finally H_2O) has decreased in enthalpy by 286 kJ. Recalling that ΔH

Compound	ΔH	Compound	ΔH	Compound	ΔH
$H_2O(l)$	-286	$LiF(s)$	-612	$H_2SO_4(aq)$	-908
$H_2O(g)$	-242	$NaF(s)$	-569	$HNO_3(l)$	-173
$H_2O_2(l)$	-187	$NaCl(s)$	-411	$HNO_3(aq)$	-207
$HF(g)$	-269	$NaBr(s)$	-360	$H_3PO_4(aq)$	-1290
$HCl(g)$	-92	$NaI(s)$	-288	$H_3PO_3(aq)$	-972
$HBr(g)$	-36	$CaO(s)$	-636	$CaCO_3(s)$	-1210
$HI(g)$	$+26$	$BaO(s)$	-558	$BaCO_3(s)$	-1220
$H_2S(g)$	-20	$Al_2O_3(s)$	-1670	$Na_2SO_4(s)$	-1380
$NH_3(g)$	-46	$Cr_2O_3(s)$	-1130	$NaHSO_4(s)$	-1130
$CH_4(g)$	-75	$CO(g)$	-110	$NaOH(s)$	-427
$C_2H_6(g)$	-85	$CO_2(g)$	-394	$Ca(OH)_2(s)$	-987
$C_2H_4(g)$	$+52$	$SiO_2(s)$	-859	$CaSO_4(s)$	-1430
$C_2H_2(g)$	$+227$	$SO_3(g)$	-395	$CaSO_4 \cdot 2H_2O(s)$	-2020

is the enthalpy of the final system minus the enthalpy of the initial system, we can write for this reaction

$$\Delta H = -286 \text{ kJ}$$

The minus sign indicates that the increase in enthalpy *of the system* is actually a decrease, the environment having gained the 286 kJ. In general for exothermic reactions (heat liberated to the environment), ΔH is negative; for endothermic reactions (heat absorbed from the surroundings) ΔH is positive.

In the preceding example the -286 kJ/mol may be referred to as the *enthalpy of formation* of liquid H_2O from the elements in their standard states at 25°C. Figure 13.2 gives representative values for the enthalpies of formation of several common compounds. Italicized abbreviations in parentheses indicate the standard state of aggregation of the substance. When the abbreviation in parentheses is *aq*, the compound's standard state is an ideal 1 *m* aqueous solution.

Once the enthalpies of formation from the elements are known, it is possible to calculate the enthalpy changes for other reactions. For instance, at 298°K and 1 atm the enthalpy change for

$$CaO(s) + CO_2(g) \longrightarrow CaCO_3(s)$$

can be found as follows:

$$\Delta H_{reaction} = \Delta H_{CaCO_3} - \Delta H_{CaO} - \Delta H_{CO_2}$$
$$= -1210 - (-636) - (-394) = -180 \text{ kJ/mol}$$

The justification for this procedure is that ΔH_{CaCO_3}, which represents the enthalpy of formation of $CaCO_3(s)$, gives the enthalpy of $CaCO_3$ relative to the elements Ca(s), C(s), and $O_2(g)$. Similarly, ΔH_{CO_2} and ΔH_{CaO} give the enthalpies of these compounds relative to the elements. The change in enthalpy between the elements as CaO plus CO_2 and as $CaCO_3$ is simply the heat of reaction. The procedure can also be justified through use of *Hess's law*, which states that *the heat of reaction is the same whether the reaction takes place in one or several steps.* The reaction above can be regarded as the sum of the following reactions:

$$CaO(s) \longrightarrow Ca(s) + \tfrac{1}{2}O_2(g) \qquad\qquad \Delta H = 636 \text{ kJ}$$

$$CO_2(g) \longrightarrow C(s) + O_2(g) \qquad\qquad \Delta H = 394 \text{ kJ}$$

$$\underline{Ca(s) + C(s) + \tfrac{3}{2}O_2(g) \longrightarrow CaCO_3(s) \qquad \Delta H = -1210 \text{ kJ}}$$

$$CaO(s) + CO_2(g) \longrightarrow CaCO_3(s) \qquad \Delta H = -180 \text{ kJ}$$

It should be noted that in the first two steps use was made of the fact that, by conservation of energy, reversal of a chemical equation causes reversal of the sign of ΔH.

Example 3

Calculate the enthalpy change on combustion of a mole of $C_2H_4(g)$ *to form* $CO_2(g)$ *and* $H_2O(g)$ *at* $298°K$ *and 1 atm.*

$$C_2H_4(g) + 3O_2(g) \longrightarrow 2CO_2(g) + 2H_2O(g)$$

$$\begin{aligned}
\Delta H_{\text{combustion}} &= 2\Delta H_{CO_2} + 2\Delta H_{H_2O} - \Delta H_{C_2H_4} - 3\Delta H_{O_2} \\
&= (2)(-394) + (2)(-242) - (52) - (3)(0) \\
&= -1324 \text{ kJ}
\end{aligned}$$

Notice that the enthalpy change for formation of an element in its standard state (for example, O_2) is zero.

■　■　■

13.5　Second law

The first law of thermodynamics concerns itself with the conservation of energy, $\Delta E = q - w$. For an isolated system the heat absorbed by the system q and the work done by the system w are both equal to zero; so $\Delta E = 0$. In other words, the energy of an isolated system is constant. Still, within this stipulation that energy be constant, changes may occur. As far as the first law is concerned, all such changes are equally possible, yet experience tells us that some changes do not occur; for example, H_2O does not freeze above its melting point; heat does not flow from low to high temperature; gases do not contract spontaneously. The second law of thermodynamics is concerned with the restrictions that are placed on what direction spontaneous processes will follow.

For our purposes the second law can be formulated as follows: *Any system when left to itself will tend to change toward a condition of greater probability.* As was noted in Sec. 7.13, a condition of greater probability is one of greater randomness, or disorder, because there are many more ways of producing disordered arrangements than an ordered one. Entropy S is the thermodynamic property that measures this increased probability of disordered systems. Hence, the second law of thermodynamics can be stated in another way: *For spontaneous* *change to occur in an isolated system, the entropy must increase; that is, ΔS must be greater than zero.*

If a system is not isolated from its surroundings, so that, for example, heat interchange may occur, then it follows that the total entropy of the system plus its surroundings must increase for a spontaneous change:

$$\Delta S_{total} = \Delta S_{system} + \Delta S_{surroundings} > 0 \qquad (1)$$

The $\Delta S_{surroundings}$ corresponds to the amount of heat transferred to the surroundings divided by the temperature at which the transfer occurs. That both heat and temperature should be involved can be appreciated by noting that heat added to the surroundings can increase thermal motion and hence increase random disorder; however, the higher the temperature, the more disordered are the surroundings already; so the less significant is the added contribution. For the situations most usually encountered, where temperature and pressure are kept constant, we can write

$$\Delta S_{surroundings} = \frac{\Delta H_{surroundings}}{T} = -\frac{\Delta H_{system}}{T}$$

where $\Delta H_{surroundings}$, the increase in enthalpy of the surroundings due to heat transfer from the system, is equal and opposite to ΔH_{system}, the increase in enthalpy of the system. Since we usually consider only what happens to the system and not what happens to its surroundings, we can substitute in Eq. (1) to get

$$\Delta S_{total} = \Delta S_{system} - \frac{\Delta H_{system}}{T} > 0 \qquad (2)$$

For the system, then, the second law states that the quantity which must increase in a spontaneous change at constant T and P is $\Delta S - \Delta H/T$. As we shall see, this is related to the free-energy change.

The definition of free energy, as given in Sec. 7.13, is

$$G = H - TS$$

The change in free energy is $\Delta G = \Delta H - T\,\Delta S - S\,\Delta T$, or at constant temperature and pressure, where $\Delta T = 0$,

$$\Delta G = \Delta H - T\,\Delta S$$

Dividing by T gives

$$\frac{\Delta G}{T} = \frac{\Delta H}{T} - \Delta S$$

or, by changing sign and rearranging,

$$-\frac{\Delta G}{T} = \Delta S - \frac{\Delta H}{T} \qquad (3)$$

According to Eq. (2), for a spontaneous change to occur in a system in contact with its surroundings the right-hand side of Eq. (3) must be greater than zero; so the left-hand side must also be greater than zero:

$$-\frac{\Delta G}{T} > 0$$

The temperature T is always positive. Hence, the only way $-\Delta G/T$ can be greater than zero is to have ΔG be negative. In other words, *for a spontaneous change to occur at constant temperature and pressure the free energy of the system must decrease.* If the free energy does not decrease, then a spontaneous change will not occur.

As an example of the foregoing principles, let us consider the reaction between one mole of liquid bromine and one mole of gaseous chlorine to form two moles of BrCl in the gaseous state:

$$Br_2(l) + Cl_2(g) \longrightarrow 2BrCl(g)$$

At 25°C and one atmosphere pressure, ΔH for the reaction as written is $+29$ kJ. This means that when one mole of liquid bromine combines with one mole of gaseous chlorine to form two moles of gaseous BrCl, 29 kJ of heat is absorbed from the surroundings. Thus the change is endothermic. Since most spontaneous reactions are exothermic ($\Delta H < 0$), it might seem that this particular reaction ($\Delta H > 0$) would not occur spontaneously. However, the second law requires that we consider also the ΔS for the change. The entropy of $Br_2(l)$ is 152 J mol^{-1} deg^{-1} at 25°C; the entropy of $Cl_2(g)$, 223 J mol^{-1} deg^{-1}; and the entropy of $BrCl(g)$, 240 J mol^{-1} deg^{-1}. Consequently, $\Delta S = 2S_{BrCl} - S_{Br_2} - S_{Cl_2} = 2(240) - 152 - 223 = +105$ J mol^{-1} deg^{-1}. At 25°C, or 298°K, the free-energy change for the reaction as written is

$$\begin{aligned}
\Delta G &= \Delta H - T\Delta S \\
&= 29 \text{ kJ} - (298 \text{ deg})(0.105 \text{ kJ/deg}) \\
&= -2 \text{ kJ}
\end{aligned}$$

Because ΔG is negative, the free energy would decrease in the course of the reaction $Br_2(l)$ plus $Cl_2(g)$ to form $BrCl(g)$. In other words, the chemical change should occur spontaneously in the direction written. We hasten to add, however, that thermodynamics does not tell how fast the change will occur but merely in what direction it should go.

What about equilibrium? As equilibrium is approached, i.e., as one chemical species converts to another, the free energy of the system with all its species decreases progressively until a minimum free energy is achieved. This state, being at a minimum, has no tendency to change and therefore represents an equilibrium state. For the BrCl example above it might seem that the free-energy minimum would be reached when complete conversion from $Br_2(l)$ and $Cl_2(g)$ has occurred. However, such a conclusion would be incorrect because it fails to take into account that the free energy of each component depends on its concentration. The dependence is such that a minimum total free energy for the components is achieved when the concentration of BrCl(g) is about 3.6 times that of Cl$_2$(g). When these equilibrium concentrations have been achieved, the conversion of a small amount of Br$_2$ plus Cl$_2$ to BrCl, or vice versa, is accompanied by no change in free energy; i.e., at equilibrium $\Delta G = 0$. To see how this comes about, we need to look separately at how the enthalpy and the entropy change during the conversion. The statement that

$$Br_2(l) + Cl_2(g) \longrightarrow 2BrCl(g) \qquad \Delta H = +29 \text{ kJ}$$

means that the enthalpy of the system progressively increases as the Br$_2$ and Cl$_2$ change into BrCl. Figure 13.3a shows schematically this progressive rise in enthalpy as a function of increasing fraction of conversion from reactants to products. The starting enthalpy of the elemental reactants has been arbitrarily set at zero. The point to note is that if there were 100 percent conversion of one mole of Br$_2$(l) plus one mole of Cl$_2$(g) into two moles of BrCl(g), all at 1 atm pressure, the enthalpy of the system would go up by 29 kJ.

What happens to the entropy during this conversion? The statement that

$$Br_2(l) + Cl_2(g) \longrightarrow 2BrCl(g) \qquad \Delta S = +105 \text{ J/deg}$$

tells us that the entropy increases by 105 J when there is complete conversion of Br$_2$ and Cl$_2$ into 2BrCl. But what about the intermediate degrees of conversion? Unlike the case for enthalpy, where the change is linear with increasing fraction of conversion, the entropy shows a more complicated dependence. The entropy of the intermediate states will be enhanced because mixtures of unlike species have many more configurations than do pure substances. Furthermore, for our purposes it will be more meaningful to look not at the entropy S but at the temperature-entropy product TS. Since our prime concern is the free energy G, which is equal to $H - TS$, a curve of TS versus fraction of conversion can be subtracted from the curve for H to find out the course of G.

Figure 13.3b shows how the TS product changes on progressive conversion of Br$_2$(l) and Cl$_2$(g) into BrCl(g). The starting entropy is 152 J mol^{-1} deg^{-1} for Br$_2$(l) and 223 J mol^{-1} deg^{-1} for Cl$_2$(g), or a total of 0.375 kJ. At 298°K, the corresponding TS product would be 112 kJ/mol. When conversion is complete, we have the entropy of two moles of BrCl(g), that is, (2)(240) J mol^{-1} deg^{-1}, which gives a TS product of 143 kJ/mol. Thus, the two end points of the curve are fixed at 112 and 143 kJ/mol, respectively.

To calculate the entropy of the intervening mixture, we need to add up the entropies of all the components in the mixture, making due allowance for the number of moles present. For the liquid phase, there is no problem; we just multiply the molar entropy of Br$_2$(l) by the number of moles of liquid bromine present. For the gas phase, we need to take account of the fact that the pressure of each gas is changing, not only because the number of moles is changing, but also because the volume of the system will have to expand if we want to carry out

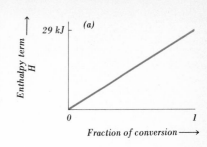

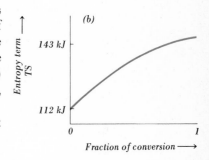

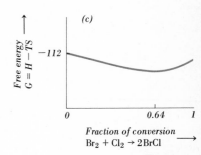

Fig. 13.3
Change of thermodynamic parameters H, TS, and G during course of reaction Br$_2$(l) + Cl$_2$(g) → 2BrCl(g). *(Scale is somewhat distorted.)*

the process at constant pressure. To keep total P constant while one mole of gas (Cl_2) converts into two moles of gas (BrCl), the volume of the system will have to double. So we will have to take account of the change of entropy with volume.

For any gas which expands from initial volume V_1 to final volume V_2 the entropy change is given by

$$\Delta S = R \ln \frac{V_2}{V_1}$$

where R is the universal gas constant and ln stands for the natural logarithm. For ideal gases, volume is inversely proportional to pressure; so the above can equally well be written as

$$\Delta S = R \ln \frac{P_1}{P_2}$$

Suppose we consider specifically the halfway point for conversion of $Br_2(l)$ and $Cl_2(g)$ into $BrCl(g)$. If we start with one mole each of $Br_2(l)$ and $Cl_2(g)$ at 1 atm and 298°K, then at the halfway point we will have $\frac{1}{2}$ mol of $Br_2(l)$, $\frac{1}{2}$ mol of $Cl_2(g)$, and one mole of $BrCl(g)$. The volume of the system would have to expand from being able to accommodate one mole of gas to being able to accommodate $\frac{3}{2}$ mol of gas; in other words, the volume goes up to three-halves of the original. Any gas contained therein would have to drop in pressure to two-thirds of its initial value, just because of this volume expansion alone. Thus, at the halfway point the pressure of $Cl_2(g)$ would become half as much, because of the disappearance of half the chlorine, times two-thirds as much, because of the volume expansion. So at the halfway point, P of $Cl_2(g)$ is $\frac{1}{3}$ atm. Similarly, P of $BrCl(g)$ at the halfway point is $\frac{2}{3}$ atm.

When chlorine gas goes from one atm to $\frac{1}{3}$ atm, its entropy *per mole* changes as follows:

$$\Delta S = R \ln \frac{P_{\text{initial}}}{P_{\text{final}}} = R \ln \frac{1}{\frac{1}{3}} = R \ln 3$$

so we can write

$$S_{Cl_2(1/3 \text{ atm})} = S_{Cl_2(1 \text{ atm})} + R \ln 3$$

Similarly, we can deduce

$$S_{BrCl(2/3 \text{ atm})} = S_{BrCl(1 \text{ atm})} + R \ln \frac{3}{2}$$

To calculate the total entropy of the system at the halfway point, we need to put in the actual moles of each constituent present:

$$S_{\text{total}} = \frac{1}{2} S_{Br_2} + \frac{1}{2} S_{Cl_2} + 1 S_{BrCl}$$

Substituting from above, we get

$$S_{\text{total}} = \frac{1}{2} S_{Br_2} + \frac{1}{2} S_{Cl_2(1 \text{ atm})} + \frac{1}{2} R \ln 3 + S_{BrCl(1 \text{ atm})} + R \ln \frac{3}{2}$$

Putting in the appropriate values and noting that $R = 8.314 \text{ J mol}^{-1}$

Fraction of conversion	H, kJ	TS, kJ	G, kJ
0.0	0.0	111.75	−111.75
0.1	2.9	116.22	−113.32
0.2	5.8	119.80	−114.00
0.3	8.7	123.37	−114.67
0.4	11.6	126.35	−114.75
0.5	14.5	129.63	−115.13
0.6	17.4	132.61	−115.21
0.7	20.3	135.59	−115.29
0.8	23.2	138.27	−115.07
0.9	26.1	140.66	−114.56
1.0	29.0	143.04	−114.04

Fig. 13.4

Thermodynamic parameters for $Br_2(l) + Cl_2(g) \rightarrow 2BrCl(g)$ as function of increasing fraction of conversion.

deg^{-1} and that $\ln x = 2.303 \log x$, we get $S_{total} = \frac{1}{2}(152) + \frac{1}{2}(223) + \frac{1}{2}(8.314)(2.303) \log 3 + 240 + (8.314)(2.303) \log \frac{3}{2} = 436$ J/deg. Multiplication by the temperature, 298°K, gives a TS product of 130 kJ. This number is intermediate between the initial value of 112 kJ corresponding to $Br_2(l)$ plus $Cl_2(g)$ and the end value of 143 kJ corresponding to two moles of $BrCl(g)$. As can be seen, the actual value for the intermediate mixture is greater than the simple average of the initial and end values. In other words, as shown in Fig. 13.3b, the TS product does not go up linearly but has a definite bulge at the intermediate mixtures.

Calculations similar to the above can be carried out for each of the mixtures encountered along the way as we convert from $Br_2 + Cl_2$ into BrCl. (See Exercise 13.25.) Figure 13.4 summarizes some of the parameters calculated for successively greater fractions of conversion. The curve for the TS results is shown schematically in Fig. 13.3b. Once the TS curve has been obtained, it can be subtracted from the enthalpy curve of Fig. 13.3a. The result is Fig. 13.3c, which shows G plotted as a function of increasing fraction of conversion. As can be seen, the free energy decreases progressively until it reaches a minimum at 64 percent conversion, whereafter it rises. Although 100 percent conversion to BrCl would give a free energy that is somewhat lower than the initial state $(Br_2 + Cl_2)$, the lowest free energy occurs for an intermediate system where some unchanged Br_2 and Cl_2 are left along with the BrCl.

At the minimum in the free-energy curve, the curve is flat. In other words, small additional conversion of $Br_2 + Cl_2$ into BrCl or a small amount of decomposition of BrCl into $Br_2 + Cl_2$ does not materially affect the free energy. Therefore, $\Delta G = 0$ at this minimum, which is the position of chemical equilibrium. We shall return to this point in Sec. 13.7.

13.6 Standard free-energy changes

The free energy of a substance (as well as its enthalpy and its entropy) depends on the state of the substance, i.e., whether it is solid, liquid, or gas. Furthermore, the free energy depends on temperature and pressure and, in the case of solutions, on concentration. Consequently, the change of free energy during a chemical reaction depends on the state, conditions, and concentrations of the reactants and products. For convenience, standard conditions have been chosen for reference. These correspond to one atmosphere pressure, a specified temperature (which is usually 25°C), and, if the substance is pure, the state s, l, or g in which the substance exists at one atmosphere and the specified temperature. For *gaseous* solutions the standard state is generally taken to correspond to a partial pressure of one atmosphere. For *liquid* solutions the standard state is sometimes taken as unit mole fraction or, more generally, one-molal concentration corrected for nonideal-solution behavior. For *solid* solutions the standard state corresponds to unit mole fraction.

As a specific example of the significance of standard states we can consider the reaction

$$Zn(s) + 2H_3O^+(aq) \longrightarrow H_2(g) + Zn^{2+}(aq) + 2H_2O$$

For this reaction at 25°C the standard free-energy change $\Delta G°$ is -148 kJ, the standard enthalpy change $\Delta H°$ is -152 kJ, and the standard entropy change $\Delta S°$ is -17.2 J mol^{-1} deg^{-1}. (Note the super-script zero, which indicates that the numbers apply to substances in their standard states only. If the standard states are other than at 25°C, a subscript may be used to show the absolute temperature at which the data apply.) $\Delta G° = -148$ kJ means that the free energy of the system at 25°C and 1 atm decreases by 148 kJ when one mole of solid zinc reacts completely with two moles of hydronium ion, from an aqueous solution in which the concentration of H_3O^+ is an ideal one molal, to produce one mole of gaseous H_2 at a pressure of 1 atm and one mole of zinc ion in aqueous solution at a concentration that is ideal one molal. "Ideal one molal" means that the concentration is one mole per kilogram of solvent except that corrections have been applied for all nonideal behavior, such as interionic attraction.

13.7   *Chemical activities and chemical equilibrium*

Most chemical reactions involve mixtures of substances. Since these substances are generally not in their standard states, we need to know how to describe quantitatively their tendency for chemical reaction under other than standard conditions. The relative tendency for chemical reaction is most often expressed in terms of *chemical activity*, which is defined as the ratio of the "effective" concentration of a species to its concentration in a standard reference state. The concentration of gases can be measured by pressure; so it is convenient to set the activity of an ideal gas as equal to its actual pressure divided by its standard-state pressure. For gases that are nonideal it is still a good approximation to measure activity as the ratio of actual pressure to standard-state pressure. For liquid solutions the activity of a solute is generally given as the ideal molal concentration of the solute divided by its molality in the standard state. It might seem that dividing by the standard-state pressure or molality is unnecessary since the numerical value in the standard state is unity. However, we must not forget units. By dividing as indicated, we arrive at activity as a *dimensionless quantity* embodying a comparison with a standard state.

To see better what is meant by activity, let us consider a mixture of gaseous hydrogen and gaseous iodine. The total free energy of the mixture G_{total} can be written as the sum of a contribution from the hydrogen plus a contribution from the iodine. Letting n be the number of moles of each component and $\overline{G}$ the free-energy contribution per mole ascribable to each component in this particular mixture, we get

$$G_{total} = n_{H_2}\overline{G}_{H_2} + n_{I_2}\overline{G}_{I_2}$$

How does $\overline{G}_{H_2}$ in the mixture differ from what it would be in the standard state of H_2? In other words, how does $\overline{G}_{H_2}$ change with H_2 pressure (i.e., concentration)? If we represent the molar free-energy contribution in the standard state by $\overline{G}_{H_2}^{\circ}$, then it turns out that

$$\overline{G}_{H_2} = \overline{G}_{H_2}^{\circ} + RT \ln a_{H_2}$$

Thus, a_{H_2}, the activity of the hydrogen, gives a measure of the change in free-energy contribution in departing from the standard state. In the standard state of hydrogen, $a_{H_2} = 1$, and hence $\overline{G}_{H_2} = \overline{G}_{H_2}^{\circ}$. For activities less than unity (that is, H_2 pressure less than one atmosphere), $\overline{G}_{H_2}$ will be less than $\overline{G}_{H_2}^{\circ}$; for activities greater than unity (that is, H_2 pressure greater than one atmosphere), $\overline{G}_{H_2}$ will be greater than $\overline{G}_{H_2}^{\circ}$. Stated differently, as the hydrogen pressure is increased, its molar free-energy contribution increases.

A mixture of H_2 and I_2 is not at equilibrium since the free energy can decrease by conversion of some H_2 and I_2 into HI. Suppose we ask what the molar free-energy change would be for the chemical reaction

$$H_2(g) + I_2(g) \longrightarrow 2HI(g)$$

in any kind of randomly selected mixture of H_2, I_2, and HI. The change in free energy ΔG is the free energy of two moles of the HI product $2\overline{G}_{HI}$ minus the sum of the free energies of one mole each of the reactants $\overline{G}_{H_2} + \overline{G}_{I_2}$:

$$\Delta G = 2\overline{G}_{HI} - \overline{G}_{H_2} - \overline{G}_{I_2}$$

The free-energy contribution of each component can now be expressed in terms of its standard-state value and its activity in the mixture. Substituting

$$\overline{G}_{HI} = \overline{G}_{HI}^{\circ} + RT \ln a_{HI}$$

$$\overline{G}_{H_2} = \overline{G}_{H_2}^{\circ} + RT \ln a_{H_2}$$

$$\overline{G}_{I_2} = \overline{G}_{I_2}^{\circ} + RT \ln a_{I_2}$$

in the above equation gives us

$$\Delta G = 2\bar{G}^\circ_{HI} + 2RT \ln a_{HI} - \bar{G}^\circ_{H_2} - RT \ln a_{H_2} \\ - \bar{G}^\circ_{I_2} - RT \ln a_{I_2}$$

In place of the terms $2\bar{G}^\circ_{HI} - \bar{G}^\circ_{H_2} - \bar{G}^\circ_{I_2}$ we can write ΔG°, which stands for what the free-energy change would be if every component had been in its standard state. Noting also that $2RT \ln a_{HI} = RT \ln a_{HI}{}^2$, we can write

$$\Delta G = \Delta G^\circ + RT \ln \frac{a_{HI}{}^2}{a_{H_2} a_{I_2}}$$

At equilibrium, the total free energy would have to be at a minimum. From the definition of a minimum, it follows that at equilibrium, $\Delta G = 0$. Therefore, for equilibrium we write

$$0 = \Delta G^\circ + RT \ln \frac{a_{HI}{}^2}{a_{H_2} a_{I_2}}$$

or

$$\Delta G^\circ = -RT \ln \frac{a_{HI}{}^2}{a_{H_2} a_{I_2}}$$

The last equation is most important. The activity ratio has just the form of the mass-action expression for the reaction $H_2 + I_2 \rightleftharpoons 2HI$; so we can set it equal to the equilibrium constant and write

$$\frac{a_{HI}{}^2}{a_{H_2} a_{I_2}} = K$$

which leads to

$$\Delta G^\circ = -RT \ln K$$

The final relation is noteworthy in that it states that the final *equilibrium state* as characterized by K is related to free-energy change from reactants to products *in the standard state*. Thus, we learn about the *equilibrium* state by considering the free-energy properties in the *nonequilibrium* standard state. The connection, of course, is that the free-energy difference between reactants and products at standard conditions tells how far away from final equilibrium the standard conditions really are.

For any chemical equation in general, the associated ΔG° may be zero, positive, or negative. If $\Delta G^\circ = 0$, then $K = 1$, which means that the final equilibrium state will be one in which reactants and products contribute equally to the activity ratio. If $\Delta G^\circ > 0$, then $K < 1$, which means that the products would have to contribute much less to the activity ratio in the equilibrium state than do the reactants; in other words, the chemical reaction would not proceed far from left to right. If $\Delta G^\circ < 0$, then $K > 1$; reaction to the right is favored, and product activities would predominate in the equilibrium state. The actual numerical value for K is dependent on the choice of standard states for the reactants and products. When the standard states are the same as those used to compute the free-energy values given in compilations such as those of the National Bureau of Standards, gases turn out to be measured in atmospheres. Hence, the equilibrium constants K_p deduced from $\Delta G° = -RT \ln K_p$ will not in general be the same as the K_c values where concentrations are expressed in moles per liter (see Sec. 11.3). Nevertheless, if the number of gas moles is the same on both sides of a chemical equation (as is true for Example 4 but not for Example 5), then its K expressed in concentrations, K_c, is numerically equal to the K expressed in pressures, K_p.

Example 4
The standard free energy of formation of HI *from* H_2 *and* I_2 *at* 490° C *is* -12.1 kJ/mol *of* HI. *Calculate the equilibrium constant for*
$H_2 + I_2 \rightleftharpoons 2HI$.

$\Delta G°_{490} =$ (2 mol of HI per mole of reaction*)(-12.1 kJ/mol)

$$= -24.2 \text{ kJ/mol}$$

$\Delta G°_{490} = -RT \ln K$

$$\ln K = -\frac{\Delta G°_{490}}{RT} = \frac{-(-24,200 \text{ J/mol})}{(8.31 \text{ J mol}^{-1} \text{ deg}^{-1})(763 \text{ deg})} = 3.82$$

$$\log K = \frac{\ln K}{2.303} = \frac{3.82}{2.303} = 1.66$$

$$K = 46$$

■ ■ ■

Example 5
For the reaction $2NO(g) + O_2(g) \rightleftharpoons 2NO_2(g)$ *at* 298° K, *the equilibrium constant* K_p *equals* 1.6×10^{12}. *Given that the standard free energy of formation of* NO(g) *is* 86.6 kJ/mol *at* 298° K, *calculate the standard free energy of formation of* $NO_2(g)$ *at* 298° K.
For $2NO(g) + O_2(g) \rightleftharpoons 2NO_2(g)$

$\Delta G° = -RT \ln K_p$

$$= -(8.31 \text{ J mol}^{-1} \text{ deg}^{-1})(298 \text{ deg})(2.303) \log (1.6 \times 10^{12})$$

$$= -69.6 \text{ kJ/mol of reaction}$$

$\Delta G° = 2\Delta G°_{NO_2} - 2\Delta G°_{NO} - \Delta G°_{O_2}$

Note that the $\Delta G°$ of formation of an element in its standard state is taken to be zero; so $\Delta G°_{O_2} = 0$.

$\Delta G°_{NO_2} = \frac{1}{2}(\Delta G° + 2\Delta G°_{NO})$

$$= \frac{1}{2}(-69.6 + 2 \times 86.6)$$

$$= 51.8 \text{ kJ/mol}$$

■ ■ ■

* The term "mole of reaction" refers to the chemical equation read in moles.

13.8 Temperature and chemical equilibrium

In the preceding sections we have considered the free-energy change for a chemical reaction at constant temperature and pressure. Specifically, we have shown that the standard free-energy change (the free energy of the products minus the free energy of the reactants, all in their standard states) is directly related to the equilibrium constant for the reaction:

$$\Delta G° = -RT \ln K$$

What happens when the temperature is changed? To find out, we first note that the definition of free energy leads to

$$\Delta G° = \Delta H° - T\,\Delta S°$$

which combines with the preceding equation to give

$$-RT \ln K = \Delta H° - T\,\Delta S°$$

Solving this for $\ln K$, we get

$$\ln K = -\frac{\Delta H°}{RT} + \frac{\Delta S°}{R}$$

Now our question becomes the following: How does $\ln K$ change with temperature? If the change of temperature is not very large, then $\Delta H°$ and $\Delta S°$ will be constant, just as R is. (In other words, the enthalpy difference and the entropy difference between pure products and pure reactants does not change much with temperature.) Hence we can rewrite the above equation as

$$\ln K = -\frac{\Delta H°}{R}\left(\frac{1}{T}\right) + \frac{\Delta S°}{R}$$

to emphasize that the variables are effectively $\ln K$ and $1/T$. The equation has the form $y = ax + b$, where a and b are constants equal to $-\Delta H°/R$ and $\Delta S°/R$, respectively. Using $1/T$ as the independent variable and $\ln K$ as the dependent variable, we get a straight-line plot with slope $-\Delta H°/R$, as is shown in Fig. 13.5. For the temperature interval between T_1 and T_2, where the respective equilibrium constants are K_1 and K_2, we can write for the slope

$$\frac{\ln K_2 - \ln K_1}{1/T_2 - 1/T_1} = -\frac{\Delta H°}{R}$$

which can also be written

$$\ln \frac{K_2}{K_1} = \frac{\Delta H°}{R}\left(\frac{1}{T_1} - \frac{1}{T_2}\right)$$

This relation represents one form of the *van't Hoff equation*. It is important not only because it provides a method for calculating K at one temperature when $\Delta H°$ and K at another temperature are known but also because it allows determination of the standard enthalpy change of a reaction $\Delta H°$ from measurements of equilibrium constants at different temperatures. It must be emphasized that the above form of the equation rests on the assumption that $\Delta H°$ and $\Delta S°$ are constant; that is, they do not change with temperature. This assumption will, in general, not be valid for large temperature intervals.

When computing with logarithms to the base 10, the above equation becomes

$$\log \frac{K_2}{K_1} = \frac{\Delta H°}{2.303R}\left(\frac{1}{T_1} - \frac{1}{T_2}\right) = \frac{\Delta H°}{19.15}\left(\frac{1}{T_1} - \frac{1}{T_2}\right)$$

where $\Delta H°$ is in joules per mole of reaction.

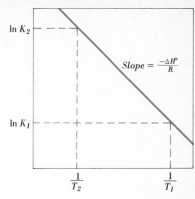

Fig. 13.5
Change of equilibrium
constant with temperature.

Example 6
For the reaction

$$2NO(g) + O_2(g) \rightleftharpoons 2NO_2(g)$$

the standard enthalpy change of reaction is $-113\ kJ/mol$ of reaction and the equilibrium constant is 1.6×10^{12} at $298°K$. Calculate K for this reaction at $373°K$, assuming $\Delta H°$ stays constant.

$$\log \frac{K_2}{K_1} = \frac{\Delta H°}{19.15}\left(\frac{1}{T_1} - \frac{1}{T_2}\right)$$

Let $K_1 = 1.6 \times 10^{12}$ at $T_1 = 298°K$; then

$$\log \frac{K_2}{1.6 \times 10^{12}} = \frac{-113,000}{19.15}\left(\frac{1}{298} - \frac{1}{373}\right)$$

$$K_2 = 1.7 \times 10^8$$

The decrease in equilibrium constant is consistent with the fact that the enthalpy change is negative, i.e., the reaction is exothermic; so a rise in temperature makes it less favored.

■ ■ ■

Example 7
The molecule NO_2 can dimerize to form N_2O_4. Calculate $\Delta H°$ for the reaction

$$2NO_2(g) \rightleftharpoons N_2O_4(g)$$

given that $K_p = 8.85$ at $298°K$ and $K_p = 0.0792$ at $373°K$.

$$\log \frac{K_2}{K_1} = \frac{\Delta H°}{19.15}\left(\frac{1}{T_1} - \frac{1}{T_2}\right)$$

$$\log \frac{0.0792}{8.85} = \frac{\Delta H°}{19.15}\left(\frac{1}{298} - \frac{1}{373}\right)$$

$$\Delta H° = -58,200\ J/mol\ of\ reaction$$

■ ■ ■

Exercises

* **13.1** *First law.* Give a concise statement of the first law. Show how your statement is equivalent to the law of conservation of energy.

* **13.2** *Terminology.* When you pump up an automobile tire, is the system (you not included) undergoing mostly an isothermal change or mostly an adiabatic change?

* **13.3** *Enthalpy.* Using a specific example, indicate the essential difference between enthalpy and energy.

* **13.4** *Enthalpy of reaction.* From the data in Fig. 13.2, calculate the enthalpy of reaction for

$$CO_2(g) + C(s) \rightleftharpoons 2CO(g)$$

at 298°K and 1 atm pressure. *Ans. 174 kJ*

* **13.5** *Standard states.* Given the reaction

$$H_2(g) + 2AgCl(s) + 2H_2O \longrightarrow 2Ag(s) + 2H_3O^+(aq) + 2Cl^-(aq)$$

for which the standard free-energy change at 25°C is −42.93 kJ, what are the standard states to which this statement refers?

** **13.6** *ΔH and ΔE.* Suppose you expand 1 mol of an ideal gas from 22.4 liters at 1.00 atm and 273°K to 44.8 liters at 0.50 atm and 273°K. What is ΔH, and what is ΔE for the process? Note that the gas is ideal.

** **13.7** *ΔH and ΔE.* Given that $\Delta H = -37.0$ kJ at 400°C and 3.00 atm total pressure for the reaction $NH_4NO_3(s) \longrightarrow N_2O(g) + 2H_2O(g)$, calculate the ΔE for the process.

Ans. $\Delta E = -53.8$ kJ

** **13.8** *Enthalpy.* Given that the molar heat of fusion of $H_2O(s)$ is 6.02 kJ and the molar heat of evaporation of $H_2O(l)$ is 40.7 kJ, calculate with the data of Fig. 13.1 the enthalpy change for each of the following processes:

a Take 5.0 g of ice from 0°C to liquid water at 25°C.
b Take 5.0 g of liquid water from 0 to 25°C.
c Take 5.0 g of liquid water from 25°C to gaseous water at 110°C and 1 atm pressure.

** **13.9** *Enthalpy of reaction.* Using the data of Fig. 13.2, figure out the order in which C_2H_6, C_2H_4, and C_2H_2 should be ranked for increasing amount of heat liberated per gram when burned to $CO_2(g)$ and $H_2O(l)$ at 298°K and 1 atm pressure.

Ans. C_2H_2, 50.0 kJ/g; C_2H_4, 50.4 kJ/g; C_2H_6, 52.0 kJ/g

** **13.10** *Enthalpy of reaction.* From the data of Fig. 13.2, which requires more energy per gram to decompose, $BaCO_3(s)$ to $BaO(s) + CO_2(g)$ or $CaCO_3(s)$ to $CaO(s) + CO_2(g)$?

** 13.11   Second law. Show that the condition for spontaneous change in an isothermal, constant-pressure system is that ΔS of the system must be greater than $\Delta H/T$ of the system.

** 13.12   Entropy. Suggest a principal reason why the entropy would increase in a reaction such as

$$Br_2(l) + Cl_2(g) \longrightarrow 2BrCl(g)$$

** 13.13   Entropy. We are given one mole of chlorine gas with an entropy of 223 J/deg at 1 atm and 298°K. To what volume would you need to expand it to increase its entropy by 10.0 percent at constant temperature?   *Ans. 357 liters*

** 13.14   Entropy. If you double the volume of a gas by expanding it at constant temperature, what do you do to its entropy?

** 13.15   Activity. Describe what is meant by the term "activity," with specific reference to the system $H_2(g) + I_2(g) \rightleftharpoons 2HI(g)$.

** 13.16 $\Delta G°$ and K. How can one prove that the standard free-energy change for a reaction is directly related to the equilibrium constant for that reaction?

** 13.17 $\Delta G°$ and K_c. Given that $K_c = 261$ at 1000°K for the reaction $2SO_2(g) + O_2(g) \rightleftharpoons 2SO_3(g)$, what is the standard free-energy change for this reaction at 1000°K? Note that $n/V = P/RT$.
                    Ans. −9.61 kJ

** 13.18 $\Delta G°$ and K. Given that at 1000°K the standard free energy of formation of $NO(g)$ is 77.95 kJ/mol while that of $NO_2(g)$ is 95.44 kJ/mol, calculate the equilibrium constant, as K_p and as K_c, for the reaction $2NO(g) + O_2(g) \rightleftharpoons 2NO_2(g)$ at 100°K.
                    Ans. $K_p = 0.0149$, $K_c = 1.22$

** 13.19   K versus T. For the reaction $PCl_5(g) \rightleftharpoons PCl_3(g) + Cl_2(g)$ the value of K_c is 0.0224 at 500°K and 33.3 at 760°K. Calculate $\Delta H°$ for the reaction.

** 13.20   K versus T. The equilibrium constant expressed with moles per liter for the reaction $COCl_2(g) \rightleftharpoons CO(g) + Cl_2(g)$ has a value of 0.0820 at 900°K and 0.329 at 1000°K. At what temperature would K_c become equal to unity?   *Ans. 1097°K*

** 13.21   K versus T. If you want an equilibrium constant to double with a 10°K rise in temperature at 298°K, what $\Delta H°$ would be required for the reaction?

** 13.22   Dissociation constants. The dissociation constant for acetic acid $HC_2H_3O_2 + H_2O \rightleftharpoons H_3O^+ + C_2H_3O_2^-$ changes from 1.657×10^{-5} at 0°C to 1.754×10^{-5} at 25°C and then to 1.633×10^{-5} at 50°C. What does this tell you about the ΔH for the above reaction? Assume ΔS is constant.
                    Ans. Changes from + 1.53 to −2.28 kJ

*** 13.23 *Second law.* Following the lines of the argument at the beginning of Sec. 13.5, show that for a constant-pressure system doing isothermal $P \Delta V$ work only, $T \Delta S_{system}$ must be greater than the heat absorbed by the system for the process to be spontaneous.

*** 13.24 *G, H, and S.* For the reaction $H_2(g) + I_2(g) \longrightarrow 2HI(g)$ at $700°K$, $\Delta H°$ is -12 kJ, and $\Delta S°$ is 3.68 J. Sketch three curves showing how H, TS, and G probably change for the system as a function of increasing percent conversion of H_2 and I_2 to HI.

*** 13.25 *Entropy.* On the basis of the information given in Sec. 13.5 for the reaction $Br_2(l) + Cl_2(g) \longrightarrow 2BrCl(g)$, calculate the entropy for the system at successively greater fractions of conversion at 0.1 intervals from 0 to 1. Start with one mole of $Br_2(l)$ and one mole of $Cl_2(g)$, and end up with two moles of $BrCl(g)$, all at 1 atm and $298°K$.

*** 13.26 *K versus T.* At $298°K$, the $\Delta G°$ of formation for $N_2O_4(g)$ is $+98.28$ kJ, whereas that of $NO_2(g)$ is 51.84 kJ. Starting with one mole of $N_2O_4(g)$ at 1 atm and $298°K$, calculate what fraction will be decomposed if the total pressure is kept constant at 1 atm and the temperature is maintained at $298°K$. If $\Delta H°$ for $N_2O_4(g) \rightleftharpoons 2NO_2(g)$ is $+58.03$ kJ, to what temperature would the system have to go to double the fraction of N_2O_4 decomposed?   *Ans. 317°K*

*** 13.27 *Temperature and equilibrium.* Given that the equilibrium constant expressed with moles per liter for $N_2(g) + 3H_2(g) \rightleftharpoons 2NH_3(g)$ has a value of 6.41 at $600°K$ and 2.37×10^{-3} at $1000°K$, calculate $\Delta G°$, $\Delta H°$, and $\Delta S°$ for the reaction at $800°K$.   *Ans. +76.6 kJ, −111.3 kJ, −235 J*

The energy changes associated with chemical reactions generally show up as heat absorbed from or emitted to the surroundings. When there is work done by the chemical system on the environment, it is usually of the mechanical, or $P \Delta V$, type. Occasionally, however, under favorable circumstances it is possible to design the chemical reaction so that it does electric work on the surroundings. In this chapter the relation between chemical energy and electric energy is explored. We consider the transport of electric energy through matter, the conversion of electric energy into chemical energy, and the conversion of chemical energy into electric energy. These topics constitute the field of *electrochemistry*.

Electric energy may be transported through matter by the conduction of electric charge from one point to another in the form of an *electric current* (see Appendix 5.5 to 5.7 for a discussion of electrical terms). In order that the electric current exist, there must be charge carriers in the matter, and there must be a force that makes the carriers move. The charge carriers can be electrons, as in the case of metals and semiconductors, or they can be positive and negative ions, as in the case of electrolytic solutions and molten salts. In the former case, conduction is said to be *electronic* or *metallic;* in the latter, *ionic* or *electrolytic.* The electric force that makes charges move is usually supplied by a battery, generator, or some similar source of electric energy. The region of space in which there is an electric force is called an *electric field.*

As pointed out in Sec. 7.6, solid metals consist of ordered arrays of positive ions immersed in a sea of electrons. For example, silver consists of Ag^+ ions arranged in a face-centered cubic pattern with the entire lattice permeated by a cloud of electrons equal in number to the number of Ag^+ ions in the crystal. The Ag^+ ions are more or less fixed in positions from which they do not move except under great stress. The electrons of the cloud, on the contrary, are free to roam throughout the crystal. When an electric field is impressed on the metal, the electrons migrate and thereby carry negative electric charge through the metal. In principle, it should be possible for an electric field to force all the loose electrons toward one end of a metal sample. In practice, it is extraordinarily difficult to separate positive and negative charges from each other without a relatively enormous increase of free energy. The only way it is possible to keep a sustained flow of charge in a wire is to add electrons to one end of the wire and drain electrons off the other end as fast as they accumulate. The metal conductor remains everywhere electrically neutral since just as many electrons move into a region per unit time as move out.

Most of the electrons that make up the electron cloud of a metal are of very high kinetic energy. Metallic conductivity would therefore be extremely high were it not for a *resistance* effect. Electric resistance is believed to arise because lattice ions vibrate about their lattice points. By creating local oscillations in the electric field and thereby interfering with the migration of electrons, the ions keep the conductivity down. At higher temperatures the thermal vibrations of the lattice increase, and therefore it is not surprising to find that as the temperature of a metal is raised, its conductivity diminishes.

In solutions the mechanism of conductivity is complicated by the fact that the positive carriers are also free to move. As pointed out in Sec. 8.4, solutions of electrolytes contain positive and negative ions. Except under highly unusual conditions, there are no free electrons in aqueous solutions. The ions are not fixed in position but are free to roam throughout the body of the solution. When an electric field is applied, as shown in Fig. 14.1, the positive ions experience a force

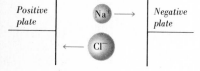

Fig. 14.1
Electrical forces on ions in solution.

in one direction, while the negative ions experience a force in the opposite direction. The simultaneous motion of positive and negative ions in opposite directions constitutes the *electrolytic current*. The current would stop if positive ions accumulated at the negative electrode and negative ions at the positive electrode. In order that the electrolytic current continue, appropriate chemical reactions must occur at the electrodes to maintain electric neutrality throughout all regions.

That ions migrate when electrolytic solutions conduct electricity can be seen from the experiment diagrammed in Fig. 14.2. The U tube is initially half filled with a deep purple solution of copper permanganate, $Cu(MnO_4)_2$, in water. The color of the blue hydrated Cu^{2+} ions is effectively masked by the purple of the MnO_4^- ions. A colorless aqueous solution of nitric acid, HNO_3, is initially floated on top of the $Cu(MnO_4)_2$ solution in each arm of the U tube. If an electric field is maintained for some time across the solution by the two electrodes, then after a while it is observed that the blue color characteristic of hydrated Cu^{2+} ions has moved into the region marked *A*, suggesting a migration of positive ions toward the negative electrode. At the same time, the purple color characteristic of MnO_4^- has moved into the region marked *B*, indicating that negative ions have migrated simultaneously toward the positive electrode.

As in the case of metallic conduction, electric neutrality must be preserved in all regions of the solution at all times. Otherwise, the current would soon cease. Figure 14.3 shows two of the possible ways by which electric neutrality can be preserved for a given region of an NaCl solution. In (*a*) one Na^+ ion enters the region to compensate for the charge of the departing Na^+ ion. In (*b*), as one Na^+ ion leaves the region, one Cl^- ion departs in the opposite direction; hence, the region shows no net change in charge. *Both* effects (*a*) and (*b*) occur simultaneously, their relative importance depending on the relative mobilities of the positive and negative ions.

Unlike metallic conduction, electrolytic conduction is usually increased when the temperature of a solution is raised.* In metals the conducting electrons are already of such high energy that a rise in temperature does not appreciably affect their kinetic energy. In solutions, not only is the average kinetic energy of the ions increased with a rise of temperature, but also the viscosity of the solvent is diminished; so the ions can migrate faster, and the solution becomes a better conductor of electricity.

14.2   Electrolysis

In order to maintain an electric current, it is necessary to have a complete circuit; i.e., there must be a closed loop whereby electric charge can return to its starting point. If the complete circuit includes,

---

* There are exceptions to this generalization. For example, with some weak electrolytes the percent dissociation (Sec. 8.5) may decrease with rising temperature. The decrease in the concentration of ions may be big enough to cause a *decrease* in conductivity with rising temperature.

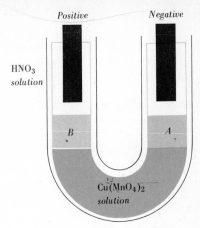

Fig. 14.2
Migration of ions in electrolytic conductivity.

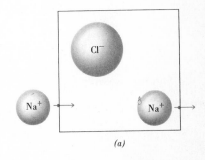

(a)

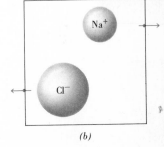

(b)

Fig. 14.3
Two ways that migrating ions could maintain electric neutrality in a region of solution.

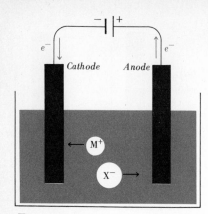

Fig. 14.4
Typical electrolysis circuit.

besides the normal complement of wires, battery, switches, etc., an electrolytic conductor, then chemical reaction must occur at the electrode interface between the metallic conductors and the electrolytic solution. Electric energy is thus used to produce chemical change; this process is called *electrolysis*.

A typical electrolysis circuit is shown in Fig. 14.4. The two vertical lines at the top of the diagram represent a battery, where the long line indicates the positive terminal and the short line, the negative one. The curved lines represent strips of connecting wire, usually copper, that join the battery to the electrodes. The electrodes, which may be made of any inert, conducting material, such as graphite or platinum, dip into the electrolytic conductor, which contains the ions M^+ and X^- that are free to move. When operating, the battery creates an electric field which pushes the electrons in the external wires in the directions shown by the arrows. Electrons are crowded onto the left-hand electrode and drained away from the right-hand electrode. The circuit is not complete, and current does not flow unless there is some way by which electrons can be used up at the left electrode and supplied at the right electrode. Chemical changes must occur. At the left electrode a *reduction* process must occur; some ion or molecule must accept electrons and thereby be reduced. An electrode at which such a reduction occurs is called a *cathode*. At the right-hand electrode electrons must be donated by some ion or molecule; in other words, an *oxidation* process must occur. The electrode at which oxidation occurs is always called an *anode*. In order for reduction to keep going at the cathode, ions will have to keep moving toward it. These ions are the positive ions; they are called *cations*. Simultaneously, negative ions move toward the anode; they are called *anions*.

For a more specific example we consider the electrolysis of *molten* NaCl. As in later cases, we assume that the electrodes themselves are inert and do not react chemically. Of the two ions present, Na^+ and Cl^-, only the Na^+ can be reduced. At the cathode, where reduction must occur, the following reaction therefore takes place:

$$Na^+(l) + e^- \longrightarrow Na(l)$$

This is called the *cathode half-reaction*. At the anode, oxidation occurs. Of the two species in the cell, only the Cl^- can be oxidized. Although the exact mechanism is not known, we can imagine that Cl^- releases an electron to the anode to form a neutral Cl atom, which then combines with another similar atom to produce the molecule Cl_2. Subsequently, the Cl_2 molecules bubble off as a gas. The net *anode half-reaction* would then be written

$$2Cl^-(l) \longrightarrow Cl_2(g) + 2e^-$$

At the cathode, electric energy has been used to convert Na^+ into liquid Na metal; at the anode, to convert Cl^- into gaseous Cl_2. By addition, the two electrode half-reactions can be combined into a single overall *cell reaction*. In order to keep electrons from accumulating anywhere in the circuit, just as many electrons must disappear at the cathode as appear at the anode. To ensure such electron balance, the half-reactions are multiplied by appropriate coefficients so that when the half-reactions are added, the electrons cancel out of the final equation. Thus, for the electrolysis of molten NaCl we have the following:

Cathode half-reaction

$$2Na^+(l) + 2e^- \longrightarrow 2Na(l)$$

Anode half-reaction

$$2Cl^-(l) \longrightarrow Cl_2(g) + 2e^-$$

Overall reaction

$$2Na^+(l) + 2Cl^-(l) \xrightarrow[electrolysis]{} 2Na(l) + Cl_2(g)$$

This reaction if left to itself would actually go spontaneously in the reverse direction. In order to emphasize that the reaction goes because of consumption of electric energy, the word "electrolysis" has been written under the arrow.

As a second example of electrolysis we consider what happens when *aqueous* NaCl is electrolyzed. Besides Na^+ and Cl^-, a new ingredient, H_2O, has been added. Under usual conditions (e.g., not too much current), it is observed that hydrogen gas is liberated at the cathode and chlorine gas is liberated at the anode. How can these observations be accounted for? The electrolysis cell contains, besides Na^+ and Cl^- ions, H_2O molecules and traces of H_3O^+ and OH^- from the dissociation of H_2O. Molecules of H_2O can be either oxidized to O_2 by removal of electrons or reduced to H_2 by addition of electrons. The H_2O must thus be considered as a possible reactant at either electrode. At the cathode, reduction must occur. Three different reactions are possible:

$$Na^+ + e^- \longrightarrow Na(s) \tag{1}$$

$$2H_2O + 2e^- \longrightarrow H_2(g) + 2OH^- \tag{2}$$

$$2H_3O^+ + 2e^- \longrightarrow H_2(g) + 2H_2O \tag{3}$$

When there are several possible reactions at a cathode, it is not easy to predict which one will occur. It is necessary to consider which reactant is reduced most *easily* and which one is reduced most *rapidly*. The strongest oxidizing agent is not necessarily the fastest. Further complications may arise when currents become very large and when concentrations of reactants become very small. The fact that hydrogen gas, not metallic sodium, is actually observed in the electrolysis of aqueous NaCl suggests that reaction (1) does not occur.* In NaCl solution, the concentration of H_3O^+ (which is only $1 \times 10^{-7}\ M$) is not large enough to make reaction (3) reasonable as a *net change*. Therefore, reaction (2) is usually written as the best description for

---

* Years ago it was thought that the metal Na was first formed by reaction (1) and that it subsequently reacted with H_2O to liberate H_2. However, there is no evidence that any intermediate Na is ever formed in this electrolysis.

the cathode reaction. [Nevertheless, in acidic solutions, the concentration of H_3O^+ may well be high enough for H_3O^+ to appear in the net electrode reaction. For example, in the electrolysis of aqueous HCl, the cathode reaction is indeed written as Eq. (3).]

In the electrolysis of NaCl solution, reaction (2) shows that OH^- accumulates in the region around the cathode; so positive ions (Na^+) must move toward the cathode to preserve electric neutrality. In addition, some of the OH^- migrates away from the cathode. Both migrations are consistent with the requirement that cations migrate toward the cathode and anions toward the anode. At the anode, oxidation must occur. Two different reactions are possible:

$$2Cl^- \longrightarrow Cl_2(g) + 2e^-$$

$$6H_2O \longrightarrow O_2(g) + 4H_3O^+ + 4e^-$$

We were told above that chlorine gas is liberated at the anode; so experiment shows that the first of these predominates. As the chloride-ion concentration around the anode is depleted, fresh Cl^- moves in, and Na^+ moves out.

In summary, the equations for the electrolysis of aqueous NaCl can be written as follows:

Cathode half-reaction

$$2e^- + 2H_2O \longrightarrow H_2(g) + 2OH^-$$

Anode half-reaction

$$2Cl^- \longrightarrow Cl_2(g) + 2e^-$$

Overall reaction

$$2Cl^- + 2H_2O \xrightarrow[electrolysis]{} H_2(g) + Cl_2(g) + 2OH^-$$

As indicated in the overall reaction, during electrolysis the concentration of Cl^- diminishes, and the concentration of OH^- increases. Since there is always Na^+ in the solution, the solution is gradually converted from aqueous NaCl to aqueous NaOH. In fact, in one process for the commercial production of chlorine by electrolysis of aqueous NaCl, solid NaOH is obtained as a by-product by evaporating H_2O from the residual solution left after electrolysis.

As a final example of electrolysis, we consider aqueous Na_2SO_4. With inert electrodes it is observed in this case that H_2 gas is formed at the cathode and O_2 is formed at the anode. At the same time, the solution near the cathode becomes basic; that near the anode, acidic. Consistent with these observations are the following electrode reactions:

Cathode

$$2e^- + 2H_2O \longrightarrow H_2(g) + 2OH^-$$

Chapter 14
Electrochemistry

Anode

$$6H_2O \longrightarrow O_2(g) + 4H_3O^+ + 4e^-$$

The overall cell reaction is obtained by doubling the cathode reaction and adding the result to the anode reaction. The four electrons cancel, and the result is

$$10H_2O \xrightarrow[\text{electrolysis}]{} 2H_2(g) + O_2(g) + 4H_3O^+ + 4OH^-$$

In this equation, both H_3O^+ and OH^- appear as products, which would seem to be a self-contradiction, except that the H_3O^+ and OH^- are formed in different regions. If the H_3O^+ from the anode region and the OH^- from the cathode region are allowed to mix, then neutralization occurs, and the above net reaction becomes

$$2H_2O \xrightarrow[\text{electrolysis}]{} 2H_2(g) + O_2(g)$$

In this electrolysis only the H_2O disappears. The Na^+ and $SO_4{}^{2-}$ initially present in the solution are also present at the conclusion of the electrolysis. Is the Na_2SO_4 necessary for the electrolysis to occur? Because of the requirement of electric neutrality, some kind of electrolytic solute must be present. Positive ions must be available to move into the cathode region to counterbalance the charge of the OH^- produced. Negative ions must be available to move to the anode to counterbalance the H_3O^+ produced.

Unlike the above cases, the electrodes themselves may also sometimes take part in the electrode reactions. In the above cells the electrodes were assumed to be inert. This would almost always be the case if the electrodes were made of graphite or the inert metal platinum. If, however, the electrode material were reactive it would have to be considered as a possible reactant. For example, copper or silver anodes themselves are frequently oxidized when no other species present is more readily oxidized.

14.3 Quantitative aspects of electrolysis

Michael Faraday empirically established early in the nineteenth century the quantitative laws of electrolysis. The Faraday laws state that the weight of product formed at an electrode is proportional to the amount of electricity transferred at the electrode and to the equivalent weight of the material. This can be accounted for by considering the electrode reactions. For example, in the electrolysis of molten $NaCl$, the cathode reaction

$$Na^+(l) + e^- \longrightarrow Na(l)$$

tells us that one Na atom is produced at the electrode when one sodium ion disappears and one electron is transferred. If the Avogadro number of electrons were transferred, one mole of Na^+ would disappear, and one mole of Na atoms would be formed. For this reaction, one equivalent of Na is 22.99 g; hence, transfer of the Avogadro number of electrons liberates 22.99 g of Na. Increasing the amount of electricity transferred increases proportionately the mass of Na produced.

The Avogadro number of electrons is such a convenient measure of the amount of electricity that it is given a special name, the *faraday*. In electrical units one faraday is equal to 96,500 coulombs. As described in Appendix 5.7, a *coulomb* is the amount of electricity that is transferred when a current of one ampere flows for one second. In other words, the current in amperes multiplied by the time in seconds is equal to the number of coulombs. The electric charge in coulombs divided by 96,500 is equal to the number of faradays.

Electrode half-reactions expressed in atoms and electrons can be read in terms of moles and faradays. Thus,

$$Na^+(l) + e^- \longrightarrow Na(l)$$

can be read either "one sodium ion reacts with one electron to form one sodium atom" or "one mole of sodium ions reacts with one faraday of electricity to form one mole of sodium atoms."

Example 1

How many grams of chlorine can be produced by the electrolysis of molten NaCl with a current of 1.00 amp for 5.00 min?

1.00 amp = 1.00 coulomb/sec

(1.00 coulomb/sec)(5.00 min)(60 sec/min) = 300 coulombs

$$\frac{300 \text{ coulombs}}{96,500 \text{ coulombs/faraday}} = 0.00311 \text{ faraday}$$

The half-reaction $2Cl^-(l) \longrightarrow Cl_2(g) + 2e^-$ tells us that we produce one mole of chlorine per two faradays of electricity.

$$(0.00311 \text{ faraday})\left(\frac{1 \text{ mol } Cl_2}{2 \text{ faradays}}\right)\left(\frac{70.9 \text{ g } Cl_2}{1 \text{ mol } Cl_2}\right) = 0.111 \text{ g } Cl_2$$

■ ■ ■

Example 2

A current of 0.0965 amp is passed for 1000 sec through 50.0 ml of 0.100 M NaCl. If the only reactions are reduction of H_2O to H_2 (at the cathode) and oxidation of Cl^- to Cl_2 (at the anode), what will be the average concentration of OH^- in the final solution?

$$\frac{(0.0965 \text{ coulomb/sec})(1000 \text{ sec})}{96,500 \text{ coulombs/faraday}} = 0.00100 \text{ faraday}$$

The cathode reaction $2e^- + 2H_2O \longrightarrow H_2(g) + 2OH^-$ shows that two faradays liberate two moles of OH^-; so 0.00100 faraday would liberate 0.00100 mol of OH^-. Assuming the final volume of solution remains at 50.0 ml, the final concentration of OH^- is

$$\frac{0.00100 \text{ mol}}{0.0500 \text{ liter}} = 0.0200 \text{ } M$$

■ ■ ■

14.4 Galvanic cells

In the cells discussed in the preceding sections, electric energy in the form of a current was used to bring about oxidation-reduction reactions. It is also possible to do the reverse, i.e., use an oxidation-reduction reaction to produce electric current. The main requirement is that the oxidizing and reducing agents be kept separate from each other so that electron transfer is forced to occur through a wire. Any device which accomplishes this is called a *galvanic*, or *voltaic*, cell after Luigi Galvani (1780) and Alessandro Volta (1800), who made the basic discoveries.

When a bar of zinc is dipped into a solution of copper sulfate, a deposit of copper forms over the zinc surface. The net reaction is

$$Zn(s) + Cu^{2+} \longrightarrow Zn^{2+} + Cu(s)$$

In this change, zinc is oxidized, and Cu^{2+} is reduced, presumably by direct transfer of two electrons from each zinc atom to each copper ion. To emphasize this transfer of electrons, the net reaction can be split into two half-reactions:

$$Zn(s) \longrightarrow Zn^{2+} + 2e^-$$

$$Cu^{2+} + 2e^- \longrightarrow Cu(s)$$

The galvanic cell operates on the principle that the two separated half-reactions can be made to take place simultaneously, with the electron transfer occurring through a wire. The galvanic cell shown in Fig. 14.5 uses the reaction

$$Zn(s) + Cu^{2+} \longrightarrow Zn^{2+} + Cu(s)$$

The dashed line represents a porous partition which separates the container into two compartments but still permits limited diffusion of ions between them. In the left-hand compartment is a solution of zinc sulfate into which a zinc bar is dipped; in the right-hand compartment is a copper bar dipping into a solution of copper sulfate. When the two electrodes are connected by a wire, electric current flows, as shown by an ammeter in the circuit. As time progresses, the zinc bar is observed to be eaten away and the copper bar plated with a fresh deposit of copper.

The cell operates as follows: At the zinc bar, oxidation occurs, thus making zinc the anode. The half-reaction

$$Zn(s) \longrightarrow Zn^{2+} + 2e^-$$

produces Zn^{2+} ions and electrons. The zinc ions migrate away from the surface of the anode into the solution, and the electrons move through the wire, as indicated by the arrow in the figure. At the copper bar, reduction occurs, thus making copper the cathode. Electrons which have come through the wire accumulate at the cathode surface, where they are picked up and used in the reaction

$$Cu^{2+} + 2e^- \longrightarrow Cu(s)$$

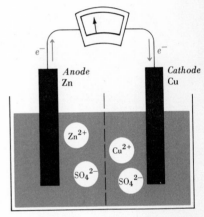

Fig. 14.5
Example of a galvanic cell.

Copper ions in the solution are thus depleted, and new copper ions have to move into the vicinity of the cathode surface. The electric circuit is complete. Consistent with previous notation, cations (Zn^{2+} and Cu^{2+}) in the solution move toward the cathode (the copper bar), and anions (SO_4^{2-}) move toward the anode (the zinc bar). Electrons are generated at the anode, flow through the wire, and are consumed at the cathode; a current is obtained from an oxidation-reduction reaction. The cell runs until either the zinc or Cu^{2+} is depleted.

In describing the operation of a galvanic cell it is not necessary to specify the relative charges of the electrodes. In fact, a simple unique assignment of charges cannot be made to account for the direction of both the electron and the ion currents. To account for the *electron current* (which goes from anode to cathode in the external wire), the anode would have to be labeled *negative* with respect to the cathode. To account for the *ion current* (negative ions move toward the anode and positive ions toward the cathode), the anode would have to be labeled *positive* with respect to the cathode. How can the anode be positive and negative at the same time? The discrepancy can be resolved by considering the electrode in detail. The Zn^{2+} ions released into the solution at the anode surface form a layer which makes the anode appear positive as *viewed from the solution*. The electrons left in the Zn bar because of Zn^{2+} formation make the anode appear negative as *viewed from the wire*.

Actually, to get a current from the cell, the Zn^{2+} ions and the Cu bar need not be initially present. Any metal support for the plating of Cu would do in place of the Cu bar. Any positive ion that does not react with Zn metal would do in place of Zn^{2+}. However, as the cell reaction proceeds, Zn^{2+} would necessarily be produced at the anode. Is the porous partition necessary? It is needed only to hinder the Cu^{2+} from easily getting over to the Zn metal, where direct electron transfer would short-circuit the whole cell. The partition must be porous in order to allow some diffusion of positive and negative ions from one compartment to the other. Otherwise, the solution in the anode compartment would soon become positively charged (owing to accumulation of Zn^{2+}), and that in the cathode compartment negatively charged (owing to depletion of Cu^{2+}). Such charge accumulation would soon cause the current to cease.

There are numerous oxidation-reduction reactions that have been made into sources of electric current. Probably the most famous example is the *lead storage battery*, or *accumulator*. Its basic features are electrodes of lead (Pb) and lead dioxide (PbO_2) dipping into a strong aqueous solution of H_2SO_4. When the cell operates, the reactions can be written as follows:

Anode

$$Pb(s) + HSO_4^- + H_2O \longrightarrow PbSO_4(s) + 2e^- + H_3O^+$$

Cathode

$$PbO_2(s) + HSO_4^- + 3H_3O^+ + 2e^- \longrightarrow PbSO_4(s) + 5H_2O$$

Overall cell reaction

$$Pb(s) + 2HSO_4^- + 2H_3O^+ + PbO_2(s) \longrightarrow 2PbSO_4(s) + 4H_2O$$

The lead sulfate ($PbSO_4$) that is formed at each electrode is insoluble and adheres to the electrode. During what is called the *charging* of the battery, the electrode reactions shown have to be reversed so as to restore the cell to its original condition. In *discharge*, as shown by the overall cell reaction, Pb and PbO_2 are depleted, and the concentration of H_2SO_4 is diminished. Since the density of the aqueous solution is chiefly dependent on the concentration of H_2SO_4, measurement of the density can be used as a simple way to tell how far the cell has discharged.*

Another familiar galvanic cell is the *Leclanché cell*, also known as the *dry cell*, which is frequently used in flashlights and similar devices. The cell consists of a Zn can containing a centered graphite stick surrounded by a moist paste of manganese dioxide (MnO_2), zinc chloride ($ZnCl_2$), and ammonium chloride (NH_4Cl). The Zn can serves as the anode, and the graphite rod as the cathode. At the anode Zn is oxidized; at the cathode MnO_2 is reduced. The electrode reactions are extremely complex and seem to vary depending on how much current is drawn from the cell. For the delivery of very small currents, the following reactions are probable:

Anode

$$Zn(s) \longrightarrow Zn^{2+} + 2e^-$$

Cathode

$$2MnO_2(s) + Zn^{2+} + 2e^- \longrightarrow ZnMn_2O_4(s)$$

Overall cell reaction

$$Zn(s) + 2MnO_2(s) \longrightarrow ZnMn_2O_4(s)$$

Other technologically important galvanic cells are the *Edison*, or *NIFE*, cell, which uses the oxidation of Fe by Ni_2O_3 in basic medium, so the overall reaction can be written

$$Fe(s) + Ni_2O_3(s) + 3H_2O \longrightarrow Fe(OH)_2(s) + 2Ni(OH)_2(s)$$

and the so-called "mercury battery," which uses the oxidation of Zn in Hg amalgam by HgO in basic medium, for which we can write

$$Zn(amalgam) + HgO(s) \longrightarrow ZnO(s) + Hg(amalgam)$$

* It would seem natural to say that when the battery is discharging, the anion HSO_4^- moves to the anode. However, as is obvious from the cathode half-reaction, some of the HSO_4^- must also move to the cathode in order to form $PbSO_4$. Thus we have the unusual but not unique situation that the anion moves simultaneously toward both electrodes, in one case because of the electric field and in the other case because of a depletion-induced concentration gradient.

14.5   Fuel cells

In principle, any oxidation-reduction reaction can be separated into half-reactions and used to drive a galvanic cell. In particular, the reaction for the oxidation of a fuel gas such as CH_4 should be so separable. Major interest has been attracted to this possibility as a means of alleviating man's energy crisis because the *direct* conversion of chemical energy to electric energy can be made considerably more efficient (i.e., up to 75 percent) than the 40 percent maximum now attainable by burning the fuel and using the heat to form steam for driving turbines.

For the oxidation of natural gas, we can write

$$CH_4(g) + 2O_2(g) \longrightarrow CO_2(g) + 2H_2O$$

which, by the methods outlined in Sec. 4.8, can be separated into two half-reactions. For acidic solution, these can be written as follows:

Anode

$$CH_4(g) + 10H_2O \longrightarrow CO_2(g) + 8H_3O^+ + 8e^-$$

Cathode

$$O_2(g) + 4H_3O^+ + 4e^- \longrightarrow 6H_2O$$

In practice, the reaction is better utilized in basic medium, where the product CO_2 exists as carbonate ion, $CO_3{}^{2-}$. For basic media, the corresponding half-reactions would be as follows:

Anode

$$CH_4(g) + 10OH^- \longrightarrow CO_3{}^{2-} + 7H_2O + 8e^-$$

Cathode

$$O_2(g) + 2H_2O + 4e^- \longrightarrow 4OH^-$$

The detailed construction of a workable cell has proved to be a challenging engineering task involving such tricks as using molten K_2CO_3 as a high-temperature electrolyte medium. However, the basic design principles must be the same for this as for any galvanic cell, viz., two electrode compartments each containing the reactants as called for by the respective half-reaction. In the present case two of the reactants are gases and must be bubbled into the cell from the exterior, as is shown schematically in Fig. 14.6. In order to make electric contact with the reactant gases, conducting, but otherwise inert, electrodes are suspended in the cell in the bubble streams. The porous partition indicates that limited ion migration between the compartments must be permitted but the reactant gases are to be kept separated from each other.

The first really workable fuel cell made use of the reaction

$$2H_2(g) + O_2(g) \longrightarrow 2H_2O$$

for which the half-reactions in basic solution are as follows:

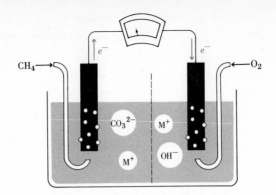

Fig. 14.6
Schematic representation of
a fuel cell.

Anode

$$H_2(g) + 2OH^- \longrightarrow 2H_2O + 2e^-$$

Cathode

$$O_2(g) + 2H_2O + 4e^- \longrightarrow 4OH^-$$

Figure 14.7 shows one design which has been used successfully. Two

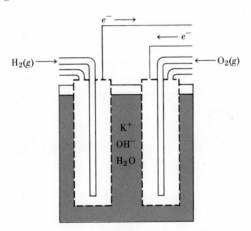

Fig. 14.7
Hydrogen-oxygen fuel cell.

chambers constructed of porous carbon dip into an aqueous solution
of KOH. Hydrogen gas is pumped into one chamber, while oxygen
gas is simultaneously introduced into the other. Because H_2 and O_2
each reacts very slowly at room temperature, suitable catalysts must
be used to accelerate the electrode reactions. These catalysts are mixed
in and pressed with the carbon. At the anode, suitable catalysts are
finely divided platinum or palladium; at the cathode, cobaltous oxide
(CoO), platinum, or silver.

14.6 *Electrode potentials*

A voltmeter connected between the two electrodes of a galvanic cell
shows a characteristic voltage, which depends in magnitude on what
reactants take part in the electrode reactions and on what their concentrations are. For example, in the Zn–Cu cell if Zn^{2+} and Cu^{2+} are
at $1\ m$ concentrations and the temperature is 25°C, the voltage meas-

ured between the Zn electrode and the Cu electrode is 1.10 V, no matter how big the cell or how big the electrodes. This voltage is characteristic of the reaction

$$Zn(s) + Cu^{2+} \longrightarrow Zn^{2+} + Cu(s)$$

The voltage measures the difference in electric potential between the two electrodes, i.e., the work done in moving unit charge from one electrode to the other. Hence, the observed voltage measures the force with which electrons would be moved around the circuit and therefore gives a quantitative measure of the relative tendency of various oxidation-reduction reactions to occur.

Figure 14.8 shows on the left a galvanic cell set up to study the reaction

$$Zn(s) + 2H_3O^+ \longrightarrow H_2(g) + Zn^{2+} + 2H_2O$$

In the anode compartment a Zn bar dips into a solution of a Zn salt. In the cathode compartment H_2 gas is led in through a tube so as to bubble over an inert electrode, made, for example, of Pt, dipped

Fig. 14.8

Comparison of zinc and copper electrodes vs. the hydrogen electrode.

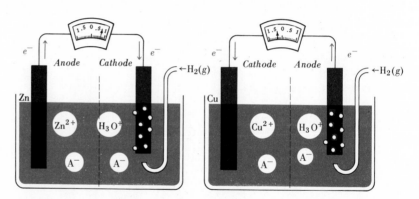

into an acidic solution. The anode reaction is

$$Zn(s) \longrightarrow Zn^{2+} + 2e^-$$

The cathode reaction is

$$2H_3O^+ + 2e^- \longrightarrow H_2(g) + 2H_2O$$

At 25°C, when the concentrations of H_3O^+ and Zn^{2+} are each at 1 *m* and when the pressure of the H_2 gas is 1 atm, the voltmeter reads 0.76 V; the deflection is in such direction as to indicate that Zn has a greater tendency to give off electrons than has H_2. In other words, the half-reaction $Zn(s) \longrightarrow Zn^{2+} + 2e^-$ has a greater tendency to occur by 0.76 V than has $H_2(g) + 2H_2O \longrightarrow 2H_3O^+ + 2e^-$.

The galvanic cell on the right in Fig. 14.8 makes use of the reaction

$$H_2(g) + Cu^{2+} + 2H_2O \longrightarrow 2H_3O^+ + Cu(s)$$

The anode reaction is

$$H_2(g) + 2H_2O \longrightarrow 2H_3O^+ + 2e^-$$

and the cathode reaction is

$$Cu^{2+} + 2e^- \longrightarrow Cu(s)$$

At 25°C, when the concentrations of H_3O^+ and Cu^{2+} are $1\,m$ and when the pressure of H_2 is 1 atm, the voltmeter reads 0.34 V; the deflection direction indicates that Cu has a smaller tendency to give off electrons than has H_2. In other words, the half-reaction $Cu(s) \longrightarrow Cu^{2+} + 2e^-$ has a smaller tendency to occur by 0.34 V than has $H_2(g) + 2H_2O \longrightarrow 2H_3O^+ + 2e^-$.

In all cells the voltage observed arises from two sources: an electric potential at the anode and an electric potential at the cathode. If either of these electrode potentials were known, the other could be obtained by subtraction. However, it is impossible to measure the electrode potential of an individual electrode since any complete circuit necessarily contains two electrodes. Conventional procedure is to select one electrode as a standard reference electrode, arbitrarily assign it a zero value of electric potential, and then refer all other electrode potentials to this arbitrarily designated zero. The procedure is equivalent to saying that any voltage measured between a given electrode and the standard reference is to be assigned completely to the given electrode. For all intents and purposes, the standard reference electrode is assumed to contribute nothing to the voltage readout observed on the meter. As reference, international convention has agreed on the *standard hydrogen electrode* (sometimes designated SHE) as being assigned a zero value at 25°C, 1 atm H_2 pressure, and $1\,m$ H_3O^+ concentration. Consequently, in a cell which contains the standard hydrogen electrode, the entire measured voltage is attributed to the half-reaction at the other electrode. Voltages thus assigned are called *oxidation-reduction potentials*, or *redox potentials*. If the half-reaction is written with the electrons on the left, the associated voltage is called a *reduction potential*; if the half-reaction is written with the electrons on the right, the associated voltage is called an *oxidation potential*. There is, unfortunately, considerable confusion as to whether the electrode potential should mean reduction potential or oxidation potential. The problem is by no means a trivial one because, as we shall see below, it means a change in sign. European chemists and most United States analytical chemists have preferred to work with reduction potentials; United States physical chemists have generally used oxidation potentials. The convention that has now been adopted internationally is that electrode potentials mean reduction potentials, not oxidation potentials. The rationale is that the polarity sign associated with the reduction potential corresponds to the sign of the electrode as seen from the outside of the cell, in other words, to the terminal of the voltmeter that it will be hooked up to.

Figure 14.9 lists some common half-reactions together with their reduction potentials. A more extensive listing is given in Appendix 7. The values given apply for the half-reaction read in the forward direction. For the reverse direction, the sign must be changed.

The forward reaction is a reduction process in which an oxidizing agent, shown just to the left of an arrow, is reduced to give the reducing agent shown to the right of an arrow. The table is so arranged that the oxidizing agents are listed in order of decreasing strength (or tendency to be reduced). In other words, there is decreasing tendency of the forward half-reaction to occur from the top of the table to the bottom. For example, of the list given, fluorine, F_2, is the best oxidizing agent, since it has the highest tendency to pick up electrons. Lithium ion, Li^+, at the bottom of the list, is the worst oxidizing agent and has the least tendency to pick up electrons.

The numerical values of the reduction potentials given in Fig. 14.9 apply to aqueous solutions at $25°C$ in which the concentration, or more exactly the activity, of each dissolved species is $1\ m$. A positive value of the reduction potential indicates that the oxidizing agent is stronger than H_3O^+; a negative value indicates that the oxidizing agent is weaker

Fig. 14.9

Some half-reactions and their standard reduction potentials.

Half-reaction	Standard reduction potential, V
$2e^- + F_2(g) \rightarrow 2F^-$	$+2.87$
$2e^- + Cl_2(g) \rightarrow 2Cl^-$	$+1.36$
$4e^- + 4H_3O^+ + O_2(g) \rightarrow 6H_2O$	$+1.23$
$2e^- + Br_2 \rightarrow 2Br^-$	$+1.09$
$e^- + Ag^+ \rightarrow Ag(s)$	$+0.80$
$2e^- + I_2 \rightarrow 2I^-$	$+0.54$
$2e^- + Cu^{2+} \rightarrow Cu(s)$	$+0.34$
$2e^- + 2H_3O^+ \rightarrow H_2(g) + 2H_2O$	zero
$2e^- + Fe^{2+} \rightarrow Fe(s)$	-0.44
$2e^- + Zn^{2+} \rightarrow Zn(s)$	-0.76
$3e^- + Al^{3+} \rightarrow Al(s)$	-1.66
$2e^- + Mg^{2+} \rightarrow Mg(s)$	-2.37
$e^- + Na^+ \rightarrow Na(s)$	-2.71
$e^- + Li^+ \rightarrow Li(s)$	-3.05

Increasing strength as oxidizing agents (left axis) — *Increasing strength as reducing agents* (right axis)

than H_3O^+. The magnitude of the potential is a quantitative measure of the relative tendency of the half-reaction to occur from left to right. (It should be noted that nothing is implied about how *fast* the reaction will go or whether the reaction will even be fast enough to be observed.)

Each oxidizing agent in Fig. 14.9 is coupled in its half-reaction with its reduced form. For example, Cu^{2+} is coupled with Cu. The reduced form is capable of acting as a reducing agent when the half-reaction is reversed. Thus, the reduction potentials in Fig. 14.9 also give information about the relative tendency of reducing agents to give off electrons. If a half-reaction, such as the one at the top of the table, has great tendency to go to the right, it is hard to reverse, and the reducing agent is a poor one. Of the reducing agents listed, F^- is the poorest, and Li(s) is the best. The half-reaction

$$Li(s) \longrightarrow Li^+ + e^- \qquad +3.05 \text{ V}$$

has greater tendency to occur than

$$2F^- \longrightarrow F_2(g) + 2e^- \qquad -2.87 \text{ V}$$

Figure 14.9 lists reducing agents (on the right) in order of increasing strength. Such a list of reducing agents arranged in decreasing order is sometimes called the *electromotive force*, or *emf, series*.

The potential of a half-reaction is a measure of the tendency of the half-reaction to occur. This potential is independent of the other half of the complete reaction. The potential of any complete reaction can be obtained by adding the potentials of its two half-reactions. The potential so obtained gives the tendency of the complete reaction to occur and gives the voltage measured for a galvanic cell which uses the reaction. For example, in the Zn–Cu cell one has the following:

Cathode

$$2e^- + Cu^{2+} \longrightarrow Cu(s) \qquad +0.34 \text{ V}$$

Anode

$$Zn(s) \longrightarrow Zn^{2+} + 2e^- \qquad +0.76 \text{ V}$$

Complete cell

$$Zn(s) + Cu^{2+} \longrightarrow Zn^{2+} + Cu(s) \qquad +1.10 \text{ V}$$

The voltage $+1.10$ V so calculated is that observed for the cell. It is positive, which indicates that the reaction tends to go spontaneously as written. It should be noted that the value 1.10 V applies when the concentrations of the ions are 1 m, since redox potentials are defined for concentrations of 1 m.

Any oxidation-reduction reaction for which the overall potential is positive has the tendency to take place as written. Whether a given reaction should take place spontaneously can be determined from the relative positions of its two half-reactions in a table of reduction potentials; e.g., in Fig. 14.9 any oxidizing agent reacts with any reducing agent below it. I_2 oxidizes Cu, H_2, Fe, etc., but does not oxidize Br^-, H_2O, Cl^-, etc. Similarly, any reducing agent reacts with any oxidizing agent above it. Zn reduces Fe^{2+}, H_3O^+, Cu^{2+}, etc., but does not reduce Al^{3+}, Mg^{2+}, Na^+, etc.

Example 3

I_2 *and* Br_2 *are added to a solution containing* I^- *and* Br^-. *What reaction will occur if the concentration of each species is 1 m?*

The half-reactions to be considered are

$$2e^- + I_2 \longrightarrow 2I^- \qquad +0.54 \text{ V}$$

$$2e^- + Br_2 \longrightarrow 2Br^- \qquad +1.09 \text{ V}$$

Method 1

From the positions in Fig. 14.9, Br_2 can oxidize I^-, whereas I_2 cannot oxidize Br^-. Therefore, the reaction is predicted to be

$$2I^- + Br_2 \longrightarrow I_2 + 2Br^-$$

Method 2

We want the overall voltage to be positive; so we pick the bigger voltage to be positive as written and have the other half-reaction inverted:

$2e^- + Br_2 \longrightarrow 2Br^-$	$+1.09$ V
$2I^- \longrightarrow I_2 + 2e^-$	-0.54 V
$2I^- + Br_2 \longrightarrow I_2 + 2Br^-$	$+0.55$ V

Reaction should occur as written since its voltage is positive.

$2Br^- \longrightarrow Br_2 + 2e^-$	-1.09 V
$2e^- + I_2 \longrightarrow 2I^-$	$+0.54$ V
$2Br^- + I_2 \longrightarrow Br_2 + 2I^-$	-0.55 V

Because this overall voltage is negative, the reaction should not occur spontaneously as written.

■ ■ ■

Further consideration of oxidation-reduction potentials and how to combine them is given in Sec. 16.2.

14.7 Nernst equation

The inherent tendency for a chemical reaction or half-reaction to occur depends not only on the chemical nature of the reactants but also upon their concentrations. The reduction potentials just discussed are *standard potentials* and as such describe the chemical nature of reactants at fixed unit concentration (more precisely, unit activity). What happens to reaction tendencies when concentrations are changed? Qualitatively, the principle of Le Châtelier predicts that increasing the concentration of a reactant favors its tendency to react and decreasing the concentration of a reactant diminishes its tendency to react. Similarly, decreasing the concentration of a product favors the tendency toward formation of that product.

Quantitatively, the change of reaction tendency with concentration is given by the *Nernst equation* as worked out by the German thermodynamicist Walther Nernst. The equation relates E, the potential for a reaction or half-reaction at nonstandard conditions, to $E°$, the standard potential for that reaction or half-reaction at unit activities. The Nernst equation for 25°C is

$$E = E° - \frac{0.0591}{n} \log Q$$

where n is the number of electrons transferred in the reaction and Q is the mass-action expression for the reaction. In writing Q, concentrations of gases are given in atmospheres of pressure, and for half-reactions, electrons are omitted from the mass-action expression. As a specific example, we consider the half-reaction

$$2e^- + 2H_3O^+ \longrightarrow H_2(g) + 2H_2O$$

for which the Nernst equation is

$$E = E^\circ - \frac{0.0591}{2} \log \frac{P_{H_2}}{[H_3O^+]^2}$$

For this half-reaction the standard potential E° is zero, and therefore

$$E = 0.00 - 0.0296 \log \frac{P_{H_2}}{[H_3O^+]^2}$$

Under standard conditions (hydrogen pressure equal to one atmosphere and hydronium-ion activity at unity), the mass-action term $P_{H_2}/[H_3O^+]^2$ is unity, the logarithm of it is zero, and $E = 0.00$ V. As a further illustration, in pure water the hydronium-ion concentration is 1.0×10^{-7} M. In such a dilute solution the molarity equals the molality, and for dilute solutions this is a good measure of activity. Substituting $P_{H_2} = 1$ and $[H_3O^+] = 1.0 \times 10^{-7}$, we get for the potential of the hydrogen electrode in pure water

$$E = 0.00 - 0.0296 \log \frac{1.0}{(1.0 \times 10^{-7})^2} = -0.41 \text{ V}$$

As a more complex example, we consider the following complete reaction:

$$2I^- + H_3AsO_4 + 2H_3O^+ \longrightarrow I_2(s) + H_3AsO_3 + 3H_2O$$

For this reaction the Nernst equation is

$$E = E^\circ - \frac{0.0591}{n} \log \frac{[I_2(s)][H_3AsO_3][H_2O]^3}{[I^-]^2[H_3AsO_4][H_3O^+]^2}$$

E° has the value $+0.02$ V [as can be calculated by taking the difference between $E^\circ = +0.54$ V for $2e^- + I_2(s) \longrightarrow 2I^-$ and $E^\circ = +0.56$ V for $2e^- + 2H_3O^+ + H_3AsO_4 \longrightarrow H_3AsO_3 + 3H_2O$ from Appendix 7]. For the overall reaction, two electrons are transferred from $2I^-$ to H_3AsO_4; so $n = 2$. In general, the activity of pure substances is unity, so both $I_2(s)$ and H_2O can be omitted from the expression, which now becomes

$$E = 0.02 - 0.0296 \log \frac{[H_3AsO_3]}{[I^-]^2[H_3AsO_4][H_3O^+]^2}$$

Under standard conditions of concentration, the mass-action fraction is unity. Its logarithm is zero, and the potential is slightly positive, thus being favorable for the reaction in the direction written. However, if the pH of the solution is raised, as by addition of base, the lowered value of $[H_3O^+]$ can make the E change sign. For example, in neutral solution with $[H_3O^+] = 1.0 \times 10^{-7}$ and assuming all other species at unit activity, we get

$$E = 0.02 - 0.0296 \log \frac{1}{(1.0 \times 10^{-7})^2} = -0.39 \text{ V}$$

The negative value for E indicates that the reaction as written should not occur but instead the reverse reaction should take place.

14.8   Free energy and cell voltage

For oxidation-reduction reactions, the net free-energy change is directly related to the voltage that would be obtained if the reaction were set up as a galvanic cell. The relationship between the overall free-energy change ΔG and the cell voltage E is

$$\Delta G = -n\mathfrak{F}E$$

where n is the number of faradays of electricity transferred between the reducing and the oxidizing agent and $\mathfrak{F}$ is the value of the faraday, which for conversion to kilojoules per mole is $96.49 \text{ kJ V}^{-1} \text{ equiv}^{-1}$. As noted in Sec. 14.7, the Nernst equation indicates that the voltage of a cell depends on the concentrations or, better, activities, of the species involved. Hence, we can substitute the general form of the Nernst equation

$$E = E° - \frac{RT}{n\mathfrak{F}} \ln Q$$

into $\Delta G = -n\mathfrak{F}E$ to get

$$\Delta G = -n\mathfrak{F}\left(E° - \frac{RT}{n\mathfrak{F}} \ln Q\right)$$
$$= -n\mathfrak{F}E° + RT \ln Q$$

The term $-n\mathfrak{F}E°$ is just equal to $\Delta G°$, the standard free-energy change for the reaction when all species are in their standard states, i.e., at unit activity. Substituting $\Delta G°$ for $-n\mathfrak{F}E°$, we get

$$\Delta G = \Delta G° + RT \ln Q$$

At *equilibrium* (that is, when the species are *not* in their standard states), ΔG equals zero. Therefore, we have at equilibrium

$$0 = \Delta G° + RT \ln Q$$

The activity quotient Q at equilibrium is equal to K, the equilibrium constant:

$$0 = \Delta G° + RT \ln K$$

Rearranging and then setting $\Delta G° = -n\mathfrak{F}E°$, we get

$$\Delta G° = -RT \ln K = -n\mathfrak{F}E°$$

Solving for $\log K$ gives

$$\log K = \frac{n\mathcal{F}E^\circ}{2.303RT}$$

At 25°C, this relation becomes

$$\log K = 16.9nE^\circ$$

For the reaction

$$Zn(s) + Cu^{2+} \rightleftharpoons Zn^{2+} + Cu(s)$$

the standard voltage is 1.10 V. The number of faradays transferred is 2. Therefore,

$$\log K = (16.9)(2)(1.10)$$

$$K = 1.5 \times 10^{37}$$

The mass-action expression for the reaction is

$$\frac{[Zn^{2+}]}{[Cu^{2+}]} = K = 1.5 \times 10^{37}$$

This means that at equilibrium, the activity of Zn^{2+} is 1.5×10^{37} times that of Cu^{2+}. It is no wonder, then, that the addition of Zn metal to a solution of copper ion results in essentially complete reduction of Cu^{2+} to Cu.

For the reaction

$$Cr_2O_7{}^{2-} + 6Fe^{2+} + 14H_3O^+ \rightleftharpoons 2Cr^{3+} + 6Fe^{3+} + 21H_2O$$

the voltage is 0.56 V, and n is six. The equilibrium constant calculated as above has a value of 6.0×10^{56}.

Example 4
We are given the reaction

$$H_2(g) + 2AgCl(s) + 2H_2O \longrightarrow 2Ag(s) + 2H_3O^+(aq) + 2Cl^-(aq)$$

At 25°C the standard free energy of formation of AgCl(s) is −109.7 kJ/mol, that of $H_2O(l)$ is −237.2 kJ/mol, and that of $(H_3O^+ + Cl^-)(aq)$ is −368.4 kJ/mol. Calculate what will be the cell voltage if this reaction is run at 25°C and one atmosphere in a cell in which the $H_2(g)$ activity is unity and the $H_3O^+(aq)$ and $Cl^-(aq)$ activities are each at 0.0100.

$$\Delta G^\circ = 2\,\Delta G^\circ_{H_3O^+,Cl^-} - 2\Delta G^\circ_{AgCl} - 2\Delta G^\circ_{H_2O}$$
$$= 2(-368.4) - 2(-109.7) - 2(-237.2) \text{ kJ}$$
$$= -43.0 \text{ kJ per mole of reaction}$$

$$E^\circ = \frac{\Delta G^\circ}{-n\mathcal{F}} = \frac{-43.0 \text{ kJ}}{2(96.49 \text{ kJ/V})} = 0.223 \text{ V}$$

$$E = E^\circ - \frac{RT}{n\mathcal{F}} \ln \frac{(a^2_{H_3O^+})(a^2_{Cl^-})}{a_{H_2}}$$

Note that the activities of *solid* Ag, of *solid* AgCl, and of H_2O are unity and, hence, can be omitted.

$$E = 0.223 - \frac{(8.31)(298)(2.303)}{2(96490)} \log \frac{(0.0100)^2(0.0100)^2}{1.00}$$

$$= 0.223 + 0.236 = 0.459 \text{ V}$$

■ ■ ■

Exercises

*14.1 *Electric conduction.* What are the essential differences between metallic and electrolytic conduction?

*14.2 *Electrolytic conduction.* How can you prove experimentally that both positive and negative ions move in electrolytic conduction? What kind of chemical system would you have to invent so that only one kind of ion would move?

*14.3 *Electrolysis.* Assuming that molten KOH is dissociated into K^+ and OH^-, sketch a cell for its electrolysis. Indicate anode, cathode, and direction of motion of all charged species in the external and internal circuits. Write the electrode reactions.

*14.4 *Electrolysis.* Explain with equations how you could use electrolysis to convert an aqueous solution of KCl into aqueous KOH.

*14.5 *Electrolysis.* For how many hours would you have to pass a current of 20.0 amp through molten NaCl to produce 1000 kg of chlorine?
Ans. 3.78 × 10⁴ h

*14.6 *Galvanic cell.* Sketch a galvanic cell that makes use of the reaction $Cu(s) + 2Ag^+ \longrightarrow Cu^{2+} + 2Ag(s)$. Label the anode and cathode; give the reaction of each. Indicate the direction of motion of all charged species in both the external and internal circuits. What voltage would be generated by such a cell if all species were at unit activity?

*14.7 *Galvanic cell.* How long could you draw 0.10 amp from a galvanic cell that starts out with 10.0 g of Zn and 0.100 liter of 1.0 M $CuSO_4$?
Ans. 193,000 sec

*14.8 *Fuel cells.* What volume of CH_4 gas at STP would you need to supply to a CH_4–O_2 fuel cell in order to get 10.0 amp for 24 h? Assume 100 percent efficiency.

*14.9 *Electrode potentials.* We are given three sticks of metal, zinc, copper, and silver, respectively. Describe how measurements of electrode potential differences would enable you to make conclusions about the relative oxidizing strength of Zn^{2+}, Ag^+, and Cu^{2+}.

*14.10 *Redox reactions.* Working from Fig. 14.9, predict what reaction would occur if Ag(s) and Cu(s) were added to a solution containing $I_2(s)$ and I^-.

** 14.11   Electrolyte conduction. How would you expect the mobility of an electrolytic ion to depend on each of the following: (*a*) increasing size of the ion, (*b*) increasing charge on the ion, (*c*) increasing temperature, and (*d*) going to a solvent of lower viscosity?

** 14.12   Electrolysis. In the electrolysis of aqueous $NaNO_3$ solution with copper electrodes, hydrogen gas is observed at the cathode and dissolution of the electrode at the anode. Diagram a suitable cell, labeling anode, cathode, and charge motion. Give equations for the most probable electrode reactions.

** 14.13   Electrolysis. Each of the following aqueous mixtures is electrolyzed with inert electrodes. What are the most probable electrode reactions in each case?
a   NaCl and NaI
b   NaCl and HCl
c   HCl and HI
d   NaOH and HNO_3

** 14.14   Electrolysis. An aqueous solution of K_2SO_4 containing litmus is electrolyzed with inert electrodes. One electrode compartment turns red; the other turns blue. Explain.

** 14.15   Electrolysis. Suppose you want to plate some copper 0.10 mm thick over a surface area of 50 cm². How long would you need to pass a current of 50 milliamp through aqueous $CuSO_4$ to accomplish this? Density of copper is 8.92 g/cm³.   *Ans. 75.3 h*

** 14.16   Lead storage battery. For a lead storage battery to be rated 200 amp-h what is the minimum amount of Pb, PbO_2, and H_2SO_4 it must contain?

** 14.17   Battery. Suggest a reason why a lead storage battery can be easily recharged whereas a Leclanché cell can be recharged only with great difficulty.

** 14.18   Edison cell. Unlike the lead storage battery, the Edison cell shows practically no drop in voltage as the cell discharges. By using half-reactions, show how this may be explained.

** 14.19   Fuel cell. A typical lead storage battery would be rated 150 amp-h. Calculate the comparative weight advantage for a corresponding H_2–O_2 fuel cell so far as consumable reactants are concerned. Assume 100 percent efficiency in each case.
Ans. 1800 g versus 50.4 g, or 35.7 : 1

** 14.20   Electrode potential. With specific reference to the Cu–Cu^{2+} case indicate the difference between oxidation potential and reduction potential.

** 14.21   Electrode potential. Under standard conditions what voltage would you expect to generate in a cell that uses the reaction $Fe + I_2(s) \longrightarrow Fe^{2+} + 2I^-$? What would the voltage become if the concentration of each ion were multiplied tenfold?

** **14.22**   *Standard reduction potentials.* If the change $PdCl_4^{2-}$ to Pd can oxidize I^- to I_2 but cannot oxidize $Ag(s)$ to Ag^+, what is the probable value of its reduction potential?

Ans. $0.54 < E° < 0.80$

** **14.23**   *Reduction potentials.* You are given that $NH_3OH^+ \longrightarrow H_2N_2O_2$ can reduce H_3AsO_4 to $HAsO_2$ but cannot reduce $Fe(CN)_6^{3-}$ to $Fe(CN)_6^{4-}$ or Sb_2O_5 to Sb_2O_4 and that $Sb_2O_4 \longrightarrow Sb_2O_5$ can reduce $H_2N_2O_2$ to NH_3OH^+ but not $Fe(CN)_6^{3-}$ to $Fe(CN)_6^{4-}$. Arrange the oxidizing agents $H_2N_2O_2$, $Fe(CN)_6^{3-}$, H_3AsO_4, and Sb_2O_5 in order of decreasing reduction potential. Write balanced half-reactions for each in acidic solution.

** **14.24**   *Nernst equation.* Show how the Nernst equation corresponds to Le Châtelier's principle for oxidation-reduction reactions.

** **14.25**   *Free energy.* Using the data of Fig. 14.9, calculate the standard free-energy change for $I_2(s) + Cu(s) \longrightarrow Cu^{2+} + 2I^-$.

** **14.26**   *Lead storage battery.* Given that $Pb^{2+} + 2e^- \longrightarrow Pb(s)$ has $E° = -0.13$ V, that K_{sp} of $PbSO_4(s)$ is 1.3×10^{-8}, and that K_{II} of H_2SO_4 is 1.3×10^{-2}, calculate the $E°$ for $PbSO_4(s) + 2e^- + H_3O^+ \longrightarrow Pb(s) + HSO_4^- + H_2O$. If $E° = 1.69$ V for $PbO_2 + HSO_4^- + 3H_3O^+ + 2e^- \longrightarrow PbSO_4(s) + 5H_2O$, what would be the voltage of a battery composed of six lead storage cells in series under standard conditions?   *Ans. -0.31, 12.0 V*

*** **14.27**   *Electrolysis.* We are given an aqueous solution that is simultaneously 0.100 M NaCl and 0.100 M NaOH. If the total volume is 250 ml, how long would you need to pass 0.100 amp through to double the OH^-/Cl^- ratio?

*** **14.28**   *Standard hydrogen electrode.* Suppose you have a galvanic cell consisting of two hydrogen electrodes hooked together. In one compartment, the hydrogen pressure is 10 atm, and the H_3O^+ concentration is 0.10 M; in the other compartment, 100 atm and 0.010 M, respectively. Sketch the cell, and indicate the motional direction of anions, cations, and electrons in the circuit. What voltage would you expect to measure between the electrodes? Which electrode would be the more positive (external circuit)?   *Ans. 0.089 V*

*** **14.29**   *Standard potentials.* The standard reduction potentials for $Fe(CN)_6^{3-} \longrightarrow Fe(CN)_6^{4-}$ and $VO^{2+} \longrightarrow V^{3+}$ are both equal to $+0.36$ V. (*a*) What will be the concentration ratio of $Fe(CN)_6^{3-}$ to $Fe(CN)_6^{4-}$ in an equilibrium mixture of all four of the above species at standard conditions? (*b*) What pH change should be made to lower the ratio 1000-fold?

*** **14.30**   *Equilibrium constant.* Given that $E° = -0.76$ V for $Zn^{2+} + 2e^- \longrightarrow Zn(s)$ and $E° = -1.12$ V for $ZnX_4^{2-} + 2e^- \longrightarrow Zn(s) + 4X^-$, calculate the equilibrium constant K for the dissociation $ZnX_4^{2-} \rightleftharpoons Zn^{2+} + 4X^-$.   *Ans. $K = 6.3 \times 10^{-13}$*

*** *14.31 Solubility product.* Under standard conditions it is observed that $E° = +0.705$ V for $M^{3+} + 3e^- \longrightarrow M(s)$ and $E° = +0.104$ V for $M(OH)_3 + 3e^- \longrightarrow M(s) + 3OH^-$. Calculate the K_{sp} of $M(OH)_3(s)$.

*** *14.32 Free energy.* Calculate the E and the ΔG for each of the following half-reactions:

a $2H_3O^+(0.10\ M) + 2e^- \longrightarrow H_2(0.10\ atm) + 2H_2O$
b $H_3O^+(0.10\ M) + e^- \longrightarrow \frac{1}{2}H_2(0.10\ atm) + H_2O$

Although it would be satisfying to be able to deduce all useful chemical information from first principles, in most cases this cannot be done. For one thing, theories sufficiently powerful to predict all observable behavior are not at hand. For another, it is not likely ever to be true that chemical facts will be obtained more easily from theory than from experiment. Therefore, it is essential that we provide ourselves with a solid base of information about the actual observed behavior of the chemical elements and compounds. The less time that is spent in looking up facts that are already known, the faster we can get on with probing the unknown.

Properties of the elements and their compounds

Part II

A systematic survey of the descriptive chemistry of the elements is best done with the aid of a periodic table, where group and period relations can be exploited for ordering information storage and retrieval. However, there is good reason to single out two elements, hydrogen and oxygen, for special preliminary consideration. The first of these, hydrogen, is unique as the only reactive element of period 1. Because period 1 has only two elements, there is no easy correlation with later periods. In particular, hydrogen cannot be exclusively or even preferentially linked to any of the groups of the periodic table; so it is best considered on its own. The other element of this chapter, oxygen, forms characteristic compounds with practically all the other elements; so its general chemistry is best discussed before that of its derivatives.

Hydrogen, $Z = 1$, has but one proton in its nucleus and one orbital electron. In the ground state, the electron is describable by a $1s$ wave function, but by the Pauli exclusion principle there can be another electron with the same wave function provided that its spin is different. This means that H atoms can bind to each other as H_2 molecules or to other atoms *via* shared electron pairs. Hence, the chemistry of hydrogen is that of an essentially simple molecule capable of reacting with many elements to form a great variety of compounds.

Oxygen, $Z = 8$, has two $1s$, two $2s$, and four $2p$ electrons. Except for fluorine, it is more electronegative than any other element and forms compounds with all elements except the lighter noble gases. Historically, the study of oxygen compounds was important in establishing the principles of chemistry and unraveling the properties of the other elements. One of these compounds, water, is the most important reaction setting in chemistry. It is of vast technological importance, it constitutes a large fraction of man's viable environment, and it is *the* solvent in which man's metabolic adventures occur.

Element	Weight %	Atom %
Oxygen	49.4	55.1
Silicon	25.8	16.3
Aluminum	7.5	5.0
Iron	4.7	1.5
Calcium	3.4	1.5
Sodium	2.6	2.0
Potassium	2.4	1.1
Magnesium	1.9	1.4
Hydrogen	0.9	15.4
Titanium	0.6	0.2

Fig. 15.1

Abundances of the most common elements in the earth's crust (atmosphere, hydrosphere, lithosphere).

15.1 Occurrence of hydrogen

In the universe, hydrogen is apparently the most abundant of all the elements. Spectral analysis of the light emitted by stars indicates that most of them are predominantly hydrogen. For example, the nearest star, the sun, has a mass composed approximately 90 percent of hydrogen. On the earth, hydrogen is much less abundant. the earth's gravitational attraction, being much less than that of stars and larger planets, is too small to hold very light molecules; so there is practically no hydrogen left in the atmosphere. There may have been some at the creation, 4.6×10^9 years ago, but even if there were, most of it would have long since escaped. If we consider only the earth's crust* (atmosphere, hydrosphere, and lithosphere), hydrogen is third in abundance on an atom basis. Of each 1000 atoms of crust, 551 are oxygen, 163 are silicon, and 154 are hydrogen. On a mass basis, hydrogen is ninth in order and contributes only 0.88 percent of the earth's crust. Figure 15.1 shows some abundance data. The specific values do not always agree from one compilation to another.

On the earth, free or uncombined hydrogen is rare. It is found occasionally in volcanic gases. Also, as shown by study of the aurora borealis, it is found in traces in the upper atmosphere. On the other hand, combined hydrogen is quite common. In water, hydrogen is bound to oxygen and makes up 11.2 percent of the total weight. The human body, two-thirds of which is water, is approximately 10% hydrogen by weight. In fossil fuels, such as coal and petroleum, hydrogen is combined with carbon in a great variety of hydrocarbons. Clay and a few other minerals contain appreciable amounts of hydrogen, usually combined with oxygen. Finally, the proteins and carbohydrates of all plant and animal matter are composed of compounds of hydrogen with oxygen, carbon, nitrogen, sulfur, etc.

15.2 Preparation of hydrogen

In producing an element for commercial use the primary consideration is usually cost. For laboratory use the important considerations are purity and convenience. For *commercial* hydrogen, chief sources are water and hydrocarbons. Hydrogen can be made inexpensively by passing steam over hot carbon:

$$C(s) + H_2O(g) \xrightarrow[1000°C]{} CO(g) + H_2(g)$$

However, the hydrogen from this source is not pure because carbon monoxide, CO, is difficult to separate. The mixture of H_2 and CO is

---

* The total mass of the earth is 5.98×10^{24} kg. Its average density is 5.52 g/cm^3. The atmosphere, which is about 1000 km thick, has a mass of 5.2×10^{18} kg, 95 percent of which is concentrated in the lower 20 km; the density at sea level is 0.00123 g/cm^3. The hydrosphere (oceans, lakes, and rivers) has a mass of 1.4×10^{23} kg, 98 percent of which is in the ocean. The lithosphere, also known as the *sial crust*, has a thickness of about 30 km in the continental areas and about 5 km in oceanic areas. Its density is 2.64 g/cm^3 at the surface and 2.87 g/cm^3 at the base.

an important industrial fuel, *water gas*. It has a very high heat of combustion.

Purer but still relatively inexpensive hydrogen can be made by passing steam over hot iron:

$$3Fe(s) + 4H_2O(g) \longrightarrow Fe_3O_4(s) + 4H_2(g)$$

The iron can be recovered by reducing the Fe_3O_4 with water gas.

The purest (99.9 percent) but most expensive hydrogen available commercially is *electrolytic hydrogen*, made from the electrolysis of water:

$$2H_2O \xrightarrow[electrolysis]{} 2H_2(g) + O_2(g) \qquad \Delta H° = +565 \, kJ$$

The reaction is endothermic and requires energy, which must be supplied by the electric current. It is the power consumption, not the raw material, that makes electrolytic hydrogen expensive. In practice, solutions of NaOH or KOH are electrolyzed in cells with iron cathodes and nickel anodes; the cells are designed to keep the anode and cathode products separate. The electrode reactions are

Anode

$$4OH^- \longrightarrow O_2(g) + 2H_2O + 4e^-$$

Cathode

$$2e^- + 2H_2O \longrightarrow H_2(g) + 2OH^-$$

Net

$$2H_2O \longrightarrow 2H_2(g) + O_2(g)$$

Considerable hydrogen is also formed as a by-product in the *chlor-alkali* industry, where Cl_2 and NaOH are produced by the electrolysis of aqueous NaCl (Sec. 14.2). Increasing use of mercury cathodes here has aggravated a serious environmental pollution problem (see Sec. 20.9).

In petroleum refineries, where gasoline is made by the catalytic cracking of hydrocarbons, hydrogen is a valuable by-product. When gaseous hydrocarbons are passed over hot catalyst, decomposition occurs to form hydrogen and other hydrocarbons. The lighter hydrocarbons, such as methane, can be economically partially oxidized by passing them with steam over a nickel catalyst to produce hydrogen:

$$CH_4(g) + H_2O(g) \xrightarrow[900°C]{Ni} CO(g) + 3H_2(g)$$

In the laboratory, pure hydrogen is usually made by the reduction of hydronium ion with zinc metal:

$$2H_3O^+ + Zn(s) \longrightarrow H_2(g) + 2H_2O + Zn^{2+}$$

In principle, such reduction should occur with any metal having an electrode potential less than zero (Sec. 14.6). For some metals, such as iron, the reaction is quite slow, even though the potential is favorable. In water, where the concentration of H_3O^+ is only $1.0 \times 10^{-7} \, M$, the reduction by metals is more difficult. The voltage for the half-reaction

$$2H_3O^+(1.0 \times 10^{-7} \, M) + 2e^- \longrightarrow H_2(g) + 2H_2O$$

is -0.41 V. In order to liberate H_2 from water, a metal must have an electrode potential more negative than -0.41 V. Thus, the element sodium reacts with water to liberate H_2 by the reaction

$$2H_2O + 2Na(s) \longrightarrow H_2(g) + 2Na^+ + 2OH^-$$

In principle, zinc too should liberate H_2 from H_2O, but the reaction is too slow at room temperature to be useful.

Laboratory hydrogen can also be made conveniently from the reaction of aluminum metal with base $2Al(s) + 2OH^- + 6H_2O \longrightarrow 2Al(OH)_4^- + 3H_2(g)$ or from the reaction of CaH_2 with water $CaH_2(s) + 2H_2O \longrightarrow Ca^{2+} + 2OH^- + 2H_2(g)$.

15.3 Properties and uses of hydrogen

Hydrogen at room temperature is a colorless, odorless, tasteless gas. Its properties are summarized in Fig. 15.2. The gas is diatomic and consists of nonpolar molecules containing two hydrogen atoms held together by a covalent bond. In order to rupture the bond, 431.8 kJ of heat must be supplied per mole. Because the dissociation is endothermic, it increases with temperature. At $4000°K$ and 1 atm total pressure, H_2 is about 60% dissociated. When H_2 reacts, one of the steps is usually the breaking of the H—H bond. Because of the high energy required for this step, the activation energy is high, and H_2 reactions are generally slow. Most hydrogen compounds contain H covalently bound since neither H^+ nor H^- is readily formed. The ionization potential (Sec. 2.7) is 13.60 eV, which is about $2\frac{1}{2}$ times the ionization potential of sodium. The electron affinity (Sec. 2.8) is 0.72 eV, which is about one-fifth that of chlorine.

Molecular hydrogen is the lightest of all gases. It is one-fourteenth as heavy as air. A balloon filled with hydrogen rises in accord with Archimedes' principle that the buoyant force on an object immersed in a fluid (such as air) is equal to the weight of fluid displaced by the object. At one time, hydrogen was used extensively to lift dirigibles, but because of major disasters arising from its combustibility it is no longer much used for this purpose. However, meteorologists still frequently send aloft weather balloons inflated with hydrogen.

The very low melting and boiling points of hydrogen indicate that the intermolecular attractions are quite small (Sec. 5.12). Because of the low boiling point, liquid hydrogen is used as a cryogenic fluid (to produce low temperatures). When the pressure above liquid hydrogen in an insulated container is reduced below 0.071 atm (the triple-point pressure), the temperature drops, and hydrogen solidifies. The critical temperature of hydrogen (above which it can exist only as a gas) is $33.2°K$.

Fig. 15.2
Properties of H_2.

Molecular weight	2.016 amu
Bond length	0.0749 nm
Bond dissociation energy $(0°K)$	431.8 kJ/mol
Approximate molecular diameter (from van der Waals equation)	0.234 nm
Normal melting point	14.1°K
Normal boiling point	20.4°K
Critical temperature	33.2°K
Critical pressure	13 atm
Density of gas (STP)	0.0899 g/liter
Density of liquid $(20°K)$	0.07 g/ml

Chemically, H_2 is able under appropriate conditions to combine directly with most elements. With oxygen, H_2 reacts to release large amounts of energy. The change

$$2H_2(g) + O_2(g) \longrightarrow 2H_2O(g) \qquad \Delta H° = -485 \text{ kJ}$$

occurs at an appreciable rate only at high temperatures or in the presence of catalyst. In the oxyhydrogen torch the above reaction occurs to produce temperatures of about 2800°C, and the reaction is self-sustaining. Mixtures of H_2 and O_2 are explosive, and especially violently so when the ratio of H_2 to O_2 is approximately 2:1. With F_2, the reaction

$$H_2(g) + F_2(g) \longrightarrow 2HF(g) \qquad \Delta H° = -537 \text{ kJ}$$

is explosive even at liquid-hydrogen temperatures.

With metals the reaction of H_2 is not nearly so violent and often requires elevated temperatures. For example, sodium hydride, NaH, is formed by bubbling H_2 through molten sodium at about 360°C. Hydrides of group II elements are just as difficult to form.

Hydrogen also reacts with certain compounds. In some cases it simply adds on to the other molecule as, for instance, in forming methyl alcohol, CH_3OH, from CO:

$$CO(g) + 2H_2(g) \xrightarrow[catalyst]{} CH_3OH(g)$$

Such addition reactions, called *hydrogenation* reactions, account for much of the industrial consumption of hydrogen. In other cases hydrogen removes atoms from other molecules, as in the reduction of tungsten trioxide, WO_3, to W.

Because of the high heat of combustion per gram of hydrogen (120 kJ/g), liquid hydrogen is a valuable rocket fuel. Its value is enhanced for propulsion purposes because the light molecular weight of the combustion products means high velocities and therefore high momentum. In terms of a long-range solution to the energy-supply crisis looming up for mankind, hydrogen offers considerable attraction. Raw material for its production (that is, H_2O) is almost inexhaustible; nuclear-reactor by-product power is likely to be available cheaply for

Section 15.3
Properties and uses of
hydrogen

419

electrolytic decomposition of water. A major unresolved problem, however, is how to transport and distribute the hydrogen safely.

15.4 Compounds of hydrogen

In its compounds H is found in the three oxidation states: $+1$, -1, and 0. In the first two cases, H forms compounds by losing a share of its lone electron or gaining a share of another electron. According to the rules for assigning oxidation numbers (Sec. 4.6), the relative electronegativity of H and the atom to which it is joined must be considered. In the general compound H_nX, the oxidation number of H is $+1$ if X is more electronegative than H and -1 if X is less electronegative than H. The oxidation state 0 represents a rather special case.

Oxidation state $+1$ This is the most important state since it includes most of the H compounds. In these compounds, H is combined with a more electronegative element, such as any element taken from the right side of the periodic table. In period 2, for example, the elements more electronegative than H are C, N, O, and F. With these elements H forms compounds such as methane, CH_4; ammonia, NH_3; water, H_2O; and hydrogen fluoride, HF. It might be noted that, even though H is thought to be more positive in these compounds, there is no uniformity in writing H first in the formula as might be expected. It should be emphasized that in all these compounds the binding of H is covalent and that none of these compounds contains simple H^+ ion.

The compounds can be formed by direct union of the elements. The reactions are often slow, sometimes requiring a large amount of activation energy; so catalysts and high temperatures may be required. For example, the reaction between N_2 and H_2 to form NH_3, an important industrial process called the *Haber process*, is usually carried out under pressure at about $500°C$ in the presence of a suitable catalyst, such as Fe.

In compounds containing more than two elements, the H is usually considered to be in a positive oxidation state. In most such compounds (for example, $NaHSO_4$) the H is bonded to an atom more electronegative than itself, usually O.

Oxidation state -1 When hydrogen is combined with an atom less electronegative than itself, the compound is said to be a *hydride*. Hydrides may be predominantly ionic, as with elements of periodic groups I and II (for example, NaH and CaH_2), or covalent, as with the lighter elements of group III (for example, B_2H_6).

In the hydrides of elements of groups I and II, the H occurs as the negative hydride ion, H^-. The compounds at room temperature are ionic solids forming cubic or hexagonal crystals. When melted, they conduct electric current and on electrolysis form H_2 *at the anode* by the reaction

$$2H^- \longrightarrow H_2(g) + 2e^-$$

The hydride ion is unstable in H_2O and is oxidized to H_2. Thus, for example, calcium hydride, CaH_2, in H_2O reacts as follows:

$$CaH_2(s) + 2H_2O \longrightarrow Ca^{2+} + 2OH^- + 2H_2(g)$$

The covalent hydrides such as silane, SiH_4, and arsine, AsH_3, are generally volatile liquids or gases. They are nonconductors and apparently contain no H^- ion. They are relatively mild reducing agents.

The term "hydride" is also applied to compounds in which H is joined to a less electronegative atom in a complex ion. Thus, for example, in the compound lithium aluminum hydride, $LiAlH_4$, the cation is Li^+, and the anion is the complex AlH_4^-. These complex hydrides are generally solids, they react with H_2O to liberate H_2, and they are of great use as reducing agents.

Oxidation state 0 Hydrogen reacts with transition metals such as uranium, copper, and palladium to form hard, brittle substances that conduct electricity and have typical metallic luster. In some cases, as with uranium hydride, UH_3, the compound is stoichiometric; i.e., the number of H atoms per metal atom is fixed and is a whole number. In other cases, as with palladium hydride, PdH_n, the compound is nonstoichiometric; i.e., the number of H atoms per metal atom is variable and can even be less than one. The state of the hydrogen in these metallic hydrides is not yet clear. It may be that the hydrogen exists as H atoms; it may be that the hydrogen is dissociated into an interstitial proton and a delocalized electron.

The dissolution of hydrogen in metals is important because metals which dissolve hydrogen are catalysts for hydrogenation reactions. The catalyst is thought to act by dissolving the hydrogen as H atoms, which react more rapidly than H_2 molecules. The catalysis by finely divided nickel of the hydrogenation of oils to give fats is explainable in this way.

When hydrogen dissolves in a metal, the H atom may go into the lattice as an interstitial or substitutional defect (Sec. 7.5) and simply expand the lattice of the metal, or it may completely alter the type of lattice. In either case the change may be significant enough to make the metal lose some of its desirable properties. This phenomenon, called *hydrogen embrittlement*, occurs even with small amounts of dissolved hydrogen, amounts that may be unavoidable in the preparation of pure metals. Thus, the large-scale industrial exploitation of the very valuable properties of the metal titanium was possible only after preparation methods that avoided hydrogen entrapment were developed.

15.5 Hydrogen bond

In some compounds a hydrogen atom is apparently bonded simultaneously to two other atoms. For example, in the compound potassium hydrogen fluoride, KHF_2, the anion HF_2^- is believed to have the structure FHF^-, in which the hydrogen acts as a bridge between the two fluorine atoms. The bridge, called a *hydrogen bond*, is unusual

Fig. 15.3
*Boiling points of some
hydrogen compounds.*

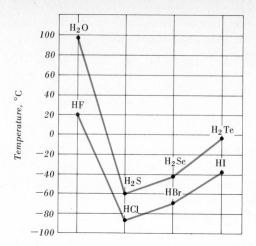

because hydrogen is limited to forming one covalent bond at a time. Hydrogen bonds seem to be formed only between small electronegative atoms such as fluorine, oxygen, and nitrogen.

Evidence in support of the existence of hydrogen bonds comes from comparing properties of hydrogen-containing substances. For example, in Fig. 15.3 are shown the normal boiling points for the hydrogen halides (lower curve) and for the hydrogen compounds of group VI elements (upper curve). It is evident that the boiling points of the members on the left, HF and H_2O, are abnormally high compared with other members of each series. In the series HF, HCl, HBr, and HI there is an increasing number of electrons per molecule; so rising boiling points would be expected because of increased van der Waals attractions (Sec. 5.12). The unexpectedly high boiling point of HF is attributed to hydrogen bonds between fluorine atoms. The hydrogen bonding makes it more difficult to detach HF from the liquid. Independent evidence for hydrogen bonding in HF comes from studies of the vapor phase, which is found to contain aggregates such as $(HF)_6$ presumed to be held together by hydrogen bonds. The unexpectedly high boiling point of H_2O in the series H_2O, H_2S, H_2Se, and H_2Te is similarly attributed to hydrogen bonding. In Sec. 15.12 the importance of hydrogen bonding in determining the structure of liquid water and ice is discussed.

What is responsible for the hydrogen bond? The simplest view is that a positively charged proton is attracted by the negative electrons of two different atoms. When a hydrogen atom is bound to a very electronegative atom, the hydrogen has such a small share of the electron pair that it is almost like a bare proton. As such, it can be attracted to another electronegative atom. Because of the tiny size of the proton, a given proton has room for only two atoms around it. This picture is consistent with the observations that hydrogen bonds are limited to very electronegative atoms and that one hydrogen atom can bridge between only two atoms.

Although relatively weak ($\approx$20 kJ) compared with most other chemical bonds ($\approx$200 kJ), hydrogen bonds are extremely important in biologic systems. Proteins, for example, which contain both $>CO$ and $>NH$ groups, owe much of their structural features and hence their properties to the existence of hydrogen bonds between $>CO$ groups in one part of the molecule and $>NH$ groups in another. (See, for example, Fig. 22.14.)

15.6 Isotopes of hydrogen

Natural hydrogen consists of three isotopes: protium ($_1^1H$); deuterium, or heavy hydrogen ($_1^2H$, or D); and tritium ($_1^3H$, or T). The protium nucleus consists of a lone proton; the deuterium nucleus, of a proton and a neutron; the tritium nucleus, of a proton and two neutrons. The protium nucleus is by far the most abundant of the three. In nature, there are 7000 times as many protium atoms (99.985 percent of the total) as deuterium atoms (0.015 percent) and only 7×10^{-14} times as many tritium atoms. The scarcity of tritium atoms in nature is due to the instability and consequent radioactivity of its nucleus.

In general, the properties of isotopes are *qualitatively* very similar. However, there may be *quantitative* differences, especially when the percentage difference in mass is appreciable. Figure 15.4 shows some of the properties of protium and deuterium.

Property	Protium	Deuterium
Mass of atom (H), amu	1.0078	2.0141
Freezing point (H_2), °K	14.0	18.7
Boiling point (H_2), °K	20.4	23.5
Freezing point (H_2O), °C	0	3.8
Boiling point (H_2O), °C	100	101.4
Density at 20°C (H_2O), g/ml	0.998	1.106

Fig. 15.4
Properties of hydrogen isotopes.

In chemical reaction, protium and deuterium show a quantitative difference both in their equilibrium and in their rate properties. Property differences arising from differences in atomic mass are called *isotope effects*. For example, the dissociation constant of ordinary water in the equilibrium

$$H_2O + H_2O \rightleftharpoons H_3O^+ + OH^-$$

is 1.0×10^{-14} at room temperature. For the corresponding dissociation of heavy water

$$D_2O + D_2O \rightleftharpoons D_3O^+ + OD^-$$

the constant is 0.2×10^{-14}, which is significantly smaller. The isotope effect on the rates of reactions is even more marked. Thus, a bond to a protium atom can be broken as much as 18 times faster than the bond to a deuterium atom. As an example, H_2 reacts with Cl_2 13.4 times as fast as D_2 does.

For elements heavier than hydrogen the isotope effect is much smaller. For example, $^{127}_{53}I$ reacts at most only 1.02 times faster than $^{129}_{53}I$, and the equilibrium properties are even more similar. The isotope effect becomes negligible for the heavier elements, where the percentage difference in mass between the isotopes is small.

The isotope effect in hydrogen is used as a basis for the separation of protium and deuterium. Since protium bonds are broken faster than deuterium bonds, electrolysis of water releases the light isotope faster than the heavy isotope. This means that there will be an enrichment of the heavy hydrogen in the residual water. By continuing the electrolysis until the residual volume is very small, practically pure deuterium oxide can be obtained. In a typical experiment, 2400 liters of ordinary water produces 83 ml of D_2O that is 99 percent pure.

15.7 Proton magnetic resonance

Just as the electron behaves like a tiny magnet (assignable to electron spin), so a proton has associated with it a small magnetic moment (assignable to nuclear spin). Because the proton is some 2000 times as massive as the electron, proton magnetism is only about one two-thousandth as great as electron magnetism. Small as it is, however, the proton magnetic moment is utilized in certain experiments to give important structural information about molecules that contain H atoms. In these experiments energy is absorbed in the hydrogen-containing sample, where it is used to change the alignment of nuclear magnetic moments from less favorable to more favorable orientations relative to an externally applied magnetic field. The energy transfer is somewhat similar to that involved in exciting sympathetic resonance as, for example, in a violin. The absorption of energy by nuclei in a magnetic field is called *nuclear magnetic resonance,** or NMR. Many nuclei other than hydrogen (for example, ^{11}B and ^{19}F) also exhibit NMR, but we shall limit our discussion to proton magnetic resonance.

Figure 15.5 is a schematic representation of one type of NMR experiment. The sample under investigation is placed in a coil between the pole pieces of an electromagnet. A radio-frequency signal is fed into the coil from an external oscillator. As the magnetic field produced by the electromagnet is gradually varied, the loss of radio-frequency power in the sample coil is measured and recorded. At a particular value of the magnetic field strength, nuclear magnetic resonance will occur and be detected as a sharp increase in power absorption by the sample.

The magnetic field strength at which absorption occurs depends on a number of factors: the identity of the nucleus being observed (e.g., whether it is ^1H or ^{19}F), the frequency of the radio-frequency signal (for example, 60 MHz, or 60×10^6 cycles/sec, as commonly

* This is a more appropriate use of the word "resonance" than that employed for certain chemical-bond descriptions (Sec. 3.7). Whereas valence-bond resonance reaches for fanciful analogies such as hybrid mermaids and hybrid jackasses, both of which are irreproducible, magnetic resonance is a clear-cut physical phenomenon capable of being reproduced.

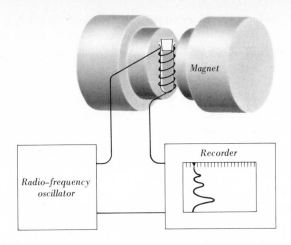

Fig. 15.5
NMR absorption.

used), and the chemical environment of the nucleus being studied. This last point, the so-called "chemical shift," is what makes NMR of such great interest to chemists, since it makes it possible to identify structurally distinct H atoms in molecules. Figure 15.6 gives a representation of the NMR absorption signal for protons in ethyl alcohol, CH_3CH_2OH. The three absorption peaks shown correspond to the three kinds of environment for the hydrogen nucleus in this molecule. The peak on the left is attributed to the hydrogen in OH, the middle peak to the hydrogens in CH_2, and the peak on the right to the hydrogens in CH_3. It turns out that the integrated areas under the three peaks are in the ratio $1:2:3$, corresponding to the relative numbers of the H atoms of the three structural types.

Why does resonance absorption occur at different magnetic field strengths for structurally different H atoms? The answer lies in the different relative shielding of the magnetic nucleus from the external magnetic field by the surrounding electrons. For example, in CH_3CH_2OH the H bound to the O is the least shielded because the very electronegative O tends to pull the bonding electrons to itself. The protonic nucleus of OH is thus more exposed to the effect of an external magnetic field than the other protons of the molecule are. The general effect of imposing an external magnetic field on an atom is to create or induce electron currents that tend to nullify the applied field. The applied field is thus partially screened in its effect on the nucleus. In the CH_3CH_2OH molecule the magnetic field "felt" by the proton is greatest at the OH site and smallest at the CH_3 site. As the external magnetic field is progressively made bigger, resonance is observed to occur first for the OH proton, then at the protons of CH_2, and finally at the most shielded ones, those of CH_3.

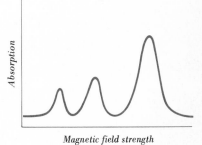

Magnetic field strength

Fig. 15.6
Plot of NMR absorption for protons in ethyl alcohol.

15.8 Occurrence of oxygen

As was noted in Fig. 15.1, oxygen is by far the most abundant element in the earth's crust both on the basis of weight and the number of atoms. Of the mass of the earth's crust 49.4 percent is due to oxygen atoms. Silicon, the next most abundant element, is only half as plentiful. On a number basis, oxygen atoms are more numerous than all other kinds of atoms combined.

In the free state, oxygen occurs in the atmosphere as O_2 molecules. Air is 20.9% oxygen by volume; i.e., for every 100 molecules in air approximately 21 are oxygen. On a mass basis, air is 23.0% oxygen; of every 100 g of air 23 g is oxygen.

In the combined state, oxygen occurs naturally in many rocks, plants and animals, and water. Of the oxygen-containing rocks, the most abundant are ones which contain silicon. The simplest of these is silica, SiO_2, the main constituent of sea sand. The most abundant rock that does not contain silicon is limestone, $CaCO_3$. In plant and animal material, oxygen is combined with carbon, sulfur, nitrogen, phosphorus, or hydrogen.

15.9 Preparation of oxygen

The industrial sources of oxygen are air and water. From air, oxygen is made by liquefaction and fractional distillation. Air, consisting of 20.946 mole percent of oxygen, 78.084 mole percent of nitrogen, 0.934 mole percent of argon, and traces of neon, carbon dioxide, and water, is first freed of carbon dioxide and water, compressed, cooled, and expanded until liquefaction results to give liquid air. On partial evaporation, the nitrogen, being lower boiling, boils away first, leaving the residue richer in oxygen. Repeated cycles of this kind give oxygen that is 99.5 percent pure.

Figure 15.7 shows the temperature-composition relations for oxygen-nitrogen mixtures. The lower curve represents the composition of the liquid; the upper curve, the gas that is in equilibrium with it. The horizontal tie lines (dashed) are constant-temperature lines joining the liquid (at the left end of a line) with the gas (at the right end of a line). As can be seen, the gas phase is at every temperature richer in nitrogen than the liquid it is in equilibrium with. The vertical dashed lines represent total evaporation lines along which a liquid of given composition is completely evaporated into the vapor phase. The staircase progression from right to left represents several consecutive equilibrium liquefactions followed by total evaporations. Starting with gas at 1 (80% N_2 and 20% O_2) we partially liquefy it to establish equilibrium liquid of composition 2 (60% N_2 and 40% O_2). We then take away this liquid and totally evaporate it to vapor (point 3). Vapor of composition 3 is then partially liquefied to establish equilibrium liquid of composition 4 (40% N_2 and 60% O_2). Again this liquid is taken off and evaporated completely to give gas at point 5. Partial liquefaction of gas 5 gives liquid of composition 6 (20% N_2 and 80% O_2); and so on. Eventually, we can end up with pure liquid oxygen. In the meantime the rejected portions are enriched in nitrogen and can be used as a source of nitrogen. To get the higher-boiling constituent (that is, O_2), we traverse the staircase from right to left; the process is called *fractional liquefaction*. To get the lower-boiling constituent (that is,

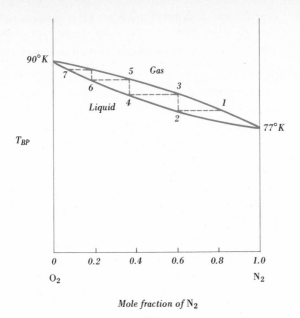

Fig. 15.7
Boiling point–composition
curve for various mixtures of
O_2 and N_2.

90°K

7

5   Gas

6

3

Liquid   4

1

2

77°K

T_{BP}

| 0 | 0.2 | 0.4 | 0.6 | 0.8 | 1.0 |

O_2

N_2

Mole fraction of N_2

N_2), we traverse the staircase left to right; the process is called *fractional distillation*. Obviously, both components can be separated from air to give liquid oxygen and liquid nitrogen.

From water, very pure oxygen can be made by electrolysis as a by-product of hydrogen manufacture. Power consumption makes electrolytic oxygen more expensive than oxygen obtained from air.

In the laboratory, oxygen is usually made by the thermal decomposition of potassium chlorate, $KClO_3$. The reaction

$$2KClO_3(s) \longrightarrow 2KCl(s) + 3O_2(g)$$

is catalyzed by the presence of various solids such as manganese dioxide, MnO_2, ferric oxide, Fe_2O_3, fine silica sand, and powdered glass. It is probable that the main function of the catalyst is to provide a surface on which the evolution of oxygen gas can occur.

15.10   Properties and uses of oxygen

At room temperature, oxygen is a colorless, odorless gas. The molecule is diatomic and, both as a liquid and gas, paramagnetic to the extent of two unpaired electrons per molecule. The bond energy of O_2, 494 kJ, lies between that of the triply bonded N_2 (941 kJ) and the singly bonded F_2 (151 kJ); it is thus in accord with a bond order of 2. The electron structure of the O_2 molecule can be rationalized, as was done in Sec. 3.9, by the molecular-orbital approach. The electronic configuration for the orbitals principally involved in the bonding can be written as $\sigma_{p_x}{}^2 \pi_{p_y}{}^2 \pi_{p_z}{}^2 \pi^*_{p_y} \pi^*_{p_z}$, where the superscript 2 indicates pairs of electrons in the three bonding orbitals (σ derived from the p_x atomic orbitals, π derived from the p_y orbitals, and π derived from the p_z orbitals). In addition, there is one electron in each of the two π^*

antibonding orbitals. The bond order, defined as half the excess of bonding over antibonding electrons, equals $\frac{1}{2}(6 - 2) = 2$. It is difficult to describe the O_2 molecule adequately from the valence-bond approach, since $:\overset{..}{O}:\overset{..}{O}:$ clearly violates the octet rule. However, as noted in the footnote on page 104, a paramagnetic, doubly bonded configuration can also be arrived at from Linnett's double-quartet approach.

When cooled to $90°K$, oxygen condenses to a pale blue liquid. At $54°K$, it solidifies to form a bluish white solid. As proposed to account for an observed reduction in the magnetism, there is probably some formation of O_4 in the liquid and solid states.

Oxygen exhibits *allotropy;* i.e., it can exist as the element in more than one form. When energy is added to diatomic oxygen, the triatomic molecule ozone, O_3, is formed by the reaction

$$3O_2(g) \longrightarrow 2O_3(g) \qquad \Delta H = +285 \text{ kJ}$$

At room temperature the equilibrium constant for this reaction is 10^{-54}. Even though K increases with temperature, the indicated equilibrium concentration of O_3 would not be appreciable at any temperature. Thus, not much O_2 can be converted to O_3 by the simple addition of heat. However, when energy is added in other forms, such as electric energy or high-energy radiation, significant amounts of O_3 can result. Once the O_3 is obtained, it only slowly reverts to O_2. In the laboratory, ozone can be made easily by passing air or oxygen through a silent electric discharge such as exists between tinfoil conductors that are connected to the terminals of an electric induction coil. About 5 percent of the oxygen is converted to ozone. Ozone is also formed in appreciable amounts by lightning bolts, by ultraviolet light, and by sparking electric motors.

Trace amounts of ozone are formed in the stratosphere, probably by absorption of ultraviolet sunlight. The maximum concentration, which is about 10^{13} molecules per cubic centimeter, is in a layer from about a 20- to 40-km altitude. It is very important as a screen against short-wavelength radiation ($\lambda < 310$ nm) coming from the sun. Much of the opposition to the supersonic transport (SST) lies in the fact that supersonic jet aircraft operate most efficiently in the ozone layer, where there is serious risk that the exhaust products (water vapor, oxides of nitrogen, and unburned hydrocarbons) may significantly deplete the ozone concentration. Organic molecules and biologic organisms are quite sensitive to light of wavelength less than 310 nm; so a significant decrease in the ozone shield may cause appreciable degradation of the quality of man's environment.

Ozone appears to be implicated also in the generation of photochemical, or Los Angeles type, smog. The mechanism is not yet understood, but it is believed that nitrogen dioxide (NO_2), which is formed as a by-product of the internal-combustion engine, is broken up by ultraviolet radiation in sunlight to give NO and O. The atomic oxygen then combines with O_2 to form O_3. Normally, the content of O_3 in the lower atmosphere would not build up very far because O_3 would be scavenged by NO to regenerate NO_2 and O_2. However, automobile exhaust emissions generate hydrocarbons that short-circuit the ozone cycle by producing free radicals (Sec. 10.7). The resultant buildup in the late afternoons of sunny days of ozone and derived compounds has produced a serious air pollution problem. Eye irritation and interference with normal respiratory processes are hallmarks of ozone pollution; further consideration of this and other forms of pollution is found in Chap. 26.

The structure of the ozone molecule is that of an isosceles triangle where the O—O—O angle is 116.8° and O—O bonds are 0.128 nm (compared with 0.121 nm in O_2). Since the ozone molecule is not paramagnetic, all its electrons must be paired. If the octet rule is followed, it is necessary to write at least two contributing resonance forms for the structure:

In the molecular-orbital description the uppermost occupied molecular orbital is a three-center one composed of side-to-side overlap of the p_z orbital on each oxygen. Taking the z axis perpendicular to the plane of the above figure, the result is to give a boomerang-shaped electron cloud above and below the plane of the figure.

Ozone gas has a sharp, penetrating odor. Its solubility in water in moles per liter is about 50 percent higher than that of oxygen, probably because O_3 is a polar molecule whereas O_2 is not. When cooled to 162°K, ozone forms a deep blue liquid that is explosive because of

Property	O_2	O_3
Molecular weight, amu	31.999	47.998
Bond length, nm	0.120	0.126
Normal melting point, °K	54.3	80.5
Normal boiling point, °K	90.2	161.7
Critical temperature, °K	154	268
Density of liquid (90°K), g/ml	1.14	1.71

Fig. 15.8
Properties of allotropic forms of oxygen.

the spontaneous tendency of O_3 to decompose to O_2. The decomposition is normally slow but increases rapidly as the temperature is increased or if a catalyst is present.

Some of the properties of oxygen and ozone are given in Fig. 15.8. Both O_2 and O_3 are good oxidizing agents, as shown by their high reduction potentials:

$$O_2(g) + 4H_3O^+ + 4e^- \longrightarrow 6H_2O \qquad E° = +1.23 \text{ V}$$

$$O_3(g) + 2H_3O^+ + 2e^- \longrightarrow 3H_2O + O_2(g) \qquad E° = +2.07 \text{ V}$$

As pointed out in Sec. 14.6, a large reduction potential indicates that the species on the left of the half-reaction is a strong oxidizing agent. Of the common oxidizing agents, ozone is second only to fluorine in oxidizing strength. In most reactions, at least at room temperature, O_2 is a slow oxidizing agent, whereas O_3 is more rapid.

Because of its cheapness and ready availability, oxygen is one of the most widely used industrial oxidizing agents. For example, in the manufacture of steel it is used to convert carbon to carbon monoxide for reducing iron oxides to iron, and to burn off residual impurities, such as carbon, phosphorus, and sulfur, that might give undesirable properties to steels. In the oxyacetylene torch, used for cutting and welding metals, temperatures in excess of $3000°C$ can be obtained by the reaction

$$2C_2H_2(g) + 5O_2(g) \longrightarrow 4CO_2(g) + 2H_2O(g)$$
$$\Delta H° = -2510 \text{ kJ}$$

Liquid oxygen is mixed with alcohol, charcoal, gasoline, powdered aluminum, etc., to give powerful explosives.

The use of oxygen in respiration of plants and animals is well known. In man, oxygen, inhaled from the atmosphere, is picked up in the lungs by the hemoglobin of the blood and distributed to the various cells, which use it for tissue respiration. In tissue respiration, carbohydrates are oxidized to provide energy required for cellular activities. Since oxygen itself is a slow oxidizing agent, catalysts (enzymes) must be present in order that reaction may proceed at body temperature. In the treatment of heart trouble, pneumonia, and shock, air oxygen is supplemented with additional oxygen.

The uses of ozone depend on its strong oxidizing properties. For example, it is used as a germicide, presumably because of its oxidation of bacteria. Inasmuch as oxidation of colored compounds often results in colorless ones, ozone is also used as a bleaching agent for wax, starch, fats, and varnishes. When added to the air in small amounts, ozone destroys odors; but it can be used safely only in low concentration because it irritates the lungs. In the laboratory, ozone aids in certain structure studies. Since it has a specific action on carbon-carbon double bonds, it can be used to determine their position in molecules.

15.11   Compounds of oxygen

Except for the oxygen fluorides, O_2F_2 and OF_2, the oxidation state of oxygen in compounds is negative. The oxidation numbers $-\frac{1}{2}$, -1, and -2 are observed.

Oxidation state $-\frac{1}{2}$ The heavier elements of group I (K, Rb, and Cs) react with oxygen to form compounds of the type MO_2, called *superoxides*. These are ionic solids containing the cation M^+ and the anion O_2^-. The solids are colored and paramagnetic; therefore, they must contain unpaired electrons.* The superoxide ion exists only in

* The anion O_2^- has 13 valence electrons. If we assume that these electrons are divided between the two atoms equally, as they most certainly are, then we must assign each atom an oxidation number of $-\frac{1}{2}$. Frequently, a fractional oxidation number is explained away by assuming that the compound contains atoms of the same element in different

the solid state. When superoxides are placed in water, O_2 and H_2O_2 are formed by the reaction

$$2MO_2(s) + 2H_2O \longrightarrow O_2(g) + H_2O_2 + 2M^+ + 2OH^-$$

Oxidation -1 Compounds which contain oxygen with oxidation number -1 are called *peroxides*. They are characterized by a direct O—O bond, which usually breaks at high temperatures. Metals such as Na, Sr, and Ba form solid peroxides which contain the peroxide ion, O_2^{2-}. This ion contains no unpaired electrons. Barium peroxide, BaO_2, is formed by heating solid barium with oxygen gas at a pressure of 3 atm. On futher heating under reduced oxygen pressure, BaO_2 decomposes to give barium oxide, BaO:

$$2BaO_2(s) \longrightarrow 2BaO(s) + O_2(g) \qquad \Delta H^\circ = 163 \text{ kJ}$$

Since this reaction is endothermic, it reverses at lower temperature; then BaO picks up O_2. At one time, this reversible process was used to extract O_2 from the air.

When solid peroxides are added to acidic solutions, hydrogen peroxide, H_2O_2, is formed. For example,

$$BaO_2(s) + 2H_3O^+ \longrightarrow Ba^{2+} + H_2O_2 + 2H_2O$$

If H_2SO_4 is used, the barium ion precipitates as insoluble barium sulfate, $BaSO_4$, leaving a dilute solution of pure H_2O_2. Commercially, most H_2O_2 is prepared by the electrolysis of cold H_2SO_4 or NH_4HSO_4 solutions followed by distillation under reduced pressure. Because H_2O_2 is unstable, owing to the reaction

$$2H_2O_2 \longrightarrow 2H_2O + O_2(g)$$

it is difficult to keep. The decomposition is slow but is catalyzed by dust and dissolved impurities such as transition metals and their compounds. It is also accelerated in the presence of light. For these reasons, solutions of H_2O_2 are stored in dark bottles with various chemicals such as diphosphate ion, $P_2O_7^{4-}$, added to tie up the catalysts by complex formation.

Pure anhydrous H_2O_2, obtained by distillation under reduced pressure, is a colorless liquid having a freezing point of -0.9°C and an estimated boiling point of 151.4°C. The structure of the gaseous molecule corresponds to a nonplanar arrangement of atoms with an O—O distance of 0.149 nm and a bond angle for H—O—O of 103°. As shown in Fig. 15.9, one H sticks out from the plane of the other three atoms at an angle of 90°. The molecule has a C_2 symmetry axis perpendicular to the O—O bond (Sec. 3.12).

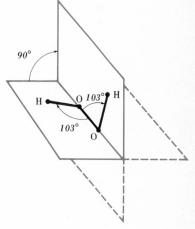

Fig. 15.9
Structure of H_2O_2 molecule.

---

oxidation states. In this case, such an explanation is unlikely since the two atoms are identical in their environment. The odd electron belongs to the O_2^- ion as a whole. The molecular-orbital description shows one electron pair and one unpaired electron in the two antibonding orbitals $\pi_{p_y}^*$ and $\pi_{p_z}^*$.

In aqueous solution hydrogen peroxide is a weak acid, dissociating as

$$H_2O_2 + H_2O \rightleftharpoons H_3O^+ + HO_2^-$$

with a dissociation constant on the order of 10^{-12}. Because oxygen also shows oxidation states of 0 and -2, compounds containing peroxide oxygen (-1) can gain or lose electrons; hence, they can act both as oxidizing agents and as reducing agents. In the decomposition

$$2H_2O_2 \longrightarrow 2H_2O + O_2(g)$$

hydrogen peroxide oxidizes and reduces itself. In the reaction

$$5H_2O_2 + 2MnO_4^- + 6H_3O^+ \longrightarrow 5O_2(g) + 2Mn^{2+} + 14H_2O$$

hydrogen peroxide is a reducing agent (goes to O_2). In the reaction

$$H_2O_2 + 2I^- + 2H_3O^+ \longrightarrow I_2 + 4H_2O$$

hydrogen peroxide is an oxidizing agent (goes to H_2O).

Oxidation state -2 Minus two is the most common oxidation state of oxygen in compounds. These compounds include the *oxides*, such as BaO, and the *oxy compounds*, such as $BaSO_4$. In none of these is there an oxygen-oxygen bond. Instead, the oxygen atoms have completed their octets by gaining a major share of two electrons from atoms other than oxygen.

All the elements except the lighter noble gases form oxides. Some of these oxides are ionic; others are covalent. In general, the more ionic ones are formed with the elements on the extreme left of the periodic table. Thus, BaO contains Ba^{2+} and O^{2-} ions and, like all ionic substances, is a solid at room temperature. It crystallizes with NaCl-type structure. It can be heated to $2000°C$ without decomposition. When placed in water, the O^{2-} ion reacts to give basic solutions:

$$O^{2-} + H_2O \longrightarrow 2OH^-$$

The ionic oxides are therefore called *basic oxides, basic anhydrides*, or, most simply, *bases*. They neutralize acids as, for example, in the reaction

$$CaO(s) + 2H_3O^+ \longrightarrow Ca^{2+} + 3H_2O$$

Elements toward the right of the periodic table do not form simple ionic oxides but instead share electrons with oxygen atoms. Many of these oxides are molecular and, like sulfur dioxide, SO_2, are gases at room temperature. They dissolve in water to give acidic solutions as, for example,

$$SO_2 + 2H_2O \rightleftharpoons H_3O^+ + HSO_3^-$$

The molecular oxides are called *acidic oxides, acidic anhydrides*, or, most simply, *acids*. They have the ability to neutralize bases as, for example, when carbon dioxide is bubbled through a basic solution:

$$CO_2(g) + OH^- \longrightarrow HCO_3^-$$

It is not possible to classify all oxides sharply as either acidic or basic. Some oxides, especially those formed by elements toward the center of the periodic table, are able to *neutralize both acids and bases*. Such oxides are called *amphoteric* (Sec. 9.5). An example of an amphoteric oxide is ZnO, which undergoes both of the following reactions:

$$ZnO(s) + 2H_3O^+ \longrightarrow Zn^{2+} + 3H_2O$$

$$ZnO(s) + 2OH^- + H_2O \longrightarrow Zn(OH)_4{}^{2-}$$

When any oxide reacts with water, the resulting compound contains OH, or *hydroxyl*, groups. If the hydroxyl group exists in the compound as the OH^- ion, the compound is called a *hydroxide*. Hydroxides are formed by the reaction of ionic oxides with water; e.g.,

$$BaO(s) + H_2O \longrightarrow Ba(OH)_2(s)$$

Barium hydroxide, $Ba(OH)_2$, is a solid which contains Ba^{2+} and OH^- ions in its lattice. It, like all hydroxides except those of group I elements, reverts to the oxide when heated. Many hydroxides, e.g., aluminum hydroxide, $Al(OH)_3$, are insoluble in water. The soluble ones give basic solutions.

Some compounds contain the OH group, not as an ion, but covalently bound to another atom. For example, in H_2SO_4 there are two OH groups and two O atoms joined to a central S atom. When placed in water, such compounds give acid solutions by rupture of an O—H bond. For this reason, they are called *oxyacids*. Most oxyacids can be dehydrated by heat to give oxides. They can also be neutralized to give *oxysalts* such as sodium sulfate, Na_2SO_4.

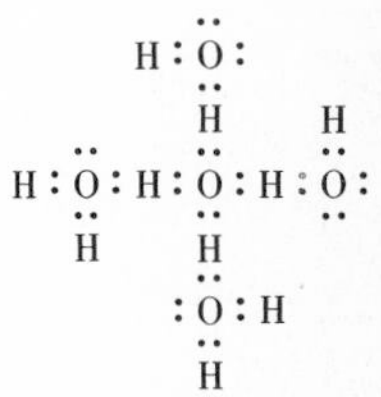

15.12   Water

The most important of all oxides, possibly the most important of all compounds, is H_2O. The H_2O molecule is nonlinear, having a C_2 axis in the plane of the molecule; the H—O—H angle is equal to $104.52°$. Because each bond is polar covalent, with the H end positive relative to the O end, the molecule has a considerable dipole moment, 1.85 debyes (Sec. 3.4). The attraction between the H atom of one molecule and the O atom of another leads to the association of H_2O molecules in both the liquid and solid states. A two-dimensional representation of the association is given in Fig. 15.10. The cluster of H_2O molecules is held together by hydrogen bonds (Sec. 15.5). The H atom, placed between two O atoms, may be considered bonded equally to both. However, the O-to-O distance is about 0.275 nm, which is more than double the normal O—H bond length, 0.0958 nm. Actually, as shown in Fig. 15.11, there are two favored positions for the H atoms separated by a low potential-energy barrier. For a given O—H—O bond, the H atom jumps back and forth from a position nearer one O atom to a position nearer the other; it is only the "average" position that is shown in Fig. 15.10. The result of hydrogen bonding is to form a giant molecule in which each O atom is surrounded by four H atoms. (The sim-

Fig. 15.10 Association of H_2O molecules. (H locations shown are midpoints of two possible positions.)

Fig. 15.11

*Potential-energy curve
showing two possible H-atom
positions in the* O ... O
hydrogen bond.

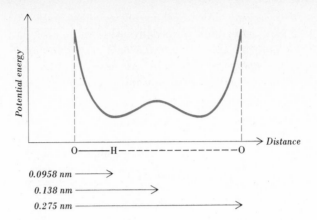

Fig. 15.12

*Tetrahedral arrangement of
O atoms in ice structure.
(Connecting regions are
occupied by H's.)*

plest formula is still H_2O, because of the four H atoms about a given O atom, only half of each H belongs to that O.) That there are four H atoms about each O is known from X-ray studies of ice. These studies do not detect the H atoms, but they do show that there are four O atoms symmetrically placed about each O. If the O atoms are joined to each other by hydrogen bonds, there must be four H atoms about each O. This can be seen by considering the central atom in Fig. 15.10.

The X-ray studies also indicate that the O atoms (of neighboring H_2O molecules) about a given O are located at the corners of a regular tetrahedron, as shown in Fig. 15.12. Because of the tetrahedral arrangement, the ice structure extends in three dimensions and is not the flat, two-dimensional representation of Fig. 15.10. Figure 15.13*a* is a better picture of the ice structure. It shows part of the crystal lattice, which extends in three dimensions. The large spheres represent O atoms, each of which is tetrahedrally surrounded by four H atoms, represented by the small circles. Every other O atom has its fourth H hidden beneath it. This hidden H joins to another O below, and so the structure continues in three dimensions. A notable feature of the structure is that it is honeycombed with hexagonal channels. Because of these holes, ice has a relatively low density.

When ice melts, the structure becomes less orderly but is not completely destroyed. In liquid water near the melting point it is thought that the O atoms are still tetrahedrally surrounded by four H atoms as in ice. However, the overall arrangement of tetrahedrons is more random and is constantly changing. An instantaneous view might be like that shown in Fig. 15.13*b*, where some of the hexagonal channels have collapsed to give a denser structure. Liquid water is denser than ice, as the data in Figs. 15.14 and 15.15 indicate.

The data in Fig. 15.14 also show that H_2O has a maximum density at 3.98°C. The maximum in the density of H_2O can be interpreted as follows: When ice is melted, the collapse of the structure leads to an increase in density. As the temperature of the liquid is raised, the collapse should continue further. However, there is an opposing effect. The higher the temperature, the greater the kinetic motion of the molecules. Hydrogen bonds are broken, and the H_2O molecules move

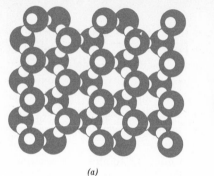
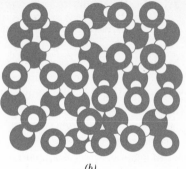

(a)   (b)

Fig. 15.13
(a) Ice structure; (b) water structure.

Temperature, °C	State	Density, g/cm³
0	Solid	0.917
0	Liquid	0.9998
3.98	Liquid	1.0000
10	Liquid	0.9997
25	Liquid	0.9971
100	Liquid	0.9584

Fig. 15.14
Density of water at various temperatures.

farther apart on the average. This effect becomes dominant at temperatures above 3.98°C. Below this temperature, collapse of structure is the more important.

In 1962, there was a considerable flurry of excitement among water chemists because of a Russian report of a strange new kind of water, which froze at −40 instead of 0°C, boiled well above 100°C, and had a considerably lower vapor pressure and higher density than ordinary water. It could only be prepared by condensing water vapor in freshly drawn glass or fused-silica capillary tubes; so for a while there was considerable activity trying to get enough of the elusive material to study. Absorption of infrared radiation suggested that the structure was an entirely new one, consisting perhaps of a network of hexagonal units in which the oxygen-oxygen distance was about 0.23 nm compared with 0.28 nm in normal water. There were even theoretical studies that purported to show that "anomalous" water should indeed be stable. Unfortunately, many attempts to reproduce the experimental findings resulted in failure, which in some cases was blithely dismissed as "improper technique," but a nagging doubt persisted that "anomalous water," which was also called *polywater*, was nothing but a specially impure silicate solution which was an artifact of the mode of preparation. Submicroscale analyses, difficult to do because of the extremely minute size of the samples available, have tended to confirm the view of the skeptics. Interest in the problem has recently died down, but only a few years ago there was great excitement because of the possibility that polywater might exist in nature, specifically in biologic systems.

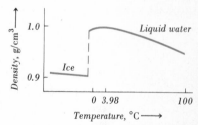

Fig. 15.15
Graph of density relations for H₂O (scales are distorted).

Section 15.12

Water

435

15.13 Water as a solvent

Water is the most common solvent, both in nature and in the laboratory. However, it is far from being a universal solvent, as is attested to by the large number of substances that are essentially insoluble in water. What are the factors that influence aqueous solubility? What predictions can be made about which substances will be soluble and which insoluble in water? The situation for water is especially complex because there is extensive hydrogen bonding that tends to keep water molecules associated to each other.

In general, water is a poor solvent for nonionic solutes. Hydrocarbons in particular, such as CH_4, C_2H_6, etc., are practically insoluble in water. In these cases, water interacts so weakly with the molecular solute that not nearly enough energy is liberated to break down the water structure. Still there are some molecular solutes which are highly soluble in water. Examples are ammonia, NH_3, and ethyl alcohol, C_2H_5OH. These substances interact strongly enough with the water molecules to break down the water structure. In the case of NH_3, hydrogen bonds are established between the N of NH_3 and the O of H_2O; in the case of ethyl alcohol, hydrogen bonds are formed between the O of C_2H_5OH and the O of H_2O. Sugars such as sucrose, $C_{12}H_{22}O_{11}$, owe their appreciable solubility in large measure to hydrogen bonding since they, like C_2H_5OH, have OH groups.

For ionic solutes, the solubility in water depends on a delicate balance between the lattice energy and the hydration energy and entropy of the ions. As the scheme below suggests

$$
\begin{array}{ccccc}
 & \xrightarrow{\ (2)\ } & M^{n+}(g) & + & X^{n-}(g) \\
 & & \downarrow{\scriptstyle(3)} & & \downarrow{\scriptstyle(4)} \\
MX(s) & \xrightarrow{\ (1)\ } & M^{n+}(aq) & + & X^{n-}(aq)
\end{array}
$$

the overall process of dissolving (1) can be resolved into a lattice breakup step (2) followed by hydration of the gaseous ions (3) and (4). The $\Delta H°$ for the lattice breakup is approximately equivalent to the lattice energy; it increases with the charge on the ions $+n$ and $-n$ and decreases with increasing size of the ions. Figure 15.16 shows some representative values. As seen for the alkali halides, with increasing anion size, the lattice energy decreases. Similarly, in going from Li to Na to K salts, the lattice energy again decreases. Ag salts would be expected to be like the K salts because the ionic radius of Ag^+ (0.126 nm) is closer to that of K^+ (0.133 nm) than it is to that of Na^+ (0.097 nm). However, the lattice energies of the Ag salts turn out to be like that of the highest Na salt. The extra-large lattice energy has been attributed to extra-large van der Waals attractions arising from the fact that Ag^+ ions contain many more electrons than do K^+ ions (Sec. 5.12). Figure 15.16 also shows that the oxides and sulfides of dipositive ions have lattice energies enormously greater than those of the $+1$ halides. The main reason for this is the fourfold increase in

LiF	1008	AgF	954	TiO	3880
LiCl	828	AgCl	904	VO	3920
LiBr	791	AgBr	895	MnO	3810
LiI	732	AgI	883	FeO	3920
				CoO	3990
NaF	904	BeO	4530	NiO	4080
NaCl	770	MgO	3920	ZnO	4030
NaBr	736	CaO	3520		
NaI	690	SrO	3310		
		BaO	3120		
KF	803				
KCl	699	BeS	3740		
KBr	674	MgS	3300		
KI	632	CaS	3040		
		SrS	2900		
		BaS	2760		

Fig. 15.16
Crystal energies in kilojoules per mole.

electric attraction expected when we double the charge on both positive and negative ions in a crystal (Sec. 7.7).

If a salt is to be appreciably soluble in water, the energy needed for breakup of the lattice has to be offset by the ion hydration process. Figure 15.17 lists values for the $\Delta H°$ and $\Delta S°$ of hydration of some common ions. The $\Delta H°$ values, which are given in kilojoules per mole, refer to the processes $M^+(g) \longrightarrow M^+(aq)$ and $X^-(g) \longrightarrow X^-(aq)$. The fact that the values are negative indicates that heat is liberated to the surroundings. $\Delta H°$ is smallest for the singly charged ions (about 400 kJ), roughly quadruples for doubly charged ions (about 1600 kJ), and becomes about 10 times as great for triply charged ions (about 4000 kJ). The larger the radius of the ion, the smaller the $\Delta H°$ of hydration.

The $\Delta S°$ of hydration applies to the process $M^+(g) \longrightarrow M^+(aq)$. The values, which are given in joules per degree-mole, are all negative, signifying that the ions in the hydrated state have a lower entropy (i.e., a smaller number of configurations) than in the gaseous state. In the gas phase, the ions are like a highly expanded gas; there are a large number of possible configurations, and the entropy is very high. When the ions go into solution, they lose much of their freedom of motion. They are more like a highly compressed gas, in that the accessible volume for movement is highly restricted; the number of possible configurations is relatively small, and the entropy is considerably lower. In the process $M^+(g) \longrightarrow M^+(aq)$ as well as in $X^-(g) \longrightarrow X^-(aq)$ we lose practically all the entropy associated with a mole of gas, which for going from 1 atm pressure to the equivalent of 1 m solution is about 50 J deg^{-1} mol^{-1}. In addition to this entropy loss due to restriction of volume, there is a further reduction of entropy to be expected from an ordering effect on the H_2O molecules in the solution. Na^+ ion, for example, when placed in water, tends to align near-neighbor H_2O molecules so that O ends point toward the Na^+

	$\Delta H°$, kJ/mol	$\Delta S°$, J deg^{-1}mol^{-1}
Li^+	−516	−118.8
Na^+	−397	−87.4
K^+	−314	−51.9
Rb^+	−289	−40.2
Cs^+	−255	−36.8
Mg^{2+}	−1910	−266.5
Ca^{2+}	−1580	−210.0
Sr^{2+}	−1430	−203.8
Ba^{2+}	−1290	−157.7
Al^{3+}	−4640	−463
Ag^+	−469	−93.7
Zn^{2+}	−203	−267.8
F^-	−506	−155.2
Cl^-	−377	−96.7
Br^-	−343	−83.3
I^-	−297	−60.7
OH^-	−502	−108.8

Fig. 15.17
Hydration parameters of some common ions.

and H ends stick out away from the central Na^+. Ca^{2+} would have an even stronger aligning effect because of its higher charge; Al^{3+} would be expected to be even more effective. In all these cases, the near-neighbor H_2O molecules lose some of their degrees of freedom, the number of possible configurations for the system goes down, and the entropy decreases. Hence, it is no surprise that $\Delta S°$ for the hydration of ions is a large negative number and, as can be seen from the data in Fig. 15.17, it gets more negative the more highly charged an ion is and the smaller its radius.

One interesting aspect about the $\Delta S°$ values shown in Fig. 15.17, which are experimental values based on temperature-solubility data, is that they are significantly smaller in absolute magnitude that one would calculate from theory for the above-mentioned processes. For example, the experimentally derived $\Delta S°$ for $Na^+(g) \longrightarrow Na^+(aq)$ is $-87.4 \text{ J deg}^{-1} \text{ mol}^{-1}$, whereas one expects about $-120 \text{ J deg}^{-1} \text{ mol}^{-1}$. One suggested explanation for the discrepancy is that there must be some breakup of the hydrogen-bonded overall structure of liquid water when ions are put into it. Forcing the O end of an H_2O molecule to face Na^+, for example, means that the H atoms have to stick out; they may not match very well the local requirements of the fairly regular liquid-water structure. Figure 15.18 is a schematic representation of what the situation may be like. Between the first hydration layer and the regular water structure there is probably a mismatch zone,

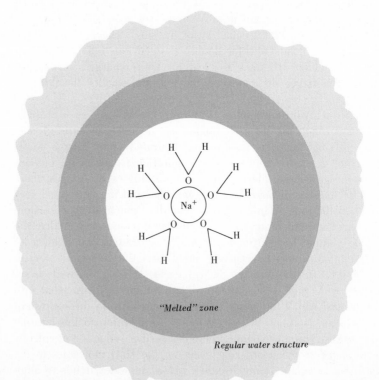

Fig. 15.18

Schematic representation of how ion hydration can produce unexpectedly high entropy because of creation of a "melted" zone.

"Melted" zone

Regular water structure

where disorder, hence entropy, is greater than in the rest of the liquid. This mismatch zone is sometimes picturesquely referred to as a "melted" zone. In it, molecular randomness is more like that of a normal liquid than like that of water with its very special hydrogen-bonded character. Attempts have been made to assign quantitatively specific degrees of structure-breaking influence to various ions, but one of the big stumbling blocks is how to split an observed effect between the cation and the anion.

No matter what the detailed interpretation is for the above $\Delta H°$ and $\Delta S°$ values, once we have the numerical values we can use them to analyze the dissolving process. For this we need to know $\Delta G°$ for

$$MX(s) \longrightarrow M^+(aq) + X^-(aq)$$

Suppose we compare AgF and AgCl in the various steps that make up the above process. We will assume that the lattice energy as tabulated in Fig. 15.16 will give us a good estimate of $\Delta H°$ for $MX(s) \longrightarrow M^+(g) + X^-(g)$. This is not quite exact, but the error is small (i.e., about 1 percent). Further, we will assume that the $\Delta S°$ for $MX(s) \longrightarrow M^+(g) + X^-(g)$ is about equal to $+200$ J deg^{-1} mol^{-1}, which is approximately what one would associate with creation of two moles of gas. Given these approximations, we can write for AgF the following:

	$\Delta H°$, kJ/mol	$\Delta S°$, kJ mol^{-1} deg^{-1}
$AgF(s) \longrightarrow Ag^+(g) + F^-(g)$	954	0.200
$Ag^+(g) \longrightarrow Ag^+(aq)$	-469	-0.094
$F^-(g) \longrightarrow F^-(aq)$	-506	-0.155
$AgF(s) \longrightarrow Ag^+(aq) + F^-(aq)$	-21	-0.049

$$\Delta G° = \Delta H° - T\Delta S°$$
$$= -21 - 298(-0.049) = -6 \text{ kJ/mol}$$

The enthalpy change is favorable for the dissolving reaction, but the entropy change is not. However, the enthalpy term is bigger; so it wins out, and the result is that AgF is quite a soluble salt. For AgCl, in contrast, we have the following situation:

	$\Delta H°$, kJ/mol	$\Delta S°$, kJ mol^{-1} deg^{-1}
$AgCl(s) \longrightarrow Ag^+(g) + Cl^-(g)$	904	0.200
$Ag^+(g) \longrightarrow Ag^+(aq)$	-469	-0.094
$Cl^-(g) \longrightarrow Cl^-(aq)$	-377	-0.097
$AgCl(s) \longrightarrow Ag^+(aq) + Cl^-(aq)$	$+58$	$+0.009$

$$\Delta G° = \Delta H° - T\Delta S°$$
$$= +58 - 298(0.009) = +55 \text{ kJ/mol}$$

Here the enthalpy change is unfavorable for dissolving, but the entropy change is favorable. Again, the enthalpy term is the bigger and wins out. The result is that AgCl is not a very soluble salt.

In analyzing the above difference between AgF and AgCl, we can see that the most important contribution is that the hydration enthalpy of Cl^- (-337 kJ/mol) is much smaller than that of F^- (-506 kJ/mol). Even though the lattice of AgCl requires less energy to break up, this is not enough. There is insufficient hydration energy from Cl^- to go with the 469 kJ/mol of Ag^+ to make it worthwhile. The smaller hydration enthalpy of Cl^- compared with F^- is, of course, due to the fact that the Cl^- is a bigger ion.

In most cases of ionic solids, not enough data are available to make the above type of analysis. In such cases, rough qualitative generalizations are sometimes useful. For example, if the charge on both anion and cation is simultaneously increased, insolubility is generally favored. Thus, for example, $BaSO_4$ (both ions doubly charged) and $AlPO_4$ (both ions triply charged) are much less soluble than NaCl (both ions singly charged). On the other hand, if the charge of only one ion is increased, the solubility is not much changed. So, for example, NaCl, $BaCl_2$, and $AlCl_3$ are all appreciably soluble. Similarly, NaCl, Na_2SO_4, and Na_3PO_4 are also soluble. Another general rule is that the more dissimilar in size the anion and cation are, the more soluble a salt is likely to be. Thus, for example, $MgCrO_4$ is very soluble, whereas $BaCrO_4$ has a $K_{sp} = 8.5 \times 10^{-11}$. Finally, it should be pointed out that there may be specific interactions with H_2O which aid solubility. Barium sulfide (BaS), for example, has a specific reaction between S^{2-} and H_2O which helps make BaS more soluble than $BaSO_4$, an insoluble salt that lacks such specific interaction. Predictions as to solubility must be made with caution, since solubility depends on a number of factors.

15.14 Hydrates

Analysis shows that many solids contain H_2O molecules. These solids, called *hydrates*, are represented by formulas such as that for nickel sulfate heptahydrate, $NiSO_4 \cdot 7H_2O$. This formula states that there are seven H_2O molecules per formula-unit but does not specify how the H_2O is bound in the crystal. For example, in $NiSO_4 \cdot 7H_2O$ all seven H_2O molecules are not equivalent. Six are bound to the Ni^{2+} ion to give $Ni(H_2O)_6^{2+}$, and the seventh is shared between $Ni(H_2O)_6^{2+}$ and neighboring SO_4^{2-} in the crystal lattice. The solid is better represented by the formula $Ni(H_2O)_6SO_4 \cdot H_2O$. In other hydrates, such as sodium carbonate decahydrate, $Na_2CO_3 \cdot 10H_2O$, H_2O molecules are not bound directly to the ions, but their principal function seems to be to improve the packing of the ions in the crystal.

Water of hydration can be driven off by heating to give *anhydrous* material. Such loss of water is usually accompanied by a change in crystal structure. However, some substances, such as certain silicate minerals called *zeolites*, and proteins, lose water on heating without much change in crystal structure. On reexposure to water they, like sponges, take up water and swell. Apparently, water taken up this way occupies semirigid tunnels within the solid.

Actually, water of hydration is more common than not in the usual salts of the chemistry laboratory. Blue copper sulfate, for example, is $CuSO_4 \cdot 5H_2O$ or, better, $Cu(H_2O)_4SO_4 \cdot H_2O$. Even acids and bases can exist as hydrates in the solid form. Examples are barium hydroxide, $Ba(OH)_2 \cdot 8H_2O$, and oxalic acid, $H_2C_2O_4 \cdot 2H_2O$.

Frequently, hydrous compounds whose composition may be known but whose structure is in doubt are encountered. Such a substance is obtained, for example, from the reaction of a base with a solution of aluminum salt. Under certain conditions, the product might have the composition AlO_3H_3. The most obvious conclusion is that the compound is the hydroxide, $Al(OH)_3$. However, it could just as well be written as the hydrated oxide, $Al_2O_3 \cdot 3H_2O$, for which the simplest formula also is AlO_3H_3. In order to distinguish the two possibilities, structure studies are needed, but they are difficult and in many cases have not been made.

Exercises

*15.1 *Comparison.* Make a tabular summary of how hydrogen and oxygen differ from each other with respect to natural occurrence, preparation, properties, and uses.

*15.2 *Preparation of hydrogen.* In terms of weight of hydrogen produced per gram of total reactants consumed, which of the methods given in Sec. 15.2 would be most efficient?

*15.3 *Hydrogen bonding.* Describe briefly what a hydrogen bond is, and tell how it might affect the boiling point or melting point of HF. What boiling point would you predict for HF if it were not a hydrogen-bonded liquid?

* 15.4 *Dissociation.* If we define $pD = -\log [D_3O^+]$ in analogy to pH, what is the pD in pure D_2O?   *Ans. 7.35*

*15.5 *Liquefaction of air.* Liquid nitrogen, boiling point $77°K$, is one of the most common cryogenic liquids used for establishing very low temperatures. Given that air is 78% N_2 and 21% O_2 show with a composition-temperature diagram how normal air can be converted by fractional liquefaction to essentially pure liquid nitrogen.

* 15.6 *Oxygen equations.* Write balanced chemical equations for each of the following processes:
  a   Thermal decomposition of $NaNO_3$ to give $NaNO_2$ and O_2
  b   Conversion of O_2 to O_3
  c   Oxidation of I^- by O_3 in acid to give IO_3^- and O_2
  d   Oxidation of CrO_2^- by O_2 in base to give CrO_4^{2-}
  e   Heating FeS_2 in air to give Fe_3O_4 and SO_2

*15.7 *Bonding.* Draw an energy-level diagram showing the relative placing and occupancy of the three highest molecular-orbital levels occupied in O_2. Sketch rough electron distributions for each of these orbitals.

*15.8 *Explosives.* One ton (2200 kg) of TNT is equivalent to 4.6×10^9 J. What weight of acetylene (C_2H_2) and oxygen would you need to match this energy output?

*15.9 *Superoxides.* What are the oxidation-state changes in the following reaction?

$$2KO_2 + 2H_2O \longrightarrow O_2 + H_2O_2 + 2K^+ + 2\ OH^-$$

Write the two half-reactions.

* 15.10 *Water.* What fraction of a water jug at 25°C should be left empty so that if the water freezes it will just fill the jug?

Ans. 8.0%

**15.11 *Elemental abundance.* As shown in Fig. 15.1, the 10 most abundant elements account for 99.2 percent of the mass of the earth's crust. Using the data from Fig. 15.1, what would have to be the atomic weight of an eleventh element that would be needed to account by itself for all the missing mass?

** 15.12 *Abundance.* The general formula for saturated hydrocarbons is C_nH_{2n+2}. What is their percent by weight hydrogen as n approaches infinity?

Ans. 14.37%

**15.13 *Heat of combustion.* Using the data of Fig. 13.2, calculate the comparative heats of combustion of water gas ($CO + H_2$) and methane (CH_4) per unit weight and per unit volume. Assume products are $CO_2(g)$ and $H_2O(g)$.

Ans. 0.35:1 for weight, 0.33:1 for volume

**15.14 *Preparation of hydrogen.* How much current would you need to pass for 100 h through aqueous NaOH to get as much hydrogen as you can get from dissolving one kilogram of aluminum in aqueous NaOH? Write the net equations for the processes.

**15.15 *Balloon.* What volume of hydrogen at STP would you need to lift a payload of 100 kg under standard conditions? The payload is not to include the weight of the hydrogen. *Ans. 83,200 liters*

**15.16 *Thermodynamics.* Given that $\Delta H° = -485$ kJ for the reaction $2H_2(g) + O_2(g) \longrightarrow 2H_2O(g)$ and $\Delta H° = -537$ kJ for $H_2(g) + F_2(g) \longrightarrow 2HF(g)$, calculate the $\Delta H°$ for the reaction $2H_2O(g) + 2F_2(g) \longrightarrow 4HF(g) + O_2(g)$. Predict whether $\Delta S°$ would be positive or negative for this reaction.

**15.17 *Hydrides.* Suppose you electrolyze molten $LiAlH_4$ with inert electrodes. Write the probable electrode reactions.

**15.18 *Isotopes.* How many deuterium atoms are there in a liter of ordinary water? In a nuclear fusion reaction such as occurs in a hydrogen bomb, deuterium nuclei fuse to give a helium nucleus. If the energy emission in this process is 5 MeV per deuterium, how many liters of gasoline (C_8H_{18}) is the liter of water equivalent to? The heat of combustion of C_8H_{18} is 5100 kJ/mol; the density of gasoline is 0.703 g/cm^3; 1 eV = 1.60×10^{-19} J.   *Ans. 243 liters*

**15.19   *Ozone.* From a perusal of the half-reactions in Appendix 7, predict what happens when ozone reacts with water. Write the equations for possible reactions.

**15.20   *Free energy.* You are given the following values of standard reduction potentials:

$$2e^- + 2H_3O^+ + H_2O_2 \longrightarrow 4H_2O \qquad\qquad E° = +1.77 \text{ V}$$

$$2e^- + 2H_3O^+ + O_2(g) \longrightarrow H_2O_2 + 2H_2O \qquad E° = +0.68 \text{ V}$$

Calculate the $\Delta G°$ for $2H_2O_2 \longrightarrow 2H_2O + O_2(g)$

**15.21   *Dissociation.* The usual commercial H_2O_2 solution is 30% by weight H_2O_2. It has a density of 1.11 g/cm^3. If the dissociation constant of H_2O_2 is 1×10^{-12}, what would be the pH of the above solution?   *Ans. 5.50*

**15.22   *Hydrogen peroxide.* When $KMnO_4$ is added to 350 ml of ostensibly 3% by weight H_2O_2 solution, 3.94 cm^3 of O_2 gas is released and collected over H_2O at a barometric pressure of 0.941 atm and 28°C. What is the actual concentration of the H_2O_2 solution?

**15.23   *Solubility thermodynamics.* From the data given in Sec. 15.13, predict a value of $\Delta G°$ and a K_{sp} corresponding to AgBr(s) $\longrightarrow$ $Ag^+(aq) + Br^-(aq)$ at 25°C.   *Ans. + 76 kJ, 4.5×10^{-14}*

**15.24   *Solution.* You are given a tiny crystal of $Ba(OH)_2 \cdot 8H_2O$, the density of which is 2.18 g/cm^3. If the crystal weighs 50.0 mg, how much water must you add to it to get a solution that is 0.120 M OH$^-$?

***15.25   *Thermodynamics.* Using the data in the first paragraph of Sec. 15.3, calculate at what temperature H_2 is 1 percent dissociated, assuming total pressure is 1 atm.

***15.26   *NMR.* In the molecule H_3PO_3, phosphorous acid, there are attached to a central P atom two OH groups, one H, and one O. In looking at the nuclear magnetic resonance predict what the power absorption vs. magnetic field trace will look like. Account for the special features of your curve.

***15.27   *Bond lengths.* The observed bond lengths in O_2^+, O_2^-, and O_2^{2-} are 0.112, 0.128, and 0.149 nm, respectively. Compare these with the value in O_2, and explain the differences.

***15.28   *Electrode potentials.* Given that the reduction potential for $O_2(g) + 4H_3O^+ + 4e^- \longrightarrow 6H_2O$ is +1.23 V and that for $O_3(g) + 2H_3O^+ + 2e^- \longrightarrow 3H_2O + O_2(g)$ is +2.07 V, calculate the $E°$ for $O_3(g) + 6H_3O^+ + 6e^- \longrightarrow 9H_2O$.

***15.29 *Hydration.* Explain how you would rationalize the following statement: In aqueous solution K^+ is more of a structure breaker than is Na^+.

***15.30 *Solubility.* Suggest a reason why a solid consisting of a small cation and a large anion might be more soluble in water than a solid composed of small cations and small anions.

Having discussed the characteristic behavior of hydrogen and oxygen in the previous chapter we can now consider the detailed descriptive chemistry of the other elements. We shall do this systematically, starting at the left side of the periodic table and moving to the right. The progression will take us from elements that are distinctly metallic in their physical behavior and are good reducing agents to elements that are nonmetallic and good oxidizing agents. In this chapter we consider the elements of groups I and II, where the chemistry is particularly simple because only one or two, respectively, valence electrons are clearly involved in the bonding. Because their chemistry is relatively simple, we can use these elements to good advantage to illustrate application of the principles established in earlier chapters.

The elements of group I are lithium ($Z = 3$), sodium ($Z = 11$), potassium ($Z = 19$), rubidium ($Z = 37$), cesium ($Z = 55$), and francium ($Z = 87$). They are usually referred to as the *alkali metals*, after the Arabic word *al-qili*, meaning "plant ashes," since the ashes of plants are particularly rich in sodium and potassium carbonate.

The elements of group II are beryllium ($Z = 4$), magnesium ($Z = 12$), calcium ($Z = 20$), strontium ($Z = 38$), barium ($Z = 56$), and radium ($Z = 88$). They are called the *alkaline-earth metals* because the alchemists referred to any nonmetallic substance insoluble in water and unchanged by fire as an "earth" and because the "earths" of this group, e.g., lime (CaO) and magnesia (MgO), give decidedly alkaline reactions.

445

The term "metal" is applied to any substance that has a silvery luster and is a good conductor of electricity and heat. Some metals, of which the alkali metals are examples, also are relatively soft, malleable (can be beaten into sheets), and ductile (can be drawn into wires). The alkaline-earth metals, on the other hand, are harder and more brittle, although they too are excellent conductors of heat and electricity. All these properties can be accounted for in terms of the metallic structure discussed in Sec. 7.6.

All the alkali elements crystallize with a body-centered cubic lattice in which the lattice points are occupied by $+1$ ions. Of the alkaline-earth elements, which form $+2$ ions, only Ba crystallizes in the body-centered cubic symmetry. The others crystallize in close-packed arrays— Be and Mg, hexagonal close-packed, and Ca and Sr, cubic close-packed. No matter what the symmetry of the cation arrangement, the valence electrons (one from each alkali atom or two from each alkaline-earth atom) make up a sea, or cloud, of free negative charges which permeates the whole lattice. Since they are not fixed in position, these electrons spread their wave functions throughout the metal and thus produce high electric conductivity (Sec. 14.1). Furthermore, it is almost invariably observed that high conductivity of electricity is accompanied by high conductivity of heat. This is not surprising because thermal energy, which is normally transported by lattice vibrations, can also be transported rapidly from one part of a metal to another by the conduction electrons.

The high luster observed in group I and group II metals is explainable by the highly mobile electrons of the metallic lattice. When a light beam strikes the surface of a metal, electric fields associated with the light wave set the electrons in the metal surface into back-and-forth oscillation. This is easy to do because the valence electrons are not tightly bound to any specific atom. However, like any moving electric charge, oscillating electrons give off electromagnetic energy as light. The net effect is that the beam of light is "reflected." In this respect the electrons in metals act like tiny radio relay stations which receive an electromagnetic signal and send it out again. Actually, nonmetals, even paper, can show high reflectivity, but only when looked at from very low angles. For nonmetals there is a critical angle beyond which the reflectivity disappears. The unusual thing about metals is that they show high reflectivity of light at all angles.

The softness, malleability, and ductility that characterize the alkali metals are accounted for by the nature of the forces holding the lattice together. For example, in metallic sodium the principal force holding the lattice together is the attraction between Na^+ ions and the valence-electron cloud. Since this attraction is uniform in all directions, there are no strongly preferred positions for the Na^+ ions. The result is that Na^+ ions can easily be moved from one lattice site to another. Under pounding, the crystal can be flattened out like a pancake with but little expenditure of energy. Also, it can be cut with a knife like soft processed cheese. All this behavior is in contrast to the case of iron or tungsten, for example, where there are strong, directed forces between adjacent positive ions because of covalent binding (Sec. 17.2). Some of this already appears in the element beryllium of group II, but the generally harder and more brittle behavior of group II relative to group I is mainly due to the +2 ions, with their greater resultant attraction for the electrons of the free-electron cloud.

16.2   Ion species and electrode potentials

The above discussion interprets metallic properties in terms of the metallic lattice, but a more fundamental question is the following: Why do the alkali elements prefer to form a crystal consisting of +1 ions and electrons? Why do the alkaline-earth elements prefer to form a crystal consisting of +2 ions and electrons? The question is complex, but it can be at least partly answered by considering the properties of the individual atoms. Figure 16.1 shows some of the properties of

Fig. 16.1
Properties of alkali atoms.

| Element | Atomic number | Electronic configuration (core e^-'s in parentheses) | Ionization potential, eV | | Ionic radius, nm (M^+) |
			First	Second	
Lithium	3	(2) $2s^1$	5.39	75.6	0.068
Sodium	11	(10) $3s^1$	5.14	47.3	0.097
Potassium	19	(18) $4s^1$	4.34	31.8	0.133
Rubidium	37	(36) $5s^1$	4.18	27.5	0.147
Cesium	55	(54) $6s^1$	3.89	25.1	0.167
Francium	87	(86) $7s^1$	....	....	(0.175)

the alkali atoms. The column headed "Electronic configuration" indicates the population according to the principal quantum number in the undisturbed neutral atom. As indicated, each of the atoms has one electron in the outermost energy level. The energy required to pull off this valence electron is given in the column of first ionization potentials. As ionization potentials go, these are relatively small values, indicating that it is relatively easy to remove this one electron. However, the second ionization potential, the energy required to pull off a second electron, is many times higher than the first. This means that, although it is relatively easy to form the M^+ ion, it is very difficult (practically impossible under ordinary conditions) to form the M^{2+} ion. All of this is consistent with the notion that a closed shell of electrons is difficult to break into. The result is that when alkali atoms come together to form liquid or solid, M^+ ions are formed.

The properties of the alkali-metal atoms shown in Fig. 16.1 are well illustrative of the general changes expected in going through a group of the periodic table. For example, the radius of the $+1$ cation* increases progressively from lithium down. This is expected because there is an increasing number of electronic shells populated. Similarly, the ionization potential shows progressive decrease in going down the group. This is consistent with increased size and resulting smaller attraction for the valence electron, as was discussed in Sec. 2.7. Actually, the change in properties in group I is so regular as to give a false sense of confidence about how well periodic-table trends can be predicted. There are traps for the unwary in later groups.

How is the situation different for group II elements from that for group I elements? Why should group II elements form M^{2+} ions whereas group I elements form M^+ ions? Figure 16.2 shows some pertinent information for the alkaline-earth atoms. The important point to note is that the electron configuration is now s^2 instead of s^1. This is reflected in the ionization potentials, where not only is the first ionization potential relatively modest but also the second. We normally expect some increase in ionization energy when we go from pulling off the first electron to a second; after all, the first is being pulled

Element	Atomic number	Electronic configuration (core e^-'s in parentheses)	Ionization potential, eV			Ionic radius, nm (M^{2+})
			First	Second	Third	
Beryllium	4	(2) $2s^2$	9.32	18.2	153.8	0.035
Magnesium	12	(10) $3s^2$	7.64	15.0	80.1	0.066
Calcium	20	(18) $4s^2$	6.11	11.9	51.2	0.099
Strontium	38	(36) $5s^2$	5.69	11.0	(43)	0.112
Barium	56	(54) $6s^2$	5.21	10.0	(36)	0.134
Radium	88	(86) $7s^2$	5.28	10.1	....	0.143

Fig. 16.2

Properties of alkaline-earth atoms.

away from a neutral atom, whereas the second is being taken from a positive ion. When we do the latter for sodium, for example, we have to break into an inner shell; so there is a relatively enormous increase in going from first to second ionization potential. For group II elements, the second electron still comes from the outermost shell; so the energy expenditure remains relatively modest. It is not until we get to the third ionization potential that we face up to a very large increase.

The above argument explains why group II elements form M^{2+} ions, not M^{3+} ions, but we still have not explained why they form M^{2+},

* From X-ray studies of ionic solids it is possible to determine the radius of an ion. There is a problem, however, in that X-ray investigations give only the distance between centers of adjacent atoms. How should this distance be apportioned? The usual procedure is to adopt one ion as a standard and to assume that it has a definite radius in all its compounds. Other radii are then assigned so that the sum of radii equals the observed spacing. A standard may be obtained from a salt such as LiI, where Li^+ is so small that the spacing can be assumed to be due to large I^- ions in contact.

not M^+. Scrutiny of the data in Fig. 16.2 shows that the second ionization potentials of the alkaline-earth atoms are almost twice as great as the first. Surely it would be energetically more favorable to pull two electrons one from each of two atoms than to pull both electrons from the same atom. It would seem that alkaline-earth elements, like group I elements, should prefer to form $+1$ ions rather than $+2$. If only the ionization potentials were involved, such would indeed be the case. For example, 6.11 eV is required to pull off the first electron from calcium, and 11.9 eV to pull off the second. This means that return of one electron to Ca^{2+} liberates 11.9 eV of energy. As a consequence, calcium gas and doubly charged calcium ions are *unstable* with respect to conversion to singly charged calcium ions:

$$Ca(g) \longrightarrow Ca^+(g) + e^- \qquad \text{requires 6.11 eV}$$

$$e^- + Ca^{2+}(g) \longrightarrow Ca^+(g) \qquad \text{liberates 11.9 eV}$$

Net

$$Ca(g) + Ca^{2+}(g) \longrightarrow 2Ca^+(g) \qquad \text{liberates 5.8 eV}$$

The net energy release is enormous (5.8 eV, or 560 kJ); hence, Ca^+ should be formed in the gas phase. It has, in fact, been detected at high temperatures. What about condensed phases, as in the solid

Element	Electrode potential, V	Density, g/cm³	Melting point, °C	Boiling point, °C
Lithium	−3.05	0.53	186	1336
Sodium	−2.71	0.97	97.5	880
Potassium	−2.93	0.86	62.3	760
Rubidium	−2.93	1.53	38.5	700
Cesium	−2.92	1.87	28.5	670

Fig. 16.3
Properties of the alkali metals.

state or in aqueous solution? Here the situation is more complicated because we now have to consider the stabilizing effect of the environment. Before we consider this for the Ca^+–Ca^{2+} case, we need to look at the general situation for any kind of metal atom. How is ionization behavior in the gas phase related to formation of species in aqueous solution? We consider first the situation for the group I elements.

The alkali metals are the most reactive metals known. Practically any oxidizing agent, no matter how weak, can be reduced by them. Even water, which is not a very good oxidizing agent, vigorously chews them up, sometimes with explosive violence. Quantitatively, the reducing strength is best described by reference to electrode potentials. Figure 16.3 lists the electrode-potential values as well as other properties that characterize the behavior of these elements. As can be seen, the electrode potentials are all large and negative. As was discussed in Sec. 14.6, large negative values mean that the half-reaction

$$e^- + M^+(aq) \longrightarrow M(s)$$

has much less tendency to go to the right than does the corresponding hydrogen-electrode half-reaction

$$e^- + H_3O^+ \longrightarrow \tfrac{1}{2}H_2(g) + H_2O$$

The electrode potential for the latter has been arbitrarily set at zero. As was also discussed in Sec. 14.6, small tendency of a half-reaction to go to the right means large tendency of that half-reaction to go to the left. In other words, the large negative numbers for the electrode potentials in Fig. 16.3 mean large tendency for the half-reaction

$$M(s) \longrightarrow M^+(aq) + e^-$$

to go to the right. Lithium (3.05 V) has the biggest tendency to do this; sodium (2.71 V) has the least. At first sight this may be surprising since the ionization potentials of Fig. 16.1 indicated that it was most difficult to pull an electron off lithium, not sodium. The apparent contradiction points up the great difference between gas-phase behavior and that involving solids and solutions. Whereas the ionization potential is concerned only with the *isolated gaseous* atom, the electrode potential is concerned with the *metal* and its ionic species in *solution*. Removal of an electron is only part of what goes on in the above half-reaction. We can see this by breaking the reaction up into its component steps:

(1) $M(s) \longrightarrow M(g)$
(2) $M(g) \longrightarrow M^+(g) + e^-$
(3) $M^+(g) \longrightarrow M^+(aq)$

In step (1) the metal is evaporated; i.e., the atoms are converted to the gaseous state, in which they are independent of each other. The energy required to do this (called the *sublimation energy*) is approximately the same for all the metals of group I. In step (2) an electron is pulled off the neutral atom to give a gaseous ion. The energy required (the ionization potential) is largest for lithium. In step (3) the gaseous ion is placed in water, i.e., hydrated. Energy (hydration energy) is liberated. The tendency of the overall change to occur depends on the net effect of all three of these steps. The fact that for lithium the tendency of the overall change to occur is greatest suggests that the relatively greater difficulty of step (2) has been more than compensated for by step (3). The hydration energy of the tiny Li^+ ion is so great that it more than makes up for the higher energy required to pull the electron off. In other words, the stabilizing effect of water on lithium ion makes the reaction

$$Li(s) \longrightarrow Li^+ + e^-$$

have a greater tendency to occur than the corresponding reaction for the other alkali elements.

The quantitative comparison of the above steps for the different alkali metals is shown in Fig. 16.4. The numbers given represent free-energy changes in kilojoules per mole. (To convert to electron volts, divide

Step	$\Delta G°$, kJ/mol				
	Li	Na	K	Rb	Cs
(1) $M(s) \longrightarrow M(g)$	+122	+78	+61	+56	+51
(2) $M(g) \longrightarrow M^+(g)$	+520	+496	+419	+403	+375
(3) $M^+(g) \longrightarrow M^+(aq)$	−480	−371	−299	−277	−244
(4) $e^- + H_3O^+ \longrightarrow \frac{1}{2}H_2(g) + H_2O$	−465	−465	−465	−465	−465
(5) $M(s) + H_3O^+ \longrightarrow M^+(aq) + \frac{1}{2}H_2(g) + H_2O$	−303	−262	−284	−283	−283
(6) $E°$ in volts for $M(s) \longrightarrow M^+(aq) + e^-$	+3.14	+2.72	+2.94	+2.93	+2.93

Fig. 16.4

Free-energy contributions to overall process
$$M(s) \longrightarrow M^+(aq) + e^-$$

by 96.49.) A positive sign indicates energy must be added to the system; a negative sign indicates energy is evolved. Steps (1) to (3) are, respectively, sublimation, ionization, and hydration. For step (2) we have simply taken the ionization potentials from Fig. 16.1 and multiplied by 96.49 kJ/eV; this does not quite give the free-energy change but is close enough to match the accuracy of the other data. Although we could make a valid comparison between the various alkali metals by simply totaling the first three steps, we have also included a fourth step which allows us to place the $\Delta G°$'s on the scale of electrode potentials. Since $E°$'s are referred to the hydrogen electrode, we need to include the other half-reaction, namely, $e^- + H_3O^+ \longrightarrow \frac{1}{2}H_2(g) + H_2O$. The $\Delta G°$ contribution for this step (4) is −465 kJ/mol. Line (5) shows the $\Delta G°$ for the overall process; it can be expressed in terms of $E°$ by using the relation $\Delta G° = -n\mathcal{F}E°$ with $n = 1$ and $\mathcal{F} = 96.49$ kJ/eV. Since under standard conditions the hydrogen electrode is assumed to contribute zero volts to the overall process, the resulting $E°$, as shown in line (6), is entirely attributed to the half-reaction $M(s) \longrightarrow M^+(aq) + e^-$. To make comparison with the actual experimental electrode potentials listed in Fig. 16.3, the signs have to be inverted, and the electrode reaction written in the reverse direction.

To see how group II elements compare with group I elements, we consider the case of calcium as a specific example. What would be the $E°$ for $Ca(s) \longrightarrow Ca^+(aq) + e^-$, or, more meaningfully, what would be the $\Delta G°$ for the overall process $Ca(s) + H_3O^+ \longrightarrow Ca^+(aq) + \frac{1}{2}H_2(g) + H_2O$? Would formation of monopositive calcium be favorable, as is the case for all the alkali metals? We again break up the overall reaction into its component steps. These are as follows:

(1)	$Ca(s) \longrightarrow Ca(g)$	$\Delta G° = +159$ kJ
(2)	$Ca(g) \longrightarrow Ca^+(g) + e^-$	$\Delta G° = +590$ kJ
(3)	$Ca^+(g) \longrightarrow Ca^+(aq)$	$\Delta G° = -300$ kJ
(4)	$e^- + H_3O^+ \longrightarrow \frac{1}{2}H_2(g) + H_2O$	$\Delta G° = -465$ kJ

For step (2), the $\Delta G°$ is estimated from the first ionization potential of calcium, which is 6.11 eV. For step (3), we have guessed a value of −300 kJ by comparison with data for known +1 ions. As can

Section 16.2

Ion species and electrode potentials

be seen, the sum of all four steps gives

$$(5) \quad Ca(s) + H_3O^+ \longrightarrow Ca^+(aq) + \tfrac{1}{2}H_2(g) + H_2O$$
$$\Delta G^\circ = -16 \text{ kJ}$$

The fact that ΔG° is negative means that the reaction should go spontaneously as written. In other words, the $+1$ aqueous ion is stable compared with the metal, just as we found above for the alkali elements. If we want to put the comparison on the basis of electrode potentials, we can use $\Delta G^\circ = -n\mathfrak{F}E^\circ$ to get $E^\circ = +0.2$ V for the half-reaction $Ca(s) \longrightarrow Ca^+(aq) + e^-$.

The fact that ΔG° for a reaction such as step (5) is negative does not necessarily mean that the final product will stop at that point since there may be other possible succeeding reactions. In the case of Na, we stop at $Na^+(aq)$ because pulling off a second electron requires so much additional energy that the process is hopeless. However, for the case of Ca we need to look seriously at the possibility of subsequent oxidation of $Ca^+(aq) \longrightarrow Ca^{2+}(aq) + e^-$. Referred to hydrogen, this reaction would be written

$$Ca^+(aq) + H_3O^+ \longrightarrow Ca^{2+}(aq) + \tfrac{1}{2}H_2(g) + H_2O$$

and the component steps would be as follows:

$$(6) \quad Ca^+(aq) \longrightarrow Ca^+(g) \qquad \Delta G^\circ = +300 \text{ kJ}$$
$$(7) \quad Ca^+(g) \longrightarrow Ca^{2+}(g) + e^- \qquad \Delta G^\circ = +1145 \text{ kJ}$$
$$(8) \quad Ca^{2+}(g) \longrightarrow Ca^{2+}(aq) \qquad \Delta G^\circ = -1518 \text{ kJ}$$
$$(9) \quad e^- + H_3O^+ \longrightarrow \tfrac{1}{2}H_2(g) + H_2O \qquad \Delta G^\circ = -465 \text{ kJ}$$

Step (6) is the reverse of step (3), which was estimated at -300 kJ. Step (7) comes from the second ionization potential of Ca. The sum of steps (6) to (9) gives the net reaction

$$Ca^+(aq) + H_3O^+ \longrightarrow Ca^{2+}(aq) + \tfrac{1}{2}H_2(g) + H_2O$$

with a free-energy change of -538 kJ. Such a large, negative free-energy change means that the reaction has a great tendency to go as written, in fact, considerably more so than the oxidation of $Ca(s)$ to $Ca^+(aq)$. Using $\Delta G^\circ = -n\mathfrak{F}E^\circ$ and taking out the hydrogen half-reaction with its $E^\circ = 0$, we get the following:

$$Ca^+(aq) \longrightarrow Ca^{2+}(aq) + e^- \qquad E^\circ = +5.6 \text{ V}$$

$$Ca(s) \longrightarrow Ca^+(aq) + e^- \qquad E^\circ = +0.2 \text{ V}$$

Since $Ca^+(aq)$ by the first half-reaction is a stronger reducing agent than $Ca(s)$ by the second, the first half-reaction will reverse the second to give the net reaction

$$2Ca^+(aq) \longrightarrow Ca(s) + Ca^{2+}(aq) \qquad E^\circ = +5.4 \text{ V}$$
$$\Delta G^\circ = -521 \text{ kJ}$$

The high negative value of ΔG° for this net reaction indicates a great tendency for $Ca^+(aq)$ to *oxidize and reduce itself*, i.e., to *dispropor-*

tionate. For all the other alkaline-earth elements the corresponding disproportionation reactions are also favored; so in no case is the $+1$ aqueous ion stable. Even in the solid state, where lattice energy plays the role that hydration energy plays in solution, the $+1$ alkaline-earth ions are not generally stable.

Once we have the $E°$ for conversion of $Ca(s)$ to $Ca^+(aq)$ and the $E°$ for conversion of $Ca^+(aq)$ to $Ca^{2+}(aq)$, we can easily calculate the $E°$ for conversion of $Ca(s)$ to $Ca^{2+}(aq)$. The important thing to remember is that $E°$ gives the free-energy change *per electron*. Hence, to add half-reactions, we first convert from $E°$ to free-energy change (i.e., multiply voltage by the number of electrons) and then add. The sum half-reaction now contains several electrons; so its free-energy change must be divided by the number of electrons to give $E°$. Following this procedure for calcium, we get

	$E°$, V	$\Delta G° = -n\mathfrak{F}E°$
$Ca(s) \longrightarrow Ca^+(aq) + e^-$	0.2	$-(1)(\mathfrak{F})(0.2)$
$Ca^+(aq) \longrightarrow Ca^{2+}(aq) + e^-$	5.6	$-(1)(\mathfrak{F})(5.6)$
$Ca(s) \longrightarrow Ca^{2+}(aq) + 2e^-$	2.9	$-(2)(\mathfrak{F})(2.9)$

The $E°$ for the last half-reaction, 2.9 V, is calculated as the sum of the first two free-energy changes, $-0.2\mathfrak{F} - 5.6\mathfrak{F}$, divided by $-n\mathfrak{F}$, where $n = 2$.

It is interesting to compare other properties of group I and group II elements. Examination of Figs. 16.1 and 16.2 shows that the ionic radius of any group II element is smaller than that of the group I element of the same period. For example, Mg^{2+} has an ionic radius of 0.066 compared with 0.097 nm for Na^+. Both these elements fall in the third period of the periodic table. Why the difference in size? Sodium ion has a nuclear charge of $+11$ and has two electrons in the K shell and eight electrons in the L shell ($1s^22s^22p^6$); magnesium ion has a nuclear charge of $+12$, and it also has two electrons in the K shell and eight electrons in the L shell ($1s^22s^22p^6$). These two ions are *isoelectronic;* i.e., they have identical electronic configurations. The difference between Na^+ and Mg^{2+} is that the latter has a higher nuclear charge. Increased nuclear charge means increased attraction for electrons, which in turn means a smaller K shell and a smaller L shell. In any isoelectronic sequence, ionic size decreases with increased nuclear charge.

Just as the *ionic size* decreases in going from group I to group II, the apparent *atomic size* (Sec. 2.6) also decreases. This smaller atomic size of the group II neutral atoms accounts for the difference between the first ionization potentials of groups I and II. Figures 16.1 and 16.2 show that in going from group I to group II there is a rather large increase in the energy required to pull off one electron (for example, 5.14 eV for Na and 7.64 eV for Mg). This, of course, is in line with the smaller size of group II atoms and their consequent tighter hold on electrons.

Fig. 16.5
Properties of alkaline-earth metals.

Element	Electrode potential (from M^{2+}), V	Density, g/cm^3	Melting point, °C	Boiling point, °C
Beryllium	-1.85	1.86	1280	2970
Magnesium	-2.37	1.74	650	1100
Calcium	-2.87	1.55	850	1490
Strontium	-2.89	2.6	770	1380
Barium	-2.90	3.6	710	1140
Radium	-2.92	5(?)	700	<1700

Finally, in comparing groups I and II there is the difference in electrode potentials. The values for group I elements are given in Fig. 16.3; the values for group II elements are shown in Fig. 16.5 (which also gives other properties). For the lighter elements, at the top of the groups, there is a distinct difference, and it is in the direction that group I elements have more negative potentials; the alkali metals therefore, are better reducing agents. For example, the electrode potential of beryllium is -1.85 V compared with that of lithium, -3.05 V. However, for the heavier elements, at the bottom of the group, there is little difference between groups I and II. Barium, for instance, has an electrode potential of -2.90 V, whereas cesium, of the same period, has -2.92 V.

Actually, it is not surprising that group I elements are stronger reducing agents than corresponding group II elements. After all, the ionization potentials of group I elements are much lower. The surprising thing is that group II elements are as good reducing agents as they are. The key to the explanation apparently lies in the hydration energy. Although it takes a fair amount of energy to pull two electrons off a group II atom, the net process $M(s) \longrightarrow M^{2+}(aq) + 2e^-$ nevertheless has a great tendency to occur because the doubly charged ion interacts strongly with water in forming the hydrated ion.

16.3 Occurrence of the elements

The alkali metals occur in nature only as $+1$ ions. Sodium and potassium are most abundant, ranking sixth and seventh of all the elements in the earth's crust. Lithium is moderately rare but is found in small amounts in practically all rocks. Rubidium and cesium are rare. Francium is essentially nonexistent since it has an unstable nucleus and is radioactive. Trace amounts of it have been prepared artificially by nuclear reactions.

Since most of the compounds of the alkali metals are water soluble, they are generally found in seawater and in brine wells. However, there are many clays which are insoluble complex compounds of the alkali metals combined with Si, O, and Al. As the result of evaporation of ancient seas, there are also large salt deposits which serve as convenient sources of the alkali metals and their compounds.

Sodium ion and potassium ion are among the indispensable constituents of animal and plant tissue. Na^+ is the principal cation of the fluids outside the cells, whereas K^+ is the principal cation inside the cells. Besides filling general physiological roles, such as aiding water retention, these ions have specific functions. For example, Na^+ depresses the activity of muscle enzymes and is required for contraction of all animal muscle. In plants, K^+, but not Na^+, is a primary requirement. As a result, more than 90 percent of the alkali content of ashes is due to potassium. Plants have such a high demand for potassium that, even in soils in which the sodium content predominates manyfold, the potassium is taken up preferentially. Since an average crop extracts from the soil about 50 lb of potassium per acre, the necessity for potassium fertilizers is obvious. The old idea of putting wood ashes in the garden makes sense because trees are plants that are rich in potassium and burning them produces a potassium-rich product.

The alkaline-earth elements are found only in compounds as $+2$ ions. As discussed in Sec. 15.12, $+2$ ions combine with -2 ions to form compounds less soluble than those of $+1$ ions. Consequently, many alkaline-earth compounds are insoluble and, unlike alkali-metal compounds, are found as insoluble deposits in the earth's crust (for example, $CaCO_3$ and $BaSO_4$). The most important of these deposits are the silicates, carbonates, sulfates, and phosphates.

Beryllium on a weight basis makes up only 0.0006 percent of the earth's crust. It is very widespread, but only in trace amounts. The only important beryllium mineral found in any quantity is a silicate, beryl, for which the formula is $Be_3Al_2Si_6O_{18}$. Enormous single crystals of beryl weighing many tons have been found. The gemstone emerald is beryl colored deep green by trace amounts of chromium.

Magnesium is the eighth most abundant element in the earth's crust, making up about 2 percent of its mass. It is widely distributed, principally as the silicate minerals such as asbestos ($CaMg_3Si_4O_{12}$) and the carbonate, oxide, and chloride. Magnesite ($MgCO_3$) and dolomite ($MgCO_3 \cdot CaCO_3$) are the principal sources of magnesium in addition to seawater and deep salt wells.

Calcium is the most abundant of the group I and group II elements on a mass basis (3.4 percent of the earth's crust), but it is outnumbered 4 to 3 on an atom basis by sodium. The principal occurrences of calcium are as the silicates, carbonate, sulfate, phosphate, and fluoride. Calcium carbonate ($CaCO_3$) as the mineral calcite, the most abundant of all nonsilicate minerals, appears in such diverse rocks as limestone, marble, and chalk. Most of these appear to be derived from the skeletons of marine animals which have been laid down on seabeds and consolidated. The mineral gypsum ($CaSO_4 \cdot 2H_2O$) is also very common. It apparently owes its origin in many cases to limestone beds which have been acted on by sulfuric acid produced from the oxidation of sulfide minerals. Phosphate rock is essentially $Ca_3(PO_4)_2$, an important ingredient of bones, teeth, and seashells.

Strontium is relatively rare and ranks twentieth in order of abundance; barium, which makes up 0.05 percent of the earth's crust, is about 2.5 times as abundant. The principal mineral of strontium is strontianite ($SrCO_3$); of barium, barite ($BaSO_4$).

Radium is very rare, but its presence is easily detected by its radioactivity. Because its nucleus spontaneously disintegrates, all the radium found is due to the nuclear breakdown of heavier elements, particularly uranium. For this reason, uranium ores such as pitchblende (impure U_3O_8) are principal sources of radium. It has been estimated that the average abundance of radium in the earth's crust is less than 1 part per million million. This makes a uranium mineral which contains $\frac{1}{4}$ g of radium per ton of ore a relatively rich source of radium.

16.4   Preparation of the elements

To prepare the alkali elements, it is necessary to reduce the $+1$ ion. This can be done chemically or electrolytically. Purely chemical methods would seem impossible since they require a reducing agent stronger than the alkali metals. However, chemical reduction can be carried out in special cases, as in the reaction of rubidium chloride with calcium at high temperature:

$$Ca(s) + 2RbCl(s) \longrightarrow CaCl_2(s) + 2Rb(g)$$

The reaction occurs in the direction indicated only because the rubidium is more volatile and escapes out of the reacting mixture, thus preventing the attainment of equilibrium. In the equilibrium state the concentration of rubidium would be very small.

In practice, the alkali metals are generally prepared by electrolysis of molten alkali halides or hydroxides. For example, sodium is made commercially in ton quantities by the electrolysis of fused NaOH, the melting point of which is $318°C$. Sodium metal is formed at the iron or copper cathodes, and oxygen at the nickel anodes. To prevent explosive oxidation of the sodium by the oxygen, the electrode compartments are separated by a bell-shaped partition. Figure 16.6 shows a schematic representation of the arrangement. A circular array of nickel anodes surrounds the cathode, and a ring of fire keeps the NaOH molten. The cathode reaction is $Na^+(l) + e^- \longrightarrow Na(l)$, and the anode reaction is $4OH^- \longrightarrow O_2(g) + 2H_2O + 4e^-$.

Since the alkaline-earth elements occur only as the $+2$ ions, preparation of the metals also requires a reduction process. Reduction can be accomplished by electrolysis of the molten halides or hydroxides or by chemical reduction with appropriate reducing agents. Beryllium, for example, is made by heating beryllium fluoride, BeF_2, with Mg and by electrolyzing a mixture of beryllium chloride, $BeCl_2$, and NaCl.

The extraction of magnesium from seawater accounts for the bulk of United States production. In the process the magnesium ion in seawater (about 0.13 percent) is precipitated as $Mg(OH)_2$ by the addition of lime, CaO. The hydroxide is filtered off and converted to $MgCl_2$ by reaction with HCl. The dried $MgCl_2$ is mixed with other salts to

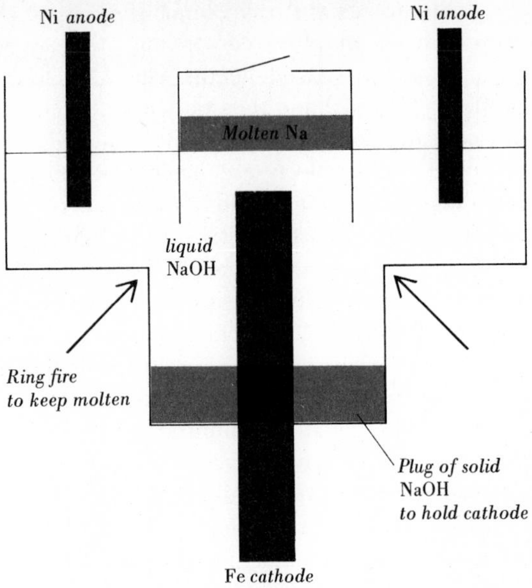

Fig. 16.6
Commercial cell for
producing sodium.

Fig. 16.6 Commercial cell for producing sodium.

lower the melting point and then electrolyzed at about $700°C$ to give metal of 99.9 percent purity.

Magnesium can also be prepared by a chemical reduction process in which magnesium oxide, obtained by heating dolomite, is reduced at high temperatures by iron and silicon. Since the reaction is carried out above $1100°C$, the boiling point of magnesium, the process produces gaseous magnesium, which escapes from the reaction mixture to condense as a very high purity product.

16.5 Properties and uses

As mentioned above the alkali metals exhibit, to a high degree, typically metallic properties. Although too chemically reactive to be generally used for metallic properties, sodium in polyethylene-encased cables is now being used in some underground, high-voltage transmission applications. Also liquid alkali metal is used to solve the difficult engineering problem of conducting heat energy from the center of a nuclear reactor to the exterior, where it can be converted into useful work. In both of these uses, the expense and difficulty involved in working with alkali metal are partially compensated for by its excellence as an electric and heat conductor.

Cesium has the distinction of being the metal from which electrons are ejected most easily by light; such light-induced emission is termed the *photoelectric effect*. One kind of *photocell*, a device for converting a light signal to an electric signal, consists of an evacuated tube containing two electrodes with a voltage difference between them. The negative electrode is coated with cesium metal, cesium oxide, or an alloy of cesium, antimony, and silver. In the absence of light the tube does not conduct electricity since there is no charge carrier to conduct current from one electrode to the other. When struck by light, the cesium-coated electrode emits electrons which are attracted to the positive electrode, and thus the circuit is completed. Television pickup devices such as the iconoscope and the image orthicon use the photocell principle. Color effects are made possible because the cesium metal has a high response to red light and a low response to blue light whereas cesium oxide is most sensitive to the blue.

Though all the alkali metals are very good reducing agents, only sodium finds extensive use for this purpose. It is used to make other metals by reducing their chlorides, and it is also used in the production of various compounds of carbon. For this latter purpose, sodium is frequently used in the form of its solution in liquid ammonia. It is a remarkable fact that sodium and the other alkali metals dissolve in the waterlike solvent ammonia to give colored solutions, which can be evaporated to give the alkali metal unchanged. In the blue solutions the alkali metal is dissociated into $+1$ ions and electrons. The electrons are associated with ammonia molecules; therefore, the anions in these solutions can be considered as solvated electrons. More concentrated solutions have a metallic, bronzelike appearance and have very high electric conductivity, indicating that the electrons are extremely mobile. Reducing properties are somewhat toned down in all these solutions compared with the pure alkali metals.

The great reactivity of the alkali metals poses a special problem in their handling. For example, water, although a relatively poor oxidizing agent, has great tendency to attack them. The tarnishing of freshly cut sodium is partially due to this oxidation by moisture in the air. To avoid such problems, alkali metals are usually stored under kerosine or other inert hydrocarbon compounds. Great care has to be taken in handling them so as not to have them accidentally come in contact with water. The prudent investigator working with the alkali metals generally keeps ready a bucket of sand to douse any fires that may occur.

The alkaline-earth metals, too, are good conductors of heat and electricity, but of them only magnesium finds any considerable use. Surprisingly, this use is based on the structural qualities of magnesium rather than on its electric properties. Lightest of all the commercially important structural metals, magnesium has relatively low structural strength, but this can be increased by alloying it with other elements. The principal elements added are aluminum, zinc, and manganese. Aluminum helps increase the tensile strength; zinc improves the working properties (machining); and manganese reduces corrosion. The use of magnesium alloys is ever increasing because of modern emphasis on weight reduction in such things as aircraft, railroad equipment, and household goods.

Too rare and costly for most large-scale uses, beryllium is important as a trace addition for hardening other metals, such as copper. It is also used as a moderator or reflector material in nuclear reactors, where it is increasingly being used to coat the uranium or plutonium fuel rods. In the finely powdered form, beryllium (and its compounds) must be handled carefully since it is extremely toxic.

Calcium, strontium, and barium are more reactive than beryllium and magnesium. The situation is complicated further by the fact that when exposed to air, they form oxides which flake off to expose fresh surface. Their great affinity for oxygen makes these elements useful as deoxidizers in steel production and as getters in the production of low-cost electron tubes. Most electronic vacuum tubes, for example, have a thin deposit of barium metal on the inner wall of the glass or metal envelope. The purpose is to scavenge any gases, such as oxygen, that were left in or might leak into the tube.

Finely divided magnesium burns rather vigorously to emit very intense light which is particularly rich in ultraviolet radiation. For this reason, magnesium is used as one of the important light sources for photography. Flashbulbs contain wire or foil of magnesium (or aluminum) packed in an oxygen atmosphere. When the bulb is fired, an electric current heats the metal and initiates the oxidation reaction.

The flame spectra of strontium salts are characteristically red, and those of barium are yellowish green. Strontium and barium salts are frequently used for color effect in pyrotechnics.

16.6 Compounds of the alkali elements

The alkali metals readily form compounds by reacting with other substances. For example, sodium metal on standing in air becomes covered with sodium peroxide, Na_2O_2. Furthermore, water vigorously attacks any of the alkali metals to liberate hydrogen:

$$2M(s) + 2H_2O \longrightarrow 2M^+ + 2OH^- + H_2(g)$$

Thus, the problem with the alkali metals is not to get them to form compounds but to keep them from doing so.

Compounds of the alkali metals, even the hydrides, are generally considered to be ionic with the alkali metal present as a $+1$ ion. Most of the compounds are quite soluble in water; hence, a convenient way to get a desired anion in solution is to use its sodium salt. The alkali-metal ions do not hydrolyze appreciably and do not form complex ions to any appreciable extent. The alkali-metal ions are colorless, and color of alkali-metal compounds is usually due to the anion.

The hydrides (Sec. 15.4) of the alkali metals are white solids prepared by heating alkali metal in hydrogen. The simple oxides, M_2O, are not so easily formed. Of the alkali metals, only lithium reacts directly with oxygen to form Li_2O. When sodium reacts with oxygen, the peroxide Na_2O_2 is formed instead. Potassium, rubidium, and cesium under similar conditions form superoxides of the type MO_2. In order to get the simple oxides, it is necessary to reduce some alkali-metal compound such as the nitrate. For example,

$$2KNO_3(s) + 10K(s) \longrightarrow 6K_2O(s) + N_2(g)$$

All the oxides are basic oxides and react with water to form hydroxides.

Commercially, however, the hydroxides of the alkali metals are made by electrolysis of aqueous alkali-chloride solutions. For example, as discussed in Sec. 14.2, sodium hydroxide, or *caustic soda*, as it is often called, is made by electrolysis of aqueous sodium chloride.

Other important compounds of the alkali metals, such as *washing soda*, Na_2CO_3, and *baking soda*, $NaHCO_3$, are discussed in later chapters in connection with the corresponding anions.

16.7 Compounds of the alkaline-earth elements

At ordinary temperatures the alkaline-earth elements form compounds only in the $+2$ oxidation state. With the exception of beryllium, all such compounds are essentially ionic. The alkaline-earth ions are colorless and except for Be^{2+} do not hydrolyze appreciably in aqueous solution. Beryllium salts hydrolyze to give acid solutions. Unlike the compounds of group I, many group II compounds are not soluble in water.

1 Hydrides. When heated in hydrogen gas, Ca, Sr, and Ba form hydrides. These are white powders which react with H_2O to liberate H_2. Calcium hydride, CaH_2, is used as a convenient, portable hydrogen supply:

$$CaH_2(s) + 2H_2O \longrightarrow Ca^{2+} + 2OH^- + 2H_2(g)$$

2 Oxides. The oxides of these elements are characteristically very high melting (*refractory*): BeO, 2530°C; MgO, 2800°C; CaO, 2580°C; SrO, 2430°C; and BaO, 1923°C. They can be made by heating the metals in oxygen or thermally decomposing the carbonates or hydroxides. For example, lime (CaO) is made from limestone ($CaCO_3$) by the reaction

$$CaCO_3(s) \longrightarrow CaO(s) + CO_2(g) \qquad \Delta H = +178 \text{ kJ}$$

Except for beryllium oxide, BeO, which is amphoteric, the oxides are basic. Both lime and magnesia (MgO) are used as linings in furnaces, sometimes specifically to counteract acidic impurities, as in steel production.

3 Hydroxides. The hydroxides are made by adding water to the oxides in a process called *slaking*. For example, the slaking of lime produces calcium hydroxide, $Ca(OH)_2$. The reaction

$$CaO(s) + H_2O \longrightarrow Ca(OH)_2(s) \qquad \Delta H = -67 \text{ kJ}$$

is exothermic and is accompanied by a threefold expansion in volume, sometimes to the consternation of building contractors whose lime supplies accidentally get wet. Lime is an important constituent of cement and is also used as an important industrial base, since it is cheaper than NaOH.

The hydroxides of the alkaline-earth elements are only slightly soluble in water; however, the solubility increases with increasing ionic size. The solubility products are given in Fig. 16.7. With the exception of $Be(OH)_2$, which is amphoteric, the other hydroxides are basic. They are assumed to be 100% dissociated in aqueous solution.

4 *Sulfates.* The sulfates of group II range from the very soluble beryllium sulfate to the practically insoluble radium sulfate. Going down the group, the solubilities decrease in regular order; for $BeSO_4$, K_{sp} is very large; $MgSO_4$, about 10; $CaSO_4$, 2.4×10^{-5}; $SrSO_4$, 7.6×10^{-7}; $BaSO_4$, 1.5×10^{-9}; and $RaSO_4$, 4×10^{-11}. This decreasing order is opposite to that observed for the hydroxides. To account for the alteration of trend, two factors need to be considered: As discussed in Sec. 15.12, solubility depends on lattice energy and on hydration energy. For the alkaline-earth sulfates the lattice energies are all about the same: the sulfate ion is so large that changing the size of the much smaller cation makes little difference. The difference in solubility must therefore be due to differences in hydration energy. From Be^{2+} to Ba^{2+}, size increases, hydration energy decreases, and the sulfates become less soluble. For the alkaline-earth hydroxides, on the other hand, the lattice energies are not the same but decrease with increasing cation size. For the group II hydroxides this is a larger effect than the change in hydration energy. Thus, the hydroxides increase in solubility down the group.

Magnesium sulfate is well known as the heptahydrate, $MgSO_4 \cdot 7H_2O$, or epsom salts. In medicine it is useful as a purgative, apparently because magnesium ions in the alimentary canal favor passage of water from other body fluids into the bowel to dilute the salt.

Calcium sulfate has already been mentioned as the mineral gypsum, $CaSO_4 \cdot 2H_2O$. When gypsum is partially dehydrated,

$$CaSO_4 \cdot 2H_2O(s) \rightleftharpoons CaSO_4 \cdot \tfrac{1}{2}H_2O(s) + \tfrac{3}{2}H_2O(g)$$
$$\Delta H = -176 \text{ kJ}$$

it forms plaster of paris, sometimes written $2CaSO_4 \cdot H_2O$. The use of plaster of paris in making casts and molds arises from the reversibility of the above reaction. On water uptake, plaster of paris sets to gypsum, and the expansion of volume results in remarkably faithful reproductions.

Barium sulfate and its insolubility have been repeatedly mentioned. Although Ba^{2+}, like most heavy metals, is poisonous, the solubility of $BaSO_4$ is so low that $BaSO_4$ can safely be ingested into the stomach and intestines. The use of $BaSO_4$ in taking X-ray pictures of the digestive tract depends on the great scattering of X rays by the Ba^{2+} ion. The scattering of X rays by atoms is proportional to the electron density of the atom. Ba^{2+} contains 54 electrons in a relatively small volume and hence scatters X rays more efficiently than do ions of lighter elements. $BaSO_4$ is also important as a white pigment.

5 *Chlorides and fluorides.* Beryllium chloride ($BeCl_2$) and beryllium fluoride (BeF_2) are unusual in that they do not conduct electricity in the molten state. For this reason, they are usually considered to be molecular rather than ionic salts. All the chlorides and fluorides of the other group II elements are typical ionic solids. Calcium fluoride, CaF_2, occurs in nature as the mineral fluorspar, and is quite insoluble

$Be(OH)_2$	*Less than 10^{-19}*
$Mg(OH)_2$	*8.9×10^{-12}*
$Ca(OH)_2$	*1.3×10^{-6}*
$Sr(OH)_2$	*3.2×10^{-4}*
$Ba(OH)_2$	*5.0×10^{-3}*

Fig. 16.7
K_{sp} for alkaline-earth hydroxides.

Section 16.7
Compounds of the
alkaline-earth elements
461

in water. The chloride, $CaCl_2$, is very soluble and, in fact, has such great affinity for water that it is used as a dehydrating agent.

6 Carbonates. All the carbonates of group II are quite insoluble and therefore are found undissolved in nature. Calcium carbonate, $CaCO_3$, or limestone, is the most common nonsilicate rock. The existence of large natural beds of $CaCO_3$ poses a special problem for water supplies because $CaCO_3$, though essentially insoluble in water, is soluble in water containing carbon dioxide. Since our atmosphere contains an average of 0.04% carbon dioxide at all times, essentially all groundwaters are solutions of carbon dioxide in water. These groundwaters dissolve limestone by the reaction

$$CaCO_3(s) + CO_2 + H_2O \rightleftharpoons Ca^{2+} + 2HCO_3^-$$

which produces a weathering action on limestone deposits and results in contamination of most groundwaters with calcium ion and bicarbonate ion, HCO_3^-.

The dissolving action of carbon dioxide-containing water explains the many caves found in limestone regions. These caves abound in weird formations produced partly by the dissolving action and partly by reprecipitation of $CaCO_3$. The optimum conditions for $CaCO_3$ deposition are slow seepage of groundwater, steady evaporation, and no disturbing air currents. In limestone caves these conditions are ideally met. Groundwater containing Ca^{2+} and HCO_3^- may seep through a fissure in the roof and hang as a drop from the ceiling. As the water evaporates along with the carbon dioxide, the above reaction reverses to deposit a bit of limestone. Later, another drop of groundwater seeps onto the limestone speck, and the process repeats. In time, a long shaft reaching down from the roof may be built up in the form of a limestone stalactite. Occasionally, drops of groundwater may drip off the stalactite to the cave floor, where they evaporate to form a spire, or stalagmite, of $CaCO_3$. The whole process of dissolving and reprecipitation of limestone is very slow and may take hundreds of years.

16.8   Hard water

Because limestone is so widespread, most groundwater contains small but appreciable concentrations of calcium ion. The presence of this Ca^{2+} (or of Mg^{2+} or Fe^{2+}) is objectionable because of formation of insoluble precipitates when such water is boiled or when soap is added. Water that behaves in this way is called *hard water.*

Hardness in water is always due to the presence of calcium, magnesium, or ferrous (Fe^{2+}) ion. However, the hardness may be of two types: (1) *temporary,* or *carbonate,* hardness, in which HCO_3^- ions are present along with the aforementioned metal ions, or (2) *permanent,* or *noncarbonate,* hardness, in which the dipositive ions but no HCO_3^- ions are in the water. In either case the hardness manifests itself by a reaction with soap (but not with detergents) to produce a scum. Soap, as discussed in Sec. 22.4, is a sodium salt of a long-chain hydrocarbon acid. The usual soap is sodium stearate, $C_{17}H_{35}COONa$, and consists of Na^+ ions and negative stearate ions. When stearate ions are added to water containing Ca^{2+}, insoluble calcium stearate forms:

$$Ca^{2+} + 2C_{17}H_{35}COO^- \longrightarrow Ca(C_{17}H_{35}COO)_2(s)$$

This insoluble calcium stearate is the familiar scum that forms on soapy water.

Hardness in water is also objectionable because boiling a solution containing Ca^{2+} and HCO_3^- deposits $CaCO_3$ just as in cave formation. In industrial boilers the formation of $CaCO_3$ is an economic headache since, like most salts, $CaCO_3$ is a poor heat conductor. Fuel efficiency is drastically cut, and boilers have been put completely out of action by local overheating due to boiler scale.

The major question is how to soften hard water effectively and economically. One way would be to add huge quantities of soap. Eventually, enough stearate ion would be added to precipitate all the objectionable Ca^{2+}, leaving the excess soap to carry on the cleansing action.

Another way (this works only for temporary hardness) is to boil the water. The reaction

$$Ca^{2+} + 2HCO_3^- \rightleftharpoons CaCO_3(s) + H_2O + CO_2(g)$$

is reversible, but the forward reaction can be made dominant by boiling off the CO_2. Boiling, however, is not practical for large-scale softening.

The third way to soften water is to precipitate the Ca^{2+} by adding washing soda, Na_2CO_3. The added carbonate ion, CO_3^{2-}, reacts with Ca^{2+} to give insoluble $CaCO_3$. If bicarbonate ion is present, the water may be softened by adding a base such as ammonia. The base deprotonates HCO_3^- to produce CO_3^{2-}, which then precipitates the Ca^{2+}. On a large scale, temporary hardness is often removed by adding limewater. The added OH^- reacts with HCO_3^- and precipitates $CaCO_3$ by the process

$$Ca^{2+} + HCO_3^- + OH^- \longrightarrow CaCO_3(s) + H_2O$$

It might seem odd that limewater, which itself contains Ca^{2+}, can be added to hard water to remove Ca^{2+}. Yet it should be noted that when $Ca(OH)_2$ is added, there are two moles of OH^- per mole of Ca^{2+}. Two moles of OH^- neutralize two moles of HCO_3^- and liberate two moles of CO_3^{2-}, thus precipitating two moles of Ca^{2+}—one that was added and one that was originally in the hard water.

A fourth way to soften water is to tie up the Ca^{2+} so that it becomes harmless. One way to do this is to form a complex containing Ca^{2+}. Certain phosphates, such as $(NaPO_3)_n$, sodium polyphosphate, act as sequestering agents by forming complexes in which the Ca^{2+} is trapped by the phosphate.

The fifth and most clever way to soften water is to replace the offending calcium ion by another ion such as Na^+. This is done by the process called *ion exchange*.

16.9 Ion exchange

An *ion exchanger* is a special type of giant molecule consisting of a three-dimensional cross-linked network containing covalently bound atoms which carry an excess of negative charge. The molecule is thus a negatively charged network with a very porous structure. The pores are filled with water molecules and enough positive ions to give an electrically neutral structure. The identity of the positive ions is not very important since their only function is to preserve electric neutrality. Consequently, one type of cation such as Ca^{2+} can take the place of another type such as Na^+ without much change in structure. It is this kind of ion exchange which is used in water softening. Hard water containing Ca^{2+} is placed in contact with an ion exchanger whose mobile ion is Na^+. Exchange then occurs; it can be represented by the equilibrium

$$Ca^{2+} + 2Na^+\ominus \rightleftharpoons 2Na^+ + \ominus Ca^{2+}\ominus$$

where the circled minus sign represents a negative site on the exchanger. The equilibrium constant for this reaction is usually on the order of 10 or less, and, therefore, in order to remove all the Ca^{2+} it is necessary to run the hard water through a large amount of ion exchanger. This is most conveniently done by pouring the water through a long tube filled with ion exchanger. Once the exchanger has given up its supply of Na^+, it cannot soften water further. However, it can be regenerated by treatment with a concentrated solution of sodium chloride, which reverses the above reaction.

The ion exchangers originally used for softening water were naturally occurring silicate minerals called *zeolites*. The giant silicate network of a zeolite is negatively charged and is composed of covalently bound silicon, oxygen, and aluminum atoms. Mobile Na^+ ions in the pores can be readily exchanged for Ca^{2+} ions. Zeolites are very closely related in structure to the clays, which also show ion exchange. Such ion exchange is important for plant nutrition since many plants receive nourishment from the soil in this fashion.

With the advent of high-polymer techniques, chemists have been able to synthesize ion exchangers superior to the zeolites. The most common synthetic exchanger consists of a giant hydrocarbon framework called a *resin* having a negative charge due to covalently bound SO_3^- groups. It has also been possible to prepare ion exchangers in which the resin network is positively charged, the charge being due to covalently bound groups of the type $N(CH_3)_3^+$. Such positively charged networks can function as *anion* exchangers; i.e., they have mobile negative ions which can be displaced by other anions.

Combination of synthetic anion exchangers with cation exchangers has made possible the removal of all ions from a salt solution. If a salt solution containing M^+ and A^- is first run through a cation exchanger whose exchangeable ions are H_3O^+, the salt solution is completely converted to a solution of an acid containing H_3O^+ and A^-. If now the acid solution is run through an anion exchanger whose

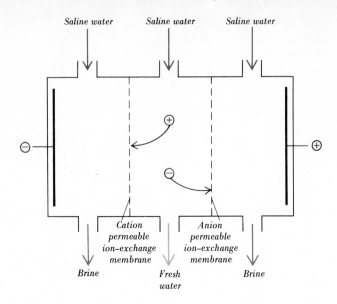

Saline water   Saline water   Saline water

⊖ — 　　　　　　　　　　　　　 — ⊕

⊕

⊖

Fig. 16.8
Electrodialysis cell for
getting fresh water from salt
water.

Cation
permeable
ion–exchange
membrane

Anion
permeable
ion–exchange
membrane

Brine　　　　　Fresh
　　　　　　　water　　　　　Brine

exchangeable ions are OH^-, the anions A^- in the solution are replaced by OH^-. Since in the original solution the number of negative charges is exactly equal to the number of positive charges, equal amounts of H_3O^+ and OH^- are exchanged into the solution. Neutralization occurs, and pure water results. Water thus deionized contains fewer ions than the most carefully distilled water.

Ion-exchange resins have recently become important as membranes for removing soluble salts from waste water. One of the great problems of water pollution (Chap. 26) is that, although solid wastes can be removed by filtration or settling and human or animal wastes can be chewed up by activated-sludge methods, there is no simple way to get out the soluble inorganic salts. Ordinary use of a water supply by an average city raises the salt content of the water by about 50 percent. This makes for problems when we attempt to recycle the water. The salt problem is of course the same as that faced by large cities on the edge of the sea which have had their populations outrace the fresh-water supply. How might one desalt water which contains too much dissolved salt?

One promising method is the *electrodialysis* method, in which positive and negative ions are separated out of a flowing current of brackish water by being made to pass through ion-exchange membranes under the influence of an electric field. Figure 16.8 shows a typical electrodialysis cell, many of which are arranged in tandem so that a small separation effect that occurs in one cell can be multiplied many times over. Salt water is fed into the three input pipes at the top of the cell. An electric field of polarity as shown tends to make the positive and negative ions drift off in opposite directions. By proper choice of resins for the ion-exchange membranes (shown by the dashed lines), we can make the left one permeable to cations and the right one permeable to anions. Specifically, the left membrane could be a cation exchanger (i.e., a resin containing SO_3^- groups), and the right membrane could be an anion exchanger [i.e., a resin containing $N(CH_3)_3^+$ groups]. Under proper flow conditions, the water coming out of the middle effluent pipe will be considerably less saline than that coming out of the two outside pipes.

Another promising technique is what is called *reverse osmosis*. In the reverse-osmosis process, instead of taking the waste out of the water, the water is squeezed out of the waste. Figure 16.9 shows a schematic representation of the setup. Salt water is fed into the top of the cell, the bottom part of which is blocked off by a semipermeable membrane. (Membrane materials can be polystyrene, cellophane, polyvinylchloride, or ethylcellulose. None of these is very satisfactory since they all go to pieces quite rapidly; so the search for an appropriate material continues.) Normally, fresh water tends to move through a semipermeable membrane toward the salty side, but by putting a sufficiently large

Fig. 16.9
Reverse-osmosis cell for getting fresh water from salt water.

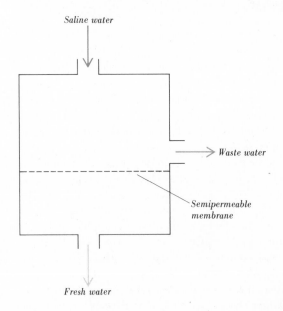

Saline water

Waste water

Semipermeable membrane

Fresh water

pressure on the inflowing saline water the normal osmotic flow can be reversed, and fresh water literally squeezed through the membrane so as to come out the bottom pipe.

Neither of the above processes is very cheap, but in water-scarce installations they are beginning to prove their value. One hopeful application for the future would be removal of nitrate ion from private water supplies. Nitrates are very soluble, and recent generous use of nitrate fertilizers has raised NO_3^- levels in some groundwaters to dangerous levels. The danger appears to be particularly great for very young infants (less than 2 months old), who are particularly susceptible to methemoglobinemia (*blue-baby syndrome*) owing to oxidation by nitrite of the oxygen-carrying iron in hemoglobin from oxidation state $+2$ to $+3$. Water intake by infants is disproportionately large, and the infant's digestive equipment is more likely to harbor the wrong kind of bacteria, which reduce nitrate to nitrite. Nitrate removal remains an unsolved problem, though ion exchange is a feasible but expensive way out.

16.10 Qualitative analysis

Because the alkali metals do not form many insoluble compounds and because the alkali ions are colorless, it is difficult to detect the presence of these elements by chemical methods. Instead, their presence is usually shown by running flame tests on the sample in question. The simplest way to run a flame test is to shape a piece of fine platinum wire into a loop, dip the loop in hydrochloric acid solution, heat to remove volatile impurities, and then use the loop to heat the sample in a burner flame. The sodium yellow is extremely intense; so even traces of it can mask other flame colors. The main reason for cleaning the platinum loop by the hydrochloric acid treatment is to help expel sodium as the relatively volatile chloride. (In general, chlorides are more volatile than most other solids.) The potassium flame is colored a delicate violet and can be observed in many cases only through cobalt glass, which filters out interfering colors such as sodium yellow. The flames of potassium, rubidium, and cesium look so similar that definite identification requires examination of the line spectra with a spectroscope. The strongest lines are lithium, 670.8 nm; sodium, 589.0 and 589.6 nm; potassium, 766.5 and 769.9 nm; rubidium, 420.2 and 421.5 nm; and cesium, 455.6 and 459.3 nm.

As a group, the alkaline-earth cations (excluding beryllium) can be distinguished from other common cations by taking advantage of the fact that, like group I elements, they form soluble sulfides but, unlike group I elements, they form insoluble carbonates.

Given a solution containing alkaline-earth cations, the barium can be precipitated as yellow $BaCrO_4$ by addition of K_2CrO_4 in the presence of an acetic acid buffer. From the residual solution (containing Sr^{2+}, Ca^{2+}, and Mg^{2+}), light yellow $SrCrO_4$ can be precipitated by subsequent addition of NH_3 and alcohol. The $BaCrO_4$ precipitates in the first step, and $SrCrO_4$ in the second, because $BaCrO_4$ ($K_{sp} = 8.5 \times 10^{-11}$) is less soluble than $SrCrO_4$ ($K_{sp} = 3.6 \times 10^{-5}$). The point of using an acetic acid buffer (Sec. 12.5) is to keep the H_3O^+ concentration around 10^{-5}, where the chromate concentration, governed by the equilibrium

$$2CrO_4{}^{2-} + 2H_3O^+ \rightleftharpoons Cr_2O_7{}^{2-} + 3H_2O$$

is too low to precipitate Sr^{2+} but high enough to precipitate Ba^{2+}. Subsequent addition of NH_3 reduces the H_3O^+ concentration, thereby increasing the $CrO_4{}^{2-}$ concentration sufficiently to precipitate $SrCrO_4$, especially in the presence of alcohol, which lowers its solubility.

Calcium ion can be separated from magnesium ion by addition of ammonium oxalate to form white, insoluble calcium oxalate, CaC_2O_4. (The K_{sp} of CaC_2O_4 is 1.3×10^{-9}, compared with 8.6×10^{-5} for

MgC_2O_4.) Finally, the presence of Mg^{2+} can be shown by adding more NH_3 and Na_2HPO_4, which precipitates white magnesium ammonium phosphate, $MgNH_4PO_4$.

Exercises

*16.1   Isoelectronic sequence. Using data from Appendix 8 for ionic radii of ions isoelectronic with Na^+ and Mg^{2+}, predict a value for the atomic radius of Ne^0.

*16.2   Chemical equations. Write balanced net reactions for each of the following processes:
  a   Electrolysis of molten barium hydroxide
  b   Oxidation of metallic sodium by air to give sodium peroxide
  c   Dissolving of metallic calcium in water
  d   Oxidation of metallic barium by carbon dioxide-containing air to give $BaCO_3$
  e   Burning of CaH_2 in oxygen

*16.3   Concentrations. Seawater contains 16,600 kg of chlorine (Cl^-), 9200 kg of sodium, and 1180 kg of magnesium per million liters of seawater. What is the molarity of each, and what concentration of charge is unaccounted for?

Ans. 0.47 M Cl^-, 0.40 M Na^+,
0.050 M Mg^{2+}, 0.03 M negative charge

*16.4   Stoichiometry. On being exposed to air, a piece of sodium weighing 5.698 g increases in weight to 5.896 g. Assuming the only reaction has been conversion to Na_2O_2, what percent of the original sodium has been oxidized?

*16.5   Thermodynamics. From data given in Sec. 16.7, calculate the $\Delta H°$ for the following process:

$$Ca(OH)_2(s) + CO_2(g) \longrightarrow CaCO_3(s) + H_2O \qquad Ans. \ -111 \ kJ$$

*16.6   Qualitative analysis. A solution possibly containing Mg^{2+}, Ba^{2+}, Sr^{2+}, and Ca^{2+} is treated with $(NH_4)_2CO_3$, and a precipitate forms. The unseparated mixture is acidified with acetic acid, and the precipitate completely dissolves. Subsequent addition of K_2CrO_4 produces a yellow precipitate. Filtered off, the residual solution produces no more precipitate on addition of NH_3 and alcohol. On the basis of these tests, state for each ion whether it is present, absent, or undetermined.

*16.7   Hard water. How would you expect the hardness of groundwater to differ in a gypsum region from a limestone region? Indicate how you could economically soften the two kinds of water. Write chemical equations for all reactions involved.

*16.8   Qualitative analysis. What single reagent or operation would enable you to distinguish between the compounds in each of the following pairs?

- *a* $BaCO_3$ and $BaSO_4$
- *b* $BaCO_3$ and $Ba(NO_3)_2$
- *c* $MgSO_4$ and $SrSO_4$
- *d* KNO_3 and $Ca(NO_3)_2$
- *e* KNO_3 and $NaNO_3$

** *16.9* *Crystal structures.* Sodium crystallizes with a body-centered cubic structure in which the unit-cell edge length is 0.424 nm. Calcium crystallizes with a face-centered cubic structure in which the unit-cell edge length is 0.556 nm. Assuming the atoms can be treated as hard spheres in contact, calculate the apparent atomic radius of sodium and calcium.

** *16.10* *Electron densities.* If sodium crystallizes in a body-centered cubic structure of unit-cell edge length 0.424 nm and calcium, face-centered cubic, 0.556 nm, what are the respective current-carrier electron densities?
Ans. 2.62 and 4.66 × 10²² electrons per cubic centimeter

** *16.11* *Ionic radii.* LiI crystallizes in the NaCl-type lattice. Its unit-cell edge length is 0.600 nm. Assuming the anions are so much bigger than the cations that the cell dimension is fixed by anion contact, calculate the apparent radius of the I^- ion. What is the maximum size of Li^+ ion that would be consistent with this? Compare with data in Appendix 8.

** *16.12* *Ionization potential.* It would appear that numerical values of ionization potential could well be predicted for the alkali atoms by using a coulomb attraction term of the form $Z'e^2/r$, where Z' is the screened nuclear charge and r is the average distance of the valence electron from the nucleus. Using the ionic radius of M^+ to estimate r, show how well this works for group I elements by plotting ionization potential versus $1/r$. Try this for group II, and suggest a reason for any observed difference.

** *16.13* *Ionization.* If you add 40.0 kJ of energy to 1.00 g of magnesium atoms in the vapor state, what is the probable makeup of the final mixture formed? *Ans. 83.7% Mg^+, 16.3% Mg^{2+}*

** *16.14* *Electrode potentials.* Calculate the $E°$ value for $2e^- + M^{2+}(aq) \longrightarrow M(s)$ from the following free-energy data: sublimation, 300 kJ; first ionization potential, 4.00 eV; second ionization potential, 8.50 eV; hydration of $M^+(g)$, -400 kJ; and hydration of $M^{2+}(g)$, -1600 kJ. *Ans. 5.31 V*

** *16.15* *Electrode potentials.* Given that $E° = +1.25$ V for $M^{3+}(aq) + 2e^- \longrightarrow M^+(aq)$ and $E° = -0.34$ V for $M^+(aq) + e^- \longrightarrow M(s)$, calculate the $E°$ for $M^{3+}(aq) + 3e^- \longrightarrow M(s)$.
Ans. +0.72 V

** *16.16* *Ion species.* On the basis of the following free-energy data for element X, predict which ionic species should be most stable in aqueous solution: sublimation, 340 kJ; first ionization potential, 7.43 eV; second ionization potential, 15.7 eV; third ionization potential, 32.0 eV; hydration of $X^+(g)$, -395 kJ; hydration of $X^{2+}(g)$, -2100 kJ; and hydration of $X^{3+}(g)$, -4500 kJ.

** 16.17   *Thermodynamics.* From data given in this chapter calculate $\Delta G°$ for each of the following processes:

 a $Li^+(g) + Na(g) \longrightarrow Li(g) + Na^+(g)$
 b $Li^+(g) + Ca(g) \longrightarrow Li(g) + Ca^+(g)$
 c $2Li^+(g) + Ca(g) \longrightarrow 2Li(g) + Ca^{2+}(g)$
 d $Mg^{2+}(g) + Ca(g) \longrightarrow Mg(g) + Ca^{2+}(g)$

Ans. -24, $+69$, $+700$, -450 kJ

** 16.18   *Corrosion.* When exposed to air, beryllium does not corrode, but barium does. One possible explanation for this is that beryllium forms a tightly protective oxide coat whereas barium does not. Given that the density of BeO is 3.01 g/cm^3 and that of BaO is 5.72 g/cm^3, calculate what happens to the volume per atom when each of these metal atoms goes from the metallic state to the oxide state.

** 16.19   *Solubility.* The K_{sp} of $CaSO_4$ is 2.4×10^{-5}. Calculate the concentration of Ca^{2+} ion in a 0.0100 M Na_2SO_4 solution that is saturated with $CaSO_4$.   *Ans. 2.0×10^{-3} M*

** 16.20   *Ion exchange.* The total ion-exchange capacity of a cation-exchange resin can be evaluated by first converting the resin completely to the acid form, washing it with NaCl solution, and titrating the wash solution with NaOH. In a given experiment 25.0 cm^3 of resin required 141.3 ml of 0.600 N NaOH solution to neutralize the wash solution. What is the capacity of the resin expressed in milliequivalents per cubic centimeter? How many liters of hard water containing 0.0015 M Ca^{2+} could be softened by 1000 cm^3 of exchanger, assuming 100 percent efficiency?

** 16.21   *Electrodialysis.* Suppose you have an electrodialysis cell which is divided into three compartments by a cation-permeable membrane and an anion-permeable membrane, respectively. Aqueous KNO_3 is placed in the middle compartment; water, in the outer two. Electrodes are placed in the outer compartments, and current is passed through the cell. Indicate what ion migrations and electrode reactions will probably occur. What chemical change will occur in the compartments?

*** 16.22   *Ionization potentials.* Suggest a reason why the second ionization potential of the alkali atoms falls off more rapidly with increasing atomic number than does the first ionization potential.

*** 16.23   *Hydration.* Using data given elsewhere in the book and given that $\Delta G° = -465$ kJ for $e^- + H_3O^+(aq) \longrightarrow \frac{1}{2}H_2(g) + H_2O$, calculate the free-energy change on hydration of $H^+(g)$. H_3O^+ has a radius of about 0.140 nm; so its hydration free energy should be about equal to that of $Rb^+(g)$. What would be an approximate value to assign to $\Delta G°$ for the process $H^+(g) + H_2O(g) \longrightarrow H_3O^+(g)$?

Ans. −1062, −785 kJ

*** **16.24** *Photoelectric effect.* How many electrons can be removed from metallic cesium with the energy required to remove one electron from an isolated cesium atom?

*** **16.25** *Solubility.* Given that $K_{sp} = 2.4 \times 10^{-5}$ for $CaSO_4$ and $K_{sp} = 7.6 \times 10^{-7}$ for $SrSO_4$, what would be the final concentrations of Ca^{2+}, Sr^{2+}, and SO_4^{2-} in a solution made by mixing 0.0100 liter of 0.300 M Na_2SO_4 with 0.0200 liter of a solution that is initially 0.100 M Ca^{2+} and 0.100 M Sr^{2+}?

Ans. 3.4 × 10⁻², 1.1 × 10⁻³, 7.1 × 10⁻⁴ M

*** **16.26** *Solubility.* Given that the K_{sp} of $CaCO_3$ is equal to 4.7×10^{-9} and the first and second dissociation constants of CO_2 in H_2O are 4.16×10^{-7} and 4.84×10^{-11}, respectively, calculate the equilibrium constant for $CaCO_3(s) + CO_2(aq) + H_2O \rightleftharpoons Ca^{2+} + 2HCO_3^-$. Using the constant, figure out how many grams of $CaCO_3$ can dissolve in one liter of H_2O that is saturated with CO_2 under normal conditions. The atmosphere generally contains about 0.04% CO_2; under 1 atm CO_2 pressure the solubility of CO_2 is 0.034 M.

*** **16.27** *Thermodynamics.* For $MgO(s)$ at 25°C the standard free energy of formation is −570 kJ/mol, and the standard enthalpy of formation is −602 kJ/mol. Calculate the ΔG for reaction of $Mg(s)$ with atmospheric oxygen at 500°C. Assume barometric pressure is 1.00 atm and air is 21% oxygen.

Ans. −514 kJ

*** **16.28** *Solubility.* If $SrCrO_4$ ($K_{sp} = 3.6 \times 10^{-5}$) is soluble in an acetic acid buffer but $BaCrO_4$ ($K_{sp} = 8.5 \times 10^{-11}$) is not, what minimum value can you set for the equilibrium constant of the reaction $2CrO_4^{2-} + 2H_3O^+ \longrightarrow Cr_2O_7^{2-} + 3H_2O$? Take 0.10 M as your criterion for solubility.

*** **16.29** *Qualitative analysis.* Mg^{2+} can be separated from Ca^{2+}, Sr^{2+}, and Ba^{2+} by addition of a solution that is 0.50 M $(NH_4)_2CO_3$ and 1.0 M NH_3. Calculate the concentration of CO_3^{2-} in this solution, and show that it is sufficient to precipitate $CaCO_3$ ($K_{sp} = 4.7 \times 10^{-9}$) but not $MgCO_3$ ($K_{sp} = 1 \times 10^{-4}$) from a solution that is 0.001 M Ca^{2+} and 0.001 M Mg^{2+}. K_{II} for $HCO_3^- + H_2O \rightleftharpoons H_3O^+ + CO_3^{2-}$ is 4.8×10^{-11}; K for $NH_3 + H_2O \rightleftharpoons NH_4^+ + OH^-$ is 1.8×10^{-5}.

Ans. .087 M

*** **16.30** *Solubility.* Suppose you shake up 0.100 mol of BaF_2 with 0.200 mol of $SrSO_4$ in one liter of water until equilibrium is established. What will be the concentrations of the various ions in the final solution? The values of K_{sp} are as follows: BaF_2, 2.4×10^{-5}; $SrSO_4$, 7.6×10^{-7}; SrF_2, 7.9×10^{-10}; and $BaSO_4$, 1.5×10^{-9}.

Intervening between groups II and III in the periodic table are subgroups of elements collectively referred to as the *transition elements*. The precise definition of "transition element" is a matter of choice, and frequently there is ambiguity as to whether a given element is included in the classification or not. Using the form of the periodic table shown in Fig. 1.4, we define transition elements as all those in the 10 subgroups intervening between main groups II and III. Thus, in the fourth period the transition elements are scandium ($Z = 21$), titanium (22), vanadium (23), chromium (24), manganese (25), iron (26), cobalt (27), nickel (28), copper (29), and zinc (30). Each of these elements heads a subgroup named after itself. The titanium subgroup, for instance, includes the elements titanium ($Z = 22$), zirconium (40), hafnium (72), and kurchatovium (104). In this and the three succeeding chapters, we consider the properties characteristic of the transition elements.

472

17.1 Electronic configuration

Transition elements owe their separate classification to belated filling of the next-to-outermost energy level of the atoms, i.e., the one with principal quantum number $n - 1$ (Sec. 2.4). The belated filling of the $3d$ orbital, for example, occurs because there is a sharp drop in the $3d$ energy level relative to the $4s$ around $Z = 21$, *after* some electrons have started to add to $4s$. Because s orbitals have appreciable probability density close to the nucleus, any increase in Z is immediately felt by s electrons. Not so for $3d$. Since d electrons have most of their probability density away from the nucleus, step-up of Z will not affect any d electrons so long as electrons are simultaneously added to screen between them and the nucleus. It is only when electrons are added outside the $3d$ probability region that they do not screen the increasing Z of the nucleus and, hence, the $3d$ level gets pulled in. Crudely put, it is as if the $3d$ orbital were concentrated on the outer periphery of the atom until about $Z = 21$, where it is more or less abruptly sucked in.

In Fig. 17.1 are given the detailed electronic configurations of the first-row transition elements. The significant difference in these configurations is that the third principal quantum level ($n = 3$) is gradually built up to 18 electrons by progressive addition to the $3d$ subshell. With the exception of chromium and copper, there are two electrons in the fourth shell (the $4s$ subshell). The apparent anomaly of chromium is due to the fact that the $3d$ and $4s$ subshells are very close in energy at this point and there is actually a decrease in energy because of loss of electron-electron repulsion on breaking up the $4s^2$ pair. In copper the dropping of a $4s$ electron to the $3d$ subshell does not lose electron-electron repulsion energy, but by this time Z has increased so much that $3d$ gets pulled lower than $4s$. Actually, minor irregularities in electron configuration may be of little practical interest since the configurations have been determined for the gaseous atoms and do not necessarily hold for other states.

When the electrons are removed from transition-metal atoms, it is not always obvious from the electronic configuration of the atom which electronic levels will be depleted. On the basis of the order of orbital

Element	Symbol	Z	Electron configuration
Scandium	Sc	21	$1s^2 2s^2 2p^6 3s^2 3p^6 3d^1 4s^2$
Titanium	Ti	22	——(18)——$3d^2 4s^2$
Vanadium	V	23	——(18)——$3d^3 4s^2$
Chromium	Cr	24	——(18)——$3d^5 4s^1$
Manganese	Mn	25	——(18)——$3d^5 4s^2$
Iron	Fe	26	——(18)——$3d^6 4s^2$
Cobalt	Co	27	——(18)——$3d^7 4s^2$
Nickel	Ni	28	——(18)——$3d^8 4s^2$
Copper	Cu	29	——(18)——$3d^{10} 4s^1$
Zinc	Zn	30	——(18)——$3d^{10} 4s^2$

Fig. 17.1

First-row transition elements.

filling in the buildup of the periodic table, it might seem that since the $3d$ electrons are added after the $4s$, they should on ionization be removed before the $4s$. However, this prediction is unwarranted because the two processes differ from each other in a major way. In the buildup of the periodic table, the number of electrons is being increased at the same time that the nuclear charge is being increased. On the other hand, in the ionization of a given atom the number of electrons is decreased at the same time that the nuclear charge stays constant. The problem is actually a very complicated one, since interelectron repulsion is also an important factor. The experimental fact is that $4s$ electrons are removed before $3d$ electrons in the ionization of transition-element atoms. The electronic configurations of the transition-element ions will be considered again later.

For the second-row transition elements, yttrium ($Z = 39$) through cadmium ($Z = 48$), the electronic expansion involves the $4d$ and $5s$ subshells, as shown in Fig. 1.18 on page 40. For the third-row transition elements, lanthanum ($Z = 57$) through mercury ($Z = 80$), a new problem arises. As shown on page 493, not only are the $5d$ and $6s$ subshells involved in the expansion, but also the $4f$ subshell is being filled to 14 electrons. The elements involved in the $4f$ expansion are called the lanthanide elements; they are discussed in Sec. 18.2. A similar problem, involving the $5f$ expansion, occurs in the last row of the periodic table, giving rise to the actinide elements. The actinides are discussed in Sec. 18.3; their electronic configurations are given on page 497.

17.2 Metallic properties

The most characteristic property of the transition elements is that they are all metals. This is not surprising, since the outermost shell contains so few electrons. However, unlike the metals of groups I and II, the transition metals are likely to be hard, brittle, and fairly high melting (see Fig. 17.2). The difference is partly due to the relatively small size of the atoms (Fig. 2.24 shows that the atomic radii are consistently small) and partly to the existence of some covalent binding between the ions. There are exceptions to this general hardness, as in the case of mercury ($Z = 80$), which is a liquid and is about as soft as a metal can be.

Figure 17.2 lists some of the characteristic properties of the transition metals. In the structure column, standard abbreviations are used, namely, fcc, face-centered cubic; hcp, hexagonal close-packed; and bcc, body-centered cubic. These structures are described in Sec. 7.4. The close-packed arrays, fcc and hcp, have a coordination number of 12, that is, 12 near neighbors for each atom; bcc has 8 for the number of near neighbors. Thus in most cases the transition metals are rather densely packed and have the high coordination numbers characteristic of good conductors. The conductivity values listed in the last column of Fig. 17.2 are given in units of ohm^{-1} cm^{-1}, that is, the reciprocal

Fig. 17.2

Metal	Structure	Density, g/cm^3	Melting point, °C	Conductivity, 0°C ohm^{-1} cm^{-1}
Sc	fcc	3	1200	
Ti	hcp	4.5	1660	1.2×10^4
V	bcc	6.0	1710	1.7×10^4
Cr	bcc	6.9	1600	6.5×10^4
Mn	bcc	7.4	1260	1.1×10^4
Fe	bcc fcc	7.9	1535	11.2×10^4
Co	fcc hcp	8.7	1490	16×10^4
Ni	fcc	8.9	1450	16×10^4
Cu	fcc	8.9	1083	64.5×10^4
Zn	hcp	7.1	419	18.1×10^4
Y	hcp	5.5	1490	
Zr	hcp bcc	6.4	1860	2.4×10^4
Nb	bcc	8.6	1950	4.4×10^4
Mo	bcc	10.2	2620	23×10^4
Tc	hcp	11.5	(2100)	
Ru	hcp	12.4	2450	8.5×10^4
Rh	fcc	12.4	1970	22×10^4
Pd	fcc	12.0	1550	10×10^4
Ag	fcc	10.5	961	66.7×10^4
Cd	hcp	8.7	321	15×10^4
La	hcp fcc	6.2	890	1.7×10^4
Hf	hcp	13.3	2200	3.4×10^4
Ta	bcc	16.6	>3000	7.2×10^4
W	bcc	19.3	3370	20×10^4
Re	hcp	20.5	3200	5.3×10^4
Os	hcp	22.7	2700	11×10^4
Ir	fcc	22.6	2450	20×10^4
Pt	fcc	21.5	1774	10.2×10^4
Au	fcc	19.3	1063	49×10^4
Hg	rhombic	14.2	−39	4.4×10^4

Properties of the transition metals.

of one ohm-centimeter,* which is the electric resistance of a one-centimeter-long wire that is one square centimeter in cross section. As illustration, the specific conductivity of sodium metal at 0°C is 23×10^4 ohm^{-1} cm^{-1}. In summary, the transition metals are generally good conductors, and silver, copper, and gold are outstandingly good.

The high electric conductivity of the transition metals can be attributed to a delocalization (i.e., spreading out over many atoms) of the *s* electrons similar to what occurs in the alkali and alkaline-earth metals. The differences in other physical properties (e.g., hardness, brittleness,

---

* It has been recommended that the reciprocal ohm as the unit of conductivity should be called the *siemens*, after Ernst Werner von Siemens, German inventor who devised an electric standard of resistance based on mercury.

and melting point) can be ascribed to covalent binding involving overlapping of partly filled d subshells of the transition-metal atoms.

17.3 Oxidation states

One of the most frequently marked characteristics of a typical transition element is the great variety of oxidation states it may show in its compounds. Figure 17.3 lists the more common states found. Included also (in parentheses) are less common states, such as those which are unstable to disproportionation in aqueous solution or which have been prepared in only a few solid-state compounds. It might be noted that there is presently considerable research activity in attempting to prepare unusual oxidation states of the transition elements; so the listing in the table should not be considered complete.

In addition to showing the large number of oxidation states, Fig. 17.3 shows several marked features. In each row there is a peaking in the numerical value of the maximum state near the middle of the row, after which the maximum state gets smaller. Thus, for the first row the maximum oxidation state increases regularly from $+3$ for Sc to $+7$ for Mn, after which there is a falloff to $+2$ for Zn. Connected with this trend is the fact that elements toward the center of the row generally show more oxidation states than those toward the ends. A careful look at the table discloses one other feature. In going down a subgroup, there is generally a trend toward favoring higher oxidation states. For example, in the Fe subgroup the $+2$ and $+3$ states, which predominate for Fe, give way to $+4$, $+6$, and $+8$, which predominate for Os.

Before making too much fuss over probable reasons for the above trends in oxidation state, we should remind ourselves that the concept of oxidation state is highly artificial and rests on a rather arbitrary assignment of shared electrons to more electronegative atoms. Nevertheless, it is helpful to associate the increasing maximum oxidation state with the increasing number of s and d electrons available for binding. The falloff after the peak in the center of the row can be related to the lowering of the d-shell energy relative to the s, and hence to the decreasing availability of those d electrons for binding. Similarly, it is helpful to connect the increasing preference for higher states down a subgroup with increasing availability of d electrons relative to s electrons as atomic size increases.

17.4 Ligand-field theory

Another frequently remarked feature of the transition elements is their relatively great tendency to form complex ions. In such complexes the transition-metal atoms are surrounded by a definite number of bound groups whose charge clouds exert a marked influence on the electronic makeup of the transition element. These bound groups, whether they are ions such as F^- or CN^- or molecules such as H_2O or NH_3, are referred to as *ligands*, and their interaction with the central atom is

Sc	Ti	V	Cr	Mn	Fe	Co	Ni	Cu	Zn
+3	(+2)	+2	+2	+2	+2	+2	+2	+1	+2
	+3	+3	+3	(+3)	+3	+3	(+3)	+2	
	+4	+4	+4	+4	(+4)	(+4)			
		+5	+6	(+6)	(+6)				
				+7					

Y	Zr	Nb	Mo	Tc	Ru	Rh	Pd	Ag	Cd
+3	+4	+3	+3	+4	+2	+3	+2	+1	+2
		+5	+4	(+6)	+3	+4	(+3)	(+2)	
			+5	+7	+4	(+6)	+4	(+3)	
			+6		(+5)				
					+6				
					(+7)				
					(+8)				

La	Hf	Ta	W	Re	Os	Ir	Pt	Au	Hg
+3	+4	(+4)	(+2)	(+3)	(+2)	(+2)	+2	+1	+1
		+5	(+3)	+4	(+3)	+3	(+3)	+3	+2
			+4	(+5)	+4	+4	+4		
			+5	+6	+6	(+6)			
			+6	+7	+8				

Fig. 17.3
Oxidation states of transition elements (less common states are given in parentheses).

the subject of *ligand-field theory*. Historically, the theory developed from a consideration of how atomic energy levels of ions in crystals are affected by the ionic surroundings. The treatment so developed, called *crystal-field theory*, forms the basis for the broader ligand-field theory. The latter may be visualized as crystal-field theory with superimposed molecular-orbital concepts.

Let us consider the crystal-field problem of a Ti^{3+} ion in an oxide lattice (e.g., in Ti_2O_3, where the Ti^{3+} occupies octahedral holes in a close-packed array of oxide ions). The electron configuration of Ti^{3+} is $1s^22s^22p^63s^23p^63d^1$. Except for $3d^1$, the electron subshells are each filled and hence spherically symmetric. As far as $3d^1$ is concerned, the electron can be accommodated in one of the five d orbitals d_{xy}, d_{yz}, d_{zx}, d_{z^2}, or $d_{x^2-y^2}$ (Sec. 2.1). Which of these orbitals is likely to be the one occupied? For an isolated ion in the gas phase, it makes no difference since all five of these orbitals are of equal energy. However, in a crystal in which there are interacting neighbors the answer depends on the disposition of these neighbors. Imagine the Ti^{3+} ion to be located at the origin of a set of axes, as shown in Fig. 17.4. For octahedral coordination there will be one ligand equally distant from the origin on each of the six axial directions. If the set of six ligands is imagined to be drawn in uniformly toward the metal ion, what effect will there be on the $3d$ electron? Specifically, which of the five $3d$ orbitals will most likely be occupied? Of the five orbitals, two, d_{z^2} and $d_{x^2-y^2}$ (see Fig. 17.5), correspond to high electron density

Fig. 17.4
Octahedral neighbors of
Ti^{3+} *in* Ti_2O_3.

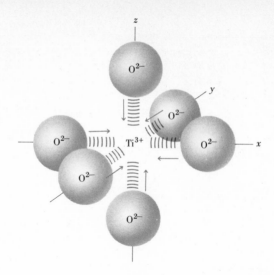

along the axes, and the other three, d_{xy}, d_{yz}, and d_{zx}, to high electron density along the 45° lines between axes.

Clearly, there will be stronger repulsion between ligand and electron when the electron is in one of the d orbitals that are concentrated along the axes (d_{z^2} or $d_{x^2-y^2}$). The repulsion will be less if the electron is in one of the d orbitals that are directed between the ligands with minimum charge density on the axes (d_{xy}, d_{yz}, and d_{zx}). The result is that, as shown in Fig. 17.5, the set of d orbitals separates into two subsets, a higher-energy one (called e_g) having high electron density on the axes and a lower-energy one (called t_{2g}) having high electron density away from the axes. For a Ti^{3+} ion the lowest-energy configuration would correspond to having the $3d$ electron in one of the t_{2g} orbitals.

The problem of calculating Δ, the energy difference between the two subsets of d orbitals in an octahedral environment, is a most difficult one. A calculation based on the assumption of only electric repulsions between charge clouds, as in the ionic model described above, is not completely satisfactory. The energy-level difference so calculated is usually considerably smaller than that actually observed by spectroscopic techniques (Sec. 17.6). The principal reason for the quantitative discrepancy between experiment and prediction from crystal-field theory lies in the initial assumption of a completely ionic model. In other words, it is not correct to assume that when a transition-element ion is surrounded by ligands, the only electronic change is within the ion. There must be considerable mixing of the charge clouds of the ligands and the central atom—covalent bonding exists. Ligand-field theory, unlike crystal-field theory, allows for such electron delocalization over the entire ligand-metal atom complex.

In its most elegant form, ligand-field theory sets up molecular orbitals which are derived from wave functions on the central atom and its surrounding ligands. For example, in the case of titanium surrounded by oxygens we can get an idea of the molecular orbitals by considering

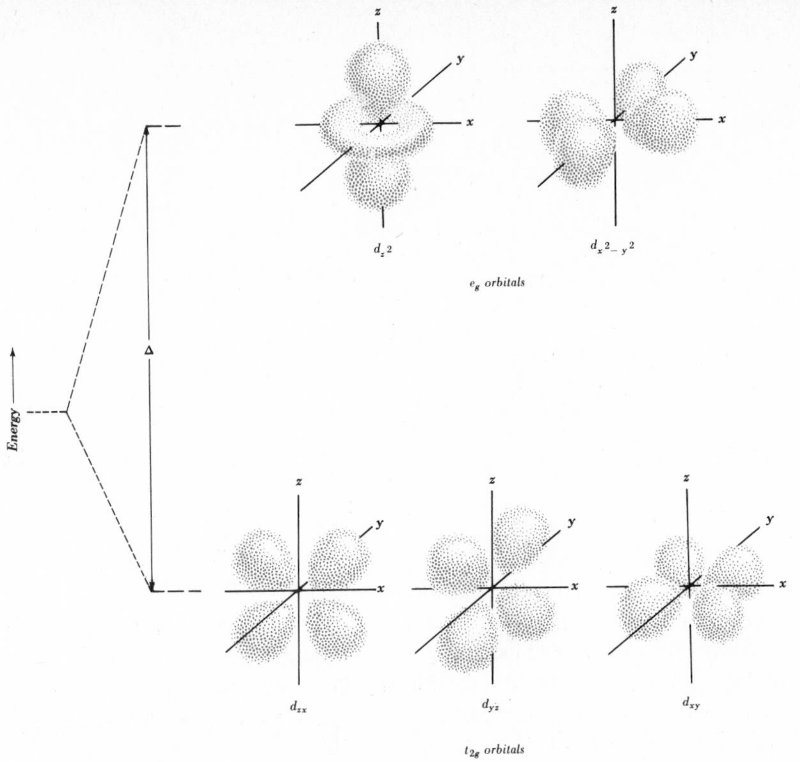

Fig. 17.5
Energy-level splitting of 3d
orbitals by octahedral
environment.

d_{z^2}

$d_{x^2-y^2}$

e_g orbitals

Δ

Energy

d_{zx}

d_{yz}

d_{xy}

t_{2g} orbitals

the overlap between d orbitals that point toward the oxygens (d_{z^2} and $d_{x^2-y^2}$) and properly oriented p orbitals of the oxygens. A schematic representation of setting up the overlap is shown in Fig. 17.6. As the ligands are brought up to their final positions, the atomic orbitals merge to give molecular orbitals. Because the electron density of the molecular orbitals is cylindrically symmetric about the bonding directions, these orbitals are said to be of the σ type (Sec. 3.9). Besides the orbitals shown in Fig. 17.6, there are other overlaps, especially with the $4s$ and the three $4p$ orbitals of the titanium. All told, from the two $3d$ (e_g), one $4s$, and three $4p$ orbitals of titanium and one $2p$ orbital from each of the six oxygen neighbors, there will be twelve molecular orbitals, six of which are bonding and six of which are antibonding. The relative disposition of these is shown in Fig. 17.7. The bonding orbitals are considerably lower in energy than either metal-atom or ligand orbitals; the antibonding orbitals are very much higher. Pairs of electrons that were initially in the ligand orbitals move down into these bonding molecular orbitals. Because oxygen is more electronegative than titanium, electrons shared in the bonding orbitals tend to be concentrated more toward oxygen than titanium. Also shown in Fig. 17.7 is the position of the relatively unaffected t_{2g} orbitals. These are labeled as nonbonding molecular orbitals. The lone $3d^1$ electron of Ti^{3+} is shown to be accommodated here.

In summary, when ligands are brought up to a transition-element ion, the d orbitals are affected differently. Those which point away from

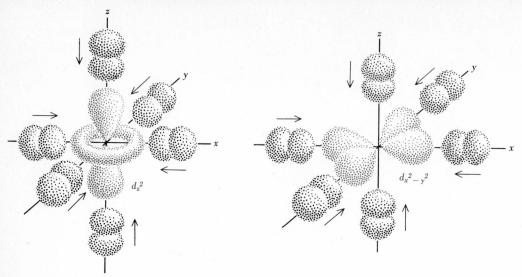

Fig. 17.6
Molecular-orbital formation from d orbitals of central atom and p orbitals of ligands.

the ligands d_{xy}, d_{yz}, and d_{zx} will be but slightly affected. Those which point toward the ligands d_{z^2} and $d_{x^2-y^2}$ merge with ligand orbitals to form molecular orbitals that bind the complex. Electron pairs from the ligands fill these bonding orbitals (effectively reducing the negative charge of the ligands and reducing the positive charge of the central ion). For the case of Ti^{3+} surrounded by six ligands, the single $3d$ electron will be accommodated in one of the d orbitals not involved in the bonding. Whether it is d_{xy}, d_{yz}, or d_{zx} is immaterial since in an octahedral environment all three are equivalent.

Finally, it must be noted that for transition-element ions having more than one d electron (for example, d^2 and d^3) the situation will be more complicated because of interelectron repulsions between the d electrons. Also, if the symmetry of the environment is not octahedral, the splitting of the d orbitals into subsets is different from that discussed above.

17.5 Magnetic properties

Because of electron spin, unpaired electrons give rise to paramagnetism, i.e., attraction into a magnetic field (Sec. 2.3). For the transition elements, where there is usually partial filling of the d subshell, paramagnetism is a likely possibility. Both in the metallic state and in compounds, transition elements may show paramagnetism resulting from presence of one or more unpaired electron spins. In a few cases (for example, Fe, Co, Ni, Fe_3O_4, and some Mn alloys) special reinforcement of paramagnetism occurs because of extensive cooperative electron-spin alignment; this gives rise to ferromagnetism (Sec. 19.1).

There are two properties of electrons in atoms that give rise to magnetism: One is the intrinsic electron spin, and the other is the moment associated with the orbital motion of the electron. In the lanthanides and actinides (in which f electrons are involved) both effects

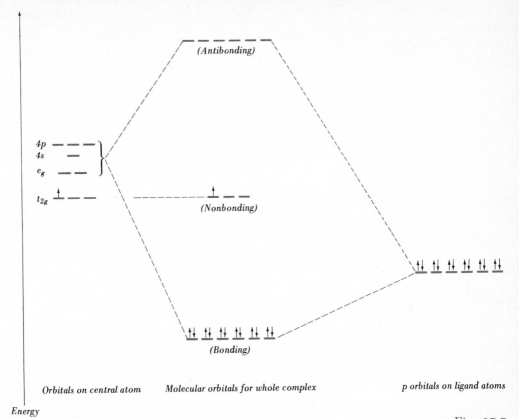

Orbitals on central atom   *Molecular orbitals for whole complex*   *p orbitals on ligand atoms*

Energy

*Fig. 17.7
Molecular-orbital diagram
for octahedral
titanium-oxygen complex.*

contribute to the measured magnetism. However, in most of the other transition elements the orbital moment contributes little to the measured magnetism, largely because interaction of d electrons with surrounding atoms prevents steady orientation of the orbital moment when a magnetic field is applied. For our purposes, the orbital moment of d electrons can be considered to be quenched, and we shall deal with the magnetic moment from the electron spin only. The magnetic moment due to electron spin can be calculated from the formula

Spin magnetic moment $= \sqrt{n(n+2)}$

where n is the number of unpaired electrons and the units are *Bohr magnetons.* In these units, a single $1s$ electron has a magnetic moment of 1.73 Bohr magnetons.

Figure 17.8 summarizes both the magnetic moments calculated from the spin-only formula and those derived from experimental data for hydrated ions of the first-row transition elements. In certain cases the data differ depending on the nature of the hydrated salt studied. Hence, there are a range of values given. As predicted by the spin-only calculation, as the number of unpaired electrons goes through a maximum at Mn^{2+} (d^5), so the magnetic moment peaks there too. In the first half of the series the agreement between calculated and experimental moments is excellent; in the second half of the series the experimental

Fig. 17.8
Calculated and observed
paramagnetic moments in
Bohr magnetons.

| Ion | e^- configuration | Unpaired e^- | Magnetic moment | |
			Calculated	Experimental
Sc^{3+}	$3d^0$	0	0	0
Ti^{3+}	$3d^1$	1	1.73	1.75
Ti^{2+}	$3d^2$	2	2.84	2.76
V^{2+}	$3d^3$	3	3.87	3.86
Cr^{2+}	$3d^4$	4	4.90	4.80
Mn^{2+}	$3d^5$	5	5.92	5.96
Fe^{2+}	$3d^6$	4	4.90	5.0–5.5
Co^{2+}	$3d^7$	3	3.87	4.4–5.2
Ni^{2+}	$3d^8$	2	2.84	2.9–3.4
Cu^{2+}	$3d^9$	1	1.73	1.8–2.2
Zn^{2+}	$3d^{10}$	0	0	0

values turn out to exceed slightly those predicted on the basis of spin only. These small discrepancies are probably due to incomplete quenching of the orbital magnetic moment.

For other compounds of the transition elements the observed magnetic moments sometimes turn out to be surprisingly small. For example, $K_4Fe(CN)_6$ does not show the paramagnetic moment of about 5 Bohr magnetons that would be predicted for Fe^{2+} (d^6—four unpaired electrons). In fact, $K_4Fe(CN)_6$ is diamagnetic and contains no unpaired electrons. How can this be accounted for? The preceding section gives a basis for understanding the phenomenon by considering how the d orbitals are affected on interaction with the environment.

For the specific case of $K_4Fe(CN)_6$ the solid consists of potassium cations, K^+, and ferrocyanide anions, $Fe(CN)_6^{4-}$. We can consider the $Fe(CN)_6^{4-}$ anion as being assembled from Fe^{2+} (d^6) and six ligand CN^- ions. If the six CN^- ions are brought up to surround the Fe^{2+} octahedrally, as in the process illustrated in Fig. 17.4, the e_g and t_{2g} orbitals of the iron atom are perturbed differently. With the approach of ligand electrons along the axes, the six electrons of the iron are repelled out of the e_g set d_{z^2} and $d_{x^2-y^2}$ into the t_{2g} set d_{xy}, d_{yz}, and d_{zx}. The only way six electrons can be accommodated in this set of three orbitals is for an electron pair to occupy each orbital. Actually, there is considerable repulsion between electrons of a pair in the same orbital. However, if the ligand-field effect is strong enough—i.e., if Δ in Fig. 17.5 is big enough—then it may be energetically preferable to put two electrons in the same orbital rather than to go to a next higher orbital. Thus, whether pairing occurs depends on the magnitude of Δ compared with electron-electron repulsion energy. The magnitude of Δ depends on several factors: the charge of the central atom, the principal quantum number of the d level involved, and the nature of the ligand. In general, Δ increases with increasing central-atom charge, with increasing principal quantum number (that is, $5d > 4d > 3d$), and with increasing polarizability of the ligand. Cyanide ion as a ligand is particularly good at producing large ligand-field splitting; water is generally not very effective. However, a notable exception is $Co(H_2O)_6^{3+}$, in which electron pairing does occur.

17.6 Spectral properties

When white light interacts with a substance giving rise to color, there is absorption of part of the visible spectrum. For example, if the blue portion of white light is absorbed, then the remainder appears red; conversely, if red frequencies are absorbed, the substance appears blue. Since most transition-element compounds are colored, there must be energy transitions which can use up some of the energy of visible light. Figure 17.9 lists some characteristic colors of first-row transition-element ions (dilute aqueous solution, no complexing, and no hydrolysis) and indicates approximate wavelengths for maximum light absorption. Because the eye is not equally sensitive to all wavelengths, it is difficult to go from the absorption maxima to the perceived color. Furthermore, the absorption does not occur sharply at a single wavelength but instead spreads out over a band of the spectrum. Since the visible region of the light spectrum extends approximately from 400 to 700 nm, some of the maxima noted in the figure occur also in the ultraviolet (below 400 nm) and some in the infrared (above 700 nm). In addition to the absorption bands noted in the figure, there is usually for any ion an additional strong absorption band in the ultraviolet, corresponding to transfer of electrons between ion and solvent.

The characteristic absorption bands of transition-element ions are attributed to electronic transitions involving the d orbitals. In the crystal-field point of view, the d orbitals are simply separated into different levels of energy. For example, in the ion Ti^{3+} surrounded by an octahedron of H_2O molecules there are the two subsets of d-orbital energies, the lower set of three, d_{xy}, d_{yz}, and d_{zx}, and the upper set of two, d_{z^2} and $d_{x^2-y^2}$. In the crystal-field picture, the minimum energy state of Ti^{3+} would correspond to having the $3d^1$ unpaired electron in one of the lower three orbitals (Fig. 17.5). As light is absorbed by the sample, the electron of a Ti^{3+} ion is raised from the lower t_{2g} set of orbitals to the upper e_g set. The absorption of energy

Ion	Configuration	Observed color	Absorption maxima, nm
$Ti(H_2O)_6^{3+}$	d^1	Violet	493
$V(H_2O)_6^{3+}$	d^2	Blue	389, 562
$V(H_2O)_6^{2+}$	d^3	Violet	358, 541, 910
$Cr(H_2O)_6^{3+}$	d^3	Violet	264, 407, 580
$Fe(H_2O)_6^{3+}$	d^5	Colorless	
$Fe(H_2O)_6^{2+}$	d^6	Pale green	962
$Co(H_2O)_6^{2+}$	d^7	Pink	515, 625, 1220
$Ni(H_2O)_6^{2+}$	d^8	Green	395, 741, 1176
$Cu(H_2O)_6^{2+}$	d^9	Blue	794

Fig. 17.9
Colors and absorption wavelengths.

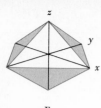

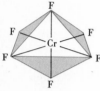

$CrF_6{}^{3-}$

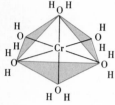

$Cr(H_2O)_6{}^{3+}$

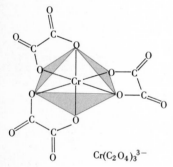

$Cr(C_2O_4)_3{}^{3-}$

Fig. 17.10
Complex ions based on
octahedral configuration.

Δ gives rise to the color. Subsequently, the stored energy has to be dissipated, apparently as heat through vibrations of the environment. In the ligand-field picture, as represented in Fig. 17.7, energy absorption occurs by excitation of the lone electron from the non-bonding molecular-orbital level (mainly t_{2g} character) to one of the antibonding states.

In cases involving more than one d electron the cause of absorption bands is qualitatively similar to the d^1 case. However, the situation can be considerably more complicated. More than one electron can be simultaneously excited, and interelectron repulsions between excited and nonexcited electrons can produce additional transitions. Hence, several absorption bands, as noted in Fig. 17.9, may be observed.

In the case of Fe^{3+} in aqueous solution no major absorption band is observed except for the electron-transfer band in the ultraviolet (which on hydrolysis shifts into the visible region so as to make the solution appear reddish brown). The reason why $Fe(H_2O)_6{}^{3+}$ lacks color is its $3d^5$ configuration. The ligand-field effect of the H_2O is so small that there is not enough energy-level splitting to bring about pairing of the d electrons; hence, there is one electron in each of the d orbitals. The only way energy could be absorbed would be for an electron from a t_{2g} orbital to be raised to an e_g orbital, where it would have to pair up. Such electron-spin change during an optical transition rarely occurs, and the transition is said to be *spin-forbidden*.

17.7 Complex ions

The availability of d orbitals participating in the chemical bonding of ligands leads to a large variety of structures for complex ions of the transition elements. For certain transition elements (especially cobalt, chromium, and platinum-like elements) the complexes formed may be extremely slow to undergo structural changes. Hence the species may be persistent enough, even in aqueous solution, to allow separation and characterization of the various complex species. The study of such complex species is a large and growing area of research in inorganic chemistry, sometimes called *coordination chemistry*. In this area, primary emphasis is on the determination of complex-ion structure and its relation to physical and chemical properties.

Most complex ions have structures that are basically octahedral arrangements of ligands about a central atom. As noted in Sec. 17.4, an octahedral arrangement of ligands favors electron overlap between the ligand and two d (e_g), one s, and three p orbitals of the central atom. This set of orbitals is the same as the one from which the set of six d^2sp^3-hybrid orbitals is derived (Sec. 3.8, Fig. 3.18). In most complex ions these orbitals are the ones used in bonding, and the complexes turn out to be octahedral. Figure 17.10 shows examples of several complex ions of chromium based on octahedral coordination of near-neighbor atoms to a central chromium atom. In $CrF_6{}^{3-}$ the octahedral corners are occupied by fluorine atoms; in $Cr(H_2O)_6{}^{3+}$ the octahedral corners are occupied by oxygen atoms, to each of which are joined the two hydrogen atoms; in $Cr(C_2O_4)_3^{3-}$ the octahedral corners are again occupied by oxygen atoms, but adjacent oxygens are now bridged *via* carbon atoms of the oxalate ions.

Any group that bridges two or more coordination positions is called a *chelating agent*, and the resulting complex is called a *chelate*. If the group binds at two points, as does oxalate, then it may also be called *bidentate*. Some chelating agents bind at more than two positions, in which case they are tridentate, tetradentate, pentadentate, or even hexadentate. An example of the last of these is ethylenediamine-tetraacetate (generally abbreviated EDTA)

$$^-OOCCH_2 \qquad\qquad H_2CCOO^-$$
$$\diagdown\qquad\qquad\qquad\qquad\diagup$$
$$N-CH_2-CH_2-N$$
$$\diagup\qquad\qquad\qquad\qquad\diagdown$$
$$^-OOCCH_2 \qquad\qquad H_2CCOO^-$$

which can wrap itself around a central atom so that binding can occur at the four negative oxygens and at the two nitrogens.

Other geometries arise if other sets of orbitals are used for binding the ligands. For example, if only one of the d orbitals is used (for example, $d_{x^2-y^2}$) and combined with an s and two of the p orbitals (for example, p_x and p_y) of the central atom, then we have a square-planar arrangement of ligands about the central atom. Figure 17.11 shows some typical examples drawn from platinum chemistry. In $PtCl_4^{2-}$ the four chlorine atoms form a square about a central platinum; in $Pten_2^{2+}$ (where *en* stands for ethylenediamine, $NH_2CH_2CH_2NH_2$), the four nitrogen atoms form a square, with bridges of CH_2CH_2 between the members of a pair. Square-planar geometry is most likely to occur for ions having d^8 configurations.

Some complexes are intermediate between octahedral and square-planar. Such is the case for $Cu(H_2O)_6^{2+}$, in which four of the H_2O molecules, arranged in a square, are nearer than the other two. The result is a considerably distorted octahedron, which can be imagined as being created from a regular octahedron (as in Fig. 17.10) by pulling outward a pair of opposite corners.

If no d orbitals are used in binding the complex, then it is likely to have tetrahedral arrangement of ligands about the central atom. In such cases it is usually one s and three p orbitals that are used for bonding, giving rise to the familiar sp^3 tetrahedral hybrids. Zn^{2+}, for example, has all its $3d$ orbitals filled. In forming a complex such as $Zn(NH_3)_4^{2+}$, the $4s$ and three $4p$ orbitals of the zinc are used. The result is a complex in which the four nitrogen atoms define a tetrahedron about the zinc.

In each of the above-considered cases all the near-neighbor atoms about a given central atom were identical. What if this were not the case? If only one near neighbor is made different from all the others, then only one species still results. For example, if one of the six F^- ions in CrF_6^{3-} is replaced by an H_2O molecule to give $CrF_5(H_2O)^{2-}$,

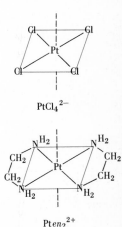

$PtCl_4^{2-}$

$Pten_2^{2+}$

Fig. 17.11
Square-planar coordination.

there is no difference in the product no matter which of the six fluorides is replaced. All the corners of a regular octahedron are equivalent. What if a second ligand is substituted? This time it matters, since two distinctly different geometries will result depending on whether the second substituted ligand goes on a corner adjacent to the first substituent or on the opposite corner.

Figure 17.12 shows the two possible configurations that result when one of the fluorides of $CrF_5(H_2O)^{2-}$ is substituted by another H_2O to give the disubstituted complex $CrF_4(H_2O)_2^-$. The product at the top has the two H_2O molecules adjacent to each other (note that the product is the same whether the second H_2O is at position 1, 2, 3, or 4); the product at the bottom has the two H_2O molecules at opposite corners of the octahedron. The product at the top and the product at the bottom are chemically different. They are said to be *isomers*—two species of the same composition differing in structural arrangement of the atoms and hence in properties. The isomer with adjacent positions

Fig. 17.12
Isomers of $CrF_4 (H_2O)_2^-$.

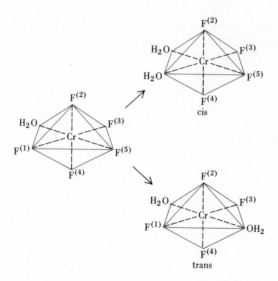

substituted is called *cis*, and that with opposite positions substituted, *trans*. Substitution of a third H_2O for F^- on a *cis*-$CrF_4(H_2O)_2^-$ gives rise to two possible products: one if the third H_2O goes into a position adjacent to both the other two H_2O's (into position 2 or 4 of the cis form in Fig. 17.12) and the other if the third H_2O goes into a position not adjacent to both the other two H_2O's (into position 3 or 5 of the cis form in Fig. 17.12). On the contrary, substitution of a third H_2O for F^- in *trans*-$CrF_4(H_2O)_2^-$ gives rise to only one possible product, which is identical with the second of the two noted above for cis.

Cis and trans substitution products are also possible in square-planar configurations, since two substituents on a square can be on either adjacent or opposite corners of the square. However, such is not the case in tetrahedral configurations. As shown in Fig. 17.13, any corner on a tetrahedron is adjacent to each of the other three. In other words, if there is a substituent at position 1 of the tetrahedron, it makes no difference whether the second substitution occurs at position 2, 3, or 4. In the square, in contrast, with a substituent at position 1, different products result if the substitution occurs at 3 (giving rise to a trans product) as opposed to substitution at either 2 or 4 (cis product).

In addition to geometric isomerism as discussed above, complex ions sometimes show a complication arising from what is called *optical isomerism*. Optical isomers occur when two species differ only in that one is the mirror image of the other. The two species have most properties identical, but they differ in that one turns the plane of polarized light clockwise and the other turns it counterclockwise. Figure 17.14 shows the relation between the two optical isomers of $Coen_3^{3+}$ [trisethylenediaminecobalt(III)]. Although in both cases cobalt is bound only to six nitrogen atoms of three *en* molecules, there are two different ways in which this is done. As with more familiar mirror-image pairs (e.g., right and left hands), no amount of turning can superimpose one of these ions on its mirror image. The pair of optical isomers are called *enantiomers*, or *enantiomorphs*. They may be individually designated *d* (for *dextro*) or *l* (for *levo*) depending on whether the plane of polarized light is rotated clockwise or counterclockwise as viewed

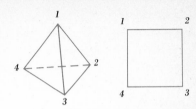

Fig. 17.13
Isomerism in tetrahedral and square-planar configurations.

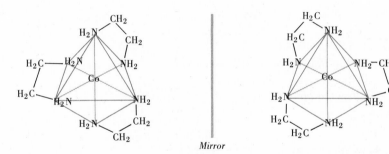

Mirror

Fig. 17.14
Optical isomers.

by one looking into the beam. The requirement for having enantiomers is that there be no center or plane of symmetry in the complex.

The general rules for naming complex ions are summarized in Appendix 2.

Exercises

*17.1   *Definition.* Sometimes transition elements are defined as those which use *d* electrons for bonding. By reference to Figs. 17.3 and 1.18, decide which elements defined as transition elements in this chapter would not fit the above criterion.

*17.2   *Oxidation states.* Taking the most simplistic view that electrons not counted with an atom are totally removed, what would be the apparent electron configuration of iron ($Z = 26$) in each of the following oxidation states: $+2$, $+3$, $+4$, and $+6$?

*17.3   *Oxidation states.* Write the formula of an oxide or a sodium oxysalt that would correspond to each of the oxidation states observed for the first-row transition elements.

* *17.4 Magnetic properties.* Arrange the following in order of increasing number of unpaired electrons: gaseous Fe^{2+}, Fe^{2+} in a strong crystal field, gaseous Cr^{3+}, Cr^{3+} in a strong crystal field, gaseous Co^{3+}, and Co^{3+} in a strong crystal field.

* *17.5 Magnetic properties.* Predict values of the spin magnetic moment for each of the following electronic configurations:
  a s^1
  b s^2
  c d^3
  d d^4 in weak ligand field
  e d^4 in strong ligand field

     Ans. 1.73, 0, 3.87, 4.90, 2.83 Bohr magnetons

* *17.6 Spectral properties.* What change in color would you expect for a blue-colored transition-metal complex ion that is subjected to high pressure?

* *17.7 Nomenclature.* Give the systematic name for each of the following: $Co(NH_3)_6{}^{3+}$, $Co(CN)_6{}^{3-}$, $Co(H_2O)(NH_3)_5{}^{3+}$, $Co(OH)(H_2O)_5{}^{2+}$, and $CoCl_4(H_2O)_2{}^-$.

* *17.8 Nomenclature.* Write the formulas for each of the following:
  a Tribromotriaquovanadium(III)
  b Tetrahydroxozincate(II)
  c Dichlorobisethylenediaminecobalt(III)
  d Pentacyanonitrosylferrate(III)
  e Hexachloromolybdate(VI)

** *17.9 Electron configuration.* By reference to Fig. 1.18 you will note that elements with $Z = 41$ to 45 have $5s^1$ instead of the expected $5s^2$. How might you account for this?

** *17.10 Metallic properties.* The data in Fig. 17.2 indicate that the two best conductors are copper and silver. What would their respective conductivities be on a per-gram basis instead of a per-cubic-centimeter basis?   *Ans. 6.4×10^4 cm^2 $ohm^{-1} \cdot g^{-1}$ for* Ag

** *17.11 Crystal-field theory.* How might you rationalize the fact that crystal-field theory works better for F^- complexes of the transition elements than it does for CN^- complexes?

** *17.12 d orbitals.* Draw sketches that distinguish the five kinds of $3d$ orbitals. What changes would you have to make to adapt your sketches to the $4d$ situation?

** *17.13 d orbitals.* Predict how the d orbitals would be split for each of the following situations:
  a Two ligands only approaching on x axis
  b Two ligands only approaching on y axis
  c Two ligands only approaching on z axis
  d Four ligands only approaching on x and y axes

** *17.14 Ligand field.* Show why in an octahedral environment a transition-metal atom is more likely to use its e_g than t_{2g} orbitals for bonding while the t_{2g} stay nonbonding.

** 17.15   *Magnetic properties.* How might you account for the fact that hydrated rhodium ion, $Rh(H_2O)_6^{3+}$, is diamagnetic?

** 17.16   *Spectral properties.* How might you account for the fact that $V(H_2O)_6^{3+}$ shows two absorption bands in the visible spectrum whereas $Ti(H_2O)_6^{3+}$ shows only one?

** 17.17   *Isomers.* How many geometric isomers are there for each of the following: $Co(H_2O)_6^{3+}$, $Co(H_2O)_5(NH_3)^{3+}$, $Co(H_2O)_4(NH_3)_2^{3+}$, $Co(H_2O)_3(NH_3)_3^{3+}$, $Co(H_2O)_2(NH_3)_4^{3+}$, $Co(H_2O)(NH_3)_5^{3+}$, and $Co(NH_3)_6^{3+}$?

** 17.18   *Cis-trans isomerism.* If in the reaction $CoCl_2(NH_3)_4^{+} + Cl^- \longrightarrow CoCl_3(NH_3)_3 + NH_3$ only one product is obtained, is the initial complex cis or trans?

** 17.19   *Optical isomers.* Which of the following can represent a pair of enantiomers?
  a $CrCl_3(NH_3)_3$
  b   cis-$Coen_2Cl_2^{+}$
  c   cis-$CoCl_2(NH_3)_4^{+}$
  d   trans-$CoCl_2en_2^{+}$
  e   trans-$CoCl_2(NH_3)_4^{+}$

** 17.20   *Ionic radii.* Using the data of Appendix 8, tabulate the ionic radius versus Z for the $+3$ and $+2$ ions of the first-row transition series. Account for any irregularities observed.

** 17.21   *Heats of reaction.* The heats of formation in kilojoules per mole of first-row transition-element oxides are as follows:

Ti_2O_3	V_2O_3	Cr_2O_3	Mn_2O_3	Fe_2O_3
1520	1238	1140	956	823

What pair of transition elements would give maximum heat evolution in a reaction of the type

$$2M + M_2'O_3 \longrightarrow 2M' + M_2O_3$$

** 17.22   *General.* You are given X, one of the first-row transition elements. It forms several oxides, but the only one you succeed in analyzing shows 38.6% oxygen. The metal is cubic and has a density between 5.0 and 8.0 g/cm^3 and a heat capacity of 0.50 $J\,g^{-1}\,deg^{-1}$. If the ΔH of fusion is 16.7 kJ/mol and the ΔS of fusion is 8.4 kJ mol^{-1} deg^{-1}, which of the transition elements is it?

*** 17.23   *Electronic configuration.* As a general rule, filled and half-filled subshells seem to have special stability. How might you explain this?

*** 17.24   *Electron configuration.* In the vicinity of $Z = 21$, the 3d-orbital energy is dropping faster with increasing Z than is the 4s.

Sketch a graph showing the probable relative course of orbital energy versus Z in this region. Account for the general character of your graph.

*** **17.25** *Metallic structure.* Chromium metal crystallizes in two forms α and β. The α form, which is the ordinary form, has a body-centered cubic structure with unit-cell edge length equal to 0.288 nm. The β form is hexagonal close packed with $a = 0.272$ nm and $c = 0.442$ nm. Assuming hard spheres in contact, what is the apparent radius of the chromium atom in each of these structures? What would be the free-electron density assuming only $4s$ ionization?

Ans. 0.125 nm and 8.4×10^{22} electrons
per cubic centimeter in bcc

*** **17.26** *Ligand field.* Construct a cube. Place Ti^{3+} on each corner of the cube and O^{2-} at each edge center. Note that the Ti^{3+} is in an octahedral field. Draw sketches to show that only one of the three p orbitals of the oxygen can contribute to σ bonding of the complex. Show also that the other p orbitals can form π bonds with the titanium but that these will be less bonding. How would Fig. 17.7 need to be modified to allow for this π bonding?

*** **17.27** *Crystal-field splitting.* Suppose you have a d^2 ion (for example, V^{3+}) in an octahedral field. Show first that in the ground state it makes no difference into which of the d_{xy}, d_{yz}, and d_{zx} orbitals each of the electrons is placed. Show also that on the absorption of energy to raise one electron into the e_g subset it makes a difference whether it goes into d_{z^2} or $d_{x^2-y^2}$. In other words, there will be two possible energy values for excitation. Show finally that if both electrons are excited, only one transition energy corresponds.

*** **17.28** *Spectral properties.* How might you account for the fact that $Mn(H_2O)_6{}^{2+}$ is almost colorless (pale pink) whereas $Mn(H_2O)_6{}^{3+}$ is deeply colored (dark red)?

*** **17.29** *Symmetry.* What symmetry elements do we have in $CrF_6{}^{3-}$ which we do not have in $Cr(C_2O_4)_3{}^{3-}$?

*** **17.30** *Symmetry.* In the complex $Co(NO_2)_6{}^{3-}$, the nitro groups, which consist of bent O—N—O units of bond angle 115°, are attached to the central cobalt through the nitrogen. Show that the symmetry of the complex is more like that of a tetrahedron than like that of an octahedron. (*Hint:* Tabulate all the symmetry elements.)

In the preceding chapter the general properties that characterize the transition elements have been discussed. In particular, the role of d orbitals in fixing observed properties has been considered. In this chapter we begin to take up the specific chemistry of individual elements. We shall do this by subgroups, starting from the left with the scandium subgroup and working toward the right. For each subgroup the emphasis will be on the top element, since in general it is both the most abundant and the most important technologically of the subgroup. The scandium subgroup is special in that it includes the lanthanides and actinides; because of this it contains more elements than any other group of the periodic table.

18.1 Scandium subgroup

The scandium subgroup contains the elements scandium ($Z = 21$), yttrium ($Z = 39$), lanthanum through lutetium ($Z = 57$ to 71), and actinium through lawrencium ($Z = 89$ to 103). The 14 elements with lanthanum are called the *lanthanides* (sometimes *lanthanoids*) or *rare-earth elements*. The 14 elements with actinium are called the *actinides* (sometimes *actinoids*). As usually displayed in the periodic table (Fig. 1.4), all the lanthanides are placed in the same position in the sixth period, below yttrium. The actinides occupy a corresponding position in the seventh period. Figure 18.1 indicates the electron configurations characteristic of the scandium-subgroup elements.

The lanthanide elements correspond to belated filling of the $4f$ subshell. Since in these elements the $4f$ subshell is third outermost, changes in its electronic population are well screened from neighboring atoms by the second-outermost and outermost shells. Consequently, all the lanthanides have properties that are remarkably alike. Similar belated filling of the $5f$ subshell occurs in the actinide series.

All the scandium-subgroup elements, including the lanthanides and actinides, are typically metallic, with high luster and good conductivity.

Fig. 18.1
Electronic configurations in scandium subgroup.

Element	Symbol	Z	Electron population
Scandium	Sc	21	2, 8, 8 + $3d^1$, $4s^2$
Yttrium	Y	39	2, 8, 18, 8 + $4d^1$, $5s^2$
Lanthanum ↓	La ↓	57 ↓	2, 8, 18, 18, 8 + $5d^1$, $6s^2$ ↓
Lutetium	Lu	71	32
Actinium ↓	Ac ↓	89 ↓	2, 8, 18, 32, 18, 8 + $6d^1$, $7s^2$ ↓
Lawrencium	Lr	103	32

They are all quite reactive, about like calcium, with electrode potentials of about -2.5 V. Many of their compounds, such as hydroxides, carbonates, and phosphates, are of low solubility. There is some slight tendency for the $+3$ ions to hydrolyze in aqueous solution to give slightly acid solutions. All these elements are not very abundant in nature.

Scandium Scandium occurs in nature only in the combined form in minerals such as monazite (a complex phosphate which can be written $CePO_4$, though much of the cerium is replaced by other M^{3+} lanthanides and Th^{4+}) and gadolinite [a complex silicate approximating yttrium iron(II) silicate]. Not much is known about the element except that it reacts vigorously with water to liberate hydrogen, has a melting point of about $1200°C$ and a boiling point of about $2500°C$, and forms compounds only in the $+3$ oxidation state. All these compounds are colorless, and none is paramagnetic. There is not a great deal of interest in the chemistry of scandium because once its $+3$ ion is formed, all the d electrons have been removed.

Yttrium Like scandium, yttrium is quite rare. It occurs in the combined form in a few rare minerals, such as gadolinite. The metal is quite vigorously reactive but only to give compounds in which its oxidation state is $+3$. As expected, these compounds are colorless and are not paramagnetic. The oxide, Y_2O_3, or yttria, is a pure white powder. Yttrium compounds, particularly the vanadate, when doped with europium, have proved to be remarkably efficient as red phosphors in color-television receivers. (The green phosphor is usually copper-activated zinc sulfide; the blue, silver-activated zinc sulfide.)

18.2 Lanthanides

The 15 elements that constitute the lanthanide series are listed in Fig. 18.2, which also gives the symbols and the most probable electron configurations. Because of the closeness of the $4f$ and $5d$ energy levels, there is considerable uncertainty in some of the electron-configuration assignments. It is not obvious in every case from the electron configurations of the neutral atoms, but all the lanthanides form $+3$ ions as their principal chemical species. It is generally assumed in forming these ions that the $6s^2$ electrons are lost along with the $5d^1$ (if present) or with one of the $4f$ (if no $5d^1$ is present).

The electron density distribution in the $4f$ orbitals is quite complicated. One of the seven orbitals can be visualized, for example, as consisting of eight lobes of high electron density extending from the center of a cube to the eight corners. The radial distribution, on the other hand, is quite simple. As one goes outward from the nucleus, there is but a single broad hump in the electron probability density, more or less as shown in Fig. 18.3. For atoms of low Z, the maximum in the $4f$ probability occurs far out from the nucleus. Stepwise buildup of

Element	Symbol	Z	Probable electron configuration	Oxidation states
Lanthanum	La	57	2, 8, 18, 18, $5s^25p^65d^16s^2$	$+3$
Cerium	Ce	58	_____$4f^26s^2$	$+3, +4$
Praseodymium	Pr	59	_____$4f^36s^2$	$+3, +4$
Neodymium	Nd	60	_____$4f^46s^2$	$+3$
Promethium	Pm	61	_____$4f^56s^2$	$+3$
Samarium	Sm	62	_____$4f^66s^2$	$+2, +3$
Europium	Eu	63	_____$4f^76s^2$	$+2, +3$
Gadolinium	Gd	64	_____$4f^75d^16s^2$	$+3$
Terbium	Tb	65	_____$4f^96s^2$	$+3, +4$
Dysprosium	Dy	66	_____$4f^{10}6s^2$	$+3$
Holmium	Ho	67	_____$4f^{11}6s^2$	$+3$
Erbium	Er	68	_____$4f^{12}6s^2$	$+3$
Thulium	Tm	69	_____$4f^{13}6s^2$	$+3$
Ytterbium	Yb	70	_____$4f^{14}6s^2$	$+3$
Lutetium	Lu	71	_____$4f^{14}5d^16s^2$	$+3$

Fig. 18.2
Lanthanide elements.

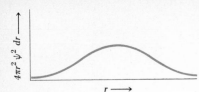

Fig. 18.3
Radial distribution function for 4f electron.

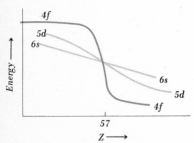

Fig. 18.4
Effect on 4f-orbital energy of atom buildup around Z = 57.

the light atoms by increasing Z and adding one electron at a time does not perturb the $4f$ subshell since the nucleus is well screened by the intervening electrons. It is only when Z exceeds about 57 that a dramatic change occurs. Stepwise increments in Z plus stepwise addition of electrons are immediately felt by the $4f$ subshell since the electrons now being added do not function well as screens to hide the increase in Z. The result is that the $4f$ subshell, instead of feeling an effective $+1$ charge pulling it to the nucleus, now feels a steadily increasing pull, and, in fact, around $Z = 57$ there occurs a pronounced shrinking in the size of the $4f$ shell. A crude picture of what happens, specifically to the energy, around $Z = 57$ is shown in Fig. 18.4. As can be seen, the $4f$, $5d$, and $6s$ are all about the same energy in this vicinity, but the $4f$ is dropping very rapidly, so it quickly changes from very high energy for $Z < 57$ to very low energy for $Z > 57$. Once the $4f$ level has been sucked in, it is buried deep inside the atom and generally is shielded from external environmental influences by electron population in $5d$ and $6s$.

Except for promethium, which has an unstable nucleus, all the lanthanides generally occur together. The richest source mineral is monazite. As a group the lanthanides are not very abundant—the most common lanthanide being cerium, which makes up but 3×10^{-4} percent of the mass of the earth's crust. Because of the great similarity in chemical properties, it is difficult to separate one lanthanide from another. Separation has been made by careful, repeated fractional crystallization and, more recently, by ion-exchange techniques. Both separations rely on slight differences of properties (e.g., solubility, complex-ion formation, and hydration) arising from size differences in the $+3$ ions. In going through the sequence from La^{3+} to Lu^{3+} the ionic radius shrinks from 0.106 to 0.085 nm as follows:

La^{3+}	Ce^{3+}	Pr^{3+}	Nd^{3+}	Pm^{3+}	Sm^{3+}	Eu^{3+}	Gd^{3+}
0.106	*0.103*	*0.101*	*0.100*	*0.098*	*0.096*	*0.095*	*0.094*

Tb^{3+}	Dy^{3+}	Ho^{3+}	Er^{3+}	Tm^{3+}	Yb^{3+}	Lu^{3+}
0.092	*0.091*	*0.089*	*0.088*	*0.087*	*0.086*	*0.085*

This shrinkage, which is called the *lanthanide contraction,* arises from the increase of nuclear charge during the progressive filling of an interior subshell. The lanthanide contraction, as we shall see below, is important in allowing separation of the lanthanide elements, but it is also significant in its effect on the relative properties of elements before and after the lanthanide sequence. For example, zirconium ($Z = 40$) and hafnium ($Z = 72$) have almost identical chemical properties because the intervening lanthanide contraction has canceled the usual increase of radius down a group and made the two kinds of atoms nearly identical in size.

The most successful technique for separating the lanthanide elements is to use ion exchange. An ion-exchange column, typically constructed from a glass tube about 1 m long and 1 cm in diameter which is packed with ion-exchange resin (Sec. 16.9), is *fixed* with the ions to be separated and then *eluted* with a complexing-agent solution. The fixing is accomplished by simply pouring a solution of the lanthanide salts onto the top of the column so that the M^{3+} ions get adsorbed. The eluting solution, which might contain, for example, EDTA (page 485) as a complexing agent, is slowly dribbled through the column so as to wash out the ions. The rate at which the M^{3+} ions come off the column depends on their relative binding to the ion exchanger compared with their relative tendency to associate with EDTA. In other words, elution depends on the equilibrium constant for the following reaction

$$M^{3+}(\text{on exchanger}) + \text{EDTA}^{4-} \rightleftharpoons \text{M(EDTA)}^-$$

The lanthanides are so similar to each other that there are but small differences in their K values, but the geometry of having many tiny beads of resin in a long column ensures many successive adsorptions on the resin and dissolving into the solution; so small equilibrium differences can be exploited. Figure 18.5 shows a typical column setup, and Fig. 18.6 a graphic record of concentration changes in the wash solution coming out of the column. As can be seen, the smaller ion, Lu^{3+}, washes out first, corresponding to its tighter binding in the EDTA complex. Actually, there is considerable overlap of the elution patterns for the different ions. Separation can be enhanced by passing a given fraction through a fresh column, perhaps modifying the pH of the EDTA solution so as to change slightly the K constants.

The lanthanide elements are soft, gray metals which need to be protected from air since they react vigorously with moisture and oxygen. The pure metals, which can now be obtained in relatively high purity (99.9 percent or better), are quite expensive (for example, \$50/g for europium). They are generally handled in closed, inert-gas chambers, particularly when used for production of sophisticated solid-state electronic and optical devices.

The electrode potentials of the lanthanides change regularly from $E° = -2.52$ V for $La^{3+}(aq) + 3e^- \longrightarrow La(s)$ to $E° = -2.25$ V for $Lu^{3+}(aq) + 3e^- \longrightarrow Lu(s)$. The trihydroxides are relatively insoluble, with K_{sp} for $M(OH)_3(s) \rightleftharpoons M^{3+} + 3OH^-$ decreasing regularly from 1.0×10^{-19} for $La(OH)_3$ to 2.5×10^{-24} for $Lu(OH)_3$.

Because the lanthanide ions are generally characterized by an incomplete $4f$ subshell, paramagnetism due to unpaired electrons is expected and observed for most lanthanide compounds. As noted in Sec. 17.5, unlike d electrons, f electrons contribute to paramagnetism through both spin moment and orbital moment. The calculation of the orbital contribution is complicated; it varies from zero for the case of Gd^{3+} $(4f^7)$ to 5.7 Bohr magnetons for the case of Ho^{3+} $(4f^{10})$. In other words, the magnetic moments of the lanthanide ions *cannot* be calculated from the spin-only formula, which was found to work reasonably well for paramagnetism arising from d electrons. The reason for this difference is that, whereas d electrons are exposed to interaction with the ligands and so are prevented from orienting freely in an external

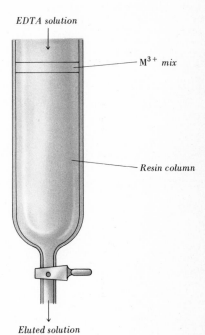

EDTA solution

M^{3+} *mix*

Resin column

Eluted solution

Fig. 18.5

Ion-exchange column for separating rare-earth elements.

Fig. 18.6
Ion-exchange separation of
rare-earth ions.

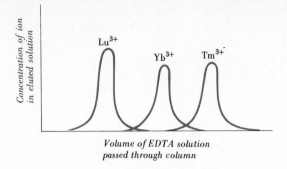

Fig. 18.6
Ion-exchange separation of
rare-earth ions.

magnetic field, f electrons are deep enough inside the ion and are shielded from the quenching effect of the environment on orbital motion.

The relatively good shielding of f electrons from interaction with the environment is important in determining the spectral characteristics of lanthanide ions. Instead of the broad absorption bands arising from d-electron transitions (Sec. 17.6), the lanthanide ions have spectra consisting generally of many sharp absorption bands. The sharpness—i.e., absorption over but a narrow range of wavelengths—is attributable to lack of interaction between the f levels and the environment. The large number (tens of hundreds) of bands is attributable to the large number of excited electron configurations possible within the partially filled f subshell.

Besides the $+3$ state, some of the lanthanides show other oxidation states. For example, cerium forms Ce^{4+} with an $E°$ of $+1.61$ V for $Ce^{4+}(aq) + e^- \longrightarrow Ce^{3+}(aq)$. In other words, Ce^{4+} is as good an oxidizing agent as MnO_4^-. Other $+4$ lanthanides (Pr^{4+} and Tb^{4+}) are even more powerful oxidizing agents. Dipositive states (Eu^{2+}, Sm^{2+}, and Yb^{2+}) also occur; of them, only Eu^{2+} is stable with respect to oxidation by water.

18.3   Actinides

The 15 actinide elements together with probable electron configurations are listed in Fig. 18.7. The electron configurations are even less certain than those given for the lanthanide elements. Not only are the energy levels close together, but because the nuclei are unstable to radioactive decay, in some cases only minute amounts of the elements have been obtained for investigation.

All these nuclei are unstable with respect to alpha emission. There is a chance that any given nucleus will spontaneously emit an alpha particle, $_2^4He$, forming thereby a new nucleus which is two charge units lower in Z and four mass units lower in mass number A. The statistical probability of radioactive decay can be described either by giving the rate constant for decay or by specifying the *half-life*, i.e., the time required for half of any assemblage of nuclei to disintegrate. Radioactive disintegration is a first-order rate process (Sec. 10.2); so we can write for the rate of decay—the number of nuclear disintegrations per

Element	Symbol	Z	Probable electron configuration	Oxidation states
Actinium	Ac	89	$2, 8, 18, 32, 18, 6s^2 6p^6 6d^1 7s^2$	$+3$
Thorium	Th	90	_____$6d^2 7s^2$	$+3, +4$
Protactinium	Pa	91	_____$5f^2 6d^1 7s^2$	$+3, +4, +5$
Uranium	U	92	_____$5f^3 6d^1 7s^2$	$+3, +4, +5, +6$
Neptunium	Np	93	_____$5f^4 6d^1 7s^2$	$+3, +4, +5, +6, +7$
Plutonium	Pu	94	_____$5f^6 7s^2$	$+3, +4, +5, +6, +7$
Americium	Am	95	_____$5f^7 7s^2$	$(+2), +3, +4, +5, +6$
Curium	Cm	96	_____$5f^7 6d^1 7s^2$	$+3, +4$
Berkelium	Bk	97	_____$5f^8 6d^1 7s^2$	$+3, +4$
Californium	Cf	98	_____$5f^{10} 7s^2$	$+2, +3$
Einsteinium	Es	99	_____$5f^{11} 7s^2$	$+2, +3$
Fermium	Fm	100	_____$5f^{12} 7s^2$	$+2, +3$
Mendelevium	Md	101	_____$5f^{13} 7s^2$	$+2, +3$
Nobelium	No	102	_____$5f^{14} 7s^2$	$+2, +3$
Lawrencium	Lr	103	_____$5f^{14} 6d^1 7s^2$	$+3$

unit time—a direct proportionality to the number of unstable nuclei present:

$$\frac{dN}{dt} = -kN$$

At any time t the number of nuclei present can be found by solving the above equation to give

$$\log \frac{N_0}{N} = \frac{kt}{2.303}$$

where N_0 is the number of nuclei at zero time. To solve for the half-life, we set N_0/N equal to 2, getting

$$t_{1/2} = \frac{2.303}{k} \log 2 = \frac{0.693}{k}$$

As with chemical reactions, k and hence $t_{1/2}$ can be determined from experiment by plotting $\log N$ versus t. Since the disintegration rate dN/dt is directly proportional to N, it is usual to plot the logarithm of the disintegration rate against time. The disintegration rate can be measured directly by use of a device such as a Geiger counter which counts electric current pulses produced by ionization tracks in gas chambers.

Of the actinide elements, thorium and uranium have rather long half-lives (1.39×10^{10} yr for $^{232}_{90}\text{Th}$ and 4.50×10^9 yr for $^{238}_{92}\text{U}$); so their specific rate constants of decay are quite small, and nuclei present at the beginning of the earth (estimated to be about 4.6×10^9 yr ago) are still with us. Even nuclei with short half-lives can be found in appreciable amounts in nature if they are generated by longer-lived parents. For example, $^{238}_{92}\text{U}$ decays eventually to the stable isotope ^{206}Pb by a sequence of consecutive disintegrations involving α (^4_2He) and β (negative electron) emissions:

Section 18.3

Actinides

$$^{238}_{92}\text{U} \xrightarrow{\alpha} {}^{234}_{90}\text{Th} \xrightarrow{\beta} {}^{234}_{91}\text{Pa} \xrightarrow{\beta} {}^{234}_{92}\text{U} \xrightarrow{\alpha} {}^{230}_{90}\text{Th} \xrightarrow{\alpha}$$

$$^{226}_{88}\text{Ra} \xrightarrow{\alpha} {}^{222}_{86}\text{Rn} \xrightarrow{\alpha} {}^{218}_{84}\text{Po} \xrightarrow{\beta} {}^{218}_{85}\text{At} \xrightarrow{\alpha} {}^{214}_{83}\text{Bi} \xrightarrow{\beta}$$

$$^{214}_{84}\text{Po} \xrightarrow{\alpha} {}^{210}_{82}\text{Pb} \xrightarrow{\beta} {}^{210}_{83}\text{Bi} \xrightarrow{\beta} {}^{210}_{84}\text{Po} \xrightarrow{\alpha} {}^{206}_{82}\text{Pb}$$

The second member in this chain, $^{234}_{90}\text{Th}$, has a rather short half-life (24.1 days); so it should disappear rapidly. However, it is constantly being regenerated by the ^{238}U decay, albeit in amounts small compared with the ^{232}Th left over from the beginning of the earth.

The actinides beyond uranium (so-called "transuranium elements") have no long-lived precursors; so they do not occur in nature in finite amounts but have to be prepared synthetically by use of particle accelerators (e.g., synchrotrons and cyclotrons) or nuclear reactors. The first of the transuranium elements to be prepared (1940) was neptunium, made by irradiation of $^{238}_{92}\text{U}$ with neutrons. The nucleus formed, $^{239}_{92}\text{U}$, decays by beta emission

$$^{239}_{92}\text{U} \longrightarrow {}^{239}_{93}\text{Np} + {}^{0}_{-1}e \qquad t_{1/2} = 23.5 \text{ min}$$

to produce $^{239}_{93}\text{Np}$, which is also beta active. It decays

$$^{239}_{93}\text{Np} \longrightarrow {}^{239}_{94}\text{Pu} + {}^{0}_{-1}e \qquad t_{1/2} = 2.3 \text{ days}$$

to produce plutonium, the second transuranium element, which is alpha active ($t_{1/2} = 24,400$ yr). Higher elements can be produced by similar irradiation of transuranium elements with neutrons or even with nuclei of a lighter element. For example, nobelium ($^{254}_{102}\text{No}$) has been made by bombarding curium ($^{242}_{96}\text{Cm}$) with carbon nuclei ($^{12}_{6}\text{C}$). Lifetimes of these higher elements get to be particularly short, partly because the probability of nuclear fission gets greater. In the fission process, a nucleus spontaneously divides not to a small chunk plus a big chunk but into two approximately equal nuclei. The process liberates enormous energies and is the basis for present nuclear reactors, most of which use ^{235}U or ^{239}Pu as their nuclear fuel. The impact of radiation hazards on quality of life and the place of nuclear fuel in man's environmental energy balance is considered in greater detail in Chap. 27.

As elements the actinides are metallic. Like the lanthanides, they generally have relatively large electrode potentials (ca. -2 V) between the first stable state and the metal. Unlike the lanthanides, the actinides (at least the early ones, which have been extensively studied) show a variety of oxidation states. Uranium, for example, forms many compounds in each of the states $+3$, $+4$, $+5$, and $+6$. In aqueous solution U^{3+} reduces water to liberate H_2 and form U^{4+}, which is slowly oxidized by air to $UO_2{}^{2+}$ (so-called "uranyl ion"). The uranyl ion is stable, but it can be reduced to form $UO_2{}^+$, which, however, is unstable to disproportionation into U^{4+} and $UO_2{}^{2+}$. Like most ions of the actinides, these species are colored. For example, $U^{3+}(aq)$ is red, $U^{4+}(aq)$ is green, and $UO_2{}^{2+}(aq)$ is yellow. As would be expected from the shielded nature of f electrons, the absorption spectra of the actinides generally consist of a large number of sharp absorption peaks. These peaks, however, are not so sharp as those arising from $4f$ transitions, primarily because $5f$ levels are not so well buried as $4f$ are.

18.4   Titanium subgroup

The elements of the titanium subgroup are titanium ($Z = 22$), zirconium ($Z = 40$), hafnium ($Z = 72$), and kurchatovium ($Z = 104$). Except for kurchatovium, they are more common than the elements of the scandium subgroup, and their chemistry is more complicated than that of scandium because of the additional oxidation state, $+4$. As Fig. 18.8 indicates, each of these elements has two s electrons in

Symbol	Z	Electronic configuration	Melting point, °C	Boiling point, °C	Ionization potential, eV	Electrode potential, V
Ti	22	$(18)\ 3d^2 4s^2$	1660	>3000	6.83	-0.9 (from TiO^{2+})
Zr	40	$(36)\ 4d^2 5s^2$	1860	>3000	6.95	-1.5 (from ZrO^{2+})
Hf	72	$(68)\ 5d^2 6s^2$	2200	>3000	5.5	-1.7 (from HfO^{2+})
Ku	104	$(100)\ 6d^2 7s^2?$	?	?	?	?

Fig. 18.8
Elements of titanium subgroup.

the outermost shell and two d electrons in the second-outermost shell. Removal of the s electrons gives the $+2$ oxidation state; further removal of one or two d electrons gives the $+3$ and the $+4$ states. Only in the case of titanium are all three of these states observed. There are some $+3$ compounds of zirconium (in nonaqueous media), but the $+4$ state is the more common. Hafnium forms compounds only in the $+4$ state. This trend of favoring the higher oxidation states in going down the group is typical of the transition subgroups.

Other characteristic properties of the Ti-subgroup elements are that the elements are metallic, have very high melting and boiling points, and are quite reactive to most oxidizing agents. Although it does not show up in their physical properties, an extraordinary similarity exists in the chemical properties of Zr and Hf. This similarity is attributed to the lanthanide contraction (Sec. 18.2), which intervenes between these two elements and makes their atoms identical in size. (Compare 0.078 nm for Hf^{4+} with 0.079 nm for Zr^{4+} in contrast to 0.068 for Ti^{4+}.)

Titanium   There was a flurry of excitement when the first moon-rock analyses indicated unusually high titanium content. However, this turned out to be a fluke of the sampling site and was not representative of the entire moon. In the earth's crust, titanium is tenth most abundant (0.58 percent by weight) and ranks ahead of such familiar elements as chlorine, carbon, and sulfur. However, it is distributed very widely, and commercially useful deposits are scarce. The principal sources are rutile (TiO_2), ilmenite ($FeTiO_3$), and iron ores. It is very difficult to prepare the pure metal because it has great affinity for carbon, nitrogen, oxygen, and hydrogen. The usual method for getting titanium is to convert the oxides with chlorine to $TiCl_4$, which is then reduced with magnesium:

$$TiO_2 \xrightarrow[C]{Cl_2} TiCl_4 \xrightarrow{Mg} \underset{99.3\%}{Ti} + MgCl_2$$

Pure titanium which has low carbon impurity and is free from hydrogen embrittlement (Sec. 15.4) is extremely strong (stronger than iron). Because it also has a high melting point and is resistant to corrosion (because of surface coatings of oxide and nitride), the metal is in great demand as a structural material, for example, in rocket engines and supersonic planes. Although the Anglo-French Concorde was able to get away with a structural skin of aluminum (mp 660°C) by limiting itself to speeds of Mach 2.2 (i.e., 2.2 times the velocity of sound), the second generation SSTs will almost inevitably have to use titanium (mp 1660°C) in order to reach the Mach 2.7 speeds that are projected. However, titanium is difficult to make and machine; so its widespread use has been delayed. Until recently, the principal use of titanium has been for hardening and toughening steel.

In its compounds, Ti exhibits oxidation states of $+2$, $+3$, and $+4$. Compounds of the first two are colored and paramagnetic because of the presence of unpaired d electrons. They are also good reducing agents. For the half-reaction $Ti^{3+} + e^- \longrightarrow Ti^{2+}$, the reduction potential is about -2 V, which indicates that Ti^{2+} is a much better reducing agent than is H_2; Ti^{2+} compounds such as $TiCl_2$ when added to H_3O^+ solutions liberate H_2.

Titanous ion, Ti^{3+}, is a violet ion which also is a convenient reducing agent. When oxidized in aqueous solution, it does not form titanic ion, Ti^{4+}, as expected, but appears in a hydrolyzed form usually written TiO^{2+} and called *titanyl ion*. From this are derived ionic salts such as $TiOSO_4$, titanyl sulfate. Like most of the $+3$ ions of the transition elements, Ti^{3+} forms an insoluble trihydroxide when base is added to its solutions. This black $Ti(OH)_3$ turns white and evolves H_2 when allowed to stand, indicating decomposition:

$$2Ti(OH)_3(s) \longrightarrow 2TiO_2(s) + 2H_2O + H_2(g)$$

The most important oxidation state of titanium is the $+4$, and probably the most important compound is TiO_2, titanium dioxide, or titania. This compound is quite inert and has good covering power; it is used extensively as a pigment in both the paint industry and the cosmetic industry. In crystalline form, it is used as a semiprecious, artificial gem. With a higher refractive index than diamond (Sec. 22.2), it has more sparkle than diamond, but, unfortunately, it is not very hard and becomes scratched.

When TiO_2 is heated with carbon in a stream of chlorine, titanium tetrachloride ($TiCl_4$) is formed. This is a colorless, fuming liquid which is used for making smoke screens. The smoke is probably an oxychloride of the type $TiOCl_2$.

Zirconium and hafnium These two elements are remarkable because they have essentially identical chemical properties. The atomic radii of Zr and Hf are 0.1454 and 0.1442 nm, respectively. Since their outer-electron configurations are the same, it is not surprising that the two elements resemble each other chemically. The resemblance is so marked that Hf atoms replace Zr atoms in crystals with ease. For this reason, all naturally occurring Zr minerals are contaminated with Hf. Zr is about 50 times as abundant as Hf, which makes up only 4.5×10^{-4} percent of the earth's crust. The principal minerals of Zr are baddeleyite (ZrO_2) and zircon ($ZrSiO_4$). In its transparent form, especially when colored, zircon finds use as a gemstone.

Zirconium and hafnium metals are extremely difficult to prepare in the pure state because, like titanium, they have such great affinity for hydrogen, carbon, oxygen, and nitrogen. One method for getting the pure metals is the thermal decomposition of the tetraiodide (ZrI_4) on a hot tungsten wire.

Practically all the known compounds of Zr and Hf correspond to the +4 oxidation state. Of these, the oxides zirconia (ZrO_2) and hafnia (HfO_2) are refractory and are used for high-temperature insulation. In the Nernst glow lamp, used in scientific work as a concentrated light source, the incandescent body is chiefly ZrO_2. The other +4 compounds readily hydrolyze to form the zirconyl (ZrO^{2+}) or hafnyl (HfO^{2+}) ions, which can also be written as $M(OH)_2^{2+}$.

Zirconium and hafnium are both important for the production of nuclear energy. Zirconium has a particularly low probability of capturing neutrons, which, coupled with its high resistance to corrosion, makes it ideal for cladding uranium or plutonium fuel rods in nuclear reactors. Hafnium, on the other hand, has a very high neutron capture probability; so every trace of hafnium has to be removed from the zirconium before the zirconium metal is used as a protective coat. In recompense, the hafnium is used to make control rods for regulating the free-neutron level that governs the rate of nuclear fission in a reactor (Chap. 27).

Like titanium, both zirconium and hafnium form complex ions, especially with fluorine. Examples of these are ZrF_6^{2-} and ZrF_7^{3-}. The latter is a rather rare illustration of seven neighbors about a central atom. The complex salts K_2ZrF_6 and K_2HfF_6 differ in solubility, and this is the basis of the separation of zirconium from hafnium by fractional crystallization.

Kurchatovium In 1964, Soviet scientists reported the synthesis of element 104 by bombardment of plutonium with neon nuclei:

$$^{242}_{94}Pu + ^{22}_{10}Ne \longrightarrow ^{260}_{104}Ku + 4^{1}_{0}n$$

As is traditionally true, they were given the privilege of naming the new element, which they promptly did after Professor Kurchatov, Soviet physicist. However, the Berkeley group which was chasing the same element could not reproduce the Soviet work but came up with another isotope of the same element, which they promptly christened rutherfordium (Rf), after Ernest Rutherford, discoverer of the nucleus. At the present time, the name as well as the discovery remain in dispute. More to the point, tracer experiments in which the radioactivity of element 104 is followed clearly indicate that the chemical properties resemble those of hafnium. Thus, element 104 is not an actinide element, corresponding to electronic expansion of the $5f$ subshell, but is a transactinide where d-orbital expansion has been resumed.

18.5   Vanadium subgroup

The elements of the vanadium subgroup are vanadium ($Z = 23$), niobium ($Z = 41$), tantalum ($Z = 73$), and hahnium ($Z = 105$). Hahnium is a recently discovered synthetic element made by heavy-ion bombardment; it is named after Otto Hahn, codiscoverer with Lise Meitner of the phenomenon of nuclear fission. The name "columbium" has sometimes been used instead of niobium. These elements are considerably less abundant than those of the titanium subgroup, and their chemistry is more complicated because of the formation of a $+5$ oxidation state. Properties are summarized in Fig. 18.9. Of these elements, vanadium shows compounds corresponding to oxidation states $+2$, $+3$, $+4$, and $+5$. Niobium shows $+5$ and some $+3$; tantalum shows $+4$ but mainly $+5$. On the basis of tracer experiments, hahnium appears to be mainly $+5$. Again because of the intervening lanthanide contraction, niobium and tantalum are very similar in chemical properties.

Vanadium   The principal minerals of vanadium are patronite (V_2S_5) and vanadinite ($Pb_5V_3O_{12}Cl$), but the element is also obtained as a valuable by-product from the uranium mineral carnotite ($KUVO_6$). The name "vanadium" comes from Vanadis, the Scandinavian goddess of beauty, and recalls the beautiful colors of the various vanadium compounds. The pure metal is very hard to prepare, and since its main use is as an additive to steel alloys, vanadium is usually made as ferrovanadium (solid solution of iron and vanadium). When added to steel, the vanadium combines with the oxygen and nitrogen, and it also dissolves in the molten iron to increase the tensile strength, toughness, and elasticity of the resulting steel.

In its compounds, vanadium shows oxidation states of $+2$, $+3$, $+4$ and $+5$. These correspond to at least partial removal of the two $4s$ electrons plus none, one, two, or all three of the $3d$ electrons. Many of the compounds are characteristically colored. Those in the lower oxidation states are good reducing agents.

Probably the most important compound of vanadium is the pentoxide, V_2O_5. This is a red or orange solid made by thermal decomposition of ammonium vanadate (NH_4VO_3). It is used as a catalyst in various oxidation reactions in which O_2 is the oxidizing agent, e.g., in the conversion of SO_2 to SO_3 for making H_2SO_4. Vanadium pentoxide is amphoteric. It dissolves in highly acid solutions to give an ion variously described as VO^{3+}, $V(OH)_2{}^{3+}$, $VO_2{}^+$, or $V(OH)_4{}^+$; in basic solution V_2O_5 dissolves to give vanadate anions such as $VO_4{}^{3-}$. When acid is gradually added to these vanadate solutions, anions which con-

Symbol	Z	Electronic configuration	Melting point, °C	Boiling point, °C	Ionization potential, eV	Electrode potential, V
V	23	$(18)\ 3d^3 4s^2$	1710	3000 (?)	6.74	-1.2 (from V^{2+})
Nb	41	$(36)\ 4d^4 5s^1$	1950	3000 (?)	6.77	-0.65 (from Nb_2O_5)
Ta	73	$(68)\ 5d^3 6s^2$	>3000	>4100	6	-0.81 (from Ta_2O_5)
Ha	105	$(100)\ 6d^3 7s^2$?	?	?	?	?

Fig. 18.9
Elements of vanadium subgroup.

tain more than one V atom per ion (so-called "polyvanadates," such as $V_2O_7^{4-}$) are formed. Eventually V_2O_5 precipitates.

Acid solutions containing vanadium in the $+5$ oxidation state go through a series of color changes when a reducing agent such as zinc is added. The solutions first turn green, then blue, and then violet, corresponding to stepwise reduction to a green mixture of yellow $+5$ and blue $+4$, a clear blue $+4$, a blue $+3$, and a violet $+2$ state. The characteristic ions are $V(OH)_2^{2+}$, V^{3+} (vanadic), and V^{2+} (vanadous). The electrode potentials relating these species are

$$V(OH)_4^+ + 2H_3O^+ + e^- \longrightarrow V(OH)_2^{2+} + 4H_2O \qquad +1.0\ V$$

$$V(OH)_2^{2+} + 2H_3O^+ + e^- \longrightarrow V^{3+} + 4H_2O \qquad +0.36\ V$$

$$V^{3+} + e^- \longrightarrow V^{2+} \qquad -0.25\ V$$

$$V^{2+} + 2e^- \longrightarrow V(s) \qquad -1.18\ V$$

Since the $E°$ for $Zn(s) \longrightarrow Zn^{2+} + 2e^-$ is $+0.76$ V, it can add to any of the above half-reactions, except the last, to give a spontaneous net reaction. In other words, $Zn(s)$ does not have the reducing strength required to convert V^{2+} to $V(s)$.

Niobium and tantalum   Both niobium and tantalum are rather rare elements, and they are almost always found together in nature. The principal minerals are columbite and tantalite, which are mixed oxides of the two metals along with those of iron and manganese. Although tantalum is rare in nature (0.00021 percent, an abundance about one-tenth that of niobium), its desirable properties have led to rather extensive use. The problem of preparation is a difficult one, and, in fact, the name "tantalum" reflects the frustrations in its first recovery. (In Greek mythology Tantalus was sent to hell, where, plagued by hunger and thirst, he was placed near food and drink which always stayed out of reach. The close relationship of niobium to tantalum is indicated by the name "niobium," after Niobe, the tragic daughter of Tantalus.) In order to separate the two elements, the mixed oxides are converted by HF and KF to K_2TaF_7 [heptafluorotantalate(V)] and K_2NbOF_5 [oxopentafluoroniobate(V)], which differ in solubility and can be separated from each other by fractional crystallization.

Tantalum is very ductile and preceded tungsten as filament material in light bulbs and electron tubes. It is also used in one kind of electrolytic *rectifier* for converting alternating current to direct current. This rectifier consists of an aqueous solution with two electrodes, one of which is Ta. When the Ta starts to act as an anode, it immediately forms an oxide coat, which cuts off the current. Ta can, however, act as a cathode and so permits flow of current in only one direction. Since Ta is very resistant to corrosion, it is used extensively for apparatus in chemical plants, especially equipment designed for handling acids. Since it is also compatible with human tissue, it finds use in surgery as for bone pins.

The most common compounds of Nb and Ta are the pentoxides, Nb_2O_5 and Ta_2O_5. These are rather inert, stable solids which can be formed by heating the finely divided metals in air or oxygen. They dissolve in concentrated bases to form niobate and tantalate anions. Little is known of these anions except that they are quite complex and can have varying numbers of metal atoms per ion. Tantalum carbide, TaC, made by heating the oxide with carbon, is extremely hard and finds use in making tools for high-speed machining of metals and wire-drawing dies.

18.6 Chromium subgroup

The chromium subgroup contains the elements chromium ($Z = 24$), molybdenum ($Z = 42$), and tungsten ($Z = 74$). They are all metals of small atomic volume, extremely high melting point, great hardness, and excellent resistance to corrosion. Their chemistry is complicated by the existence of several oxidation states ranging from $+2$ to $+6$ and by the formation of many complex ions, including oxyanions. Some properties are given in Fig. 18.10. It might be noted in the electronic configuration that the s^1 characteristic of chromium and molybdenum gives way to s^2 for tungsten. This is tied to the fact that the $6s$ orbital, being larger than $5s$ or $4s$, has in it less electron-electron repulsion and so can tolerate $6s^2$ configuration. All these elements form compounds in which they show the $+6$ oxidation state. In addition, chromium commonly shows $+2$ and $+3$ states, molybdenum $+3$, $+4$, and $+5$ states, and tungsten $+4$ and $+5$.

Chromium   Cr is one of the less abundant metals (0.037 percent of the earth's crust), but still it is approximately 50 times as abundant as Mo and W. Its principal mineral is chromite ($FeCr_2O_4$), some of which is reduced directly by heating with C in order to provide ferrochromium (solid solution of Cr in Fe) for addition to alloy steels. Low-chrome steels (up to 1% Cr) are quite hard and strong; high-chrome steels (up to 30% Cr), or *stainless steels*, are very resistant to corrosion. Most of the remaining chromite is converted to sodium chromate (Na_2CrO_4) by heating it with Na_2CO_3 in air:

$$8Na_2CO_3(s) + 4FeCr_2O_4(s) + 7O_2(g) \longrightarrow$$
$$2Fe_2O_3(s) + 8Na_2CrO_4(s) + 8CO_2(g)$$

Symbol	Z	Electronic configuration	Melting point, °C	Boiling point, °C	Ionization potential, eV	Electrode potential, V
Cr	24	$(18)\ 3d^5 4s^1$	1600	2500 (?)	6.76	-0.91 (from Cr^{2+})
Mo	42	$(36)\ 4d^5 5s^1$	2620	>3700	7.18	-0.2 (from Mo^{3+})
W	74	$(68)\ 5d^4 6s^2$	3370	5900	7.98	-0.12 (from WO_2)

Fig. 18.10
Elements of chromium subgroup.

The sodium chromate is leached out with acid to form $Na_2Cr_2O_7$, an important oxidizing agent.

Chromium metal is very hard and, although quite reactive in the powdered form, in the massive form is quite resistant to corrosion. Furthermore, it takes a high polish, which lasts because of formation of an invisible, self-protective oxide coat. Consequently, chromium finds much use as a plating material, both for its decorative effect (0.00005 cm thick) and for its protective effect (0.0075 cm thick). The plate is usually put on by electrolyzing the object in a bath made by dissolving $Na_2Cr_2O_7$ and H_2SO_4 in water. The mechanism of the electroplating is complicated; since plating will not occur unless the sulfate is present, the sulfate must be involved in some intermediate formed during the electrolysis.

All the compounds of chromium are colored, a fact which suggested the name "chromium," from the Greek word for color, *chroma*. The characteristic oxidation states are $+2$, $+3$, and $+6$, represented in acid solution by Cr^{2+} (chromous), Cr^{3+} (chromic), and $Cr_2O_7{}^{2-}$ (dichromate) and in basic media by $Cr(OH)_2$, $CrO_2{}^-$ or $Cr(OH)_4{}^-$ (chromite), and $CrO_4{}^{2-}$ (chromate).

The chromous ion, Cr^{2+}, is a beautiful blue ion obtained by reducing either Cr^{3+} or $Cr_2O_7{}^{2-}$ with Zn metal. However, it is rapidly oxidized in aqueous solution by air. The electrode potential for $Cr^{3+} + e^- \longrightarrow Cr^{2+}$ is -0.41 V, which means that Cr^{2+} should be able to reduce H_3O^+, but the latter reaction is very slow. When base is added to solutions of chromous salts, chromous hydroxide precipitates. On exposure to air, $Cr(OH)_2$ is oxidized by O_2 to give $Cr(OH)_3$ (also written $Cr_2O_3 \cdot xH_2O$).

Many chromic salts, such as chromic nitrate, $Cr(NO_3)_3$, and chromic perchlorate, $Cr(ClO_4)_3$, dissolve in water to give violet solutions, in which the violet color is due to the hydrated chromic ion, $Cr(H_2O)_6{}^{3+}$. This is one of the best characterized hydrated ions in solution. To show that it is really a hexaaquo ion, an ingenious method has been developed, known as the *isotope dilution method*. H_2O containing the ^{18}O isotope is added to an ordinary solution containing Cr^{3+} and normal H_2O. The solvent H_2O is then sampled at intervals (after the Cr^{3+} has been precipitated out) and examined for its $H_2{}^{18}O/H_2{}^{16}O$ ratio. It is found that six H_2O per Cr^{3+} do not participate in the instantaneous dilution of the added $H_2{}^{18}O$. This method works for Cr^{3+} because Cr^{3+} is very slow in exchanging its bound H_2O for H_2O that comes from the solvent. If one waits awhile before sampling the solvent H_2O

after addition of $H_2^{18}O$, there will be a slow exchange:

$$Cr(H_2^{16}O)_6^{3+} + H_2^{18}O \rightleftharpoons Cr(H_2^{16}O)_5(H_2^{18}O)^{3+} + H_2^{16}O$$

The ratio of $H_2^{18}O/H_2^{16}O$ can be obtained by analysis with a mass spectrometer. The sample is injected into an evacuated chamber where electron bombardment causes the molecules to be ionized. These are accelerated and bent in a magnetic field, the heavier molecules following a curve of bigger radius (Fig. 1.12). Monitoring the ion-collector current tells the relative number of $H_3^{18}O$ and $H_2^{16}O$ molecules in the original sample.

If to a solution containing violet $Cr(H_2O)_6^{3+}$ a high concentration of chloride ion is added, some of the hydrate water is replaced, and the solution slowly turns green because of formation of a chloro complex. Solutions of chromic salts can be kept indefinitely exposed to the air without oxidation or reduction. In general, they are slightly acid because of hydrolysis of the chromic ion. The hydrolysis reaction can be written in either of the following ways:

(1) $Cr^{3+} + H_2O \longrightarrow CrOH^{2+} + H^+$

 $H^+ + H_2O \longrightarrow H_3O^+$

 $\overline{Cr^{3+} + 2H_2O \longrightarrow CrOH^{2+} + H_3O^+}$

which corresponds to abstraction of OH^- from H_2O by Cr^{3+}, leaving H^+ which immediately attaches to another H_2O.

(2) $Cr(H_2O)_6^{3+} + H_2O \longrightarrow Cr(H_2O)_5OH^{2+} + H_3O^+$

which corresponds to proton transfer from the aquocomplex to solvent H_2O, a process we have previously called dissociation of a weak acid.

When a base is gradually added to chromic solutions, a green, slimy precipitate, which can be formulated either as $Cr(OH)_3 \cdot xH_2O$ or $Cr_2O_3 \cdot xH_2O$, first forms but then disappears as excess OH^- is added. A deep green color characteristic of chromite ion, written as CrO_2^- or $Cr(OH)_4^-$, is produced. The precipitation and redissolving associated with this amphoteric behavior can be described as follows:

$Cr^{3+} + 3OH^- \longrightarrow Cr(OH)_3(s)$

$Cr(OH)_3(s) + OH^- \longrightarrow CrO_2^- + 2H_2O$

The green species in the final solution is certainly more complicated than CrO_2^- and probably contains more than one Cr atom per ion. When filtered off and heated, the insoluble hydroxide loses water to form Cr_2O_3, chromic oxide or chromic sesquioxide. This is an inert, green powder much used as artists' pigment, chrome green.

Chromic ion forms a great number of complex ions. In all of these the chromium atom is surrounded by six other atoms arranged at the corners of an octahedron. Typical octahedral complexes are CrF_6^{3-}, $Cr(NH_3)_6^{3+}$, $Cr(H_2O)_6^{3+}$, $Cr(H_2O)_5Cl^{2+}$, and $Cr(NH_3)_4Cl_2^+$. It is characteristic of chromic complexes that they form and dissociate very slowly. In potassium chrome alum, $KCr(SO_4)_2 \cdot 12H_2O$, the $Cr(H_2O)_6^{3+}$ complex occurs as a unit occupying some of the crystal lattice sites.

In the $+6$ oxidation state, chromium is known principally as the chromates and dichromates. The chromate ion, CrO_4^{2-}, can be made quite easily by oxidizing chromite ion, CrO_2^-, in basic solution with a moderately good oxidizing agent such as hydrogen peroxide. The reaction is

$$2CrO_2^- + 3HO_2^- \longrightarrow 2CrO_4^{2-} + H_2O + OH^-$$

where the peroxide is written as HO_2^- in basic solution. The chromate ion is yellow and has a tetrahedral structure with four oxygen atoms bound to a central chromium atom.

When solutions of chromate salts are acidified, the yellow color is replaced by a characteristic orange, the result of formation of $Cr_2O_7^{2-}$, dichromate ion:

$$2CrO_4^{2-} + 2H_3O^+ \rightleftharpoons Cr_2O_7^{2-} + 3H_2O$$

The change is reversed by adding base. The structure of the dichromate ion, shown in Fig. 18.11, consists of two tetrahedra sharing an oxygen atom at a common corner. Each of the chromium atoms is at the center of a tetrahedron bound to four oxygen atoms. The dichromate ion is a very good oxidizing agent, especially in acid solution. The half-reaction

$$Cr_2O_7^{2-} + 14H_3O^+ + 6e^- \longrightarrow 2Cr^{3+} + 21H_2O$$

has an electrode potential of $+1.33$ V; therefore, $Cr_2O_7^{2-}$ is able to oxidize all but the very poorest reducing agents. It will, for example, oxidize hydrogen peroxide to form oxygen gas:

$$Cr_2O_7^{2-} + 3H_2O_2 + 8H_3O^+ \longrightarrow 3O_2(g) + 2Cr^{3+} + 15H_2O$$

It might seem strange that in basic solution hydrogen peroxide oxidizes chromium whereas in acid solution chromium oxidizes hydrogen peroxide. The reason for this is that in going from the $+3$ to the $+6$ state of chromium (Cr^{3+} to $Cr_2O_7^{2-}$), oxygen atoms have to be added, whereas in going from the $+6$ to $+3$ state ($Cr_2O_7^{2-}$ to Cr^{3+}) oxygen is removed. In acid solution H_3O^+ helps in removing oxygen by forming water; in basic solution the scarcity of H_3O^+ facilitates addition of oxygen. In the general case of preparing compounds, the change to compounds of higher oxidation state is usually most easily done in basic solution; to go to compounds of lower oxidation state it is best to work in acid solution.

When solutions of dichromate ion are made very acid, especially in the presence of a dehydrating agent such as concentrated H_2SO_4, the uncharged species CrO_3 is formed. This deep red solid is chromium trioxide, or, as it is sometimes called, chromic anhydride. It is a very powerful oxidizing agent and is used extensively in preparing organic compounds. Suspensions of CrO_3 in concentrated H_2SO_4 are frequently

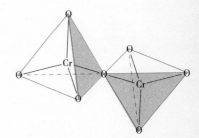

Fig. 18.11
Dichromate ion.

used as "cleaning solution" for glass equipment in laboratories. The cleaning action is due to oxidation of grease.

Molybdenum Molybdenum is found in nature as the mineral molybdenite, MoS_2, a beautiful blue-gray material of metallic luster, frequently confused with graphite. Like graphite, MoS_2 has a layered structure and acts as a lubricant, but it has the added advantage of not losing its lubricating properties when subject to a vacuum, as in outer space. When heated in air, MoS_2 is oxidized to the trioxide, MoO_3, which can then be reduced to the metal by heating with hydrogen. Because of its very high melting point (2620°C), molybdenum is obtained as a powder. This is pressed into bars and heated, so as to sinter the particles together to give sheet or wire for use as supports in X-ray tubes, electron tubes, electric furnaces, etc., where high temperatures may develop locally.

The major part of the Mo metal produced goes into Fe alloys, where it acts as a toughening agent, favoring fine-grained structure. The Mo is added as ferromolybdenum (55 to 75% Mo in Fe), made by reducing mixed oxides of Mo and Fe.

The most important oxidation state of Mo is +6, as, for example, in molybdenum trioxide, MoO_3. This trioxide is acidic and dissolves in basic solutions to form a complicated series of oxyanions called the *molybdates*, the simplest of which is MoO_4^{2-}. More complicated molybdates, having Mo—O—Mo bridges and containing up to 24 Mo atoms per ion, are known. These are called *polymolybdates* but are not well characterized. Neither MoO_3 nor the molybdates are particularly good oxidizing agents. When reduced, MoO_3 can form a deep blue oxide of variable composition, $MoO_{2.5-3.0}$, which is apparently some sort of nonstoichiometric compound with a defect structure (Sec. 7.5). Further reduction of MoO_3 or the molybdates usually forms the metal, but under appropriate conditions Mo^{3+} can be formed in aqueous solution.

Molybdenum is important as a trace element in the soil, where it plays an important role in nitrogen fixation by bacteria. It is believed to be essential for the growth of plants, but only in trace amounts. Large amounts, as present in molybdenum-rich soils, can be taken up to make the plants toxic for animals that forage on them.

Tungsten The element tungsten is frequently also called *wolfram*, whence the symbol W. It occurs in nature principally as the tungstates, e.g., as a calcium tungstate, $CaWO_4$, or scheelite, and as a mixture of iron and manganese tungstates, $(Fe,Mn)WO_4$, or wolframite. In order to get the metal, tungstates are treated with acid to precipitate insoluble tungstic acid, H_2WO_4, which is then dehydrated by ignition to tungstic oxide, WO_3. Hydrogen reduction of WO_3 produces W.

Since W has such a high melting point (3370°C),* it is obtained, like Mo, as a fine powder, which has to be sintered into workable

* At room temperature W has an extremely small vapor pressure. It has been calculated to correspond to one W atom per universe. *E pluribus unum.*

form. Fine W wire for lamp filaments can be produced by heating sintered W powder in H_2 and subjecting the specimen to prolonged, vigorous pounding in a swaging machine (a machine for beating wire from large cross section down to small). Finally, an electric current through the wire heats it to a very high temperature, at which the microparticles coalesce.

The metal is rather inert to common oxidizing agents such as O_2 and HNO_3. However, it can be dissolved in a mixture of concentrated HNO_3 and HF. Most W goes into steel production, especially to make *high-speed steel* for cutting tools. Addition of W increases the ability to hold hardness at high temperatures and slows down the tearing off of small particles that causes the dulling of fast-finishing tools.

In its compounds tungsten usually occurs as the +6 oxidation state. The oxide WO_3 is an acidic oxide and dissolves in basic solutions or in molten basic oxides to form tungstates and polytungstates. Like the molybdates, the tungstates are not particularly good oxidizing agents. Unusual are the nonstoichiometric compounds called *tungsten bronzes*, $M_xWO_3 (0 < x < 1)$, in which metal atoms such as sodium fit interstitially into a WO_3 matrix. When x is small, the materials are rather poor semiconductors; but when x exceeds about 0.25, they become metallic, with high reflectivity, good electric conductivity, and other free-electron symptoms. The unusual thing is that, even though they appear to contain sodium dissolved in a solid matrix, the bronzes are actually extraordinarily inert to chemical attack by acids. Only if the whole WO_3 matrix is destroyed, as by alkaline attack, do the tungsten bronzes go into solution. They are considered to be promising as electrode materials for fuel cells, especially when doped with platinum.

18.7   Manganese subgroup

The elements of the manganese subgroup are manganese $(Z = 25)$, technetium $(Z = 43)$, and rhenium $(Z = 75)$. Manganese is by far the most important of the group; technetium is radioactive and does not occur in nature; rhenium is so rare as to constitute a chemical curiosity. Some of the properties of the subgroup are given in Fig. 18.12.

Manganese   Mn is not a very common element (abundance, 0.08 percent), but in the earth's crust it is as abundant as C and more so than S. The most important minerals are the oxides: MnO_2, or pyrolusite, Mn_2O_3, or braunite (usually contaminated with iron oxide), and Mn_3O_4, or hausmannite. Probably the best way to prepare the metal is by reduction with powdered Al. However, since most metallic Mn goes into steel production, alloys of Mn are used instead. Two such alloys are ferromanganese (about 80% Mn in Fe) and spiegeleisen (about 20 to 30% Mn and about 5% C in Fe); they are made by reducing mixed oxides of Fe and Mn in a blast furnace with C or CO acting as reducing agent. When added to steel, Mn has two functions: In low amounts, it acts as a scavenger by combining with O and S in the molten Fe to form easily removable substances. In high

Symbol	Z	Electronic configuration	Melting point, °C	Boiling point, °C	Ionization potential, eV	Electrode potential, V
Mn	25	$(18)\ 3d^5 4s^2$	1260	1900	7.43	-1.18 (from Mn^{2+})
Tc	43	$(36)\ 4d^6 5s^1$?	. . . .	. . . .	. . . .	$+0.4(?)$ (from Tc^{2+})
Re	75	$(68)\ 5d^5 6s^2$	3200	. . . .	7.87	$+0.25$ (from ReO_2)

Fig. 18.12

Elements of manganese subgroup.

amounts (up to 14 percent), it imparts special hardness and toughness such as is needed for resistance to battering abrasion.

In its chemical compounds manganese shows oxidation states of $+2$, $+3$, $+4$, $+6$, and $+7$. Most of these compounds are colored and paramagnetic. In the $+2$ state, manganese exists as the manganous ion, Mn^{2+}. Although Mn^{2+} solutions are essentially colorless, many manganous salts, such as manganous sulfate, $MnSO_4$, and manganous chloride, $MnCl_2$, have pink coloration. Unlike Cr^{2+}, Mn^{2+} is a very poor reducing agent, and neutral or acid solutions of manganous salts can be kept indefinitely exposed to oxygen or other oxidizing agents. When base is added to Mn^{2+}, a white precipitate of $Mn(OH)_2$ is formed. This solid, unlike manganous salts, is promptly oxidized by air to the $+3$ state.

In the $+3$ state, manganese can exist as the manganic ion, Mn^{3+}, but for long periods and high concentrations only in solids and complex ions. Unlike Cr^{3+}, Mn^{3+} is a very powerful oxidizing agent and can even oxidize water to liberate oxygen. The electrode potential of $Mn^{3+} + e^- \longrightarrow Mn^{2+}$ is $+1.51$ V, high enough that manganic ion can oxidize itself to the $+4$ state:

$$Mn^{3+} + 6H_2O \longrightarrow MnO_2(s) + 4H_3O^+ + e^- \qquad E° = -0.95 \text{ V}$$

$$Mn^{3+} + e^- \longrightarrow Mn^{2+} \qquad\qquad\qquad\qquad E° = +1.51 \text{ V}$$

$$2Mn^{3+} + 6H_2O \longrightarrow Mn^{2+} + MnO_2(s) + 4H_3O^+ \quad E° = +0.56 \text{ V}$$
$$\Delta G° = -54 \text{ kJ}$$

This disproportionation can be prevented and the $+3$ state stabilized by (1) complexing the manganese, for example, with cyanide, CN^-, to give $Mn(CN)_6{}^{3-}$ or with oxalate, $C_2O_4{}^{2-}$, to give $Mn(C_2O_4)_3{}^{3-}$; (2) forming an insoluble salt, such as $MnPO_4$, manganic phosphate, or the hydroxide, written as $Mn(OH)_3$, $MnOOH$, or even Mn_2O_3; or (3) adding a large excess of Mn^{2+} and H_3O^+.

In the $+4$ state the principal compound of manganese is manganese dioxide, MnO_2. It is not a simple stoichiometric compound because no matter how careful the preparation, the product always contains fewer than two (generally ≈ 1.97) oxygen atoms per manganese atom. As mentioned in Sec. 14.4, MnO_2 is the oxidizing agent in the dry cell.

When MnO_2 is heated with basic substances in air, it is oxidized from its original black color to a deep green, the result of conversion to manganate ion, $MnO_4{}^{2-}$. Though stable in alkaline solution, this ion (which represents Mn in the $+6$ state) disproportionates when the solution is acidified:

$$3MnO_4^{2-} + 4H_3O^+ \longrightarrow MnO_2(s) + 2MnO_4^- + 6H_2O$$

The MnO_4^- ion is the permanganate ion and shows manganese in its highest oxidation state. It is a very good oxidizing agent, especially in acid solution, where it is usually reduced all the way to manganous ion. For the half-reaction

$$MnO_4^- + 8H_3O^+ + 5e^- \longrightarrow Mn^{2+} + 12H_2O$$

the electrode potential is $+1.51$ V. This means that MnO_4^- is one of the strongest common oxidizing agents. Solutions of permanganate salts are frequently used in analytical chemistry to determine amounts of reducing agents by titration. Titration is simplified by using the disappearance of the deep violet color of MnO_4^- as the end-point indicator. The usual procedure is to make the solutions acid, so as to ensure complete reduction to Mn^{2+}. In neutral or alkaline solutions, MnO_2 is formed instead; in very basic solutions, MnO_4^{2-}. When $KMnO_4$ is treated with concentrated H_2SO_4, a violently explosive oil, Mn_2O_7, or manganese heptoxide, is formed.

The chemistry of manganese illustrates well how the characteristics of compounds change in going from low oxidation state to high. In the low oxidation states manganese exists as a cation which forms basic oxides and hydroxides. In the higher oxidation states it exists as anions derived from acidic oxides.

Technetium Since the technetium nucleus is unstable to radioactive decay, it does not occur in nature. The name is derived from the Greek *technetos,* for artificial, and draws attention to the fact that technetium is made synthetically by bombardment of molybdenum with neutrons. The longest-lived isotope of technetium is $^{97}_{43}Tc$, having a half-life of 2.6×10^6 yr; so, once obtained, it can be worked with, though special shielding precautions against the radiation must be taken. Technetium does not resemble manganese nearly so much as it resembles the following element, rhenium.

Rhenium Rhenium, named after the Rhine River, is a very rare element (abundance, 10^{-7} percent by mass) occurring in trace amounts in molybdenite, columbite, and pyrolusite. The element is extracted by oxidation to perrhenic acid, $HReO_4$, and precipitation as slightly soluble potassium perrhenate, $KReO_4$. In many of its properties, rhenium is much like tungsten. It has a high melting point (about $3200°C$) and a high density (21 g/cm^3), and it is not particularly reactive. When heated in air, it gives off clouds of pale yellow rhenium heptoxide, a volatile solid which, unlike Mn_2O_7, is not explosive. Like WO_3, Re_2O_7 is an acidic oxide and not expecially reactive. Rhenium is too scarce to be much used, but it does have remarkable catalytic properties for hydrogenation reactions.

18.8 Qualitative analysis

Of the elements discussed in this chapter, only manganese and chromium are included in the common schemes of qualitative analysis. Their confirming tests depend on the characteristic colors of compounds of these elements.

If chromium and manganese are present in the original unknown as chromate (or dichromate) and permanganate, they will be reduced by H_2S (as from thioacetamide) in acid solution to Cr^{3+} and Mn^{2+}, forming finely divided sulfur in the process. When the solution, still containing sulfide, is made basic, $Cr(OH)_3$ and MnS will precipitate. Treatment with acid and an oxidizing agent serves to remove the sulfide, and subsequent addition of excess $NaOH$ precipitates manganese as a hydroxide and converts the chromium to soluble chromite ion. The appearance of a green color in the solution at this point is a strong indication of the presence of chromium. It can be confirmed by treating the green, basic solution with H_2O_2 to oxidize chromite to yellow chromate, which can be precipitated as yellow $BaCrO_4$ by the addition of barium ion in an acetic acid buffer.

The hydroxide precipitate suspected of containing manganese on treatment with acid and a very strong oxidizing agent such as sodium bismuthate, $NaBiO_3$, or sodium periodate, $NaIO_4$, produces the characteristic violet color of MnO_4^- if manganese is present.

Exercises

*18.1 *Lanthanides*. Using the orbital-filling diagram of Fig. 2.22, show the electron configuration at the $4f$ level for each of the lanthanide elements. What can you say about the relative levels of the $5d$ and $4f$ orbitals in this series?

*18.2 *Nuclear equation*. In a balanced nuclear equation, the sum of the subscripts on the nuclear symbols (which gives the positive charge) must be conserved as is the sum of the superscripts (which gives the number of mass units). Using ^4_2He for the alpha particle and $^0_{-1}e$ for the beta particle, write balanced nuclear equations for each of the following:

a Alpha decay of $^{238}_{92}U$

b Beta decay of $^{234}_{90}Th$

c Alpha decay of $^{239}_{94}Pu$

d Bombardment of curium 246 by carbon 12 to give nobelium 254 and four neutrons

e Bombardment of californium 98 with boron 11 to give lawrencium 259 and two neutrons

*18.3 *Uranium equations*. Write balanced chemical equations for each of the following reactions:

a Metallic uranium dissolves in acidic solution to liberate H_2 and form U^{3+}.

b U^{3+} reacts with neutral water to form $UO_2(s)$ and liberate hydrogen gas.

c An aqueous acidic solution of U^{4+} on exposure to air is oxidized to $UO_2{}^{2+}$.

d $UO_2{}^+$ disproportionates in acidic solution to give U^{4+} and $UO_2{}^{2+}$.

** 18.4 *Stoichiometry.* A 1.000-g sample of $Sc_{2-x}Y_x(C_2O_4)_3 \cdot 5H_2O$ is able to reduce 25.0 ml of 0.100 M $MnO_4{}^-$ by the reaction in acid solution $H_2C_2O_4 + MnO_4{}^- \longrightarrow CO_2 + Mn^{2+}$. What is the value of x? *Ans. 0.818*

** 18.5 *Lanthanide contractions.* What is meant by the lanthanide contraction? How does it manifest itself chemically? Where would you look for evidence of an analogous actinide contraction?

** 18.6 *Ion-exchange separation.*

a If you had to predict where the elution peaks for Er^{3+} and La^{3+} fall in Fig. 18.6, where would you put them? Justify.

b If you did an ion-exchange separation of Co^{2+} and Ni^{2+}, which would you probably wash out of the column first? Explain.

** 18.7 *Solubility.* What would be the concentrations of La^{3+} and Lu^{3+} in a solution that is simultaneously saturated with $La(OH)_3(s)$ and $Lu(OH)_3(s)$? The K_{sp}'s are 1.0×10^{-19} and 2.5×10^{-24}, respectively.

** 18.8 *Actinides.* Why do many of the actinide elements show a greater variety of oxidation states than do the corresponding lanthanide elements?

** 18.9 *Actinide elements.* Draw curves showing the probable relation of $5f$, $6d$, and $7s$ levels at the beginning of the actinide series.

** 18.10 *Radioactive decay.* The half-lives of $^{232}_{90}Th$ and $^{238}_{92}U$ are 1.39×10^{10} and 4.50×10^9 yr, respectively. Suppose you start at time $t = 0$ with equal numbers of these nuclei. What will be the abundance ratio after 1.00×10^9 yr? After 5.00×10^9 yr? *Ans. 1.11, 1.68*

** 18.11 *Lunar rocks.* One of the first reports of lunar-rock analysis (from Mare Tranquilitatis) reported elemental composition as equivalent to 7.6 wt % TiO_2 and 17.1 wt % FeO. Assuming that the Fe and Ti were actually present as ilmenite ($FeTiO_3$), what weight percent of ilmenite would the above analysis have corresponded to, and what weight percent of Fe or Ti would not have been accounted for?

** 18.12 *Lunar rocks.* Analysis of the lunar soil brought back by Apollo 11 from Mare Tranquilitatis was reported (all in weight percent) as follows: 42.3% SiO_2, 7.3% TiO_2, 14.1% Al_2O_3, 15.8% FeO, 7.9% MgO, 12.0% CaO, and 0.5% Na_2O. Translate these data into elemental abundances by atom weight, and compare with the earth's crust analysis given in Fig. 15.1.

Exercises

** *18.13 Vanadium potentials.* From the vanadium potentials given in Sec. 18.5, calculate the $E°$ for $V^{3+}(aq) + 3e^- \longrightarrow V(s)$.

Ans. -0.87 *volt*

** *18.14 Vanadium stoichiometry.* When $NH_4VO_3(s)$ is heated, it gives off $NH_3(g)$ plus $H_2O(g)$ and leaves $V_2O_5(s)$. If a given sample of $NH_4VO_3(s)$ is heated so as to lose 10.0 percent of initial weight, what percent of the NH_4VO_3 has been decomposed?

** *18.15 Tantalum thermodynamics.* A common way to prepare metals is to heat their oxides with powdered aluminum so as to reduce the oxide to metal and form Al_2O_3. Given that the free energy of formation of Ta_2O_5 is equal to -1969 kJ/mol and that of Al_2O_3 is -1575 kJ/mol, would you expect to be able to make tantalum by heating aluminum with Ta_2O_5? Justify your answer.

** *18.16 Chromium stoichiometry.* Write the equation for the reduction of chromite ore to ferrochromium by C, assuming CO as the other product. To end up with a 30 wt % high-chrome steel (assume Fe and Cr as the only constituents), how much Fe_2O_3, calculated as weight percent of the chromite, would you need to add to the original chromite ore?

** *18.17 Dichromate solution stoichiometry.* The amount of iron in an iron ore can be determined by dissolving the ore in acid, treating the solution to ensure the iron is Fe^{2+}, and then titrating with dichromate in acid to give Fe^{3+} and Cr^{3+}. For a particular ore composed mainly of Fe_3O_4, it was found that a 1.000-g sample required 9.27 ml of 0.150 M $K_2Cr_2O_7$. What is the apparent wt % of Fe_3O_4 in the ore?

Ans. 64.4%

** *18.18 Hydrolysis.* Given that $K = 1 \times 10^{-10}$ for $CrOH^{2+} \rightleftharpoons Cr^{3+} + OH^-$, what pH would you expect for a solution that is 0.10 M Cr^{3+}?

** *18.19 Chromium chemistry.* Write chemical formulas for the chromium-containing species in each of the materials written in italics: *Chrome alum* is dissolved in water to give a *violet solution*. Slow addition of NaOH gives a slimy, *green precipitate*, which then dissolves to give a clear *green solution* when excess NaOH is added. Subsequent addition of H_2O_2 and heating gives a *yellow solution*, which on acidification eventually turns *orange*.

** *18.20 Molybdenum stoichiometry.* When a 1.500-g sample of MoS_2 was heated in air so as to be oxidized partially to MoO_3 and $SO_2(g)$, the observed weight loss was 75.0 mg. What fraction of the MoS_2 was oxidized?

Ans. 0.497

** *18.21 Manganese chemistry.* Conserving the Mn atoms, how could you make the following conversions? Write balanced net equations for each reaction involved.

a $KMnO_4$ to $MnSO_4$

b MnO_2 to K_2MnO_4

c $MnSO_4$ to $Mn(OH)_3$

*** **18.22**   *Scandium.* If the first three ionization potentials of scandium are 6.56, 12.89, and 24.75 eV, respectively, and the potential for $Sc^{3+}(aq) + 3e^- \longrightarrow Sc(s)$ is -2.08 V, what must be the minimum amount of heat evolved on hydration of $Sc^{3+}(g)$? Why is the actual value probably greater than this?

*** **18.23**   *4f electron.* Sketch the $4f$ orbital mentioned as an example in Sec. 18.2. From comparison of your result with what you know about $3d$, $2p$, and $1s$, what regularity can you detect? What prediction might you make for a $5g$ orbital?

*** **18.24**   *5f electrons.* Besides being farther from the nucleus and of higher energy, how would the electron probability distribution of $5f$ electrons compare with that of $4f$? (*Hint:* Consider $1s$, $2s$, and $3s$.)

*** **18.25**   *Radioactive decay.* When $^{238}_{92}U$ ($t_{1/2} = 4.50 \times 10^9$ yr) decays by alpha activity, it produces $^{234}_{90}Th$ ($t_{1/2} = 24.1$ days). The only reason ^{234}Th persists in nature is because it is constantly regenerated by ^{238}U decay. In the steady state, what fraction of a ^{238}U sample is ^{234}Th?   *Ans. Th/U = 1.47 × 10^{-11}*

*** **18.26**   *Vanadium spectra.* Using ligand-field theory suggest a reason why the absorption spectrum of aqueous V^{3+} contains two broad bands whereas that of aqueous V^{4+} contains but one.

*** **18.27**   *Vanadium nonstoichiometry.* Samples of vanadium pentoxide are generally oxygen deficient, corresponding to the formula V_2O_{5-x}. The value of x can be determined by dissolving the sample in sulfuric acid and titrating with $Ce^{4+} + e^- \longrightarrow Ce^{3+}$ to oxidize any apparent V^{4+} to V^{5+}. A given sample weighing 2.6487 g requires 41.28 ml of 0.100 M Ce^{4+} to accomplish this. What is the value of x in the sample?

*** **18.28**   *Isotope dilution method.* Suppose you add 1.00 g of $H_2{}^{18}O$ to 100.0 g of 1.50 m $Cr(ClO_4)_3$ solution in normal water, where 0.204% of the oxygen is ^{18}O. If after mixing you immediately extract 1.00 g of water, what $H_2{}^{18}O/H_2{}^{16}O$ ratio would you expect to see in it:

a Assuming $6H_2O$ per Cr^{3+} are not involved in the mixing.

b Assuming all the water is involved in the dilution.

      Ans. 0.0181, 0.0156

*** **18.29**   *Technetium radioactivity.* If the half-life of $^{97}_{43}Tc$ is 2.6×10^6 yr, what is the probability that a given nucleus in a particular sample will disintegrate within one year?

      Ans. 27 chances out of 100 million

*** **18.30**   *Manganese free-energy change.* From the information given in Appendix 7 calculate the $\Delta G°$ for the following reaction:

$$3MnO_4{}^{2-} + 4H_3O^+ \longrightarrow MnO_2(s) + 2MnO_4{}^- + 6H_2O$$

The first five subgroups of the transition elements were discussed in Chap. 18 as individual vertical subgroups of the periodic table. In the next three subgroups, the chemical resemblance along the horizontal sequence is more pronounced than the chemical resemblance down the subgroup. For this reason, it is convenient to consider the next three subgroups in terms of horizontal triads. In the top transition period, the elements iron ($Z = 26$), cobalt ($Z = 27$), and nickel ($Z = 28$) make up the *iron triad*; in the middle period, ruthenium ($Z = 44$), rhodium ($Z = 45$), and palladium ($Z = 46$) are the *light platinum triad*; in the bottom period, osmium ($Z = 76$), iridium ($Z = 77$), and platinum ($Z = 78$) are the *heavy platinum triad*. All these elements are metals of low atomic volume with high melting points and high densities. The elements of the first triad are moderately reactive; those of the platinum triads are fairly inert. In the original Mendeleev periodic table all these elements were lumped into what was called group VIII, and they are still occasionally referred to by that designation. The platinum elements (together with gold and silver) are sometimes referred to as the *noble metals*.

19.1 Iron triad

In progressing from left to right in the first transition period, there is a progressive rise in maximum oxidation state from $+3$ for scandium to $+7$ for manganese. As discussed in Sec. 17.3, the following iron-triad elements show maximum oxidation states that are less than those of the preceding elements. Furthermore, the compounds containing the maximum oxidation states are such strong oxidizing agents that they are not usually encountered. Specifically, iron shows a maximum oxidation state of $+6$, but only the $+2$ and $+3$ states are common; cobalt shows only $+2$ and $+3$; nickel, usually only $+2$. Figure 19.1 indicates schematically the electron population of the neutral atoms. Removal of the two $4s$ electrons is relatively easy in each case; hence a $+2$ state is formed. Additional removal of a $3d$ electron gives a $+3$ state. In the case of iron this happens easily because a half-filled $3d$ level is left; in cobalt and nickel it does not happen so readily.

Sublevel	Iron	Cobalt	Nickel
1s	2	2	2
2s and 2p	8	8	8
3s and 3p	8	8	8
3d	↑↓ ↑ ↑ ↑ ↑	↑↓ ↑↓ ↑ ↑ ↑	↑↓ ↑↓ ↑↓ ↑ ↑
4s	↑↓	↑↓	↑↓

Fig. 19.1
Electron population of iron-triad elements.

Property	Fe	Co	Ni
Melting point, °C	1535	1490	1450
Boiling point, °C	2700	2900	2700
Ionization potential, eV	7.90	7.86	7.63
Electrode potential (from M^{2+}), V	−0.44	−0.28	−0.25
Density, g/cm³	7.9	8.7	8.9

Fig. 19.2
Elements of the iron triad.

The $+3$ state of cobalt must be stabilized as by the ligand field (Sec. 17.4) of a complex ion; the $+3$ state of nickel is very rare, and compounds of $+3$ nickel are powerful oxidizing agents.

The properties of the iron-triad elements are very similar, as shown in Fig. 19.2. The melting points and boiling points are uniformly high; the energies required to pull an electron off the gas atom are nearly the same for the three elements; and all the electrode potentials are moderately more negative than the electrode potential of hydrogen. In addition to the properties listed, these elements are alike in that all are *ferromagnetic*; i.e., they are strongly attracted into a magnetic field and show permanent magnetization when removed from such a field. That these elements are magnetic is not surprising. The electronic configurations in Fig. 19.1 would lead us to expect that there would be unpaired electrons in the $+2$ ions such as might exist in the metal lattice. However, it is surprising that the magnetization is so large and so persistent. The explanation is that in these metals there are *domains* of magnetization, regions of a million or so ions, all of which cooperatively direct their individual magnetic effects along the same direction. In an unmagnetized piece of metal the individual domains have their directions of magnetization randomly distributed in space so that, in sum, the magnetic effect cancels. When placed in a magnetic field, the domain directions are turned so that all point in the same way, giving rise to a large magnetic effect. If the piece of metal is now removed from the field, it remains permanently magnetized unless the domain orientation is disorganized, as by heating or pounding. Of all the elements, only iron, cobalt, and nickel show this kind of magnetism at room temperature. Apparently they are the only ones that satisfy the conditions necessary for domain formation. These conditions are that the ions contain unpaired electrons and that the distance between ions be exactly right, in order that the interaction for lining up all the ions to form a domain may be effective. Manganese metal has most of the properties needed to be ferromagnetic, but the ions of the metal are too close; addition of copper to manganese increases this average spacing, and the resulting alloy is ferromagnetic.

In their compounds the iron-triad elements behave like typical transition elements. Many of the compounds are colored and paramagnetic, and frequently they contain complex ions.

19.2 Iron

The element Fe has an industrial importance which exceeds that of any other element. It is very abundant, ranking fourth in the earth's crust (after O, Si, and Al); it is very common, being an essential constituent of several hundred minerals; it is easy to make by simply heating some of its minerals with C; it has many desirable properties, especially when impure. For all these reasons, Fe has become such a distinctive feature of civilization that it marks one of the ages in archaeological chronology.

About 5 percent of the earth's crust is iron. Some of this iron is meteoritic in origin and occurs in the uncombined, metallic state. However, most of it is combined with oxygen, silicon, or sulfur. The important source minerals are hematite (Fe_2O_3), limonite ($Fe_2O_3 \cdot H_2O$), magnetite (Fe_3O_4), and siderite ($FeCO_3$), usually contaminated with complex iron silicates from which these minerals are produced by weathering. Iron sulfides, such as iron pyrites (FeS_2), or *fool's gold*, are also quite abundant, but they cannot be used as sources of iron because sulfur is an objectionable impurity in the final product. Recently, improvements in extractive technology have opened up vast deposits of iron-containing ore in the Lake Superior region of the United States that were previously considered unworkable. These deposits consist of taconite, a hard, fine-grained material composed mainly of SiO_2, Fe_3O_4, and Fe_2O_3. Though the problem of conserving natural resources in the face of exponential population growth remains a serious one, development of the taconite reserves has expanded the critical

United States reserves-to-annual consumption ratio to almost 200 yr. Further consideration of the resource depletion problem is given in Chap. 28.

In addition to abundant iron in the earth's crust, there is a possibility that the center of the earth may be iron. Indirect evidence based on the study of earthquake waves and tidal action indicates that the core of the earth is liquid and has a density corresponding to that of liquid iron at high pressure.

Iron is practically never produced in a pure state, since it is difficult to make and too expensive for most purposes. Furthermore, impure iron, i.e., steel, has desirable properties, especially when the specific impurity is carbon in carefully controlled amounts. The industrial production of steel is carried out on a massive scale in the well-known blast furnace, in which occur complicated high-temperature reactions involving iron ore, limestone, and carbon. The iron ore, limestone, and coke are added at the top of the huge vertical structure, and preheated air or oxygen is blown in at the bottom. As the molten iron forms, it trickles down to a pit at the bottom, from which it is periodically drawn off. All told, it takes about 12 h for material to pass through the furnace.

The actual chemical processes which occur in the blast furnace are still obscure. It is generally agreed, however, that the active reducing agent is not carbon, but carbon monoxide. As the charge settles through the furnace, the coke is oxidized by the incoming oxygen by the reaction

$$2C(s) + O_2(g) \longrightarrow 2CO(g) \qquad \Delta H^\circ = -220 \text{ kJ}$$

thus forming the reducing agent carbon monoxide and liberating large amounts of heat. As the carbon monoxide moves up the furnace, it encounters oxides of iron in various stages of reduction, depending on the temperature of the particular zone. At the top of the furnace, where the temperature is lowest (250°C), the iron ore (mostly Fe_2O_3) is reduced to Fe_3O_4 by the reaction

$$3Fe_2O_3(s) + CO(g) \longrightarrow 2Fe_3O_4(s) + CO_2(g)$$

As the Fe_3O_4 settles, it gets reduced further to FeO:

$$Fe_3O_4(s) + CO(g) \longrightarrow 3FeO(s) + CO_2(g)$$

Finally, toward the bottom of the furnace, FeO is eventually reduced to iron:

$$FeO(s) + CO(g) \longrightarrow Fe(s) + CO_2(g)$$

Since the temperature at the lowest part of the furnace (1500°C) is above the melting point of the impure iron, the solid melts and drips down into the hearth at the very bottom. The net equation for the reduction of Fe_2O_3 is

$$Fe_2O_3(s) + 3CO(g) \longrightarrow 2Fe(l) + 3CO_2(g)$$

In addition to the foregoing reactions, there occurs the combination of carbon dioxide with hot carbon

$$C(s) + CO_2(g) \longrightarrow 2CO(g)$$

and the thermal decomposition of limestone by the reaction

$$CaCO_3(s) \longrightarrow CaO(s) + CO_2(g)$$

Both of these reactions are helpful: The former raises the concentration of the reducing agent carbon monoxide, and the latter facilitates the removal of silica-containing contaminants present in the original ore. Lime (CaO), being a basic oxide, reacts with the acidic oxide SiO_2 to form calcium silicate ($CaSiO_3$). In the form of a lavalike *slag*, calcium silicate collects at the bottom of the furnace, where it floats on the molten iron and protects it from oxidation by incoming oxygen.

Four times a day the liquid iron and molten slag are drawn off through tapholes in the bottom of the furnace. About 1000 tons of impure iron can be produced per day from one furnace. For each ton of iron, there is also produced approximately $\frac{1}{2}$ ton of slag. Since slag is essentially calcium aluminum silicate, some of it is put to good use in making cement (Sec. 22.6).

Pig iron, the crude product of the blast furnace, contains about 4% carbon, 2% silicon, a trace of sulfur, and up to 1% of phosphorus and manganese. Sulfur is probably the worst impurity (making steel break when worked) and must be avoided, since it is hard to remove in refining operations. The refining operations are specialized and of great variety. Only a few of the basic ideas will be discussed.

When pig iron is remelted with scrap iron and cast into molds, it forms *cast iron*. This can be either *gray* or *white*, depending on the rate of cooling. When cooled slowly (as in sand molds, where heat loss is slow), the carbon impurity separates out almost completely in the form of tiny flakes of graphite (Sec. 22.2), giving gray cast iron which is relatively soft and tough. When cooled rapidly (as in water-cooled molds), the carbon does not have a chance to separate out but remains combined in the form of the compound iron carbide, Fe_3C, also called *cementite*. White cast iron is as much as 75% cementite and is extremely hard and brittle.

Most pig iron is refined into steel by burning out the impurities to leave small controlled amounts of carbon. In the *open-hearth* process (which accounts for over half of the United States production), some of the carbon is removed by oxidation with air and iron oxide, the latter being added as hematite and rusted scrap iron. The process is usually carried out on a shallow hearth so arranged that a hot-air blast can play over the surface. In the *basic* open-hearth process, limestone is added to provide CaO for converting oxidation products, such as acidic P_2O_5, into slag. Since it takes about 8 h to refine a batch of steel by this process, there is ample time for continuous testing to maintain quality control.

The *bessemer process* (which has practically disappeared from the United States but still accounts for much of the European production) is much more rapid (10 to 15 min) but gives a less uniform product.

In this process, molten pig iron taken directly from the blast furnace is poured into a large pot having blowholes at the bottom, and a blast of air or commercial oxygen is swept through the liquid mix to burn off most of the carbon and silicon. Frequently, the bessemer and open-hearth processes are combined to take advantage of the good points of each: A preliminary blowing in the bessemer converter gets rid of most of the carbon and silicon; a following burn-off in an open-hearth furnace gets rid of phosphorus.

Increasingly important (more than one-third of United States production) is the *oxygen process*, in which pure oxygen is played over the top of the melt through a water-cooled lance.

A fourth process, generally reserved for specialty steels, is the *electric-crucible process*, in which carefully regulated heating under controlled atmospheres is brought about by induction of high-frequency electric currents. Elements such as chromium, vanadium, or manganese can be added to produce desired properties. In order to prevent the formation of blowholes (as in Swiss cheese) when molten steel is poured into ingots, it is generally necessary for finished steel to contain some manganese. The function of this manganese is apparently to combine with the oxygen and keep the oxygen from bubbling out as the steel solidifies.

The properties of Fe in the form of steel are very much dependent on the percentage of impurities present, on the heat treament of the specimen, and even on the working to which the sample has been subjected. For these reasons, the following comments about Fe properties do not necessarily apply to every given sample. Compared with most metals, Fe is a fairly good reducing agent, but it is not so good as the preceding transition elements. With nonoxidizing acids it reacts to liberate H_2 by the reaction $Fe(s) + 2H_3O^+ \longrightarrow Fe^{2+} + H_2(g) + 2H_2O$. It also has the ability to replace less active metals in their solutions. For example, a bar of Fe placed in a solution of $CuSO_4$ immediately is covered with a reddish deposit of Cu formed by the reaction $Fe(s) + Cu^{2+} \longrightarrow Fe^{2+} + Cu(s)$. In concentrated HNO_3, Fe, like many other metals (Cr, Mo, Co, Ni, etc.), becomes *passive;* i.e., it loses the ability to react with H_3O^+ and Cu^{2+} as above and appears to be inert. When scratched or subjected to shock, however, reactivity is restored. It may be that passivity is due to formation of a submicroscopic surface coating of oxide, which slows down the rates of oxidation below the limits of detectability. When the film is broken, reactivity is restored. Passivity is important in some methods of preventing corrosion of Fe.

19.3   Compounds of iron

The two common oxidation states of iron are +2 (ferrous) and +3 (ferric). Under vigorous oxidizing conditions it is also possible to get compounds such as $BaFeO_4$, barium ferrate, but in general the +6 state is rare. Compounds in which the oxidation state is fractional, as in Fe_3O_4, can be thought of as mixtures of two oxidation states. Fe_3O_4

is a spinel (Sec. 7.4) in which Fe^{2+} and Fe^{3+} ions occur in interstices of the close-packed oxide structure.

In the $+2$ state iron exists as ferrous ion, Fe^{2+}. This in water is a pale green, almost colorless ion which, except in acid solutions, is rather hard to keep since it is easily oxidized to the $+3$ state by oxygen in the air. However, since the rate of oxidation by oxygen is inversely proportional to H_3O^+ concentration, acid solutions of ferrous salts can be kept for long periods. When base is added to ferrous solutions, a nearly white precipitate of ferrous hydroxide, $Fe(OH)_2$, is formed. On exposure to air, $Fe(OH)_2$ turns brown, owing to oxidation to hydrated ferric oxide, $Fe_2O_3 \cdot xH_2O$. For convenience, the latter is often written as $Fe(OH)_3$, ferric hydroxide, and the oxidation can be written as

$$4Fe(OH)_2(s) + O_2(g) + 2H_2O \longrightarrow 4Fe(OH)_3(s)$$

However, pure ferric hydroxide has never been prepared.

In the $+3$ state, iron exists as the colorless ferric ion, Fe^{3+}. The aqueous solutions of ferric salts are generally acid, indicating that appreciable hydrolysis must take place. This can be written as

$$Fe^{3+} + 2H_2O \rightleftharpoons FeOH^{2+} + H_3O^+$$

Apparently, the yellow-brown color so characteristic of ferric solutions is mainly due to $FeOH^{2+}$. By addition of an acid such as HNO_3, the color can be made to disappear. (However, the color will not disappear on addition of HCl because $FeCl^{2+}$, which is yellow, forms.) On addition of base, a slimy, red-brown, gelatinous precipitate forms; it may be written as $Fe(OH)_3$. This can be dehydrated to form a yellow or red Fe_2O_3.

In both the $+2$ and $+3$ states, iron shows a great tendency to form complex ions. For example, ferric ion combines with thiocyanate ion, NCS^-, to form $FeNCS^{2+}$, which has such a deep red color that it can be detected at concentrations of $10^{-5}\ M$. The formation of this complex is the basis of one of the most sensitive qualitative tests for the presence of Fe^{3+}. With cyanide ion, CN^-, both Fe^{2+} and Fe^{3+} form complexes; these are so stable to dissociation that they can be thought of as single units such as SO_4^{2-} ion. As shown in Fig. 19.3, both ferrocyanide, $Fe(CN)_6^{4-}$, and ferricyanide, $Fe(CN)_6^{3-}$, are octahedral with six CN groups joined through carbon to the central iron. If, as is usually done in assigning oxidation states, we give cyanide a -1 charge, then the iron has to be assigned a $+2$ charge in ferrocyanide and a $+3$ charge in ferricyanide. We can then interpret the observed diamagnetism of ferrocyanide as arising from strong crystal-field splitting of the $3d$ orbitals to force pairing of the six electrons in the t_{2g} set. Similarly, the observed one-spin paramagnetism of ferricyanide can be explained by five electrons in t_{2g} (two pairs and one unpaired). Actually, a ligand-field model such as was shown in Fig. 17.7 might be more appropriate in that it treats $Fe(CN)_6$ as a single unit. In any case, these ions take part in some very complicated reactions, among which are the formation of prussian blue and Turnbull's blue.

Fig. 19.3
*Octahedral complex ions of
iron and cyanide.*

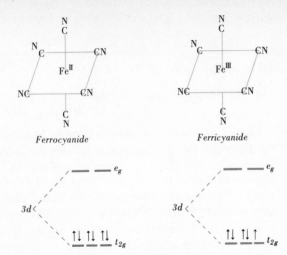

Ferrocyanide

Ferricyanide

Prussian blue is a deep blue precipitate obtained by mixing a solution of a ferric salt with a solution of potassium ferrocyanide, $K_4Fe(CN)_6$. It is used extensively as a dye for blueprint paper, for ink, and for bluing, in the last case because its color counteracts the yellow color of fabrics and makes them appear white. Turnbull's blue is a similar deep blue precipitate, obtained by mixing a solution of a ferrous salt with a solution of potassium ferricyanide, $K_3Fe(CN)_6$. The surprising thing about prussian blue and Turnbull's blue is that apparently they are the same, having identical overall compositions, corresponding to $KFeFe(CN)_6$. However, despite claims to the contrary no one has yet been able to carry out a definitive experiment to prove that they are the same or, for that matter, that they are different.

19.4 Corrosion of iron

Corrosion is a general term applied to the process in which uncombined metals change over to compounds. In the special case of iron the corrosion process is called *rusting*. Economically, rusting is a serious problem, and it has been estimated that one-seventh of the annual production of iron goes simply to replace that lost by rusting. Still, despite much study, corrosion is a mysterious process, and its chemistry not well understood.

Rust appears to be a hydrated ferric oxide with a chemical composition corresponding approximately to $2Fe_2O_3 \cdot 3H_2O$, that is, three moles of water per two moles of ferric oxide. However, since the water content is variable, it is preferable to write $Fe_2O_3 \cdot xH_2O$. Because iron will not rust in dry air or in water that is completely free of air, it would seem that both oxygen and water are required for rust formation. Furthermore, it is observed that rusting is generally speeded up by the presence of acids, by strains in the metal, by contact with less active metals, and by the presence of rust itself (autocatalysis).

In order to account for the observed facts, the following steps have been proposed as the mechanism by which rusting occurs:

$$Fe(s) \longrightarrow Fe^{2+} + 2e^- \tag{1}$$

$$e^- + H_3O^+ \longrightarrow H + H_2O \tag{2}$$

$$4H + O_2(g) \longrightarrow 2H_2O \tag{3}$$

$$4Fe^{2+} + O_2(g) + (12 + 2x)H_2O \longrightarrow$$
$$2(Fe_2O_3 \cdot xH_2O)(s) + 8H_3O^+ \tag{4}$$

In step (1) ferrous ions are produced by loss of electrons from neutral Fe. However, this process cannot go very far unless there is some way to get rid of the electrons which accumulate on the residual Fe. One way to do this is by step (2), in which H_3O^+ ions, either from the water or from acid substances in the water, pick up the electrons to form neutral H atoms. (Normally, we would expect these H atoms to pair to form H_2 molecules; however, H_2 gas is usually not observed in rust formation.) Since Fe is known to be a good catalyst for hydrogenation reactions in general, it is believed that step (3) now occurs to use up the H atoms. In the meantime, the ferrous ion reacts with O_2 gas by step (4) to form the rust and restore H_3O^+ required for step (2). The net reaction, obtained by adding all four steps, is

$$4Fe(s) + 3O_2(g) + 2xH_2O \longrightarrow 2(Fe_2O_3 \cdot xH_2O)(s)$$

Since H_3O^+ accelerates step (2) and is replenished in step (4), it is a true catalyst for the reaction and explains the observation that acids speed up the rate of rust formation. (A remarkable example of this is observed when iron pipes are located so as to be in contact with cinders. Such pipes corrode much more rapidly than they normally would, apparently because weathering of S compounds in the cinders forms H_2SO_4 that catalyzes the rusting.)

The above mechanism also accounts for many other observations and, in particular, for the process often called *electrolytic corrosion*. For example, when iron pipes are connected to copper pipes, the iron is observed to corrode much faster than normally. The explanation lies in step (1). Residual electrons accumulating from the dissolution of iron flow from the iron to the copper, where their energy is lower. This removes the excess negative charge from the iron and allows more Fe^{2+} to leave the metal. A complicating feature which also accelerates the reaction is that H atoms, which now form on the negative copper surface instead of the iron, detach themselves more readily from copper than from iron, thus accelerating step (3). Electrolytic corrosion used to be very common, for example, when new copper plumbing was hooked up directly to an existing system composed of iron pipes. It can now be prevented by simply inserting a junction that consists of an alloy which does not conduct electric current very well.

One of the strongest supports for the above stepwise rusting mechanism comes from the observation that the most serious pitting of a rusting iron bar occurs in that part of the bar where the oxygen supply is restricted. The reason for this is that where the oxygen supply is unrestricted, step (4) promptly occurs to deposit rust before the Fe^{2+} formed by step (1) can move very far away. This, of course, makes it more difficult for more iron to dissolve, and the reaction is self-stopping. However, if the oxygen supply is restricted, especially in an aqueous environment, Fe^{2+} may have a chance to diffuse away before encountering enough oxygen to form rust. This means that the rust may deposit some distance away from the point where pitting occurs. Common examples of this are observed at the edges of overlapping plates or around rivet heads. In the latter case, as shown in Fig. 19.4a, the rivet shank, although protected from air, is eaten away, but the rust forms where the rivet head overlaps the plate. Apparently, moisture that seeps in allows Fe^{2+} to diffuse out to the surface, where it can react with oxygen. Another slightly different example is found in the well-known waterline rust which forms around partially immersed steel posts. When a new post is placed in water, pitting usually starts where there are strains in the metal, but the rust forms, as shown in Fig. 19.4b, near the waterline, where the dissolved oxygen supply tends to be plentiful. This makes the situation go from bad to worse since the waterline rust now acts as a curtain to keep any oxygen from reaching the iron. Self-protection is no longer possible, and severe pitting can now occur where the oxygen supply is restricted. The common practice of undercoating car bottoms to prevent rusting can actually do more harm than good if the covering job is poorly done.

Although there are still many unanswered questions about rusting, it is clear what must be done to prevent it. The most direct approach is to shut off the reactants oxygen and water. This can be done by smearing grease over the iron to be protected, painting it either with an ordinary paint or better with an oxidizing paint so as to make the iron passive, or plating the iron with some other metal. All these methods are used to some extent. Painting or greasing is probably the cheapest, but it must be done thoroughly; otherwise rusting may only be accelerated by partial exclusion of oxygen. Plating with another metal is more common when appearance is a factor. Chrome plating, for example, is usually chosen because of its dressy look. Zinc plating, or galvanizing, though it does not look so good, is actually more permanent. Tin plating looks good and is relatively cheap, but it is not reliable.

The relative merits of metals used for plating depend on the activity of the metal relative to Fe and the ability of the metal to form a self-protective coat. Zn, for example, is a self-protecting metal which reacts with O_2 and CO_2 in the air to form an adherent coating of hydroxycarbonate, $Zn_2(OH)_2CO_3$, which prevents further corrosion. Furthermore, it has a bigger electrode potential than does Fe; so if a hole is punched in a Zn plating so that both Zn and Fe are exposed to oxidation, it is the Zn that is preferentially oxidized. The Zn compound forms a plug to seal the hole. Sn also forms a self-protective coat, but Sn has a smaller electrode potential than Fe; so if a tin coating is punctured, it is the underlying Fe that is preferentially oxidized.

One of the most elegant ways to protect iron from corrosion is by *cathodic protection*. In this method, iron is charged to a negative voltage compared with its surroundings. This tends to make the iron

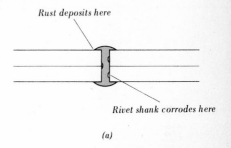

Rust deposits here

Rivet shank corrodes here

(a)

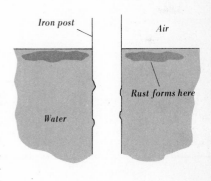

Iron post

Air

Rust forms here

Water

(b)

Fig. 19.4
Iron corrosion as enhanced by restricted oxygen supply.

act as a cathode instead of the anode required for oxidation and effectively stops corrosion. Actually, zinc plating is a method of cathodic protection, since zinc has a larger electrode potential than does iron and forces electrons onto the iron. In practice, for pipelines and standpipes cathodic protection is obtained by driving stakes of zinc or magnesium, for example, into the ground and connecting them to the object to be protected. In salt water, where rusting is unusually severe, the steel plates of ships have been protected by strapping blocks of magnesium to the hulls. These preferentially corrode (since they are acting as anodes) but can easily be replaced, while the iron is essentially untouched.

Cathodic protection also explains why tin plating (as in the ordinary tin can) is so unlasting. As long as the tin coating is unpunctured, there is no corrosion, since tin is a rather inert metal and can be exposed indefinitely to the atmosphere. Once the coating is punctured (and this happens very easily because it is very thin), there is real trouble, and the iron is worse off than if the tin plating were not there. The reason for this is that iron, being more active than tin, acts as an anode in *setting up cathodic protection for the tin*. This, of course, accelerates the dissolving of iron and the formation of rust; hence the rust spreads very rapidly.

19.5   Cobalt and nickel

The other elements of the iron triad, cobalt and nickel, although less important than iron, do have some interesting aspects. Both are much less abundant than iron (cobalt at 0.002% and nickel at 0.008%) and are harder to extract from their minerals. The name "cobalt" reflects this difficulty, since it comes from the German word *Kobold*, meaning "goblin." Cobalt minerals look very much like copper minerals and were occasionally worked by mistake as sources of copper. Furthermore, since arsenic is usually present with cobalt, poisonous fumes were inevitably present, all obviously due to black magic. Similar troubles with nickel minerals led to their being named after *Nickel*, a mischievous underground spirit in German superstition. Cobalt minerals and nickel minerals frequently occur together, associated with those of iron and copper. The principal compounds in these minerals are oxides, sulfides, and arsenides.

Cobalt   The more important minerals of cobalt are cobalt glance ($CoAsS$), linnaeite (Co_3S_4), and smaltite, or cobalt speiss ($CoAs_2$). The extraction of cobalt is very complex and involves roasting in a blast furnace, dissolving with sulfuric acid, and precipitating by addition of sodium carbonate. The hydroxide so produced is dehydrated to the oxide, which can then be reduced with hydrogen.

The properties of the metal have been given in Fig. 19.2. The electrode potential of -0.28 V for the reaction $Co^{2+} + 2e^- \longrightarrow Co(s)$ indicates that Co metal should dissolve in acids with the liberation of H_2. The reaction with H_2SO_4 or HCl is slow and does not occur at all in concentrated HNO_3 where Co becomes passive. The ferromagnetism of Co is very strong, higher than that of Fe, and accounts for its extensive use in magnets, especially in alloys such as the *alnico* alloys (Co, Ni, Al, and Cu). Other alloys such as stellite (55% Co, 15% W, 25% Cr, and 5% Mo) are important for their extreme hardness and resistance to corrosion. They are used, for example, in high-speed tools and surgical instruments.

In its compounds cobalt shows oxidation states of $+2$ (cobaltous) and $+3$ (cobaltic). Unlike ferrous ion, the cobaltous ion is quite stable to oxidation, and solutions of cobaltous salts can be kept indefinitely exposed to the air. Most cobaltous solutions are pink and presumably contain the hydrated ion $Co(H_2O)_6^{2+}$. Addition of base precipitates dark blue, insoluble hydroxide, which in the absence of oxygen can be dehydrated to give grayish brown cobaltous oxide, CoO. This is a basic oxide much used to produce blue color in pottery and enamel. When heated in the presence of air, cobaltous hydroxide dehydrates to give Co_3O_4, cobaltocobaltic oxide, reminiscent of Fe_3O_4. Cobaltous forms many complex ions, which, however, are easily oxidized.

In aqueous solution the cobaltic ion Co^{3+} is a very powerful oxidizing agent. The potential for $Co^{3+} + e^- \longrightarrow Co^{2+}$ is $+1.84$ V, which means that Co^{3+} is strong enough to oxidize water to form oxygen. Only a few simple cobaltic salts such as CoF_3 and $Co_2(SO_4)_3 \cdot 18H_2O$ have been made, and these decompose in aqueous solution:

$$4Co^{3+} + 6H_2O \longrightarrow 4Co^{2+} + O_2(g) + 4H_3O^+$$

Unlike the simple ion, the complex ions of Co^{3+} are quite stable to reduction. There are a tremendous number of these, ranging from the simple $Co(CN)_6^{3-}$ and $Co(NH_3)_6^{3+}$ to complicated *polynuclear* complexes, in which several cobalt atoms are bridged together by shared complexing groups. Vitamin B_{12} is an organic complex in which cobalt occurs in a distorted octahedral environment provided by a cyanide group and five other ligand nitrogens which are interconnected with each other by various hydrocarbon links. Vitamin B_{12} is synthesized by many bacteria; it is required by green algae and has recently been implicated in the mechanism by which supposedly harmless mercury wastes get converted to deadly dimethylmercury, $Hg(CH_3)_2$.

Nickel The principal minerals of nickel are pentlandite and nickeliferous pyrrhotite, both of which are mixed sulfides of iron and nickel, and garnierite, which is a mixed silicate of magnesium and nickel. Since most ores are very poor in nickel content, they have to be concentrated before smelting, usually by *flotation*. In this process the ore is ground up and then agitated briskly with water to which oil and wetting agents have been added. Earthy particles (*gangue*) are wet by the water and hence sink, whereas the fine particles of minerals get carried off with the froth. The concentrate is then roasted in air to get rid of some of the sulfur as SO_2, burned in a furnace (smelted) to form oxide, and finally reduced with carbon. To get pure nickel, the final product must be refined, either electrolytically or by taking advantage of the instability of volatile nickel carbonyl. In the Mond process, carbon monoxide at 80°C is passed over impure nickel to form volatile $Ni(CO)_4$, nickel tetracarbonyl. This is distilled off, purified, and then heated to about 200°C, where it decomposes into nickel and carbon monoxide.

The properties of nickel metal are much like those of cobalt, except that nickel is less ferromagnetic and more inert to chemical oxidation. More than 65% of nickel production goes into iron alloys to increase their strength and corrosion resistance. The rest of it goes into nickel-copper alloys, e.g., into nickel coinage (which can vary anywhere from 10 to 100% nickel), or is used as the pure metal. In the latter case it is used for plating steel and as a catalyst for hydrogenation reactions.

The chemistry of nickel compounds is essentially that of the $+2$ state. In aqueous solution it exists either as the green nickelous ion, Ni^{2+}, or as a complex ion. On treatment with base, nickelous ion precipitates as the light green nickelous hydroxide, $Ni(OH)_2$, which can be dehydrated thermally to black NiO. The complex ions of nickel are almost as numerous as those of cobalt. However, unlike cobaltous complexes, they are quite stable to air oxidation. Furthermore, not all of them are octahedral [as in the blue $Ni(NH_3)_6^{2+}$] but may be planar [for example, yellow $Ni(CN)_4^{2-}$].

In basic solution nickelous hydroxide can be oxidized by the powerful oxidizing agent hypochlorite, ClO^-. The product is a dark-colored oxide of indefinite composition variously described as NiO_2, Ni_2O_3, or Ni_3O_4. No matter what it is, it is a very good oxidizing agent and contains nickel in an oxidation state higher than $+2$. As such, it forms the cathode material in the Edison storage battery. On discharge, the cathode reaction can be written as

$$Ni_2O_3(s) + 2e^- + 3H_2O \longrightarrow 2Ni(OH)_2(s) + 2OH^-$$

Since $Ni(OH)_2$ sticks to the cathode and since the reaction is reversible, the discharged cell can be recharged by the application of an external voltage. The anode in the Edison cell is usually of iron, with NaOH as electrolyte. The anode reaction is probably

$$Fe(s) + 2OH^- \longrightarrow Fe(OH)_2(s) + 2e^-$$

Although it has a lower voltage, the Edison cell (1.4 V) has an advantage over the lead storage cell (2.0 V) because the OH^- produced at the Edison cathode is used up at the anode; hence, there is no concentration change in the electrolyte as the cell runs down; so there is no change in its output voltage. Also, the materials are less dense and stronger than those of the lead storage cell; consequently, a battery of Edison cells is more easily portable and more rugged. Unfortunately, the chemical components are more expensive; so its general usefulness is limited.

19.6 Light platinum triad

In the second transition period the three elements following technetium are ruthenium ($Z = 44$), rhodium ($Z = 45$), and palladium ($Z = 46$).

Fig. 19.5
Light platinum elements.

Property	Ru	Rh	Pd
Atomic number	44	45	46
Electronic configuration	$4d^7 5s^1$	$4d^8 5s^1$	$4d^{10}$
Melting point, °C	2450	1970	1550
Boiling point, °C	>2700	>2500	2200(?)
Ionization potential, eV	7.5	7.7	8.3
Electrode potential (from M^{2+}), V	+0.5	+0.6	+1.2
Density, g/cm^3	12.4	12.4	12.0

Since these elements chemically resemble platinum but have only about half the density of platinum, they are called the light platinum elements. Their properties are summarized in Fig. 19.5. As can be seen, these elements are high-melting and high-boiling with rather high densities. They are not very reactive, and their electrode potentials indicate that, unlike the iron-triad elements, they are much poorer reducing agents than hydrogen. For this reason they are difficult to oxidize and, in fact, occur in nature as the uncombined elements. Because of great similarity in chemical properties, they usually occur together.

Ruthenium Ruthenium occurs as a natural alloy of ruthenium, osmium, and iridium. To prepare the pure metal, the alloy is heated with alkaline oxidizing agents (e.g., a mixture of KOH and KNO_3) to form potassium ruthenate (K_2RuO_4). After being dissolved and acidified, the mixture is boiled to eliminate osmium as volatile osmium tetroxide (OsO_4), and then after the solution has been made basic, ruthenium is distilled off as ruthenium tetroxide, RuO_4. Reduction with hydrogen gives the metal. It is quite inert and thus far has found use only as a hardening agent for platinum and for making other hard, inert alloys such as are required for fountain-pen tips.

In its principal compounds, ruthenium shows oxidation states corresponding to +2, +3, +4 and +6. Some of these compounds are simple, such as the volatile, toxic ruthenium tetroxide (RuO_4); others, such as $K_3Ru(C_2O_4)_3$, are complex. In general, they are rarely encountered.

Rhodium Rhodium is a rather rare element, amounting to only 10^{-7} percent of the earth's crust. It occurs principally with platinum, from which it can be separated by fusion with $KHSO_4$. Rhodium dissolves to give $KRh(SO_4)_2$, a rose-colored salt, which can be leached out and recrystallized from water. The metal is rather inert and finds use for plating scientific instruments (e.g., precision weights and optical reflectors). An alloy of rhodium and platinum is used to make high-temperature scientific apparatus such as crucibles and thermocouples. The principal oxidation state is +3 and is represented by many simple salts, such as $RhCl_3$, as well as by complex ones such as K_3RhCl_6.

Palladium Palladium is the most abundant (10^{-6} percent) of the platinum elements. Since it alone of the platinum elements forms an insoluble cyanide, it can be separated from the others by precipitation of $Pd(CN)_2$. On ignition, this decomposes to give pure metal. Like the other platinum elements, palladium is inert, but not so much so that it cannot dissolve in concentrated nitric acid. One of its most remarkable properties is that it has the ability to absorb hydrogen. At dull red heat, a piece of palladium can absorb about 1000 times its own volume of hydrogen. When the temperature is raised further, the hydrogen is expelled. Apparently in this absorption process the H_2 molecule is ripped apart into H atoms, which can then fit into the palladium lattice. It may be that this dissociation $H_2 \longrightarrow 2H$ also explains the powerful catalytic effect of palladium on hydrogenation reactions. Because it is not corroded in air and is capable of taking a high polish, palladium finds use for making optical mirrors and for jewelry. The compounds, both simple and complex, are essentially those of the $+2$ oxidation state.

19.7   Heavy platinum triad

The elements of the heavy platinum triad, osmium ($Z = 76$), iridium ($Z = 77$), and platinum ($Z = 78$), resemble very closely the elements just above them. Their properties are shown in Fig. 19.6. They are all very high-melting, of extraordinarily high density, and generally quite unreactive.

Fig. 19.6
Heavy platinum elements.

Property	Os	Ir	Pt
Atomic number	76	77	78
Electronic configuration	$5d^6 6s^2$	$5d^7 6s^2$	$5d^9 6s^1$
Melting point, °C	2700	2450	1774
Boiling point, °C	>5300	>4800	4100
Ionization potential, eV	8.7	9.2	8.96
Electrode potential (from M^{2+}), V	+0.9	+1	+1.2
Density, g/cm³	22.7	22.6	21.5

Osmium   The chief source of osmium is naturally occurring osmiridium, an alloy of osmium and iridium which is usually contaminated with ruthenium. As described in Sec. 19.6, this alloy can be dissolved by strong, alkaline oxidizing agents, and osmium tetroxide can be driven off as a volatile material from the acidified product. Metallic osmium forms when OsO_4 is reduced with almost any reducing agent, the half-reaction

$$OsO_4(s) + 8H_3O^+ + 8e^- \longrightarrow Os(s) + 12H_2O \qquad E° = +0.85 \text{ V}$$

indicating that OsO_4 is about as good an oxidizing agent as is nitrate ion. In the massive state, osmium metal is quite inert, even to aqua regia. It is very hard, especially when alloyed with iridium, and is used as the alloy for the tips of fountain pens.

Compounds of osmium are known in the $+2$, $+3$, $+4$, $+6$, and

$+8$ oxidation states. The most important of these is OsO_4, osmium tetroxide, sometimes called osmic acid. Though solid at room temperature (its melting point is $40°C$, and its boiling point is $130°C$), it is very volatile and hence very dangerous since it is corrosive to animal tissue, especially the eyes. An aqueous solution of OsO_4 can be used to stain fats in the preparation of microscopical slides since unsaturated fatty acids easily reduce OsO_4 to black metal or lower oxides. OsO_4 is also of importance in synthetic organic chemistry since it is a specific catalyst for the addition of OH to carbon-carbon double bonds by hydrogen peroxide.

Iridium Iridium is obtained from natural osmiridium by driving off the osmium and ruthenium as mentioned above. The iridium is usually separated as the slightly soluble ammonium hexachloroiridate(IV), $(NH_4)_2IrCl_6$. Thermal decomposition produces the metal, which in the massive state is quite inert. The chief use of iridium metal is for additive hardening of platinum, as for making electric contacts and the fine-hole plates used in spinning rayon. Compounds, corresponding principally to the $+3$ and $+4$ states, are known but are not commonly encountered.

Platinum Of all the platinum elements, Pt is the most useful. Although not very abundant (5×10^{-7} percent), it occurs in concentrated deposits, a fact which makes its separation feasible. However, the demand for Pt is so great that increasingly rather poor deposits have to be worked. In order to isolate Pt from the other platinum metals, the naturally occurring alloys are treated with aqua regia. Pt and Pd dissolve, and from the resulting solution Pt is precipitated as insoluble ammonium hexachloroplatinate(IV), $(NH_4)_2PtCl_6$. Thermal decomposition produces the metal.

The metal is quite inert to many kinds of chemical attack, and for this reason, especially when hardened with a few percent of iridium, it is used in making jewelry and laboratory equipment. In using platinum ware (e.g., crucibles), fused alkalis, such as NaOH, must be avoided because platinum dissolves in molten bases to form platinates. Also to be avoided are phosphorus, silicon, arsenic, antimony, lead, etc., with which platinum forms alloys. Industrially, probably the most important use of platinum is as a catalyst. For example, it catalyzes the oxidation of ammonia in the manufacture of nitric acid; it acts as a catalyst for the automobile-exhaust afterburner, to cut down air pollution. Finely divided platinum supported on aluminum oxide is also much used as a catalyst for processes that upgrade the octane rating of gasoline.

Although rather inert, platinum occurs in many chemical combinations, principally in the $+2$ and $+4$ oxidation states. Many of these compounds contain complex ions, and practically all of them are unstable with respect to thermal decomposition.

19.8 Qualitative analysis

All three iron-triad elements precipitate as black sulfides insoluble in basic solution. (If ferric ion were present, it would be reduced by H_2S in acid solution to ferrous ion.) FeS can be separated from CoS and NiS because it dissolves fairly quickly in Na_2SO_4–$NaHSO_4$ buffer whereas CoS and NiS are slow to dissolve. Separation of iron from cobalt and nickel can also be achieved by making use of the fact that Fe^{2+} plus an excess of NH_3 forms in air-insoluble ferric hydroxide whereas Co^{2+} and Ni^{2+} form soluble ammonia complexes. The presence of iron can be confirmed by adding thiocyanate, after oxidation of Fe^{2+} to Fe^{3+} with H_2O_2, if necessary. The deep red color of $FeNCS^{2+}$ shows that iron is present.

To distinguish cobalt from nickel, the sulfides CoS and NiS can be dissolved in acid solution, boiled with bromine water to destroy H_2S, and treated with potassium nitrite. The appearance of insoluble, yellow potassium hexanitritocobaltate(III), $K_3Co(NO_2)_6$, shows the presence of cobalt. Nickel can be identified by adding a special reagent, dimethylglyoxime, which from basic solution precipitates the reddish orange, voluminous solid nickel dimethylglyoxime, $Ni[CH_3C(NO)C(NOH)\text{-}CH_3]_2$. The latter is a square-planar complex of the following structure:

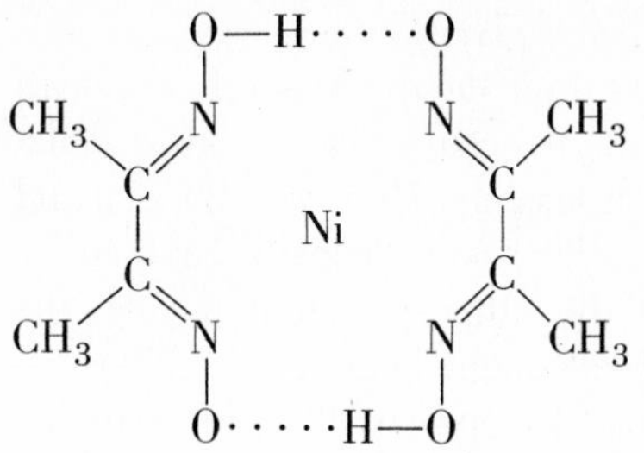

Exercises

* 19.1   Oxidation states. What is the principal reason for the decrease in maximum oxidation state observed in the compounds of the iron-triad elements?

*19.2   Ferromagnetism. From data given in this chapter estimate the number of domains in an iron needle 3 cm long and 0.5 mm in diameter. What would be the approximate diameter of each domain, assuming spherical shape?   *Ans. 5 × 10¹⁴, 3 × 10⁻⁶ cm*

*19.3   Stoichiometry. You are given an iron ore sample that is either pure hematite, limonite, magnetite, or siderite. If it analyzes to 72.4% iron by weight, which is it?

*19.4   Stoichiometry. In principle, how many tons of coke should be added to a blast furnace per ton of Fe_2O_3? What change would you make in your figure to allow for the fact that the exhaust gas contains typically 25% CO and 12% CO_2 by volume? (*Compare:* A typical medium-sized furnace which produces 600 tons of iron per day consumes 1160 tons of ore and 580 tons of coke.)

*19.5 Steel. How many tons of air (21% O_2 by volume) would you need to supply to burn to CO_2 the C from 1 ton (1000 kg) of cast iron (4% C by weight) to make a mild steel (1.8% C by weight)?

Ans. 0.25 ton

*19.6 Stoichiometry. If all the carbon in a 3.85% C (by weight) steel is present as Fe_3C, what percent of the steel is Fe_3C?

*19.7 Structure. The solid-state structure of $KFe(II)Fe(III)(CN)_6$ can be visualized as composed of a simple cubic array of iron atoms with CN groups along the cube edges. What fraction of the cubes should contain K^+ at their centers so as to come out with the right stoichiometry?

*19.8 Rusting. Suppose you have a steel trash can you want to protect from corrosion. Explain in detail why each of the following would be effective:
a Painting it with lead chromate
b Washing it with aqueous ammonia
c Keeping it thoroughly dry
d Hooking it up to the lead electrode of a car battery
e Daubing it with tar

*19.9 Nonstoichiometry. Nickelous oxide is likely to be nonstoichiometric with a formula that can be written NiO_{1+x}. If a given sample analyzes to give 78.75% nickel, what is its value of x?

*19.10 Chemical equations. Write balanced net equations for each of the following processes:
a Metallic nickel is dissolved in acid to give nickelous ion.
b Aqueous NaOH is added to the solution from (a) to precipitate nickelous hydroxide.
c The precipitate from (b) is cooked with hypochlorite ion to give Ni_2O_3 and chloride ion.
d Acid is added to (c) whereupon the Ni_2O_3 promptly oxidizes the Cl^- to Cl_2.

*19.11 Storage battery. Calculate how much Ni_2O_3 and Fe you would need in an Edison storage battery to deliver 150 amp-h. Compare with the weight of Pb and PbO_2 you would need to get the equivalent from a lead storage battery.   Ans. 619 versus 1250 g

*19.12 Palladium. If one cubic centimeter of palladium can absorb 1000 cm^3 of hydrogen gas at STP, what will be the formula of the PdH_x compound formed?   Ans. $PdH_{0.79}$

**19.13 Structures. Iron crystallizes with body-centered cubic structure (unit-cell edge length equal to 0.286 nm); cobalt and nickel can be obtained face-centered cubic with unit cells 0.355 and 0.352 nm, respectively. Assuming the metals are composed of M^{2+} ions and using data from Appendix 8, calculate the ratio of interatomic distance to ionic radius for each of these metals. Suggest a reason for any trend observed.

**19.14   Thermal effects. A quick, efficient way to cool molten iron is to spray it with cold water. To cool a ton (1000 kg) of molten iron from 1500 to 80°C, how many tons of 10°C water will be required? The heat capacity of iron is about 0.68 J g^{-1} deg^{-1}, that of water is about 4.2 J g^{-1} deg^{-1}. The heat of fusion of iron is 16.2 kJ/mol; the heat of vaporization of water is 40.7 kJ/mol.   *Ans. 0.5 tons*

**19.15   Solution stoichiometry. A mixed sample of Fe_3O_4 and Fe_2O_3 is dissolved in sulfuric acid and titrated with $KMnO_4$. What percent (by weight) of the original sample is Fe_3O_4 if it takes 35.00 ml of 0.100 M $KMnO_4$ to titrate a 5.00-g sample?

** 19.16   Electrode potentials. Given that the K_{sp} of $Fe(OH)_2$ is 1.8×10^{-15} and that of $Fe(OH)_3$ is 6×10^{-38}, calculate the $E°$ for $Fe(OH)_3(s) + e^- \longrightarrow Fe(OH)_2(s) + OH^-$. The $E°$ for $e^- + Fe^{3+} \longrightarrow Fe^{2+}$ is $+0.77$ V.

**19.17   Dissociation.   Given   that $K = 5.0 \times 10^{-9}$ for $Fe(CN)_6{}^{4-} \rightleftharpoons Fe(CN)_5{}^{3-} + CN^-$, calculate the equilibrium concentration of CN^- in 0.10 M $K_4Fe(CN)_6$ solution. Qualitatively, how would you expect this to compare with the analogous CN^- in 0.10 M $K_3Fe(CN)_6$? Explain.

** 19.18   Stoichiometry. The iron content of rust can be determined by dissolving the rust in acid, reducing the iron to Fe^{2+}, and then titrating to Fe^{3+} with dichromate in acid. A given sample of $Fe_2O_3 \cdot xH_2O$ weighing 1.786 g requires 22.0 ml of 0.150 M $K_2Cr_2O_7$ in such a titration. Calculate the value of x.   *Ans. 1.15*

**19.19   Rusting. Normally, the big struggle is to prevent rusting, but suppose you wanted to accelerate it. Indicate three things you could do, and explain why each of them would work.

**19.20   Iron. Explain why in preparing an aqueous solution of ferrous ion you would be better off using $Fe(NH_4)_2(SO_4)_2$ than $FeSO_4$.

**19.21   Cobalt complexes. Draw structural formulas, and give systematic names for each of the following:
  a $Co(H_2O)(NH_3)_5{}^{3+}$
  b   *cis*-$Co(H_2O)_2(NH_3)_4{}^{3+}$
  c $Co(C_2O_4)_3{}^{3-}$
  d $CoCl_3(NH_3)_3$
  e   *trans*-$Co(CN)_4(H_2O)_2{}^{2-}$
Which of these represent pairs of isomers?

**19.22   Magnetism. In a tetrahedral environment the crystal-field splitting of t and e orbitals is inverted from what it is in an octahedral environment. Given that $Ni(CO)_4$ has a tetrahedral structure using sp^3 bonding of neutral CO to a central Ni, predict the magnetism of the complex.

**19.23   Solubility. The K_{sp} of $Ni(OH)_2$ is 1.6×10^{-16}. Given 1.00 g of $Ni(OH)_2(s)$ in contact with 100 ml of water that is saturated with respect to $Ni(OH)_2$, calculate the pH of the solution. To what value must the pH be brought to dissolve all the $Ni(OH)_2$?

Ans. 8.84, 6.59

**19.24 *Osmium.* For OsO_4, the ΔH at the melting point ($40°C$) is 14.3 kJ, and the ΔH at the boiling point ($130°C$) is 39.5 kJ. Calculate the ΔS at each of these transitions, and suggest an important reason for the difference.

**19.25 *Platinum.* Platinum dissolves in aqua regia (a mixture of concentrated HNO_3 and HCl) because of the high oxidizing ability of nitrate ion and the great complexing ability of Cl^-. Assuming formation of $PtCl_6^{2-}$ and NO_2, write the two half-reactions and the balanced net equation for dissolving platinum in aqua regia.

**19.26 *Qualitative analysis.* You have a black solid which is either CoS or NiS. What tests can you run to distinguish which you have? Write balanced equations for the possible reactions.

**19.27 *Conversions.* How would you go about making the following conversions to conserve the element marked by the asterisk? Indicate specific reagents, and write equations for any reactions involved.

a   From $\overset{*}{Co}S$ to $\overset{*}{Co}SO_4$

b   From $\overset{*}{Fe}_3O_4$ to $\overset{*}{Fe}S$

c   From $\overset{*}{Ni}(CO)_4$ to $\overset{*}{Ni}(OH)_2$

d   From $\overset{*}{Fe}CO_3$ to $\overset{*}{Fe}_2O_3$

***19.28 *Thermodynamics.* Given that the free energies of formation of CO, CO_2, Fe_2O_3, Fe_3O_4, and FeO are -137, -394, -740, -1014, and -244 kJ/mol, respectively, calculate the $\Delta G°$ for each of the stepwise reduction reactions by CO from Fe_2O_3 to Fe. Suggest a reason for any discrepancies observed.

***19.29 *Free energy.* Under standard conditions iron dissolves in acid solution to give Fe^{2+} and H_2. Calculate the standard free-energy change for dissolving iron in:

a $1\ M\ H_3O^+$

b $1\ M\ OH^-$

The K_{sp} of $Fe(OH)_2$ is 1.8×10^{-15}. Assume hydrogen pressure is 1 atm.   *Ans. -84.9, $-9.6\ kJ$*

***19.30 *Magnetism.* In terms of crystal-field theory account for the fact that removal of an electron from $Co(H_2O)_6^{2+}$ decreases the magnetic moment by 3.87 Bohr magnetons whereas removal of an electron from $Co(NH_3)_6^{2+}$ decreases the magnetic moment by 1.73 Bohr magnetons.

The chemical similarity between the members of the iron triad (iron, cobalt, and nickel) is more pronounced horizontally along the period than vertically down the subgroup, as indicated in Chap. 19. With the elements of the next two subgroups the situation is reversed, and it is more convenient to compare them vertically than to compare them horizontally. The head elements of the next subgroups are copper ($Z = 29$) and zinc ($Z = 30$). The copper subgroup also includes silver ($Z = 47$) and gold ($Z = 79$); the zinc subgroup, cadmium ($Z = 48$) and mercury ($Z = 80$). With these elements, the sequence we have referred to as the transition elements is completed.

536

20.1 Copper subgroup

The elements of the copper subgroup, copper, silver, and gold, have been known to man since antiquity, for, unlike most of the preceding elements discussed, they are sometimes found in nature in the uncombined, or native, state. Originally decorative in function, they soon were adapted to use in coins because of their relative scarcity and resistance to corrosion. Originally, only silver and gold were used as coins, but then someone discovered the happy coincidence that copper could be added not only to make the coins cost less but also to increase their life in circulation because of increased hardness. Since then, copper, silver, and gold have been called the *coinage metals,* even though their principal uses are quite different.

Some of the important properties of these elements are shown in Fig. 20.1. All the elements are typically metallic, with relatively high melting points and rather high boiling points. The positive electrode potentials indicate that the metals are not very reactive. According to the electronic configurations, there is in the ground state of these atoms one electron in the outermost energy level. When this electron is removed, the $+1$ ion results. This is all that we expect since the second-

Symbol	Z	Electronic configuration	Melting point, °C	Boiling point, °C	Ionization potential, eV	Electrode potential (from M^+), V
Cu	29	$(18)\ 3d^{10}4s^1$	1083	2300	7.72	$+0.52$
Ag	47	$(36)\ 4d^{10}5s^1$	961	1950	7.57	$+0.799$
Au	79	$(68)\ 5d^{10}6s^1$	1063	2600	9.22	$+1.7$

Fig. 20.1
Elements of copper subgroup.

outermost shell is filled and presumably is hard to break into. In this respect these elements resemble the alkali metals (Chap. 16) and consequently are sometimes classified as a group IB. However, the d electrons in the second-outermost shell are close enough in energy to the outermost electrons that they can be removed with little additional energy, especially if there is some way to stabilize the resulting $+2$ or $+3$ ions. Apparently, this is exactly what happens. Copper forms $+1$ and $+2$ compounds; silver forms $+1$, $+2$, and $+3$ (although the $+2$ and $+3$ are rare); and gold forms $+1$ and $+3$ compounds. However, even with variable oxidation states, the chemistry of these elements is simpler than that of the preceding transition elements.

20.2 Copper

Considering its usefulness and familiarity, it is surprising that copper is such a small fraction (0.0001 percent) of the earth's crust. Fortunately, its deposits are concentrated and easily worked. Still, the reserves-to-annual consumption ratio is small enough (about 40 yr) that the world needs to worry about copper resource exhaustion in the near

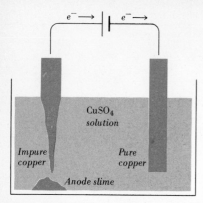

Fig. 20.2

Electrorefining of copper.

future. Besides native copper, which occurs mainly in the Lake Superior region of the United States and is 99.9 percent pure, the element occurs as two principal classes of minerals: sulfide ores (accounting for about 90 percent of the occurrence) and oxide ores. The principal sulfide ores are chalcocite (Cu_2S), chalcopyrite, or copper pyrites ($CuFeS_2$), and covellite (CuS); the principal oxide ores are cuprite (Cu_2O), malachite [$CuCO_3 \cdot Cu(OH)_2$], and tenorite (CuO).

In order to make the metal, the sulfide minerals are first concentrated by flotation (Sec. 19.5), roasted in air, and then smelted. The roasting and smelting process, represented, for example, by the simplified overall equation

$$2CuFeS_2(s) + 5O_2(g) \longrightarrow 2Cu(s) + 2FeO(s) + 4SO_2(g)$$

produces tremendous quantities of sulfur dioxide, much of which is converted on the spot into sulfuric acid. Still, industrial smelters in the United States pour something like 4 million tons/yr of sulfur dioxide into the air, thereby contributing a serious pollution problem.

Raw copper product is about 97 to 99 percent pure and must be refined (purified) for most uses. This can be done best in a $CuSO_4$ electrolysis cell, such as that sketched in Fig. 20.2. In the electrolysis cell the impure copper is made the anode, and pure copper the cathode. By careful control of the electrolysis voltage, the copper can be transferred from the anode to the cathode. The principle of operation can be seen from the following example, in which we consider the purification of a typical bar of copper containing iron and silver as impurities: The iron represents an impurity that is more easily oxidized than copper; the silver, an impurity that is less easily oxidized than copper. The pertinent half-reactions are as follows:

$$Fe(s) \longrightarrow Fe^{2+} + 2e^- \qquad E° = +0.44 \text{ V}$$

$$Cu(s) \longrightarrow Cu^{2+} + 2e^- \qquad E° = -0.34 \text{ V}$$

$$Ag(s) \longrightarrow Ag^+ + e^- \qquad E° = -0.80 \text{ V}$$

By keeping the cell voltage at an appropriate value, only the iron and copper are oxidized and go into the solution as ions. Silver, oxidizing with more difficulty, simply drops off to the bottom of the cell as the anode dissolves away. At the cathode, where reduction must occur, the high concentration of Cu^{2+} and the fact that Cu^{2+} is more readily reduced than Fe^{2+} combine to bring about deposition of pure copper. The Fe^{2+} remains in solution, and the solid silver stays at the bottom of the cell. Some common impurities in crude copper are iron, nickel, arsenic, antimony, and bismuth (all of which, like iron, are oxidized and remain oxidized in the refining solution), and silver, gold, and traces of platinum metals (all of which, like silver, are not oxidized and collect at the bottom of the cell). The residue at the bottom of the cell beneath the anode is called the *anode slime*. With efficient operation, the recovery of noble metals from the anode slime pays for the whole refinery operation, leaving the copper as profit.

Metallic copper is malleable, ductile, and a very good conductor of heat and electricity. Except for silver, it has the lowest electric resistance of any metal (page 475) and is used extensively in wires and switches that carry current. Technological advances in understanding electric power transmission have recently cut copper requirements from about 90 kg per megawatt of generated power to 25 kg/MW. Approximately 25 percent of the copper going into the electrical industry is recycled copper.

Chemically, copper is a poorer reducing agent than hydrogen and does not dissolve in acids unless they contain oxidizing anions. When exposed to the air, it slowly tarnishes, with the formation of a green hydroxy carbonate, but this adheres to the metal and protects it from further corrosion. (The green patina observed on bronze statues is copper hydroxy carbonate.) Copper is an important constituent of thousands of alloys ranging from simple *brasses* (copper plus zinc) and *bronzes* (copper plus tin) to more complex and specialized alloys such as Monel metal (copper, nickel, iron, and manganese).

The compounds of copper correspond to oxidation states of $+1$ (cuprous) and $+2$ (cupric). The $+1$ state is easily oxidized and is stable only in very insoluble compounds or in complex ions. The $+2$ state is the one commonly observed in most situations, especially in an aqueous environment. The simple cuprous ion, Cu^+, cannot exist in aqueous solution since it oxidizes and reduces itself by the reaction

$$2Cu^+ \longrightarrow Cu^{2+} + Cu(s)$$

A comparison of the electrode potentials

$$Cu^+ + e^- \longrightarrow Cu(s) \qquad E° = +0.52 \text{ V}$$

$$Cu^{2+} + e^- \longrightarrow Cu^+ \qquad E° = +0.15 \text{ V}$$

indicates that Cu^+ is a better oxidizing agent than is Cu^{2+}. This means that when Cu^+ ions are placed in aqueous solutions, some of the Cu^+ ions take electrons away from other Cu^+ ions. Disproportionation (self-oxidation-reduction) occurs, with the formation of solid copper and cupric ion. This reaction takes place, for example, when cuprous oxide, Cu_2O, is placed in a solution of sulfuric acid. The net reaction

$$Cu_2O(s) + 2H_3O^+ \longrightarrow Cu(s) + Cu^{2+} + 3H_2O$$

can be considered to be the sum of two steps:

$$Cu_2O(s) + 2H_3O^+ \longrightarrow 2Cu^+ + 3H_2O$$

$$2Cu^+ \longrightarrow Cu(s) + Cu^{2+}$$

However, the cuprous condition can be stabilized by formation of insoluble substances or complex ions. For instance, in the presence of chloride ion, cuprous ion can form insoluble cuprous chloride:

$$CuCl(s) \rightleftharpoons Cu^+ + Cl^- \qquad K_{sp} = 3.2 \times 10^{-7}$$

The above electrode potentials then become

$$CuCl(s) + e^- \longrightarrow Cu(s) + Cl^- \qquad E° = +0.14 \text{ V}$$

$$Cu^{2+} + Cl^- + e^- \longrightarrow CuCl(s) \qquad E° = +0.54 \text{ V}$$

which indicates that CuCl is not a good enough oxidizing agent to oxidize itself (i.e., reverse the second half-reaction). Thus, cuprous chloride can be obtained as a stable, white solid in contact with aqueous solutions. If there is a high concentration of chloride ion in the aqueous phase, then an additional complication appears in the formation of complex ions such as $CuCl_2^-$, called dichlorocuprate(I). This is a nonlinear polymeric complex which can be prepared by boiling $CuCl_2$ with copper turnings in concentrated hydrochloric acid. A deep brown color is first formed, probably because of a complex containing both cuprous and cupric copper. As all the cupric state becomes reduced, the solution turns colorless. If the chloride-ion concentration of the colorless solution is decreased, as by dilution, white CuCl precipitates.

The cuprous state is also found in cuprous oxide, Cu_2O, a reddish, insoluble solid. It can be formed by addition of base to a solution of a cuprous complex (for example, $CuCl_2^-$) followed by dehydration. The reddish color observed on metallic copper that has been heated in air is apparently due to a surface coating of Cu_2O. In the classic test for reducing sugars (e.g., glucose, which, unlike sucrose, acts as a mild reducing agent), Cu_2O is formed as a red precipitate when a reducing sugar is heated with an alkaline solution of a cupric salt.

Although many anhydrous cupric salts are white, hydrated cupric salts and their aqueous solutions are blue, owing to the presence of hydrated cupric ion. This may be written $Cu(H_2O)_6^{2+}$, but two oppositely located H_2O molecules are farther away than the other four. In general, aqueous solutions of cupric salts are acidic because of hydrolysis

$$Cu^{2+} + 2H_2O \rightleftharpoons CuOH^+ + H_3O^+$$

but the hydrolysis is not very extensive ($K = 4.6 \times 10^{-8}$). When base is added to these solutions, light blue cupric hydroxide, $Cu(OH)_2$, is formed. The hydroxide is slightly soluble in excess base, and so it might be called slightly amphoteric. When treated with aqueous ammonia solution, $Cu(OH)_2$ dissolves to give a deep blue solution. The color is usually attributed to a copper-ammonia complex ion $Cu(NH_3)_4^{2+}$:

$$Cu(OH)_2(s) + 4NH_3 \longrightarrow Cu(NH_3)_4^{2+} + 2OH^-$$

Like many other complexes of cupric ion, $Cu(NH_3)_4^{2+}$ is paramagnetic owing to an unpaired electron. It has a planar structure (but with H_2O above and below the plane, it could be called highly distorted octahedral), and it can be destroyed by heat or by addition of acid. Heat is effective because it boils the NH_3 out of the solution

$$Cu(NH_3)_4^{2+} \longrightarrow Cu^{2+} + 4NH_3(g)$$

and thus favors dissociation of the complex. Addition of acids results in neutralization of the NH_3 and similarly favors breakup of the complex:

$$Cu(NH_3)_4^{2+} + 4H_3O^+ \longrightarrow Cu^{2+} + 4NH_4^+ + 4H_2O$$

It is interesting to note that addition *of an acid* to a basic solution containing $Cu(NH_3)_4^{2+}$ *can produce* $Cu(OH)_2$ *precipitation.* As acid is added, the concentration of Cu^{2+} rises to compensate for the gradual neutralization of NH_3 until eventually the K_{sp} of $Cu(OH)_2$, 1.6×10^{-19}, is exceeded.

One of the least soluble of cupric compounds is cupric sulfide, CuS. This is the black precipitate which is easily prepared by bubbling hydrogen sulfide through a solution of cupric salt. The very low K_{sp} of CuS (8×10^{-37}) indicates that not even very concentrated H_3O^+ can dissolve appreciable amounts of it. For instance, in $10\,M\,H_3O^+$ the relation from Sec. 12.9 $[H_3O^+]^2[S^{2-}] = 1 \times 10^{-22}$ indicates that the S^{2-} concentration is $1 \times 10^{-22}/(10)^2$, or 1×10^{-24}, M. From the K_{sp} of CuS, the copper concentration would be $8 \times 10^{-37}/1 \times 10^{-24}$, or $8 \times 10^{-13}\,M$. Thus, no appreciable amount of CuS can dissolve in this fashion. It is possible, however, to dissolve appreciable amounts of CuS by heating it with nitric acid. Dissolving occurs, not because H_3O^+ reacts with the S^{2-}, but because hot nitrate ion (especially in acid solution) is a very good oxidizing agent and oxidizes the sulfide ion to elementary sulfur. The net reaction is

$$3CuS(s) + 2NO_3^- + 8H_3O^+ \longrightarrow 3Cu^{2+} + 3S(s) + 2NO(g) + 12H_2O$$

Probably the best-known cupric compound is copper sulfate pentahydrate, $Cu(H_2O)_4SO_4 \cdot H_2O$. In this material each cupric ion is surrounded by a distorted octahedron of O atoms; four of these lie in a square and belong to four H_2O molecules, and the other two belong to neighboring sulfate groups. The odd H_2O molecule, the fifth one, is not directly bound to the cupric ion but forms a bridge between SO_4^{2-} and other H_2O groups. The pentahydrate, or blue vitriol, as it is sometimes called, is used extensively as a germicide and fungicide since the cupric ion is toxic to lower organisms. Its application to water supplies for controlling algae and its use on grapevines to control molds depend on this toxicity. Although trace copper is essential to all organisms, e.g., as a constituent of metalloenzymes, it is very toxic to algae, fungi, and seed plants.

20.3 Silver

Silver is a rather rare element (10^{-8} percent of the earth's crust), occurring principally as native silver, argentite (Ag_2S), and horn silver (AgCl). Only about one-fifth of current silver production comes from silver ores; the rest is mainly a by-product of copper and lead production. The main problem in extracting silver from its ores is to get the rather inert silver (or the very insoluble silver compounds) to go into solution. This can be accomplished by blowing air for a week or two through a suspension of the ore in dilute aqueous sodium cyanide (NaCN) solution. With native silver, the reaction can be written

$$4Ag(s) + 8CN^- + 2H_2O + O_2(g) \longrightarrow 4Ag(CN)_2^- + 4OH^-$$

Were it not for the presence of cyanide ion, the oxygen would not oxidize the silver to a higher oxidation state. This can be seen from a comparison of the following potentials:

$$Ag(s) \longrightarrow Ag^+ + e^- \qquad\qquad E^\circ = -0.799 \text{ V}$$

$$Ag(s) + 2CN^- \longrightarrow Ag(CN)_2^- + e^- \qquad E^\circ = +0.31 \text{ V}$$

In the absence of cyanide ion, metallic silver is a rather poor reducing agent, and hence it is difficult to oxidize it to Ag^+. In the presence of cyanide ion, Ag^+ forms a strongly associated complex ion and is thus stabilized. What this means is that when silver reacts to form the silver-cyanide complex ion, it acts as a fair reducing agent and is rather easily oxidized. Similar reasoning applies to the dissolving of argentite (Ag_2S). This sulfide is very insoluble ($K_{sp} = 5.5 \times 10^{-51}$), and air oxidation of the sulfur by itself is not sufficient to get it into solution. However, in the presence of cyanide ion, solution does occur. In fact, the stability of the complex $Ag(CN)_2^-$ is so great that with high concentrations of cyanide ion the reaction

$$Ag_2S(s) + 4CN^- \longrightarrow 2Ag(CN)_2^- + S^{2-}$$

can be made to proceed to a useful extent even without invoking air oxidation to oxidize the S^{2-}. To recover the silver from the residual solutions, it is necessary to use a rather strong reducing agent, such as aluminum metal or zinc metal in basic solution. A possible reaction is

$$Zn(s) + 2Ag(CN)_2^- + 4OH^- \longrightarrow 2Ag(s) + 4CN^- + Zn(OH)_4^{2-}$$

where some of the zinc in the final solution is also present as a cyanide complex, $Zn(CN)_4^{2-}$.

Massive silver appears almost white because of its high luster. It is too soft to be used pure in jewelry and coinage and is usually alloyed with copper for these purposes. Because of expense it cannot be used much for its best property, its electric and thermal conductivity, which is second to none. In the finely divided state silver usually appears black because the haphazard arrangement of tiny crystalline faces reflects light in all directions with very little probability of sending it to the eye of the observer. Also with smaller particles (of colloidal dimensions), metallic reflection of the type discussed in Sec. 16.1 cannot occur.

The compounds of silver are essentially all of the $+1$ state, although $+2$ and $+3$ compounds have been prepared under extreme oxidizing conditions. For example, an oxide believed to be AgO is formed when ozone is passed over elementary silver. The compound is not very stable toward decomposition to silver and oxygen and, in general, behaves as a very strong oxidizing agent. In the $+1$ state silver forms the ion Ag^+, sometimes called *argentous ion*, after the Latin word for silver, *argentum*. It does not hydrolyze appreciably in aqueous solution; it is a good oxidizing agent; and it forms many complex ions [for example, $Ag(NH_3)_2^+$, $AgCl_2^-$, and $Ag(CN)_2^-$, all of which are linear]. When base is added to solutions of silver salts, a brown oxide, which shows little sign of being amphoteric, is formed:

$$2Ag^+ + 2OH^- \longrightarrow Ag_2O(s) + H_2O$$

However, the oxide does dissolve in an aqueous solution of ammonia because of formation of the colorless complex ion $Ag(NH_3)_2{}^+$, diamminesilver(I):

$$Ag_2O(s) + 4NH_3 + H_2O \longrightarrow 2Ag(NH_3)_2{}^+ + 2OH^-$$

Solutions containing $Ag(NH_3)_2{}^+$ are frequently used as sources of silver for silver plating. They have the advantage of providing low concentrations of Ag^+; so reduction by mild reducing agents, such as glucose, slowly deposits a compact silver plate. Care should be taken in disposing of the waste solutions since evaporation leaves dangerous solid residues which may be violently explosive. The composition of the solids is not known but has been described both as silver amide, $AgNH_2$, and as silver nitride, Ag_3N.

Probably the most interesting of all the silver compounds are the silver halides, AgF, AgCl, AgBr, and AgI. Except for silver fluoride, which is very soluble in water (up to 14.3 mol per 1000 g of water), these halides are quite insoluble. The solubility products, 1.7×10^{-10} for AgCl, 5.0×10^{-13} for AgBr, and 8.5×10^{-17} for AgI, indicate a decrease in solubility from AgCl to AgI. The low solubility is rather surprising, because salts of $+1$ cations and -1 anions are usually soluble. In this respect AgF is normal; it dissolves much like NaF or KF. The abnormal insolubility of the other silver halides is attributed to the fact that their lattice energies (Sec. 15.12) are higher than expected and cannot be compensated for by hydration effects. The principal reason for the higher lattice energy is that there are strong van der Waals attractions between Ag^+ ions and the halide ions which are superposed on the ordinary ionic attractions. Suppose we compare AgCl with KCl. Since Ag^+ (ionic radius 0.126 nm) and K^+ (ionic radius 0.133 nm) have about the same size, we would expect the ionic attractions in the solid to be about the same. However, the Ag^+ has 46 electrons, whereas K^+ has only 18. In general, the more electrons an atom has, the more easily it can be polarized, and hence the stronger its van der Waals attraction to neighboring atoms (Sec. 5.12). Consequently, the lattice of AgCl should be held together more strongly than that of KCl. In fact, the lattice energy of AgCl is 904 kJ/mol; that of KCl, 699 kJ/mol. Since more energy is required to break up the AgCl lattice than the KCl, AgCl should be less soluble. In further support of this picture is the observed decrease in solubility from AgCl to AgBr to AgI. As the anion comes to contain more electrons, the van der Waals attraction increases, and the lattice energy becomes greater.

Except for AgF, the silver halides are sensitive to light. For this reason, they find use in making photographic emulsions. The chemistry of the photographic process is not well understood; it is complicated and apparently involves enhanced reactivity of defect structures (see

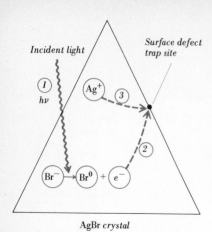

Incident light

Surface defect trap site

(1) hν

Ag⁺ (3)

Br⁻ → Br⁰ + e⁻

(2)

AgBr crystal

Fig. 20.3
Formation of latent image in photographic process. Sequence shown is repeated about 50 times.

Sec. 7.5). The basic steps are usually described as (1) exposure, (2) development, and (3) fixing:

1 Exposure. When photographic film, consisting of a dispersion of silver bromide in gelatin, is exposed to light, grains of silver bromide are activated, depending on the intensity of the incident light. This is not a visible change (in fact, it is called formation of a *latent image*) and, according to one theory (the Gurney-Mott theory), involves the following sequence of events (see Fig. 20.3):

(a) Incoming $h\nu$ kicks an electron out of a Br^- to form Br^0 and e^-.

(b) The e^- wanders through the crystal of AgBr and gets trapped at a surface defect, which might, for example, be a speck of Ag_2S.

(c) An interstitial Ag^+ diffuses to the trap site, where Ag^+ and e^- combine to give Ag^0.

(d) A second quantum of light energy $h\nu$ comes along and ejects a second e^-, which then migrates to the Ag^0 and converts it to Ag^-.

(e) A second interstitial Ag^+ subsequently diffuses over and combines:

$$Ag^+ + Ag^- \longrightarrow Ag_2$$

(f) The process continues repeatedly until a clump of Ag_n atoms is built up which activates the whole AgBr grain for subsequent attack by a mild reducing agent (developer). It is estimated that a clump of about 50 silver atoms is needed for this purpose.

2 Development. Grains that have been sufficiently activated can now be reduced with a mild reducing agent. Either hydroquinone (HOC_6H_4OH, a benzene ring with two OH's substituted for oppositely located H atoms) or sodium sulfite (Na_2SO_3) is typical. In neutral solution the potential for the conversion

$$HO\langle\bigcirc\rangle OH \longrightarrow O=\langle\bigcirc\rangle=O + 2e^- + 2H^+$$

Hydroquinone   *Quinone*

is about 0.12 V, which is just about what is required for

$$AgBr(s) + e^- \longrightarrow Ag(s) + Br^- \qquad E^\circ = 0.03 \text{ V}$$

The differential blackening in a photograph is due to formation of elemental silver in the activated grains, there being more of them where the light was strongest.

3 Fixing. Since AgBr slowly turns black when exposed to light, the whole film would turn black eventually. However, the photographic image can be fixed by washing out the nonactivated AgBr grains. Although very insoluble ($K_{sp} = 5.0 \times 10^{-13}$), AgBr will dissolve in solutions containing high concentrations of thiosulfate ion, $S_2O_3^{2-}$, by the reaction

$$AgBr(s) + 2S_2O_3^{2-} \longrightarrow Ag(S_2O_3)_2^{3-} + Br^-$$

Thus, the final step involves soaking the film in a fixing bath, the essential component of which is $Na_2S_2O_3$. The result is a fixed negative image of the exposure. To get a positive image, the whole process is repeated. By shining light through the negative onto another emulsion, developing, and then fixing it, the light and dark areas can be inverted.

In color photography the processes are much more involved. Fundamentally, they depend on having film coated with three emulsion layers, each of which is sensitive to one of three primary colors. On exposure and development, images are formed in each of the three layers. By appropriate choice of dyes and other chemicals, these three images can be colored separately to reproduce by superposition the original multicolored pattern.

20.4 Gold

Gold is a very rare element, being about one-tenth as abundant as silver. It occurs naturally as *native gold* (where it is usually alloyed with silver) and less frequently as compounds of tellurium, such as $AuTe_2$ (gold telluride, or calaverite). The recovery is generally a mechanical process which makes use of the very high density (19.3 g/cm^3) of the metal. Chemical extraction is usually by a cyanide process like that used for silver:

$$4Au(s) + 8CN^- + O_2(g) + 2H_2O \longrightarrow 4Au(CN)_2^- + 4OH^-$$

Although gold is the most malleable and ductile of all metals and is a very good conductor of heat and electricity, its principal use is for currency and jewelry. It is rather inert; so it finds some use as electrode material, lining for reaction chambers, and plating to protect more active metals.

The compounds of gold correspond to +1 and +3 oxidation states; these are called aurous and auric, respectively, after the Latin word for gold, *aurum*. The aurous ion, Au^+, cannot exist in aqueous solution because it is such a good oxidizing agent. The electrode potential for $Au^+ + e^- \longrightarrow Au(s)$ is $+1.7$ V, thus making Au^+ about as strong an oxidizing agent as permanganate ion. It will, in fact, oxidize itself to Au^{3+}. However, the +1 state of gold can be stabilized by complexing with cyanide ion to form $Au(CN)_2^-$. The electrode potential for $Au(CN)_2^- + e^- \longrightarrow Au(s) + 2CN^-$ is -0.60 V.

In the +3 state, gold exists as complex ions. The most common of these is the chloraurate ion, $AuCl_4^-$, which is obtained when gold is dissolved in aqua regia. Aqua regia, which consists of one part of concentrated nitric acid and about three parts of concentrated hydrochloric acid, can dissolve gold, whereas concentrated HNO_3 or concentrated HCl alone cannot. The reason for this is that HNO_3 cannot oxidize gold unless the chloride ion is present to complex the product. In other words, the half-reaction $Au(s) + 4Cl^- \longrightarrow AuCl_4^- + 3e^-$, which has $E° = -1.0$ V, is easier to carry out than the half-reaction $Au(s) \longrightarrow Au^{+3} + 3e^-$, which has $E° = -1.42$ V.

The net equation for dissolving gold in aqua regia can be written

$$Au(s) + 3NO_3^- + 4Cl^- + 6H_3O^+ \longrightarrow$$
$$AuCl_4^- + 3NO_2(g) + 9H_2O$$

though, in fact, a variety of nitrogen products are obtained.

20.5 Zinc subgroup

The elements of the zinc subgroup are zinc, cadmium, and mercury. They are more active than the elements of the copper subgroup, but their chemistry is somewhat simpler. The zinc-subgroup elements have a characteristic oxidation state of $+2$, except for mercury, which also forms $+1$ compounds. Some of the more important properties are listed in Fig. 20.4.

Fig. 20.4
Elements of zinc subgroup.

Symbol	Z	Electronic configuration	Melting point, °C	Boiling point, °C	Ionization potential, eV	Electrode potential (from M^{2+}), V
Zn	30	(18) $3d^{10}4s^2$	419	907	9.39	-0.76
Cd	48	(36) $4d^{10}5s^2$	321	767	8.99	-0.40
Hg	80	(68) $5d^{10}6s^2$	-38.9	357	10.43	$+0.85$

As seen from the electronic configurations, each of these elements has two electrons in the outermost energy level. The situation is reminiscent of that found for the alkaline-earth elements (Sec. 16.1). The low melting points may at first sight be surprising, but they are not entirely unexpected. In progressing from left to right through the transition sequence, the low point in the atomic volume has been passed (see Fig. 2.24), and the atoms get bigger from there on. As the atoms get bigger, they are farther apart, and forces of attraction are smaller. Thus, it becomes easier to melt the elements. Probably of greater importance is the fact that the d shells of the second-outermost shells are filled; therefore, there is little chance for d-orbital covalent binding between ions as found in other transition elements. In mercury, the interatomic forces are so weak that the melting point is below room temperature.

There is a striking difference between these elements and the analogous group II elements. In Fig. 20.5 the element zinc is compared with calcium, a typical element of group II. Although both of these elements have but two electrons in the outermost shell and have no partially filled shells, their properties are quite different. For example, as shown by the electrode potentials, calcium is a very powerful reducing agent, whereas zinc is only moderately strong. The fundamental reason for the change in properties is the decreased size of the zinc atom. Since the nuclear charge has increased by 10 units in going from $Z = 20$ to $Z = 30$ but the number of electron shells stays the same, there is a greater attraction for all electrons in zinc, and the shells are pulled in. The atom is thus smaller. The valence electrons are fourth-shell electrons in both cases and are held more tightly in the zinc atom.

Property	Calcium	Zinc
Atomic number	20	30
Electronic configuration	2, 8, 8, 2	2, 8, 18, 2
Atomic volume, cm³/mol	26	9
Size of M^{2+} (radius), nm	0.094	0.070
First ionization potential, eV	6.11	9.39
Second ionization potential, eV	11.87	17.89
Third ionization potential, eV	51.21	40.0
Density, g/cm³	1.55	7.14
Melting point, °C	810	419
Boiling point, °C	1300(?)	907
Electrode potential, V	−2.87	−0.76

Fig. 20.5
Comparison of calcium and zinc.

20.6 Zinc

About 100 times as abundant as copper, zinc occurs principally as the mineral sphalerite (ZnS), also called zinc blende. The metal is prepared by roasting the sulfide in air to convert it to oxide and then reducing the oxide with finely divided carbon. The reactions are

$$2ZnS(s) + 3O_2(g) \longrightarrow 2ZnO(s) + 2SO_2(g)$$

$$ZnO(s) + C(s) \longrightarrow Zn(g) + CO(g)$$

Since the second reaction is carried out at about 1200°C, above the boiling point of zinc, the metal forms as a vapor and must be condensed. Very rapid condensation produces the fine powder known as zinc dust.

Massive Zn has fairly good metallic properties except that it is rather brittle, especially at 200°C, where it can even be ground up into a powder. It is a moderately active metal and can reduce H_2O to H_2, but only when heated. With acids, ordinary Zn gives the well-known evolution of H_2. Strangely enough, this is very rapid when the Zn is impure but almost too slow to be observed when the Zn is very pure. Impurities (especially arsenic and antimony) apparently speed dissolving by serving as centers from which H_2 gas can evolve.

In air, zinc tarnishes but slightly, probably because it forms a self-protective coat of oxide, hydroxide, or carbonate. Because it itself withstands corrosion so well and because it can give cathodic protection (Sec. 19.4) to iron, zinc is often used as a coating on iron to keep it from rusting. Iron protected in this way, called galvanized iron, can be made by dipping the iron into molten zinc or by plating zinc on it from an electrolytic bath. The other important use of zinc is in alloys such as the brasses, which are essentially copper-zinc alloys.

In all its compounds zinc shows only a $+2$ oxidation state. The zinc ion, Zn^{2+}, is colorless and not paramagnetic. In aqueous solutions it hydrolyzes to give slightly acid solutions. The hydrolysis, usually written

$$Zn^{2+} + 2H_2O \rightleftharpoons Zn(OH)^+ + H_3O^+ \qquad K = 1.1 \times 10^{-9}$$

does not proceed so far toward the right as does that of Cu^{2+} ion. Thus, for equal concentrations solutions of zinc salts are somewhat less acid than those of cupric salts. When base is added to solutions of zinc salts, white zinc hydroxide, $Zn(OH)_2$, is precipitated. This hydroxide is amphoteric, and therefore further addition of base dissolves it to give zincate ion, $Zn(OH)_4{}^{2-}$. The concentration of Zn^{2+} in equilibrium in basic solution is very small. This means that the half-reaction

$$Zn(s) \longrightarrow Zn^{2+} + 2e^- \qquad E° = +0.76 \text{ V}$$

has greater tendency to go to the right in basic solution than in acid solution. Consequently, zinc metal is a stronger reducing agent for basic solutions than for acid solutions. For basic solutions, the half-reaction can be written as

$$Zn(s) + 4OH^- \longrightarrow Zn(OH)_4{}^{2-} + 2e^- \qquad E° = +1.22 \text{ V}$$

Like other transition elements, zinc has great tendency to form stable complex ions. For example, zinc hydroxide is easily dissolved in aqueous ammonia because of the formation of a tetrahedral complex, $Zn(NH_3)_4{}^{2+}$, tetraamminezinc(II). The hydroxide can also be dissolved in cyanide solutions because of the formation of $Zn(CN)_4{}^{2-}$, tetracyanozincate(II), which is also tetrahedral. As shown by the following equilibrium constants

$$Zn(NH_3)_4{}^{2+} \rightleftharpoons Zn^{2+} + 4NH_3 \qquad K = 3.4 \times 10^{-10}$$

$$Zn(CN)_4{}^{2-} \rightleftharpoons Zn^{2+} + 4CN^- \qquad K = 1.2 \times 10^{-18}$$

the cyanide complex is less dissociated than is the ammonia complex. The greater stability of the cyanide complex is reflected in the fact that zinc metal is a stronger reducing agent in cyanide solutions than in ammonia solutions. The potentials are

$$Zn(s) + 4CN^- \longrightarrow Zn(CN)_4^{2-} + 2e^- \qquad E° = +1.26 \text{ V}$$

$$Zn(s) + 4NH_3 \longrightarrow Zn(NH_3)_4^{2+} + 2e^- \qquad E° = +1.04 \text{ V}$$

When hydrogen sulfide is passed through solutions of zinc salts which are not too acid, white zinc sulfide precipitates. Although the solubility product of ZnS (1×10^{-22}) is rather small, so that ZnS is essentially insoluble in neutral solutions, addition of acid lowers the sulfide-ion concentration sufficiently that ZnS becomes soluble. The enhanced solubility of ZnS in acid solution gives a method for separating it from other sulfides such as CuS, Ag_2S, and CdS.

Zinc sulfide is used extensively in the white pigment lithopone, an approximately equimolar mixture of ZnS and $BaSO_4$. Unlike white lead paints, it is not toxic. ZnS is also used in making fluorescent screens because impure ZnS acts as a phosphor; i.e., it can convert energy such as that of an electron beam into visible light. The action of phosphors is very complex and is closely related to the properties of defects in solid-state structures (Sec. 7.5). The simplest view is that an electron beam impinging on impure ZnS uses its energy to detach electrons from the impurity centers to which they are bound. An electron so removed moves through the crystal until it encounters some other center to which it can be bound by giving off some energy, usually as a flash of light.

Zinc oxide is probably the most important of the zinc compounds. It can be made by oxidizing zinc vapor in air, either by boiling the metal or by heating zinc ore with carbon. ZnO has many specialized uses (e.g., filler and vulcanization activator in rubber tires, pigment, ointment base, and cement), but its most interesting use is as a photoconductor in copying machines. It conducts electric current when illuminated; when the light is shut off, it reverts to an insulator. In the Xerox method of electrostatic printing, a photoconductive plate having an electrically conducting backing is uniformly charged by exposure to a silent electric discharge. The photoconductive material is then discharged by exposure to light, but only in those places where light is reflected from white parts of a document to be copied. The residual electric image is then used to attract negatively charged black powder (i.e., carrier plus resin pigment) for subsequent transfer to paper where it is fused by heat to fix a permanent image.

20.7   Cadmium

The properties of cadmium are so similar to those of zinc that the two elements invariably occur together. There are no important minerals of cadmium, which is only about one-thousandth as abundant as zinc. The principal source of cadmium is the flue dust from the purification of zinc by distillation. Since cadmium is more volatile than zinc, it evaporates first and concentrates in the first distillates. The principal use of cadmium is as a plate on other metals, such as steel. It is particularly good as a protective coat for alkaline conditions because, unlike zinc, it is not amphoteric and does not dissolve in base. The other principal use of cadmium is in making low-melting alloys, such as Wood's metal (mp 70°C).

In its compounds the usual oxidation state of cadmium is +2. It exists in aqueous solutions as colorless Cd^{2+} ion. With H_2S, it forms insoluble, yellow CdS ($K_{sp} = 1.0 \times 10^{-28}$), which is used as the pigment *cadmium yellow*. Like zinc, cadmium forms a variety of complex ions, including $Cd(NH_3)_4{}^{2+}$, $Cd(CN)_4{}^{2-}$, $CdCl_4{}^{2-}$, and $CdI_4{}^{2-}$, all of which are tetrahedral. Some of the salts of cadmium are unusual in the sense that they do not dissociate completely into ions in aqueous solution as practically all other salts do. Cadmium sulfate, for example, has a dissociation constant of 5×10^{-3} and so can be called a weak salt.

Although cadmium is a relatively rare element, it is now recognized as a major environmental contaminant. Natural geochemical processes appear to be the major source of pollution, but man is increasingly contributing in a localized way. With a world production on the order of 15 million kg/yr (40 percent in the United States) and given the high volatility of cadmium, it is inevitable that there be some leakage into the environment. The most notorious example of cadmium poisoning comes from Japan in the form of *Itai-itai disease*. (An equivalent English name would be "ouch-ouch.") Symptoms include pain in the bones and joints, a waddling gait caused by bone deformation, and, in the later stages, susceptibility to multiple fractures by as little a disturbance as coughing. Contamination of water, food, and air all contribute as pathways for ingestion into the human body. The cadmium accumulates in the liver and kidneys and apparently acts by inactivating sulfur-containing enzymes. Excretion *via* urine and the gastrointestinal tract is very slow, which makes the cadmium hazard particularly serious. Recently, tobacco smoke (which contains cadmium) has been implicated as a cadmium pollutant to smoker and nonsmoker alike. Low-dosage cadmium poisoning may be associated with high blood pressure and some chronic diseases such as bronchitis and emphysema.

20.8 Mercury

The only common mineral of mercury is cinnabar (HgS), from which the element is produced by roasting in air:

$$HgS(s) + O_2(g) \longrightarrow Hg + SO_2(g)$$

Unlike any other metal,* mercury is a liquid at room temperature, and its symbol emphasizes this, since it comes from the Latin *hydrargy-*

* Cesium metal has a melting point of 28.5°C, or 83.3°F, and gallium metal has a melting point of 29.8°C, or 85.6°F. Thus, the uniqueness of mercury as a liquid metal disappears on hot days.

rum, meaning "liquid silver." The liquid is not very volatile (vapor pressure is 0.0000024 atm at 25°C), but the vapor is very poisonous, and *prolonged exposure even to the liquid should be avoided.*

Liquid mercury has a high metallic luster, but it is not a very good metal in that it has a higher electric resistance than any of the other transition metals. However, for some uses, as in making electric contacts, its fluidity is such a great advantage that its mediocre conductivity can be tolerated. Furthermore, its inertness to air oxidation, its relatively high density, and its uniform expansion with temperature lead to special uses as in barometers and thermometers.

Liquid mercury dissolves many metals, especially the softer ones such as copper, silver, gold, and the alkali elements. The resulting alloys, which may be solid as well as liquid, are called *amalgams.* Probably their most distinctive property is that the reactivity of the metal dissolved in the mercury is thereby lowered. For example, the reactivity of sodium in sodium amalgam is so low that the amalgam can be kept in water with only slow evolution of hydrogen.

In its compounds mercury shows both +1 (mercurous) and +2 (mercuric) oxidation states. In this respect, it is unlike the other members of the zinc subgroup. The mercurous compounds are unusual because they all contain two mercury atoms bound together. In aqueous solutions, the ion is a double ion corresponding to Hg_2^{2+}, in which there is a covalent σ bond between the two mercury atoms. Experimental evidence for this is the lack of paramagnetism of mercurous compounds. The ion Hg^+ would have one unpaired electron in its $6s$ orbital and would be paramagnetic, whereas the ion Hg_2^{2+} would have the two electrons paired in a σ bonding molecular orbital and would not be paramagnetic. Further experimental evidence for Hg_2^{2+} comes from a study of the equilibrium between liquid mercury, mercuric ion, and mercurous ion. There are two possible ways of writing this equilibrium, depending on whether mercurous mercury exists as Hg_2^{2+} or Hg^+:

$$Hg(l) + Hg^{2+} \rightleftharpoons Hg_2^{2+} \qquad K = \frac{[Hg_2^{2+}]}{[Hg^{2+}]}$$

$$Hg(l) + Hg^{2+} \rightleftharpoons 2Hg^+ \qquad K' = \frac{[Hg^+]^2}{[Hg^{2+}]}$$

If the amounts of mercurous mercury and mercuric mercury in solution are determined for various equilibrium solutions, it is found that the ratio of mercurous to mercuric is constant, but the ratio of mercurous squared to mercuric is not. In other words, K is found to be a true constant (1.7×10^2) for all experiments, but K' is not. Apparently, there is little, if any, Hg^+ in solution.

Except for the doubling, mercurous ion behaves much like Ag^+; for example, it reacts with chloride ion to precipitate white mercurous chloride, Hg_2Cl_2, also known as calomel. When exposed to light, calomel darkens by partial disproportionation into Hg and $HgCl_2$.* Just as the silver halides decrease in solubility in going from AgF to AgI, so do the mercurous halides. Mercurous fluoride, Hg_2F_2, is quite soluble in water, but the solution immediately decomposes to form HF and insoluble, black Hg_2O. For the other halides the solubility products are as follows:

$$Hg_2Cl_2(s) \rightleftharpoons Hg_2{}^{2+} + 2Cl^- \qquad K_{sp} = 1.1 \times 10^{-18}$$

$$Hg_2Br_2(s) \rightleftharpoons Hg_2{}^{2+} + 2Br^- \qquad K_{sp} = 1.3 \times 10^{-22}$$

$$Hg_2I_2(s) \rightleftharpoons Hg_2{}^{2+} + 2I^- \qquad K_{sp} = 4.5 \times 10^{-29}$$

Unlike Ag^+, mercurous ion does not form an ammonia complex. When aqueous ammonia is added to Hg_2Cl_2, the solid turns black because of formation of finely divided mercury:

$$Hg_2Cl_2(s) + 2NH_3 \longrightarrow HgNH_2Cl(s) + Hg + NH_4{}^+ + Cl^-$$

The compound $HgNH_2Cl$, mercuric ammonobasic chloride, is white, but its color is obscured by the intense black of the mercury. This difference in behavior toward NH_3 provides a simple test for distinguishing AgCl from Hg_2Cl_2.

In the $+2$ state, mercury is frequently represented as the simple ion Hg^{2+}, although it is usually found in the form of complex ions, insoluble solids, or weak salts. For example, in a solution of the weak salt mercuric chloride, the concentration of Hg^{2+} is much smaller than the concentration of undissociated $HgCl_2$ molecules. With excess chloride ion the complexes $HgCl_3{}^-$ and $HgCl_4{}^{2-}$ are also formed. In ammonia solutions complex ions containing one, two, three, and four NH_3 molecules are known. For complete dissociation of $Hg(NH_3)_4{}^{2+}$ the constant is 5.2×10^{-20}. The complex $Hg(CN)_4{}^{2-}$ is even more stable ($K = 4 \times 10^{-42}$).

Although mercuric sulfide as found in nature is red, when H_2S is passed through a mercuric solution, a black precipitate of HgS is obtained. The color difference may be due to differences in crystal structure. The solubility product of black HgS is very low (1.6×10^{-54}), but not so low as that of platinum sulfide, PtS ($K_{sp} = 8 \times 10^{-73}$). In order to dissolve these very insoluble sulfides, drastic measures are required. HgS, for example, will not dissolve even in boiling nitric acid. Aqua regia, however, which supplies both nitrate for oxidizing the sulfide and chloride for complexing the mercuric, does take it into solution.

The electrode potentials

$$2Hg^{2+} + 2e^- \longrightarrow Hg_2{}^{2+} \qquad E° = +0.92 \text{ V}$$

$$Hg_2{}^{2+} + 2e^- \longrightarrow 2Hg(l) \qquad E° = +0.79 \text{ V}$$

* Calomel was once used in medicine as a purgative. However, when exposed to light there is possibility of forming $HgCl_2$, which is a deadly poison. Because of the toxicity hazard, calomel is no longer used in internal medicine. *Judex damnatur cum nocens absolvitur.*

are so close that any reducing agent which is able to reduce mercuric ion to mercurous ion is also able to reduce mercurous ion to mercury. Thus, if a limited amount of reducing agent such as Sn^{2+} (stannous ion) is added to a mercuric solution, only Hg_2^{2+} is formed: but if Sn^{2+} is added in excess, the reduction goes all the way to Hg.

20.9 Mercury in the environment

Although nature has tried to lock mercury away as the very insoluble HgS, man has opened the Pandora's box by extracting the metal and by burning fossil fuels. It is estimated that each year 5,000,000 kg of mercury is added to the atmosphere by the combustion of coal. The supposition until recently was that since elemental mercury is rather inert and eventually goes to the very insoluble sulfide, no great harm was done. We know now that the mercury threat to the environment was vastly underrated.

Metallic mercury is not highly toxic. In a classic suicide attempt 2 g of liquid mercury was injected into the veins but had no effect—the liquid mercury simply collected as a puddle in the heart. The vapor of mercury, on the other hand, is a hazard, particularly over long exposure. The yellow sulfur that is sometimes seen sprinkled around in a laboratory where there is spilled mercury represents an attempt to convert the mercury to HgS and thereby prevent buildup of even small amounts of vapor.

Inorganic compounds of mercury, when soluble, are toxic, but only moderately so. Mercuric chloride, for example, when taken orally damages the intestinal tract (leading to bloody diarrhea) and the kidneys (leading to suppression of urine and ultimately to uremic death). The organic compounds, particularly dimethylmercury, $(CH_3)_2Hg$, are extremely toxic, and their continued use as slimicides, fungicides, mildew killers, and germ sprays poses a continual threat to the environment.

The most famous case of environmental contamination by mercury occurred in Japan as *Minamata disease*. Minamata is a small fishing village on the southwest coast of Kyushu, and there in 1953 the cats and crows began to do strange things. Soon the people began to show signs of a strange illness characterized progressively by soreness of hands and face, tunnel vision, dizziness, loss of control of body movements, mental disorder, and finally death. Health authorities correlated the symptoms with fish consumption and by analysis of silt in the harbor (which contained 2000 ppm mercury) implicated a nearby factory which had used $HgCl_2$ as a catalyst for making polyvinyl chloride, a common plastic. In 1965 a similar disease struck Niigata, afflicting 120 people (5 of whom died) and leading to the birth of 27 defective children. The culprit in both cases was believed to be $(CH_3)_2Hg$ stored in the body tissue of the fish and produced by bacterial action in the mud slimes on mercury wastes.

A related case involved Sweden in the 1950s, where there was noted a significant drop in the population of seed-eating birds (e.g., pheasant, partridge, and pigeons). Neutron activation analysis, an ultrasensitive technique in which neutron bombardment changes an isotope of an element into a radioactive one that can be counted by its radioactive disintegration, showed excessively high mercury contents in the livers of the birds. The ultimate source appeared to be organic mercury fungicides (for example, CH_3—Hg—CN) used as seed dressings in agriculture.

Where does most of the man-made mercury pollution come from? This is a difficult question to answer because it depends on (1) the amount of mercury a particular usage represents and (2) the leakage from that use into the environment. The biggest user of mercury in the United States is the chlor-alkali industry, which uses mercury for the cathode in the electrochemical decomposition of aqueous sodium chloride. Total United States annual consumption of mercury is about 2.7 million kg; about a quarter of this goes to chlor-alkali. Figure 20.6 shows a typical electrolysis cell. As can be seen, the process is a continuous one in which aqueous sodium chloride is made to flow into

Fig. 20.6
Chlor-alkali electrolytic cell.

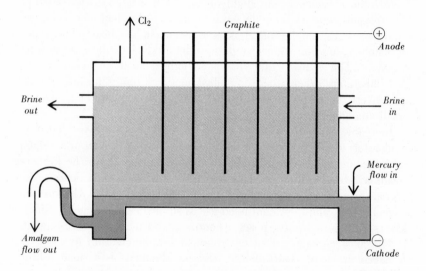

the cell and is electrolyzed between graphite anodes and a flowing-mercury-pool cathode. At the cathode, the Na^+ is reduced to Na^0, which dissolves in the mercury to give a sodium amalgam. [The other possible cathode reaction $e^- + H_2O \longrightarrow \frac{1}{2}H_2(g) + OH^-$ does not occur because of the great difficulty kinetically of releasing hydrogen at a mercury surface.] The actual cathode reaction is

$$Na^+(aq) + e^- \longrightarrow Na(amalgam)$$

and the amalgam produced is siphoned off into a water-containing vessel made of iron. Hydrogen discharge off an iron surface, unlike mercury, is quite rapid; so the sodium amalgam promptly reacts with the water to form aqueous NaOH and regenerate the mercury. In principle, there is no mercury loss; so there should be no problem for the environment. However, in practice there is loss (to the products, to the effluent water, and to air ventilation) amounting to about 250 g of mercury per ton of chlorine produced. Until recently, the total loss was about 400 tons/yr, but this has recently been reduced by almost 90 percent by recirculating the waste water into settling ponds. The damage, however, has been done. Waste mercury in many cases was simply washed out to accumulate in river and lake bottoms. Unless decontamination procedures are undertaken, anaerobic bacteria in the mud slimes will feast on the mercury and generate dimethylmercury for absorption into the food chain. Marine diatoms with mercury compounds adsorbed on their surfaces get eaten by higher organisms, fish eat the organisms, man eats the fish, and so on up the food chain. Each step concentrates the mercury contamination.

Fish in contaminated water have been found to contain appreciable amounts of mercury. For example, in Lake Erie the following average mercury levels (in parts per million) have been observed: walleyed pike, 1.4 to 3.57; sucker, 0.88; Northern pike, 0.64; and white bass, 0.53 to 0.80. These numbers are to be compared with FDA (Food and Drug Administration) established maximum permissible levels of 0.5 ppm for food and 0.005 ppm for water. Pertinent also is the fact that, unlike the usual situation where permitted levels are set at one-hundredth of the concentration where poison symptoms appear, for the case of mercury this is set at one-tenth. The problem is that if permitted levels are made lower than 0.5 ppm, they approach the background level in food of 0.2 ppm!

Other sources of mercury pollution are the paper and pulp industry, commercial laundries, seed dressings in agriculture, coal combustion, and sewage effluents. In the paper industry, PMA, or phenyl mercury acetate ($C_6H_5HgCOOCH_3$), has been used for in-process slime control. In the United States the practice has been given up because of a government directive not to use PMA for paper that would contact food. Since the paper use cannot be guaranteed, use of PMA for any paper was abandoned. Commercial laundries, especially those with diaper service, sometimes use PMA to suppress mold. Mercury seed dressings, although suspended for interstate commerce in the United States, continue to produce tragedies elsewhere. In several cases, seed grains that had been mercury-treated and had been dyed red to show this were diverted from normal use, washed to remove the dye, fed to cattle, and eventually fed to people *via* sick animals that were rushed to the slaughterhouse.

Coal contains about 1 ppm Hg. Worldwide combustion of coal amounts to about 5×10^9 tons/yr, which means about 5000 tons of Hg into the atmosphere. This airborne contamination may be part of the explanation why fish in isolated mountain lakes have sometimes shown surprisingly large Hg content. Hg is excreted by fish only very slowly. The half-life in freshwater fish is about 200 days; in humans it is about 70 days.

Further discussion of the environmental pollution problem is given in Chap. 26.

20.10 Qualitative analysis

Mercurous ion and silver ion can be separated from the other cations by adding HCl to precipitate white, insoluble Hg_2Cl_2 and AgCl. If NH_3 is added to a mixture of these chlorides, a black color appears, owing to formation of Hg and $HgNH_2Cl$. Since NH_3 converts AgCl into soluble $Ag(NH_3)_2^+$ and Cl^-, the filtrate contains the silver, and AgCl can be reprecipitated by addition of HNO_3.

If H_2S is added to an acidic solution containing Cd^{2+}, Hg^{2+}, Cu^{2+}, and Zn^{2+}, the first three precipitate as insoluble sulfides (yellow CdS, black HgS, and black CuS). If the residual solution is then made basic with NH_3, white ZnS is formed. A confirmatory test for ZnS would be to dissolve it in HCl plus HNO_3, evaporate to dryness, and reprecipitate by addition of H_2S in a SO_4^{2-}–HSO_4^- buffer.

The separation of CdS, HgS, and CuS makes use of the fact that CdS and CuS are soluble in boiling HNO_3 whereas HgS is not. Residual HgS can be confirmed by dissolving it in aqua regia and reducing with $SnCl_2$ to give Hg_2Cl_2 and Hg. Addition of NH_3 to a solution containing Cu^{2+} and Cd^{2+} gives the blue color characteristic of $Cu(NH_3)_4^{2+}$. Cd can be detected by first precipitating out the Cu^{2+} with H_2S in acid solution in the presence of high-concentration chloride ion (which keeps Cd^{2+} in solution as $CdCl_4^{2-}$) and then adding $NaC_2H_3O_2$ and H_2S. The added acetate ion serves to reduce the H_3O^+ concentration, thereby raising the S^{2-} concentration sufficiently to precipitate yellow CdS.

Exercises

*20.1 *Periodic table.* Why is the vertical relation in the copper and zinc subgroups closer than the horizontal relation, whereas in the three preceding subgroups the opposite is the case?

*20.2 *Structure.* Copper, silver, and gold all crystallize with face-centered cubic structures. The unit-cell edge lengths are 0.3608, 0.4078, and 0.4070 nm, respectively. Calculate the theoretical density for each of these elements. The observed values are 8.92, 10.5, and 19.3 g/cm³, respectively. *Ans. 8.987, 10.57, 19.41 g/cm³*

*20.3 *Copper ores.* Without doing any calculations tell whether CuS or $CuFeS_2$ has higher percent by weight sulfur. Indicate your reasoning.

*20.4 *Electrolysis.* How long would it take to refine a metric ton (1000 kg) of copper with a current of 100 amp?

*20.5 *Photography.* Explain in terms of the Gurney-Mott theory why the sensitivity of a photographic film can be enhanced by partial substitution of AgI for AgBr.

*20.6 *Entropy of fusion.* Given that Zn, Cd, and Hg require, respectively, 102, 54.4, and 11.8 J/g to melt, how great is the molar entropy change for each of the fusion processes?
Ans. 9.64, 10.3, 10.1 J deg⁻¹ mol⁻¹

*20.7   *Mercury.* Explain why solutions of mercurous nitrate generally are given a pool of liquid mercury at the bottom of the bottle.

**20.8   *Electrode potential.* Show that $E° = +0.14$ V for $CuCl(s) + e^- \longrightarrow Cu(s) + Cl^-$ given that $E° = +0.52$ V for $Cu^+ + e^- \longrightarrow Cu(s)$ and $K_{sp} = 3.2 \times 10^{-7}$ for CuCl.

**20.9   *Thermodynamics.* From data given in Sec. 20.2, calculate $\Delta G°$ for each of the following processes:

$2Cu^+ \longrightarrow Cu(s) + Cu^{2+}$

$2CuCl(s) \longrightarrow Cu(s) + Cu^{2+} + 2Cl^-$

**20.10   *Equilibrium constant.* If the K_{sp} of $Cu(OH)_2$ is 1.6×10^{-19} and total K_{diss} for $Cu(NH_3)_4{}^{2+}$ is 5×10^{-15}, what is K for $Cu(OH)_2(s) + 4NH_3 \rightleftharpoons Cu(NH_3)_4{}^{2+} + 2OH^-$?

Ans. 3.2 × 10⁻⁵

**20.11   *Precipitation.* A solution containing 0.10 M Cu^{2+} and 0.10 M Ag^+ with 0.30 M H_3O^+ is saturated with H_2S until both Ag_2S and CuS precipitate as much as possible. What will be the final concentration of Cu^{2+} and Ag^+ in the residual solution?

**20.12   *Conversions.* Indicate the reagents and reactions necessary for the following conversions:
  a Cu_2O to CuO
  b $CuCO_3$ to $CuSO_4$
  c   CuS to $CuSO_4$
  d   Cu to CuS
  e   CuS to Cu

**20.13   *Symmetry.* In the vapor phase cuprous chloride exists as a trimer $(CuCl)_3$, the molecules of which are six-membered rings with C_3 and σ_v as their principal symmetry elements. Sketch a possible structure for the molecule.

**20.14   *Nomenclature and structure.* In the ion tetrachlorocuprate(II) the Cu–Cl distance is 0.222 nm, and the geometry is approximately tetrahedral; in the ion *trans*-diaquotetrabromocuprate(II) the Cu–Br distance is 0.246 nm, and the geometry is approximately octahedral. Calculate the interhalogen distance in these two complexes. Compare with the radii in Appendix 8, and explain.

**20.15   *Dissociation constant.* From the data given in Sec. 20.3, calculate K_{diss} for the total dissociation of $Ag(CN)_2{}^-$.

**20.16   *Chemical equations.* Write net equations for each of the following processes:
  a   Aqueous NH_3 is added dropwise to a solution of silver nitrate until a precipitate of Ag_2O appears.
  b   Excess NH_3 is added to (*a*) until the precipitate redissolves.
  c   Half of the solution from (*b*) is treated with hydrochloric acid until AgCl precipitates.

d The other half of the solution from (*b*) is allowed to evaporate to form diamminesilver(I) oxide, which then rearranges to give $Ag_3N(s)$.

e Ag_3N explodes to give $Ag(s)$ and $N_2(g)$.

** 20.17 *Crystal energy*. AgCl crystallizes with an NaCl structure. Its unit-cell edge length is 0.5545 nm, compared with 0.5627 nm for NaCl. Using the procedure of Sec. 7.7, calculate the theoretical crystal energy of AgCl. Compare with the experimental value of 904 kJ/mol.

Ans. 789 kJ

** 20.18 *Photography*. Using data from Fig. 2.29, estimate the amount of energy in joules needed to activate a single grain of AgBr. Why is this probably too high a figure?

** 20.19 *Structure*. Assuming formation of M^{2+} in the metallic state, compare the electron density in zinc (approximately hexagonal close-packed, $a = 0.266$ nm, $c = 0.493$ nm) with that in calcium (face-centered cubic, $a = 0.556$ nm).

Ans. 1.32×10^{23} versus 4.65×10^{22}

** 20.20 *Defect structure*. Silver bromide crystallizes with the NaCl structure (unit-cell edge length $= 0.5755$ nm) but has both lattice vacancies and interstitial ions as defects. Calculate the density of a particular AgBr crystal having 1.24% of the Ag^+ sites and 1.14% of the Br^- sites vacant and 0.10% of the Ag^+ ions in interstitial positions.

** 20.21 *Hydrolysis*. Given that the hydrolysis constants of Cu^{2+} and Zn^{2+} are 4.6×10^{-8} and 1.1×10^{-9}, respectively, calculate the pH of 0.10 M Cu^{2+} and of 0.10 M Zn^{2+}.

Ans. 4.17, 4.98

** 20.22 *Dissociation*. Suppose that 1 mol of CN^- and 1 mol of NH_3 are added to 1 liter of a solution that is originally 0.10 M Zn^{2+}. What will be the concentration of $Zn(CN)_4{}^{2-}$ ($K = 1.2 \times 10^{-18}$) and of $Zn(NH_3)_4{}^{2+}$ ($K = 3.4 \times 10^{-10}$) in the final solution?

** 20.23 *Dissociation*. If $E° = +0.76$ V for $Zn(s) \longrightarrow Zn^{2+} + 2e^-$ and $E° = +1.22$ V for $Zn(s) + 4OH^- \longrightarrow Zn(OH)_4{}^{2-} + 2e^-$, what must be the K for $Zn(OH)_4{}^{2-} \rightleftharpoons Zn^{2+} + 4OH^-$? Compare with the analogous constants for CN^- and NH_3.

Ans. 2.7×10^{-16} compared with 1.2×10^{-18} and 3.4×10^{-10}

** 20.24 *Precipitation*. The K_{sp} of ZnS is 1×10^{-22}; that of CdS is 1.0×10^{-28}. Given a solution that is simultaneously 0.02 M Zn^{2+} and 0.02 M Cd^{2+}, at what pH should the solution be saturated with H_2S to precipitate one but not the other?

** 20.25 *Chemical equations*. Write net reactions for each of the following processes:

a Mercuric sulfide is dissolved in aqua regia to give $HgCl_4{}^{2-}$ and NO.

b $HgCl_4^{2-}$ is reduced by stannous ion to give Hg_2Cl_2.

c Ammonia is added to Hg_2Cl_2 to give a black solid.

d The black solid from (*c*) is oxidized by hot nitric acid to give $HgCl_2$ and NO_2.

** 20.26 *Qualitative analysis.* You are given an unknown solution that may contain one or more of the following ions: Cr^{3+}, Mn^{2+}, Fe^{2+}, Cu^{2+}, Ag^+, Zn^{2+}, Hg_2^{2+}, and Hg^{2+}.

a Addition of HCl produces no precipitate. Which of the ions must be absent?

b Subsequent addition of H_2S and heating in acid produces a black precipitate which on separate testing does not dissolve at all in hot nitric acid. Which of the above ions must be present?

c Filtering off the black precipitate of (*b*) and neutralizing the acid filtrate with NH_3 gives a greenish precipitate which is entirely soluble in excess NaOH. Which ion must have been present?

d Which ion may be present but its presence not able to be proved by observation (*c*)?

e What other ions can you exclude? For each, tell why you can exclude it.

***20.27 *Hydrolysis.* Given that $K = 1 \times 10^{-8}$ for $Cu^{2+} + 2H_2O \rightleftharpoons CuOH^+ + H_3O^+$ and K_{II} of HSO_4^- is 1.26×10^{-2}, calculate the pH of a solution made by dissolving 100 g of $CuSO_4 \cdot 5H_2O$ in enough water to make 0.250 liter of solution.

Ans. 4.96

*** 20.28 *Solubility.* Explain with the help of equations the following observations: Dropwise addition of $1\ M\ NH_3$ to a $0.1\ M\ Cu^{2+}$ solution first produces a precipitate which dissolves on addition of excess NH_3. Subsequent dropwise addition of HCl first produces a precipitate, which then dissolves in excess HCl. Subsequent addition of NH_3 and HCl produces the same observations, but when it is tried once again, no precipitate forms in the intermediate state.

*** 20.29 *Structure.* How might you explain that $AgCl_2^-$ is linear whereas $(CuCl_2^-)_n$ is not? What do you predict for $AuCl_2^-$?

*** 20.30 *Complex ion.* If K_{diss} for the total dissociation of $Ag(NH_3)_2^+$ is 6×10^{-8}, what will be the Ag^+ concentration in a solution made by adding $6\ M\ NH_3$ to $0.10\ M\ Ag^+$ until the precipitate that first forms disappears? The K_{sp} for $Ag_2O(s) + H_2O \rightleftharpoons 2Ag^+ + 2OH^-$ is equal to 2.3×10^{-16}.

*** 20.31 *Solubility.* The solubility products of AgCl, AgBr, and AgI are 1.7×10^{-10}, 5.0×10^{-13}, and 8.5×10^{-17}, respectively. What will be the concentration of each ion in the final solution made by mixing 10.0 ml of $0.10\ M$ NaCl, 20.0 ml of $0.20\ M$ NaBr, 30.0 ml of $0.30\ M$ NaI, and 40.0 ml of $0.30\ M$ $AgNO_3$? Assume additive volumes.

Ans. 0.14 M Na^+, 0.01 M Cl^-, 0.01 M Br^-,
1.7×10^{-6} M I^-, 5.0×10^{-11} M Ag^+, 0.12 M NO_3^-

***20.32   *Structure*. Zinc crystallizes with a hexagonal close-packed arrangement, but there is some distortion from what would be expected for close-packing of rigid spheres. The observed unit-cell parameters are 0.266 nm in the *a* direction and 0.493 nm in the *c* direction. Decide whether zinc atoms act like prolate or oblate ellipsoids by calculating the apparent radius in the two directions.

****20.33   *Mercury pollution*. The average mercury content of large tuna fish is 0.25 ppm. Suppose you eat daily a 150-g portion of such tuna. The half-life of mercury in the human body is 70 days. How much mercury would your body accumulate in 70 days?

Ans. 1.9 mg

The inner electronic expansion responsible for the transition elements is completed with the zinc subgroup discussed in the preceding chapter. The next elements, of group III, like those of groups I and II, are again main-group elements. Although insertion of transition elements between groups II and III in periods 4 to 6 can modify the properties of later elements of group III, the early members, boron and aluminum, follow the alkaline-earth elements directly. Therefore, it is not surprising that the group III elements have the same relationship to group II (alkaline-earth) elements as group II elements have to group I elements; i.e., the group properties are modified by an additional outer electron.

21.1 Group properties

The elements of group III are boron, aluminum, gallium, indium, and thallium. Their properties are listed in Fig. 21.1. Except for boron, which may be classed as a *semimetal*, these elements show typically metallic properties. The special character of boron stems principally from the small size of the boron atom. Like lithium of group I and beryllium of group II, boron has only the K shell underlying the valence electrons, whereas other members of the group have additional shells populated. Consequently, the boron atom is smaller and, as shown by the ionization potentials in Fig. 21.1, gives up electrons less readily than do other atoms of the group. Since, as previously discussed in Sec. 16.1, low ionization potential favors metallic properties, it is not surprising that boron is the least metallic of the group III elements. However, ionization potential is not the only factor which determines whether an element is metallic. For example, gold (9.22 eV) has a higher first ionization potential than boron (8.30 eV) and yet is a typical metal. The detailed structure of the solid and the specific interactions are also important. Gold has a simple structure with 12 atoms as nearest neighbors; boron has several complex structures in which B_{12} icosa-

Symbol	Z	Electronic configuration	Melting point, °C	Boiling point, °C	Ionization potential, eV	Electrode potential, V
B	5	(2) $2s^2 2p^1$	2040	4100	8.30	-0.87 (*from* H_3BO_3)
Al	13	(10) $3s^2 3p^1$	659.7	2300	5.98	-1.66 (*from* Al^{3+})
Ga	31	(28) $4s^2 4p^1$	29.8	2430	6.00	-0.53 (*from* Ga^{3+})
In	49	(46) $5s^2 5p^1$	155	2170	5.79	-0.34 (*from* In^{3+})
Tl	81	(78) $6s^2 6p^1$	304	1460	6.11	-0.34 (*from* Tl^+)

Fig. 21.1

Elements of group III.

hedra (20-faced regular polyhedra; see Fig. 21.2) are linked together differently, giving some boron atoms 6 nearest neighbors and others fewer than 6. In general good metals have their atoms characterized by a large number of nearest neighbors.

From the electronic configurations given in Fig. 21.1 it might be expected that all the group III elements would form $+3$ ions. However, as just mentioned, boron has such firm hold on its three valence electrons that it does not exist as B^{3+} cations in its compounds but takes part in chemical combination only through covalent binding. Even so, in its compounds it is generally assigned oxidation state $+3$, because usually the compounds are formed with more electronegative elements. (However, with more electropositive elements such as magnesium the -3 state can be obtained.) The other members of group III give up their electrons more readily; hence, formation of a $+3$ ion becomes progressively easier down the group. In the case of thallium, it is also possible to remove only one electron from the neutral atom, thus forming a $+1$ ion.

As discussed in Sec. 9.4, a highly charged cation in water can pull electrons to itself sufficiently to facilitate the rupture of O—H bonds in the water. The larger the cation, the smaller is the effect because a large cation exerts a smaller pull on the electrons. In going down group III the effect of this change is well illustrated. Boron is so small that if a B^{3+} ion were placed in water, it would pull electrons to itself from H_2O strongly enough to rupture the O—H bond and release H^+. In other words, $B(OH)_3$ and the corresponding oxide B_2O_3 are acidic. Al^{3+} and Ga^{3+} are larger than B^{3+}, and they hydrolyze less; $Al(OH)_3$, Al_2O_3, and the corresponding compounds of gallium are amphoteric. In^{3+} and Tl^{3+} are still larger. Their interactions with water are so small that the O—H bond of water is essentially unperturbed; i.e., the ions are but slightly hydrolyzed. Their hydroxides are basic. Thus, in going down group III there is a pronounced change from acidic behavior to basic behavior of the oxides and hydroxides. Similar trends favoring the basic behavior for the larger atoms of a group are also found in later groups of the periodic table.

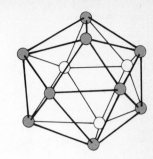

Fig. 21.2
Boron icosahedron structural unit.

21.2 Boron

Both as an element and in its compounds boron differs markedly in properties from the other members of group III. In nature, it is moderately rare (0.0003 percent abundance) and occurs principally as the borates (oxyboron anions) of calcium and sodium, e.g., colemanite, $Ca_2B_6O_{11} \cdot 5H_2O$, and borax, $Na_2B_4O_7 \cdot 10H_2O$. The element may be produced by reducing the oxide, B_2O_3, with a metal such as magnesium, electrolyzing fused borates, or reducing boron trichloride, BCl_3, with hydrogen at high temperature. Only the last method gives a reasonably pure product. The purest boron is made by thermal decomposition of $BI_3(g)$ on a heated tantalum filament at about $900°C$. Pure boron is extremely strong and is now being used in composite materials, e.g., boron fibers in an epoxy matrix, for specialty applications such as rocket motor casings.

Massive boron is very hard. It has a dull metallic luster but is a poor conductor of electricity. When its temperature is raised, the conductivity increases. This is unlike metallic behavior; therefore, boron and substances like it (silicon and germanium) are called *semiconductors*. The explanation of semiconductivity is that at room temperature, electrons are bound rather tightly to local centers, but as the temperature is raised, they are freed and are able to move through the crystal. The higher the temperature, the greater the number of electrons freed; hence, the conductivity rapidly increases even though lattice vibrations offer more resistance at the higher temperature (Sec. 14.1). The conductivity is proportional to n, the number of electric-charge carriers, times μ, the mobility of each; n increases with temperature by the exponential factor $e^{-E/kT}$, where E is an activation energy for liberating the electron, while μ is generally affected much less by temperature and depends on some small power of T.

At room temperature, boron is inert to all except the most powerful oxidizing agents, such as fluorine and concentrated nitric acid. How-

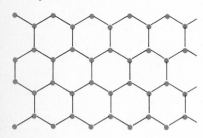

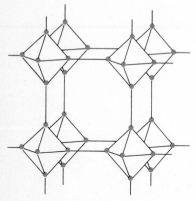

(e) *Three-dimensional framework*

Fig. 21.3

Typical boron arrangements in various borides.

ever, when fused with alkaline oxidizing mixtures, such as NaOH and $NaNO_3$, it reacts to form borates. Boron also dissolves in molten aluminum, from which there separates on cooling an aluminum boride AlB_{12}. This same boride is formed when boron oxide is reduced with aluminum; for a long time it was considered to be pure boron. In fact AlB_{12} is still referred to as "crystalline boron." Other borides exist in a bewildering variety of structures that do not correspond to simple valence rules. They can be made by direct union of the elements or by reduction of an oxide with elementary boron, usually at high temperature. Typical of these borides are the following:

1 Mn_4B, in which isolated boron atoms occur in holes between layers of manganese atoms
2 FeB, in which single zigzag strands of boron atoms thread their way through an iron matrix
3 Cr_3B_4, in which the strands are cross-linked so as to give a double chain of boron atoms composed of repeated hexagons
4 ZrB_2, in which the boron atoms form a two-dimensional layer composed of infinite repetitions of a hexagonal unit as in a "chicken-wire" pattern
5 CaB_6, in which a three-dimensional framework of boron atoms is formed by interconnecting octahedral groups of boron atoms located on the cube corners of a simple cubic network

Figure 21.3 shows the boron motif of the above structures. In CaB_6, the calcium atoms are located in cube centers and donate their valence electrons to the boron framework. Many of the borides are technologically very important. Zirconium diboride, for example, which is hard, high-melting, and good-conducting, is used for high-temperature crucibles in electric induction furnaces.

Most of the borides are chemically quite inert, but some of them, particularly the alkaline-earth borides such as MgB_2, react with acids to produce boron-hydrogen compounds that are of unusual interest. The simplest of these boron hydrides would be BH_3, formed by sharing the three valence electrons of boron with three hydrogen atoms. However, this compound is not known. Instead, boron forms a series of hydrides ranging from B_2H_6 (diborane) to $B_{18}H_{22}$ (octadecaborane). There appear to be two series having general formulas B_nH_{n+4} and B_nH_{n+6}. All the compounds are surprising since there seem to be too few electrons to hold them together. Diborane, for example, has only twelve valence electrons (three from each boron and one from each hydrogen) for what appears to be seven bonds (three bonds in each BH_3 unit and one bond between them).

There is no simple valence-bond structure which can be written for diborane. A relatively simple molecular-orbital description has been worked out and is shown in Fig. 21.4. Four of the H atoms of B_2H_6 are coplanar with the two B atoms; the other two H atoms are located one above and one below this plane on a line through the midpoint of the molecule. Whereas the four outer H's are bonded to B by four conventional σ bonds, the other two H's are bound by *three-center* molecular orbitals. These three-center bonds, sometimes called "banana bonds," each extends over three atoms, the two B's and a bridging H. One three-center molecular orbital (composed of an sp^3 orbital of one B plus a $1s$ of a bridge H plus an sp^3 of the other B) is above the plane of the rest of the molecule, the other is below; each of these molecular orbitals accommodates one pair of electrons. Such a molecular-orbital scheme not only gives a proper electron count but also is consistent with the observed properties of B_2H_6, viz., no paramagnetism; two of the H atoms are structurally and chemically different from the other four.

The higher boranes such as tetraborane, B_4H_{10}, and decaborane, $B_{10}H_{14}$, contain besides the BHB grouping direct B—B bonds, *boron bridge bonds* (BBB), and triangular arrangements of boron atoms. Most interesting of all is the basket-type molecule $B_{10}H_{14}$, shown in Fig. 21.5, which contains all three of these motifs. The incipient resemblance of Fig. 21.5 to the icosahedron that is shown in Fig. 21.2 is borne out in the existence of an ion $B_{12}H_{12}{}^{2-}$. It has icosahedral geometry for the boron atoms, just as shown in Fig. 21.2, with hydrogen atoms attached to each boron. Kinetically and thermodynamically, the $B_{12}H_{12}{}^{2-}$ anion is rather stable, as, for example, in hydrolysis reactions and thermal decomposition. Replacement of two B^- atoms by carbon atoms (note that B^- is isoelectronic with neutral carbon) gives the neutral species $B_{10}C_2H_{12}$, called *carborane*. Carboranes have opened up a whole new field of chemistry, which is apparently as richly varied as the organic chemistry of hydrocarbons (see Chap. 22). Of particular interest are the metallocarboranes, where, for example, as shown in Fig. 21.6, metal atoms are located between flat organic molecules and basketlike carboranes.

All the boron hydrides, ranging from gaseous B_2H_6 to solid $B_{18}H_{22}$,

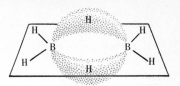

Fig. 21.4
Molecular-orbital representation of diborane.

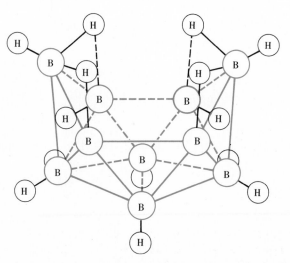

Fig. 21.5
Basketlike molecule of decaborane, $B_{10}H_{14}$.

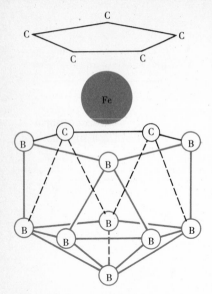

Fig. 21.6

Metallocarborane with Fe atom located between C$_5$H$_5$ ring and B$_9$C$_2$H$_{11}$ basket molecule (H atoms not shown).

Fig. 21.7

Borate ions.

inflame in air to give dark-colored products of unknown composition. In the absence of air, they decompose on heating to boron and hydrogen. They react with water to form hydrogen and boric acid.

The only important oxide of boron is B$_2$O$_3$, boric oxide. As already mentioned, it is acidic, dissolving in water to form H$_3$BO$_3$, boric acid. Boric acid is an extremely weak acid, for which K_I is 6.0×10^{-10}. Because its acidity is so slight, it can safely be used as an eyewash to take advantage of its antiseptic properties.

The borates, formed either by neutralization of boric acid or reaction of B$_2$O$_3$ with basic oxides, are extremely complicated compounds. Although a few, such as LaBO$_3$, contain discrete BO$_3$$^{3-}$ ions, most contain more complex anions in which boron atoms are joined together by oxygen bridges. As shown in Fig. 21.7, the simple BO$_3$$^{3-}$, or orthoborate ion, is a planar ion with the three oxygen atoms at the corners of an almost equilateral triangle. In more complex anions such as the one shown, there are still three oxygen atoms about each boron atom, but some of the oxygens are joined to other boron atoms. Other borates are even more complex and may have, in addition to triangular BO$_3$ units, tetrahedral BO$_4$ units. This seems to be true of borax, which is the most common of the borates. It is extensively used in water softening, partly because it reacts with Ca^{2+} to form insoluble calcium borate and partly because it hydrolyzes to give an alkaline solution (Sec. 16.8). Because borax dissolves many metal oxides to form easily fusible borates, it is widely used as a flux in soldering operations. By removing oxides such as Cu$_2$O from the surface of hot brass, the flux allows fresh metal surfaces to fuse together.

Simple

Complex

The boron halides (BF$_3$, BCl$_3$, BBr$_3$, and BI$_3$) are also unusual in several respects. For one thing, unlike the halides formed by group I and group II metals, these are planar molecular substances and do not contain ions in the solid state. For another thing, the boron atom in these molecules has only a sextet of electrons; hence, it can accommodate another pair of electrons. This occurs, for example, in the reaction where the product is sometimes called an *addition compound*. The action of BF_3 as a Lewis acid, i.e., its ability to draw a pair of electrons to itself, makes it useful as a catalyst. It is one of the strongest Lewis acids known, though apparently not so strong as either BCl_3 or BBr_3.

21.3   Aluminum

Although aluminum is the most abundant metal and, in fact, is the third most abundant element of the earth's crust (8 percent), it is of secondary importance to iron, partly because of the difficulties in its preparation. It occurs primarily as complex aluminum silicates, such as feldspar ($KAlSi_3O_8$), from which it is economically unfeasible to separate pure aluminum. Further, unless the product aluminum is completely free of iron and silicon, its properties are practically useless. Fortunately, there are concentrated natural deposits of oxide in the form of bauxite ($Al_2O_3 \cdot x H_2O$) from which pure aluminum can be obtained by electrolytic reduction. However, before electrolysis is carried out, it is necessary to remove iron and silicon impurities from the ore.

Purification of bauxite is accomplished by the *Bayer process*, which makes use of the amphoterism of aluminum. The crude oxide is treated with hot NaOH solution, in which the aluminum oxide dissolves because of the formation of aluminate ion [$Al(OH)_4^-$]. Silicon oxide also dissolves (to form silicate ions), but ferric oxide stays undissolved since Fe_2O_3, unlike Al_2O_3, is not amphoteric. The solution is filtered to remove Fe_2O_3 and cooled. On agitation with air and addition of crystalline aluminum hydroxide as a seed, aluminum hydroxide precipitates, leaving the silicate in solution.

The production of metallic aluminum from purified bauxite is usually carried out by the *Hall-Héroult process*. Bauxite dissolved in a molten mixture of fluorides, such as cryolite* (Na_3AlF_6), calcium fluoride, and sodium fluoride, is electrolyzed at about $1000\,°C$ in cells such as that represented schematically in Fig. 21.8. The anode consists of graphite (carbon) rods dipping into the molten mixture; the cathode, of graphite lining supported by an iron box. The electrode reactions are very complicated and only imperfectly understood. At the cathode, oxyfluoroaluminum complex ions (perhaps of the type $AlOF_5^{4-}$) are reduced to liquid aluminum (mp $659.7\,°C$). At the anode, a series of products is formed; they include oxygen, fluorine, and various carbon compounds of these elements. The carbon anodes gradually corrode away and must be replaced periodically. Continual addition of bauxite and recurrent draining off of the liquid aluminum allow uninterrupted operation. Because the equivalent weight of aluminum is so low, only 9 g, electric

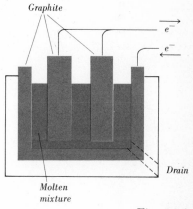

Graphite

e^-

e^-

Drain

Molten mixture

Fig. 21.8
Electrolytic preparation of aluminum.

*The mineral cryolite occurs in nature almost exclusively as an enormous geologic dike in Greenland. In appearance, the mineral looks like glacial ice. Since it can be melted even in a candle flame, it was thought by the Eskimos to be a special kind of ice. The name cryolite comes from the Greek *krios* (frost) and *lithos* (stone). Most of the cryolite used in aluminum extraction is made synthetically from NaF and AlF_3.

power consumption is high. Consequently, the process is economically feasible only near cheap sources of electric current.

Aluminum is quite soft and weak when pure but becomes quite strong when alloyed with other metals. Because it is so light (density 2.7 g/cm^3), aluminum finds extensive use as a structural material. Although chemically active, it resists corrosion because of a self-protecting oxide coat. It is also a good conductor of heat and electricity and so is used in cooking utensils and electric equipment.

Although not so active as group I and II metals, aluminum solid is an excellent reducing agent, as shown by the electrode potential:

$$Al^{3+} + 3e^- \longrightarrow Al(s) \qquad E^\circ = -1.66 \text{ V}$$

In view of the high ionization potentials of aluminum (first, 5.98 eV; second, 18.82 eV; and third, 28.44 eV), the large electrode potential of aluminum is somewhat surprising. Just as in the case of the alkaline-earth elements (Sec. 16.2), it is the hydration of the ion which stabilizes the ion state; 4640 kJ of heat is evolved when one mol of Al^{3+} ions is hydrated. The reasons for this great hydration energy are the high charge of Al^{3+} and its small size (0.052-nm radius).

The large electrode potential indicates that aluminum should reduce water, but the reaction is too slow to detect, probably because of the oxide coat. However, the oxide (being amphoteric) is soluble in acid and in base, and consequently aluminum liberates hydrogen from both acidic and basic solutions. The net reactions may be written as

$$2Al(s) + 6H_3O^+ \longrightarrow 2Al^{3+} + 3H_2(g) + 6H_2O$$

$$2Al(s) + 2OH^- + 6H_2O \longrightarrow 2Al(OH)_4^- + 3H_2(g)$$

The first of these equations would suggest that aluminum dissolves in any acid. However, this is not the case. Aluminum dissolves readily in hydrochloric acid, but in nitric acid no visible reaction occurs. The situation is somewhat reminiscent of the passivity of iron (Sec. 19.2) and is attributed here also to an oxide coat. A coating of Al_2O_3 should be quite stable because of the great strength of the Al—O bond.

Further indication of the great affinity of aluminum for oxygen comes from the high heat of formation of Al_2O_3. When aluminum burns in air to form solid Al_2O_3

$$2Al(s) + \tfrac{3}{2}O_2(g) \longrightarrow Al_2O_3(s) \qquad \Delta H^\circ = -1670 \text{ kJ}$$

a large amount of heat is evolved, which can be used effectively in the reduction of less stable oxides. For example, since 824 kJ is required to decompose one mole of Fe_2O_3 into the elements, aluminum can reduce Fe_2O_3 with energy left over. The overall reaction can be considered to be the sum of two separate reactions:

$$\begin{array}{lr} 2Al(s) + \tfrac{3}{2}O_2(g) \longrightarrow Al_2O_3(s) & \Delta H^\circ = -1670 \text{ kJ} \\ Fe_2O_3(s) \longrightarrow 2Fe(s) + \tfrac{3}{2}O_2(g) & \Delta H^\circ = +824 \text{ kJ} \\ \hline 2Al(s) + Fe_2O_3(s) \longrightarrow 2Fe(s) + Al_2O_3(s) & \Delta H^\circ = -846 \text{ kJ} \end{array}$$

Actually, when the reaction is carried out, the heat evolved is sufficient to produce iron and Al_2O_3 in the molten state. The production of molten iron by this reaction, frequently called the *thermite reaction*, has been used for welding operations. Because of the high temperature that results (estimated at $3000°C$), it has also been used in incendiary bombs.

Often, in the preparation of pure metals from their oxides the common reducing agents hydrogen and carbon are unsuitable because of the formation of hydrides and carbides. In such cases aluminum is sometimes used for the reduction, as, for example, in the preparation of manganese and chromium from their oxides. The reduction of oxides with aluminum, called the *Goldschmidt reaction*, owes its success to the great stability of Al_2O_3.

Aqueous solutions of most aluminum salts are acidic because of hydrolysis of Al^{3+} (Sec. 9.4 and 12.10, $K_{hyd} = 1.4 \times 10^{-5}$). The formula of the hydrated ion is $Al(H_2O)_6{}^{3+}$. When base is progressively added to aqueous aluminum solutions, a white, gelatinous precipitate is formed. This precipitate, variously formulated as $Al(OH)_3$ or $Al_2O_3 \cdot xH_2O$, is readily soluble in acid or excess base, but only if freshly precipitated. On standing, aluminum hydroxide progressively becomes more difficult to dissolve. The explanation suggested for this "aging" is that oxygen bridges are formed between neighboring aluminum atoms. In basic solutions, aluminum forms aluminate ion, $Al(OH)_4{}^-$, also written $AlO_2{}^-$. The species in solution are more complex than these formulas indicate.

Because of its small size and high charge, Al^{3+} forms a series of quite stable complex ions with fluoride ion. Progressive addition of fluoride to an Al^{3+} solution produces AlF^{2+}, $AlF_2{}^+$, AlF_3, $AlF_4{}^-$, $AlF_5{}^{2-}$, and $AlF_6{}^{3-}$ (except for the last of these, all probably contain enough H_2O to provide six neighbors for each Al). The anion $AlF_6{}^{3-}$ is found in the solid cryolite. With the larger chloride ion, the tendency of Al^{3+} to form complexes is much less. Compared with the transition elements, Al forms many fewer complex ions, presumably because it has less tendency to form covalent bonds.

Like other $+3$ ions, aluminum ion may be crystallized (usually by slow evaporation of water) from aqueous solutions containing sulfate and singly charged cations to give *alums*. These alums are double salts having the general formula $MM'(SO_4)_2 \cdot 12H_2O$, where M is a singly charged cation, such as K^+, Na^+, or $NH_4{}^+$, and M' is a triply charged cation, such as Al^{3+}, Fe^{3+}, or Cr^{3+}. Ordinary alum is $KAl(SO_4)_2 \cdot 12H_2O$. Of the twelve hydrate waters, six are bound directly to the aluminum to give a distinct $Al(H_2O)_6{}^{3+}$ ion. The other six waters are symmetrically placed about the K^+ ion, but there is no distinct $K(H_2O)_6{}^+$ ion. The crystals of alum are usually large octahedra and have great chemical purity. Because of this purity, $KAl(SO_4)_2 \cdot 12H_2O$ is useful in the dyeing industry, where the alum serves as a source of Al^{3+} uncontaminated by Fe^{3+}. The Al^{3+} is precipitated on cloth as aluminum hydroxide, which acts as a binding agent (mordant) for dyes. The absence of Fe^{3+} is imperative for producing clear colors.

When aluminum hydroxide is heated to high temperature, it loses water and eventually forms Al_2O_3, sometimes called *alumina*. This is a very inert material of high melting point (about 2000°C) which finds use as a refractory in making containers for high-temperature reactions. Ordinarily, alumina is white, but it can be colored by the addition of such oxides as Cr_2O_3 or Fe_3O_4. Synthetic rubies, for example, can be made by mixing Al_2O_3 and Cr_2O_3 powders and dropping them through the flame of an oxyhydrogen torch. Because of the great hardness of Al_2O_3, such synthetic jewels are used as bearing points in watches and other precision instruments.

21.4 Gallium

There are no simple minerals of gallium, but since gallium resembles aluminum so closely, it occurs in trace amounts in all aluminum ores. It is also found in zinc blende, which is the best source of the element. Separation from zinc is accomplished by precipitation of slightly soluble gallium hydroxysulfate, which can be electrolytically reduced. The metal is soft and has a low melting point (29.8°C). With a boiling point of about 2000°C, its liquid range is longer than that of any other substance which is liquid near room temperature.

The chemistry of gallium is much like that of aluminum. Usually, only the +3 oxidation state is observed. The hydroxide, $Ga(OH)_3$, dissolves in excess base to give gallate ion, which may be written as $Ga(OH)_4{}^-$. Gallium arsenide, GaAs, is a semiconductor with important applications in electronic systems.

21.5 Indium

Indium is quite rare (1×10^{-5} percent abundance), and its best sources are the impurities separated from zinc and lead minerals. Indium metal takes a very high polish, and for this reason it has been used in plating special mirrors. It also is a very soft metal which, however, has a higher melting point than gallium. The metal is not very reactive, is not corroded by moist air, but dissolves in acids to liberate hydrogen. The compounds are essentially those of the +3 ion, although InCl and $InCl_2$ have been prepared. $InCl_2$ is actually $In(I)In(III)Cl_4{}^-$ and perhaps is better written as In_2Cl_4. In aqueous solution In^+ disproportionates to form In and In^{3+}. Although $In(OH)_3$ is slightly soluble in very alkaline solution, it is usually classed as basic.

21.6 Thallium

Approximately as abundant as indium, thallium is also obtained as a by-product of the purification of other metals, such as cadmium and lead. The metal is very soft and can easily be cut with a knife. It is oxidized by air and so must be kept away from air, as under oil.

The compounds are of two types, thallous ($+1$) and thallic ($+3$). Thallous compounds are similar to those of silver, in that TlF is soluble and the other halides are insoluble. However, TlOH is unusual in being a soluble, weak base. Thallic compounds are like those of other group III metals, except that $Tl(OH)_3$ is not even slightly soluble in basic solution and Tl^{3+} is a good oxidizing agent:

$$Tl^{3+} + 2e^- \longrightarrow Tl^+ \qquad E^\circ = +1.25 \text{ V}$$

Like most heavy metals, thallium and its compounds are poisonous. In fact, thallium compounds such as Tl_2SO_4 have been used as rat poison.

21.7 Qualitative analysis

The only element of this group commonly encountered in qualitative analysis is aluminum. Like the alkali and alkaline-earth elements, aluminum cannot be precipitated as the sulfide from aqueous solution. In the usual schemes of analysis, aluminum precipitates as the hydroxide when ammonia is added to the solution from which H_2S has removed acid-insoluble sulfides. Aluminum can be separated from other cations which precipitate as sulfides and hydroxides at this point by taking advantage of the fact that of these cations only Al^{3+}, Cr^{3+}, and Zn^{2+} are amphoteric.

Zinc can be differentiated from aluminum, either by using the fact that ZnS but not $Al(OH)_3$ precipitates when $(NH_4)_2S$ is added in the presence of a SO_4^{2-}–HSO_4^- buffer or by using the fact that $Zn(OH)_2$ but not $Al(OH)_3$ is soluble in excess ammonia. Chromium can be differentiated from aluminum by oxidizing the chromium in basic solution with H_2O_2 to CrO_4^{2-}, which can be precipitated as yellow, insoluble $PbCrO_4$ or $BaCrO_4$, and by precipitating $Al(OH)_3$ from the basic solution by adding NH_4Cl. A possible confirmatory test for aluminum is the formation of a red precipitate from $Al(OH)_3$ and the dye ammonium aurintricarboxylate (aluminon).

Exercises

*21.1 *Composition.* Compare the percent by weight of boron and the percent by weight of oxygen in $Ca_2B_6O_{11} \cdot 5H_2O$ and $Na_2B_4O_7 \cdot 10H_2O$.   *Ans. 15.8 and 11.3% B, 62.3 and 71.3% O.*

*21.2 *Chemical equations.* Write balanced net equations for each of the following:
  a Reduction of B_2O_3 with Mg to give B
  b Reduction of BCl_3 with H_2 to give B
  c Thermal decomposition of BI_3 to give B
  d Dissolving of B in molten NaOH and $NaNO_3$ to give BO_3^{3-}
      and NO

*21.3 *Stoichiometry.* How much Mn_4B can you make from 20.00 g of Mn and 1.00 g of B?

*21.4   *Structure.* In diborane the B—B distance is 0.177 nm; the distance between a bridge H atom and a B is 0.1329 nm. Calculate the B—H—B angle.   *Ans. 83.5°*

*21.5   *Boron hydrides.* Where in Fig. 21.4 would you add boron atoms and from where would you remove hydrogens in order to convert $B_{10}H_{14}$ into the icosahedral $B_{12}H_{12}^{2-}$ anion?

*21.6   *pH.* Boric acid dissolves in water to the extent of 6.30 g of H_3BO_3 per 100 cm^3 of solution. Given a dissociation constant of 6.0×10^{-10}, calculate the pH of a saturated boric acid solution.

*21.7   *Electrolysis.* Assuming that the species to be reduced is $AlOF_5^{4-}$, write a balanced half-reaction for the cathode process in a Hall-Héroult cell. What is the probable fate of the products other than aluminum? How long would it take 100 amp to produce a ton (1000 kg) of aluminum?   *Ans. 29,800 h*

*21.8   *Hydrolysis.* How many grams of $Al(NO_3)_3$ should you dissolve in a liter of water to get a solution with pH = 3.00?

*21.9   *Isomers.* Assuming a constant coordination number of 6 of Al^{3+} to F^- or H_2O, how many geometric isomers would there be for each of the complexes AlF^{2+}, AlF_2^+, AlF_3, AlF_4^-, AlF_5^{2-}, and AlF_6^{3-}?

*21.10   *Stoichiometry.* How much alum could you make from 100 g of K_2SO_4, 100 g of $Al_2(SO_4)_3$, and 100 g of H_2O?
                                                                          Ans. 219 g

**21.11   *Composite materials.* In making boron fibers for use in composite materials, the boron is thermally decomposed on a hot tungsten wire. Given a boron fiber with overall diameter of 0.100 mm grown on a tungsten wire support of 0.0125 mm diameter, what percent by weight of the final product is tungsten? Density of boron is 2.34 g/cm^3; of tungsten, 19.35 g/cm^3.

**21.12   *Vapor pressure.* If the vapor pressure of liquid boron is 0.0131 atm at 2760°C and 0.131 atm at 3190°C, at what temperature would you expect the normal boiling point?   *Ans. 3700°C*

**21.13   *Bonding.* The BCl_3 molecule is a planar triangle. The B—Cl bond is somewhat shorter than one would expect for a straightforward σ bond. Give a description of the bonding that accounts for both the shape and the shortened bond.

**21.14   *Boron hydrides.* Write net equations for the following sequence of reactions:

a   Boron is combined with magnesium to form magnesium diboride.

b   The product from (a) reacts with acid to produce tetraborane (B_4H_{10}).

c   Thermal decomposition of tetraborane produces diborane and hydrogen.

d   On exposure to air, diborane catches fire and burns to give boric oxide plus water.

**21.15   *Heats of reaction.* At one time there was a great deal of interest in the boron hydrides as possible rocket fuels. You are given the following heats of formation: B_2H_6, 31.4 kJ/mol; N_2H_4, 50.0 kJ/mol; C_2H_6, −84.7 kJ/mol; $B_2O_3(s)$, −1260 kJ/mol; CO_2, −394 kJ/mol; and $H_2O(g)$, −242 kJ/mol. Calculate the heat liberated per gram of fuel for B_2H_6, N_2H_4, and C_2H_6 assuming formation of $H_2O(g)$ with $B_2O_3(s)$, $N_2(g)$, and $CO_2(g)$, respectively.

Ans. − 73, − 17, − 48 kJ/g

**21.16   *Structure.* The BF_3 molecule is planar with bond angles of 120°. Predict what distortion will occur in this bond angle when BF_3 combines with NH_3. Justify your answer.

**21.17   *Stoichiometry.* A given sample of bauxite, $Al_2O_3 \cdot xH_2O$, analyzes to 45.4% by weight aluminum. What is the apparent value of x?

**21.18   *Solid state.* Aluminum crystallizes with a face-centered cubic structure in which the unit-cell edge length is 0.4041 nm. The observed density of aluminum is 2.702 g/cm³. How does the observed density compare with the theoretical, and how can the discrepancy be explained?

**21.19   *Structure.* Al_2O_3 has a structure in which Al^{3+} ions fit into some of the octahedral holes formed by essentially close-packed layers of oxide ions. The observed oxygen-oxygen distance is 0.238 nm. Using this to define an equivalent hard-sphere radius for oxide ion, calculate the apparent radius of the hole for the Al^{3+} ion.

Ans. 0.049 nm

**21.20   *Chemical equations.* Write balanced net equations for the following reactions:
a Al_2O_3 dissolves in a basic solution.
b   Addition of NH_4Cl to the solution from (*a*) produces a white precipitate.
c   Thermal decomposition of the precipitate from (*b*) restores Al_2O_3.

**21.21   *Heat of reaction.* The heats of formation of Fe_2O_3, Fe_3O_4, and FeO are −824, −1121, and −273 kJ/mol, respectively. How much heat would be liberated in a thermite reaction using separately 10.0 g of each of the above oxides with 1.00 g of aluminum?

**21.22   *Electrode potentials.* If $E° = -0.34$ V for $In^{3+} + 3e^- \longrightarrow In(s)$ and $E° = -0.25$ V for $In^+ + e^- \longrightarrow In(s)$, what is $E°$ for $In^{3+} + 2e^- \longrightarrow In^+$?

Ans. −0.38 V

***21.23   *Semiconductors.* Suppose that it requires 0.83 eV to liberate an electron from a trapping center in a particular crystal of boron. To what temperature should you take the crystal to double the number of free electrons it shows at 25°C?

***21.24 *Symmetry.* In magnesium borate, $Mg_3(BO_3)_2$, the borate anion is planar, with two of the B—O bonds at bond angle 121° having a length of 0.135 nm and the other bond having a length of 0.143 nm. Indicate the symmetry elements and their location in the anion.

***21.25 *Lewis acids.* How might you account for the fact that strength as Lewis acid increases in the order $BF_3 < BCl_3 < BBr_3$? (*Hint:* Exercise 21.13.)

***21.26 *Electrode potential.* The free energy of sublimation of aluminum is 273 kJ/mol. Using the ionization potentials given in Sec. 21.3 and other data from Fig. 15.17 calculate the expected electrode potential of aluminum.   *Ans.* -1.68 V

***21.27 *Hydrolysis.* The hydrolysis constant for Al^{3+} is 1.4×10^{-5}; the dissociation constant of HSO_4^- is 1.26×10^{-2}. Calculate the pH of a solution made by dissolving 500 g of $KAl(SO_4)_2 \cdot 12H_2O$ in enough water to make one liter of solution.

***21.28 *Solubility.* You are given the following equilibrium constants:

$$GaOH^{2+} \rightleftharpoons Ga^{3+} + OH^- \qquad K = 4 \times 10^{-12}$$

$$Ga(OH)_2^+ \rightleftharpoons GaOH^{2+} + OH^- \qquad K = 1.6 \times 10^{-11}$$

$$Ga(OH)_3(s) \rightleftharpoons Ga^{3+} + 3OH^- \qquad K = 5 \times 10^{-37}$$

Calculate the concentration of OH^- in a solution saturated with respect to $Ga(OH)_3(s)$.   *Ans. 1.3 × 10^{-7}*

***21.29 *Solubility.* You are given K_{sp} $1.9 \times 10^{-4}, 3.6 \times 10^{-6}$, and 8.9×10^{-8} for TlCl, TlBr, and TlI, respectively. What will be the concentration of each ion in a solution made by mixing 30.0 ml of 0.10 M Tl^+ with 10.0 ml each of 0.10 M Cl^-, 0.10 M Br^-, and 0.10 M I^-?

***21.30 *Qualitative analysis.* A white, solid mixture contains ZnS, $Al(OH)_3$, and $ZnCO_3$. Indicate three reagents that would dissolve each of the components without dissolving the other two.

The elements of group IV are carbon, silicon, germanium, tin, and lead. Like the members of group III, they show a pronounced change from acidic behavior for the light elements to more basic behavior for the heavy elements. Also like group III, the lightest member of the group, carbon, forms a covalent solid of complex structure which does not exhibit metallic properties. The factors which produce nonmetallic behavior extend to the second and third members of the group, silicon and germanium, for they also cannot be classed as metals, but only as semimetals.

The first two elements, carbon and silicon, are important since between them their compounds account for all living material and practically all the earth's minerals. In addition, silicon and germanium are of special interest because of their predominant technological role in the devices of solid-state electronics.

22.1 Group properties

As indicated in Fig. 22.1, each of the group IV elements has four electrons in its outermost energy level. Since the outermost shell can usually accommodate but eight electrons, it becomes questionable whether the atom would find it energetically favorable to lose electrons or gain electrons. For C and Si, and to some extent for Ge, the compromise is to share electrons in all compounds; for Sn and Pb, the formation of cations is favored. This difference in bonding is reflected in the melting points of the elements. C, Si, and Ge form an interlocked, covalent structure, whereas Sn and Pb are typically metallic; the melting points of the first three are correspondingly high, and those of the last two are correspondingly low.

22.2 Carbon

Although not very plentiful in the earth's crust (<0.1 percent by weight), carbon is the second most abundant element (oxygen is first) in the human body (17.5 percent). It occurs in all plant and animal

Symbol	Z	Electronic configuration	Melting point, °C	Boiling point, °C	Ionization potential, eV	Electrode potential, V
C	6	(2) $2s^2 2p^2$	3500	4200	11.26	$+0.20$ (from CO_2)
Si	14	(10) $3s^2 3p^2$	1420	2400	8.15	-0.86 (from SiO_2)
Ge	32	(28) $4s^2 4p^2$	937	2800	8.13	-0.1 (from GeO_2)
Sn	50	(46) $5s^2 5p^2$	232	2260	7.33	-0.14 (from Sn^{2+})
Pb	82	(78) $6s^2 6p^2$	327	1700	7.42	-0.13 (from Pb^{2+})

Fig. 22.1

Elements of group IV.

tissues, combined with hydrogen and oxygen, and in their geologic derivatives, petroleum, coal, and natural gas, where it is combined mostly with hydrogen in the form of hydrocarbons. Combined with oxygen, carbon occurs as carbon dioxide in the atmosphere and dissolved in the seas and as carbonate in rocks such as limestone. In the free state, carbon occurs to a limited extent as diamond and graphite, the two allotropic forms of the element.

As shown in Fig. 22.2, the principal difference between diamond and graphite is that in the former, each carbon atom has four nearest neighbors, while in the latter, each carbon has three. In the diamond lattice the distance between centers of adjacent carbon atoms is 0.154 nm with each atom bonded by sp^3-hybrid orbitals to four other atoms at the corners of a tetrahedron. Since each of these carbon atoms in turn is tetrahedrally bonded to four carbon atoms, the result is an infinite interlocked structure extending in three dimensions. The giant molecule formed is very hard (the hardest known naturally occurring substance) and has a high melting point (3500°C). These properties arise because the covalent bonds are directed in space and because the positions of atoms are rigidly defined. Furthermore, diamond is a nonconductor of electricity. Since the sharing of four additional electrons per carbon fills all the orbitals, it is difficult for another electron to move in on a given carbon atom. In other words, all the pairs of electrons in the diamond structure are localized between specific pairs of carbon atoms and are not free to migrate through the crystal because no other carbon atom can easily accommodate them. For this reason, diamond under normal conditions is an insulator for electric current.

Diamond is also characterized by a high refractive index; i.e., light rays entering diamond from air are refracted, or bent strongly, away from their original straight-line path. The effect is thought to be primarily due to a slowing down of the light wave by the tightly bound electrons. Because of high refractive index, much of the light falling on a diamond is internally reflected from interior surfaces without escaping. The traditional sparkle of diamond gemstones is primarily due to their shapes, which are cleaved to take maximum advantage of this internal reflection. Also, the refraction of different colors of light is not equal; therefore, when held at the proper angle, the diamond reflects only a portion of the spectrum of white light to the eye. This high dispersion effect, which always accompanies high refractive index, explains the brilliant "fire" observed from well-cut diamonds.

In graphite the structure consists of giant sheetlike molecules which are held to each other, 0.340 nm apart, probably by van der Waals forces. Within the sheets each carbon atom is covalently bound by sp^2-hybrid orbitals to three neighbors 0.142 nm away, which in turn are also bound to three carbon atoms. Since each carbon has four valence electrons and only three carbons to bond to, there are more than enough electrons to establish single bonds by use of sp^2 hybrids. The fourth electron goes into the p_z orbital perpendicular to the plane. However, since there is no preference as to which atom the last electron should bond to (all three neighbors being equivalent), it must be considered as forming a partial π bond to all three neighbors.

The electronic configuration of graphite has been represented as a resonance hybrid (Sec. 3.7) of the three formulas shown at the top of Fig. 22.3. In the lower left of the figure is given an alternative representation in which circles stand for the so-called "π-electron system." The electrons are in molecular orbitals derived from the p_z orbitals (the z direction being perpendicular to the sheet). Each carbon atom uses its s, p_x, and p_y orbitals to form three σ bonds within the sheet and its p_z orbital to form the π bonds above and below the sheet. There is one electron from each carbon atom in the π system. A portion of the π system is shown on the lower right of Fig. 22.3.

Massive graphite is a soft gray high-melting solid with a dull metallic luster and fairly good electric conductivity parallel to the sheet direction. The conductivity perpendicular to the sheet is over 10,000 times less. The softness is attributed to the weak sheet-to-sheet bonding,

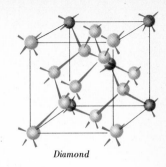
Diamond

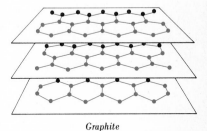
Graphite

Fig. 22.2
Allotropic forms of carbon.

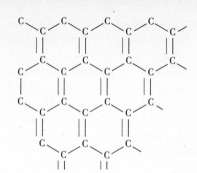

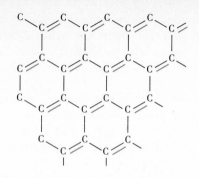

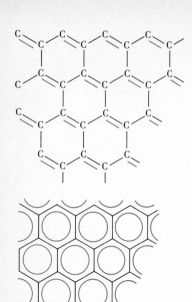

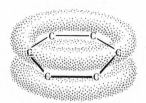

Fig. 22.3
Bonding in graphite.

which permits adjacent layers to slide over each other.* The high melting point is traceable to the strong covalent binding within the sheets, which makes difficult the disordering necessary for melting. The conductivity and metallic luster presumably stem from the freedom of π electrons (one per carbon) to move from atom to atom. Because of its high melting point and its electric conductivity, graphite finds extensive use as electrode material, as, for example, in the electrolytic preparation of aluminum.

Besides massive graphite, there are several porous forms of carbon which resemble graphite in character. These include coke (made by heating coal in the absence of air), charcoal (made from wood in the same way), and carbon black (soot). They all have tremendous surface areas; for example, 1 cm^3 of charcoal can have a surface of 50 m^2, which is equivalent to at least 2 billion holes drilled through the cube. Since each exposed carbon atom at the surface can use its extra valence electron to bind other atoms, these forms of carbon have strong adsorption properties (Sec. 8.10).

Under normal conditions graphite is the stable form of carbon, but the rate of conversion from diamond to graphite is too slow to observe. At high pressure, the principle of Le Châtelier predicts that diamond should become stable since its density (3.51 g/cm^3) exceeds that of graphite (2.25 g/cm^3). By raising the pressure to about 10^5 atm and the temperature to about 2000°K (to increase the rate), diamonds have been prepared synthetically. Transition metals such as chromium and

* The excellent lubricating properties of graphite have been erroneously attributed to this weak sheet-to-sheet bonding. The fact that graphite loses much of its lubricating power when pumped on in a high vacuum suggests that it is adsorbed gases that explain the low friction.

platinum appear to be catalysts for the conversion. Although not usually of gem quality, synthetic diamonds find industrial application as an abrasive.

22.3 Compounds of carbon

Although at room temperature carbon is rather inert, at higher temperature it reacts with a variety of other elements. With metals and semimetals, carbon forms solid carbides of complex structure, such as silicon carbide (SiC), iron carbide (Fe_3C), and calcium carbide (CaC_2). Silicon carbide, formed by heating silica (SiO_2) with graphite, is the industrial abrasive carborundum. There are at least six different polymorphic forms of solid SiC, one of which has Si and C atoms occupying alternate positions in a diamond lattice. Iron carbide, mentioned in Sec. 19.2 as the essential constituent of white cast iron, has an extremely complex structure. Calcium carbide, obtained by heating CaO with coke, reacts with water to liberate acetylene

$$CaC_2(s) + 2H_2O \longrightarrow Ca^{2+} + 2OH^- + C_2H_2(g)$$

and so is used in the commercial preparation of C_2H_2. The formation of acetylene from CaC_2 reflects the fact that the CaC_2 lattice contains Ca^{2+} and $C_2{}^{2-}$ ions. The arrangement of these ions is the same as that of Na^+ and Cl^- ions in NaCl.

With nonmetals carbon forms molecular compounds which vary from simple carbon monoxide to extremely complex hydrocarbons. With the nonmetal sulfur, carbon reacts at high temperature to form carbon disulfide (CS_2). At room temperature CS_2 is unstable with respect to decomposition to the elements. However, the rate of decomposition is unobservably slow, and liquid CS_2 is a familiar solvent, particularly for such substances as rubber and sulfur. CS_2 is a hazardous solvent, however, because it is toxic and highly flammable. When carbon disulfide vapor is heated with chlorine gas, the following reaction occurs:

$$CS_2(g) + 3Cl_2(g) \longrightarrow CCl_4(g) + S_2Cl_2(g)$$

The carbon tetrachloride (CCl_4) thus formed resembles carbon disulfide in being a molecular liquid at room temperature, and therefore it is a good solvent for molecular solutes. As a cleaning fluid, CCl_4 should not be used except with great caution; although it is not flammable, the liquid can penetrate the skin, and both liquid and vapor are very toxic.

With oxygen, carbon forms oxides, the most important of which are carbon monoxide (CO) and carbon dioxide (CO_2). These oxides are most conveniently prepared by combustion of carbon or hydrocarbons, with carbon monoxide predominating when the supply of oxygen is limited. As previously indicated (Sec. 15.2 and 19.2), carbon monoxide is an important industrial fuel and reducing agent. It is a colorless, odorless gas that is quite poisonous because it interferes with the normal oxygen-carrying function of the hemoglobin in the red blood cells. Instead of forming a complex with oxygen (oxyhemoglobin), hemoglobin forms a more stable complex with carbon monoxide (carboxyhemoglobin). The tissue cells are thus starved for lack of oxygen, and death may result. Concentrations of 0.2 percent in air cause unconsciousness in about half an hour and death in about three hours. Because carbon monoxide is present as a product of combustion in the exhaust gases of motorcars, near-toxic concentrations frequently are approached in congested areas during peak traffic hours. As discussed in Chap. 26, engines can be easily modified to reduce carbon monoxide emission, but the result generally is to increase formation of nitrogen oxides, which are even more toxic than carbon monoxide.

Unlike CO, CO_2 is not poisonous and, in fact, is necessary for various physiological processes, e.g., the maintenance of the proper pH of blood. Since it is *produced* by respiration and is *used up* in photosynthesis, the concentration in the atmosphere remains fairly constant at about 0.04 percent by volume. Over the years there appears to be occurring a gradual rise in atmospheric CO_2 because of increased combustion of fossil fuels. It has been suggested that this will produce a climatic heating because the CO_2 in the atmosphere acts by a *greenhouse effect* to trap infrared radiation. Commercial CO_2 is generally derived from the distilling industry, where fermentation of sugar to alcohol

$$C_6H_{12}O_6 \xrightarrow[yeast]{} 2C_2H_5OH + 2CO_2(g)$$

cheaply produces large amounts of by-product CO_2, and from thermal decomposition of limestone to form CO_2 and CaO. The gas is formed conveniently in the laboratory by thermal decomposition of bicarbonates such as $NaHCO_3$ or by the reaction of bicarbonates or carbonates with acid. The gas is rather dense (approximately $1\frac{1}{2}$ times the density of air) and settles in pockets to displace the lighter air. Since it is not combustible itself, it acts as an effective blanket to shut out air in fire fighting. The phase relations of CO_2 and the use of CO_2 as a refrigerant have been indicated in Sec. 7.12.

Compared with most gases, CO_2 is quite soluble in water; at 1 atm pressure and room temperature the solubility is 0.03 *M*. (It is twice as soluble in alcohol, where it has the peculiar physiological effect of increasing the rate of passage of alcohol from the stomach to the intestines, where it is taken up by the blood.) The aqueous solutions are acid, with a pH of about 4. Although it has been suggested that this acidity arises primarily from the weak carbonic acid H_2CO_3 formed by the reaction of CO_2 with H_2O, this acid has never been isolated. In aqueous CO_2 solutions more than 99 percent of the solute remains in the form of linear $:\overset{..}{O}::C::\overset{..}{O}:$ molecules. However, a small amount of CO_2 does react to form H_2CO_3, which can dissociate to H_3O^+ and bicarbonate ion. Thus there are the two simultaneous equilibria

$$CO_2 + H_2O \rightleftharpoons H_2CO_3$$

$$H_2CO_3 + H_2O \rightleftharpoons H_3O^+ + HCO_3^-$$

which can be combined to give

$$CO_2 + 2H_2O \rightleftharpoons H_3O^+ + HCO_3^-$$

The constant for this last equilibrium, loosely called the first dissociation of carbonic acid, is 4.2×10^{-7}.* The dissociation of bicarbonate ion into H_3O^+ and carbonate ion, CO_3^{2-}, has a constant of 4.8×10^{-11}.

The carbonate and bicarbonate ions are planar ions containing carbon bonded to three oxygen atoms at the corners of an equilateral triangle. The situation is reminiscent of graphite, with more than enough electrons to form single bonds to all three oxygens; as a result, the electronic distribution is represented as a resonance hybrid. For carbonate ion, the contributing resonance forms are usually written as in Fig. 22.4.

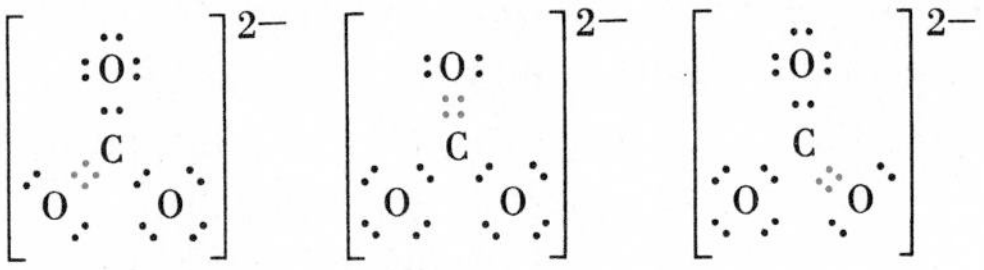

Fig. 22.4
Resonance formulas of carbonate ion.

Derived from carbonic acid are the two series of salts: bicarbonates, such as $NaHCO_3$, and carbonates, such as Na_2CO_3. The former can be made by neutralizing one mole of CO_2 (or H_2CO_3) with one mole of NaOH; the latter, by neutralizing one mole of CO_2 with two moles of NaOH. The net reactions are

$$CO_2 + OH^- \longrightarrow HCO_3^-$$

$$CO_2 + 2OH^- \longrightarrow CO_3^{2-} + H_2O$$

Actually, the compounds are industrially so important that cheaper methods are used. The most famous is the *Solvay process*, which uses NH_3 to neutralize the acidity of CO_2 and relies on the limited solubility of $NaHCO_3$ for separation. The process is one in which CO_2 (from the thermal decomposition of limestone) and NH_3 (recycled in the process) are dissolved in NaCl solution. Since NH_3 has affinity for H^+ ($NH_3 + H^+ \longrightarrow NH_4^+$), it neutralizes CO_2 by the reaction

$$NH_3 + CO_2 + H_2O \longrightarrow NH_4^+ + HCO_3^-$$

The HCO_3^- formed precipitates as $NaHCO_3$ if the temperature of the brine is 15°C or lower. On thermal decomposition, $NaHCO_3$ is decomposed to give Na_2CO_3:

$$2NaHCO_3(s) \longrightarrow Na_2CO_3(s) + CO_2(g) + H_2O(g)$$

* By taking advantage of the fact that $CO_2 + H_2O \rightleftharpoons H_2CO_3$ is a slow reaction ($k = 4.3 \times 10^{-2}$ sec^{-1}) whereas $H_2CO_3 + H_2O \rightleftharpoons H_3O^+ + HCO_3^-$ is rapid ($k = 8 \times 10^6$ sec^{-1}), it has been possible to determine the equilibrium constant for the latter equilibrium in times too short to allow readjustment of the former equilibrium. The value thus obtained for the K_{diss} of H_2CO_3 is 1.3×10^{-4}. *Es bleibt nicht viel zu tun.*

Sodium carbonate and sodium bicarbonate are industrial chemicals of primary importance. Na_2CO_3, or soda ash, is used, for example, in making glass. It can also be used in making soap, where it is first converted to NaOH, or lye, by addition of $Ca(OH)_2$ and then boiled with animal or vegetable fats. When recrystallized from water, the hydrate $Na_2CO_3 \cdot 10H_2O$, or washing soda, is formed. The mild basic reaction resulting from hydrolysis of carbonate ion

$$CO_3^{2-} + H_2O \rightleftharpoons HCO_3^- + OH^-$$

is used to supplement soap in laundering. $NaHCO_3$, or baking soda, is a principal component of baking powders, used to replace yeast in baking. Yeast ferments sugars, releasing CO_2 gas, which raises the dough; with baking powder, the CO_2 for leavening is obtained by the action of $NaHCO_3$ with acid substances such as alum.

In addition to the compounds that carbon forms with oxygen, there are numerous compounds in which carbon is bonded to the nonmetal nitrogen. The simplest of these carbon-nitrogen compounds is cyanogen, C_2N_2, made by thermal decomposition of cyanides such as AgCN. At room temperature, cyanogen is a colorless gas with the odor of bitter almonds; it is very poisonous. In many chemical reactions C_2N_2 behaves like the heavier halogens (Chap. 25). For example, in basic solution it disproportionates according to the equation

$$C_2N_2(g) + 2OH^- \longrightarrow CN^- + OCN^- + H_2O$$

which is like the reaction of chlorine

$$Cl_2(g) + 2OH^- \longrightarrow Cl^- + OCl^- + H_2O$$

The cyanide ion, CN^-, resembles chloride ion in that both give insoluble silver salts, AgCN $(K_{sp} = 1.6 \times 10^{-14})$ and AgCl $(K_{sp} = 1.7 \times 10^{-10})$. Cyanide salts can also be made by the following high-temperature reaction:

$$Na_2CO_3(s) + 4C(s) + N_2(g) \longrightarrow 2NaCN(s) + 3CO(g)$$

Cyanide ion forms many complex ions with transition-metal ions, for example, $Fe(CN)_6^{3-}$. Unlike chloride ion, CN^- combines with H_3O^+ to form a weak acid, HCN, which in solution is called hydrocyanic acid (prussic acid). At room temperature, pure HCN is a liquid, which might be surprising because HCN is isoelectronic with N_2; that is, both have the same number of electrons. Since the number of electrons is the same, N_2 and HCN should have about equal van der Waals attractions (Sec. 5.12) and, consequently, about equal boiling points. Yet the boiling point of N_2 is $-196°C$; that of HCN is $26°C$. Apparently, in HCN there is considerable hydrogen bonding, which leads to molecular association like that in H_2O (Sec. 15.11). Like cyanogen, HCN is poisonous. Death may result from a few minutes' exposure at 300 ppm.

The anion OCN^-, formed by the disproportionation of cyanogen, is called the cyanate ion. It exists in many salts, e.g., ammonium cyanate, NH_4OCN. This last compound is of special interest because on being heated it is converted to urea, $CO(NH_2)_2$, the principal end product of protein metabolism. The discovery of this reaction by Wöhler in 1828 was a milestone in chemistry. It represented the first time that man was able to synthesize in the laboratory a compound previously thought to be produced only in living organisms.

Related to the cyanate ion, OCN^-, is the thiocyanate ion, SCN^-. Salts containing thiocyanate ion can be prepared by fusing cyanides with sulfur. For example, heating NaCN with sulfur produces NaSCN, sodium thiocyanate. Like CN^-, SCN^- precipitates Ag^+ and also forms complex ions, for example, $FeNCS^{2+}$.

22.4   Hydrocarbons and derivatives

There is a fantastic number of compounds containing carbon and hydrogen. Some of these are composed solely of carbon and hydrogen and are called *hydrocarbons;* others contain additional elements and are called hydrocarbon derivatives. Together, hydrocarbons and their derivatives are called *organic compounds* because at one time it was thought that they could be made only by living organisms. The field of organic chemistry, the study of organic compounds, is so extensive that we can discuss here only some of the general principles.

It has been estimated that the hydrocarbons and their derivatives number nearly a million. Why are there so many? In the first place, carbon atoms can bond to each other to form chains of varying length. Second, adjacent carbon atoms can share one, two, or three pairs of electrons; therefore, a carbon chain of given length can have different numbers of attached hydrogen atoms. Third, the more atoms a molecule contains, the more ways they may be arranged in space to give compounds having the same composition but differing in structure. Finally, different atoms or groups of atoms can be substituted for hydrogen atoms to yield a large number of derivatives. In the following paragraphs, these four factors are briefly discussed and illustrated.

The carbon atom has four valence electrons, and, as discussed in Sec. 3.8, it is expected to use its sp^3-hybrid orbitals to form four covalent bonds directed toward the corners of a tetrahedron. It matters little whether the bonds are formed to other carbon atoms or to hydrogen atoms because carbon and hydrogen atoms are of about the same electronegativity. This means that, instead of being restricted to the simplest hydrocarbon, CH_4 (methane), a whole series of compounds is possible; examples are C_2H_6 (ethane), C_3H_8 (propane), and C_4H_{10} (butane). The structural formulas of these are usually written as in the top part of Fig. 22.5, although it should be remembered that the molecules are three-dimensional, as shown in the lower part of the figure. Excellent sources of hydrocarbons are natural gas and petroleum, the former consisting of the light hydrocarbons (mostly methane and ethane) and the latter of heavier hydrocarbons all the way up to molecules containing 90 carbon atoms.

Since the hydrocarbons differ in boiling points, they can be separated from each other by distillation. At room temperature they all are

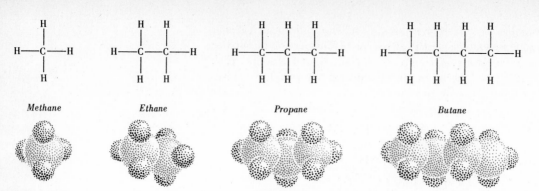

Methane   Ethane   Propane   Butane

Fig. 22.5
Some hydrocarbons.

Ethylene

Acetylene

Butadiene

Fig. 22.6
*Some unsaturated
hydrocarbons.*

chemically inert, but at higher temperature they can be burned in air to form CO, CO_2, and H_2O and thus are used as fuels. The mixture of hydrocarbons ranging from C_7H_{16} (heptane) to $C_{10}H_{22}$ (decane) is gasoline; $C_{10}H_{22}$ to $C_{16}H_{34}$, kerosine; $C_{15}H_{32}$ to $C_{22}H_{46}$, fuel oil; and $C_{19}H_{40}$ to $C_{35}H_{72}$, lubricating oil. In order to improve the yield of gasoline from petroleum, large molecules (with more than ten C atoms) can be broken down by *thermal* and *catalytic cracking*, or small molecules (with fewer than seven C atoms) can be combined by *polymerization* and by *catalytic re-forming*. Paraffin wax and asphalt contain hydrocarbons ranging from $C_{36}H_{74}$ to $C_{90}H_{182}$.

Unsaturated hydrocarbons contain double or triple bonds. Examples are ethylene, C_2H_4, acetylene, C_2H_2, and butadiene, C_4H_6, for which the structural formulas are shown in Fig. 22.6. In general, they are more reactive than corresponding saturated hydrocarbons. For example, they undergo addition reactions, in which hydrogen or other atoms add on to double or triple bonds. An example of such an addition reaction is found in the conversion of vegetable oils to synthetic fats by catalytic hydrogenation. Unsaturated hydrocarbons may also undergo *polymerization* reactions, in which small molecules couple together to form extended chains. For example, ethylene polymerizes at high temperatures to form the plastic polyethylene in a manner which can be visualized as follows:

The double bond can be thought of as opening to form an unstable intermediate, which then joins with other molecules to produce a high polymer. The term *high polymer* is applied to any large molecule which contains recognizable repeating units.

The third factor mentioned above as contributing to the large number of organic compounds is *isomerism,* i.e., the existence of more than one compound with the same molecular formula. As illustration, C_2H_6O can signify either

$$H-\underset{\underset{\displaystyle H}{|}}{\overset{\overset{\displaystyle H}{|}}{C}}-\underset{\underset{\displaystyle H}{|}}{\overset{\overset{\displaystyle H}{|}}{C}}-O-H \qquad \text{or} \qquad H-\underset{\underset{\displaystyle H}{|}}{\overset{\overset{\displaystyle H}{|}}{C}}-O-\underset{\underset{\displaystyle H}{|}}{\overset{\overset{\displaystyle H}{|}}{C}}-H$$

The first is ethyl alcohol, and the second is dimethyl ether; they have different properties (for example, bp = 78.5 and $-23.7°C$, respectively) owing to the changed position of the oxygen. There are many examples of isomerism in the saturated-hydrocarbon series. All the members from butane on have two or more isomers. ($C_{40}H_{82}$ has been calculated to have more than 61 trillion isomers.) Butane has only two: normal (*n-*) butane and isobutane. Their conventional (two-dimensional) structural formulas are

n-Butane   *Isobutane*

Though it might seem possible that there would be other isomers of C_4H_{10}, these are the only ones. Other two-dimensional formulas can be written, but they can be shown to be equivalent to one or the other of the above. The problem arises because the two-dimensional display formulas do not allow for rotation in space about the various bonds. In a saturated hydrocarbon each carbon is tetrahedrally surrounded by four groups, and the molecule can assume various configurations by rotation about the individual bonds. The spatial relations can best be seen by the use of molecular models such as those diagrammed in Fig. 22.7. Of the five configurations shown, the first four correspond to the same molecule (*n*-butane) twisted into different shapes; the fifth corresponds to a different molecule (isobutane), and no amount of twisting can convert it to the normal isomer. Isomers always differ in properties, but sometimes the differences are so slight as to make separation difficult.*

For higher hydrocarbons (and other organic compounds as well) there is, besides the *structural isomerism* just described, the possibility of

---

* Recent large demand for normal hydrocarbons to make biodegradable detergents (hydrocarbon-digesting bacteria can eat straight chains but give up on branched chains) has been met by use of *molecular sieves,* ingenious silicate compounds (Sec. 22.6) which sustain perfusion by molecules of small cross section but exclude those of large cross section.

Normal butane　　　　　　　　　　Isobutane

Fig. 22.7
Various configurations of butane molecules.

optical isomerism as well. (Recall Sec. 17.7.) This happens, for example, with the hydrocarbon

$$H-\overset{\overset{\displaystyle H}{|}}{\underset{\underset{\displaystyle H}{|}}{C}}-\overset{\overset{\displaystyle H}{|}}{\underset{\underset{\displaystyle H}{|}}{C}}-\overset{\overset{\displaystyle CH_3}{|}}{\underset{\underset{\displaystyle H}{|}}{C^*}}-\overset{\overset{\displaystyle H}{|}}{\underset{\underset{\displaystyle H}{|}}{C}}-\overset{\overset{\displaystyle H}{|}}{\underset{\underset{\displaystyle H}{|}}{C}}-\overset{\overset{\displaystyle H}{|}}{\underset{\underset{\displaystyle H}{|}}{C}}-H$$

which can be abbreviated

$$C_2H_5-\overset{\overset{\displaystyle CH_3}{|}}{\underset{\underset{\displaystyle H}{|}}{C^*}}-C_3H_7$$

The carbon with the asterisk has attached to it four different groups; so there is no center of symmetry or plane of symmetry in the molecule. Hence, as indicated in Sec. 17.7, the molecule can exist in two forms, one of which is the mirror image of the other. Optical isomers are very difficult to separate; they have identical melting and boiling points and identical chemical properties except when reacting with molecules that also show optical isomerism. As a consequence, they can sometimes be separated by taking advantage of slight solubility differences in a solvent that is composed of a single optical isomer.

A complicating feature which increases the number of hydrocarbon isomers is the possibility of having atoms arranged in rings. The most common of these cyclic compounds is benzene, C_6H_6, which consists of 6 carbon atoms at the corners of a hexagon with a hydrogen atom attached to each carbon atom. All 12 atoms are in one plane. The carbon-carbon bonds are all equivalent, and therefore the molecule can be considered as a resonance hybrid. The contributing forms may be written as shown in Fig. 22.8a; ordinarily only one of the forms is shown and that concisely as a simple hexagon with nonreacting carbon and hydrogen atoms omitted as in Fig. 22.8b. Alternatively, both forms may be merged into a single representation as shown in Fig. 22.8c. Other groups may be substituted for one or more of the hydrogen atoms to give derivatives.

The fourth reason for the large number of organic compounds is the formation of derivatives. These derivatives differ in properties from the parent hydrocarbons, and, in fact, their properties are mainly determined by the nature of the substituent. Because the hydrocarbon residue usually remains intact throughout chemical reactions, it is convenient to consider such hydrocarbon derivatives as combinations of hydrocarbon residues and substituents. Each substituent, called a *functional group*, imparts characteristic properties to the molecule. A simple example of a functional group is the *alcohol*, or OH, group. In methyl alcohol (CH_3OH) and ethyl alcohol (C_2H_5OH), derived, respectively, from methane (CH_4) and ethane (C_2H_6), the substitution of OH for one of the hydrogens has the effect of bestowing waterlike properties on what was originally a volatile but inert hydrocarbon. In the higher alcohols (containing many C atoms) the influence of OH in changing properties is less pronounced, and the properties are more hydrocarbonlike. The general formula for any alcohol can be written ROH, where R stands for a hydrocarbon residue. In methyl alcohol, R is the methyl group (CH_3); in ethyl alcohol, R is the ethyl group (C_2H_5). Figure 22.9 shows the common classes of organic compounds based on different functional groups.

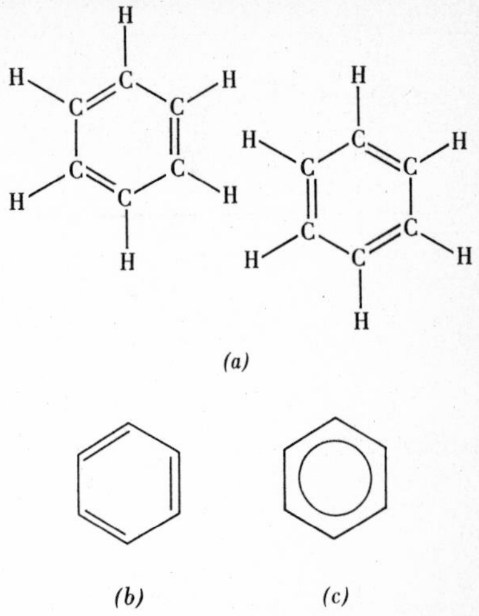

The functional group consisting of a carbon with a doubly bonded oxygen and a singly bonded OH, usually written

$$\overset{O}{\underset{\|}{}}$$

COOH or $-\overset{O}{\underset{\|}{C}}-OH$, imparts acid properties to organic molecules. Compounds containing the COOH group are called *carboxylic acids* and have the general formula RCOOH. If R stands for CH_3, the acid is CH_3COOH, acetic acid, which we have earlier written as $HC_2H_3O_2$. (The grouping CH_3CO, which is called the *acetyl group*, is frequently represented by Ac; so the formula of acetic acid is sometimes also

(a)

(b)

(c)

Fig. 22.8
Representations of benzene.

Fig. 22.9
Classes of organic compounds.

Functional group	General formula	Name	Examples
$-O-H$	ROH	*Alcohols*	CH_3OH, *methanol* C_2H_5OH, *ethanol*
$-\overset{O}{\underset{\|}{C}}-H$	RCHO	*Aldehydes*	HCHO, *formaldehyde* CH_3CHO, *acetaldehyde*
$-\overset{O}{\underset{\|}{C}}-$	RCOR'	*Ketones*	CH_3COCH_3, *acetone*
$-\overset{O}{\underset{\|}{C}}-O-H$	RCOOH	*Acids*	HCOOH, *formic acid* CH_3COOH, *acetic acid*
$-\overset{O}{\underset{\|}{C}}-O-$	RCOOR'	*Esters*	CH_3COOCH_3, *methyl acetate*
$-O-$	ROR'	*Ethers*	$C_2H_5OC_2H_5$, *diethyl ether*
$-\overset{H}{\underset{\|}{N}}-H$	RNH_2	*Amines*	CH_3NH_2, *methylamine*

written as HOAc.) Like acetic acid ($K_{diss} = 1.8 \times 10^{-5}$), other organic acids are generally weak electrolytes. As acids, they undergo the usual neutralization reactions with bases. They also react with alcohols in an entirely different reaction called *esterification*. Esterification reactions can be described by the general equation

$$\underset{Acid}{R-\overset{\overset{\displaystyle O}{\|}}{C}-O-H} + \underset{Alcohol}{R'-O-H} \longrightarrow \underset{Ester}{R-\overset{\overset{\displaystyle O}{\|}}{C}-O-R'} + H_2O$$

which shows the splitting out of water and the formation of an ester, RCOOR'. As shown by isotope-tracer experiments, the acid contributes the OH and the alcohol the hydrogen atom to the product water. In that they are slow and in that they involve only parts of molecules, esterification reactions are typical of most organic reactions.

Esters are quite common in nature. For example, animal *fats* (e.g., lard, suet, and tallow) and vegetable *oils* (e.g., olive oil and cottonseed oil) are composed of mixtures of esters such as stearin. As shown in Fig. 22.10, stearin is a polyfunctional ester containing three ester groups. When boiled with sodium hydroxide, it undergoes a *saponification* reaction, which breaks it down to a polyalcohol and a sodium salt of stearic acid (sodium stearate, $NaOOCC_{17}H_{35}$, or soap). The general equation for saponification is

$$\underset{Ester}{RCOOR'} + OH^- \longrightarrow \underset{Ion\ of\ acid}{RCOO^-} + \underset{Alcohol}{R'OH}$$

Soap (which we shall consider as sodium stearate, even though it can be the salt of any long-chain acid) is remarkable because it gives ions which are largely hydrocarbon modified by a charged group at one end. When placed in water, these ions do not really dissolve because hydrocarbons are insoluble in polar solvents. Instead, *micelles* are formed in which the hydrocarbon parts of the stearate ions cluster together as shown schematically in Fig. 22.11. The negative charges at the surface of the micelle are dissolved in the water (they are *hydrophilic*, or "water-loving"); the hydrocarbon chains in the interior are dissolved in each other (they are *hydrophobic*, or "water-hating"). X-ray investigations of soap suspensions show that at low concentrations the micelles are approximately spherical, with a diameter of about 5 nm. The cleansing action of soap is thought to stem from the dissolving of grease (essentially hydrocarbon in nature) in the hydrocarbon clusters.

In addition to the fats, the two other major groups of substances of which living material is composed are the *carbohydrates* and *proteins*. The carbohydrates were so named because the common ones can be represented by formulas $C_x(H_2O)_y$; they contain H and O in a 2:1 ratio, just as in water. However, they are not hydrates in any sense of the word but consist of rather complex ring structures in which the C atoms have H atoms and OH groups attached to them. One of the simpler carbohydrates is glucose, $C_6H_{12}O_6$, which occurs in many fruits and the blood of many animals. As shown in Fig. 22.12a, the

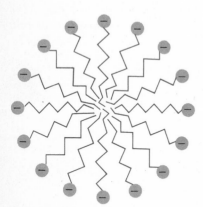

Fig. 22.10
Stearin molecule. (The hydrocarbon chains are probably coiled up and intertwined; so the molecule is rather spherical in shape.)

Fig. 22.11
Soap micelle.

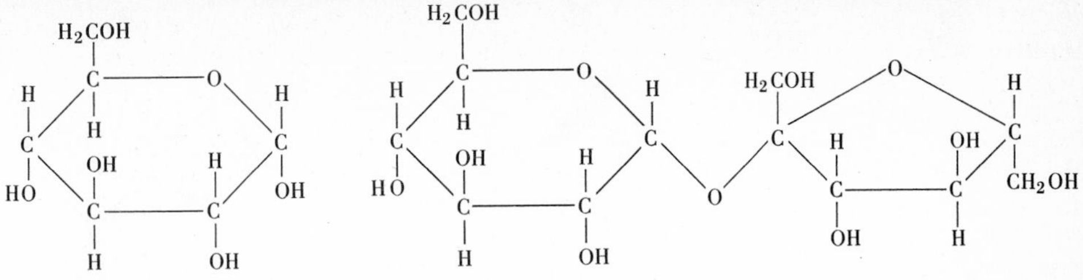

Fig. 22.12

Simple carbohydrates: (a) glucose, (b) sucrose.

molecule contains a six-membered ring consisting of five C atoms and one O atom. Sucrose, $C_{12}H_{22}O_{11}$, shown in Fig. 22.12*b*, is the most important commercial sugar; it contains two rings (six-membered and five-membered) in the molecule. Starch and cellulose are natural high-polymer carbohydrates consisting of long chains of glucose rings hooked together by O bridges.

The proteins are extremely complex molecules which are also natural high polymers. When boiled in acid or base, they undergo hydrolysis to form relatively simple *amino acids* (carboxylic acids containing the amino, or NH_2, group). Some 26 such amino acids have been identified, and all proteins, as in hair, fingernails, skin, muscles, tendons, and blood, are considered to be condensation products of two or more of these acids. The characteristic feature of all proteins is the group

$$\begin{matrix} H & O \\ | & \| \\ -N & -C- \end{matrix}$$

, called the *peptide link*. Figure 22.13 shows how the peptide link might be established between two amino acid molecules by splitting out H_2O. Further polymerization is possible since there is a free NH_2 group (on the far left) and a free acid group (on the far right). From various combinations of the 26 amino acids a large variety of high-molecular-weight (10^4 to 10^7 amu) proteins is possible. A structural feature common to many of these proteins is the helix shown in Fig. 22.14. The helical conformation is maintained by hydrogen bonds between amino acid groups in successive turns of the helix.

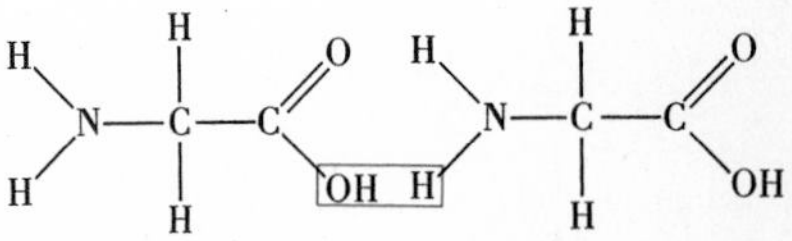

Fig. 22.13

Formation of a peptide link.

22.5 Organic reactions

There are two important features which, in general, characterize the reactions between organic compounds: One of these features is the relative slowness of most organic reactions compared with many familiar inorganic reactions. For example, whereas the reaction between hydrochloric acid and sodium hydroxide is practically instantaneous, the esterification between acetic acid and ethyl alcohol takes hours and occurs even then only if the reaction mixture is heated and a catalyst such as sulfuric acid is added. The other characteristic feature of organic reactions is that, in general, the greater part of the reacting molecule remains relatively unchanged during the course of the reaction. In other words, many of the atoms are undisturbed by the reaction and maintain their arrangement relative to their neighbors even though somewhere else in the molecule fairly extensive changes are going on.

Fig. 22.14
Protein helix.

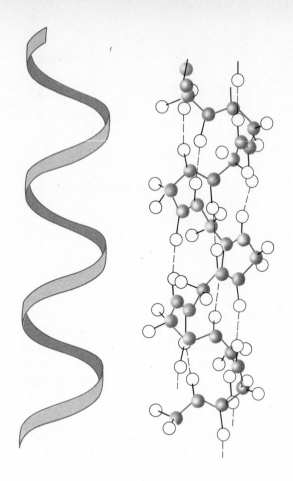

Both the slowness of the reactions and the retention of the major part of a molecule's identity can be exceedingly exasperating at times, but in general they prove very useful. For one thing, they allow the organic chemist to focus attention on a small portion of the molecule with the confident expectation that the rest of the molecule will not change much while the reaction is being carried out. For another thing, the slowness of the reactions allows the changes to be stopped well before equilibrium is established. As a result, it is frequently possible to isolate compounds which if allowed to remain in the reaction mixture would react further to give different products. Even so, it is not practicable to eliminate completely side reactions which lead to other than the desired products. Consequently, the organic chemist must often be content with considerably less than 100 percent of the theoretically possible yield of the compound desired.

Of the many types of organic reactions, some of the most commonly encountered are *addition, substitution,* and *polymerization:*

1 A simple example of an addition reaction is the adding of HBr to C_2H_4, ethylene. The overall change is

$$CH_2=CH_2 \text{ (Ethylene)} + H-Br \longrightarrow H-CH_2-CH_2-Br \text{ (Bromoethane)}$$

Ethylene   *Bromoethane*

The mechanism of the reaction is believed to be stepwise. HBr dissociates to give a proton and a bromide ion. The proton, being positively charged, adds to the double bond, which is an electron-rich region. This pulls electrons away from one of the carbon atoms so that it looks positive, thereby attracting the negative bromide ion. The sequence looks like this:

$$CH_2=CH_2 + H^+ \longrightarrow CH_3-\overset{\oplus}{C}H_2$$

$$CH_3-\overset{\oplus}{C}H_2 + Br^- \longrightarrow CH_3-CH_2Br$$

The positively charged ion, containing the carbon with only six instead of the usual eight electrons, is called a *carbonium ion*. It is highly reactive and frequently appears as an intermediate in organic reactions. Carbonium ions can be primary ($1°$), secondary ($2°$), or tertiary ($3°$) depending on the number of other carbon atoms joined to the positive carbon. Thus, we can have

$$CH_3-\overset{\oplus}{\underset{H}{C}}H \qquad CH_3-\overset{\oplus}{\underset{CH_3}{C}}H \qquad CH_3-\overset{\oplus}{\underset{CH_3}{C}}CH_3$$

Primary ($1°$)   *Secondary ($2°$)*   *Tertiary ($3°$)*

In general, the ease of formation decreases in the order $3° > 2° > 1°$. This decreasing order helps to explain a famous generalization in organic chemistry known as Markovnikov's rule: *When HX adds to a double bond, the hydrogen atom goes to the carbon that already has more hydrogen.* To illustrate, addition of HCl to isobutylene proceeds as follows:

$$CH_3-\underset{CH_3}{C}=CH_2 + H^+ \longrightarrow CH_3-\overset{\oplus}{\underset{CH_3}{C}}-CH_3$$

$$CH_3-\overset{\oplus}{\underset{CH_3}{C}}-CH_3 + Cl^- \longrightarrow CH_3-\underset{\underset{Cl}{|}}{\underset{CH_3}{C}}-CH_3$$

The alternative path, where the proton adds to the other carbon of the double bond, is not nearly so well favored because it would lead to an intermediate

$$CH_3-\underset{\underset{H}{|}}{\overset{\overset{CH_3}{|}}{C}}-\overset{\oplus}{CH_2}$$

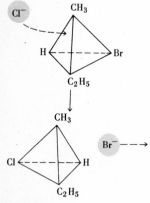

Fig. 22.15
Substitution of Cl *for* Br *by a displacement mechanism. (C in center of tetrahedron is not shown.)*

which is a primary carbonium ion, not a tertiary. The tertiary one would be favored because the positive charge on a carbon can be more readily relieved when there are several attached methyl groups to feed in their electrons.

2 Substitution reactions are somewhat more complicated than addition reactions because they involve removal of an attached atom or group of atoms and the addition of a different atom or group. One way in which this can occur is by *displacement*, where an entering group approaches one side of the molecule while the leaving group departs from the other side. As indicated in Fig. 22.15, the tetrahedral arrangement of groups around the central carbon is inverted in the process. If the attached groups are all unlike so that the molecule is optically active, the substitution inverts the optical activity from dextro to levo, or vice versa. This kind of substitution reaction by displacement is known as a *Walden inversion*.

Another important class of substitution reactions is those involving benzene and its derivatives. We take as an example the chlorination of benzene:

$$\text{Benzene} \quad + \text{Cl—Cl} \longrightarrow \quad \text{Chlorobenzene} + \text{HCl}$$

Benzene   *Chlorobenzene*

Although benzene appears to contain double bonds, it does not add molecules such as Cl_2 or HCl but instead undergoes substitution reaction because to do otherwise would destroy the resonance stabilization of the ring system. Hence when a positive ion such as H^+ (or Cl^+) adds to the C_6H_6 ring to form a carbonium ion, the following step is not the addition of a negative ion (say Cl^-) but the elimination of a proton so as to preserve the maximum resonance possibility. For the above reaction $FeCl_3$ is a catalyst, and the reaction is believed to occur as follows:

$$FeCl_3 + Cl_2 \longrightarrow FeCl_4^- + Cl^+ \quad \text{(fast)}$$

$$Cl^+ + C_6H_6 \longrightarrow C_6H_6Cl^+ \quad \text{(very slow)}$$

$$C_6H_6Cl^+ \longrightarrow C_6H_5Cl + H^+ \quad \text{(fast)}$$

The function of the $FeCl_3$ is to form the cation Cl^+, which adds to the ring system and brings about elimination of H^+. The net effect is to substitute a chlorine for one of the six equivalent H atoms of C_6H_6. The overall reaction is fairly difficult to bring about; it is nothing so easy as either the addition to the double bond or the Walden inversion mentioned above. Long heating with the catalyst is required, and even then the yield of product is not very good.

As noted, all the H atoms in the benzene molecule are equivalent; therefore it makes no difference which H atom is replaced by Cl. Only one kind of chlorobenzene molecule is produced. However, in chlorobenzene not all the H atoms are equivalent (since some are closer than others to the Cl atom), and when a second Cl atom is substituted for a second H, three different products are possible. These are

Orthodichloro	Metadichloro	Paradichloro
benzene	benzene	benzene
o-$C_6H_4Cl_2$	m-$C_6H_4Cl_2$	p-$C_6H_4Cl_2$

They differ from each other in that the ortho isomer has the two substituents attached to adjacent C atoms; the meta isomer, to C atoms separated by a single CH group; the para isomer, to C atoms separated by two CH groups. All three isomers are formed when monochlorobenzene is converted to dichlorobenzene, but the ortho and para isomers predominate.

3 · The third type of reaction mentioned above, polymerization, can be of two very different kinds. In one of these, called *addition polymerization*, molecules add together to form giant molecules; in the other, called *condensation polymerization*, small molecules are split out as the giant molecule is built up. The formation of polyethylene is a typical addition polymerization. Polymerization can be initiated by heating ethylene at high pressure or by use of initiators such as organic peroxides. Peroxide-initiated polymerization is rather common for the production of polymers other than polyethylene such as polyvinyl chloride (PVC, one of the most common plastics, as in Saran wrap) and is thought to proceed in the following manner:

$$R—O—O—R \longrightarrow R—O\cdot + \cdot O—R$$

Organic peroxide   *Peroxide free radicals*

Vinyl chloride

$$\underset{\substack{H \\ H}}{\overset{\substack{Cl \\ H}}{R-O-\overset{|}{\underset{|}{C}}-\overset{|}{\underset{|}{C}}\cdot}} + \underset{\substack{H \\ H}}{\overset{Cl}{C}}=\underset{\substack{H}}{\overset{H}{C}} \longrightarrow \underset{\substack{H \ H \ H \ H}}{\overset{Cl \ H \ Cl \ H}{R-O-\overset{|}{\underset{|}{C}}-\overset{|}{\underset{|}{C}}-\overset{|}{\underset{|}{C}}-\overset{|}{\underset{|}{C}}\cdots}}$$

In the first step the peroxide is assumed to have broken a covalent bond so as to leave an unpaired electron (shown by the dot) on each R—O· residue. The residue is a *free radical* (Sec. 10.7), is very reactive, and initiates a chain reaction. As shown in the second step the free radical combines with a vinyl chloride molecule to form a new free radical which is one —CHClCH$_2$— unit longer than the original. Chain propagation thus continues, and a giant molecule is progressively built up until the free radical is destroyed, either by combining with another free radical or by reacting with some chemical added as an inhibitor. Such a mechanism accounts satisfactorily for the observation that the final polymer product contains giant molecules of different molecular weights. It also explains the fact that a small amount of initiator can produce a great deal of polymerization. It should be noted that in the final giant molecules, the RO end groups are an insignificant fraction of the molecules and so the polymers are essentially aggregations of CHClCH$_2$ units added together.

In condensation polymerization, the buildup of the polymer occurs not by the addition of a whole molecule to a lengthening chain but by reaction of a lengthening chain with a small molecule accompanied by elimination of a simpler species such as H$_2$O. In order that polymerization may occur, two conditions must be met: One end of molecule A must be able to interact with the other end of molecule B so as to split out a small group such as H$_2$O. Second, both molecules A and B must contain two functional groups so that after A and B combine the free ends can continue to react to extend the polymer. An example of condensation polymerization is the formation of Dacron polyester. Here the reaction is between methyl terephthalate, CH$_3$OOCC$_6$H$_4$COOCH$_3$, and ethylene glycol, HOCH$_2$CH$_2$OH, and involves elimination of methyl alcohol, CH$_3$OH. The reaction can be pictured as follows:

The first step is then followed by reactions at both ends of the molecule to give giant polymers.

The chemistry of silicon, the second member of group IV, resembles that of carbon in several respects. For example, silicon forms a tetrahedral SiH_4 and a few higher hydrosilicons which contain chains of silicon atoms. However, since Si—O bonds are formed preferentially to Si—H or Si—Si bonds, the chemistry of silicon is primarily concerned with oxygen compounds rather than with hydrosilicons. Furthermore, unlike the smaller carbon atom, which forms multiple bonds, silicon invariably forms single bonds. As a result, the oxygen-silicon compounds contain Si—O—Si bridges in which oxygen is bonded by single bonds to two silicon atoms instead of being bonded by a double bond to one silicon atom. This is unlike the case of carbon, where oxygen is frequently found bonded to a single carbon atom as the

$$\begin{array}{c}\diagdown\\\diagup\end{array}\hspace{-2pt}C{=}O \text{ group.}$$

Silicon is the second most abundant element in the earth's crust (25.8 wt %) and is about as important in the mineral world as carbon is in the organic. As SiO_2 (silica) it accounts for most beach sands, quartz, flint, and opal; as complex oxysilicates of aluminum, iron, magnesium, and other metals it accounts for practically all rocks, clays, and soils.

The preparation of elemental Si is quite difficult. It can be accomplished by the reduction of SiO_2 with Mg or by the reduction of the chloride with Na. Since it is mainly used for addition to steel, it is more usually prepared as ferrosilicon by reduction of mixtures of SiO_2 and iron oxides with coke. The element is a semimetal with a crystal structure like that of diamond. At room temperature it is inert to most reagents but will dissolve in basic solutions to liberate H_2. At elevated temperatures it reacts with many metals, such as Mg, to form silicides (such as Mg_2Si).

For the electronics industry, the need is for ultrapure silicon, i.e., silicon that contains less than 0.001 percent foreign elements. As mentioned in Sec. 7.5, some solids such as silicon and germanium are electrical insulators in the pure state but become good conductors when doped with group III or group V elements. The enhanced conductivity increases with the amount of impurity and rapidly increases with increasing temperature. Many important devices such as transistors and solar batteries depend on such electric properties. To get ultra-high-purity materials, the starting materials such as $SiCl_4$ and zinc have to be prepurified by the best techniques available, e.g., vacuum distillation and selective adsorption. As a last step in preparing the ultrapure material, an ingot of high-purity silicon is subject to a *zone-refining* process. In this process, the high-purity ingot is slowly drawn through a long, inert-gas-filled quartz tube. Induction heating coils wound in narrow strips around the tube, as shown in Fig. 22.16, melt the silicon in narrow zones. At the interface of the solid and liquid there is an equilibrium Si(s) $\rightleftharpoons$ Si(l). Impurities, which are generally more solu-

Fig. 22.16

Zone-refining technique to purify silicon.

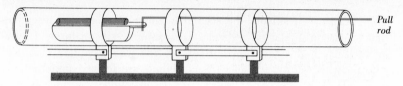

Quartz tube   *Heater loops*   Pull rod

ble in the liquid phase than in the solid phase, remain in the molten zones and get swept to the end of the ingot, where they can be cut off and discarded. Zone refining has made it possible to reduce impurity concentrations to one part per billion. For use in solar cells a thin wafer of ultrapure silicon is doped with boron (a group III element), giving it a *p*-type character. One side of the wafer then has a group V element such as phosphorus or arsenic diffused into it, giving it *n*-type character. When the *n*-type surface is exposed to the sun, photons of solar energy penetrate to about the depth of the *p-n* junction and give up their energy in a burst by creating an electron and a hole (recall Sec. 7.5). When either the electron or the hole, depending where the electron-hole pair is created, migrates across the *p-n* junction, a voltage is created. A single cell can generate only about 0.4 V, but with thousands of them placed in series large power production can be achieved. Unfortunately, the cells are rather expensive; so even modest power demands are very costly.

Almost all the compounds of silicon are oxy compounds. However, other compounds which are unstable with respect to conversion to the oxy compounds can be prepared. Thus, the hydrosilicons, which are prepared by reaction of silicides with acid and are analogous to the hydrocarbons, are unstable in oxygen with respect to rapid conversion to SiO_2. Silane (SiH_4), for example, is oxidized as follows:

$$SiH_4(g) + 2O_2(g) \longrightarrow SiO_2(s) + 2H_2O(g) \qquad \Delta G° = -1220 \text{ kJ}$$

which is to be compared with

$$CH_4(g) + 2O_2(g) \longrightarrow CO_2(g) + 2H_2O(g) \qquad \Delta G° = -800 \text{ kJ}$$

Disilane (Si_2H_6), trisilane (Si_3H_8) and tetrasilane (Si_4H_{10}) have been prepared, but they are progressively even less stable as the silicon-silicon chain length increases. Derivatives of the silanes such as silicon tetrachloride ($SiCl_4$) are also known, but they are unstable with respect to air oxidation to SiO_2.

The silicates (oxy compounds of silicon) have been extensively investigated, and in practically every case the silicon atom is found to be tetrahedrally bonded to four oxygen atoms. As shown in Fig. 22.17, four valence electrons ($3s^2 3p^2$) from silicon and six valence electrons from each oxygen are insufficient to complete the octets of all the atoms. Consequently, to produce a stable compound the oxygen atoms may obtain electrons from other atoms and become negative in the process. This produces the discrete orthosilicate anion, SiO_4^{4-}, found, for example, in the mineral zircon ($ZrSiO_4$). Alternatively, the oxygen atoms may complete their octets by sharing electrons with other silicon atoms.

Fig. 22.17

Tetrahedral SiO_4 *unit.*

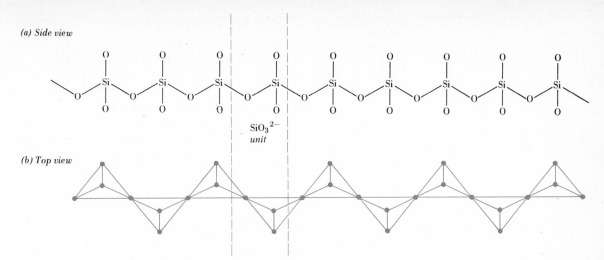

(a) Side view

SiO_3^{2-}
unit

(b) Top view

Fig. 22.18
Pyroxene chain. In (b) the
view is along the Si—O
bond sticking out of the
plane of the paper.

Since one, two, three, or four of the oxygen atoms can thus bridge to other silicon atoms, increasingly complex structures are possible. One bridge oxygen per silicon atom gives $Si_2O_7^{6-}$, which is analogous to $Cr_2O_7^{2-}$. Two bridge oxygens per silicon lead to formation of extended chains called *pyroxene* chains. The pyroxenes are a class of minerals— spodumene, $LiAl(SiO_3)_2$, is an example—which are second only to the feldspars as the most common constituents of igneous rock. As shown by the two views of Fig. 22.18, the pyroxene chains are strings of corner-sharing tetrahedra where the repeat unit is SiO_3^{2-}. Each of the oxygen atoms that is not a bridge oxygen has picked up an electron to complete its octet; so each pyroxene chain is a negatively charged anion. In the compounds, cations such as Li^+ and Al^{3+} hold the solid together by ionic attractions. Pyroxene chains of varying length occur in the material called *water glass*, which can be made by dissolving silica in NaOH solution. The equation for formation can be written

$$SiO_2 + 2NaOH \longrightarrow \text{``}Na_2SiO_3\text{''} + H_2O$$

although, as indicated by the quotation marks, the ratio of sodium to silicon can be highly variable. Water glass is used as an egg preservative (where it presumably acts by sealing the pores), fabric fireproofer, adhesive for cardboard cartons, and in the children's toy called "Chemical Garden." In the latter, bits of colored salt are dropped into a sodium silicate solution to produce frondlike growths. The mechanism is not completely understood but apparently involves hydrolysis of the colored ions to produce acid, which reacts with the silicate to form a membrane of $SiO_2(H_2O)_x$. Osmosis causes water to flow into the membrane making it expand, thus simulating growth.

If two pyroxene strands are laid parallel to each other and cross-linked by shared O bridges, the result is to form a band anion known as an *amphibole* chain (Fig. 22.19). Amphibole chains are found in a variety of common minerals, including, for example, the hornblendes, which are hydroxysilicates containing Ca, Mg, and Fe as cations to

Fig. 22.19
Amphibole chain.

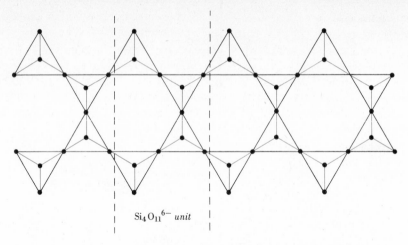

$Si_4O_{11}^{6-}$ unit

balance the charge of the anion. The anion charge is variable because approximately every fourth tetrahedron contains Al instead of Si.

If the SiO_4 tetrahedra are cross-linked indefinitely in two dimensions, the result is to form infinite sheets that give characteristic layering to some silicate minerals. A portion of an extended two-dimensional sheet built up of SiO_4 tetrahedra with three bridge oxygen atoms per silicon is shown in Fig. 22.20. The colored circles represent oxygen

Fig. 22.20
Silicate sheet.

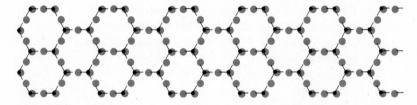

atoms below the plane of the paper, and the black circles represent silicon atoms in the plane with oxygen atoms above the plane. The oxygen atoms sticking out above the plane are negatively charged and are attracted to positive ions, which in turn are attracted to other, similar sheet silicate ions. If the sheets were perfect, they would have the general formula $(Si_2O_5^{2-})_n$; so if the counterion were Mg^{2+}, the formula of the compound would be $MgSi_2O_5$. Although it is not needed for structural reasons, there is almost always association with hydroxide sandwiched in between layers of Mg^{2+}. Thus, we have the compound $2MgSi_2O_5 \cdot Mg(OH)_2$, or $Mg_3(Si_2O_5)_2(OH)_2$, otherwise known as *talc*. A schematic representation of the layer stacking in talc is shown in Fig. 22.21a. The sandwich formed by the assemblage is neutral, and there is only weak van der Waals attraction to parallel structures; hence, talc and substances like it are very soft. On *Mohs' hardness scale*, which rates diamond, the hardest natural substance known, as 10, talc is given the lowest rating, 1. Since Mohs' scale is logarithmic, this means that talc is about 10 powers of 10 less hard than diamond.

The Si in the SiO_4 tetrahedra is commonly replaced by Al, usually in the ratio of about 1 out of 4. Since Al is tripositive whereas Si is tetrapositive, substitution of Al for Si in the layer structure above results in a net negative charge for the whole sandwich. This attracts positive ions such as K^+. Thus, for example, we can explain the compound $KMg_2AlSi_3O_{10}$ as being derived from $2(MgSi_2O_5) = Mg_2Si_4O_{10}$ with replacement of one Si by Al and K. If combined with $Mg(OH)_2$, the above compound comes out to be $KMg_3AlSi_3O_{10}(OH)_2$, a form of mica. As shown in Fig. 22.21b, the stacking in mica is repetitive with rather strong binding of adjacent sandwiches by layers of K^+ ions attracted to either side. Although mica shows easy cleavage parallel to the plane direction, it has lost the softness characteristic of talc. Clay minerals have analogous layer structures.

In the limit there can be four bridge O atoms per Si. This leads to the three-dimensional structures as found in silica (e.g., quartz, SiO_2), feldspars (e.g., orthoclase, $KAlSi_3O_8$), and zeolites (e.g., ultramarine, $Na_3Al_2Si_6O_{12}S$). In *silica* the framework contains only Si and O atoms and is electrically neutral. If the framework is an ordered one, the silica is crystalline, as in quartz; if it has been disordered, as by supercooling molten SiO_2, the silica is noncrystalline. Crystalline silica has a very high melting point ($1700\,^\circ$C), but noncrystalline (or vitreous) silica can be softened at a considerably lower temperature ($\approx 1200\,^\circ$C). Thus softened, it can be blown into various forms, such as laboratory ware, which take advantage of its desirable properties. SiO_2 transmits both visible and ultraviolet light, has a low thermal coefficient of expansion (only about one-twentieth that of glass or steel), and is inert to most chemical reagents. However, it is dissolved by solutions of HF to form complex fluosilicate ions, SiF_6^{2-}, and, to a limited extent, by basic solutions to form various silicate ions.

In *feldspars*, the common constituents of igneous rocks (e.g., granite, gneiss, and basalt), replacement of part of the Si by Al gives the framework a negative charge, which must be compensated for by cations, usually K^+, Na^+, or Ca^{2+}. Orthoclase, $KAlSi_3O_8$, can be considered to be derived from $4SiO_2 = Si_4O_8$ with replacement of one out of four Si by K and Al. The Al sits in a tetrahedron of O atoms, just as does the Si, but the K^+ is found in largish cavities in the framework. In the *zeolites* the cavities are very much larger so that besides cations, H_2O can be accommodated too. Figure 22.22 shows a portion of the framework that characterizes the ultramarine structure. As can be seen, the structural unit contains an enormous cavity in the cube center which accommodates the counterion Na^+ but also may contain anions such as S_2^{2-}, Cl^-, or even SO_4^{2-}. The beautiful blue color of the gemstone *lapis lazuli* is believed to be due to light absorption by the polysulfide ion in an ultramarine framework. Action of zeolites as ion exchangers was mentioned in Sec. 16.9, but their use in that application has been largely supplanted by synthetic resins. Synthetic zeolites have been tailor-made to create cavities and passage cross sections of desired dimensions for use as molecular sieves. Figure 22.23 shows a portion of the structure of the synthetic zeolite *Linde A*. Each of the globular clusters at the cube corners is like the entire cluster shown in Fig. 22.22. Thus there are eight small cavities (each

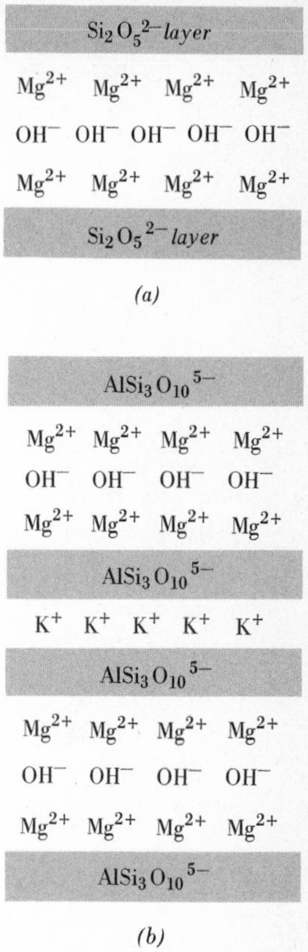

$Si_2O_5{}^{2-}$ *layer*

Mg^{2+} Mg^{2+} Mg^{2+} Mg^{2+}

OH^- OH^- OH^- OH^- OH^-

Mg^{2+} Mg^{2+} Mg^{2+} Mg^{2+}

$Si_2O_5{}^{2-}$ *layer*

(a)

$AlSi_3O_{10}{}^{5-}$

Mg^{2+} Mg^{2+} Mg^{2+} Mg^{2+}

OH^- OH^- OH^- OH^-

Mg^{2+} Mg^{2+} Mg^{2+} Mg^{2+}

$AlSi_3O_{10}{}^{5-}$

K^+ K^+ K^+ K^+ K^+

$AlSi_3O_{10}{}^{5-}$

Mg^{2+} Mg^{2+} Mg^{2+} Mg^{2+}

OH^- OH^- OH^- OH^-

Mg^{2+} Mg^{2+} Mg^{2+} Mg^{2+}

$AlSi_3O_{10}{}^{5-}$

(b)

Fig. 22.21
(a) *Edge view of layer stacking in talc,* $Mg_3(Si_2O_5)_2(OH)_2$; (b) *edge view of layer stacking in mica,* $KMg_3AlSi_3O_{10}(OH)_2$.

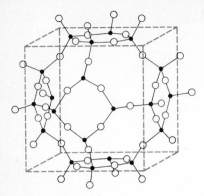

Fig. 22.22
Ultramarine. (Filled circles are Si or Al atoms. Open circles are O atoms. Central cavity accommodates Na+ and various anions such as S_2^{2-}, Cl−, or even SO_4^{2-}.)

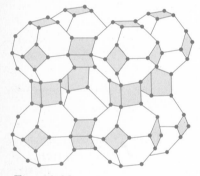

Fig. 22.23
Linde A. (Circles represent positions of Si or Al atoms. O bridges not shown.)

about 0.66 nm in diameter) at the cube corners and a very large cavity (about 1.14 nm) in the cube center. The cavities are connected by openings of 0.20 and 0.42 nm, respectively. Molecules that can squeeze through the openings can be absorbed in the cavities and held there by various attractive forces such as van der Waals forces. Even hydrocarbons can be absorbed. As already mentioned on page 585, molecular sieves can be used to separate straight-chain from branched hydrocarbons.

Derived from SiO_2 are other silicate systems of practical importance, e.g., glass and cement. Glass is made by fusing SiO_2 (as quartz sand) with basic substances such as CaO and Na_2CO_3. Special glasses such as Pyrex glassware contain other acidic oxides (B_2O_3) substituted for some of the SiO_2. Like silica, glass will dissolve in solutions of HF and is slowly etched by basic solutions. As a consequence of the latter reaction, it is frequently observed that glass stoppers stick fast in reagent bottles containing basic solutions such as NaOH and Na_2CO_3. Cement, a complex aluminum silicate, is made by sintering limestone and clay at high temperature and grinding the product to a fine powder. When mixed with water and allowed to stand, it sets to a hard, rigid solid by a series of complex reactions. Although these reactions are only imperfectly understood, they seem to involve slow hydration of silicates to form some sort of interlocking structure. The hydration is accompanied by the evolution of considerable heat, which may cause cracking unless provision is made for its removal.

The high thermal stability of Si—O—Si chains has been exploited in the *silicones*, compounds in which organic residues are bonded to Si atoms in place of negatively charged silicate oxygens. A typical example of a silicone is the chainlike methyl silicone shown in Fig. 22.24. Thanks to the methyl groups, this silicone has lubricating properties characteristic of hydrocarbon oils, but, unlike hydrocarbons, it is unreactive even at high temperatures. More complicated silicone polymers are made possible by having oxygen or hydrocarbon bridges between chains. These rubbery materials are used as electrical insulators at elevated temperatures.

22.7 Germanium

The least abundant (0.0007 percent) of the group IV elements is germanium, the principal source of which is zinc ores. The element may be prepared by the reduction of GeO_2 with carbon or hydrogen. In many respects, compounds of germanium resemble those of silicon. For example, magnesium germanide (Mg_2Ge) reacts with acid to produce hydrogen compounds, but only GeH_4, Ge_2H_6, and Ge_3H_8 are known. These compounds, like the hydrosilicons, are unstable with respect to oxidation to GeO_2. Germanates, derived from GeO_2 and analogous to silicates, have been but slightly investigated. Like silicon, germanium forms a volatile tetrachloride, which fumes in air because of hydrolysis to GeO_2.

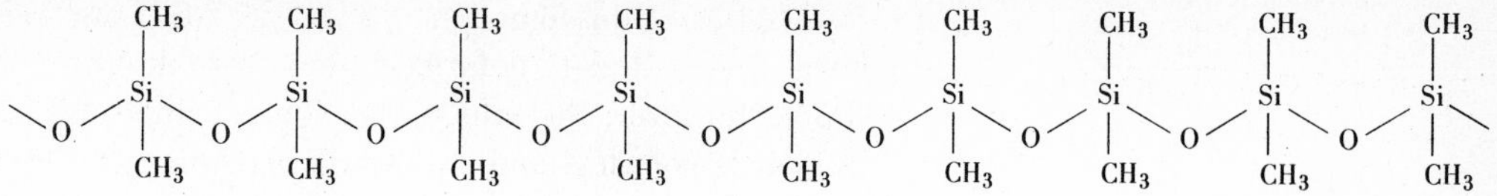

Fig. 22.24
Methyl silicone chain.

The element has taken on unusual industrial importance because of the special properties of slightly impure germanium. Like silicon, germanium has the diamond structure and is a semiconductor; by incorporating traces of a group III or group V element in the lattice, the conductivity can be greatly increased. This increase in conductivity comes about because the substituted element creates a deficiency or excess of valence electrons compared with the four demanded by the diamond structure. As a result, there is a mobility of electrons in the lattice either because the excess electron from a substituted group V atom can easily be dissociated from the impurity center and so may move through the entire structure or because a deficiency (hole) due to a substituted group III atom is filled by an electron from a neighboring bond, which in turn allows further electron motion of the germanium valence electrons. By joining two crystals of germanium, one containing group III impurity and the other containing group V impurity, an *n-p* junction, which passes electricity more easily in one direction than in the other, is formed. This junction can be used to convert alternating current to direct current. Combination of two junctions produces an *n-p-n* or *p-n-p* transistor, where the middle section can be attached to a weak electric signal and controlled as a valve to regulate electric current flow between the two end sections. Transistors thus can be used as detectors and amplifiers for electric signals.

22.8 *Tin*

The principal source of tin is cassiterite (SnO_2), from which the element is prepared by carbon reduction. Although usually considered a metal, the element also exists in a nonmetallic form (α, or gray, tin), which is stable below 13°C.* Ordinary tin (β, or white, tin) is a rather inert metal, which resists corrosion because of an oxide coat. Because of its inertness, it is widely used as a protective plating for steel, especially in making tin cans. The steel is coated, either by being dipped in molten tin or by being made the cathode in an electrolytic bath which contains dissolved tin salts. For reasons mentioned in Sec. 19.4, tin-plated steel does not corrode until the tin coat is punctured, whereupon corrosion of the steel is accelerated by the presence of the tin.

---

* The conversion of shiny metallic tin to powdery gray tin was first observed on the tin organ pipes in early European cathedrals. At the low temperatures prevalent in these unheated churches the metallic pipes slowly developed grotesque, cancerous "growths." The phenomenon, called *tin plague* or *tin disease*, was first blamed on the devil, then on microorganisms, and finally on the more prosaic conversion of β to α tin. The tin swells because the density of α tin (5.75 g/cm³) is less than that of β tin (7.31 g/cm³). *Damnant quod non intelligunt.*

Two series of tin compounds, stannous ($+2$) and stannic ($+4$), are known. The $+2$ state is formed when metallic tin is dissolved in acid solution; however, the rate of reaction is rather slow. In solution, the Sn^{2+} ion is colorless and hydrolyzes according to the reaction

$$Sn^{2+} + 2H_2O \rightleftharpoons SnOH^+ + H_3O^+$$

for which the equilibrium constant is about 0.01. Thus, Sn^{2+} is about as strong an acid as HSO_4^-. Gradual addition of base to a solution of a stannous salt precipitates a white solid usually described as stannous hydroxide, $Sn(OH)_2$. Further addition of base dissolves the precipitate to form stannite ion, which can be written, for example, as either $Sn(OH)_3^-$ or $HSnO_2^-$. Stannite ion is a powerful reducing agent. Furthermore, on standing, solutions of stannite disproportionate to give 0 and $+4$ oxidation states:

$$2Sn(OH)_3^- \longrightarrow Sn(s) + Sn(OH)_6^{2-} \qquad \Delta G^\circ = -4.6 \text{ kJ}$$

In acid solution stannous ion is frequently complexed with anions. For example, in chloride solutions the whole series $SnCl^+$, $SnCl_2$, $SnCl_3^-$, and $SnCl_4^{2-}$ has been identified. Solutions of stannous chloride are frequently used as convenient, mild reducing agents. The reducing species in these solutions is usually represented as Sn^{2+} and assigned an electrode potential

$$Sn^{4+} + 2e^- \longrightarrow Sn^{2+} \qquad E^\circ = +0.15 \text{ V}$$

Chloride complexing of Sn^{2+} would make the potential slightly more positive, but the effect is canceled since chloride ion also forms complexes with Sn^{4+}.

In the stannic state, tin is often represented as the simple Sn^{4+} ion. However, because of its high charge, Sn^{4+} probably does not exist as such in aqueous solutions. When base is added to stannic solutions, a white precipitate forms; it may be $Sn(OH)_4$ or, more probably, a hydrated oxide $SnO_2 \cdot xH_2O$. The precipitate is soluble in excess base to give stannate ion, usually written $Sn(OH)_6^{2-}$ or SnO_3^{2-}.

Both stannous and stannic sulfides are insoluble in water and can be precipitated by H_2S in acid solution. Stannic sulfide, SnS_2, is a yellow solid which is soluble in high concentrations of sulfide ion. The reaction can be written

$$SnS_2(s) + S^{2-} \longrightarrow SnS_3^{2-}$$

where the complex ion SnS_3^{2-} is called the thiostannate ion and is analogous to stannate ion, SnO_3^{2-}. The dissolving of SnS_2 in excess S^{2-} can be used to distinguish it from another yellow sulfide CdS. Owing to the stability of the thiostannate ion, brown-black, insoluble stannous sulfide, SnS, can be oxidized by the relatively poor oxidizing agent S_2^{2-}, polysulfide ion:

$$SnS(s) + S_2^{2-} \longrightarrow SnS_3^{2-}$$

When solutions of thiostannate are acidified, SnS_2 is precipitated.

22.9 Lead

Conforming to the general trend of increasing metallic character down a group, lead is the most metallic of the group IV elements. Like tin, it shows oxidation states of $+2$ and $+4$, but the $+4$ state is more highly oxidizing. It might be pointed out here that on the right side of the periodic table, the heavier elements show a maximum oxidation state corresponding to the group number and a second state two units lower, e.g., in group III, Tl^{3+} and Tl^+, and in group V, Bi^{5+} and Bi^{3+}. This presumably stems from leaving a pair of s electrons (the so-called "inert pair") in the outer shell.

Lead occurs principally as the mineral galena, PbS, from which the element is produced in several different ways. In one of them the sulfide ore is roasted in air until it is completely converted to the oxide, which is then reduced with carbon in a small blast furnace:

$$2PbS(s) + 3O_2(g) \longrightarrow 2PbO(s) + 2SO_2(g)$$

$$2PbO(s) + C(s) \longrightarrow 2Pb(l) + CO_2(g)$$

In an alternative process the sulfide ore is only partially oxidized by air, the product containing a mixture of PbO, PbS, and $PbSO_4$. This mixture is then smelted in the absence of air, with the result that the PbS reduces PbO and $PbSO_4$ to lead:

$$PbS(s) + 2PbO(s) \longrightarrow 3Pb(l) + SO_2(g)$$

$$PbS(s) + PbSO_4(s) \longrightarrow 2Pb(l) + 2SO_2(g)$$

The crude lead may contain impurities such as antimony, copper, and silver. The silver is generally recovered by extracting it with molten zinc since zinc and lead are only slightly soluble in each other whereas silver is 270 times more soluble in molten zinc than in molten lead. If lead of high purity is required, it can be refined by an electrolytic process analogous to that used for copper (Sec. 20.2). Pure lead is a soft, low-melting metal which, when freshly cut, has a silvery luster that rapidly dulls and turns blue-gray on exposure to air. The tarnishing is due to the formation of a surface coat of oxides and carbonates. Primary uses of lead are in the manufacture of lead storage batteries (Sec. 14.4), alloys such as type metal and solder, and "white-lead" paint (hydrated lead hydroxycarbonate). The latter has been indicted as a source of chronic lead poisoning in ghetto children (lead salts generally have a sweet taste, which may contribute a physiological component to the urge to teethe on an old lead-painted surface). In spite of its toxicity, "white lead" has excellent adhering and covering ability; so it continues in wide use.

Practically all the common lead compounds correspond to lead in the $+2$ state. This state is called plumbous, from the Latin name for the element, *plumbum*. In aqueous solutions of plumbous salts, for example, $Pb(NO_3)_2$, the lead is usually formulated as Pb^{2+} ion. However, like stannous ion, Pb^{2+} forms many complex ions. The series of equilibria

$$PbCl^+ \rightleftharpoons Pb^{2+} + Cl^- \qquad K = 0.8$$

$$PbBr^+ \rightleftharpoons Pb^{2+} + Br^- \qquad K = 0.07$$

$$PbI^+ \rightleftharpoons Pb^{2+} + I^- \qquad K = 0.03$$

shows that, although the complexes are not especially stable, there is increasing stability in going from chloride to iodide. This trend of stability is the same as that found for other heavy-metal ions on the right side of the periodic table, for example, Hg^{2+}, Cd^{2+}, and Sn^{2+}. When the halide concentration of plumbous solutions is increased, insoluble plumbous halides form. In excess halide ion the precipitates redissolve, presumably because of the formation of complex ions of the type $PbCl_3^-$ and $PbBr_4^{2-}$. Unlike the two other common insoluble chlorides, $AgCl$ and Hg_2Cl_2, $PbCl_2$ can also be dissolved by raising the temperature.

Plumbous ion hydrolyzes somewhat less than stannous ion. When base is added, white $Pb(OH)_2$ is precipitated. Being amphoteric, it dissolves in excess base to form plumbite ion $[Pb(OH)_3^-$, or $HPbO_2^-]$. Unlike stannite ion, plumbite ion is stable in solution. The potential

$$Pb(s) + 3OH^- \longrightarrow Pb(OH)_3^- + 2e^- \qquad E° = +0.54\ V$$

indicates that lead in basic solution is a stronger reducing agent than it is in acid solution:

$$Pb(s) \longrightarrow Pb^{2+} + 2e^- \qquad E° = +0.13\ V$$

With most -2 anions Pb^{2+} forms insoluble salts, for example, $PbSO_4$, $PbCO_3$, PbS, $PbCrO_4$, and $PbHPO_4$. Lead sulfide is the least soluble of these, and the others convert to it in the presence of sulfide ion.

The principal compound of lead in the $+4$, or plumbic, state is PbO_2, lead dioxide. This compound, used in large amounts for the cathode of lead storage batteries (Sec. 14.4), can be made by oxidation of plumbite with hypochlorite ion in basic solution. The reaction can be written

$$Pb(OH)_3^- + ClO^- \longrightarrow Cl^- + PbO_2(s) + OH^- + H_2O$$

With acid solutions PbO_2 is a potent oxidizing agent:

$$PbO_2(s) + 4H_3O^+ + 2e^- \longrightarrow Pb^{2+} + 6H_2O \qquad E° = +1.46\ V$$

which is made even more potent in the presence of concentrated acid and anions which precipitate Pb^{2+}. In very concentrated solutions of base PbO_2 dissolves to form plumbates, such as PbO_4^{4-}, PbO_3^{2-}, and $Pb(OH)_6^{2-}$. Red lead, Pb_3O_4, much used as an undercoat for painting structural steel, can be considered to be plumbous plumbate, Pb_2PbO_4. Its use in preventing corrosion depends on the fact that, as a strong oxidizing agent, it renders iron passive (Sec. 19.2 and 19.4), perhaps by forming an Fe_3O_4 coat.

Like most heavy metals, lead and its compounds are poisonous. For example, the decline of the Roman Empire has been attributed in part to chronic lead poisoning brought about by use of lead pipes in the water system. Being reserved to the wealthy, there was a selective decimation of the aristocracy. More recently, lead has come under attack as an environmental pollutant because of its emission in automobile exhausts. As noted in Chap. 26, tetraethyl lead, $Pb(C_2H_5)_4$, is one of the most effective antiknock agents in the internal-combustion engine. Its use is being phased out, not just because of toxicity effects but because lead in an exhaust deactivates afterburner catalysts. Fairly large doses of lead are required for toxicity, but the danger is amplified because the lead tends to accumulate in the body (central nervous system). The toxicity may be due to the fact that lead and other heavy metals are powerful inhibitors of enzyme reactions.

22.10   Qualitative analysis

Lead and tin precipitate as sulfides in $0.3\ N$ acid solution, although much of the lead may precipitate as white $PbCl_2$ along with $AgCl$ and Hg_2Cl_2 when HCl is added to the original unknown. $PbCl_2$ can be separated from the other two by leaching with hot water. Addition of K_2CrO_4 and acetic acid to the leach solution gives the confirmatory yellow precipitate $PbCrO_4$.

Lead sulfide (black) can be separated from tin sulfide, either SnS (brown-black) or SnS_2 (yellow), by treatment with ammonium polysulfide, which converts SnS and SnS_2 to $SnS_3{}^{2-}$ but leaves the PbS undissolved. PbS can be dissolved with hot HNO_3 (unlike black HgS) and reprecipitated as a white sulfate with H_2SO_4 (unlike Cd^{2+}, Bi^{3+}, and Cu^{2+}). To confirm, the $PbSO_4$ is dissolved in ammonium acetate and precipitated as $PbCrO_4$.

If to a solution containing $SnS_3{}^{2-}$ HCl is added in excess, the tin (unlike As_2S_3) stays in solution, probably as a chloride complex. Evaporation (to drive off H_2S) in the presence of iron followed by $HgCl_2$ addition confirms tin if a precipitate of white Hg_2Cl_2 or black mercury is observed.

The carbonate and bicarbonate anions can be detected easily by adding acid to an unknown and allowing any escaping gas to come in contact with $Ba(OH)_2$ solution. A white milkiness develops, owing to formation of $BaCO_3$.

Exercises

*22.1   Structure. How many carbon atoms are there in the unit cell shown for diamond in Fig. 22.2?

*22.2   Combustion. In oxidizing CH_4 to CO_2 and CO, what relative volumes of air (21 mole % O_2) and fuel would you need to ensure conversion 90% to CO_2 and 10% to CO? Assume kinetic factors are equal.   Ans. 9.3:1

*22.3   Stoichiometry. A given sample of $NaHCO_3$ is heated until it loses 10.0 percent of its weight. What percent of the $NaHCO_3$ has been decomposed to Na_2CO_3?

*22.4   *Isomers.* Draw the carbon skeletons for all the possible geometric isomers of heptane, C_7H_{16}.   *Ans. There are nine*

*22.5   *Optical isomerism.* What is the smallest hydrocarbon that exhibits optical isomerism? What is the smallest alcohol that exhibits optical isomerism?

*22.6   *Soap micelle.* The density of a liquid hydrocarbon generally is on the order of 0.75 g/cm^3. Figure out a crude estimate of how many soap molecules, $NaOOCC_{17}H_{35}$, there are in a typical micelle.
Ans. About 100

*22.7   *Polymerization.* Suppose the R group of the organic peroxide used to initiate addition polymerization is C_6H_5CO. How long a polymer chain of vinyl chloride would you need so that the end group as above accounts for less than 1.0 percent of the total weight?

*22.8   *Equations.* Write chemical equations for the following:
a   Reduction of silica to elemental Si by Mg
b   Chlorination of elemental Si to silicon tetrachloride
c   Hydrolysis of silicon tetrachloride to form silica
d   Dissolving of silica in hydrofluoric acid to give hexafluorosilicate(IV)

*22.9   *Ultrapure germanium.* Make a flow sheet showing the conversion of germanium dioxide to ultrapure germanium. Go successively through impure Ge, $GeCl_4$, GeO_2, pure Ge, and ultrapure Ge. Write chemical equations for any reactions involved, and describe any other processes required.

*22.10   *Silicates.* Make a tabular summary of the classification of silicates starting with the orthosilicate anion and ending up with SiO_2. What is the essential basis of your classification?

**22.11   *Structure.* Show that the diamond structure can be represented as two interpenetrating face-centered cubic lattices displaced relative to each other by one-fourth the body diagonal of the cube shown in Fig. 22.2.

**22.12   *Bonding.* Explain why the near carbon-carbon bond in graphite is shorter than that in diamond.

**22.13   *Surface area.* Suppose you have 1 cm^3 of charcoal that has an adsorbing area of 50 m^2. If you want to get the equivalent surface from 1 cm^3 of diamond, into what sized cubes should you grind it? If an H_2S molecule needs an area of 0.060 nm^2 to be adsorbed, how many moles of H_2S can you adsorb on this surface?
Ans. 1.2×10^{-5} cm, 0.0014 mol

**22.14   *Stoichiometry.* What would be the pH of a solution made by dissolving 5.00 g of CaC_2 in enough water to make 250 ml of solution?

**22.15   Thermodynamics. You are given $\Delta H°$ of formation as follows: $CS_2(g)$, $+115$ kJ; $CCl_4(g)$, -107 kJ; and $S_2Cl_2(l)$, -60 kJ. (a) Calculate $\Delta H°$ for the process $CS_2(g) + 3Cl_2(g) \longrightarrow CCl_4(g) + S_2Cl_2(l)$. (b) Would you expect $\Delta G°$ to be greater or less than this? Explain.   *Ans. -282 kJ, less negative*

**22.16   Heat of combustion. Calculate the heat liberated on combustion to CO_2 and H_2O of 1 kg of CH_4, C_2H_6, and C_3H_8, respectively, given the following heats of formation: $CO_2(g)$, -394 kJ/mol; $H_2O(g)$, -242 kJ/mol; $CH_4(g)$, -74.8 kJ/mol; $C_2H_6(g)$, -84.7 kJ/mol; and C_3H_8, -103.8 kJ/mol.

**22.17   Thermodynamics. The $\Delta G°$ of formation for $CO_2(g)$ is -394 kJ/mol; that of $CO(g)$ is -137 kJ/mol. Calculate K_p for the reaction $CO_2(g) + C(s) \rightleftharpoons 2CO(g)$ at 25°C. If $\Delta H°$ for this process is $+177$ kJ, at what temperature would K_p become equal to 1.00?
Ans. 9.33×10^{-22}, 927°K

**22.18   Dissociation. Air generally contains 0.04% by volume of CO_2. If the solubility of CO_2 in water at 1 atm pressure is 0.03 M, what would be the equilibrium concentration of CO_2 dissolved in water in contact with air? What pH would this give rise to?

**22.19   Hydrolysis. Calculate the pH of a solution made by dissolving 100 g of $Na_2CO_3 \cdot 10H_2O$ in enough water to make 500 ml of solution.   *Ans. 12.08*

**22.20   Symmetry. Write the structural formulas of the three different isomers of trichlorobenzene. Indicate the symmetry elements for each.

**22.21   Mechanism. Give structural formulas for the two possible products resulting from the addition of HCl to $CH_3CH{=}CH_2$. Indicate the probable mechanism of the reaction and the probable major product.

**22.22   Polymer. Saran is a copolymer of about 85% by number of $CH_2{=}CCl_2$ and 15% of $CH_2{=}CHCl$. How many grams of HCl would you liberate to the atmosphere on incinerating one kilogram of such a polymer? Ignore the end groups.

**22.23   Silicates. Identify each of the following, and indicate the main structural feature that characterizes it:
 a   Pyroxene chain
 b   Amphibole chain
 c   Talc
 d   Mica
 e   Feldspar

**22.24   Silicates. How are the zeolites related to silica? Why are they useful as ion exchangers and molecular sieves?

**22.25   Water glass. Write chemical equations and account for the formation of brown-colored fronds when ferric nitrate crystals are dropped into an aqueous solution of water glass.

Exercises

22.26 *Tin.* Indicate how you would carry out the following conversions. Write equations where possible.

  a Cassiterite to elemental tin
  b Elemental tin to stannous hydroxide
  c Stannous hydroxide to sodium stannate
  d Sodium stannate to stannic sulfide
  e Stannic sulfide to sodium thiostannate

22.27 *Hydrolysis.* Calculate the percent hydrolysis of Sn^{2+} in a solution that is initially $0.10\ M\ Sn^{2+}$. *Ans. 27%*

22.28 *Solubility.* Given that $K_{sp} = 1.6 \times 10^{-5}$ for $PbCl_2(s) \rightleftharpoons Pb^{2+} + 2Cl^-$ and $K = 0.8$ for $PbCl^+ \rightleftharpoons Pb^{2+} + Cl^-$, what would be the K'_{sp} for $PbCl_2(s) \rightleftharpoons PbCl^+ + Cl^-$? Which predicts a higher solubility, K_{sp} or K'_{sp}?

22.29 *Unit cell.* If the carbon-carbon distance in the diamond unit cell shown in Fig. 22.2 is equal to 0.154 nm, what is the edge length of the unit cell? *Ans. 0.356 nm*

22.30 *Symmetry.* Indicate all the symmetry elements, and tell where they are located in the diamond unit cell shown in Fig. 22.2.

22.31 *Thermodynamics.* The standard state of carbon at 25°C and 1 atm pressure is graphite. Under these same conditions diamond has an enthalpy 1.88 kJ/mol higher and an entropy 3.26 J mol^{-1} deg^{-1} lower. (*a*) Calculate $\Delta G°$ under standard conditions for the process graphite $\longrightarrow$ diamond. (*b*) Assuming no change of ΔH or ΔS with T or P calculate (1) the temperature at 1 atm pressure and (2) the pressure at 25°C that would be required to make diamond stable.

Ans. 2.85 kJ, $-576°K$, 14,800 atm

22.32 *Bond length.* The anion in CaC_2 is C_2^{2-}. Account for the fact that the bond length in C_2^{2-} (0.120 nm) is longer than that in C_2 (0.131 nm) but about the same as that in C_2H_2 (0.120 nm). (*Hint:* Use MO theory.)

22.33 *Simultaneous equilibria.* Calculate the pH of a solution made by dissolving 100 g of $NaHCO_3$ in enough water to make 1500 ml of solution.

22.34 *Mechanism.* Suppose you have a sample of optically active C_4H_9OH that is pure dextro. You now carry out a reaction with HBr, the result of which is substitution of Br for OH. If the optical activity has disappeared, what could you conclude about the probable mechanism of the reaction?

22.35 *Electrode potential.* If the K_{sp} of $PbSO_4$ is 1.3×10^{-8}, what would be the electrode potential for $PbO_2(s)$ acting as an oxidizing agent in the presence of $1\ M\ H_2SO_4$? *Ans. 1.64 V*

In group V there is a complete change of properties from nonmetallic to metallic in going down the group. The lighter members of the group, nitrogen and phosphorus, are typical nonmetals and form only acidic oxides; the middle members, arsenic and antimony, are semimetals and form amphoteric oxides; the heaviest member, bismuth, is a metal and forms mostly basic oxides. The terms "pnicogen" and "pnictide," derived from the Greek word *pnigmos,* meaning suffocation, are sometimes used to designate the group V elements and their simple compounds with electropositive elements.

609

23.1 Group properties

Figure 23.1 summarizes some of the specific properties of the group V elements. As shown, each of the atoms has five valence electrons (ns^2np^3) in its outermost principal quantum level. Since the outer octet is only slightly more than half filled, complete ionization by loss of five electrons or gain of three electrons is unlikely. Sharing electrons with more electronegative atoms would correspond to a maximum oxidation state of $+5$; sharing with less electronegative atoms, to a minimum oxidation state of -3. Both of these states are observed for all the group V elements, though the stability of the -3 state decreases down the group. In addition, a $+3$ state corresponding to leaving an unshared pair of electrons on the group V atom is common to all. Nitrogen and phosphorus are unusual in that they show all the oxidation states from -3 to $+5$, inclusive.

The pronounced change from nonmetallic to metallic behavior down the group is due principally to increasing size of the atoms. As the ionization potentials of Fig. 23.1 indicate, it is much more difficult to pull an electron off the small nitrogen atom (first IP 14.5 eV) than off the larger bismuth atom (first IP 8 eV). Furthermore, the nitrogen

Symbol	Z	Electronic configuration	Melting point, °C	Boiling point, °C	Ionization potential, eV	Electrode potential, V
N	7	(2) $2s^2 2p^3$	-210.0	-195.8	14.5	$+1.25$ (*from* NO_3^-)
P	15	(10) $3s^2 3p^3$	44.1	280	11.0	-0.50 (*from* H_3PO_3)
As	33	(28) $4s^2 4p^3$	Sublimes	Sublimes	10	$+0.23$ (*from* As_4O_6)
Sb	51	(46) $5s^2 5p^3$	631	1380	8.6	$+0.21$ (*from* $Sb(OH)_2^+$)
Bi	83	(78) $6s^2 6p^3$	271	1500	8	$+0.32$ (*from* $Bi(OH)_2^+$)

Fig. 23.1

Elements of group V.

atom, being small and holding its electrons tightly, can, like the preceding element, carbon, form multiple bonds to other atoms. One of the results is that nitrogen forms simple diatomic molecules whereas under ordinary conditions other members of the group do not. However, phosphorus, arsenic, and possibly antimony do form discrete tetratomic molecules [P_4, As_4, and $Sb_4(?)$] in at least some of their allotropic forms, indicating that the tendency to form covalent bonds persists down the group. Bismuth, which holds its electrons least tightly and hence is most metallic, still retains some of this covalent character. It shows up, for example, in the fact that bismuth is not a very good metal—rather brittle, it has even greater electric resistance than mercury. Also, elemental bismuth has an extraordinarily high diamagnetism (repulsion out of magnetic fields) compared with other metals. This behavior indicates that the electron cloud in bismuth metal is not like that of a typical metal but that the electrons are somewhat restricted in their motion.

The increasing basicity of oxides going down the group is also primarily due to increasing size. As pointed out in Sec. 9.5, the action of an oxide or hydroxide as an acid or base depends on the extent of hydrolysis, which in turn depends on the charge and on the size of the atom. Since N^{3+} would be much smaller than Bi^{3+}, it would interact with water more strongly and be more likely to result in acid properties. Thus, it is not surprising that N_2O_3 is an acidic oxide, dissolving in water to give H_3O^+ and neutralizing bases, whereas Bi_2O_3 is a basic oxide, dissolving to give OH^- and neutralizing acids. Of the intermediate elements, phosphorus forms the acidic oxide P_2O_3, and arsenic and antimony form the amphoteric oxides As_2O_3 and Sb_2O_3.

Except for white phosphorus, none of the group V elements is particularly reactive in the elemental state. This is partly because of slowness of the reactions and partly because of the small electrode potentials. In the $+5$ state all the elements except phosphorus form compounds that are powerful oxidizing agents.

23.2 Nitrogen

Nitrogen is about one-third as abundant as carbon and occurs principally *free* as diatomic N_2 in the atmosphere and *combined* as Chile saltpeter ($NaNO_3$). In plants and animals N is found combined in the form of proteins, which average in composition 51% by weight C, 25% O, 16% N, 7% H, 0.4% P, and 0.4% S.

Elemental nitrogen is usually obtained by fractional distillation of liquid air. Since N_2 has a lower boiling point ($77.4°K$) than O_2 ($90.2°K$), it is more volatile and evaporates preferentially in the first fractions of gas (Sec. 15.9). Very pure N_2 can be made by thermal decomposition of some nitrogen compounds, such as ammonium nitrite, NH_4NO_2:

$$NH_4NO_2(s) \longrightarrow N_2(g) + 2H_2O(g) \qquad \Delta H° = -220 \text{ kJ}$$

It is interesting to note that pure nitrogen obtained from decomposition of compounds such as NH_4NO_2 was the key that led to the discovery of the noble gases. Lord Rayleigh, in 1894, was the first to note that nitrogen from the decomposition of compounds was of slightly lower density (1.2505 g/liter at STP) than the residual gas obtained from the atmosphere by removal of oxygen, carbon dioxide, and water (1.2572 g/liter at STP). In conjunction with Sir William Ramsay, Rayleigh removed the nitrogen from the air residue by various reactions, such as the combination of nitrogen with hot magnesium to form solid magnesium nitride, Mg_3N_2. After removal of the nitrogen, there was still some remaining gas which, unlike any gas known at that time, was completely unreactive. It was christened "argon" from the Greek word *argos* meaning "lazy." Later spectroscopic investigations showed that crude argon, and hence the atmosphere, contains the other noble-gas elements helium, neon, krypton, and xenon. Including the noble gases, the average composition of the earth's atmosphere is as shown in Fig. 23.2. In addition to the noble gases listed, there are traces of radon, Rn, in the atmosphere. The concentration

Fig. 23.2

Components of dry air.

Component	Percent by volume	Boiling point, °K
Nitrogen (N_2)	78.09	77.4
Oxygen (O_2)	20.95	90.2
Argon (Ar)	0.93	87.4
Carbon dioxide (CO_2)	0.023–0.050	Sublimes
Neon (Ne)	0.0018	27.2
Helium (He)	0.0005	4.2
Krypton (Kr)	0.0001	121.3
Hydrogen (H_2)	0.00005	20.4
Xenon (Xe)	0.000008	163.9

is very low and variable because radon is produced near radioactive deposits by nuclear disintegration of other elements and is itself unstable to nuclear decomposition. As can be seen from the data, nitrogen is by far the predominant constituent of the atmosphere.*

The N_2 molecule contains a triple bond, and the bonding molecular orbitals may be designated $\sigma_{p_x}^2 \pi_{p_y}^2 \pi_{p_z}^2$. Although very stable with respect to dissociation into single atoms, N_2 is thermodynamically unstable with respect to oxidation by O_2 in the presence of water to form nitrate ion, NO_3^-. It is fortunate that this reaction is very slow; otherwise, atmospheric N_2 and O_2 would combine with the oceans to form solutions of dilute nitric acid. In practice, nitrogen is frequently used when an inert atmosphere is required, as, for example, in incandescent lamp bulbs to retard filament evaporation.

The compounds of nitrogen, though not so numerous as those of carbon, are just as varied. In many respects, their chemical reactions are more complicated because there are usually no residues which retain identity throughout a reaction. Only a few of the compounds and their reactions can be considered here.

The principal compound of nitrogen is probably ammonia, NH_3. It occurs to a slight extent in the atmosphere, primarily as a product of the putrefaction of nitrogen-containing animal or vegetable matter. Commercially it is important as the most economic pathway for nitrogen *fixation*, i.e., the conversion of atmospheric N_2 into useful compounds. In the Haber process, synthetic ammonia is made by passing a nitrogen-hydrogen mixture through a bed of catalyst consisting of iron with oxides such as Al_2O_3 added. By using a temperature of about 500°C (a compromise between the requirements of kinetics and equilibrium) and a pressure of about 1000 atm, there is about 50 percent conversion of N_2 to NH_3.

$$N_2(g) + 3H_2(g) \longrightarrow 2NH_3(g) \qquad \Delta H° = -92 \text{ kJ}$$

*The genesis of the atmosphere is still a mystery. It is partly ascribed to degassing of the earth because of radioactive warming of the interior and because of exsolution due to solidification of the crust. The early atmosphere was probably reducing, but most of the lighter molecules (for example, H_2 and CH_4) were lost because the earth's gravity is not great enough to hold high-velocity molecules. The oxygen is surmised to have been added after photosynthetic plants developed. *Magnus ab integro saeculorum nascitur ordo.*

NH$_3$ is a polar molecule, pyramidal in shape, with the three H atoms occupying the base of the pyramid and an unshared pair of electrons, the apex. The structure leads to a compound which is easily condensed (liquefaction temperature of $-33\,^{\circ}$C) to a liquid of great solvent power. In many respects, liquid ammonia is as versatile a solvent as water, and, like water, it can dissolve a great variety of salts. In addition, it has the rather unique property of dissolving alkali and alkaline-earth metals to give solutions which contain solvated electrons. Thus, for example, when sodium is placed in liquid ammonia, it smoothly dissolves to give either a deep blue (dilute) or bronze (concentrated) solution. The dilute solutions are best described as containing ammoniated sodium ions of the type Na(NH$_3$)$_n{}^+$ and ammoniated electrons, e(NH$_3$)$_n{}^-$. The ammoniated electron, which can be visualized as an electron trapped in a cavity dug for itself in the midst of the ammonia molecules, acts very much like a large anion, at least in dilute solutions. In concentrated solutions, the electron is best described as a delocalized wave spread over the whole sample, as in a metal. When cooled to very low temperatures, the metal–ammonia solutions produce some very highly unusual solid compounds. These include, for example, Li(NH$_3$)$_4{}^0$ and Ca(NH$_3$)$_6{}^0$. In the former, four NH$_3$ molecules tetrahedrally surround a Li atom; in the latter, six NH$_3$ octahedrally surround a Ca atom. There is no anion in the crystal structure, just an electron gas that suffuses the whole structure.

Ammonia gas is very soluble in water, which is easily explained by the fact that both NH$_3$ and H$_2$O are polar molecules. Not so easy to explain is the basic character of the aqueous solutions formed. At one time it was thought that the NH$_3$ molecules react with H$_2$O to form molecules of the weak base ammonium hydroxide

$$\text{H:N:H:O:}$$

with H atoms below N and O

which could then dissociate into ammonium ions (NH$_4{}^+$) and hydroxide ions. However, NMR experiments indicate that in aqueous NH$_3$ solutions, protons jump back and forth so rapidly between N and O atoms that the distinction between NH$_3$ plus H$_2$O and NH$_4$OH is arbitrary. Thus, the basic nature of aqueous NH$_3$ can be represented by either of the equilibria

$$\text{NH}_3 + \text{H}_2\text{O} \rightleftharpoons \text{NH}_4{}^+ + \text{OH}^-$$

$$\text{NH}_4\text{OH} \rightleftharpoons \text{NH}_4{}^+ + \text{OH}^-$$

and K for either is 1.8×10^{-5}. By neutralizing NH$_3$ with acids, ammonium salts can be formed; these contain the tetrahedral NH$_4{}^+$ ion. They resemble potassium salts, except that they give slightly acid solutions. This can be interpreted either as a hydrolysis

$$\text{NH}_4{}^+ + 2\text{H}_2\text{O} \rightleftharpoons \text{NH}_4\text{OH} + \text{H}_3\text{O}^+$$

or as a dissociation of a Brønsted-Lowry acid

$$NH_4^+ + H_2O \rightleftharpoons NH_3 + H_3O^+$$

and K for either is 5.5×10^{-10}. Some ammonium salts, such as ammonium nitrate, NH_4NO_3, and ammonium dichromate, $(NH_4)_2Cr_2O_7$, are thermally unstable because they undergo autooxidation. As illustration, NH_4NO_3 sometimes explodes when heated to produce nitrous oxide, N_2O, by the reaction

$$NH_4NO_3(s) \longrightarrow N_2O(g) + 2H_2O(g) \qquad \Delta H^\circ = -37.0\ kJ$$

Whereas ammonia and ammonium salts represent nitrogen in its lowest oxidation state (-3), the highest oxidation state of nitrogen ($+5$) appears in the familiar compounds nitric acid (HNO_3) and nitrate salts. Nitric acid is one of the most important industrial acids, and large quantities of it are produced, principally by the catalytic oxidation of ammonia. In the process, called the *Ostwald process*, the following steps are important:

$$4NH_3(g) + 5O_2(g) \xrightarrow{Pt} 4NO(g) + 6H_2O(g)$$

$$2NO(g) + O_2(g) \longrightarrow 2NO_2(g)$$

$$3NO_2(g) + 3H_2O \longrightarrow 2H_3O^+ + 2NO_3^- + NO(g)$$

In the first step a mixture of ammonia and air is passed over a platinum catalyst heated to about $800^\circ C$. On cooling, the product nitric oxide (NO) is oxidized to nitrogen dioxide (NO_2) which disproportionates in solution to form nitric acid and NO. By keeping a high concentration of O_2, the remaining NO is converted to NO_2, and the last reaction is driven to the right. To get 100 percent acid, it is necessary to distill off volatile HNO_3.

Pure nitric acid is a colorless liquid which on exposure to light turns brown because of slight decomposition to brown NO_2:

$$4HNO_3 \longrightarrow 4NO_2(g) + O_2(g) + 2H_2O$$

It is a strong acid in that it is 100% dissociated in dilute solutions to H_3O^+ and nitrate ion, NO_3^-. Like carbonate ion (Fig. 22.4), nitrate ion is planar with a C_3 axis perpendicular to the plane and C_2 axes in the plane; it is sometimes represented as a resonance hybrid of three contributing formulas. The ion is colorless and forms a great variety of nitrate salts, most of which are quite soluble in aqueous solutions.* Owing to the low complexing ability of nitrate ion, practically all these salts are dissociated in aqueous solution.

In acid solution nitrate ion is a good oxidizing agent. By proper choice of concentrations and reducing agents it can be reduced to compounds of nitrogen in all the other oxidation states. The possible half-reactions and their electrode potentials are

* Because of the solubility of the nitrates it is not usual to find solid nitrates occurring naturally as minerals. The extensive deposits of $NaNO_3$ in Chile occur in a desert region where there is insufficient rainfall to wash them away. These deposits probably originated from the decomposition of nitrogenous deposits of marine organisms which were cut off from the sea. *Aquae nisi faillit augur annosa cornix.*

$$NO_3^- + 2H_3O^+ + e^- \longrightarrow NO_2(g) + 3H_2O$$
$$E° = +0.79 \text{ V}$$

$$NO_3^- + 3H_3O^+ + 2e^- \longrightarrow HNO_2 + 4H_2O$$
$$E° = +0.94 \text{ V}$$

$$NO_3^- + 4H_3O^+ + 3e^- \longrightarrow NO(g) + 6H_2O$$
$$E° = +0.96 \text{ V}$$

$$2NO_3^- + 10H_3O^+ + 8e^- \longrightarrow N_2O(g) + 15H_2O$$
$$E° = +1.12 \text{ V}$$

$$2NO_3^- + 12H_3O^+ + 10e^- \longrightarrow N_2(g) + 18H_2O$$
$$E° = +1.25 \text{ V}$$

$$NO_3^- + 8H_3O^+ + 6e^- \longrightarrow NH_3OH^+ + 10H_2O$$
$$E° = +0.73 \text{ V}$$

$$2NO_3^- + 17H_3O^+ + 14e^- \longrightarrow N_2H_5^+ + 23H_2O$$
$$E° = +0.83 \text{ V}$$

$$NO_3^- + 10H_3O^+ + 8e^- \longrightarrow NH_4^+ + 13H_2O$$
$$E° = +0.88 \text{ V}$$

Since all the electrode potentials are quite positive, nitrate ion is a much better oxidizing agent than H_3O^+ by itself. This presumably is responsible for the observation that some metals such as copper and silver, which are too poor as reducing agents to dissolve in HCl, for example, will dissolve in HNO_3. Both of these acids contain the oxidizing agent H_3O^+, but only nitric acid has the additional oxidizing agent NO_3^-. Some metals, such as gold, which are insoluble in HCl and in HNO_3 are soluble in a mixture of the two acids. This mixture, called *aqua regia*, usually consists of one part of concentrated HNO_3 to three parts of concentrated HCl. As mentioned in Sec. 20.4, the dissolving power of aqua regia is due to the oxidizing ability of nitrate ion in strong acid plus the complexing ability of chloride ion.

Reduction of NO_3^- usually produces a mixed product. Since the various electrode potentials of nitrate shown above are very roughly the same, the reduction may yield any of several species. The actual composition of the product depends on the rates of the different reactions. These rates in turn are influenced by the concentration of NO_3^-, the concentration of H_3O^+, the temperature, and the reducing agent used. Thus, for example, in *concentrated* nitric acid, copper reacts to give brown NO_2 gas, but in *dilute* nitric acid, copper reacts to form colorless NO gas. However, since NO is easily oxidized by air to NO_2, some brown fumes may also appear when dilute nitric acid is used.

As can be seen from the above list of reduction products of NO_3^-, compounds of nitrogen are possible in the +4, +3, +2, +1, −1, and −2 states, as well as +5 and −3. Some of the more common representative species of these states are discussed below:

1 The +5 state. In addition to nitric acid and the nitrates, nitrogen corresponding to the +5 state is found in nitrogen pentoxide, N_2O_5.

This material, which is the acid hydride of HNO_3, can be produced by treating concentrated nitric acid with a very strong dehydrating agent such as phosphoric oxide, P_4O_{10}. At room temperature, N_2O_5 is a white solid which decomposes slowly into NO_2 and oxygen. At slightly elevated temperatures, it may explode. With water it reacts quite vigorously to form HNO_3.

   2   *The +4 state.* When concentrated nitric acid is reduced with metals, brown fumes are evolved. The brown gas is NO_2, nitrogen dioxide. It has a bent molecule ONO with a bond angle of 134°. Since the molecule contains an odd number of valence electrons (five from the nitrogen and six from each of the oxygens), it should be, and is, paramagnetic. When the brown NO_2 gas is cooled, its color fades, and the paramagnetism diminishes. These observations are interpreted as indicating that two NO_2 molecules pair up (dimerize) to form a single molecule of N_2O_4, nitrogen tetroxide. The equilibrium

$$2NO_2(g) \rightleftharpoons N_2O_4(g) \qquad \Delta H° = -61.1 \text{ kJ}$$

is such that at 60°C and 1 atm pressure half the nitrogen is present as NO_2 and half as N_2O_4. As the temperature is raised, decomposition of N_2O_4 is favored. The NO_2–N_2O_4 mixture is poisonous and is a strong oxidizing agent. As already mentioned in connection with the Ostwald process, NO_2, or, more correctly, a mixture of NO_2 and N_2O_4, dissolves in water to form HNO_3 and NO.

   3   *The +3 state.* The most common representatives of the +3 oxidation state are the salts called nitrites. Nitrites such as $NaNO_2$ can be made by heating sodium nitrate above its melting point:

$$2NaNO_3(l) \longrightarrow 2NaNO_2(l) + O_2(g)$$

They can also be made by chemical reduction of nitrates with such substances as C and Pb. Nitrites are important industrially in the manufacture of azo dyes, which contain the —N=N— group as in azobenzene, C_6H_5—N=N—C_6H_5. The azo dyes are intensely colored and account for more than half of synthetic dyes. When acid is added to a solution of nitrite, the weak acid HNO_2, nitrous acid ($K_{diss} = 4.5 \times 10^{-4}$), is formed. It is unstable and slowly decomposes by several complex reactions, including

$$3HNO_2 \longrightarrow H_3O^+ + NO_3^- + 2NO(g)$$

$$2HNO_2 \longrightarrow NO(g) + NO_2(g) + H_2O$$

   4   *The +2 state.* The oxide NO, nitric oxide, is, like NO_2, an "odd" molecule in that it contains an uneven number of electrons. However, unlike NO_2, NO is colorless and does not dimerize appreciably in the gas phase. In the liquid phase, as shown by a decrease of paramagnetism, some dimerization occurs to form N_2O_2. For simple NO molecules in the gas phase, there is magnetic-resonance evidence that the odd electron spends half its time with the N and half with the O. In the molecular-orbital description, there are three bonding pairs as in N_2 plus the odd electron which is accommodated in a π^* antibonding

orbital. Thus, the NO molecule is intermediate between N_2 and O_2 (Sec. 3.9).

Nitric oxide can be made in several ways:

$$4NH_3(g) + 5O_2(g) \longrightarrow 4NO(g) + 6H_2O$$

$$3Cu(s) + 8H_3O^+ + 2NO_3^- \longrightarrow 3Cu^{2+} + 2NO(g) + 12H_2O$$

$$N_2(g) + O_2(g) \longrightarrow 2NO(g)$$

The first of these reactions is the catalytic oxidation that is the first step of the Ostwald process for making HNO_3. The second is observed with dilute nitric acid but not with concentrated. The third is extremely endothermic (by 180 kJ) and can occur only at high temperature or when large amounts of energy are added. Apparently, this last reaction occurs when lightning bolts pass through the atmosphere and is one of the paths by which atmospheric nitrogen is made available to plants. It also occurs in the internal-combustion engine. The higher the temperature an engine operates at or the higher its air-fuel ratio (up to a certain limit), the more NO is formed in the exhaust gases. The role of nitrogen oxides in air pollution is considered further in Chap. 26. In air, NO is rapidly oxidized to brown NO_2:

$$2NO(g) + O_2(g) \longrightarrow 2NO_2(g)$$

Nitric oxide also combines with many transition-metal cations to form complex ions. The most familiar of these complexes is $FeNO^{2+}$, the nitroso ferrous ion, which forms in the brown-ring test for nitrates. When concentrated sulfuric acid is carefully poured into a solution containing ferrous ion and nitrate, a brown layer appears at the junction of the H_2SO_4 and the nitrate-containing solution. The NO for the complex is formed by reduction of NO_3^- by Fe^{2+}.

5 *The +1 state.* When solid ammonium nitrate is gently heated, it melts and undergoes autooxidation according to the following equation:

$$NH_4NO_3(l) \longrightarrow N_2O(g) + 2H_2O(g)$$

The compound formed, N_2O, called nitrous oxide, or *laughing gas*, has a linear molecule with the oxygen at one end. Although rather inert at low temperature, N_2O decomposes to N_2 and O_2 at higher temperatures. Perhaps because of this decomposition, substances which burn briskly in air actually burn more vigorously in nitrous oxide. Compared with the other oxides of nitrogen, nitrous oxide is considerably less poisonous. However, small doses are mildly intoxicating; large doses produce general anesthesia and in dentistry are frequently used for this purpose. Nitrous oxide has an appreciable solubility in fats, a property which has been exploited in making self-whipping cream. Cream is packaged with nitrous oxide under pressure to increase its solubility. When the pressure is released, the nitrous oxide escapes to form tiny bubbles, which produce whipped cream.

6 *The −1 state.* Hydroxylamine, NH_2OH, is representative of nitrogen with oxidation number −1. It can be considered to be derived

from NH_3 by substituting a hydroxyl group for one of the hydrogen atoms. However, the preparation of NH_2OH involves not NH_3 but rather the reduction of nitrates or nitrites by appropriate reducing agents such as sulfur dioxide (SO_2) or tin. Pure hydroxylamine is a solid at room temperature and is unstable, especially at higher temperatures. The decomposition, which is sometimes explosive, produces a mixture of products including NH_3, H_2O, N_2, and N_2O. In dilute aqueous solution the decomposition is slow. Like NH_3, NH_2OH has an unshared pair of electrons and so can pick up a proton to form NH_3OH^+:

$$
\begin{array}{c}
\text{H} \\
\overset{..}{\text{H}:\overset{..}{\text{N}}:\overset{..}{\text{O}}:\text{H}} + \text{H}_2\text{O} \rightleftharpoons \left[\begin{array}{c} \text{H} \\ \text{H}:\overset{..}{\text{N}}:\overset{..}{\text{O}}:\text{H} \\ \text{H} \end{array} \right]^+ + \text{OH}^- \quad K = 6.6 \times 10^{-8}
\end{array}
$$

Thus, hydroxylamine solutions are slightly basic, but less so than NH_3 solutions. Analogous to ammonium salts, such as NH_4Cl, there are hydroxylammonium salts, such as NH_3OHCl. Since hydroxylamine and its salts correspond to nitrogen in an intermediate oxidation state, they can act both as oxidizing agents and as reducing agents.

7 *The -2 state.* In many ways similar to ammonia is the compound hydrazine, N_2H_4. This compound can be made by bubbling chlorine through a solution of ammonia:

$$ \text{Cl}_2(g) + 4\text{NH}_3 \longrightarrow \text{N}_2\text{H}_4 + 2\text{NH}_4^+ + 2\text{Cl}^- $$

When pure, N_2H_4 is a colorless liquid at room temperature. Like ammonia, it is a good solvent for many salts and even for the alkali metals. Hydrazine is unstable with respect to disproportionation

$$ 2\text{N}_2\text{H}_4(l) \longrightarrow \text{N}_2(g) + 2\text{NH}_3(g) + \text{H}_2(g) \qquad \Delta H^\circ = -193 \text{ kJ} $$

and is violently explosive in the presence of air or other oxidizing agents. It is quite poisonous. In aqueous solution it acts as a base, since it can add one or two protons to the unshared pairs of electrons:

$$
\begin{array}{c}
\text{H} \ \text{H} \\
\overset{..}{\text{H}:\overset{..}{\text{N}}:\overset{..}{\text{N}}:\text{H}} + \text{H}_2\text{O} \longrightarrow \left[\begin{array}{c} \text{H} \ \text{H} \\ \text{H}:\overset{..}{\text{N}}:\overset{..}{\text{N}}:\text{H} \\ \text{H} \end{array} \right]^+ + \text{OH}^- \quad K = 9.8 \times 10^{-7}
\end{array}
$$

Salts of the type N_2H_5Cl and $N_2H_6Cl_2$ are known. In aqueous solution hydrazine and its salts are good oxidizing and reducing agents, though reaction is sometimes slow.

Hydrazine has become important as a rocket propellant. For example, the reaction

$$ \text{N}_2\text{H}_4(l) + 2\text{H}_2\text{O}_2(l) \longrightarrow \text{N}_2(g) + 4\text{H}_2\text{O}(g) \qquad \Delta H^\circ = -643 \text{ kJ} $$

which takes place in the presence of Cu^{2+} ion as catalyst, is strongly exothermic and is accompanied by a large increase in volume. The heat liberated expands the gases still further and adds to the thrust.

8 *The -3 state.* In addition to ammonia and the ammonium salts, nitrogen forms other compounds in which it is assigned an oxidation

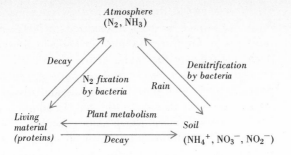

Fig. 23.3
Nitrogen cycle.

state of -3. These include the nitrides, such as Na_3N, Mg_3N_2, and TiN, many of which can be formed by direct combination of the elements. Some of these, for example, Na_3N and Mg_3N_2, are quite reactive and combine with water to liberate ammonia. Others, for example, TiN, are very inert and can be used to make containers for high-temperature reactions. The compound nitrogen triiodide (NI_3) might also be included with the -3 oxidation state of nitrogen, since nitrogen is more electronegative than iodine. At room temperature, NI_3 is a solid which is violently explosive and is well known for the fact that even a fly's landing on it can set it off.

The above list of nitrogen compounds is by no means exhaustive, but it does serve to indicate the great complexity of nitrogen chemistry. Even more complexity is found in the proteins, the nitrogen compounds which are essential constituents of all living matter. As was described in Sec. 22.4, the proteins are natural high polymers containing the

peptide link, or $-\overset{\overset{\displaystyle H}{|}}{N}-\overset{\overset{\displaystyle O}{\|}}{C}-$ group. There are a great variety of protein molecules, most of which are of extraordinarily high molecular weight, sometimes as high as a million amu. The structure of these is only now being worked out. Their synthesis by organisms remains incompletely understood, but it seems to involve amino acids as intermediates (Sec. 22.4). In nature there is constant interconversion between animal and plant proteins. However, the interconversion is not without loss because the decay of protein material produces some elemental nitrogen which escapes to the atmosphere. Living organisms, with the exception of some bacteria, are unable to utilize elemental nitrogen for the production of proteins. Thus, in order to maintain life nitrogen must somehow be restored to a biologically useful form.

The *nitrogen cycle*, which traces the path of nitrogen atoms in nature, is shown in simplified form in Fig. 23.3. When plant and animal proteins are broken down, as in digestion and decay, the principal end products are NH_3 and N_2, which are released to the atmosphere, and various nitrogen-containing ions, which are added to the soil. NH_3 in the atmosphere can be returned to the soil by being dissolved in rain. Elemental nitrogen can be returned by two paths: (1) Nitrogen-fixing bacteria which live on the roots of leguminous plants, such as clover, convert N_2 to proteins and other nitrogen compounds. (2) Lightning discharges initiate the otherwise slow combination of N_2 and O_2 to form NO, which in turn is oxidized to NO_2. The NO_2 dissolves in rainwater to form nitrates and nitrites, which are washed into the soil. As a final step of the cycle, plants absorb nitrogen compounds from the soil and convert these to plant proteins. Ingested as food, the plant proteins are broken down by animals and reassembled as animal proteins or excreted as waste to the soil. In addition, there are some forms of denitrifying bacteria which convert some of the nitrogen compounds in the soil directly to atmospheric nitrogen.

The nitrogen cycle as outlined above is in precarious balance. Frequently, the balance is locally upset, as, for example, by intensive cultivation and removal of crops. In such cases, it is necessary to replenish the nitrogen by addition of synthetic fertilizers, such as NH_3, NH_4NO_3, or KNO_3. Too much fertilization, however, can result in environmental pollution. Runoff of excess nutrient into streams can encourage plant life to the stage that too rich a growth (eutrophication) can occur. Also, as mentioned in Sec. 16.9, excess nitrate ion in groundwater can prove toxic to infants.

23.3 Phosphorus

The second element of group V, phosphorus, is considerably more abundant than nitrogen. Its principal natural form is *phosphate rock*, which is mostly $Ca_5(OH,F)(PO_4)_3$, *apatite*, where the OH^- and F^- substitute freely for each other, giving either hydroxyapatite or fluorapatite. Like nitrogen compounds, phosphorus compounds are essential constituents of all animal and vegetable matter. Bones, for example, contain about 60% $Ca_2(PO_4)_2$; nucleic acids such as DNA, which is responsible for transfer of genetic information, and RNA, which participates in biosynthesis of proteins, are polyester chains of sugars and phosphates.

Elemental phosphorus can be made by reduction of calcium phosphate with coke in the presence of silica sand. The reaction can be represented by the equation

$$Ca_3(PO_4)_2(s) + 3SiO_2(s) + 5C(s) \longrightarrow 3CaSiO_3 + 5CO(g) + P_2(g)$$

Since the reaction is carried out at high temperature, the phosphorus is formed as a gas, which is condensed to a solid by running the product gases through water. This condensation serves not only to separate the phosphorus from the carbon monoxide but also to protect it from reoxidation by air.

There are several forms of solid phosphorus, but only the white and red forms are important. White phosphorus consists of discrete tetrahedral P_4 molecules, as shown on the left of Fig. 23.4. The bond angle of $60°$ corresponds to a great deal of strain, consistent with the very high reactivity of the molecule. The structure of red phosphorus has not yet been completely determined, but there is evidence that it is polymeric and consists of chains of P_4 tetrahedra linked together, possibly in the manner shown in Fig. 23.4. At room temperature the

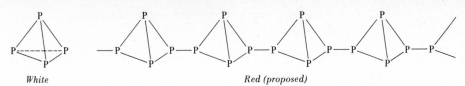

White   Red (proposed)

Fig. 23.4

White and red phosphorus.

stable modification of elemental phosphorus is the red form. Because of its highly polymerized structure it is less volatile, less soluble (especially in nonpolar solvents), and less reactive than white phosphorus. The white form must be handled with care because it ignites spontaneously in air and is extremely poisonous.

At relatively low temperatures (below $800°C$) phosphorus vapor consists primarily of P_4 molecules. At higher temperatures there is considerable dissociation to give P_2 molecules. Thus, only at elevated temperature does elemental phosphorus resemble elemental nitrogen in being diatomic. The favoring of tetratomic molecules over diatomic may be attributed to the larger size of the phosphorus atom. In general, large atoms have more difficulty than small atoms in forming multiple bonds, which would be required in $P\equiv P$.

At room temperature ordinary red phosphorus is not especially reactive, but at higher temperatures it reacts with many other elements to form a variety of compounds. For example, when heated with calcium, it forms solid calcium phosphide, Ca_3P_2. With chlorine it can form either liquid phosphorus trichloride, PCl_3, or solid phosphorus pentachloride, PCl_5, depending on the relative amount of chlorine supplied. The three compounds just mentioned illustrate the three most important oxidation states of phosphorus, -3, $+3$, and $+5$.

When Ca_3P_2 is placed in water, it reacts vigorously to form phosphine, PH_3, a toxic gas:

$$Ca_3P_2(s) + 6H_2O \longrightarrow 2PH_3(g) + 3Ca^{2+} + 6OH^-$$

In structure PH_3 resembles NH_3 in being a pyramidal molecule. Like NH_3, PH_3 can add a proton to form a phosphonium ion, PH_4^+, which, however, is found only in solid salts, such as PH_4I. Compared with NH_3, PH_3 is practically insoluble in water and is much less basic. In air, PH_3 usually bursts into flame, apparently because it is ignited by spontaneous oxidation of the impurity P_2H_4.[*]

When phosphorus is burned in a limited supply of oxygen, it forms the oxide P_4O_6 (phosphor*ous* oxide). (Note the spelling: The element ends in -*us*, and the $+3$ compounds end in -*ous*.) Below room temperature this compound is a white solid which melts at $23.8°C$. Its structure, shown in Fig. 23.5, can be visualized as derived from a P_4 tetrahedron by insertion of an oxygen atom between each pair of phosphorus atoms. P_4O_6 is the anhydride of phosphorous acid, and when cold water is added to it, H_3PO_3 is formed. Phosphorous acid is peculiar because

---

[*] The will-o'-the-wisp, or faint, flickering light, sometimes observed in marshes may be due to spontaneous ignition of impure PH_3. The PH_3 might be formed by reduction of naturally occurring phosphorus compounds.

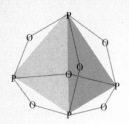

Fig. 23.5
Phosphorous oxide (P_4O_6).

although it contains three hydrogen atoms per molecule, only two of them can dissociate:

$$H_3PO_3 + H_2O \rightleftharpoons H_3O^+ + H_2PO_3^- \qquad K_I = 1.6 \times 10^{-2}$$

$$H_2PO_3^- + H_2O \rightleftharpoons H_3O^+ + HPO_3^{2-} \qquad K_{II} = 7 \times 10^{-7}$$

It has been suggested that the reason the third hydrogen does not dissociate is that it is attached directly to the phosphorus instead of to an oxygen. The structure of H_3PO_3 would then be $HPO(OH)_2$ instead of $P(OH)_3$. Phosphorous acid can also be made by the hydrolysis of phosphorus trichloride. The reaction

$$PCl_3 + 6H_2O \longrightarrow H_3PO_3 + 3H_3O^+ + 3Cl^-$$

is quite vigorous and liberates considerable heat, partly because of the high heat of hydration of the hydrogen ion liberated to the solution. Neutralization of H_3PO_3 by bases can produce two series of salts, the dihydrogen phosphites, for example, NaH_2PO_3, and the monohydrogen phosphites, for example, Na_2HPO_3. The phosphites, especially in basic solutions, are very strong reducing agents. Even in acid solution (where they immediately are converted to H_3PO_3) they are moderately good reducing agents

$$H_3PO_3 + 3H_2O \longrightarrow H_3PO_4 + 2H_3O^+ + 2e^- \qquad E^\circ = +0.28 \text{ V}$$

being slightly better than nickel metal.

In the $+5$ state, phosphorus exists as several oxy compounds of varying complexity. In contrast with the oxy compounds of nitrogen, none of the $+5$ compounds of phosphorus is an especially good oxidizing agent. The least complicated is the oxide, P_4O_{10}, called phosphoric oxide, phosphorus pentoxide, or phosphoric anhydride. This is the white solid which is usually formed when red phosphorus is burned in an unlimited supply of oxygen or when white phosphorus spontaneously catches fire in air. Though called a pentoxide (because of its simplest formula, P_2O_5), this material both in the vapor and in the most stable solid modification is known to consist of discrete P_4O_{10} molecules. The structure can be visualized as being derived from the molecule shown in Fig. 23.5 by addition of an oxygen atom sticking out from each phosphorus. Consistent with the molecular nature of the solid, P_4O_{10} is quite volatile and can be readily sublimed. At 360°C the vapor pressure of the solid is about 1 atm. It is remarkable that further heating of P_4O_{10} to about 500°C converts it not to a liquid but to a highly polymerized solid. Apparently some of the P—O—P bonds in the P_4O_{10} unit are broken and reestablished to adjacent P_4O_{10} units.

When exposed to moisture, P_4O_{10} turns gummy as it picks up water. The affinity for water is so great that P_4O_{10} is frequently used as an efficient dehydrating agent. With a large amount of water, the acid H_3PO_4, or orthophosphoric acid, is formed. This is a triprotic acid for which the stepwise dissociation is as follows:

$$H_3PO_4 + H_2O \rightleftharpoons H_3O^+ + H_2PO_4^- \qquad K_I = 7.5 \times 10^{-3}$$

$$H_2PO_4^- + H_2O \rightleftharpoons H_3O^+ + HPO_4^{2-} \qquad K_{II} = 6.2 \times 10^{-8}$$

$$HPO_4{}^{2-} + H_2O \rightleftharpoons H_3O^+ + PO_4{}^{3-} \qquad K_{III} = 10^{-12}$$

(Like $SO_4{}^{2-}$ and $CrO_4{}^{2-}$, $PO_4{}^{3-}$ is tetrahedral in structure.) From H_3PO_4, three series of salts are possible: the dihydrogen phosphates, the monohydrogen phosphates, and the normal phosphates. When dissolved in water, salts such as NaH_2PO_4 (monosodium dihydrogen phosphate) give slightly acid solutions. The slight acidity results from the fact that the dissociation of $H_2PO_4{}^-$ to produce H_3O^+ and $HPO_4{}^{2-}$ ($K_{II} = 6.2 \times 10^{-8}$) slightly exceeds the hydrolysis of $H_2PO_4{}^-$ to produce OH^- and H_3PO_4 ($K_{hyd} = K_w/K_I = 1.3 \times 10^{-12}$). Solutions of Na_2HPO_4 are slightly basic because the hydrolysis of $HPO_4{}^{2-}$ to produce OH^- and $H_2PO_4{}^-$ ($K_{hyd} = K_w/K_{II} = 1.6 \times 10^{-7}$) slightly exceeds the dissociation of $HPO_4{}^{2-}$ to produce H_3O^+ and $PO_4{}^{3-}$ ($K_{III} = 10^{-12}$). Solutions of Na_3PO_4 are quite basic because there is no acid dissociation to counterbalance the strong hydrolysis of $PO_4{}^{3-}$ to produce OH^- and $HPO_4{}^{2-}$ ($K_{hyd} = K_w/K_{III} = 10^{-2}$). Since $H_2PO_4{}^-$ in water gives an acid reaction, $Ca(H_2PO_4)_2$ is used with $NaHCO_3$ in some baking powders to produce carbon dioxide. The reaction may be written as

$$H_2PO_4{}^- + HCO_3{}^- \longrightarrow CO_2(g) + H_2O + HPO_4{}^{2-}$$

but it does not occur until water is added to the baking powder. Since $PO_4{}^{3-}$ in water gives a basic reaction and since $Ca_3(PO_4)_2$ is rather insoluble, trisodium phosphate has been used in water softening though it is under vigorous attack by the environmentalists for contributing to water pollution *via* eutrophication. (See Chap. 26.)

H_3PO_4 is only one of a series of phosphoric acids that may be formed by the hydration of P_4O_{10}. To distinguish it from other phosphoric acids, H_3PO_4 is called orthophosphoric acid, and its salts are called orthophosphates. Among the other phosphoric acids are pyrophosphoric acid, $H_4P_2O_7$, and metaphosphoric acid, HPO_3, both of which can be made by heating H_3PO_4. Unlike H_3PO_4 and $H_4P_2O_7$, which are discrete molecules, HPO_3 is polymeric; i.e., several HPO_3 groups are bound together. On standing in water, all the phosphoric acids convert to orthophosphoric acid. Perhaps more important than pyrophosphoric and metaphosphoric acids are their salts, a great variety of which are known. The pyrophosphates are relatively simple. Two series of salts are known: the normal pyrophosphates (for example, $Na_4P_2O_7$) and the dihydrogen pyrophosphates (for example, $Na_2H_2P_2O_7$). The structure of the normal pyrophosphate ion, shown in Fig. 23.6, consists of two PO_4 tetrahedra sharing a corner. In the dihydrogen pyrophosphate ion a proton is bound to one of the oxygen atoms on each of the tetrahedra. Pyrophosphates are used for water softening and as complexing agents in electroplating baths.

The metaphosphates, with simplest formula MPO_3, exist in a bewildering variety of complex salts. They are all polymeric in structure and can be thought of as being built up of $PO_3{}^-$ units in such a way that each phosphorus atom remains tetrahedrally associated with four oxygen atoms. In other words, there must be oxygen bridges between

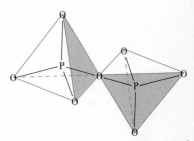

Fig. 23.6
Pyrophosphate ion.

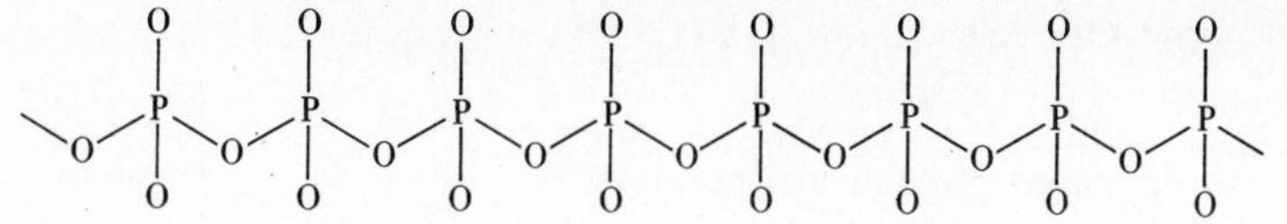

Fig. 23.7
Metaphosphate chain.

phosphorus atoms, and, furthermore, two of the four oxygen atoms about a given phosphorus atom must be bridge oxygens. Thus, the situation in the metaphosphates (illustrated in Fig. 23.7) is in some respects comparable with that of the silicate chains (shown in Fig. 22.18). Of the many metaphosphates reported, some of which are certainly not pure substances but are mixtures instead, we might mention the trimetaphosphate $Na_3P_3O_9$. This material is a white, crystalline solid which is produced by heating NaH_2PO_4 for several hours at about 550°C. The reaction can be written

$$3NaH_2PO_4 \longrightarrow Na_3P_3O_9(s) + 3H_2O(g)$$

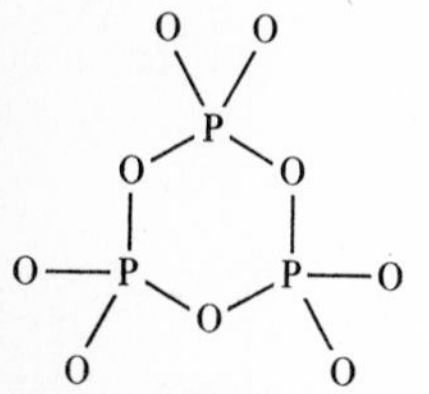

The product is quite readily soluble in water but, unlike many of the other metaphosphates, does not precipitate Pb^{2+} or Ag^+ out of solutions of their salts. It is generally believed that the trimetaphosphate ion is a cyclic polymer with a structure like that shown in Fig. 23.8. When heated above 620°C, $Na_3P_3O_9$ and, indeed, all other forms of metaphosphate melt to a clear, colorless liquid. If this liquid is cooled suddenly (quenched), it does not crystallize but instead forms a glass (sometimes called Graham's salt). The glass is quite soluble in water and in solution can precipitate Ag^+ and Pb^{2+} but not Ca^{2+}. In fact, it seems to form a complex with Ca^{2+} which makes it impossible to precipitate Ca^{2+} with the usual reagents such as carbonate. Because of the sequestering action on Ca^{2+}, the material has been used extensively in water softening under the trade name Calgon. At one time it was believed that Graham's salt was the hexametaphosphate $Na_6P_6O_{18}$, but more recent investigations indicate that it is a much higher polymer of the type $(NaPO_3)_n$, where n can be as high as 1000. It no doubt consists of a mixture of chains of varying length made up of PO_3 units. The two ends of the chains might be terminated by —OH groups.

Like nitrogen, phosphorus is an essential constituent of living cells. It occurs as phosphate groups in complex organic molecules. One of the principal functions of these phosphate groups is to provide a means for storing energy in the cells. For example, when water splits a phosphate group off adenosine triphosphate (ATP)

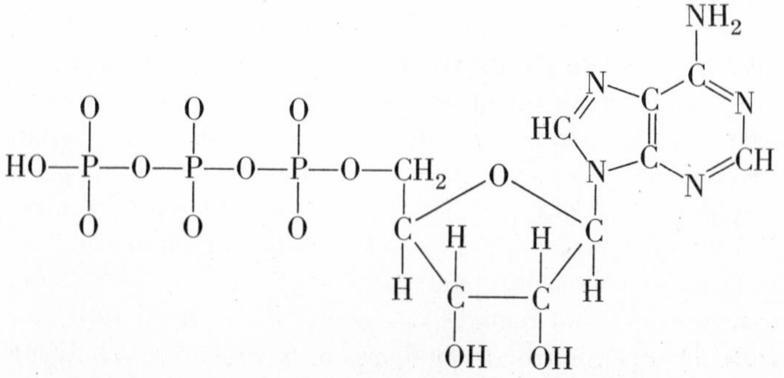

Fig. 23.8
Trimetaphosphate ion.

to form adenosine diphosphate (ADP), approximately 33 kJ of heat is liberated per mole. This energy can be used for the mechanical work of muscle contraction. Further discussions of this interesting subject are found in textbooks on biochemistry.

23.4 Arsenic

Conforming to the increasing metallic character going down group V, elemental arsenic exists as a metallic modification (gray arsenic) as well as a nonmetallic one (yellow arsenic). The metallic form is the stable modification at room temperature, and it can be made by carbon reduction of arsenious oxide, As_4O_6, or by the thermal decomposition to arsenic and FeS of naturally occurring arsenical pyrites (FeAsS). Yellow arsenic, the analog of white phosphorus, can be made by sudden cooling of arsenic vapor. Like white phosphorus, it consists of tetra-tomic molecules, As_4, and is volatile and soluble in nonpolar solvents.

The principal oxidation states of arsenic are $+3$ and $+5$. The -3 state, represented by the poisonous compound AsH_3 (arsine), is even less stable to air oxidation than the corresponding state for phosphorus. In the $+3$ state arsenic forms arsenious oxide (As_4O_6), commonly called *white arsenic*. When treated with water, As_4O_6 gives a slightly acid solution which is thought to contain the hydroxide $As(OH)_3$, or H_3AsO_3 (also written $HAsO_2$). This hydroxide is amphoteric; it can neutralize acids to give solutions containing $As(OH)_2^+$, and it can neutralize bases to give solutions containing arsenite ions [variously written as $H_2AsO_3^-$, AsO_2^-, or $As(OH)_4^-$]:

$$As(OH)_3(s) + H_3O^+ \longrightarrow As(OH)_2^+ + 2H_2O$$

$$As(OH)_3(s) + OH^- \longrightarrow H_2AsO_3^- + H_2O$$

When H_2S is bubbled into an arsenious solution, a yellow precipitate is formed. It is usually described as As_2S_3, but it probably has the molecular formula As_4S_6. As discussed in Sec. 8.8, arsenious sulfide has great tendency to form colloids stabilized by adsorption of negative ions. These colloids can be coagulated by addition of H_3O^+ or other positive ions. Like stannous sulfide, arsenious sulfide can be oxidized by polysulfide, S_2^{2-}.

In the $+5$ state the principal compounds of arsenic are arsenic acid and its derivatives, the arsenates. Arsenic acid is primarily orthoarsenic acid, H_3AsO_4, a triprotic acid with successive dissociation constants of 2.5×10^{-4}, 5.6×10^{-8}, and 3×10^{-13}. Salts of arsenic acid, especially lead arsenate and calcium arsenate, are much used as insecticides. Arsenate ion is considerably better as an oxidizing agent than is phosphate. For example, the electrode potential for the half-reaction

$$H_3AsO_4 + 2H_3O^+ + 2e^- \longrightarrow As(OH)_3 + 3H_2O$$

$$E° = +0.56 \text{ V}$$

indicates that H_3AsO_4 will oxidize I^- to I_2 ($I_2 + 2e^- \longrightarrow 2I^-$, $E° = +0.54$ V), though the reaction can be reversed at low H_3O^+

concentration. When H_2S is bubbled through a solution containing arsenic acid, a yellow precipitate, As_2S_5 (or perhaps As_4S_{10}), is formed. This sulfide is dissolved by excess sulfide ion to produce thioarsenate ions, $AsS_4{}^{3-}$. Addition of acid lowers the sulfide-ion concentration and reprecipitates As_2S_5.

The compounds of arsenic are among the most important of the systemic poisons. Because they are practically tasteless, they were great favorites in the Middle Ages for homicidal purposes. They are no longer popular because there are sensitive chemical tests for traces of arsenic compounds. Useful antidotes for arsenic poisoning are limewater $[Ca(OH)_2]$ and epsom salts $(MgSO_4 \cdot 7H_2O)$ because they precipitate oxyanions of arsenic.

23.5   Antimony

The element antimony is not very abundant (0.0001 percent, about one-fifth as abundant as arsenic), but it occurs in concentrated form as stibnite, Sb_2S_3. Its symbol, Sb, comes from *stibium*, the Latin name for the element. In order to prepare the element, stibnite can be heated with scrap iron:

$$Sb_2S_3(s) + 3Fe(s) \longrightarrow 3FeS(s) + 2Sb$$

The element exists in several allotropic forms, the stable one at room temperature being gray antimony. Yellow antimony, which is presumably the analog of yellow arsenic, is stable below $-90°C$. Explosive antimony, prepared by electrolysis of antimony trichloride, is a black material which on being scratched converts to the gray form with considerable violence. Ordinary gray antimony has a metallic appearance but is a rather poor metal. It is used principally in alloys with lead, as in making battery plates and shrapnel.

In the -3 state antimony forms the very unstable compound SbH_3 (stibine). This, like arsine, is quite poisonous and easily oxidized to the metal. In the $+3$ state antimony forms the oxide Sb_2O_3 (antimony trioxide, or antimony sesquioxide), which, at least in one crystal modification, exists as Sb_4O_6 molecules. It is an amphoteric oxide, dissolving in acid to give $Sb(OH)_2{}^+$ (or SbO^+) ion and dissolving in base to give antimonite anions, usually written $SbO_2{}^-$ or $Sb(OH)_4{}^-$. When solutions of antimonites (such as $NaSbO_2$) are gradually acidified, a white precipitate which has the composition $Sb_2O_3 \cdot xH_2O$ is first formed. Apparently, no simple $Sb(OH)_3$ is formed. The sulfide, Sb_2S_3, is orange when freshly precipitated. In many respects its chemical reactions are like those of arsenious sulfide.

In the $+5$ state antimony forms the pentoxide Sb_2O_5, which is a slightly stronger oxidizing agent than H_3AsO_4. It is practically insoluble in acid but does dissolve in base to give antimonate ion, usually written $Sb(OH)_6{}^-$. The fact that this ion has six oxygens about the central atom, rather than four as in arsenate, is ascribed to the larger size of antimony. Unlike As_2S_5, Sb_2S_5 is soluble in acid, but solution is accompanied by reduction of antimony to the $+3$ state.

23.6 Bismuth

Since the bismuth minerals, bismuth glance (Bi_2S_3) and bismuth ochre (Bi_2O_3), are rather rare, most commercial bismuth is produced as a by-product of lead production and electrolytic refining of copper. It is a rather poor metal which is used principally to make easily fusible alloys such as Wood's metal (50% by weight bismuth, 25% lead, 12.5% tin, and 12.5% cadmium), for which the melting point is 70°C, and is added in small amounts to harden lead plates for storage batteries.

Like antimony, bismuth forms an unstable hydrogen compound, BiH_3 (bismuthine), and a sesquioxide, Bi_2O_3. However, Bi_2O_3 is basic and not amphoteric like Sb_2O_3. Although insoluble in water, it dissolves in acid solution to give hydrolyzed bismuth ion, which may be BiO^+, $BiOH^{2+}$, or $Bi(OH)_2^+$. Two series of salts are known: simple bismuth salts, for example, $Bi(NO_3)_3 \cdot 5H_2O$, and oxysalts, e.g., bismuthyl nitrate, $BiONO_3$.

When fused with strong oxidizing agents such as Na_2O_2 in the presence of NaOH, Bi_2O_3 is converted to a compound with remarkable oxidizing power. For example, it even oxidizes Mn^{2+} to MnO_4^-. Though called sodium bismuthate and given the formula $NaBiO_3$, it is insoluble and probably is not a definite compound but a mixture of oxides, which may include Bi_2O_5.

The only known sulfide of bismuth is Bi_2S_3. It is formed as a black precipitate when H_2S is passed through bismuth-containing solutions. Bi_2S_3 is insoluble in dilute acids but dissolves in hot, concentrated nitric acid as a result of oxidation of sulfide to elemental sulfur. Unlike the corresponding sulfides of arsenic and antimony, bismuth sulfide is not dissolved by either sulfide ion or polysulfide ion.

Bismuth telluride, $Bi_2Te_{3\pm x}$, is an interesting nonstoichiometric compound, which is being applied to the problem of direct conversion of heat into electricity. Depending on the value of x, the material can be an n-type or p-type semiconductor, but in addition it acts as a thermoelectric element. When a bar of such thermoelectric material is placed in a thermal gradient so that one end is warmer than the other, a voltage is generated which can be used to do electric work. Conversely, if an electric current is passed through the material, one end gets hot, and the other end gets cold. Thus, thermoelectric materials can also be used for cooling.

23.7 Qualitative analysis

The ammonium ion is easily detected by adding NaOH to the unknown and heating to expel NH_3, which can be detected by its characteristic odor or by allowing the vapor to turn moist litmus blue. Nitrate ion may be detected by the brown-ring test described in item (4) of Sec. 23.2.

The anions phosphate and arsenate can be recognized by their formation of white, insoluble magnesium ammonium salts ($MgNH_4PO_4$ and $MgNH_4AsO_4$). Arsenate in the presence of phosphate can be distinguished by addition of $AgNO_3$ and acetic acid, which converts $MgNH_4AsO_4$ to red Ag_3AsO_4. Phosphate in the presence of arsenate is more difficult to detect and requires preliminary removal of arsenic as insoluble As_2S_3. This is accomplished by reducing the arsenic from the $+5$ oxidation state to the $+3$ with I^- in acid solution and then adding H_2S. From the filtrate, yellow, insoluble ammonium phosphomolybdate, $(NH_4)_3PMo_{12}O_{40}$, can be precipitated by treatment with hot ammonium molybdate solution, $(NH_4)_2MoO_4$.

The cations of arsenic, antimony, and bismuth precipitate as As_2S_3 (red-yellow), Sb_2S_3 (black-red), and Bi_2S_3 (brown-black) when H_2S is added in $0.3\ N$ acid. Bi_2S_3 differs from Sb_2S_3 and As_2S_3 in being insoluble in $(NH_4)_2S$ solution. The presence of bismuth can be confirmed by reducing $Bi(OH)_3$ to black bismuth metal with stannite ion in basic solution. From the solution containing AsS_3^{3-} and SbS_3^{3-} obtained by the $(NH_4)_2S$ treatment, As_2S_3 precipitates on addition of $6\ M$ HCl. (The arsenic can be confirmed by reducing to AsH_3 with aluminum metal in strong base and allowing the AsH_3 vapor to blacken paper wet with silver nitrate solution. The black color results from the formation of finely divided silver metal.) The presence of dissolved antimony can be shown by the formation of a characteristic orange-red color (possibly an oxysulfide) on addition of $Na_2S_2O_3$.

Exercises

*23.1 *Acidic oxides.* How does the acidic character of the X_2O_3 oxides change going down group V? Using data from Appendix 8, account for this trend.

*23.2 *Metallic behavior.* How does metallic character change in going down group V? Why should the most metallic of the group still be a rather poor metal?

*23.3 *Protein composition.* From the average composition of proteins given in Sec. 23.2, calculate the number of atoms of each kind per atom of nitrogen.   *Ans. 3.7 C, 1.4 O, 6 H, 0.01 P, 0.01 S*

*23.4 *Protein composition.* The simplest amino acid is glycine, H_2NCH_2COOH, which can be thought of as being derived from acetic acid by replacement of a methyl hydrogen by an amino group. How does the percent composition of this amino acid compare with the average composition of proteins given in Sec. 23.2?

*23.5 *Combustion stoichiometry.* In a situation where $N_2O(g)$ supports combustion, what volume of $N_2O(g)$ would you need to take to substitute for one volume of air?   *Ans. 0.42*

*23.6 *Phosphorus bonding.* Give a valence-bond description of the bonding in P_4 and in P_2. Show how your description accounts for the shapes of the molecules. Account also for an observed difference in bond lengths: The P—P distance in P_4 is 0.221 nm; in P_2, 0.189 nm.

*23.7 *Conversions.* How would you go about carrying out the following conversions? Indicate special conditions, and write balanced equations.

   a As_4O_6 to As_4
   b As_4O_6 to As_2S_3
   c As_4O_6 to AsH_3
   d As_4O_6 to Na_3AsO_4

** 23.8   *Vapor pressure*. The heat of vaporization of $N_2(l)$ is 200 J/g. If the normal boiling point of liquid nitrogen is $77.4°K$, what would it be at 0.010 atm?

** 23.9   *Density of air*. Given the composition of dry air shown in Fig. 23.2 and taking CO_2 at 0.04% by volume, calculate the density expected for dry air at 0.965 atm and 25°C.

Ans. 0.00114 g/cm^3

** 23.10   *Density*. Assuming that Rayleigh's observed density difference between pure nitrogen and impure nitrogen was entirely attributable to argon, calculate his molar ratio of argon-to-nitrogen in the impure sample, and compare it with the present figure for air.

** 23.11   *Molecular orbital*. Draw sketches of the three bonding orbitals populated in N_2. Predict what would happen to the bond length on (*a*) removal of one electron to give N_2^+ and (*b*) addition of one electron to give N_2^-.

** 23.12   *Bond angle*. Given that the H-to-H distance in NH_3 is 0.1624 nm and the N—H distance is 0.101 nm, calculate the bond angle H—N—H.   *Ans. 107.0°*

** 23.13   *Ammonia*. For the process $N_2(g) + 3H_2(g) \rightleftharpoons 2NH_3(g)$, $\Delta H°$ is -92 kJ, and $\Delta G°$ is -33.3 kJ. Calculate K_p for the process (*a*) at 25°C and (*b*) at 500°C.

** 23.14   *Ammonia*. At $-33°C$, the density of pure liquid NH_3 is 0.683 g/cm^3; the density of Na is 0.97 g/cm^3. Given that the density of $1\ M$ Na in NH_3 is 0.660 g/cm^3, calculate the volume expansion per mole of Na when $Na(s) + NH_3(l) \longrightarrow 1\ M$ solution. If we attribute all the volume expansion to the electrons, what would be the apparent radius of the electron?   *Ans. 43 cm^3, 0.26 nm*

** 23.15   *Solid state*. The compound $Ca(HN_3)_6{}^0$ crystallizes in a body-centered cubic arrangement. The unit-cell edge length is 0.912 nm. Calculate the theoretical density.

** 23.16   *Solubility*. At 25°C and 1 atm pressure, 790 cm^3 of NH_3 gas can dissolve in 1 cm^3 of H_2O to give a solution of density 0.8805 g/cm^3. Calculate (*a*) the molarity and (*b*) the pH of the final solution.   *Ans. 18.4 M, 12.25*

** 23.17   *Electrode potentials*. Given the electrode potentials on page 615, calculate the $E°$ for (*a*) $NO_2(g)$ to $NO(g)$, (*b*) $NO(g)$ to $\frac{1}{2}N_2O(g)$, and (*c*) $N_2O(g)$ to $N_2(g)$.

** **23.18** *Combustion stoichiometry*. What air-fuel ratio (by weight) do you need to oxidize C_8H_{18} to CO_2 and H_2O? What would this ratio be if the air were replaced by nitrous oxide?   *Ans. 15.1, 9.63*

** **23.19** *Symmetry*. What are the symmetry elements possessed by the P_4O_6 molecule shown in Fig. 23.5? What difference would it make if you had P_4O_{10} instead?

** **23.20** *Structure*. Predict which has the bigger P-to-P distance, P_4O_6 or P_4O_{10}. Check your prediction from the following data: P—O—P bond angle is $127.5°$ in P_4O_6 and $124.5°$ in P_4O_{10}. The distance from the bridge oxygen to phosphorus is 0.165 nm in P_4O_6 and 0.160 nm in P_4O_{10}.

** **23.21** *Dissociation*. Given $K_I = 1.6 \times 10^{-2}$ and $K_{II} = 7 \times 10^{-7}$ for H_3PO_3, calculate the H_3O^+ and HPO_3^{2-} concentrations in 1.0 M H_3PO_3 solution.   *Ans. 0.12 and 7×10^{-7} M*

** **23.22** *Equations*. Write balanced net equations for each of the following:
  a Oxidation of HPO_3^{2-} in base by O_2 to give PO_4^{3-}
  b Hydrolysis of $AsCl_3$ to give $As(OH)_3$
  c Oxidation of I^- by H_3AsO_4 in acid to give I_2 and $As(OH)_3$
  d Oxidation of Mn^{2+} by $NaBiO_3(s)$ in acid to give MnO_4^- and BiO^+

** **23.23** *Simultaneous equilibria*. Suppose you make a solution by dissolving 0.590 mol of H_3PO_4 and 0.210 mol of NaH_2PO_4 in enough water to make a liter of solution. What will be the concentration of all species in the final solution?

*** **23.24** *Thermodynamics*. The free energy of formation of $HNO_3(aq)$ is -110.5 kJ/mol; that of $H_2O(l)$ is -237 kJ/mol. Calculate $\Delta G°$ for forming one mole of $HNO_3(aq)$ from $N_2(g)$, $O_2(g)$, and $H_2O(l)$. Repeat the calculation, allowing for the fact that the N_2 pressure is 0.78 atm and the O_2 pressure is 0.21 atm. Assuming 100 percent dissociation into H_3O^+ and NO_3^-, how much HNO_3 concentration do you need to make $\Delta G = 0$?   *Ans. 8.0 and 13.1 kJ, 0.070 M*

*** **23.25** *Molecular structure*. In the trans form of nitrous acid, all the atoms are in the same plane, but the HO is directed away from the other O. The O—H distance is 0.098 nm; the distance from the central N to the hydroxyl O is 0.146 nm; the distance to the other O is 0.120 nm. If the H—O—N bond angle is $105°$ and the O—N—O bond angle is $118°$, how far is the H from the other O?

*** **23.26** *Molecular structure*. In PH_3, the P—H distance is 0.1421 nm, and the H—P—H bond angle is $93.5°$. How far is the P away from the center of a regular tetrahedron defined by the plane of the three H atoms?   *Ans. 0.0347 nm farther away from the base*

***23.27 *Buffer solution.* You are given a $1.00\ M$ NH_3–$1.00\ M$ NH_4^+ buffer solution. How much of this solution would you need to take in order to be able to absorb 0.010 mol of added H_3O^+ without changing pH by more than 0.01 unit?

***23.28 *Gaseous equilibrium.* A bulb contains a mixture of $NO_2(g)$ and $N_2O_4(g)$. If the gas density is 1.53 g/liter at 0.965 atm and $100°C$, what is the numerical value of the equilibrium constant K_p for the equilibrium $2NO_2(g) \rightleftharpoons N_2O_4(g)$ at $100°C$? If $\Delta H° = -61.1$ kJ for the reaction as written, what are $\Delta G°$ and $\Delta S°$ at $100°C$? *Ans. 0.062 atm^{-1}, +8.6 kJ, $-190\ J$*

***23.29 *Phosphorus.* The standard state of phosphorus at $25°C$ and 1 atm pressure is white, crystalline phosphorus. With respect to this standard state, $P_4(g)$ has free energy 24.4 kJ/mol higher and enthalpy 54.9 kJ/mol higher. With respect to the same standard state $P_2(g)$ has free energy 102.9 kJ/mol higher and enthalpy 141.5 kJ/mol higher. From these data calculate K_p for $P_4(g) \rightleftharpoons 2P_2(g)$ at $800°C$, and estimate what percent of $P_4(g)$ would be dissociated at a total pressure of 1.00 atm.

***23.30 *Titration.* Sketch a titration curve analogous to Fig. 12.3 corresponding to progressive addition of 0.03 mol of solid NaOH to one liter of $0.010\ M$ H_3PO_4.

The most important element of group VI, oxygen, has already been discussed in Chap. 15. The other members of the group, sulfur, selenium, tellurium, and polonium, differ markedly from oxygen, especially in the formation of positive oxidation states. Although oxygen can also show a positive oxidation state, its highest state is $+2$, and this is extremely rare. On the other hand, all the other elements of the group form compounds in which oxidation numbers $+4$ and $+6$ are assigned. The $+2$ state is common to all. In the earth's crust, selenium, tellurium, and polonium are extremely rare, and sulfur is much less plentiful than the very abundant oxygen.

The elements of the group are sometimes known as *chalcogens*, from the Greek word *chalkos* for copper and *genēs* for born. Most copper minerals are either oxygen or sulfur compounds and frequently contain the other members of the group.

24.1 Group properties

On the far right of the periodic table, the elements have characteristically high ionization potentials, and metallic properties are hard to find. However, in going down the group electrons are held less tightly; hence, there is some suggestion of metallic behavior in the heavier elements of group VI. Ionization potentials and other properties are given in Fig. 24.1.

Fig. 24.1
Elements of group VI.

Symbol	Z	Electronic configuration	Melting point, °C	Boiling point, °C	Ionization potential, eV	Electrode potential, V (X to H_2X)
O	8	(2) $2s^2 2p^4$	−219	−183.0	13.61	+1.23
S	16	(10) $3s^2 3p^4$	119	444.6	10.36	+0.14
Se	34	(28) $4s^2 4p^4$	220	685	9.75	−0.40
Te	52	(46) $5s^2 5p^4$	450	1390	9.01	−0.72
Po	84	(78) $6s^2 6p^4$	....	....	8.43	−1.0

Oxygen stands alone from the group in being a diatomic gas at room temperature. The other elements are solids with structural units more complex than diatomic molecules. All the elements of the group show allotropy. Just as oxygen can exist both as diatomic oxygen and triatomic ozone, so the other elements can be obtained in more than one form, the forms differing either in the number of atoms per molecule or in the arrangement of molecules in the solid. There is in going down the group increasing tendency toward formation of long strings of atoms held together by covalent bonds, except that the bottom element, polonium, appears to be typically metallic. The increasing complexity of structure is principally due to the increasing atomic size down the group. In general, the larger the atom, the less the tendency to form multiple bonds, and the greater the tendency of each atom to be bound to more than one other atom.

Because of the increasing number of electronic shells populated, we would expect an increase of the atomic size from O to Po. This increase is reflected in the values assigned to the radii of the −2 ions. From X-ray studies of crystal structures the following radii have been assigned: O^{2-}, 0.140 nm; S^{2-}, 0.184 nm; Se^{2-}, 0.198 nm; and Te^{2-}, 0.221 nm. These values are quite high compared with those of positive ions having the same electronic configuration. For example, Na^+, which is isoelectronic with O^{2-}, has an ionic radius of 0.097 nm. The comparison can be extended by noting the ionic radii of the other alkali-metal ions given in Fig. 16.1. The fact that the nuclear charges of the alkali-metal ions are greater than those of isoelectronic group VI ions is apparently the main reason for the difference of size.

Perhaps the most striking variation in these elements is the decreasing oxidizing strength from oxygen to polonium. As the electrode potentials in the last column of Fig. 24.1 indicate, there is much greater

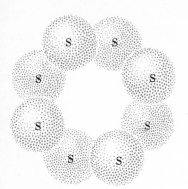

Fig. 24.2
S_8 *molecule.*

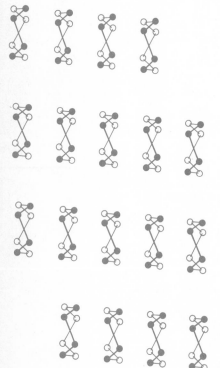

Fig. 24.3

Arrangement of S_8 rings in rhombic sulfur.

tendency for oxygen to form H_2O than for polonium to form H_2Po. In fact, unlike H_2O and H_2S, the compounds H_2Se, H_2Te, and H_2Po are better reducing agents than hydrogen.

When bound to more electronegative atoms, the elements of group VI show positive oxidation states. Positive oxidation states of oxygen are found only in compounds with fluorine (for example, OF_2) since fluorine is the only element more electronegative than oxygen. All the other elements of group VI form oxy compounds in which the elements, being less electronegative than oxygen, are assigned positive oxidation numbers. An examination of these compounds shows that the $+4$ and $+6$ states are the most common. There is thus a difference of two units between the most common positive states of group VI; this is also true in group V ($+3$ and $+5$), and in group VII, at least for chlorine and iodine ($+5$ and $+7$). This difference of two units is consistent with the notion that electrons in molecules generally exist as pairs.

24.2 Sulfur

Although not very abundant (0.05 percent), sulfur is readily available because of its occurrence in large beds of the free element. These beds, usually located several hundred feet underground, are thought to be due to bacterial decomposition of calcium sulfate. They are exploited by the *Frasch process*, pumping superheated water (at about $170°C$) down to the beds to melt the sulfur and blowing it to the surface with compressed air. Since the product is about 99.5 percent pure, it can be used without purification for most commercial purposes. Besides being found as the free element, sulfur occurs naturally in many sulfide and sulfate minerals, such as $CuFeS_2$, Cu_2S, and $CaSO_4 \cdot 2H_2O$. It is also important in some biologic molecules such as the amino acid cysteine, $HSCH_2CH(NH_2)COOH$. S—S links also occur as bridges between various parts of protein molecules (e.g., insulin).

There are several allotropic modifications of sulfur, the most important being rhombic (also known as α) and monoclinic (β) sulfur, which differ from each other in the symmetry of their crystals. In the rhombic form, which is the stable one at room temperature, sulfur atoms are linked to each other as shown in Fig. 24.2 to give puckered, eight-membered rings having bond angles $105°$ and bond lengths 0.207 nm. The rings are stacked next to each other as in Fig. 24.3 to form pipes that line up parallel to one another in one plane but are crisscross relative to the pipes in an adjacent plane. There are 16 S_8 rings per rhombic unit cell. Above $96°C$, monoclinic sulfur is stable; the arrangement of sulfur atoms in it is not known.

When heated above the melting point, sulfur goes through a variety of changes. Starting as a mobile, pale yellow liquid, it gradually thickens above $160°C$ and then becomes less viscous as the boiling point is approached. If the thick liquid, which may be dark red if impurities are present, is poured into water, amorphous, or plastic, sulfur is produced. X-ray analysis of amorphous sulfur shows that it contains long strings of sulfur atoms. In accordance with this, the change in viscosity with temperature has been attributed to opening of S_8 rings, which then couple up to form less mobile long chains. These in turn are broken into fragments as their kinetic energy is increased.

The phase relationships of sulfur are shown in the phase diagram given in Fig. 24.4. Because sulfur can exist in two solid modifications, the diagram contains four regions, corresponding to two solid states, one liquid and one gas. At any temperature and pressure lying within the triangle, monoclinic sulfur is the stable form. Thus, if rhombic sulfur is heated at 1 atm pressure to about 100°C and held there, it slowly converts by atomic rearrangement in the solid to monoclinic sulfur. This is a very slow process, and under usual conditions of heating the transformation is not observed. For the usual rapid melting of sulfur, the effective phase diagram is that delineated by the dashed lines instead of the solid ones. In other words, if heating is so rapid that equilibrium is not attained, solid rhombic sulfur superheats without changing to monoclinic and melts at a temperature (112.8°C) below the melting point of monoclinic sulfur (119.25°C).

Although much of the sulfur produced is used directly in insecticides, fertilizers, paper and pulp fillers, and rubber, most of it is converted to industrially important compounds, especially sulfuric acid. Sulfuric acid is produced from sulfur dioxide, SO_2, usually made by burning sulfur in air:

$$S(s) + O_2(g) \longrightarrow SO_2(g)$$

Sulfur dioxide is also a by-product of the preparation of various metals from their sulfide ores. For example, SO_2 is formed in the roasting of the copper ore chalcocite, or Cu_2S:

$$Cu_2S(s) + O_2(g) \longrightarrow 2Cu(s) + SO_2(g)$$

In the contact process, which accounts for nearly all the H_2SO_4 production, the SO_2 is oxidized by air in the presence of catalysts such as vanadium pentoxide (V_2O_5) or platinum:

$$2SO_2(g) + O_2(g) \longrightarrow 2SO_3(g)$$

The product, SO_3, or sulfur trioxide, is the anhydride of H_2SO_4, and we would expect the final step in preparing sulfuric acid to be the dissolving of SO_3 in water. However, SO_3 reacts with water to form a fog of H_2SO_4, and the uptake of SO_3 by water is extremely slow. The usual method for circumventing this difficulty is to dissolve the SO_3 in pure H_2SO_4 in a reaction which goes smoothly to produce $H_2S_2O_7$, pyrosulfuric acid. On dilution with water, 100% H_2SO_4 results:

$$SO_3(g) + H_2SO_4 \longrightarrow H_2S_2O_7$$

$$H_2S_2O_7 + H_2O \longrightarrow 2H_2SO_4$$

Pure H_2SO_4 is a liquid at room temperature; it freezes at 10°C. In many respects liquid H_2SO_4 resembles water. For example, it con-

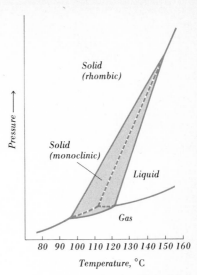

Fig. 24.4

Phase diagram for sulfur (pressure axis distorted).

ducts electricity slightly, presumably because, like water, it is dissociated into ions:

$$2H_2SO_4 \rightleftharpoons H_3SO_4^+ + HSO_4^- \qquad K = 2.9 \times 10^{-4}$$

Furthermore, like water, it dissolves many substances, even ionic solids. However, H_2SO_4 differs from water in that its extent of dissociation is considerably greater and in that H_2SO_4 may force a proton on any dissolved species. For instance, when acetic acid is placed in pure H_2SO_4, the following reaction occurs:

$$CH_3COOH + H_2SO_4 \longrightarrow CH_3COOH_2^+ + HSO_4^-$$

Pure H_2SO_4 has great affinity for water and forms several compounds, or hydrates, such as $H_2SO_4 \cdot H_2O$ and $H_2SO_4 \cdot 2H_2O$. Ordinary commercially available, concentrated sulfuric acid is approximately 93% H_2SO_4 by weight and can be thought of as a solution of H_2SO_4 and $H_2SO_4 \cdot H_2O$. The monohydrate may be H_3O^+ and HSO_4^-, and the large heat observed to be liberated when concentrated sulfuric acid is added to water may be due to formation of H_3O^+ and subsequent hydration of it and of HSO_4^-. Frequently, concentrated H_2SO_4 is used as a dehydrating agent, as, for example, in desiccators to keep substances dry. It is also used in reactions to favor splitting off of water. As an example of the latter, H_2SO_4 is used in the manufacture of ethers from alcohols:

$$2C_2H_5OH \xrightarrow{\;H_2SO_4\;} C_2H_5OC_2H_5 + H_2O$$

In aqueous solutions, H_2SO_4 is a strong acid, but only for dissociation of one proton. The dissociation constant for the second proton is 1.26×10^{-2}. Because of the dissociation

$$HSO_4^- + H_2O \longrightarrow H_3O^+ + SO_4^{2-}$$

solutions of HSO_4^-, such as solutions of sodium hydrogen sulfate ($NaHSO_4$), are acid. Because of the reverse reaction, solutions of SO_4^{2-}, such as solutions of sodium sulfate (Na_2SO_4), are slightly basic. However, the extent of hydrolysis is extremely small ($K_{hyd} = 7.94 \times 10^{-13}$), and these solutions of sulfate salts are essentially neutral.

For the half-reaction

$$HSO_4^- + 3H_3O^+ + 2e^- \longrightarrow SO_2 + 5H_2O$$

the electrode potential is $+0.11$ V. This means that HSO_4^- at $1\ m$ concentration is a mild oxidizing agent. However, its action as an oxidizing agent in dilute solution at room temperature is not observed because reaction is so slow. With hot, concentrated solutions of sulfuric acid, oxidation is observed. For example, sodium bromide plus hot H_2SO_4 produces some bromine by oxidation of Br^- to Br_2. Furthermore, some of the less active metals such as copper are soluble in hot, concentrated sulfuric acid, presumably because of oxidation by sulfate.

Although sulfate is not an especially good oxidizing agent, there is a closely related derivative which is an extremely powerful oxidizing agent. This derivative is produced by electrolytic oxidation of cold, concentrated sulfuric acid and has been assigned the formula $H_2S_2O_8$. The acid is called peroxydisulfuric acid, and its salts are called peroxy compounds since they contain an oxygen-oxygen bond. The structural relation between H_2SO_4 and $H_2S_2O_8$ is shown in the following half-reaction:

$$\text{H—O—}\overset{\displaystyle O}{\underset{\displaystyle O}{\overset{|}{\underset{|}{S}}}}\text{—O—O—}\overset{\displaystyle O}{\underset{\displaystyle O}{\overset{|}{\underset{|}{S}}}}\text{—O—H} + 2H_3O^+ + 2e^- \;\rightleftharpoons$$

$$2\left(\text{H—O—}\overset{\displaystyle O}{\underset{\displaystyle O}{\overset{|}{\underset{|}{S}}}}\text{—O—H}\right) + 2H_2O$$

The forward reaction (for which $E° = +2.01$ V) applies to the action of $H_2S_2O_8$ as an oxidizing agent; the reverse reaction, to the electrolytic preparation. Both the acid and its salts, such as ammonium peroxydisulfate, $(NH_4)_2S_2O_8$, are very strong oxidizing agents and can, for example, oxidize manganous salts to permanganate. On a relative scale, the peroxydisulfates $(E° = +2.01$ V) are about as good oxidizing agents as ozone $(E° = +2.07$ V). In addition to their use as oxidizing agents, the peroxydisulfates are important as intermediates in the preparation of hydrogen peroxide:

$$K_2S_2O_8(s) + 2H_3O^+ \longrightarrow 2H_2SO_4 + H_2O_2 + 2K^+$$

In addition to the oxy compounds that sulfur forms in the $+6$ oxidation state, there are important oxy compounds corresponding to the $+4$ state of sulfur. The simplest of these is the dioxide, SO_2, which is formed either by burning sulfur in air or by reducing sulfates. At room temperature, SO_2 is a gas, but it is quite easily liquefied (bp $-10°$C). The easy liquefaction reflects the fact that the molecule is polar because it has a nonlinear arrangement of atoms (shown in Sec. 3.7). Sulfur dioxide has a disagreeable, choking odor and is somewhat poisonous. Its emission from copper and zinc smelters and from power plants that burn fossil fuels (e.g., coal contains an average 2.7% by weight sulfur) makes SO_2 a serious environmental pollutant. It is especially toxic to lower organisms such as fungi and for this reason is used for sterilizing dried fruit and wine barrels. With water, SO_2 dissolves to give acid solutions, which contain about 5% of the sulfur as sulfurous acid, H_2SO_3. The compound H_2SO_3 has never been isolated pure; any attempt to concentrate the solution, as by heating, simply expels SO_2. H_2SO_3 is a weak diprotic acid, for which the principal equilibria are

$$SO_2 + H_2O \rightleftharpoons H_2SO_3 \qquad\qquad K = 0.05$$

$$SO_2 + 2H_2O \rightleftharpoons H_3O^+ + HSO_3^- \qquad K = 1.25 \times 10^{-2}$$

$$HSO_3^- + H_2O \rightleftharpoons H_3O^+ + SO_3^{2-} \qquad K = 5.6 \times 10^{-8}$$

It forms two series of salts. The sulfites, for example, Na_2SO_3, give slightly basic solutions owing to hydrolysis of SO_3^{2-}; the hydrogen sulfites, for example, $NaHSO_3$, give slightly acid solutions because the dissociation of HSO_3^- outweighs its hydrolysis. Addition of concentrated acids to either solid sulfites or solid hydrogen sulfites liberates SO_2 and is a convenient way of making sulfur dioxide in the laboratory. Sulfites, hydrogen sulfites, and sulfurous acid are mild reducing agents and are relatively easily oxidized to sulfates, though sometimes the reaction is quite slow.

When solutions containing sulfite ion are boiled with elemental sulfur, the solid sulfur dissolves in a reaction

$$S(s) + SO_3^{2-} \rightleftharpoons S_2O_3^{2-}$$

which is easily reversed by addition of acid. The ion formed, $S_2O_3^{2-}$, is called thiosulfate ion, where the prefix *thio-* indicates substitution of a sulfur atom for an oxygen atom. Apparently, $S_2O_3^{2-}$ contains two different kinds of sulfur atoms, as found by the following experiment: Solid sulfur containing a radioactive isotope of sulfur was boiled with a solution containing nonradioactive sulfite ions. The thiosulfate ions formed were found to be radioactive, but after acid was added so as to reverse the above reaction, all the radioactivity was recovered as precipitated solid sulfur. The implication is that the same sulfur atom which adds to SO_3^{2-} to form $S_2O_3^{2-}$ is dropped off when acid is added. This can be true only if the added sulfur atom is bound in $S_2O_3^{2-}$ in a way that is unlike the binding of the sulfur atom already in SO_3^{2-}. Otherwise, the two sulfur atoms in $S_2O_3^{2-}$ would be identical, and the addition of acid would not preferentially drop off the added radioactive sulfur atom but would have a 50-50 chance of retaining it and its activity in the complex SO_3^{2-}. The structure proposed for $S_2O_3^{2-}$ has one sulfur atom at the center of a tetrahedron with the other sulfur atom and the three oxygen atoms at the four corners of the tetrahedron. The decomposition of thiosulfate by acid is quite slow, at least so far as formation of visible solid sulfur is concerned. Indeed, when acid is added to thiosulfate solutions, nothing is observed at first. Then a white milkiness develops as colloidal sulfur is produced by gradual agglomeration of sulfur atoms. Thiosulfate ion acts as a mild reducing agent

$$2S_2O_3^{2-} \longrightarrow S_4O_6^{2-} + 2e^- \qquad E° = -0.08 \text{ V}$$

and has, for example, the ability to reduce iodine, I_2, to iodide ion, I^-. The reaction, which produces tetrathionate ion, $S_4O_6^{2-}$, is frequently used in determining the amount of iodine in a solution. It also makes possible the quantitative analysis of many oxidizing agents. The unknown oxidizing agent is reduced with an excess of I^-, and the liberated I_2 is titrated with a thiosulfate solution. Thiosulfate ion also has the ability to form complex ions with the ions of some metals, especially Ag^+. The silver-thiosulfate complex ion is so stable

$$Ag^+ + 2S_2O_3{}^{2-} \longrightarrow Ag(S_2O_3)_2{}^{3-} \qquad K = 1.6 \times 10^{13}$$

that thiosulfate solutions can dissolve the insoluble silver halides (see Sec. 20.3 for the use of thiosulfate in the photographic fixing process).

Besides occurring in the positive oxidation states, sulfur forms compounds corresponding to negative oxidation states, especially -2. The most familiar of these compounds is probably hydrogen sulfide, H_2S, notorious for its rotten-egg odor. Not so well known is the fact that hydrogen sulfide is as poisonous as hydrogen cyanide and four times as poisonous as carbon monoxide. The presence of hydrogen sulfide in sewer gas is due to putrefaction of sulfur-containing organic material. The pure compound can be made by bubbling hydrogen gas through molten sulfur. In the laboratory, it is conveniently prepared by interaction of some sulfide such as FeS with acid

$$FeS(s) + 2H_3O^+ \longrightarrow Fe^{2+} + H_2S(g) + 2H_2O$$

or by the warming of a solution of thioacetamide

Thioacetamide   Acetamide

This latter reaction is frequently used as an easily controlled laboratory source of hydrogen sulfide for qualitative analysis. Like water, hydrogen sulfide has a bent molecule and is polar; however, it is considerably harder to liquefy (bp $-61°C$), presumably because of the lack of hydrogen bonding (Sec. 15.5) in the liquid. Gaseous hydrogen sulfide burns to produce water and either sulfur or sulfur dioxide, depending on the temperature and the oxygen supply. It is a mild reducing agent and can, for example, reduce ferric ion to ferrous ion:

$$2Fe^{3+} + H_2S(g) + 2H_2O \longrightarrow 2Fe^{2+} + S(s) + 2H_3O^+$$

During the course of the reaction, the solution becomes milky from the production of colloidal sulfur. In aqueous solution hydrogen sulfide is a weak diprotic acid for which the dissociation constants are $K_I = 1.1 \times 10^{-7}$ and $K_{II} = 1 \times 10^{-14}$. A detailed consideration of the equilibria in aqueous hydrogen sulfide solutions is given in Sec. 12.9.

Derived from H_2S are the sulfides, such as Na_2S and HgS. In regard to their solubility in water, the sulfides vary widely from those which, like Na_2S, are quite soluble in water to those which, like HgS, require drastic treatment to be brought into solution.* In Fig. 24.5 are listed various representative sulfides and methods required to dissolve them. The alkali-metal and alkaline-earth-metal sulfides dissolve readily in

---

* HgS can be "dissolved" by heating it with aqua regia, but it is questionable whether this is a good description of the process. Although the mercury dissolves as $HgCl_4{}^{2-}$, the sulfur does not dissolve but is oxidized to insoluble elemental sulfur.

Fig. 24.5

Solubilities of sulfides.

Soluble in water	Na_2S, K_2S, $(NH_4)_2S$, BaS
Soluble in 0.3 M H_3O^+	ZnS, FeS, MnS, CoS
Soluble in hot HNO_3	CuS, Ag_2S, PbS, SnS
Soluble in aqua regia	HgS

water to give basic solutions, with extensive hydrolysis of sulfide ion:

$$Na_2S(s) + H_2O \longrightarrow 2Na^+ + HS^- + OH^-$$

This equation represents the principal net reaction; about 10% of the sulfide ion remains unhydrolyzed. Because the sulfides of the group I and group II metals are so soluble, they cannot be precipitated by bubbling H_2S through solutions of their salts.

As already discussed in Sec. 12.9, some sulfides that are insoluble in water can be dissolved simply by raising the H_3O^+ concentration. ZnS, for example, is soluble in 0.3 M H_3O^+ because the H_3O^+ serves to lower the concentration of sulfide ion (in equilibrium with solid ZnS) by combining with it to form H_2S. The net equation can be represented as

$$ZnS(s) + 2H_3O^+ \longrightarrow Zn^{2+} + H_2S + 2H_2O$$

The sulfides in the third row of Fig. 24.5 are so insoluble that they cannot be dissolved by H_3O^+ alone. However, hot nitric acid oxidizes sulfide to sulfur and hence lowers the sulfide-ion concentration sufficiently to permit solubility. For CuS, the net reaction can be written as

$$3CuS(s) + 8H_3O^+ + 2NO_3^- \longrightarrow$$
$$3Cu^{2+} + 3S(s) + 2NO(g) + 12H_2O$$

The least soluble of the sulfides shown in Fig. 24.5, mercuric sulfide, is not appreciably soluble in hot HNO_3. In order to "dissolve" it, aqua regia must be used in order that oxidation of the sulfide ion may be accompanied by complexing of the mercuric ion. The net reaction may be written as

$$HgS(s) + 2NO_3^- + 4Cl^- + 4H_3O^+ \longrightarrow$$
$$HgCl_4^{2-} + 2NO_2(g) + S(s) + 6H_2O$$

although the reduction product of nitrate is probably a mixture of oxides rather than just NO_2.

The differences in solubility behavior of metal sulfides can be used to great advantage in the separation and identification of various elements.

In addition to sulfides, sulfur forms polysulfides, in which two or more sulfur atoms are bound together in a chain. These polysulfides can be made, for example, by boiling a solution of a soluble sulfide with elemental sulfur. With Na_2S and sulfur, the product is usually described as Na_2S_x and is thought to consist of Na^+ ions and S_x^{2-} ions. The polysulfide chains are of varying length formed by progressive addition of sulfur atoms to sulfide ion:

$$\left[:\ddot{S}:\right]^{2-} + \ddot{S}: \longrightarrow \left[:\ddot{S}:\ddot{S}:\right]^{2-}$$

$$\left[:\ddot{S}:\ddot{S}:\right]^{2-} + \ddot{S}: \longrightarrow \left[:\ddot{S}:\ddot{S}:\ddot{S}:\right]^{2-} \cdots$$

The simplest of the polysulfide chains is the disulfide, S_2^{2-}; it is found in the mineral FeS_2, iron pyrites, or fool's gold. Solid FeS_2 has an NaCl-like structure consisting of an array of alternating Fe^{2+} and S_2^{2-} ions. In acid solution disulfides (and other polysulfides) break down to form solid sulfur and H_2S. In some respects, disulfides resemble peroxides. They are, for example, oxidizing agents, especially for metal sulfides. Thus, a solution of Na_2S_2 can oxidize stannous sulfide, SnS ($+2$ state of tin), to SnS_3^{2-} ($+4$ state of tin):

$$SnS(s) + S_2^{2-} \longrightarrow SnS_3^{2-}$$

Tin is assigned oxidation number $+4$ in the thiostannate ion, SnS_3^{2-}, because acidification produces stannic sulfide, SnS_2. The SnS_2 is identical with that precipitated from stannic solutions by H_2S.

In addition to the above compounds, which are mainly inorganic, there are many important organic compounds that contain sulfur. These include, for example, the sulfonates, which are of interest as synthetic detergents. Actually, the term "detergent" includes soap, but by common usage it has come to mean a soap substitute. Prime among these are the benzenesulfonates such as

$$CH_3(CH_2)_9-\overset{\overset{\displaystyle H}{|}}{\underset{\underset{\displaystyle CH_3}{|}}{C}}-\langle\bigcirc\rangle-\overset{\overset{\displaystyle O}{\|}}{\underset{\underset{\displaystyle O}{\|}}{S}}-O^{\ominus}\,Na^+$$

which is sodium dodecylbenzenesulfonate. As can be seen, it resembles soap in having a long hydrocarbon tail and a charged anionic group at the end. As with soap (Fig. 22.11) the hydrocarbon ends cluster to give a micelle which can dissolve oils and give desirable surface-tension properties to the solution. The above molecule is an example of a linear alkylsulfonate (LAS) and is to be contrasted with an alkyl-benzenesulfonate (ABS) of the type

$$CH_3-\overset{\overset{\displaystyle CH_3}{|}}{CH}-CH_2-\overset{\overset{\displaystyle CH_3}{|}}{CH}-CH_2-\overset{\overset{\displaystyle CH_3}{|}}{CH}-CH_2-\overset{\overset{\displaystyle H}{|}}{\underset{\underset{\displaystyle CH_3}{|}}{C}}-\langle\bigcirc\rangle-\overset{\overset{\displaystyle O}{\|}}{\underset{\underset{\displaystyle O}{\|}}{S}}-O^{\ominus}\,Na^+$$

The alkylbenzenesulfonates are very good surfactants (surface-active agents); they reduce surface tension and are good wetting agents, but because of excessive chain branching they are not biodegradable. Hence, they accumulate in the water environment and, even at 1 ppm levels, lead to serious foaming problems. Although they are not harmful even up to 50 ppm, there is a certain lack of appeal in drinking water with a "head" on it. At present, at least in the United States most ABS detergents have been replaced by LAS detergents. To make the LAS detergents, kerosine-type hydrocarbons are separated by molecular-sieve screening to give normal hydrocarbon isomers, which are then chlorinated and reacted directly with benzene in the presence of $AlCl_3$, or dehydrohalogenated, to give unsaturated hydrocarbons which are subsequently reacted with benzene. The final step is to sulfonate the hydrocarbon-benzene chain by SO_3 or H_2SO_4. The detergent industry is big business, and a search for suitable soap substitutes continues. Most detergent formulations include about 35% LAS as surfactant, 45% phosphate as sequestering agent, and 15% bleaches, perfumes, etc. The role of phosphates in particular is being questioned because they contribute to degradation of the water environment by overstimulating plant growth (eutrophication).

24.3 Selenium

Selenium is about as rare as gold. It occurs principally with sulfur, both as elemental selenium in native sulfur and as selenides in various sulfide minerals. Most commercial selenium is obtained as a by-product of the electrolytic refining of copper. The element exists in several allotropic forms, the most stable of which at room temperature is hexagonal, or "metallic," selenium. In this form, selenium atoms are joined in extremely long spiral chains arranged parallel to each other. When dissolved in molten sulfur, these chains form Se_8 molecules, as is indicated by the freezing-point depression of the sulfur.

"Metallic" selenium is a poor conductor of electricity in the dark, but its conductivity increases proportionally to its illumination; i.e., it is a photoconductor. This property is utilized in photocopying the same way as ZnO (Sec. 20.6) and in the selenium photocell, used in exposure meters for measuring light intensity. The basic feature of a selenium photocell is a sandwich consisting of a copper plate, a selenium coating, and a thin, translucent gold film. Electric leads go to the copper and to the gold film, but the circuit is not complete until light falls on the selenium. In the selenium rectifier, for converting alternating current to direct current, a similar cell is used; it differs in that the gold film of the photocell is replaced by an alloy such as Wood's metal. The selenium-alloy junction acts as a barrier to current, but in only one direction. In spite of these interesting properties, most selenium goes to the glass industry, where it is added in small amounts to molten glass to counteract the objectionable green color due to iron impurities. In large amounts it gives red glass.

The chemistry of selenium is similar to that of sulfur. It reacts with metals to form selenides, for example, Al_2Se_3, which decompose in acid to give gaseous hydrogen selenide, H_2Se. This, like H_2S, is toxic and burns to give selenium or SeO_2, but it is a stronger reducing agent than H_2S. Selenium dioxide, SeO_2, is a colorless solid which dissolves in water to give the weak selenious acid, H_2SeO_3 ($K_I = 2.7 \times 10^{-3}$ and $K_{II} = 2.5 \times 10^{-7}$). Selenic acid, H_2SeO_4, the analog of sulfuric acid, can be formed by oxidation of selenious acid with hydrogen peroxide or chlorine. It is a colorless solid which is not so strong an acid as H_2SO_4 but is a stronger oxidizing agent.

24.4 Tellurium

Nearly as abundant as selenium, tellurium is one of two elements—the other one is antimony—with which gold occurs chemically combined in nature. It is also present as tellurides in copper and lead minerals and, in fact, is obtained principally as a by-product of their refining. The most stable modification of the element is hexagonal, or "metallic," tellurium. Its structure is much like that of selenium, but its photo-conductivity is only slight. Because it is a semiconductor, its principal but limited use is in making rectifiers.

Hydrogen telluride, H_2Te, is a vile-smelling gas, unstable with re-spect to decomposition to the elements. In water it acts as a moderately weak acid ($K_I = 2.3 \times 10^{-3}$), slightly stronger than H_2Se ($K_I = 1.9 \times 10^{-4}$), which in turn is stronger than H_2S. Bi_2Te_3 is an impor-tant solid-state material for direct conversion of heat to electric current. Tellurium dioxide (TeO_2) and tellurous acid (H_2TeO_3) resemble the corresponding selenium compounds. Unlike selenic acid, telluric acid (H_6TeO_6) contains six oxygen atoms bound to the central tellurium atom, presumably because of the larger size of the tellurium atom. It is a very weak acid but a strong oxidizing agent.

Investigation of the chemical behavior of tellurium compounds, and to some extent of selenium compounds, has been retarded because of the foul-smelling nature of the compounds. These are taken up by the body and given off in the perspiration and breath. Elimination is slow, and the stench of "tellurium breath" may linger for months.

24.5 Polonium

Work on polonium has been retarded because of its high radioactivity. The nucleus is unstable and, like the radium nucleus, decomposes with the emission of alpha particles: ^{208}Po, $t_{1/2} = 2.9$ yr; ^{209}Po, 100 yr; and ^{210}Po, 138 days. Since alpha radiation is damaging to the human body, the element is extremely dangerous. Amounts greater than 4×10^{-12} g cannot be tolerated in the body.

Polonium occurs naturally in uranium minerals such as pitchblende, but only to the extent of 5×10^{-9} percent of the mineral. It is con-stantly being produced by the radioactive decay of other elements, but since it itself decays, the concentration stays constant.

X-ray studies of trace amounts indicate that the element exists in two forms which appear to be metallic, one of which seems to resemble lead. Little is known of its compounds, but it does form PoH_2, Po^{2+}, PoO_2, and PoO_3.

24.6 Qualitative analysis

An unknown solution might contain sulfur as sulfide, sulfate, or sulfite. The sulfite, on addition of acid, generates H_2S gas, which can be detected either by its odor or by its blackening (due to PbS formation) of filter paper wet with a solution of lead acetate. Sulfate, on addition of barium nitrate and acid, produces white, insoluble $BaSO_4$. If sulfite is present, it will not precipitate with Ba^{2+} in acid solution; however, if Br_2 is added, the sulfite will be oxidized to sulfate, and $BaSO_4$ forms.

Exercises

*24.1   *Bonding.* Given the ns^2np^4 electron configuration of the group VI elements, what bonding angle do you predict? Justify your prediction.

*24.2   *Group properties.* Suggest a reason why the melting point of oxygen is so much lower than that of the other group VI elements.

*24.3   *Ionization.* To detach an electron from an isolated sulfur atom, what wavelength radiation needs to be used?   *Ans. 120 nm*

*24.4   *Structure.* Draw a picture (side view) of the S_8 molecule showing the unshared pairs of electrons.

*24.5   *Solid state.* The dimensions of the unit cell for rhombic sulfur are 1.046, 1.287, and 2.449 nm, respectively. What is the theoretical density of rhombic sulfur? (The experimental value is 2.07 g/cm^3.)

*24.6   *Heating curves.* Draw two heating curves (temperature vs time for uniform addition of heat) corresponding to fast and slow heating of sulfur across the midlevel of the phase diagram in Fig. 24.4.

*24.7   *Pollution.* The world burns about 5×10^9 tons (1 ton $\approx$ 1000 kg) of coal per year. Assuming an average sulfur content of 2.7 percent by weight, what would be the resulting parts-per-million level of SO_2 in the atmosphere assuming global, uniform distribution in a 20-km layer? See footnote, Sec. 15.1.   *Ans. 0.05 ppm*

*24.8   *Detergents.* Describe what is meant by a micelle. Why do micelles form? Explain how development of molecular-sieve silicates helped to remove foam from rivers.

**24.9   *Ionic sizes.* Calculate for each of the group VI elements the ratio of the radius of its dinegative ion to the radius of the following alkaline-earth ion. (See Appendix 8.) Suggest a reason for the trend.

**24.10   *Group properties.* Given the electrode potentials of Fig. 24.1, which of the H_2X compounds of group VI should be able to reduce neutral water to hydrogen gas?

**24.11   *Superheated water.* What pressure probably is needed to have the liquid water at 170°C for the Frasch process? The heat of vaporization of H_2O is 40.7 kJ/mol. *Ans. 7.95 atm*

** *24.12 Structure.* Given 0.207 nm for the S—S distance and 105° for the S—S—S bond angle, calculate the S-to-S distance across the S_8 ring.

** *24.13 Bonds.* Using molecular-orbital theory, predict the relative bond energies for S_2 and P_2. What should happen to the bond strength on removal of one electron from each of these?

** *24.14 Entropy.* Which has a larger entropy change on melting, rhombic sulfur or monoclinic sulfur? The ΔH for the process rhombic $\longrightarrow$ monoclinic is positive. Justify your answer.

** *24.15 Thermodynamics.* The standard free energy of formation of $SO_2(g)$ is -300 kJ/mol; that of $SO_3(g)$ is -370 kJ/mol. Calculate the equilibrium constant for $2SO_2(g) + O_2(g) \rightleftharpoons 2SO_3(g)$ at standard conditions. If $\Delta H°$ for the above process is -197 kJ, what will be the value of K_p at 500°C? *Ans. $3 \times 10^{24}, 2 \times 10^3$*

** *24.16 pH.* What concentration of H_2SO_4 solution do you need to have pH = 2.00? What concentration of $NaHSO_4$ solution do you need to achieve the same thing? *Ans. 0.0064 and 0.018 M*

** *24.17 Dissociation.* Selenic acid, H_2SeO_4, is a strong electrolyte for dissociation of the first proton and has $K_{diss} = 8.9 \times 10^{-3}$ for the dissociation of the second. What will be the concentration of H_3O^+, $HSeO_4^-$, and SeO_4^{2-} in the final solution made by mixing 100 ml of 0.200 M HCl with 100 ml of 0.200 M Na_2SeO_4? Assume volumes are additive.

** *24.18 Equations.* Write balanced net equations for each of the following reactions:
  a Oxidation of $S_2O_3^{2-}$ in base by HO_2^- to give SO_4^{2-}
  b Oxidation of Mn^{2+} in acid by $HS_2O_8^-$ to give MnO_4^- and HSO_4^-
  c Dissolving of PbS(s) in hot nitric acid to form HSO_4^- and NO(g)
  d Oxidation of disulfide ion in base by chromate ion to give elemental sulfur and $Cr(OH)_4^-$

** *24.19 Conversions.* Indicate reagents and reactions required to carry out the following conversions so as to conserve the sulfur atoms:
  a FeS_2 to H_2SO_4
  b SO_2 to S
  c Na_2SO_3 to Na_2SO_4
  d H_2S to $NaHSO_3$

** *24.20 Symmetry.* What are the symmetry elements of sulfate ion? Which of these are lost in going from sulfate to thiosulfate? What new symmetry elements appear?

** *24.21 Toxicity.* The maximum allowable concentration of hydrogen sulfide is 20 ppm (by weight). What pressure does this correspond to at STP? How many grams of FeS would you need to dissolve in acid to infect an average room (50 m³) to this toxic concentration?

Ans. 1.7 × 10⁻⁵ atm, 3.3 g

** 24.22   Hydrolysis. Calculate the percent hydrolysis of sulfide ion in 0.50 M Na_2S.

** 24.23   Solubility. The K_{sp} of ZnS is $1 × 10^{-22}$; that of CdS is $1.0 × 10^{-28}$. What concentration of hydronium ion should you maintain in a solution of 0.50 M Zn^{2+} and 0.50 M Cd^{2+} so that addition of H_2S precipitates one but not the other?

** 24.24   Selenium. Write equations for each step in the progressive conversion Se $\longrightarrow$ Al_2Se_3 $\longrightarrow$ H_2Se $\longrightarrow$ SeO_2 $\longrightarrow$ H_2SeO_3 $\longrightarrow$ H_2SeO_4.

** 24.25   Tellurium. Suggest a reason why the +6 acid of tellurium is a weak electrolyte whereas the +6 acid of sulfur is a strong electrolyte.

*** 24.26   Hydrolysis. Calculate the percent hydrolysis of SO_4^{2-} in 0.10 M Na_2SO_4.

Ans. 2.6 × 10⁻⁴%

*** 24.27   pH. Assuming that only the first dissociation $SO_2 + 2H_2O \rightleftharpoons H_3O^+ + HSO_3^-$ needs to be considered, calculate the pH of the final solution made by mixing 25.0 ml of 0.100 M SO_2 with 10.0 ml of 0.100 M NaOH. Ignore the presence of H_2SO_3, and assume volumes are additive.

*** 24.28   Thermodynamics. The standard free energy of formation of $S_2O_3^{2-}(aq)$ is -532 kJ/mol, and that of $SO_3^{2-}(aq)$ is -497 kJ/mol. Calculate $\Delta G°$ for $S_2O_3^{2-}(aq) \longrightarrow S(s) + SO_3^{2-}(aq)$ and for $S_2O_3^{2-}(aq) + H_3O^+(aq) \longrightarrow S(s) + HSO_3^-(aq) + H_2O$.

Ans. +35 and −6 kJ

*** 24.29   Simultaneous equilibria. What is the pH of 0.500 M $NaHSO_3$? What happens to the pH when you add 24.0 ml of 0.500 M NaOH to 280 ml of 0.500 M $NaHSO_3$? Assume volumes are additive.

*** 24.30   Polonium. How many alpha particles per second would be emitted from $4 × 10^{-12}$ g of ^{210}Po ($t_{1/2} = 138$ days)?

Ans. 700

Although the chemistry of the group VII elements is somewhat complex, similarities within the group are more pronounced than in any of the other groups except I and II. The elements fluorine, chlorine, bromine, iodine, and astatine are collectively called *halogens* (from the Greek *halos*, salt, and *genēs*, born), or "salt producers," because they all have high electronegativity and form negative halide ions such as are found in ionic salts. Except for fluorine, they also show positive oxidation states.

647

25.1 Group properties

Because of their high electronegativity, the halogens show practically no metallic properties, though solid iodine has a somewhat metallic appearance. Astatine, the heaviest member of the group, may also have some metal properties, but it is a short-lived radioactive element (even the longest-lived isotope has a half-life of only 8.3 h), and not enough of it has been accumulated to see whether the solid is metallic. Other properties of the group are shown in Fig. 25.1. Although the bond between the halogen atoms in the X_2 molecules is fairly strong (see Fig. 25.2), the attraction between X_2 molecules is quite weak and due only to van der Waals forces. In going down the group there is an increasing number of electrons per X_2 molecule, and we would expect van der Waals attraction to increase. Thus, it is not surprising that the boiling points increase in going from F_2 to I_2. At room temperature fluorine and chlorine are gases, bromine is a liquid, and iodine is a solid; all are volatile, however, so even for bromine and iodine, vapors are present at room temperature.

Fig. 25.1
Elements of group VII.

Symbol	Z	Electronic configuration	Melting point, °C	Boiling point, °C	Ionization potential, eV	Electrode potential, V (X_2 to X^-)
F	9	(2) $2s^2 2p^5$	−223	−187	17.42	+2.87
Cl	17	(10) $3s^2 3p^5$	−102	−34.6	13.01	+1.36
Br	35	(28) $4s^2 4p^5$	−7.3	58.78	11.84	+1.09
I	53	(46) $5s^2 5p^5$	114	183	10.44	+0.54
At	85	(78) $6s^2 6p^5$	....	....	....	+0.2

As indicated by the relatively high values of the ionization potentials, it is fairly difficult to remove an electron from a halogen atom. In fact, more energy is required to remove an electron from a halogen atom than from any other atom in the same period except for the noble gas (compare values in Fig. 2.28). Within the group itself there is, of course, a decrease in the ionization potential going down the group; the larger the halogen atom, the less firmly bound are the outermost electrons, and the lower is the energy required to remove an electron from the neutral atom.

Of greater significance chemically are the electrode potentials, given in the last column of Fig. 25.1. The potentials show that fluorine gas is the best oxidizing agent of the group. The reason for this, however, is not so simple as it appears. The overall half-reaction

$$\tfrac{1}{2}X_2(g) + e^- \longrightarrow X^-(aq)$$

can be constructed from the set of consecutive steps

$$\tfrac{1}{2}X_2(g) \xrightarrow{(1)} X(g) \xrightarrow{(2)} X^-(g) \xrightarrow{(3)} X^-(aq)$$

Step (1) corresponds to breaking up a half mole of diatomic X_2 mole-

Step	Fluorine	Chlorine	Bromine	Iodine
$\frac{1}{2}X_2(g) \longrightarrow X(g)$	$+63$	$+105$	$+79$	$+59$
$X(g) + e^- \longrightarrow X^-(g)$	-333	-348	-324	-295
$X^-(g) \longrightarrow X^-(aq)$	-460	-348	-318	-279
$\frac{1}{2}X_2(g) + e^- \longrightarrow X^-(aq)$	-730	-591	-563	-515

Fig. 25.2
Free-energy data for halogens in kilojoules per mole.

cules to give monatomic gas; in step (2) each gaseous X atom picks up one electron to form a mononegative X^- ion; in step (3) the X^- ions go from the gaseous state to the aqueous state; i.e., they become hydrated. The free-energy changes for the various steps are listed in Fig. 25.2. In the case of fluorine, for example, the sum of the three processes shown has a net free-energy change of -730 kJ for $\frac{1}{2}X_2(g) + e^- \longrightarrow X^-(aq)$. The electron, however, as discussed in Sec. 16.2, is referred to a hydrogen-electrode standard, which amounts to having an additional change of $+465$ kJ in the free energy:

$$\frac{1}{2}X_2(g) + e^- \longrightarrow X^-(aq) \qquad \Delta G^\circ = -730 \text{ kJ}$$

$$\frac{1}{2}H_2(g) + H_2O \longrightarrow H_3O^+ + e^- \qquad \Delta G^\circ = +465 \text{ kJ}$$

$$\frac{1}{2}X_2(g) + \frac{1}{2}H_2(g) + H_2O \longrightarrow X^-(aq) + H_3O^+ \quad \Delta G^\circ = -265 \text{ kJ}$$

By using the relation $\Delta G^\circ = -n\mathcal{F}E^\circ$, we see that the final ΔG° corresponds to a voltage of $+2.75$ V. In other words, the calculated electrode potential for fluorine is $+2.75$ V. Similar calculations for chlorine, bromine, and iodine give E° values of $+1.31$, $+1.0$, and $+0.52$ V, respectively. For the case of bromine and especially for the case of iodine, it is necessary to correct for the fact that the final state of X_2 is not gaseous but liquid and solid, respectively. This correction amounts to $+1.7$ kJ for bromine and $+9.6$ kJ for iodine.

Comparing the halogens with each other, we see from Fig. 25.2 that the most significant difference between the elements is the hydration step. Except for chlorine, it requires about the same amount of energy to break up the X_2 molecule. It is about equally favorable to add an electron to any gaseous halogen atom [actually, most favorable in the case of $Cl(g)$]. Hence, if only the sum of the first two steps were considered, there would not be much difference between the various halogens, and the electrode potentials would be within 0.35 V of each other. The biggest single factor in making the E°'s differ greatly from each other is the hydration free energy. In going from F^- to I^- there is a large increase in ionic radius (0.133 nm for F^-, 0.181 nm for Cl^-, 0.196 nm for Br^-, and 0.220 nm for I^-), which makes for a large decrease in affinity for water. The reason why F_2 is such a good oxidizing agent [and $F^-(aq)$ is such a poor reducing agent] is that the fluoride ion is so strongly hydrated.*

* Although this discussion has been confined to aqueous solution, the uniquely great oxidizing strength of fluorine shows up even when water solutions are not involved. Thus, for example, dry F_2 can oxidize solid NaCl to Cl_2 and NaF. In these cases oxidation results because the smaller size of the fluoride ion leads to more favorable lattice energies.

Fluorine is the most electronegative of all the elements; therefore, it can show only a negative oxidation state. The other halogens, however, also show positive oxidation states in compounds with more electronegative elements. Most of these compounds contain oxygen, which has electronegativity between that of fluorine and chlorine. In their oxy compounds chlorine, bromine, and iodine show a maximum oxidation number of +7. In addition, chlorine, bromine, and iodine form compounds in which the halogen atom is assigned oxidation numbers +1 and +5.

25.2   Fluorine

Fluorine is about half as abundant as chlorine and is widely distributed in nature. It occurs principally as the minerals fluorspar, CaF_2; cryolite, Na_3AlF_6; and fluorapatite, $Ca_5F(PO_4)_3$. Because none of the ordinary chemical oxidizing agents is capable of extracting electrons from fluoride ions, elemental fluorine is prepared only by electrolytic oxidation of molten fluorides, such as KF–HF mixtures. At room temperature fluorine is a pale yellow gas which is extremely corrosive and reactive. With hydrogen it forms violently explosive mixtures because of the reaction

$$H_2(g) + F_2(g) \longrightarrow 2HF(g) \qquad \Delta H° = -537 \text{ kJ}$$

On the skin it causes severe "burns," which are quite slow to heal.

Hydrogen fluoride is usually made by the action of sulfuric acid on fluorspar. Because of hydrogen bonding (Sec. 15.5), liquid HF has a higher boiling point (19.5°C) than any of the other hydrogen halides. Hydrogen bonding is also present in the gas phase and accounts for polymeric species, $(HF)_x$, where x is some small number such as 6 or less. In aqueous solutions HF is called hydrofluoric acid and is unique among the hydrogen halides in being a weak, rather than a strong, acid ($K_{diss} = 6.7 \times 10^{-4}$) and in being able to dissolve glass. The latter reaction is attributed to the formation of fluosilicate ions as in the equation

$$SiO_2(s) + 6HF \longrightarrow SiF_6{}^{2-} + 2H_3O^+$$

where glass is represented for simplicity as SiO_2. Other complex ions of fluorine are known, for example, $AlF_6{}^{3-}$, $ZrF_7{}^{3-}$, and $TaF_8{}^{3-}$, in which the small size of fluoride ion permits relatively large numbers of them to be attached to another atom.

In general, most simple fluoride salts formed with +1 cations are soluble (for example, KF and AgF) and give slightly basic solutions because of the hydrolysis of F^- to HF. With +2 cations, however, the fluorides are usually insoluble (for example, CaF_2 and PbF_2), but their solubility is somewhat increased in acid solution. The formation of insoluble, inert fluorides as surface coatings is apparently the reason why fluorine and its compounds can be stored in metal containers such as copper.

With oxygen, fluorine forms two compounds, oxygen difluoride (OF_2) and dioxygen difluoride (O_2F_2). The first can be prepared by passing fluorine very rapidly through dilute NaOH solution:

$$2F_2(g) + 2OH^- \longrightarrow 2F^- + OF_2(g) + H_2O$$

It is somewhat less reactive than F_2 and slowly reacts with H_2O to form HF and O_2. O_2F_2 results as a red liquid when an electric discharge is passed through a mixture of fluorine and oxygen below $-100°C$. It is unstable with respect to decomposition to the elements.

Most amazing of the fluorine compounds are the fluorocarbons. These are materials which can be considered to be derived from the hydrocarbons (Sec. 22.4) by substitution of fluorine atoms for hydrogen atoms. Thus, the fluorocarbon corresponding to methane, CH_4, is tetrafluoromethane, CF_4. This compound is typical of the saturated (i.e., containing no double bonds) fluorocarbons in being extremely inert. For example, unlike methane, it can be heated in air without burning. Furthermore, it can be treated with boiling nitric acid, concentrated sulfuric acid, and strong oxidizing agents such as potassium permanganate with no change. Reducing agents such as hydrogen and carbon do not affect it even at temperatures as high as $1000°C$. Because of their inertness, the fluorocarbons find application for special uses. For example, $C_{12}F_{26}$ is an ideal insulating liquid for heavy-duty transformers that operate at high temperature. Just as ethylene, C_2H_4, can polymerize to form polyethylene (Sec. 22.4), so tetrafluoroethylene, C_2F_4, can polymerize to form polytetrafluoroethylene. The polymerization can be imagined to proceed by the opening up of the double bond to form an unstable intermediate, which joins with other molecules to produce a high polymer:

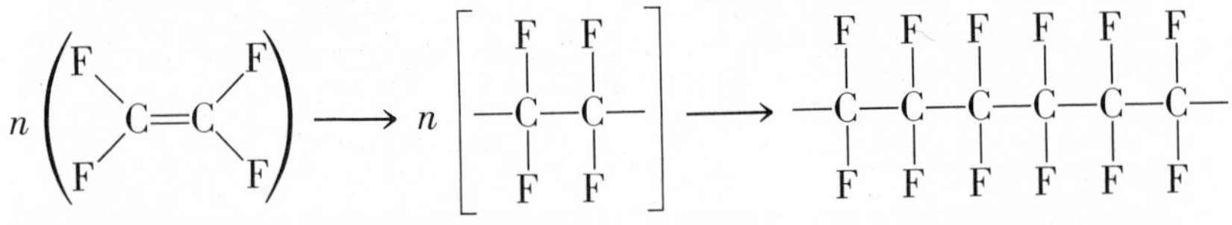

The high polymer is a plastic known commercially as Teflon and, like the other saturated fluorocarbons, it is inert to chemical attack. It is unaffected even by boiling aqua regia or ozone. Though rather expensive, fluorocarbon polymers find use as structural materials where corrosive conditions are extreme, as in chemical plants. They are also familiar for coating "greaseless" frying pans—a boon to brides, bachelors, and calorie counters.

Finally, as noted in the following section, fluorine was the key element in finally cracking the problem of making the noble-gas elements combine chemically with other elements.

25.3 Noble-gas compounds

The search for compounds of the group 0 elements (helium, neon, argon, krypton, xenon, and radon) has been carried on for many years. Until recently, the best that could be done were the *clathrate*, or "enclosure," compounds, in which noble-gas atoms were trapped in cagelike holes in a crystal lattice. The synthetic breakthrough came as a result of an observation by Neil Bartlett that O_2 reacts with PtF_6 to form a solid compound O_2PtF_6. X-ray diffraction data indicated that the orange-colored compound consists of O_2^+ and PtF_6^-. Recognizing that the ionization potential of the O_2 molecule (12.2 eV) is very close to that of xenon (12.1 eV), Bartlett tried the reaction between xenon and PtF_6. Reaction occurred readily at room temperature to produce a red solid, $XePtF_6$, the first real compound of a noble-gas element.

Since this discovery in 1962, a number of other compounds have been prepared. These are principally compounds of xenon (with fluorine or oxygen), but a few compounds of krypton and radon have also been prepared. Some of the compounds are complex. For example, xenon reacts with PtF_6, PuF_6, and RhF_6 to form compounds of the type $XePuF_6$, but not with the more stable hexafluorides such as UF_6. This observation, coupled with the fact that RuF_6 reacts with xenon to produce $Xe(RuF_6)_2$, suggests that the MF_6 may act as a fluoridating agent and not just as an electron acceptor. In any case PtF_6 acts as one of the most powerful oxidizing agents known.

The most striking example of simple-compound formation is the reaction of xenon with fluorine. For instance, heating of a 1:5 mixture of Xe and F_2 gases in a nickel container at 400°C followed by quenching produces a colorless solid product corresponding to XeF_4. Variation in the initial Xe-to-F_2 ratio leads to the compounds XeF_2 and XeF_6, also colorless solids. Some properties of these compounds are shown in Fig. 25.3. Chemically, they all react with H_2 to produce HF and

Fig. 25.3
Properties of xenon fluorides.

Formula	Melting point, °C	Vapor pressure at 20°C, atm	ΔH_{evap}, kJ/mol	Bond energy, kJ/mol
XeF_2	*ca. 130*	*0.004*	*51.5*	*160*
XeF_4	*100*	*0.004*	*64.0*	*130*
XeF_6	*46*	*0.036*	*37.7*	*130*

Xe. All dissolve in liquid HF, and in the case of XeF_6 there is about 30 percent conversion to ions, presumably XeF_5^+ and HF_2^-. In water, a variety of reactions occurs. XeF_2 oxidizes H_2O to O_2.

$$2XeF_2 + 2H_2O \longrightarrow O_2(g) + 2Xe(g) + 4HF$$

XeF_4 disproportionates to give Xe gas and some Xe(VI) species. After evaporation of the solution, the compound XeO_3 is obtained. This is a white, nonvolatile compound that is tremendously explosive (about like TNT). It can be detonated by simple rubbing, pressing, or slight heating. Because XeO_3 is a frequent product of Xe-compound reactions, there is an element of added excitement in the work with Xe compounds. XeF_6 hydrolyzes in aqueous solutions to produce various products depending on the pH of the solution. In acid solution, Xe(VI) species are most stable [such as $XeOF_4$ and perhaps $Xe(OH)_6$*]; in

* This can also be written H_6XeO_6 and has been called xenic acid. It is hoped that the corresponding Kr compound can be formed so that we can also have "kryptic acid."

base, Xe(VIII) species predominate. From NaOH solution, a solid of the formula $Na_4XeO_6 \cdot 8H_2O$ precipitates. X-ray studies show it to contain the octahedral ion XeO_6^{4-}, called the perxenate ion.

Because the noble-gas compounds were so late in arriving, there was awaiting an impressive array of elegant instruments for structural studies. Within one year of their preparation, many of the compounds became among the best characterized compounds in chemistry. Among the techniques brought to bear on the problem were X-ray, electron, and neutron diffraction as well as all types of spectroscopy, including NMR, infrared, ultraviolet, and microwave (radar).

The shapes of the various molecules (shown in Figs. 25.4 and 25.5) are as follows: XeF_2 is linear, XeF_4 is square-planar, XeO_3 is a pyramid, $XeOF_4$ is a square pyramid. The Xe—F bond length is about 0.2 nm, and the bond energy is about 130 to 160 kJ/mol. For Xe—O the bond length is about 0.18 nm. None of these values is unusual. Hence, xenon forms normal chemical bonds. The bonding can be accounted for by a valence-bond scheme which utilizes the $5d$ orbitals of xenon. In XeF_2, for example, one of the $5d$ orbitals can be hybridized with the $5s$ and the three $5p$ orbitals to form a set of five sp^3d hybrids. These are directed, as in Fig. 25.4 (left), toward the corners of a trigonal bipyramid. (This can be visualized as two triangular pyramids sharing a common base.) Within the five sp^3d orbitals there are ten electrons (one from each of the bound fluorine atoms plus eight from xenon) to give two ordinary covalent bonds and three unshared pairs. If the two fluorine atoms are as far apart as possible, we get the structure shown on the left in Fig. 25.4. For XeF_4 we need to use two of the $5d$ orbitals of xenon. The combination (two $5d$, one $5s$, and three $5p$) leads to six sp^3d^2-hybrid orbitals directed toward the corners of an octahedron. The twelve electrons (one from each of four fluorines and eight from xenon) are disposed as four covalent bonds and two unshared pairs. With the four fluorine atoms in a plane, the structure is square-planar, as is shown on the right in Fig. 25.4.

In $XeOF_4$, shown on the left in Fig. 25.5, sp^3d^2-hybrid orbitals again lead to a basically octahedral disposition of twelve electrons (four from four F atoms, eight from Xe, and none from the O atom). The O atom, two short of a full subshell, can be thought of as adding on to an unshared pair of the XeF_4 structure. In similar fashion, we can rationalize XeO_3. The three O atoms of XeO_3 can be imagined to accept a share of three of the four electron pairs in the outer shell of an Xe atom. If the four pairs constitute an sp^3 set, the geometry is nearly tetrahedral, with three bonding pairs directed toward three vertexes and the unshared pair going to a fourth. The observed O—Xe—O bond angle of 103° is reasonably close to the value of 109° expected for the tetrahedral angle.

For XeF_6 the simple valence-bond theory would lead to a prediction of a distorted octahedral structure. The reasoning goes as follows: Each of the six fluorine atoms makes available one electron, to give a total of six. The xenon atom has eight valence electrons. Thus, the hybrid orbitals need to accommodate fourteen electrons, or seven pairs. Using

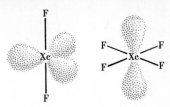

Fig. 25.4
Xenon fluorides.

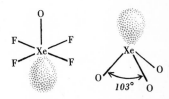

Fig. 25.5
Oxyxenon compounds.

the 5s and three 5p orbitals of the xenon, we need three more orbitals from the 5d set of xenon. This would give us the hybrid set sp^3d^3. As we have repeatedly seen, sp^3d^2 leads to an octahedron. The use of an additional d orbital for a pair of unshared electrons introduces a distortion, the nature of which depends on which d orbital is used. The reason that the prediction of the expected structure is less definite for XeF_6 than for the other compounds is that in the case of XeF_6 the structure has not yet been experimentally determined. At first, the data suggested XeF_6 was a regular octahedron; later experiments showed it not to be precisely octahedral.

Finally, it might be noted that molecular-orbital methods provide alternative descriptions for the Xe compounds without using d orbitals. This is appealing because in Xe the $5d$ orbitals are considerably higher in energy than are the $5p$. (Note that in previously discussed transition-metal complexes the octahedral orbitals were d^2sp^3 instead of sp^3d^2; i.e., the d orbital used was 1 lower in principal quantum number than the s and p.)

To account for XeF_2, for example, molecular-orbital theory forms molecular orbitals by combining p orbitals of xenon with p orbitals of the two fluorines. The result is a linear three-center orbital somewhat analogous to the three-center "banana bonds" used for B_2H_6 (Sec. 21.2).

Figure 25.6 is a schematic version of how one of the molecular orbitals can be set up. With the three atoms being brought together

Fig. 25.6
Molecular orbital for XeF_2.
(Xe 5p orbital is shown schematically since it has radial nodes and is more complex than the 2p orbital of F.)

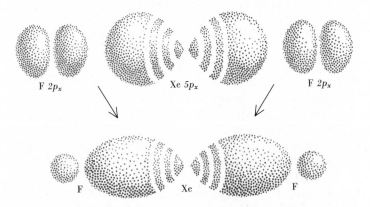

along the x axis, the $2p_x$ orbitals of the F atoms can be considered to coalesce with the $5p_x$ orbital of Xe to give the bonding orbital shown. This orbital accommodates two electrons; it is bonding because of the relatively high electron density in the two Xe–F internuclear regions.

Actually, from three atomic orbitals three molecular orbitals must result. Figure 25.6 shows only the lowest-energy one of the three. Of the other two, one is nonbonding and corresponds to high electron density on the two F atoms, with little contribution from the xenon. This orbital also accommodates two electrons. The third molecular orbital, of very high energy, corresponds to considerable depletion of electron density from the internuclear region; hence, it is antibonding.

In the above set of molecular orbitals for XeF_2, we have four electrons to accommodate—one from each of the F atoms and two that were originally in the $5p_x$ orbital of Xe. Of these four electrons, two can be placed in the bonding orbital, and the remaining two in the nonbonding one. Since the antibonding orbital is unused, the net result is a bonding situation.

The above molecular-orbital construction can be extended by using successively the $5p_y$ and the $5p_z$ orbitals of the Xe to form additional molecular orbitals perpendicular to the set discussed above. In such fashion, XeF_4 and XeF_6 have been rationalized. However, there are serious difficulties, and the actual situation must be more complex than implied by successively combining the p orbitals as above. The final solution of the theoretical problems posed by the bonding in noble-gas compounds may be different from either of the above descriptions, but in any case, the final solution should have general applicability beyond these particular compounds.

25.4   Chlorine

Chlorine is the most abundant (0.2 percent) of the halogens and occurs as chloride ion in seawater, salt wells, and salt beds, where it is combined with Na^+, K^+, Mg^{2+}, and Ca^{2+}. On a small scale, the element can be made by chemical oxidation as with MnO_2:

$$MnO_2(s) + 2Cl^- + 4H_3O^+ \longrightarrow Mn^{2+} + Cl_2(g) + 6H_2O$$

On a commercial scale, chlorine is more economically prepared by electrolytic oxidation of either aqueous or molten NaCl (see Sec. 14.2). The element is a greenish yellow gas (in fact, it gets its name from the Greek, *chloros,* green) and has a choking odor. Although not so reactive as fluorine, it is a good oxidizing agent and explodes with hydrogen when mixtures of H_2 and Cl_2 are exposed to ultraviolet light. The affinity for hydrogen is so great that it reacts with hydrogen-containing compounds such as turpentine ($C_{10}H_{18}$) to form HCl and carbon. Most of the commercial chlorine is used as a bleach for paper and wood pulp and for large-scale disinfecting of public water supplies. Both of these uses depend on its oxidizing action.

The most important compounds of chlorine are those which correspond to the oxidation states -1, $+1$, $+5$, and $+7$, although there also are compounds of chlorine in the other positive states $+3$, $+4$, and $+6$. The -1 state is familiar as the one assigned to chlorine in HCl and chloride salts. Although HCl can be produced by direct combination of the elements, a more convenient method of preparation is the heating of NaCl with concentrated H_2SO_4:

$$NaCl(s) + H_2SO_4 \longrightarrow NaHSO_4(s) + HCl(g)$$

Hydrogen chloride gas is very soluble in water, and it is the aqueous solutions that are properly referred to as hydrochloric acid. Commercially available, concentrated hydrochloric acid is 37% by weight HCl, or 12 M. Unlike HF, HCl is a strong acid and is essentially completely dissociated into ions in 1 M solution. Why is HCl so much stronger as an acid than HF? The higher hydration energy of the fluoride ion would tend to favor the dissociation of HF. The fact that HCl is more highly dissociated arises because the bond in HCl (427 kJ/mol) is weaker than that in HF (575 kJ/mol). Inasmuch as HCl is a strong acid, there is no appreciable tendency for chloride ion to hydrolyze in aqueous solution. Thus, solutions of NaCl and KCl, for example, are neutral. Of the common chlorides, silver chloride (AgCl), mercurous chloride (Hg_2Cl_2), and lead chloride ($PbCl_2$) are rather insoluble.

The +1 oxidation state of chlorine is represented by hypochlorous acid, HOCl, and its salts, the hypochlorites. Hypochlorous acid is produced to a limited extent when chlorine gas is dissolved in water. Disproportionation of the dissolved chlorine occurs according to the equation

$$Cl_2 + 2H_2O \rightleftharpoons Cl^- + H_3O^+ + HOCl \qquad K = 4.7 \times 10^{-4}$$

The yield of products can be greatly increased by tying up the Cl^- and H_3O^+, as by adding silver oxide (Ag^+ to precipitate AgCl and oxide to neutralize H_3O^+). The formula of hypochlorous acid is usually written HOCl, instead of HClO, to emphasize the fact that the proton is bonded to the oxygen and not directly to the chlorine. The acid is weak, with a dissociation constant of 3.2×10^{-8}, and exists only in aqueous solution. Even in solution it slowly decomposes with evolution of oxygen:

$$2HOCl + 2H_2O \longrightarrow 2H_3O^+ + 2Cl^- + O_2(g)$$

HOCl is a powerful oxidizing agent, as shown by the electrode potential for the half-reaction:

$$2HOCl + 2H_3O^+ + 2e^- \longrightarrow Cl_2(g) + 4H_2O \qquad E° = +1.63 \text{ V}$$

The value is more positive than that for permanganate ion in acid solution (+1.51 V), indicating that the HOCl is a stronger oxidizing agent than MnO_4^-. Hypochlorites, such as NaClO, can be made by neutralization of HOCl solutions, but they are produced more economically by the disproportionation of chlorine in basic solution:

$$Cl_2 + 2OH^- \longrightarrow Cl^- + ClO^- + H_2O$$

Commercially, the process is efficiently carried out by electrolyzing cold, aqueous sodium chloride solutions and stirring vigorously. The stirring serves to mix chlorine produced at the anode

$$2Cl^- \longrightarrow Cl_2 + 2e^-$$

with hydroxide ion produced at the cathode

$$2e^- + 2H_2O \longrightarrow H_2(g) + 2OH^-$$

so that reaction can occur. Solutions of hypochlorite ion so produced are sold as laundry bleaches, e.g., Clorox. Another common household bleach which owes its action to the oxidizing power of hypochlorite ion is bleaching powder, or chlorinated lime. It is largely $4Ca(ClO)_2 \cdot Ca(OH)_2$ and is prepared by treating calcium hydroxide with chlorine.

In aqueous solution, hypochlorite ion is unstable with respect to self-oxidation and, when warmed, disproportionates by the equation

$$3ClO^- \longrightarrow 2Cl^- + ClO_3^-$$

to produce chloride ion and chlorate ion (ClO_3^-). Chlorate ion contains chlorine in oxidation state $+5$. Its structure is pyramidal with the three oxygen atoms forming the base of the pyramid and the chlorine atom the apex. Probably the most important chlorate salt is $KClO_3$, used as an oxidizing agent in matches, fireworks, and some explosives. Since $KClO_3$ is only moderately soluble in water, it can be precipitated by addition of KCl to chlorate-containing solutions. The chlorate solutions can be produced by electrolyzing hot chloride solutions that are vigorously stirred. Steps in the production can be summarized as follows:

$$2Cl^- + 2H_2O \xrightarrow{electrolyze} Cl_2 + 2OH^- + H_2(g)$$

$$3Cl_2 + 6OH^- \xrightarrow{stir, \; heat} 5Cl^- + ClO_3^- + 3H_2O$$

$$K^+ + ClO_3^- \longrightarrow KClO_3(s)$$

As seen from the equation for the second step, only one-sixth of the chlorine is converted to ClO_3^-, which makes the process seem rather inefficient. However, on continued electrolysis the chloride produced in the second step is reoxidized in the first step.

Unlike hypochlorite ion, chlorate ion is the anion of a strong acid. The parent acid, $HClO_3$, chloric acid, has not been prepared in the pure state since it is unstable. When attempts are made to concentrate chloric acid solutions, as by evaporation, violent explosions occur. The principal reaction is

$$4HClO_3 \longrightarrow 4ClO_2(g) + O_2(g) + 2H_2O(g)$$

but the chlorine dioxide, ClO_2, produced may decompose further. In acid aqueous solutions, chlorate ion, like hypochlorite ion, is a good oxidizing agent. The electrode potential for the half-reaction

$$2ClO_3^- + 12H_3O^+ + 10e^- \longrightarrow Cl_2(g) + 18H_2O$$

$$E^\circ = +1.47 \text{ V}$$

indicates that ClO_3^- is almost the equal of MnO_4^-.

When $KClO_3$ is heated, it can decompose by two reactions

$$2KClO_3(s) \longrightarrow 2KCl(s) + 3O_2(g)$$

$$4KClO_3(s) \longrightarrow 3KClO_4(s) + KCl(s)$$

the first of which is catalyzed by surfaces, such as powdered glass or MnO_2, from which oxygen can readily escape. In the absence of such catalysts, especially at lower temperatures, the formation of potassium perchlorate ($KClO_4$) is favored. A more efficient method of preparing perchlorates is to use electrolytic oxidation of chlorate solutions. Since $KClO_4$ is only sparingly soluble in water (less than $KClO_3$), it can be made by addition of K^+ to perchlorate solutions. The perchlorate ion has a tetrahedral configuration with the chlorine atom at the center of the tetrahedron and the four oxygen atoms at the corners. In aqueous solutions perchlorate ion is potentially a good oxidizing agent

$$2ClO_4^- + 16H_3O^+ + 14e^- \longrightarrow Cl_2(g) + 24H_2O$$
$$E^\circ = +1.39 \text{ V}$$

especially in acid solution, but its reactions are so very slow that they are usually not observed. For example, a solution containing ClO_4^- and the very strong reducing agent Cr^{2+} (chromous ion) can be kept for weeks without any appreciable oxidation to Cr^{3+} (chromic ion).

Like chlorate ion, perchlorate ion is an anion of a strong acid. Consequently, in aqueous solution there is practically no transfer of protons from H_3O^+ to ClO_4^-. However, when perchlorate salts are treated with sulfuric acid, pure hydrogen perchlorate ($HClO_4$) may be distilled off under reduced pressure. The anhydrous compound is a liquid at room temperature and is extremely dangerous because it may explode spontaneously. With water, $HClO_4$ forms a series of hydrates. The monohydrate, $HClO_4 \cdot H_2O$, is a crystalline solid which actually contains H_3O^+ and ClO_4^- at the lattice points. Like the anhydrous material or the concentrated solutions, the hydrate should be treated with respect because of the possibility of explosions. The danger is especially great in the presence of reducing agents such as organic material (e.g., wood, cloth, etc.). Dilute aqueous solutions of $HClO_4$ are safe and are useful reagents for the chemist. For one thing, perchlorate ion has less tendency to form complex ions with metal cations than any other anion. For another thing, perchloric acid is probably the strongest of all common acids and in aqueous solution is more completely dissociated than the usual strong acids hydrochloric, sulfuric, and nitric.

Why is perchloric a stronger acid than the other oxyacids of chlorine? The dissociation of an oxyacid involves breaking a hydrogen-oxygen bond to form a hydrated hydronium ion and a hydrated anion. The bigger the anion, the less its hydration energy. Consequently, since ClO_4^- is obviously bigger than ClO^-, for example, we might expect $HClO_4$ to be less dissociated than $HOCl$. Since the reverse is true, the bond holding the proton to OCl^- must be stronger than the bond holding H^+ to ClO_4^-. That this is reasonable can be seen by noting that oxygen is more electronegative than chlorine; therefore, addition of oxygen atoms to the chlorine of an $HOCl$ molecule pulls electrons away from the H—O bond and tends to weaken it. This picture is supported by observing that the oxyacid $HOClO$ (chlorous acid) has a dissociation constant, $K_{diss} = 1.1 \times 10^{-2}$, larger than that of $HOCl$, $K_{diss} = 3.2 \times 10^{-8}$. In general, for any series of oxyacids the acid corresponding to highest oxidation number (i.e., the most oxygen) is the most highly dissociated. Thus, HNO_3 is stronger than HNO_2; H_2SO_4 is stronger than H_2SO_3; etc.

The above discussion of the oxy compounds of chlorine primarily concerned the oxidation states $+1$, $+5$, and $+7$. Brief mention was made of two compounds which represent two other states, ClO_2 ($+4$ state) and $HClO_2$ ($+3$ state). The first of these compounds, chlorine dioxide, is produced when $HClO_3$ explodes, but a safer method of preparing it involves reduction of acid chlorate solutions with sulfur dioxide, SO_2, or oxalic acid, $H_2C_2O_4$. Pertinent equations are

$$2ClO_3^- + SO_2(g) + H_3O^+ \longrightarrow 2ClO_2(g) + HSO_4^- + H_2O$$

$$2ClO_3^- + H_2C_2O_4 + 2H_3O^+ \longrightarrow 2ClO_2(g) + 2CO_2(g) + 4H_2O$$

Chlorine dioxide is a yellow gas at room temperature but is easily condensed to a red liquid (bp $11\,°C$). The gas is paramagnetic, indicating that the molecule contains an unpaired electron. Despite the fact that ClO_2 is explosive, it is produced in large quantities for bleaching flour, paper, etc. Its use depends on the fact that it is both a strong and a rapid oxidizing agent. When placed in basic solution, chlorine dioxide disproportionates:

$$2ClO_2(g) + 2OH^- \longrightarrow ClO_2^- + ClO_3^- + H_2O$$

One product is chlorate ion, previously discussed; the other is chlorite ion, ClO_2^-, which is frequently prepared commercially by this reaction. Chlorite ion is the anion of the moderately weak acid $HClO_2$, or chlorous acid. Like $HOCl$, it exists only in solution, and even in solution it decomposes. The principal reaction seems to be

$$5HClO_2 \longrightarrow 4ClO_2(g) + H_3O^+ + Cl^- + H_2O$$

Chlorites are important industrial bleaching agents because they can bleach without appreciably affecting other properties of the substance bleached. Like other oxy compounds of chlorine, they must be used with caution because the dry salts may explode when in contact with organic material.

25.5 Bromine

Bromine, from the Greek word *bromos* for stink, occurs as bromide ion in seawater, brine wells, and salt beds, and is less than a hundredth as abundant as chlorine. The element is usually prepared by chlorine oxidation of bromide solutions, as by sweeping chlorine gas through seawater. Since chlorine is a stronger oxidizing agent than bromine, the reaction

$$Cl_2(g) + 2Br^- \longrightarrow Br_2 + 2Cl^-$$

occurs as indicated. Removal of the bromine from the resulting solution can be accomplished by sweeping the solution with air because bromine is quite volatile. At room temperature pure bromine is a mobile but dense red liquid of pungent odor. It is a dangerous substance since it attacks the skin to form slow-healing sores.

Although less powerful an oxidizing agent than chlorine, bromine readily reacts with other elements to form bromides. Hydrogen bromide, like HCl, is a strong acid but is more easily oxidized than HCl is. Whereas HCl can be made by heating the sodium salt with H_2SO_4, HBr cannot. The hot H_2SO_4 oxidizes HBr to Br_2, and a nonoxidizing acid such as H_3PO_4 must be used instead.

In basic solution, bromine disproportionates to give bromide ion and hypobromite ion (BrO^-). The reaction is quickly followed by further disproportionation

$$3BrO^- \longrightarrow 2Br^- + BrO_3^-.$$

to give bromate ion, BrO_3^-. Bromic acid, $HBrO_3$, has never been prepared pure. In aqueous solution, it is a strong acid and a good oxidizing agent. The electrode potential

$$2BrO_3^- + 12H_3O^+ + 10e^- \longrightarrow Br_2 + 18H_2O$$
$$E° = +1.50 \text{ V}$$

indicates that bromate is a slightly stronger oxidizing agent than chlorate. It has the added virtue of being faster in its action. Bromine in the $+7$ state as perbromate, BrO_4^-, can be made by electrolytic oxidation of BrO_3^- or by action of XeF_2 or F_2 on BrO_3^-. The BrO_4^- is a sluggish oxidizing agent but is potentially stronger than BrO_3^-. The electrode potential for

$$BrO_4^- + 2H_3O^+ + 2e^- \longrightarrow BrO_3^- + 3H_2O \qquad E° = +1.76 \text{ V}$$

indicates BrO_4^- is stronger than MnO_4^-.

One of the most important uses of bromine is in making silver bromide for photographic emulsions (Sec. 20.3). However, the principal use of bromine has been in making dibromoethane ($C_2H_4Br_2$) for addition to gasolines which contain tetraethyllead. Tetraethyllead, $(C_2H_5)_4Pb$, added to gasoline as an antiknock agent, decomposes on burning to form lead deposits. The dibromoethane prevents accumulation of lead deposits in the engine but, of course, increases the amount of lead exhausted into the air, principally as $PbClBr$.

25.6 Iodine

Of the halogens, iodine is the only one which occurs naturally in a positive oxidation state. In addition to its occurrence as I^- in seawater and salt wells, it is found as sodium iodate ($NaIO_3$), small amounts of which are mixed with $NaNO_3$ in Chile saltpeter. The Chilean ore is processed by the reduction of $NaIO_3$ with controlled amounts of $NaHSO_3$. The principal reaction is

$$2H_2O + 5HSO_3^- + 2IO_3^- \longrightarrow I_2 + 5SO_4^{2-} + 3H_3O^+$$

Excess hydrogen sulfite must be avoided, for it would reduce I_2 to I^-. In the United States most of the iodine is produced by chlorine oxidation of I^- from salt wells.

At room temperature iodine crystallizes as black leaflets with metallic luster. Although, as shown by X-ray analysis, the solid consists of discrete I_2 molecules, its properties are different from those of usual molecular solids. For example, its electric conductivity, though small, increases with increasing temperature like that of a semiconductor. Furthermore, liquid iodine also has perceptible conductivity, which decreases with increasing temperature like that of a metal. Thus, feeble as they are, metallic properties do appear even in the halogen group.

When heated, solid iodine readily sublimes to give a violet vapor which consists of I_2 molecules. The violet color is the same as that observed in many iodine solutions, such as those in CCl_4 and in hydrocarbons. However, in water and in alcohol the solutions are brown, presumably because of unusual interactions between I_2 and the solvent. When iodine is brought in contact with starch, a characteristic deep blue color results; the color has been attributed to a starch–I_2 complex. The formation of the blue color is the basis for using starch–potassium iodide mixtures as a qualitative test for the presence of oxidizing agents. Oxidizing agents convert I^- to I_2, which with starch forms the colored complex. With very strong oxidizing agents the color may fade with oxidation of I_2 to a higher oxidation state.

Iodine is only slightly soluble in water (0.001 M), but the solubility is vastly increased by the presence of iodide ion. The color changes from brown to deep red because of the formation of the triiodide ion, I_3^-. The triiodide ion is also known in solids such as NH_4I_3, X-ray investigations of which indicate that the I_3^- ion is linear. No electronic formula conforming to the octet rule can be written for this ion. Apparently an iodine atom, perhaps because of its large size, can accommodate more than eight electrons in its valence shell. A possible model would be like that shown in Fig. 25.7. Hybridization of the $5d$ with the $5s$ and three $5p$ orbitals of the central iodine would give sp^3d orbitals directed to the vertexes of a trigonal bipyramid. The outer iodines could then form ordinary σ bonds along the threefold axis, leaving three unshared pairs in the equatorial plane. In basic solutions, I_2 disproportionates to form iodide ion and hypoiodite ion (IO^-):

$$I_2 + 2OH^- \longrightarrow I^- + IO^- + H_2O$$

Further disproportionation to give iodate ion (IO_3^-) is hastened by heating or by addition of acid. Iodate ion in acid solution is a weaker oxidizing agent than either bromate ion or chlorate ion. This is shown by the electrode potential:

$$2IO_3^- + 12H_3O^+ + 10e^- \longrightarrow I_2 + 18H_2O \qquad E^\circ = +1.20 \text{ V}$$

Since IO_3^- is a weaker oxidizing agent than ClO_3^-, iodates can be made by oxidizing I_2 with ClO_3^-. Furthermore, iodate salts are not quite so explosive as chlorates or bromates. The greater stability of iodates also is evident in the fact that HIO_3, unlike $HClO_3$ and $HBrO_3$, can be isolated pure (as a white solid). The latter acids detonate when attempts are made to concentrate them.

In the +7 state, the oxysalts of iodine are called periodates, but there are several kinds of periodates. There are those derived from HIO_4 (metaperiodic acid), those derived from H_5IO_6 (paraperiodic acid), and possibly others. In the metaperiodates the iodine is bonded

Fig. 25.7
Possible electron configuration of triiodide ion.

tetrahedrally to four oxygen atoms (this ion is analogous to ClO_4^-); in the paraperiodates there are six oxygen atoms bound octahedrally to the iodine atom. The fact that there are paraperiodates but no paraperchlorates is apparently due to the larger size of the iodine atom. As in I_3^-, it is necessary to assume that in the paraperiodates the valence shell of iodine is expanded to contain more than eight electrons. Paraperiodic acid, H_5IO_6, is moderately weak ($K_I = 5.1 \times 10^{-4}$), but metaperiodic acid, HIO_4, seems to be strong. H_5IO_6 can be prepared by anodic oxidation of HIO_3; if heated at low pressure, H_2O can be driven off to form HIO_4.

In going down the halogen group the atoms of the elements get progressively larger, and it becomes easier to oxidize the halide ion to the free halogen. This shows up in the instability of iodide solutions to air oxidation. The oxidation is slow for basic and neutral solutions but becomes appreciably faster for acid solutions.

So far as uses are concerned, iodine is less widely used than other halogens. It finds limited use for its antiseptic properties, both as tincture of iodine (solution of I_2 in alcohol) and as iodoform (CHI_3). Since small amounts of iodine are required in the human diet, traces of sodium iodide (10 ppm) are frequently added to table salt.

25.7 Astatine

Since astatine does not occur to an appreciable extent in nature, all that is known about it is based on experiments done with trace amounts of artificially produced element. It can be made by bombarding bismuth nuclei with alpha particles

$$^{209}_{83}Bi + {}^4_2He \longrightarrow {}^{211}_{85}At + 2{}^1_0n$$

and its chemistry is studied by observing whether radioactive astatine is carried along with iodine through the course of chemical reactions. On the basis of such tracer studies, it is concluded that astatine forms an astatide ion (At^-) and compounds in two positive oxidation states, probably $+1$ and $+5$.

25.8 Interhalogen compounds

In view of the fact that halogen atoms combine with each other to form diatomic molecules, it is not surprising that an atom of one halogen can combine with an atom of another halogen. Thus, we have compounds such as ICl, iodine monochloride, which can be prepared by direct union of the elements. ICl can also be made by reaction of iodate with iodide in concentrated HCl:

$$6H_3O^+ + IO_3^- + 2I^- + 3Cl^- \longrightarrow 3ICl + 9H_2O$$

Pure ICl exists as a low-melting, red solid (mp 27°C) which when melted can be electrolyzed to produce I_2 at the cathode and Cl_2 at the anode. Since molten ICl conducts electric current and since I_2 is discharged at the cathode, it has been assumed that the liquid contains

I^+ ions. In dilute acid solution, ICl hydrolyzes to give chloride ion and probably hypoiodous acid (HOI), but the latter disproportionates, giving the net reaction

$$5ICl + 9H_2O \longrightarrow 5Cl^- + IO_3^- + 2I_2 + 6H_3O^+$$

ICl is sometimes used for adding iodine to organic molecules.

In addition to ICl there are other interhalogen compounds of the type XY. In fact, all combinations except IF are known. More surprising than these compounds of type XY is the existence of more complex interhalogens of types XY_3, XY_5, and XY_7. Although there are three examples of XY_3 (ClF_3, BrF_3, and ICl_3) and two of XY_5 (BrF_5 and IF_5), there is only one XY_7 (IF_7). All of these compounds are of interest as examples of failure of the octet rule. As shown in Fig. 25.8, their structures are highly unusual. BrF_3 is T-shaped; BrF_5 is a square pyramid; and IF_7 is a pentagonal bipyramid.

25.9 Qualitative analysis

Fluoride differs from Cl^-, Br^-, and I^- in forming an insoluble magnesium salt but also a soluble silver salt.

Iodide can be distinguished from Br^- and Cl^- in that its silver salt is insoluble in excess NH_3. Its presence can be confirmed by oxidizing I^- to I_2, which imparts a violet color to CCl_4.

Both Br^- and Cl^- form insoluble silver salts, but AgBr is more yellowish than the pure white of AgCl. Furthermore, AgBr dissolves with greater difficulty in excess NH_3 than does AgCl. Finally, when Br^- is oxidized to Br_2 in the presence of CCl_4, the Br_2 in CCl_4 solution is brown, whereas Cl_2 in CCl_4 is yellow.

The oxyanions ClO_3^-, BrO_3^-, and IO_3^- can be detected by reducing them in acid (by adding sulfite or nitrite) and then analyzing for the corresponding halide ion as above.

Perchlorate ion, once encountered only rarely but now becoming increasingly common, can be recognized easily because it is one of the few anions that forms a white, sparingly soluble potassium salt.

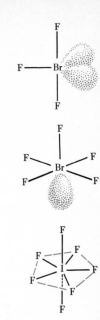

Fig. 25.8
Structures of interhalogen compounds.

Exercises

*25.1 Precipitation. The K_{sp} of CaF_2 is 1.7×10^{-10}. Should precipitation occur on mixing equal volumes of 0.010 M $Ca(NO_3)_2$ with 0.010 M HF? Justify your answer.

*25.2 Equivalent. What would be the weight of 1 equiv of OF_2 for its reaction with H_2O to give HF and O_2? What would be the oxidizing agent, and what would be the reducing agent in this reaction?
Ans. 27.0 g

*25.3 Stoichiometry. A 5.000-g sample of impure MnO_2 when heated with HCl generates 0.746 liter of chlorine gas at 298°K and 0.965 atm pressure. What is the purity of the sample?

*25.4 Dilution. How many milliliters of 1.50 M HCl must you add to 250 ml of 12.0 M HCl to prepare a solution that is 6.00 M? Assume volumes are additive.   *Ans. 333 ml*

25.5   Chemical equations. Write balanced chemical equations for each of the following:

a   Preparation of $Cl_2(g)$ by electrolysis of aqueous NaCl
b   Conversion of $Cl_2(g)$ in base to Cl^- and ClO^-
c   Heating of ClO^- to produce Cl^- and ClO_3^-
d   Reduction of ClO_3^- to Cl_2 by $Mn^{2+} \longrightarrow MnO_2$ in acid

25.6   Molality. How many moles of $HClO_4 \cdot H_2O$ should you dissolve in a liter of water to make a solution that is 0.250 m $HClO_4$?

25.7   Bromine. How many grams of dibromoethane would you have to add to a gallon of gasoline to scavenge all the lead if the gasoline contains 3.85 ml of tetraethyllead (density 1.65 g/ml) per gallon?
Ans. 3.68 g

**25.8   Electrode potential.* Estimate the electrode potential for $\frac{1}{2}At_2(g) + e^- \longrightarrow At^-(aq)$ by using the following estimated values:

$$At_2(g) \longrightarrow 2At(g) \qquad \Delta G° \approx +80 \text{ kJ}$$
Electron affinity of At $EA \approx 2.7$ eV
$At^-(g) \longrightarrow At^-(aq) \qquad \Delta G° \approx -250 \text{ kJ} \qquad$ *Ans. +0.05 V*

**25.9   Heat of reaction.* If $\Delta H°$ of formation of $HF(g)$ is -13.4 kJ/g and that of $HCl(g)$ is -2.53 kJ/g, calculate the heat of reaction for $2HCl(g) + F_2(g) \longrightarrow 2HF(g) + Cl_2(g)$.

**25.10   Dissociation.* What is the percent dissociation of HF in 0.10 M HF? $K_{diss} = 6.7 \times 10^{-4}$. Calculate the pH of this solution.
Ans. 8.2%, 2.09

**25.11   Hydrofluoric acid.* Just as it can dissolve glass, $HF(aq)$ can dissolve cement. Writing cement as $Ca_4Al_2SiO_9$, write a balanced net equation for its dissolving in HF.

**25.12   Fluorocarbons.* How might you account for the fact that the double bond in tetrafluoroethylene is a bit shorter than the double bond in ethylene?

**25.13   Fluorocarbons.* CF_4 is used as a low-temperature refrigerant under the name Freon-14. Its normal boiling point is $-127.8°$C, and its heat of vaporization is 12.62 kJ/mol. At what pressure should you pump on liquid Freon-14 to set a temperature of $-135°$C?

**25.14   Noble-gas-type compounds.* Why is it plausible that O_2 (12.2 eV) should have a lower ionization potential than O (13.61 eV)?

**25.15   Noble-gas-type compounds.* Describe the bonding and electron distribution in O_2PtF_6. Predict the extent of the magnetism.

**26.16   Symmetry.* Indicate the order of the principal rotation symmetry axis and its location in each of the following: XeF_2, XeF_4, XeO_3, and $XeOF_4$.

25.17 *Bond strengths.* Traces of water can be extremely hazardous in the study of xenon chemistry because of possible formation of XeO_3, which is violently explosive. Given that the bond energy of Xe—O is about 110 kJ/mol and that of O_2 is 494 kJ/mol, predict a value of $\Delta H°$ for $XeO_3(s) \longrightarrow Xe(g) + \frac{3}{2}O_2(g)$. *Ans. −411 kJ*

25.18 XeF_2 *bonding.* Draw sketches showing the electron distribution in the orbitals involved in bonding XeF_2 by (*a*) the valence-bond method and (*b*) the molecular-orbital method.

25.19 *Chlorine.* The solubility of chlorine in water is 310 cm³ (STP) of gas in 100 g of water. Assume this much chlorine has been dissolved and equilibrated as follows:

$$Cl_2 + 2H_2O \rightleftharpoons Cl^- + H_3O^+ + HOCl \qquad K = 4.7 \times 10^{-4}$$

Calculate the equilibrium concentration of HOCl formed.

25.20 *Hydrolysis.* If K_{diss} of HOCl is 3.2×10^{-8}, what percent of ClO^- in 0.25 *M* NaClO is hydrolyzed? What is the pH of the solution? *Ans. 0.11%, 10.44*

25.21 *Electrode potential.* If $E° = +1.63$ V for $2HOCl + 2H_3O^+ + 2e^- \longrightarrow Cl_2(g) + 4H_2O$ and K_{diss} of HOCl is 3.2×10^{-8}, what would the $E°$ become for 1 *m* basic solution?

25.22 *Conversions.* How would you go about making the following conversions so as to conserve the atoms with the asterisks?

a $\overset{*}{N}aCl$ to $\overset{*}{N}aClO_3$

b $K\overset{*}{B}r$ to $\overset{*}{B}r_2$

c $\overset{*}{I}_2$ to $Na\overset{*}{I}O_3$

d $Na\overset{*}{I}O_3$ to $\overset{*}{I}_2$

25.23 *Electrode potentials.* If $E°$ for $ClO_3^- \longrightarrow Cl_2$ is $+1.47$ V and that for $Cl_2 \longrightarrow Cl^-$ is $+1.36$ V, what is $E°$ for $ClO_3^- \longrightarrow Cl^-$?

25.24 *Solubility.* The solubility of $KClO_4$ in water is 7.5 g per liter of solution at 0°C and 218 g per liter of solution at 100°C. Calculate the K_{sp} at these two temperatures, and derive a value for $\Delta H°$ of solution. *Ans. 2.93×10^{-3}, 2.46, 57.0 kJ/mol*

25.25 I_3^-. The solubility of iodine in pure water is 0.0279 g per 100 g of water. In a solution that is originally 0.10 *M* KI, it is possible to dissolve 1.14 g of iodine per 100 g of solution. What is K for the reaction $I_2 + I^- \rightleftharpoons I_3^-$? *Ans. 710*

25.26 *Mixing problem.* What will be the final concentration of I^- in a solution made by mixing 35.0 ml of 0.100 *M* $NaHSO_3$ with 10.0 ml of 0.100 *M* KIO_3? Assume volumes are additive.

**25.27   *Periodic acid.* Suggest a reason why the paraperiodic acid H_5IO_6 is weaker as an acid than is the metaperiodic acid HIO_4.

***25.28   *Radioactivity.* One curie is defined as the amount of material, which is equivalent to one gram of radium, that gives 3.7×10^{10} nuclear disintegrations per second. The half-life of ^{210}At is 8.3 h. How many grams of astatine would be one curie?

***25.29   *Electrode potential.* The standard E° for $\frac{1}{2}F_2(g) + e^- \longrightarrow F^-(aq)$, which is $+2.87$ V, applies when the fluoride concentration is $1\ m$. What would the corresponding E° be in $1\ m\ H_3O^+$, i.e., for the electrode reaction $\frac{1}{2}F_2(g) + e^- + H_3O^+ \longrightarrow HF(aq) + H_2O$? The K_{diss} of HF is 6.7×10^{-4}.

***25.30   *Thermodynamics.* From data given in Fig. 25.3 and 25.1 and elsewhere in the book, estimate the ΔH° for the reaction

$$2XeF_2 + 2H_2O(g) \longrightarrow O_2(g) + 2Xe(g) + 4HF(g)$$

Ans. $-200\ kJ$

The planet earth has been in existence for about 4.6×10^9 yr. Man has been on it only relatively briefly. Recently, the steep part of exponentially expanding population growth has taken over, accompanied by enormous multiplication in technological capability of harming the environment. Fortunately, as man has become more clever at doing damage to his planet, he has also become more skilled at detecting that damage. As man gets more ingenious at getting into trouble, he becomes even more ingenious at figuring ways of getting out of it.

In this part we look at the chemical aspects of the human environment, particularly as they affect man's survival and his search for an improved quality of life.

Man and his
chemical environment

Part III

We examine in this chapter chemical influences on the three minimal inputs necessary for life: air, water, and food. Of first concern is the pollution of the atmosphere, especially by the ubiquitous internal-combustion engine. Then we look at water quality and how it is degraded by man's activities. Finally, we consider the problem of food additives and the residues that come from attempts to ensure an adequate food supply.

26.1 Pollutants in the atmosphere

Air is a loose term used to describe the mixture of gases in a relatively thin layer around the earth. The total is anywhere from 400 to 1500 km thick, but 95 percent of it is in a 20-km layer. The composition is variable, with the biggest variation in CO_2 content (rises near decay, burning, or breathing and falls over water and growing vegetation) and in H_2O content (which ranges from 0.1 to 5 percent by volume, depending on the weather and the temperature).

The definition of a normal atmosphere depends on what point in geologic time we are talking about. The primitive atmosphere was probably a reducing one, a supposition based on observation of other planets and the fact that there are enough $+2$ compounds of iron in the upper lithosphere to have extracted all the oxygen from the atmosphere in a geologically short time. Whether the present oxygen in the atmosphere came later as a result of plant photosynthesis or photolysis of H_2O by ultraviolet radiation remains to be established, but the present belief is that the early reducing atmosphere, containing perhaps NH_3, CH_4, and H_2O, was the source from which compounds of biologic interest developed. Electric discharges and ultraviolet radiation could have led to formation of amino acids, aldehydes, organic acids, and urea. Gradual enrichment of this primeval "soup" would have favored polymerization of amino acids to give polypeptides and primitive proteins. Adsorption of the macromolecules on inorganic minerals, perhaps semi-organized in a solid-state structure such as a zeolite, could have led to large organic complexes with transient forms exhibiting some of the properties associated with life. There may have been a primitive photosynthesis in which photons of wavelength shorter than 200 nm reduced CO_2 and H_2O in a single step to form carbohydrates and O_2.* However, such a process would have been self-stopping since some of the evolved O_2 would have been converted photochemically into O_3, which acts as a natural filter for absorbing sun's radiation shorter than 300 nm. At $\lambda > 300$ nm, photons are not energetic enough to drive photosynthesis in a single step. Photosynthesis as we know it now, i.e., a chain of consecutive low-energy steps, did not occur until suitable light-gathering pigment molecules (e.g., chlorophyll) appeared. At that stage, the atmosphere could well have changed from reducing to oxidizing.

The present average composition of clean, dry air is represented in Fig. 26.1. The actual values determined depend on the method of sampling and the averaging time over which the sampling is integrated. For most normal gases (for example, N_2 and O_2), concentrations vary but little with time; for contaminants that originate at a point or in an area, the maximum and minimum readings tend to converge as the averaging time lengthens. CH_4 occurs as a natural part of the carbon cycle; certain bacteria generate CH_4 in their metabolic processes. N_2O and NO_2 are part of the nitrogen cycle previously discussed (Sec. 23.2).

* It is interesting to note that the first emission of oxygen into the primitive atmosphere probably constituted a serious air pollutant for the early organisms then flourishing.

Fig. 26.1
*Average composition of
clean, dry air in parts per
million by volume.*

Component	Average concentration, ppm	Component	Average concentration, ppm
N_2	780,900	Kr	1
O_2	209,500	N_2O	0.5
Ar	9,300	H_2	0.5
CO_2	300	Xe	0.08
Ne	18	NO_2	0.02
He	5.2	O_3	0.01
CH_4	1.1	NH_3	Trace

The ozone content arises from production in the upper atmosphere by solar radiation ($O_2 \longrightarrow 2O$ accompanied by $O + O_2 \longrightarrow O_3$) followed by partial transport into the lower atmosphere; some natural ozone also comes from lightning and forest fires.

Air is never found completely clean in nature. Natural pollutants such as SO_2, H_2S, and CO arise from volcanic activity, vegetative decay, and forest fires. However, our concern here is with man-made pollutants, which in the United States are estimated to amount to between 2 and 3 kg *per person per day*. Given the vast mass of air over the United States (3×10^{18} kg), this might seem like a trivial amount, but dilution is not uniform. There is very little diffusion of pollutants into layers above 1 km of altitude, and geologic and man-made barriers can limit lateral mixing so that pollutant concentrations of 50 to 100 ppm are not uncommon.

The major air pollutants (accounting for more than 90 percent of the total) are CO, oxides of nitrogen, hydrocarbons, oxides of sulfur, and particulates. The CO generally results from incomplete combustion of carbon or carbon-containing compounds; the oxides of nitrogen (primarily NO and NO_2, but frequently lumped together as NO_x) mainly come from combination of air N_2 and O_2 induced by high-temperature combustion. Transport, through evaporation and exhaust emission of unburned gasoline, accounts for the major part of hydrocarbon pollution. Combustion again of sulfur-containing fuels and smelting operations contributes the oxides of sulfur (mostly SO_2, some SO_3, generally described as SO_x) as well as particulates, small, solid particles and liquid droplets as in fly ash from coal combustion. Figure 26.2 gives some representative data for relative amounts of different air pollutants and their primary sources. As can be seen, approximately half of the total is CO, and over half of that comes from the automobile.

Actually, it is somewhat distorting to consider pollutants on a mass basis since they vary greatly in their relative toxicity. For example, the tolerance level of CO is 32 ppm by volume, which corresponds to 40 mg/m^3; the tolerance level of NO_x is 0.5 mg/m^3. Hence, on a mass basis NO_x is 80 times as toxic as CO. When adjusted for such differences in relative toxicity, transportation, which accounts for 42 percent of the total air pollutants in Fig. 26.2 on a mass basis, drops to 15 percent of the total mass times toxicity correction. The chief culprit on an

Fig. 26.2
Air pollutant emissions, 10^9
kg/yr, United States.

	CO	NO$_x$	Hydro-carbons	SO$_x$	Particulates
Transportation					
Gasoline automobiles	53.6	6.0	13.8	0.2	0.5
Diesels	0.2	0.5	0.4	0.1	0.3
Aircraft	2.2	Negl	0.3	Negl	Negl
Railroads	0.1	0.4	0.3	0.1	0.2
Vessels	0.3	0.2	0.1	0.3	0.1
Nonhighway	1.6	0.3	0.3	0.1	0.1
	58.0	7.4	15.2	0.8	1.2
Fuel combustion (stationary)					
Coal	0.7	3.6	0.2	18.3	7.4
Fuel oil	0.1	0.9	0.1	3.9	0.3
Natural gas	Negl	4.4	Negl	Negl	0.2
Wood	0.9	0.2	0.4	Negl	0.2
	1.7	9.1	0.7	22.2	8.1
Industrial processes	9.2*	0.2	4.2	6.6	6.8
Solid-waste disposal	7.1	0.5	1.5	0.1	1.0
Miscellaneous					
Forest fires	6.5	1.1	2.0	Negl	6.1
Structural fires	0.2	Negl	0.1	Negl	0.1
Coal refuse burning	1.1	0.2	0.2	0.5	0.4
Agricultural burning	7.5	0.3	1.5	Negl	2.2
	15.3	1.6	7.7†	0.5	8.8
Total	91.3	18.8	29.3	30.2	25.9

* Steel (5.4), petroleum (2.5), paper (0.7), other (0.6).
† Includes organic solvent evaporation (2.8), gasoline marketing (1.1).

adjusted pollutant-level basis becomes stationary fuel combustion (e.g., power plants), accounting for 42 percent of the total.

Why are these air pollutants harmful? In the case of CO, CO combines with hemoglobin

$$CO + hemoglobin\ (Hb) \longrightarrow carboxyhemoglobin\ (COHb)$$

with an association constant approximately 200 times as great as that for the binding of O_2. Hence the CO is bound preferentially, and the COHb is not available for oxygen transport in the blood. Less than 1.0% COHb in the blood has no effect; the normal level is 0.5%, partly as a result of CO produced in the body during destructive metabolism of heme. From 2 to 5% COHb, effects appear on the central nervous system with impaired time-interval discrimination and degradation of visual acuity. When COHb > 5%, cardiac and pulmonary functional changes appear. To put these in perspective, the percent COHb in the blood is approximately equal to 0.16 times the CO concentration in the air (in parts per million) $+0.5$. Thus, 10 ppm CO in the air means 2.1% COHb in the blood; 30 ppm means 5.3%; etc. For comparison, a light cigarette smoker (10 cigarettes per day) who does not inhale typically shows 2.3% COHb; a heavy smoker (40 cigarettes per day) who does inhale, 6.9% COHb.

As far as oxides of nitrogen are concerned, both NO_2 and NO are toxic at fairly low concentrations, with NO_2 being about five times as toxic as NO. Threshold limit values are 5 and 25 ppm, respectively. At present levels NO is not a major health hazard, but it does convert into NO_2 by the reaction

$$NO + \tfrac{1}{2}O_2 \longrightarrow NO_2$$

The main toxic effects of NO_2 are on the lungs, and concentrations greater than 100 ppm are generally fatal, usually because of pulmonary edema. The more disturbing aspect of NO_x emission is that the oxides of nitrogen appear to be involved as intermediates in the generation of serious air pollutants such as ozone and peroxy organic compounds, which are toxic at 0.1 ppm levels and are present in photochemical smog. The proposed mechanism of photochemical-smog generation is considered in Sec. 26.3. Hydrocarbons *per se* are generally not very harmful, but, as we shall see later, they too are key factors in multiplying the photochemical-smog problem.

Sulfur dioxide constitutes the bulk of the SO_x contamination of the atmosphere. (There is also some oxidation to SO_3, which subsequently reacts with water vapor to form an aerosol of H_2SO_4.) SO_2 can be detected by odor at 3 ppm levels. Throat irritation sets in at about 10 ppm; eye irritation and coughing, 20 ppm. Maximum allowed exposure for short intervals (for example, 30 min) is 50 to 100 ppm. The main effect of SO_2 is on the respiratory system, and it seems to hit particularly severely the aged and the chronically ill. Mortality studies of the infamous London smogs before pollution controls were instituted disclosed detectable increases in the death rate when the SO_2 level exceeded 0.25 ppm. During the killer smog of January 1956, when SO_2 levels rose to 0.40 ppm, the death rate among the aged jumped from 130 to 180 day^{-1}. Besides the effect on health, SO_x pollution leads to increased corrosion rates of metals as well as accelerated degradation of marble, limestone, and mortar.

26.2 Internal-combustion engine

One of the prime villains in air pollution is the automobile, and to understand the problem of control it is necessary to know something about the workings of the internal-combustion engine. The dilemma, as we shall see, is that increasing the air-fuel ratio from 13 to 16 dramatically reduces hydrocarbon and CO emissions but raises NO_x emission. Unfortunately, the maximum power of a conventional internal-combustion engine comes at maximum flame speed, and this corresponds to an air-fuel ratio of 13.

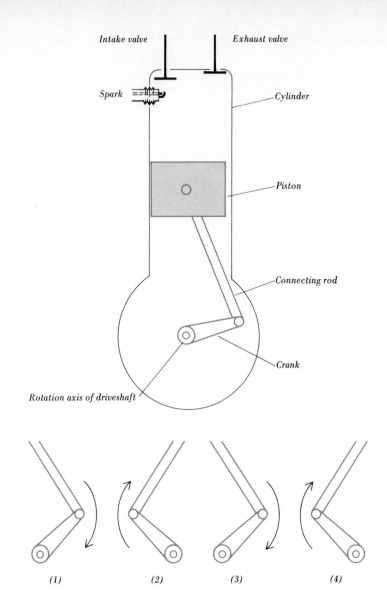

Fig. 26.3
Four-stroke SI (spark-ignition) engine.

Intake valve

Exhaust valve

Spark

Cylinder

Piston

Connecting rod

Crank

Rotation axis of driveshaft

(1) *(2)* *(3)* *(4)*

Jean Joseph Lenoir invented the idea of spark ignition of hydro-carbon-air mixtures in 1860, but the idea was a failure until 1876 when Nikolaus Otto suggested that the mixture be compressed before spark ignition. Such compression raises the temperature and increases the work per stroke. Figure 26.3 gives a schematic diagram of the so-called "Otto" or "four-stroke SI engine." (SI stands for spark ignition and is in contrast to CI, which stands for compression ignition as in a diesel engine. In a diesel engine, oxidation of hydrocarbon is initiated explosively by high compression of a fuel-air mixture.) The main components are a cylinder, a piston, a connecting rod, and a crank to turn the drive shaft. As indicated at the bottom of the figure, there are four strokes for each power cycle, corresponding to two full rotations of the crank clockwise about the rotation axis of the drive shaft:

1   *Intake stroke.* With the intake valve open and the exhaust valve closed, a mixture of volatile hydrocarbon and air is sucked into the cylinder chamber as the piston moves on a downstroke.

2   *Compression stroke.* With both valves closed, the piston moves on an upstroke, compressing the fuel-air mixture by a factor ranging from 7 to 12. The compression ratio is defined as the (displacement volume + clearance volume) divided by the clearance volume. Spark ignition occurs near the end of this stroke to initiate combustion.

3   *Power, or work, stroke.* With both valves still closed, combustion occurs with pressure and temperature increases that drive the piston down. The exhaust valve opens at the end of this stroke.

4   *Exhaust stroke.* With the exhaust valve open and the intake valve closed, the upstroke of the piston pushes the exhaust products out of the cylinder chamber.

The carburetor is a device located in front of the intake valve for providing a fairly definite and homogeneous mixture of fuel and air (approximately 15 parts by weight of air per 1 part of fuel). On the intake stroke, a drop in pressure draws liquid fuel through a nozzle, which sprays the fuel into an air mixing chamber. The hydrocarbon-air mix is then sucked into the piston chamber.

Liquid fuels are a complex mix of hydrocarbons, averaging 85% by weight C and 15% H together with a trace of S. The corresponding empirical formula would be C_8H_{17}. Ideal as a fuel would be iso-octane, C_8H_{18}, otherwise known as 2,2,4-trimethylpentane. Its structural formula is

$$
\begin{array}{ccc}
& CH_3 & CH_3 \\
& | & | \\
CH_3-C-CH_2-CH-CH_3 \\
& | \\
& CH_3
\end{array}
$$

Using iso-octane as a typical fuel, the overall combustion reaction would be written

$$C_8H_{18} + 12\tfrac{1}{2}O_2(+46.6N_2) \longrightarrow 8CO_2 + 9H_2O(+46.6N_2)$$

The 46.6 mol of N_2 given parenthetically takes account of the fact that the oxidant is generally air. On a mass basis, the one mole of C_8H_{18} corresponds to 114 g; the $12\tfrac{1}{2}$ mol of O_2 and the 46.6 mol of N_2 (plus associated argon) come to 1725 g. Thus, on a mass basis the theoretical air-fuel ratio is 1725:114, or 15.1:1. Air-fuel ratios greater than this are called *lean*; air-fuel ratios less than this are called *rich*.

Given the heat of combustion of C_8H_{18} to CO_2 plus H_2O as about 500 kJ/mol and taking account of the varying enthalpy of the products (CO_2, H_2O, and N_2) at various temperatures, one can calculate that the theoretical flame temperature resulting from the above reaction would be about 2700°C. Actually, the equilibrium flame temperature (ca. 1800°C) is much less than this because of limitations due to CO_2 dissociation into CO and O_2 (would be 18 percent at 2700°C), heat loss, and the fact that the above process is not at constant volume. Also, most gas-phase reactions are more complex than indicated by the stoichiometric equation since there may be opposing reactions, competing reaction paths, and possible chain mechanisms (Sec. 10.7). It appears that *all* combustion reactions go by chain mechanisms. In the internal-combustion engine, the problem is even more complicated because the flame is not a *diffusion* flame, as with compression ignition, but a *premixed* flame. Spark ignition produces reactive intermediates, and the reaction becomes self-sustaining when the energy released by combustion is slightly greater than the heat loss. Propagation of the flame, i.e., propagation of the chemical reaction into the unburned mixture ahead of the flame front, arises from transfer of heat and reactive particle diffusion. (Using the Arrhenius equation $k = Ae^{-\Delta E/RT}$ one can see that with an activation energy of about 125 kJ k increases from 10^{-20} at 300°K to 10^{-3} at 3000°K.) Also, it is sometimes possible to have self-ignition of the fuel-air mix ahead of the flame front. This might occur, for example, when the rise in pressure that normally accompanies combustion compresses the "end gas" so that its temperature rises above the self-ignition temperature. Spontaneous ignition may spread from several point sources, unlike a smooth flame front, and the result is *knock*. It may range from a thud to a small ping; a little bit of pinging is considered desirable because one wants to hasten the final stages of combustion before the piston gets far into its expansion stroke.

As chain reactions, combustion reactions are very complex. Reacting radicals have to be generated and regenerated to act as chain carriers. Reaction velocity depends intricately on the concentration of chain carriers and on the length of the chain. Also, as was discussed in Sec. 10.7, chain-branching and chain-breaking can dramatically influence the observed rates. For hydrocarbon combustion, the following sequence of steps seems to be involved:

1 Radicals are produced in some unknown fashion to initiate the chain. One possible mechanism is cracking of a hydrocarbon, as in the following reaction:

$$C_8H_{18} \longrightarrow C_7H_{15}\cdot + CH_3\cdot$$

2 The radicals generated then proceed to strip H atoms from hydrocarbon molecules, forming new radicals in the process, e.g.,

$$CH_3\cdot + C_8H_{18} \longrightarrow CH_4 + C_8H_{17}\cdot$$

Some of these large radicals may break up into smaller radicals, leaving unsaturated hydrocarbons as products:

$$C_8H_{17}\cdot \longrightarrow CH_3\cdot + C_7H_{14}$$

3 The radicals may react with O_2 to form peroxide radicals:

$$R\cdot + O_2 \longrightarrow R-O-O\cdot$$

4 Several paths now arise. The peroxide radical might decompose, or it might strip H from another hydrocarbon to set up a radical-peroxide chain:

$$R\cdot \xrightarrow{O_2} ROO\cdot \xrightarrow{RH} ROOH + R\cdot$$

Depending on how the peroxide decomposes, whether by breaking the R—O bond or the O—O bond, a variety of products is possible. Typical products include organic peroxides, H_2O_2, formaldehyde, higher aldehydes, and ketones. Only in relatively cool flames are the final products CO_2, CO, and H_2O. Hot flames, i.e., self-ignition flames, are excessively rich in aldehydes and ketones. Engine misfiring as well as excessively high compression ratios generally leads to aggravated air pollution because of the highly reactive species created in the preflame and self-ignition reactions. As will be discussed in Sec. 26.4, some of the remedies adopted to mitigate these problems have created bigger problems of their own, e.g., lead poisoning.

26.3 *Photochemical smog*

There are two kinds of smog. The oldest, which is a mixture of coal smoke and fog, has plagued mankind since the fourteenth century. Particles of smoke from coal combustion act as condensation nuclei on which fog droplets condense. The word "smog" comes from an elision of "smoke" and "fog," best typified by such classic examples as the London smog of December 1952, which killed 3500, and the Donora, Pennsylvania, smog which killed 20 and made hundreds ill. The fog part is largely $SO_2(+ SO_3)$ and humidity; it is generally worst in the early morning hours and appears to worsen shortly after sunrise (perhaps because of photochemically induced oxidation $SO_2 + \frac{1}{2}O_2 \longrightarrow SO_3$ followed by reaction with humidity to give H_2SO_4 aerosol).

The other predominant kind of smog is called photochemical or Los Angeles–type smog. It has no relation to smoke or fog; it is worst in the sunshine and, unlike London–type smog which peaks early in the day, it peaks in the afternoon. Also, whereas London smog is characterized by bronchial irritation, Los Angeles smog tends to produce eye irritation (i.e., lacrimation) and plant damage (e.g., metallic sheen on leaves).

The first clues to the cause of photochemical smog came in 1950 when it was noticed that articles made of rubber showed severe cracking in periods of high smog. Laboratory tests suggested the culprit is ozone. The second clue came in 1952 when it was noted that ozone plus double-bonded hydrocarbons give products which yield the same kind of damage to plants as does Los Angeles smog. In the same year Professor Haagen-Smit of Cal Tech discovered that double-bonded hydrocarbons plus NO_2 give products that crack rubber. Originally, the products were believed to be organic peroxides, R—O—O, but subsequently they were identified with ozone. The fourth clue came from statistical studies which suggested a correlation with the automobile, specifically that the automobile exhaust is a main source of double-bonded hydrocarbons and a major source of oxides of nitrogen.

The chemical reactions that are involved in the creation of photochemical smog are complex and still only incompletely understood. They start with the combination of N_2 and O_2 in the air. At ordinary temperatures these do not react with each other, but when the temperature rises to the levels found in hydrocarbon combustion flames, they react to form nitric oxide

$$N_2(g) + O_2(g) \longrightarrow 2NO(g)$$

Some of this NO reacts further to form nitrogen dioxide

$$2NO(g) + O_2(g) \longrightarrow 2NO_2(g)$$

but this, being a third-order reaction, does not go very far before diffusion away from the exhaust source reduces the temperature and concentration to levels where conversion to NO_2 becomes negligible. Hence, as pointed out above, only part of the exhaust nitrogen oxides is NO_2, the rest being NO. Actually, man-made NO_x is but 10 percent of the total in the environment, the rest coming from natural bacterial action. Typical background levels in the United States are 0.002 ppm NO and 0.004 ppm NO_2; in urban areas these often reach 100 times these values.

Nitrogen dioxide is part of the *photolytic* NO_2 *cycle*, which can be written as follows:

1 $NO_2 \xrightarrow{h\nu} NO + O$

2 $O + O_2 + M \longrightarrow O_3 + M$

3 $O_3 + NO \longrightarrow NO_2 + O_2$

In step (1), NO_2, which is an efficient absorber of ultraviolet light, is cleaved to give NO and atomic oxygen. The cutoff radiation for this step is 400 nm. For $\lambda > 400$ nm, the photons are too weak to cause dissociation; the molecules are simply excited to higher states. For $\lambda < 400$ nm (about the violet edge of the visible spectrum), dissociation results. Step (2) shows a combination of the monatomic oxygen with an ordinary diatomic O_2 to form ozone, M being a third body such as N_2, O_2, Ar, or CO_2, which can carry off the large excess energy. In step (3), O_3 recombines with the NO formed in step (1) to regenerate NO_2 and O_2. The cycle is a closed one and would be balanced except for competing reactions with hydrocarbons. Hydrocarbons, such as those emitted in automobile exhausts, are believed to react by a free-radical mechanism either with O or O_3 to form secondary pollutants such as aldehydes, ketones, and organic peroxides. In the process, NO

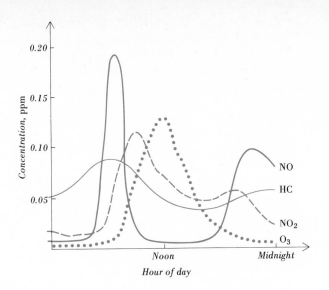

Fig. 26.4
Concentrations of NO, NO$_2$,
O$_3$, *and hydrocarbons*
during a smog day. (Scale
for hydrocarbon
concentrations should be
multiplied by 100.)

is oxidized to NO$_2$, and lack of the NO scavenger makes the O$_3$ content rise.

The components of a typical smog are CO (2 ppm), NO (0.15 ppm), NO$_2$ (0.20 ppm), unsaturated hydrocarbons (0.05 ppm), aromatics (0.20 ppm), O$_3$ (0.10 ppm), aldehydes (0.20 ppm), and organic peroxides (0.03 ppm). How some of these change with time during a representative smog day is shown in Fig. 26.4. As can be seen, they correlate differently with the time of day, partly because of changes in automobile traffic and partly because of change in sunlight. Before 6 A.M., the NO and NO$_2$ concentrations are steady. From about 6 to 8 A.M., there is a sharp increase in NO and hydrocarbons, which correlates with rise in automobile traffic. As the sun comes up, the NO$_2$ content goes up, and the NO goes down, presumably because of photochemically induced oxidation of NO to NO$_2$. As the NO drops below about 0.1 ppm, the O$_3$ content starts to build up, reaching a peak in the early afternoon. Again in the period 5 to 8 P.M., the NO rises as the sun goes down and traffic picks up. A belated rise in NO$_2$ may be attributed to late oxidation of the NO by O$_3$ built up during the day.

The dynamics of the above conversions are only imperfectly understood. Some of the critical rate constants have not yet been unambiguously evaluated, and almost half of the total pollution mass balance cannot yet be accounted for. Attempts to simulate the problem in artificial chambers have proved to be frustrating because of a large dependence on the nature, pretreatment, and area of the enclosing surface. This is typical of chain reactions and points up the probable role of free radicals in the smog reactions. How the hydrocarbons upset the photolytic NO$_2$ cycle is subject to vigorous debate. Most proposed schemes go through the unsaturated hydrocarbons and have them attacked by monatomic oxygen or, less probably, by ozone. Hydrocarbon free radicals are generated as in

$$CH_3CH{=}CHCH_3 + O \longrightarrow CH_3CHO + CH_3CH\cdot$$

and these go on to react with O_2 or NO to produce other unstable intermediates as well as various by-products. Often a chain of 20 or more consecutive steps is needed to account for all the products, and even then the scheme may be unacceptable because it may disagree with the observed rates. Currently, a great deal of attention is being given to the possible role of *singlet oxygen,* an excited state of the O_2 molecule in which the two electron spins are not oriented parallel to each other as in ordinary paramagnetic O_2 but are antiparallel even though in separate molecular orbitals. Apparently, when singlet oxygen reacts with organic molecules, it produces rather unusual reaction products; it is hoped that an understanding of their mechanism of formation will go far to clear up the mystery of photochemical-smog generation.

So far as we are concerned, the important point is that hydrocarbons in the smog cycle generate free radicals which use up some of the ambient NO, thereby depriving the O_3 its normal route in the photolytic NO_2 cycle. O_3 itself is an extremely irritating pollutant even at low concentrations, but it also reacts to produce other highly reactive organic oxidants. One such is peroxyacetylnitrate (sometimes abbreviated PAN), for which the formula is

$$CH_3-\overset{\displaystyle O}{\overset{\displaystyle \|}{C}}-O-O-NO_2$$

It has a deleterious effect on plants even at levels as low as 0.01 ppm. Other similar organic peroxides are if anything even more injurious.

The photochemical-smog problem is by no means unique to Los Angeles. Most big cities, especially if they have lots of sunshine, are getting periods of smog alert. Los Angeles, as usual, seems to have run into this problem of civilization ahead of everybody else and more spectacularly. Part of the reason is the urban sprawl, with its enormous number of motor-vehicle point sources of pollution; the other part of the reason is abundant sunshine and a geographic situation generally ideal for generating and trapping photochemical pollutants. In the next two sections we look at some of the problems in air pollution control.

26.4 Octane dilemma

In 1970 the U.S. Congress passed public law 91-604, known as the Clean Air Amendments of 1970, or, more popularly, the "Muskie bill," calling for a 90 percent reduction of pollutant emissions from their 1970 levels. Implementation was turned over to the Environmental Protection Agency (EPA). Figure 26.5 shows the emission standards as set for a defined cycle of driving in various modes (idle, accelerate, and decelerate). Because of problems in devising test procedures and technical modifications to meet new standards, the precise figures and their scheduling cannot be regarded as rigid. They are useful only as a relative guide to the chase for a clean exhaust, i.e., one in which you can put your nose and breathe.

	Prior to control	1970	1976	Goal for 1980
Hydrocarbons	11	2.2	0.5	0.25
CO	80	23	11	4.7
NO_x	4	....	0.9	0.4
Particulates	0.3	....	0.1	0.03

Fig. 26.5

Emission standards for control of pollution by motor vehicles, given in grams per mile.

The auto pollutants that are the easiest to control are hydrocarbons and carbon monoxide. As shown in Fig. 26.6, increasing the air-fuel ratio (i.e., making the mixture more lean) dramatically reduces hydrocarbon and carbon monoxide emission. This is how most automobile manufacturers were able to meet projected standards for hydrocarbons and carbon monoxide so quickly. Minor modification of present engines was all that was needed. However, as Fig. 26.6 makes abundantly clear, the result is to aggravate the NO_x problem. Running engines at higher air-fuel ratios means higher combustion temperatures, and that means more generation of NO and NO_2. Two solutions have been suggested: One is to keep the air-fuel ratio low (this would make the "hot rodders" happy since the maximum power corresponds to air-fuel $\approx$ 13) and add an afterburner where exhaust gases could be more completely burned after addition of more air. The other suggestion is to add a catalytic reactor where the exhaust gases could be passed over or through a bed of catalyst to convert to less noxious products. The problem is that most catalysts are deactivated (i.e., "poisoned") rather rapidly by the lead added to gasoline to improve its antiknock qualities. However, removal of lead from gasoline, where it is generally added as tetraethyllead, $Pb(C_2H_5)_4$, to the extent of about 1 cm^3/liter, is not easy. High-compression engines tend to knock rather badly because of self-ignition brought about by the compression. Tetraethyllead (TEL) helps to cut this down by easily generating a smooth supply of free radicals for maintaining an orderly flame front. What can be done to get rid of the lead? It appears to be needed for smooth burning but negates the possibility of being able to clean up the exhaust. To appreciate the full dimensions of the problem, we need to look more specifically at the problem of petroleum fuels.

Crude petroleum consists of an almost infinite variety of hydrocarbons, e.g., straight chains, branched chains, unsaturated hydrocarbons, rings, etc. Besides traces of S, O, N, H_2O, and SiO_2, there is generally 83 to 87% by weight C and 11 to 14% H. The hydrocarbons comprise four families: (1) the *paraffins*, or alkanes, which are saturated hydrocarbon chains of general formula C_nH_{2n+2}; (2) the *naphthenes*, such as cyclopentane, which are saturated ring compounds of general formula C_nH_{2n}; (3) the *olefins*, or unsaturated chains of formula C_nH_{2n}; and (4) the *aromatics*, unsaturated ring compounds of formula C_nH_{2n-6} In the refining of petroleum, the various volatility fractions are separated from each other by distillation. The groups generally differ by the number of C's per molecule and can be classified as follows: gases (C_1 to C_4), gasoline (C_5 to C_{12}), kerosine (C_{10} to C_{16}), fuel and gas

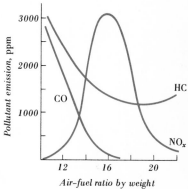

Fig. 26.6

Effect of air-fuel ratio on pollutant emissions. (Vertical scale needs to be multiplied by 40 for CO, by 0.4 for HC.)

Section 26.4

Octane dilemma

681

oil (C_{15} to C_{22}), and lubricating oils (C_{19} to C_{35}). The residue of the distillation gives paraffin wax and asphalt (C_{36} to C_{90}).

The problem with petroleum is that the demand-supply ratio for the gasoline fraction exceeds that of the other fractions. To enhance the supply of gasoline, four recourses are possible:

1 Cracking. Thermal or catalytic decomposition of higher hydrocarbons to lower ones, for example, $C_{14}H_{30} \longrightarrow C_7H_{16} + C_7H_{14}$. The products are generally *iso*paraffins rather than *normal;* so they have good antiknock properties.

2 Polymerization. Olefinic hydrocarbons of the type C_4 and C_5 are polymerized to give 100% olefinic C_8 and C_9 compounds. Again these have high antiknock properties.

3 Alkylation. Small isoparaffins such as isobutane (C_4H_{10}) are made to combine with a small olefin such as butene (C_4H_8) to give a large isoparaffin (C_8H_{18}).

4 Reforming. High-knock compounds are converted to low-knock compounds by dehydrogenation. Approximately 40 percent of present fuels are made this way.

The isohydrocarbons and the unsaturated hydrocarbons are better for low-knock combustion than are the normal saturated hydrocarbons. The arbitrarily chosen references for low-knock combustion are iso-octane (2,2,4-trimethylpentane), C_8H_{18}, which is assigned an *octane number* of 100 and *n*-heptane, C_7H_{16}, which knocks badly and is assigned an octane number of 0. Other fuels are then rated by comparison in a standard engine with mixtures of iso-octane and *n*-heptane that give the same knock. Thus, for example, octane number 70 means a fuel that knocks like 70 parts (by volume) of iso-C_8H_{18} and 30 parts of *n*-C_7H_{16}. Octane numbers above 100 are possible (e.g., isodecane, $C_{10}H_{22}$, is 113); such fuels are usually rated by finding the number of milliliters of $Pb(C_2H_5)_4$ required per gallon of iso-octane to give the same antiknock. Octane numbers less than 0 are also possible (for example, *n*-C_8H_{18} is -20).

Although tetraethyllead is one of the best fuel additives, there are other compounds that have antiknock qualities. Some of these are listed in Fig. 26.7. As can be seen, there is nothing quite like $Pb(C_2H_5)_4$, though the aromatics represent possible inexpensive substitutes. Actually, there is a compound that might be included in Fig. 26.7 to rival, even surpass, tetraethyllead. It is called AK33X and on a weight basis is twice as efficient as $Pb(C_2H_5)_4$. The proper chemical name of this superadditive is methylcyclopentadienylmanganese-tricarbonyl. It has the structure shown in Fig. 26.8. Three carbonyl (CO) groups are attached to Mn on one side; a five-membered flat ring, C_5H_5, with one of the H's replaced by a methyl group, sits on the other side of the Mn and joins to it by use of its π-electron system. Unfortunately, AK33X is very expensive to make and is light sensitive.

So where is the dilemma? The antilead people say take the lead out of gasoline so that you can make catalytic reactors practical: You will decrease lead pollution of the environment, you will cut down on

Compound	Weight for a given effect, g	Relative amount needed
Tetraethyllead	0.0295	1
Aniline, $C_6H_5NH_2$	1.00	34
Ethyl iodide, C_2H_5I	1.55	53
Ethanol, C_2H_5OH	4.75	161
Xylene, $C_6H_4(CH_3)_2$	8.00	271
Toluene, $C_6H_5CH_3$	8.8	298
Benzene, C_6H_6	9.8	332

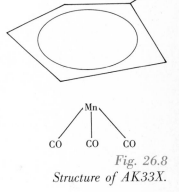

Fig. 26.7
Relative efficiencies of antiknock additives.

Fig. 26.8
Structure of AK33X.

metal corrosion in engine and exhaust components, and you will be able to dispense with other gasoline additives such as ethylene dibromide you now need to add to scavenge the lead. The prolead people counter with their own arguments: If you take the lead out, the octane rating will go down, you will have power loss, you will need other additives, and these will cost money. Furthermore, you will need new refining equipment to get the additives, and this also will cost money. Also, you will need for your additives blending components dependent on petroleum, which is already in short supply. Finally, if you take the lead out, the fuel efficiency will not be so great, fuel economy may go down by 5 percent, and you will lose the advantage of having those lead deposits lubricate your valve seats.

Clearly there is a dilemma, even excluding the economic question. Addition of aromatics, which is now one of the main ways low-lead gasoline is doctored to keep up its octane rating, tends to raise the content of unsaturated hydrocarbons in the engine exhaust. As we saw above, unsaturated hydrocarbons are key reactants in triggering the photochemical-smog cycle; so we are back where we started from!

26.5   Control of air pollution

The above discussion centered primarily on the motor vehicle as a source of air pollution. One reason for this is the excessive burden automobile exhausts add to the air pollution problem; another is the ubiquity of the motor-vehicle problem and its portent for the future.*
Still, major steps are being taken to minimize vehicular production of air contaminants. For CO, significant reduction has been achieved by adjusting carburetors to avoid excessively low air-fuel ratios and by modifying carburetor and combustion-chamber design so as to ensure good air-fuel mixing and minimal quenching of high-temperature equilibrium CO. Projected improvements are to add thermal exhaust reactors or catalytic reactors, in which air is added to the exhaust gases for a final afterburn or catalytic conversion to CO_2. Other suggestions have been made that we try to find substitute fuels for gasoline. Liquid

* In some places motor-vehicle numbers are rising faster than the population. It has been noted that if present trends continue, by 1980 one out of four cars on the New Jersey Turnpike will be driverless!

natural gas, which is mostly methane, and liquid petroleum gas (mainly propane and butane) could be used as clean-burning fuels, but they are running into short supply for other purposes, and there are transport problems. Finally, it has been suggested that the internal-combustion engine as motive power in transport be replaced by steam, electricity, or gas turbines. So far, none of these suggestions is practical as a simple, flexible, economic, convenient, and acceptable substitute.

For the NO_x problem, no final answer is in sight. Two approaches are being examined: (1) Decrease the generation of NO_x by decreasing the flame temperature, and (2) remove the NO_x after formation by catalytic decomposition. For the former, two-stage combustion has been suggested wherein the first stage is carried out with less than stoichiometric air-fuel ratio (*ergo* lower flame temperature), and the second stage, after partial cooling and more air injection, with greater than stoichiometric air-fuel ratio. To remove NO_x *after* combustion, two interesting proposals have been made: One is to decompose NO catalytically into N_2 and O_2, thus generating harmless residuals. The equilibrium K for the decomposition $2NO \rightleftharpoons N_2 + O_2$ is 9×10^{10} at $500°C$; so NO is unstable thermodynamically. However, it is kinetically stable, and even with catalysts such as $Co_3O_4 > CuO > Cr_2O_3 > ZnO$ the reaction is discouragingly slow. The other cute suggestion was to bring about catalytically $NO + CO \longrightarrow \frac{1}{2}N_2 + CO_2$, thus getting rid of two pollutants at the same time.

The reduction of O_3 and organic peroxide levels is essentially a secondary pollutant problem. There the solution lies in cutting back on hydrocarbon and NO_x emissions. If that is done, O_3 and PAN will automatically drop.

The SO_x problem is more a problem of stationary power sources and industrial installations than roving point polluters such as the automobile. In the old days, the smell of SO_2 indicated that the devil was within spitting distance; now, it suggests a high-sulfur fuel or a smelting operation. To remedy the problem, low-sulfur coal or fuel oil can be substituted for high-sulfur fuels. Two-thirds of the United States coal reserve contains 1.0% by weight sulfur, or less, compared with an average 2.7% in the current consumption. Unfortunately, the low-sulfur stocks are mainly west of the Mississippi, away from the major markets, and they generally are of lower heat content. Proposals have been made to remove sulfur from fuels before combustion, but this is not easy to do. In one method, powdered coal is treated with steam and controlled oxygen to form CH_4, H_2, CO, H_2S, and CO_2, after which the CO_2 and H_2S are removed and the gases burned as fuel.

A most promising solution to the SO_x problem is to inject powdered limestone into the combustion zone so as to get the reaction

$$2CaCO_3(s) + 2SO_2(g) + O_2(g) \longrightarrow 2CaSO_4(s) + 2CO_2(g)$$

and pass the flue gas through a slurry of lime. The SO_2 removal has been shown to be 90 percent efficient even on a full plant scale.

26.6 Odors

One of the most direct assaults on the senses arising from environmental pollution comes from the stinks of civilization. The biggest offenders, involved in both air pollution and water pollution, are the *mercaptans* (organic compounds of the type R—S—H, which have a putrid odor), the *amines* (for example, RNH_2, R_2NH, and R_3N, which are described as having fishlike or "animal" odors), the low-molecular-weight *aldehydes* (for example, HCHO and CH_3CHO), and the *sulfides* (e.g., oil of garlic, $CH_2{=}CHCH_2{-}S{-}S{-}CH_2CH{=}CH_2$). The reasons for odors and how one goes about masking them are largely shrouded in mystery, but fascinating glimmers of understanding are beginning to appear.

The sense of smell depends on a group of highly specialized receptor cells located far back and up in the nose in what is called the *olfactory cleft*. Figure 26.9 shows the relative location of the olfactory cleft. The receptor area has a total of 6 to 7 cm^2 and is yellow, whereas the rest of the lining is red. In keen-scented animals, such as the dog, the area is much larger and is brown instead of yellow. The receptor area, as shown in Fig. 26.10, is covered with tiny hairlike growths (olfactory hairs) projecting from the ends of sensory cells squeezed into a matrix of supporting cells. The short hairs, which are the receptor centers, are probably normally immersed in a film of mucus; so to be smelled not only must a substance be volatile (to get into the nasal passage) but also soluble in the mucous layer. Microscopical studies indicate that the olfactory hairs are stained by osmic acid (OsO_4), which is taken as an indication of the presence of unsaturated fatty acids. Fat or lipoid solubility is usually taken as a criterion for odoriferousness; water solubility also seems to help.

Odor classification is highly subjective, and no classification scheme is universally accepted. Most interesting is the scheme proposed by Henning, in which six basic odors (fruity, flowery, resinous, spicy, burnt, and foul) were arranged on an olfactory prism as shown in Fig. 26.11. Odors were described as being on a corner and therefore having one component (e.g., geranium was classified as flowery), on an edge (e.g., vanillin had two components, spicy and flowery), or on a face where it was possible to have three or four components (e.g., rose was flowery, spicy, burnt, and foul). No odors could have five or six components. Certain combinations (e.g., foul, fruity, and spicy) were not possible. Other combinations demanded certain components; thus, if something smelled foul, fruity, and resinous, it would also smell burnt. More to the point, Henning associated with his six basic odors certain chemical groupings, as illustrated in Fig. 26.12. The para arrangement of substituents on the benzene ring went with spicy odors; the ortho arrangement, with flowery ones. Branched chains, as in citral, went with fruity odors, even though unsaturation generally gives rise to an irritating acrid odor. (Acraldehyde, $CH_2{=}CHCHO$, for example, smelled when a candle is blown out, has double bonds and an aldehyde group as in citral, but citral, found in oil of orange and oil of lemon,

Olfactory cleft

Septum

Front view of interior of nose

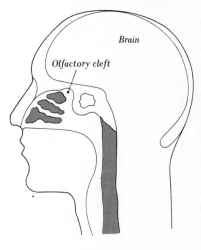

Brain

Olfactory cleft

Side view

Fig. 26.9
Olfactory cleft for odor detection.

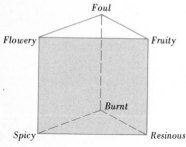

To olfactory bulb in
lower part of brain

Membrane

Sensory cells

Supporting cells

Olfactory hairs

Mucous layer

Fig. 26.10
Schematic representation of disposition of olfactory receptors in the active area of the nasal passage.

Foul

Flowery

Fruity

Burnt

Spicy

Resinous

Fig. 26.11
Olfactory prism.

has a characteristic fruity smell.) To resinous, as shown in Fig. 26.12*d*, Henning associated bridge groups stretched across a ring, as in pinene. Burnt went with smooth rings, as in pyridine, and foul was characteristic of sulfur compounds. According to Henning, intermediate odors came about when several characteristic groupings occurred simultaneously in the same molecule. Thus, vanillin, which has —OH and —CHO groups para to each other on a benzene ring and an —OCH$_3$ group that is simultaneously ortho to the —OH group, is both spicy and flowery.

The relation of odor to chemical structure, however, is not simple. Dissimilar structures often smell the same. Thus, for example, HCN, nitrobenzene ($C_6H_5NO_2^-$), and benzaldehyde (C_6H_5CHO) all smell like bitter almonds. Similarly, the five molecules shown in Fig. 26.13 all smell like camphor. Attempts have been made to correlate odors with changes in series of organic compounds, but with little success. In the alcohol series, ROH, the early members such as C_2H_5OH have a rather sweet smell, but they get more disagreeable, until a peak is reached at nonyl alcohol ($C_9H_{19}OH$), after which the disagreeable character decreases. In the case of the organic acids, the early ones such as formic acid (HCOOH, irritating), acetic acid (CH_3COOH, vinegary), and butyric acid (C_3H_7COOH, sweaty) tend to be disagreeable, but the later ones (e.g., palmitic acid, $C_{15}H_{31}COOH$) are odorless. Similarly, the aldehydes go from pungent formaldehyde (HCHO) to sharp, disagreeable acetaldehyde (CH_3CHO), to suffocating propionaldehyde (C_2H_5CHO), and then they become pleasant. Decreasing volatility due to lengthening hydrocarbon chain certainly affects odor, but so does solubility. The solubility problem is illustrated by sugar-type molecules. Generally, the presence of two or more hydroxy groups as in glycols, glycerols, and the sugars makes for high aqueous solubility, but all these materials are odorless. Apparently, the factor that makes for high water solubility decreases lipoid solubility, and it may be that to have an odor a substance must dissolve in the fatty material of the receptor olfactory hairs.

A further puzzle is the great variation in threshold limits for detection. NH$_3$ needs to be at a concentration of 0.037 mg/liter to be detected by the average person, SO$_2$ at 0.009 mg/liter, HCN at 0.001 mg/liter, and H$_2$S at 0.00018 mg/liter. Ethyl mercaptan (C_2H_5SH), however, can be detected at 6.6×10^{-7} mg/liter, which corresponds to about 10^9 molecules per cubic centimeter. Considering how little of a sniff gets into the olfactory cleft and how little is the chance of interacting with a receptor, investigators have concluded that as few as eight molecules are all that are needed to trigger the receptor action.

How does odor come about? There appear to be three master theories: (1) *molecular interaction*, involving chemical reaction between odorant and receptor; (2) *vibration interaction*, in which infrared-frequency bond vibrations stimulate the receptor; and (3) *radiant-energy interaction*, where photons of energy, perhaps in the infrared range, stimulate a receptor somewhat as visible light stimulates the eye. At present,

Class	Example	Structural formula
(a) Spicy	Anisaldehyde	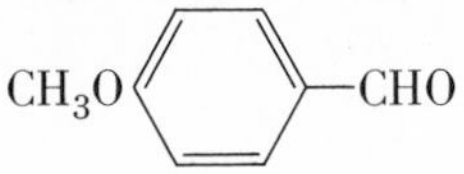 CH_3O—⬡—CHO
(b) Flowery	Coumarin	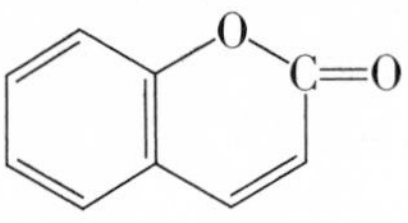 (benzene ring fused with O—C=O ring)
(c) Fruity	Citral	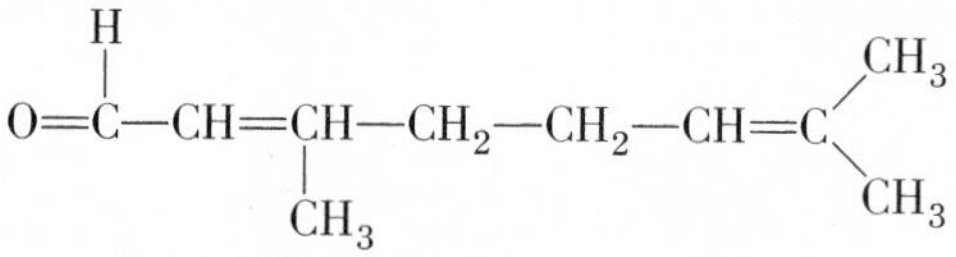 $O=\overset{\underset{\mid}{H}}{C}-CH=CH-\underset{\underset{CH_3}{\mid}}{CH}-CH_2-CH_2-CH=C\overset{CH_3}{\underset{CH_3}{\diagdown}}$
(d) Resinous	Pinene	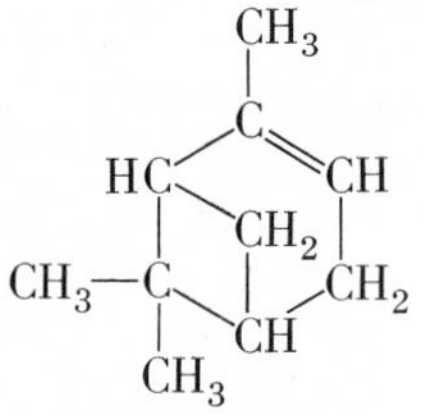 (bicyclic pinene structure)
(e) Burnt	Pyridine	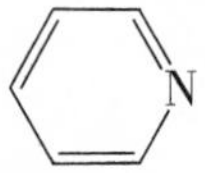 (pyridine ring, N)
(f) Foul	Mercaptans	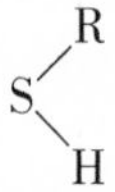 $S\overset{R}{\underset{H}{\diagup}}$

Fig. 26.12
Odor class examples.

no theory satisfactorily accounts for all the observations and explains the subtle regularities that have been noted. Attempts to blend the master theories usually involve some kind of odorant-receptor binding *via* weak chemical interaction (e.g., van der Waals and hydrogen bonding) in a specific site where normal infrared absorption and emission can be interfered with.

Most interesting of the current theories is the stereochemical theory of odor proposed by John Amoore, which is based on the geometry of molecules. It suggests that corresponding to the different basic odors there are at the olfactory nerve endings different kinds of receptor sites, each of which will accept only appropriately shaped molecules. Thus, the stereochemical theory of odor is much like the "lock-and-key" theory of enzyme action. Figure 26.14 shows the shape classification proposed. Spherical or globular molecules such as camphor (cf. Fig. 26.13) are supposed to fit into matching hemispheric receptor sites and, in some as yet not understood fashion, produce the sensation we know as mothball smell. One far-out but probably correct suggestion is that the receptor site acts as a resonance cavity for long-wave infrared radiation but this action is destroyed when the odorant molecule binds into the cavity. In any event, appropriate signals are sent to the brain, which are there interpreted as camphor smell. Other typical odors are suggested to have their own unique receptor-site geometries. The pictures in Fig. 26.14 are intended to be illustrative rather than definitive.

Fig. 26.13
Molecules with camphorlike odor.

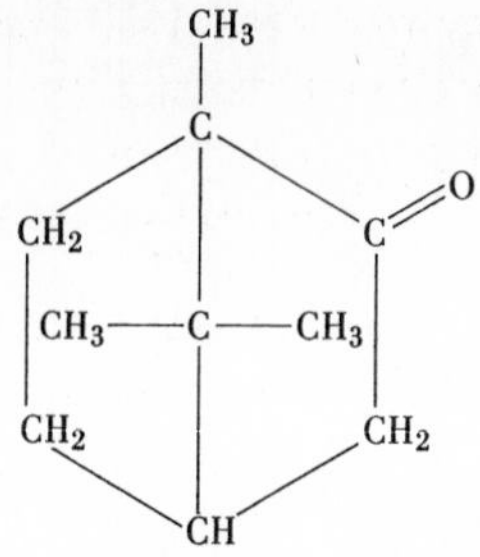

Camphor 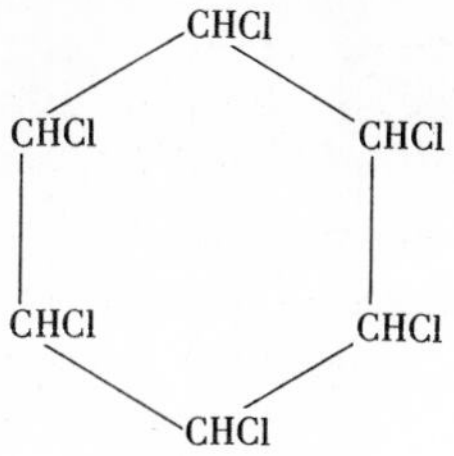   Hexachlorohexane

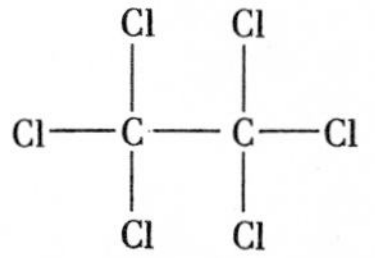

Hexachloroethane 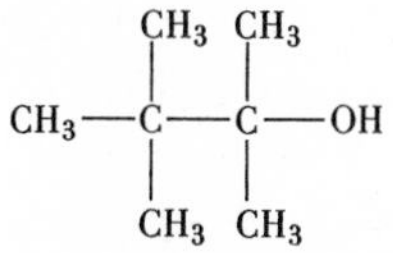   Pentamethyl ethanol 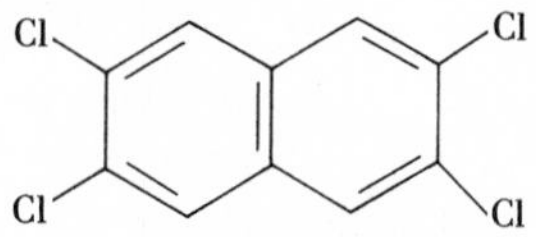   Tetrachloronaphthalene

At the bottom of Fig. 26.14 are shown two kinds of odors that are generally related to rather simple-shaped molecules and which perhaps depend not on shape but on electric charge perturbation. Amoore suggests that pungent odors such as that of formic acid arise from the electrophilic, or electron-seeking, nature of the molecules. The carbon atom of a carboxylic acid (COOH) grouping is usually regarded as positive because of the electron-withdrawing action of the oxygen. Such

Fig. 26.14

Classification of molecules and receptor sites according to stereochemical theory of odor. (See "Annals New York Academy of Sciences," vol. 116, p. 458.)

Primary odor	Chemical example	Molecular shape	Receptor site
Camphoraceous	Camphor	Spherical	
Musky	Pentadecano-lactone	Disk	
Floral	Phenylethyl methyl-ethyl-carbinol	Disk and tail	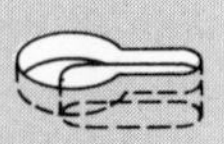
Pepperminty	Menthol	Wedge	
Ethereal	Diethyl ether	Rod	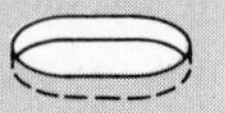
Pungent	HCOOH	Simple (electrophilic)	
Putrid	C₂H₅SH	Simple (nucleophilic)	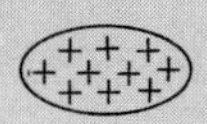

a positive carbon would bind to electron-rich sites or at least attract electrons to itself. Both effects are probably associated with producing the sensation of a pungent odor. In like fashion, putrid odors are associated with electron-rich (nucleophilic) molecules such as the sulfides, which bind to electron-deficient sites.

Regardless of what odor theory eventually triumphs, certain facts will need to be explained: All normal people can smell. Brain lesions, severed olfactory nerves, and obstructed nasal passages reduce the ability to detect odors. Preferential anosmia (i.e., inability to perceive certain odors) seems not to be well established. Some substances are odorous, others are not. Substances can be smelled at a distance. Substances of different chemical constitution may smell the same. Substances of similar constitution generally have the same odor, but isomers and stereoisomers may smell quite unlike each other. Molecules of high molecular weight are generally nonodorous. The quality as well as the intensity of an odor depends strongly on dilution. The sense of smell is generally rapidly fatigued, though fatigue for one kind of odor usually does not kill the ability to perceive dissimilar odors.

26.7 Water pollution

Most of the large rivers of the world are nothing but open sewers. Little attention had been given to the problem because water is a renewable resource, globally abundant, which is constantly recycled through natural distillation *via* solar evaporation, cloud condensation, and rain. When there were no neighbors upstream, one could watch the pollution problems flow away, but with increasing population and increasing technological waste, that becomes more difficult. Also, some of the dumped chemicals that were safely flushed away (e.g., mercury and polychlorinated biphenyls) are now coming back to haunt us. What happens, for example, when our water pollutants gathered in the sea reach the threshold levels of toxicity at which phytoplanktons, which account for 50 percent of our photosynthetic oxygen, cease to function?

Water quality is a relative property dependent on the use to which the water is put. Generally, it is a function of dissolved oxygen (DO), dissolved solids, biochemical oxygen demand (BOD), suspended sediments, pH, and temperature.

Dissolved oxygen is required by all aquatic plant and animal life. Fish require the highest levels, vertebrates next, and bacteria the least. Warm-water biota, including game fish, require 5 ppm; cold-water biota at or near saturation, no lower than 6 ppm. Figure 26.15 shows how the solubility of oxygen varies with temperature; it drops from a high of 15 ppm at 0°C to about 6 ppm at 40°C. Levels below saturation arise from the decay of oxygen-demanding wastes. Most of these are organic, and if we represent them as carbon, we can write

$$C \; + \; O_2 \; \longrightarrow \; CO_2$$
$$12\,g \qquad 32\,g$$

which indicates that a 9-ppm level of dissolved oxygen would be totally

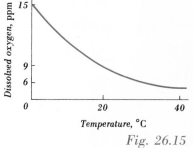

Fig. 26.15
Dissolved oxygen as a function of temperature.

exhausted by $(\frac{12}{32})(9)$, or about 3 ppm of carbon waste. This is equivalent to about a drop of oil in 10 liters of water.

The BOD indicates the amount of dissolved oxygen used up during oxidation of oxygen-demanding waste. It is measured by incubation of the water sample for five days at 20°C, with dissolved oxygen measured before and after. For nearly pure water, BOD runs about 1 ppm; fairly pure, 3 ppm; doubtful, 5 ppm. The U.S. Public Health Service limits water effluents into streams to BOD values less than 20 ppm. For comparison, untreated municipal sewage generally has BOD values in the range 100 to 400 ppm; barnyard and cattle feedlot runoffs, in the range 100 to 10,000 ppm; and food-processing waste, 100 to 10,000 ppm. Domestic animal wastes in the United States are equivalent to a population of 1.9×10^9 people. As far as BOD equivalence is concerned, on a scale where man $= 1$, sheep would be 2.5, pig 1.9, chicken 0.1, cow 16.4, and horse 11.3. Paper and pulp processing, notorious for dumping SO_2 and like compounds into our water resources, are equivalent to 216×10^6 people in BOD.

What happens as dissolved oxygen gets depleted by BOD? Plant and animal life disappear, either because of death or migration. Bacterial decomposition shifts from aerobic (O_2-requiring) to anaerobic (not requiring O_2). The products of metabolism change. Whereas under aerobic conditions $C \longrightarrow CO_2$, $N \longrightarrow NH_3 + HNO_3$, $S \longrightarrow H_2SO_4$, and $P \longrightarrow H_3PO_4$; under anaerobic conditions $C \longrightarrow CH_4, C_2H_4$, etc., $N \longrightarrow NH_3 +$ amines, $S \longrightarrow H_2S$, and $P \longrightarrow PH_3$ and other lower-valent phosphorus compounds. The main point to note is that under anaerobic conditions, the decomposition products tend to be more odoriferous and more likely to be toxic.

Water pollution can be *chemical, physical, physiological,* or *biological.* If *chemical,* it may be classified as organic or inorganic. Organic pollutants include proteins (domestic sewage, waste from creameries, canneries, and slaughterhouses), fats (sewage, wool processing, soap production, and food processing), carbohydrates (sewage, textile mills, and paper mills), resins, coal, and oil. Inorganic pollutants might be acids, alkalies, heavy metal cations, and certain anions. Acid mine drainage is a primary source of stream pollution, especially in coal-producing regions. The actual pollutants are H_2SO_4 and soluble iron salts formed as a result of reaction between air, water, and pyrites present in the coal seams. Certain types of bacteria also appear to be involved, but their role is not understood. It is estimated that about 4×10^9 kg of H_2SO_4 per year goes into United States streams, 60 percent of which originates in abandoned mines. Acid stream pollution is one of the primary causes of fish kill in the United States. Other sources of acid pollution are chemical plants, iron and copper production, and the wood-pulp and paper industry. Alkalies generally come from chemical plants, textile operations, and tanning operations.

Figure 26.16 indicates water pollutant limits as recommended by the U.S. Public Health Service and as observed on a national average in the public water supplies. The observed average is based on 2500 samples, and, of course, there is a wide range in the actual values

Fig. 26.16
Water pollutant limits in
milligrams per liter.

Substance	USPHS (1962)	U.S. average (1969)	Remarks
Ag	0.05	0.008	Limit set for cosmetic reasons, leads to discoloration.
As	0.01	0.0001	Serious systemic poison, cumulative.
Ba	1.0	0.034	Not common, serious toxic effect on heart.
Cd	0.01	0.003	Seepage from electroplating, 15 ppm in food causes illness.
Cr	0.05	0.0023	Not natural, suggests plating or tannery pollution.
Cu	1	0.13	Essential and beneficial, adult needs 1 mg/day, detectable taste at 1–5 ppm, large doses may cause liver damage, used for algae control.
Pb	0.05	0.013	Serious, cumulative body poison.
^{226}Ra	3	2.2	Bone-seeking α emitter, destroys bone marrow.
^{90}Sr	10	<1.0	Bone-seeking β emitter.
Zn	5	0.19	Essential and beneficial, milky at 30 ppm, metallic taste at 40 ppm.
Cl	250	27.6	Limit set for taste reasons, salty if too much.
CN	0.01	0.00009	Rapid fatal poison, safety factor 100.
F	0.7–1.2	0.32	Prevents dental caries in small amounts, mottling of enamel above 1.2 ppm.
NO_3	45	6.3	Fertilizer runoff, can cause methemoglobinemia in infants.
SO_4	250	46	Laxative effect above 750 mg/liter, often the cause of traveler's diarrhea (e.g., Montezuma's revenge, Delhi belly).
ABS	0.5	0.05	Not highly toxic, safety factor 15,000.

found. The limits are quoted in milligrams per liter, which is essentially the same as parts per million by weight. For radium and strontium, units are picocuries per liter, where one curie is the radiation equivalent of one gram of radium (that is, 3.7×10^{10} disintegrations per second). The designation ABS at the bottom of Fig. 26.16 stands for alkylbenzenesulfonate, one of the components of some detergents. The detergent problem as a factor in water pollution is discussed in the next section. As can be noted, phosphate is not included in the listing of

Fig. 26.16. It has not been considered a water pollutant in public water supplies in the same way as the toxic materials mentioned, but increasing runoff from fertilizer and detergent use greatly affects biologic activity in streams and lakes so that, as also discussed in the next section, phosphate pollution is a problem.

Physical pollution of water generally refers to color, turbidity, elevated temperature (i.e., thermal pollution), and suspended matter. Color *per se* is not harmful unless it is associated with a toxic chemical, but it may change the quality of light that penetrates to a given depth and hence inhibit plant growth. Turbidity, which arises from soil erosion and colloidal wastes, can be corrected by addition of coagulants such as $FeCl_3$, alum, or $Fe_2(SO_4)_3$. Colloid particles (e.g., clay in natural waters, and proteins, fats, and carbohydrates in waste waters) are usually stabilized by having negative charges at their surfaces, and these can be neutralized by addition of ions. Thermal pollution usually arises from use by manufacturing and power plants of streams for cooling. The result is decreased dissolved oxygen (Fig. 26.15) and increased rate of biochemical activity.

Physiological pollution of water comes from bad taste and objectionable odor. These usually go together, and the most often mentioned contaminants are the mercaptans and the amines. Special mention should be given to phenol (hydroxybenzene), which the USPHS limits to 0.002 ppm. It is not that bad by itself, but the usual chlorination of water supplies converts it to ortho- or parachlorophenol, which tastes like medicine.

Biological pollution of water may include bacteria, viruses, protozoa, parasites, and plant toxins. Infections of the intestinal tract (e.g., cholera, typhoid, and dysentery), polio, and infectious hepatitis have frequently been traced to contaminated water supplies. Generally, no check is made for these pathogenic contaminants because it is a 24-h problem to detect them and that is usually too late. Instead, one looks for a benign indicator such as *coliform* bacteria, the presence of which alerts to fecal contamination.

Waste-water treatment can be classified into three successive stages: primary, secondary, and tertiary. About one-fourth of the United States waste water gets only primary treatment; 5 percent of it gets none! In primary treatment the waste water is passed (1) through screens to take out the large solids, (2) successively into grit and sedimentation tanks, where the smaller sediments are allowed to settle, and then (3) finally through a chlorine treatment to destroy the bacteria. Most of the solids, about a third of the BOD, and a few percent of the persistent organic compounds and potential plant nutrients are removed in this way. In secondary treatment, further pollutant reduction is achieved by adding one of two possible processes: a *trickling filter* or *activated-sludge treatment*. For the trickling filter, a one- to three-meter bed of gravel and rocks is provided through which the sewage is passed slowly enough that bacteria multiply on the stones and consume most of the organic matter. The process is about 75 percent effective. In the activated-sludge method, the incoming sewage is inoculated with some activated sludge (from recirculation), passed into an aeration tank, where it is mixed with air, and then passed into a sedimentation tank, where the recirculate separates out from the effluent, which goes on to chlorine treatment. The process takes several hours but is 90 percent effective at removing organic wastes. Use of pure oxygen instead of air significantly raises the efficiency.

Tertiary treatments are mostly only exploratory. They are used only when water supply is so critical that drinking-quality water needs to be produced in a completely recycled water system or from naturally contaminated sources. They are also used when it is necessary to remove refractory organic compounds (sometimes known as biorefractories) that do not yield to secondary treatment. One such method is to treat the nearly purified water with activated charcoal, filter off the adsorbing solid, and regenerate it with steam distillation of the refractory organics. To remove phosphate, precipitation of the highly insoluble phosphates can be achieved by adding CaO, $Fe(OH)_3$, or $Al(OH)_3$. Other inorganic salts, which a typical city about doubles in content when it uses its water supply, are very difficult to remove. However, even such ions as nitrate can be removed by the admittedly expensive electrodialysis and reverse-osmosis methods previously described (Figs. 16.8 and 16.9).

26.8 Detergents and eutrophication

The term "detergent" has generally come to mean a soap substitute which generally has three ingredients: a surfactant, a builder, and a miscellaneous variety of additives. The *surfactant* is a surface-active agent, which because of its dual hydrocarbon and polar character dissolves partly in organic material and partly in aqueous phases. Soap (Sec. 22.3) is a surfactant, as are the alkylbenzenesulfonates (Sec. 24.2). The *builder* is usually a sodium phosphate of the type $Na_5P_3O_{10}$ or $Na_4P_2O_7$, which acts as a sequestering agent on dipositive cations and raises the pH by hydrolysis. *Additives* may include bleaches, perfumes, enzymes, etc.

Initially, the principal surfactants were alkylbenzenesulfonates made from propylene tetramer and benzene:

$$4CH_3CH=CH_2 \xrightarrow[\textit{150 to 250°C}]{H_3PO_4} \underset{\textit{Propylene tetramer}}{\overset{\overset{\displaystyle CH_3}{|}\quad\overset{\displaystyle CH_3}{|}}{CH_3CH-(CH_2CH)_2-CH_2CH=CH_2}}$$

Propylene

The tetramer adds to benzene in the presence of HF, $AlCl_3$, or H_2SO_4; the product can be then sulfonated with SO_3 or H_2SO_4, and finally neutralized with NaOH to give

$$\overset{\overset{\displaystyle CH_3}{|}\qquad\overset{\displaystyle CH_3}{|}\qquad\overset{\displaystyle CH_3}{|}\qquad\overset{\displaystyle CH_3}{|}}{CH_3CH-CH_2CH-CH_2CH-CH_2CH}-\left\langle\!\!\!\bigcirc\!\!\!\right\rangle-SO_3^-Na^+$$

Although very effective as surfactants, ABS detergents were not popular. Their sales did not boom until formulations were changed to include the phosphate builders. Unfortunately, bacteria did not like the branched chains, and the ABS accumulated in the water environment. Even 1 ppm can lead to excessive foam formation in a river.

The search for suitable substitutes for ABS was based on the known fact that ordinary straight-chain soaps are readily biodegradable as are the long-chain alcohol sulfates. However, the search was not successful until synthetic molecular sieves (Sec. 22.6) were developed. With pore diameters of 0.5 nm, straight chains of cross section 0.49 nm can be separated and utilized as follows:

$$\text{Kerosine} \xrightarrow[\text{sieve}]{\text{molecular}} n\text{-paraffin} \xrightarrow{\text{Cl}_2} \text{monochloroparaffin}$$

$(C_{10}\ to\ C_{16})$

olefin $\quad$ alkylbenzene

(via $-\text{HCl}$, benzene; and benzene/AlCl_3)

The products have linear carbon chains and when sulfonated give linear alkylbenzenesulfonates (LAS), which are completely biodegradable. (The mechanism of biodegradation is not well understood, but it appears to go at least partly by enzymatic fatty acid degradation in which $-\text{CH}_2\text{CH}_2-$ pieces are progressively chewed up.) Most detergent formulations now use LAS instead of ABS.

Not all detergent surfactants are sulfonates. Some are derived from long-chain alcohol sulfates. These became possible when Ziegler invented his ethylene polymerization process. C_2H_4 is built up on $Al(C_2H_5)_3$ to form an aluminum trialkyl-containing long-chain hydrocarbon, and this is subsequently oxidized and hydrolyzed to give long, straight-chain alcohols. Thus, for example, one can get lauryl alcohol, $n\text{-CH}_3(\text{CH}_2)_{10}\text{CH}_2\text{OH}$, which on sulfation with H_2SO_4 followed by neutralization with NaOH gives sodium lauryl sulfate, $n\text{-CH}_3(\text{CH}_2)_{10}\text{CH}_2\text{OSO}_3^-\text{Na}^+$. Lauryl sulfate is an excellent surfactant and, like LAS, is also biodegradable.

What about the phosphates? There is no degradability problem with them. The main builder in detergents, $Na_5P_3O_{10}$, simply undergoes slow hydrolysis to orthophosphate:

$$P_3O_{10}{}^{5-} + 2H_2O \longrightarrow 2HPO_4{}^{2-} + H_2PO_4{}^-$$

The problem is that phosphates are important nutrients for growth and their excessive presence in domestic waste water can nourish biologic processes beyond what are desirable rates. This phenomenon, known as *eutrophication* (from the Greek work *eutrophos*, meaning "well nourished"), can quickly degrade an aquatic environment. Progressive enrichment of water with nutrients is a natural process, observed, for example, in the aging of lakes. Streams bring in nutrients; plants and animals grow; organic deposits accumulate on the bottom; the lake gets shallower, warmer, and richer in nutrient; plants take root at the bottom; the lake becomes a marsh, a field, and eventually a forest. Normally, the process takes hundreds of years; man with his phosphate wastes risks accelerating it.

However, the eutrophication problem is not simply a phosphate problem. Liebig's law of the limiting factor in ecology states that the rate of growth in activity of some processes in an organism is controlled by some limiting environmental factor. If the limiting factor is removed, another factor becomes limiting; and so on. Algae, for example, need a variety of elements (in descending order C, H, O, N, P, S, K, Mg, Ca, Na, Fe, Mn, Cu, Zn, B, V, Cl, Mo, Co, and Si), but it is not always obvious which element supply is limiting. Reduction of phosphates, say, from 10 to 5 ppm, may not check growth at all if in a given environment 0.01 ppm is the limiting-factor level. To slow down eutrophication, it is the limiting nutrient that has to be reduced. Frequently, it is phosphorus; so, as with Lake Tahoe in California and Lake Washington in Seattle, total removal of sewage effluent produces marked reversal of eutrophication. That does not mean, unfortunately, that phosphate-removal treatment will always work equally well.

In seeking a phosphate substitute, the question often asked is the following: Why not go back to soap? One reason is that fats and oils may no longer be available in the amounts needed. Another is that soap is generally not suitable for automatic dishwashers and washing machines as presently designed.

Early in 1970 there was a flurry of excitement in the discovery of a substitute builder for phosphate in the form of NTA, nitrilotriacetate, $N(CH_2COONa)_3$. This is a good sequestering agent but was banned from detergents late in 1970 because of possible teratogenetic (monster-creating) effects when combined with Hg or Cd pollution. Apparently, complex formation between NTA and these metals increases the possibility of transmission across the placental barrier into a fetus, thereby increasing the likelihood of birth defects. Normally NTA is degraded in waste treatment systems, but under special conditions, such as the anaerobic conditions that exist in some septic tanks, the NTA may persist and get back to a water-well system.

A third approach to phosphate removal is to use, instead of LAS, surfactants which do not need phosphate builders. Incidentally, detergents now account for only 13 percent of United States phosphate consumption. Most phosphate goes into the fertilizer industry, but the form there is less soluble and less mobile than what goes down the kitchen sink.

Attempts to control eutrophication by reduction of nitrogen nutrients do not seem practical. The problem is compounded by overzealous agricultural use of NH_3 and nitrates as fertilizers, where the excesses, being highly soluble, quickly get into the stream runoffs. The problem is that many nuisance species of algae have the ability to fix atmospheric N_2 directly, hence being independent of dissolved nitrogen compounds.

26.9 Chemistry of flavor

Flavor is a complex sensation comprising taste, odor, roughness (e.g., gritty chocolate), hotness (e.g., ginger), coldness (e.g., peppermint), pungency, and blandness. Odor is by far the factor of greatest influence. Without odor we are left primarily with bitter, sweet, sour, and salt. How many different tastes are there? Linnaeus, the Swedish botanist, in 1754, came up with eleven: sweet, sour, sharp, salty, bitter, fatty, insipid, aqueous, astringent, viscous, and nauseous. Psychologists say that people generally categorize sensations (e.g., sound, pitch, and odor) into about seven groups. If so, we can take these to be the four primary tastes—sour, bitter, salty, and sweet—plus perhaps metallic, alkaline or soapy, and astringent. Many texts say there are four types of taste buds, responding to the primary tastes, but, as we shall see below, this idea is no longer tenable.

The taste receptors are located on the tongue and the soft palate. It is estimated there are 9000 such receptors, called taste buds, in man. In a child, the whole upper surface of the tongue as well as the insides of the cheeks is covered; in an adult, there are no taste buds in the middle of the tongue or on the cheeks. Figure 26.17 shows what the taste bud looks like. It is a goblet-shaped cluster of long, slender sensory cells (10 to 15) arranged like the segments of an orange in a matrix of supporting cells. Tiny hairlike projections from the sensory cells reach into the taste pore and pick up receptor signals for transmission to the brain. The sensory cells are constantly renewed, being formed by differentiation at the edge of the bud and then moving into the center, where in 4 to 5 days they die and are absorbed. When the taste nerves are cut, the taste buds disappear, but they regenerate when the nerve grows back. The taste buds are located in the papillae, pimplelike projections distributed on the surface of the tongue.

The taste sensations for sweet are greatest at the tip of the tongue; for sour, at the sides; for bitter, at the back; and for salty, relatively homogeneously around the edge. It used to be that these regional differences were ascribed to differences in the relative density of four different kinds of taste receptors, each responding to but one of the primary tastes. It now appears that sites of different specificity occur on the cell membrane of a single cell. Sensory cells that respond to two basic tastes are most numerous; those responding to one or three tastes, next most numerous; and those responding to all four, least numerous.

The sensitivity to the basic tastes varies greatly. For the *sour* taste, the threshold is at $0.0005\ M$ HCl. Sourness is proportional to the concentration of H_3O^+, but the weak organic acids taste more sour than predicted from their values of K_{diss}. The threshold for the *salty* taste is $0.01\ M$, and except for NaCl there are usually other qualities such as bitterness or sourness. Low-molecular-weight salts are predominantly salty; high-molecular-weight salts, bitter. Lead salts and beryllium salts (both of which are poisonous) taste sweet. In decreasing order of contribution to saltiness, some common cations are NH_4^+,

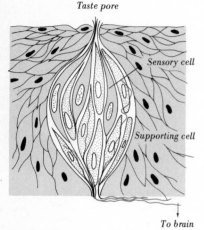

Taste pore

Sensory cell

Supporting cell

To brain

Fig. 26.17
Taste bud.

K+, Ca2+, Na+, Li+, and Mg2+. One theory says this corresponds to decreasing order of binding to surface proteins. The sensation of *sweetness* is associated mainly with organic compounds, especially the alcohols, glycols, sugars, and their derivatives. The threshold for sucrose is 0.1 M; that for saccharin (Fig. 26.18) is one seven-hundredth as great. Slight changes in molecular structure can make enormous differences in taste. As shown in Fig. 26.19, the molecule on the left, 2-amino-4-nitropropoxybenzene, is sweet (actually 4000 times as sweet as sucrose). The molecule in the middle, where the NH_2 and NO_2 groups have been interchanged, is tasteless; that on the right, where NH_2 is replaced by another NO_2, is bitter. *Bitterness*, incidentally, is often associated with sweetness. There are many chemical classes that taste bitter. In general, increasing molecular weight or increasing chain length increases the bitterness. Many of the bitter compounds are toxic (e.g., the alkaloids such as nicotine, strychnine, and morphine). Quinine, which is the classic example for a bitter taste, has a threshold of 0.000008 M.

In an attempt to unravel the mechanism of taste reception, microelectrodes have been inserted into single taste cells and the cells examined for response to various standard solutions (for example, NaCl,

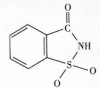

Fig. 26.18
Saccharin. (The commercial sweetener is actually sodium saccharinate, where the amino H is replaced by Na.)

Fig. 26.19
Effect of structure changes on taste.

OCH2CH2CH3	OCH2CH2CH3	OCH2CH2CH3
NH2	NO2	NO2
NO2	NH2	NO2
Sweet	Tasteless	Bitter

sucrose, quinine, and HCl). All the cells tested responded to H_3O^+ with a big signal. With sucrose, most cells showed no response, and the others but a small one. With NaCl, a fair response came from all the cells. The biggest variation, some cells responding hardly at all while others generated big signals, came with the quinine. Clearly, the receptors are not highly specific; their coding must be exceedingly complex. The fact that the nerve fibers for taste touch several taste cells strongly suggests preliminary processing of information before it gets to the brain, but how this is done is not understood. Neither is it understood how the mechanism of signal generation proceeds. One suggestion is that chemical reaction between the taste stimulant and the molecules of the sensory-cell surface leads to a small, temporary breakdown of the cell membrane near an associated nerve axon. As shown schematically in Fig. 26.20, cell membranes are thought to consist of a double layer of phospholipid molecules with protein molecules more or less dissolved in it. Breakdown in the molecular geography of the surface could allow K+ ions, which are in excess inside the cell, to seep out. There would be an instantaneous change of electric potential across the membrane, which could generate an electric pulse for transmission to the brain. The frequency and intensity of the pulses,

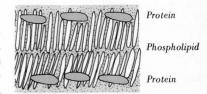

Protein

Phospholipid

Protein

Fig. 26.20
Schematic representation of cell membrane.

presumably dependent on the concentration and nature of stimulant, would then be the way information is transmitted.

The microelectrode experiments, most of which have been done by Lloyd Beidler, suggest that the sensory-cell response is related to concentration of stimulant as follows:

$$\frac{C}{R} = \frac{C}{R_s} + \frac{1}{KR_s}$$

Here C is the concentration of taste stimulant, R is the electric response, R_s is the saturation-response value (when C gets very big, R levels out), and K is an equilibrium constant for the stimulus plus receptor reaction. Plots of C/R versus C enable one to deduce values of K, and these turn out to be on the order of 10. Such K values are about like those for ion-exchange equilibria or hydrogen-bond formation. Sodium saccharinate, however, shows a K value of about 1500.

One of the most intriguing problems in understanding the chemistry of flavor is to explain the action of certain substances which act as taste inhibitors or as taste enhancers. As an example of a *taste inhibitor*, we have gymnemic acid, a glucoselike acid which occurs as the potassium salt in the leaves of an Indian plant *Gymnema sylvestre* (milkweed family). When the leaves of this plant are chewed for a minute of two, one loses the ability to sense sweet tastes (also bitter) for an hour or two. Salty or sour tastes are not affected. Another example of a taste inhibitor is the berry known as *miraculous fruit* of Nigeria. It alters sour taste so that, for example, a sour lemon tastes like a sweet orange. The action mechanism of these taste inhibitors is not understood. It has been suggested that they act by blocking receptor sites, being adsorbed without themselves creating a stimulus.

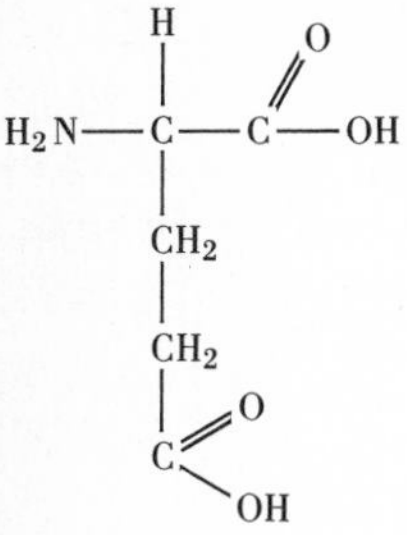

Even more mysterious is the action of flavor enhancers, also known as *taste potentiators*. The most famous of these is MSG, monosodium glutamate, the sodium salt of glutamic acid (Fig. 26.21). First identified as the flavor component of sea tangle (a kind of seaweed), MSG is now generally made by hydrolysis of vegetable proteins, casein, soybeans, and beet molasses. Large amounts of it are used in cooking, where among other things it has given rise to what has been called the "Chinese restaurant syndrome." The symptoms are burning sensation, chest pain, facial pressure, and, sometimes, characteristic headaches. The symptoms, which appear after ingestion of about 3 g of MSG, do not last very long (i.e., about 45 min) and do not affect most people except at higher threshold values. Another taste potentiator, much used in Japan but now also being introduced in the United States, is 5'-IMP, first identified as the flavor component of dried bonito (a kind of fish). 5'-IMP stands for 5'-inosinic acid, where the 5' indicates the position of a phosphate group on a ribose (sugar) part of the molecule. Figure 26.22 shows the structure. 5'-IMP is but one of a whole series of compounds known as the 5'-nucleotides (derived from the sugar phosphates). Some of these act as flavor enhancers; some

Fig. 26.21
Glutamic acid.

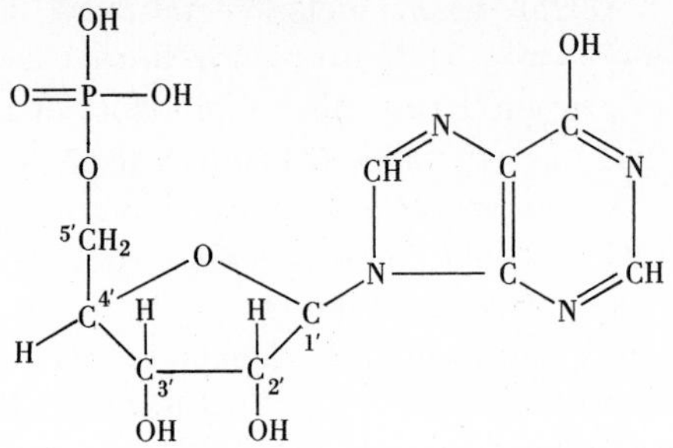

Fig. 26.22
5'-inosinic acid (5'-IMP).

do not. How the flavor enhancers work is not understood. Microelec-trode cell-response studies with MSG yield the surprising result that there is no enhancement of the electric signals. It may be that MSG does not affect taste at all but acts only by enhancing odor. Meat flavor, for example, is very largely a matter of odor, and the often-given description of MSG as imparting meat flavor to foods may be connected to this most important aspect of flavor. A wry commentary on our priorities in setting the quality of life is that until recently MSG was being added to baby foods. Since babies primarily respond to sweet (which is not enhanced by MSG), they probably got no benefit from the MSG in their food; the adults, however, must have thought it tasted better! Though questions have recently been raised about the absolute safety of MSG, it remains on the GRAS (generally recognized as safe) list. Legally, as discussed in the next section, it is not even considered a food additive.

26.10 Food additives

According to the National Academy of Sciences, "a food additive is a substance or mixture of substances, other than a basic foodstuff, that is present in food as a result of any aspect of production, processing, storage, or packaging." So defined, there are about 2500 food additives. These fall into five classes: flavors, colors, preservatives, texture agents, and miscellaneous:

Flavors, both natural and synthetic, account for roughly half of food additives. Many are on the GRAS list. In some cases, not much is known about the toxicologic effects.

Colors are often added to increase palatability. Some are natural; most are synthetic coal-tar derivatives. The World Health Organization (WHO) has tested out some 140 that seem all right. Most countries allow only a dozen or so.

Preservatives are added to prevent spoilage, prolong shelf life, and inhibit rancidity. Some of the most widely used ones, such as BHT (butylated hydroxytoluene) and BHA (butylated hydroxyanisole), act as antioxidants in retarding oxidative breakdown of fats and oils. Such oxidative breakdown frequently produces objectionable flavors and odors. The antioxidants act as preferential oxygen acceptors.

Texture agents include emulsifiers, stabilizers, thickening agents, and surfactants. They are used in bread, pastry, and ice cream, for example, to prevent phase separation. Most of the texture agents are plant extracts or vegetable derivatives (e.g., lecithin from soybeans).

In the *miscellaneous* category, we would include *bleaches* such as ClO_2 in flour milling, *buffers* such as citric acid and citrates in soft drinks, and *sequestering agents*. Typical of the sequestering agents are citric acid, sodium hexametaphosphate, and EDTA. They act by tying up trace metals such as Fe, Cu, Pb, and Zn which can catalyze oxidative breakdown of food and sometimes markedly reduce shelf life.

The GRAS list was set up in the Food Additives Amendment of 1958 to the United States Federal Food, Drug, and Cosmetic Act of 1938. The amendment states that no additive can be used in food unless the FDA (Food and Drug Administration), after a careful review of test data, agrees that the compound is safe at its intended level of use. Exception is made for all additives that because of widespread use are generally regarded as safe by experts in the field. Included in the GRAS list are substances such as pepper, cinnamon, citric acid, MSG, and about 575 others. They are not even considered food additives. For most of the items on the GRAS list, the FDA does not establish tolerance limits; for non-GRAS items it does. Limits may vary with intended use. For example, the sequestrant CaNaEDTA can be up to 25 ppm in beer (where it chelates trace Fe and prevents sudden release of CO_2 when the container is opened), 100 ppm in pecan pie, and 275 ppm in cooked, canned crab meat. No judgment is made by the FDA whether the additive is desirable or necessary. Also it can change its mind. In 1963, FDA approved the use of cobaltous salts in beer to stabilize foam; in 1966, after some unexplained deaths by heavy beer drinkers, the approval was canceled.

Most controversial in the 1958 law is the so-called "cancer clause," which states that no chemical can be added to foods if in any amount it produces cancer when ingested by man or animals. This clause was tacked on in the final hours of debate as an amendment and is often referred to as the *Delaney clause*. It is hotly controversial because the proponents argue that we know so little about cancer that we cannot risk adding carcinogens to the food supply in any amount whereas the opponents say there must be a no-effect level, otherwise we would all get cancer. As discussed further in the next section, permissible limits for carcinogenic compounds are difficult to establish because of the sometimes very long delay (for example, 30 yr for aniline-derivative dyes) between exposure and disease.

Besides food additives of an intentional sort, such as discussed above, there may be in foods contaminants that arise from nonintentional sources. One of these, mercury, has already been discussed in Sec. 20.9. Another, cadmium, was considered in Sec. 20.7. Of a quite different sort are the organochlorine compounds that occur as persistent residues in the environment from use as pesticides. Some of the aspects of the problem are considered in Sec. 26.12.

26.11 Carcinogens

In 1775, a London physician, Sir Percival Pott, noted the high incidence of scrotal cancer in chimney sweeps. Correctly, he attributed it to soot. It is now known that soot, smoke, and other products of incomplete combustion are rich in carcinogenic (cancer-causing) compounds. The compounds in this case are polycyclic aromatic compounds, which means they consist of several benzene rings fused together (i.e., sharing atoms in common). The most potent are the five- or six-ring compounds, two examples of which are shown in Fig. 26.23. Benzopyrene is the type example. It constitutes up to 1.5 percent of coal tar, a by-product in the destructive distillation of coal. The molecule is very susceptible to substitution reactions at the 5 position, which may be related to its biologic activity. The carcinogenic character is lost when OH groups are substituted for H at the 10 or 8 position. This kind of detoxification occurs in metabolism. Benzopyrene occurs generally as an air pollutant, being about 10 times as prevalent in city air as in rural air. Cigarette smokers are also exposed to elevated levels of benzopyrene. There is some correlation between benzopyrene exposure and the incidence of lung cancer, but the relation cannot be a simple one since transport workers, who generally have an extra-high exposure, do not have correspondingly high cancer rates.

Tantalizing regularities have been pointed out in cancer rates, and attempts have been made to correlate these with nutrition, environment, genetics, chemical exposure, etc., but with only limited success. The United States used to have a high stomach cancer rate; now the rate is relatively low; on the other hand, Japan, Iceland, and Chile have high rates. In Poland, the rate of stomach cancer is very high, and that of the colon is very low; yet when Polish emigrants come to the United States, their stomach cancer rate falls; the colon rate increases. In China, the Chinese are very susceptible to cancer of the nasopharynx; after two or three generations in the United States, the rate goes way down.

Figure 26.24 shows what a great variation there is in the annual cancer incidence in the various sites. It has been suggested that despite some probable influence of genetic factors, the main reason for the different cancer rates lies in differences in the air, water, and food environment. Certainly the chemical carcinogen content of these must be fairly decisive. However, it is not all that easy to recognize what is carcinogenic. Animal testing is usually considered essential, especially on rats and mice. Strong carcinogens are relatively easy to detect because a large number of tumors appear in the tested population; weak carcinogens, however, generally produce few tumors, and these can easily be confused with spontaneous ones, which appear in any population. Another problem in the testing is that there may be marked species differences. 2-Naphthylamine, for example, a dye intermediate, produces bladder tumors in man and dog, liver tumors in the mouse, and breast cancer in the rat, but it is inactive in the rabbit. Testing on but one species could be misleading. Finally, we have to recognize that

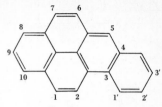

3,4–Benzopyrene
Also known as Benzo[a]pyrene

1,2:5,6–Dibenzanthracene
Also known as Dibenz[a,h]anthracene

Fig. 26.23
Examples of polycyclic aromatic carcinogens.

Fig. 26.24

Annual incidence of cancer
per 100,000 population
(ages 35 to 64).

	Stomach	Esophagus	Lungs	Breast
United States	16	6	155	95
England	32	4	130	100
Japan	160	20	22	30

induction of cancer is a slow process. Even in rodents it may take several months; in man it may take from 5 to 40 yr.

What types of compounds are likely to be carcinogenic? The list grows as we know more about structural relations, but the following is representative:

1. Polycyclic aromatic hydrocarbons
2. Aromatic amines
3. Azo dyes
4. Biological alkylating agents
5. N-Nitroso compounds

We shall look at examples of each of these as well as at two other specific carcinogens: aflatoxin and diethylstilbestrol (DES).

1 We have already seen two examples of *polycyclic aromatics* in Fig. 26.23. The molecules are generally planar or near planar. They usually have four to six aromatic rings. Their potency is sensitive to small changes in structure (e.g., methyl groups in some positions amplify the action, in others they do not). The activity is generally destroyed by OH substitution, strongly enhanced by methoxy (CH_3O) substitution. They are soluble in benzene, insoluble in water, but solubilized by caffeine and DNA. Their mechanism of carcinogenic action is not understood.

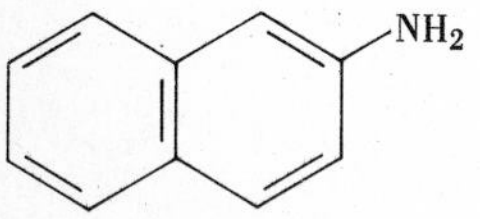

2–Naphthylamine

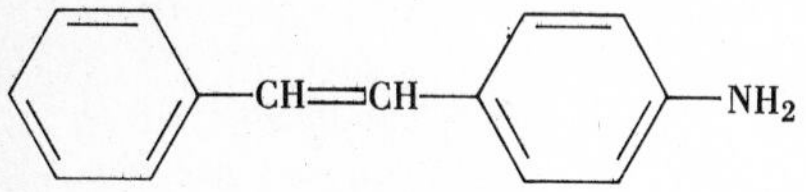

4–Stilbenamine

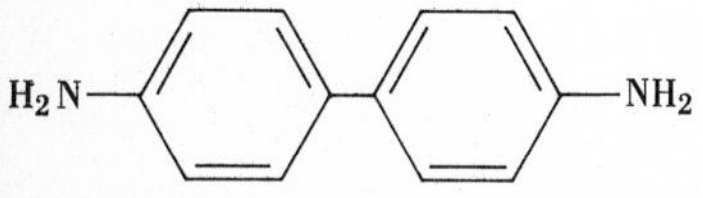

Benzidine

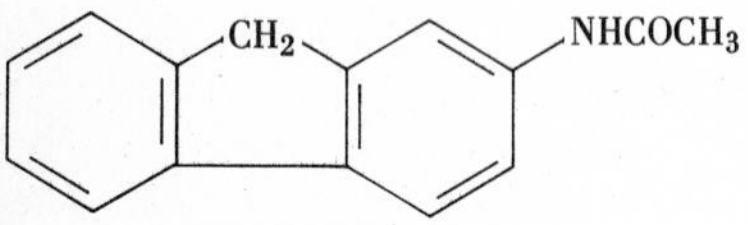

N–2–Fluorenylacetamide

Fig. 26.25
Some carcinogenic aromatic amines.

2 The *aromatic amines,* some examples of which are shown in Fig. 26.25, are potent carcinogens for the human urinary bladder. They used to be an occupational hazard in the dye industry (in some countries, they still are) and in the rubber and electrical industries, where they find use as antioxidants. 2-Naphthylamine was used extensively in the manufacture of dyes, but its use is now prohibited in some states of the United States. It is one of the most dangerous carcinogens known, producing bladder cancer at a 50 percent rate, sometimes after being latent for 30 yr. In one studied group of 15 exposed persons, 100 percent of them became afflicted. Benzidine, used also in the manufacture of dyes, is a useful reagent to test for H_2O_2 in milk and for detection of blood.

3 *Azo dyes* are compounds containing a doubly bonded pair of nitrogen atoms attached to aromatic rings. They represent the most numerous class of dyes. Most of these are innocuous, but some, such as the one shown in Fig. 26.26 (butter yellow), are carcinogenic. Butter yellow was used as a food colorant but has now been replaced by the diethyl analog, which is harmless.

4 *Biological alkylating agents* are organic compounds that readily introduce alkyl (hydrocarbon) groups at reactive sites in proteins and nucleic acids. It is a paradox that they have often proved to be tumor growth inhibitors but often they themselves are carcinogenic. (This points up a caution that should be made about chemical carcinogens. Sometimes it is a metabolite, i.e., a product of metabolism, rather than the compound being investigated that is the real carcinogen. Experimental indication of this comes, for example, when a tumor appears not at the site of ingestion but some place else in the test animal, say the liver.) A number of such anticancer drugs have been derived from the chemical-warfare agents mustard gas, $S(CH_2CH_2Cl)_2$, and the related nitrogen mustards, for example, $CH_3N(CH_2CH_2Cl)_2$.

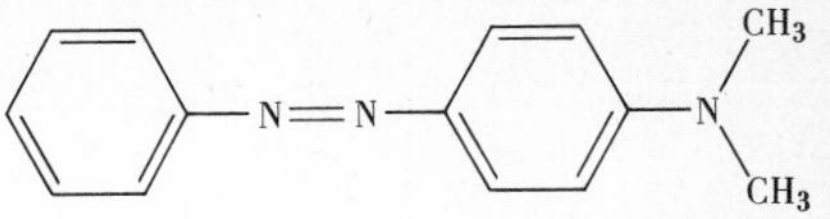

5 The *N-nitroso compounds* are organic compounds in which a nitroso group (—N=O) is bonded to another nitrogen. If the other nitrogen has two organic residues, R and R′, attached to it as in

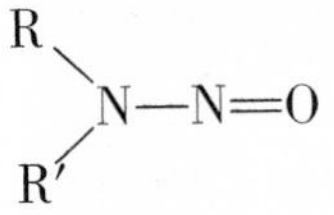

$$\begin{array}{c} R \\ \diagdown \\ \diagup \\ R' \end{array} N-N=O$$

then the resulting compound is called a *nitrosamine*. About a hundred of these have been tested for carcinogenicity; some 75 percent have produced lesions. Unlike the polycyclic aromatics, which produce their tumors at the site of application, the nitrosamines rarely produce cancer at the site of injection. It is therefore probable that they are only precursors to the real carcinogen. Carcinogenic activity is strong when the R and R′ groups are *n*-alkyl, isopropyl, vinyl, or cyclohexyl; the compounds are inactive when R and R′ are $(CH_3)_3C—$ or $C_6H_5—$. On a molar basis, the nitrosamines are more potent than the aromatic amines or the azo dyes.

Nitrosamines are in the environment in unexpected places. They are used as gasoline and lubricant additives, where they function as antioxidants. They are also fungicides and insecticides. More startling, however, is the fact that they have been found in processed meat (e.g., dried beef and cured pork). Although not present in raw bacon, they have been found at 0.03- to 0.11-ppm levels after cooking. It is surmised that reaction between $NaNO_2$ and amines in the food produces the nitrosamines. ($NaNO_2$ is often used in preserving certain foods, e.g., sausage, luncheon meat, and canned fish, partly to improve color and partly to inhibit the growth of *Clostridium botulinum*, with its deadly toxin.) Questions are now being raised as to the advisability of continuing to allow $NaNO_2$ to be used as a food additive. Because nitrate is a possible precursor of nitrite, its use is also being questioned even though there is no evidence NO_3^- is harmful *per se*. Man's primary intake of NO_3^- comes from water or from vegetables (spinach, beets, radishes, eggplant, celery, and lettuce)—some of which go up to 3000 ppm. Why are the levels so high? Some species accumulate nitrate; sometimes the growth environment is overrich because of too much fertilization; sometimes nutrient deficiencies (e.g., molybdenum) trigger nitrate overaccumulation. In normal cases, nitrate is rapidly excreted

Fig. 26.26
Carcinogenic azo dye, butter or methyl yellow.

Fig. 26.27
Aflatoxin B₁.

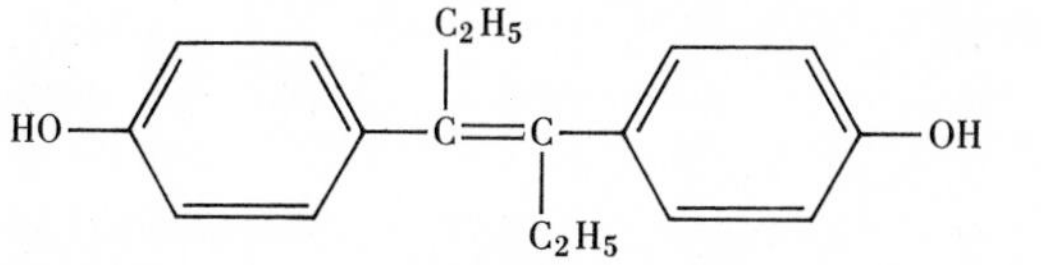

Fig. 26.28
Diethylstilbestrol (DES).

in the urine. The hazard comes from possible reduction $NO_3^- \longrightarrow NO_2^-$ in the microbial environment of the intestine or in infants less than 4 months of age even in the stomach since their gastric acidity is not yet fully developed.

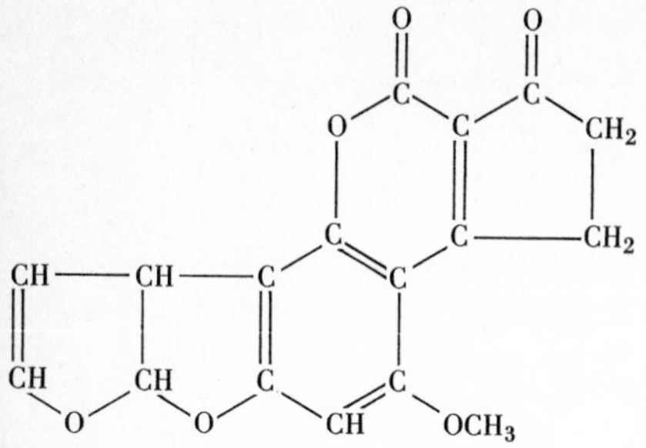

Besides the above, two other carcinogen hazards need to be mentioned. One of these, aflatoxin, is a natural hazard, stemming from mold-contaminated food; the other, DES, is a synthetic hormone, added as a growth accelerator in meat production. Aflatoxin, one form of which is shown in Fig. 26.27, is a pentacyclic organic compound produced by a common fungus in grain, *Aspergillus flavus*. It was identified as the causative agent of a strange disease known as *turkey X*, which appeared around 1960 in such diverse places as turkeys in England, ducks in Kenya, pheasant in Uganda, and trout in the United States. It is most commonly found in grain, coconuts, cottonseed, and peanuts after improper harvest or storage. Toxic exposure, as in contaminated feeds, may lead to liver damage; chronic exposure, to malignant liver tumors. Subcutaneous injection of 2×10^{-6} g in rats produced tumors in 5 out of 16 cases. One of the strongest carcinogens known, aflatoxin is believed to act by binding to DNA (deoxyribonucleic acid) and interfering with transcription of information for protein synthesis.

Aflatoxins are a relatively rare problem in the United States but a major one in Africa and the Far East.

Diethylstilbestrol (DES) is a female sex hormone with the structure shown in Fig. 26.28. It used to be fed to chickens as a chemical way to make capons, and there is a classic story of a New York male restaurant worker who developed female-sized breasts from eating an excessive number of chicken necks. Until banned in 1973, DES was also being fed in the United States to roughly 75 percent of the 30 million cattle marketed annually. DES made the animals fatten faster on less grain; so it saved about $90 million/yr in feed costs. This was in spite of the fact that DES had been known since 1940 to cause cancer in animals and had been banned in 22 countries. The DES controversy boiled up in 1970 when a teenage girl was admitted to Massachusetts General Hospital with the rarest of all cancers, cancer of the vagina (which is virtually unknown in women under 30). Six such cases rapidly turned up and were traced back to the fact that the mothers had been given relatively large doses of DES during pregnancy to prevent miscarriage. DES was known to have a long latency period, but here we have the effect of a specific carcinogen being transferred *to the next generation*. In January 1973, DES was banned in the United States as a feed supplement, although permitted as an implant in animals being fattened for market. The latter was believed to be safe because release was slow and in such tiny amounts that no residue was detected in the marketed product. It is now totally banned. It is interesting to note that in 1962 a hole was punched in the Delaney amendment expressly for DES to allow carcinogenic drugs to be fed to animals provided none of the residue is detected in food derived from such animals. Also it should be emphasized that there is no evidence that *trace* amounts of DES have harmful effects. The controversy, which spills over into the whole carcinogen problem, is between those who say "microscopic amounts will not hurt anybody" and those who say "we must do everything we can to reduce the carcinogenic burden in the environment."

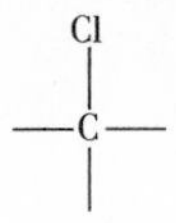

26.12 Insecticides

There are three main classes of organic insecticides: organochlorine, organophosphorus, and carbamate. The important functional groups of these and representative examples are sketched in Fig. 26.29. The phosphate shown, DDVP, is of interest as the active ingredient in Vapona or No-Pest Strip insecticide.

DDT is the classic example of the chlorinated hydrocarbon insecticides that have generated so much controversy because of their persistence in the environment. DDT was first synthesized in 1874 as part of a doctoral thesis by Othmar Zeidler of Germany. It was rediscovered in 1939 by the Swiss entomologist Paul Mueller, who received a Nobel Prize for uncovering its powerful insecticide properties. The Allies in World War II put it to good use as a delousing agent to replace the scarce pyrethrum. After the war, DDT showed spectacular success in controlling diseases such as typhus, malaria, and yellow fever, which are transmitted by insect carriers. Two problems developed: (1) Certain insects developed immunity. Whereas in 1948 there were 12 immune species, by 1967 this number had risen to 165. (2) Because of its long persistence time (half-life in the environment is 2 to 4 yr compared with 1 to 10 weeks for organophosphorus and 1 week for carbamate), DDT has ample opportunity to move around in the environment. It accumulates upward in the food chain, where it appears to have special deleterious effects on fish and birds of prey. Disquietingly large residues of DDT are now appearing throughout the environment.

There is no evidence that any human has ever been harmed by DDT through normal use. DDT concentrations in man average 11.0 ppm in the United States and range from 2.2 ppm in England to 31.0 ppm in India. The lethal dose (expressed as values of LD_{50}, the dose that is lethal to 50 percent of a given tested population) varies for oral ingestion from 10 ppm (milligrams per kilogram of body weight) for the cockroach to 400 ppm for the rat. Unfortunately, the mechanism of DDT action is not known. It is thought to dissolve in the fatty membrane surrounding nerve fibers and interfere with ion transport in and out of the nerve membrane. The mode of action appears to be quite different from that of the organophosphorus and carbamate insecticides, which, like most other drugs, interfere with signal trans-

Organochlorine   *Organophosphorus*

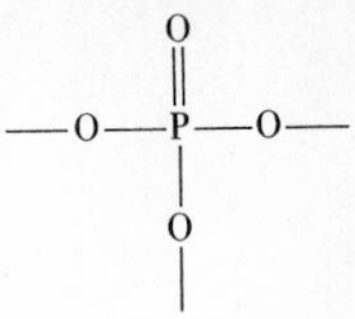

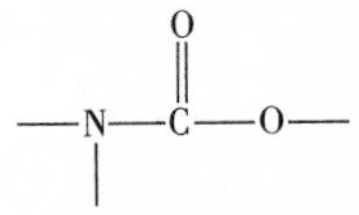

Carbamate

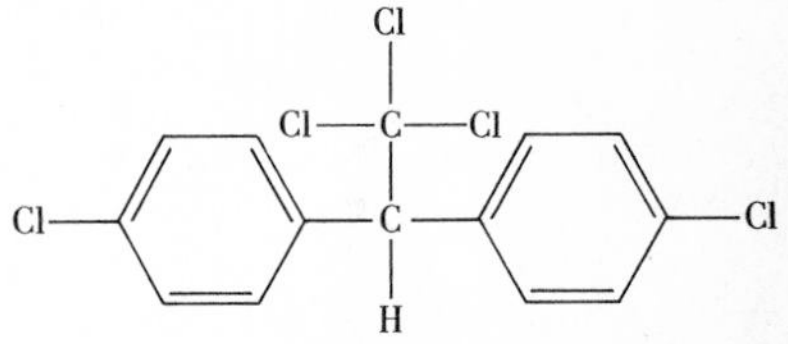

Dichlorodiphenyltrichloroethane (DDT)

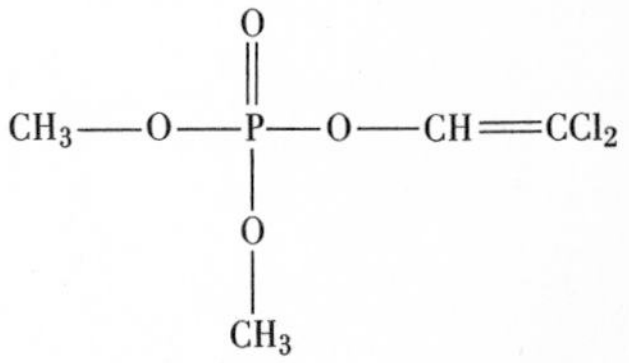

Dimethyl dichlorovinyl phosphate (DDVP)

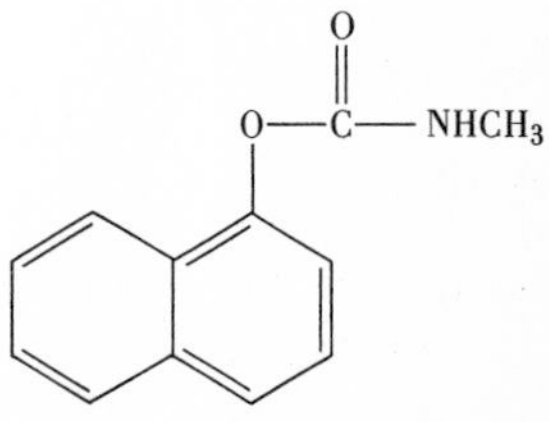

1-Naphthyl N-methylcarbamate (Sevin)

Fig. 26.29
Some organic insecticides.

mission at the synaptic gap (discussed further in Sec. 27.2). One of the worries about DDT for man is that since DDT stores in the fatty tissue, there is risk of its being massively released into body fluids when the fat cells are metabolized, as under starvation conditions.

The tendency of DDT to concentrate up the food chain is rather impressive. Oysters living in water with 0.001 ppm DDT, for example, show 700 ppm in their bodies. Similarly, the DDT content of seawater at 0.000001 ppm typically rises to 0.0003 ppm in the planktons to 0.5 ppm in marine fish and eventually to 10 ppm in birds of prey. For birds, DDT has been implicated as the cause of thin eggshells, perhaps because of inhibition of the enzymes that control Ca^{2+} metabolism.

Exercises

(*Note:* There are no free answers for the problems of "Man and His Chemical Environment.")

*26.1   *Air pollution.* Starting with the figure of 3 kg per person per day for man-made air pollution in the United States, estimate how many weight parts per million this would correspond to if accumulated over one year. Assume uniform dilution of the pollutants below 10 km of altitude, which would correspond to 85 percent of atmospheric mass. United States population is about 200×10^6. Why does your result represent an impractical figure for evaluating air pollution?

*26.2   *Air pollution.* What are the main air pollutants, where do they come from, and how can they be controlled? Which of the pollutants would you consider to be the biggest problem? What solution would you recommend?

*26.3   *Smog.* What is smog? How does London smog differ from Los Angeles smog? Describe the conditions that lead to formation of each, and suggest ways to prevent each.

*26.4   *Water pollution.* Discuss the pros and cons of considering phosphate as a water pollutant. How would you go about treating water so as to remove phosphate?

*26.5   *Detergents.* What are the three usual components of a commercial detergent? What is the function of each? Why was it necessary to change the formulation from alkylbenzenesulfonates to linear alkylsulfonates? Why is it likely formulations will have to be changed again?

*26.6   *Eutrophication.* What is meant by eutrophication? Why has it become a recent problem? What recommendations would you make to help alleviate the problem in general?

*26.7   *Flavor.* Examine the labels on various prepared foods in a supermarket, and make a list of those containing added MSG. Why should the addition of MSG to baby food be discouraged?

*26.8   *Food additives.* Identify each of the following: FDA, GRAS, Delaney clause, EDTA, and BHA.

*26.9   *Carcinogens.* What classes of compounds are likely to be carcinogenic? Give a structural formula for an example of each.

*26.10   *DDT.* Sales of DDT in the United States have been banned, but developing countries believe this will hurt them. What is the argument here? What recommendation would you make?

**26.11   *Clean air.* Given the data of Fig. 26.1 for the composition of clean, dry air in parts per million by volume, calculate the weight percent composition for the four most abundant components.

**26.12   *Internal-combustion engine.* What would be the ideal air-fuel ratio if air were replaced by pure oxygen? Estimate how far a typical United States car could go on a standard cylinder of high-pressure oxygen gas. Assume average fuel consumption is 15 mi/gal. One gallon = 3.8 liters; density of octane is 0.70 g/ml. A large standard cylinder of oxygen contains about 50 liters of oxygen at 150 atm.

**26.13   *Octane dilemma.* Why is tetraethyllead added to gasoline? Discuss the probable effect on air pollution of the current move to low-lead gasolines.

**26.14   *Odors.* What properties of sucrose might help explain its lack of odor? How might you change the molecule so as to give it an odor?

**26.15   *Odors.* Suggest a molecular mechanism by which an odor killer such as Air-Wick deodorant might act. How might you test out your hypothesis?

**26.16   *Water pollution.* Comment on this oft-repeated saying: Water in a stream purifies itself in 7 mi. For what sort of pollution might this well be true? For what kind of pollution is it nonsense?

**26.17   *Water treatment.* What recommendations for water treatment would you probably make to get drinkable water from the following sources: (*a*) deep well contaminated with fluoride, (*b*) stream below a sausage factory, (*c*) stream below the opening of an abandoned mine shaft, (*d*) New York Harbor, and (*e*) spring in Death Valley?

**26.18   *Detergents.* What do molecular sieves have to do with detergents? How might molecular sieves help resolve the phosphate problem?

**26.19   *Food additives.* Discuss critically the following statement: If it were not for food additives, the world's food supply would have to be increased by 10 percent.

**26.20   *Carcinogens.* Check the prepared-meat labels in a supermarket, and estimate what fraction show addition of $NaNO_2$ or $NaNO_3$. What is the argument for discouraging this?

***26.21 *Wankel engine.* The Wankel engine is a compression engine in which the back-and-forth motion of a piston has been replaced by a rotating motion of a triangular rotor in a specially designed chamber. Because the surface-volume ratio is greater than normal, there is faster quenching of combustion products. The result is higher emission of unburned hydrocarbons and lower combustion temperatures. Show how this might be used to advantage to solve some air pollution problems. (*Hint:* The Wankel engine is smaller than a conventional engine.)

***26.22 *Internal-combustion engine.* Suppose you design a four-cylinder automobile in which the exhaust gases from three of the cylinders are mixed with air and fed into the fourth cylinder. How might this be used to help solve the NO_x pollution problem?

***26.23 *Photochemical smog.* What are the main components of photochemical smog? Why is sunlight necessary for its formation? Explain why raising the air-fuel ratio in automobile engines might be expected simultaneously to help alleviate the problem but also make it worse.

***26.24 *Chemistry of flavor.* How might you explain the observation that weak organic acids taste more sour than corresponding HCl solutions of the same pH?

In the preceding chapter we considered the influence on the quality of life of chemical factors in the air, water, and food supply. In this chapter we look at two other perturbing influences on life: drugs and radiation. Both of these hold the potential for major disaster as well as for improved well-being of all mankind.

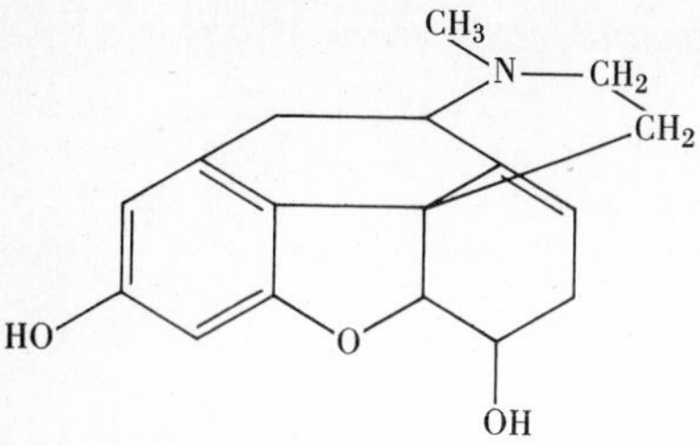

Morphine

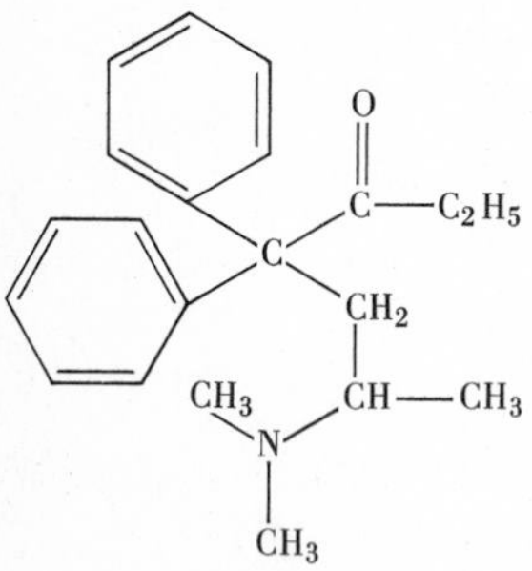

Methadone

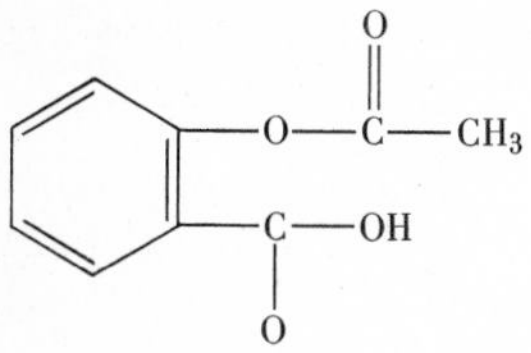

Aspirin

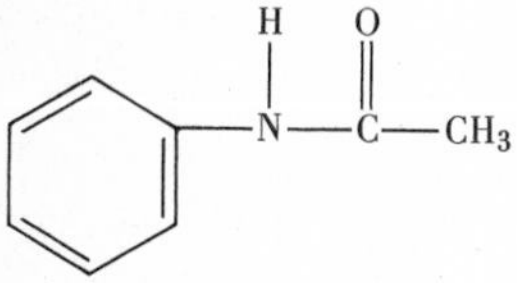

Acetanilid

Fig. 27.1
Examples of some analgesics (pain relievers).

27.1 Drugs

The term "drug" is applied to any chemical substance that modifies the function of living tissue so as to produce physiological or behavioral change. Most famous of the drugs are the *narcotics*, which act to diminish the awareness of sensory impulses, particularly pain, by the brain. When the drugs are used primarily for the relief of pain, then they are generally called *analgesics*. The prime example of a narcotic analgesic is morphine, long used for the relief of severe pain, as in terminal cancer. As discussed further in Sec. 27.3, morphine is a derivative of opium, long famous for its induction of a state of euphoria. Its chemical makeup and mode of action will be considered later. Other important classes of drugs are the barbiturates (sedatives and hypnotics), the stimulants (i.e., "speed"), the tranquilizers, and the hallucinogens (e.g., marihuana and LSD).

In describing the action of drugs, pharmacologists find it useful to distinguish between *pain* and *suffering*. Pain is a specific sensation from a sensory input, as from an injury or a physiological malfunction. Suffering, on the other hand, which has been identified as mental anguish, derives from a physiological reaction to sensory pain. Morphine relieves pain by depressing the pain reception centers of the brain; it also relieves suffering through disruption of the association pathways of the cortex. Typical of the effect of morphine is the state of euphoria, where anxiety and tension are replaced by calmness and equanimity. This state of euphoria may be a valuable component in its pain-killing effect. Unfortunately, most of the narcotic analgesics are complicated by a number of undesirable side effects. These include depressed respiratory activity, nausea and vomiting, inhibited defecation and urination, increased perspiration, etc. As we shall see later, all these effects form a related package involving changes in the central nervous system.

Before looking at the way drugs may affect the nervous system, it is well to have in mind the chemical makeup that characterizes the major drug classes. Fig. 27.1 shows some representative analgesics (pain relievers).

The *barbiturates* are depressant drugs which decrease the activity level and sensitivity to the environment. With increasing dose, the effect generally goes from sedation to hypnosis to anesthesia. The important type compound is shown in Fig. 27.2, where when $R_1 = C_2H_5$ and $R_2 = C_6H_5$, we have the well-known phenobarbital.

The *stimulant* drugs, which include the amphetamines (e.g., Benzedrine), cocaine, and caffeine, are stimulants of the central nervous system. They lead to elevated blood pressure, increased heart action, and enhanced alertness and activity. Figure 27.3 shows the structure of amphetamine itself and caffeine. Caffeine is found in coffee, tea, cocoa, and cola. The amount in coffee (90 to 125 mg per cup) (150 cm^3) is generally recognized as being quite large, as is the amount in tea (30 to 70 mg per 150 cm^3). Few people, however, realize there is 22 mg of caffeine in a 30-g piece of chocolate. Cola drinks, where caffeine is an essential ingredient by law, contain 30 to 45 mg per bottle (400 cm³). Taken orally, caffeine is rapidly absorbed in the stomach. Concentration in the blood reaches a peak in about 30 to 60 min and remains appreciable for 6 to 12 h.

The *tranquilizers*, two examples of which are shown in Fig. 27.4, are antianxiety and antidepression drugs. They are generally available only by physician's precription and are used in the treatment of mental disorders. Phenothiazines are used to treat schizophrenia; meprobamate is a more mild tranquilizer, used to treat neuroses or emotional problems.

The *hallucinogens* are also known as psychotomimetic or psychedelic drugs. Their most famous example is LSD (lysergic acid diethylamide), which is discussed in greater detail in Sec. 27.5. Marihuana, a much less potent drug than LSD, is also hallucinogenic. It is described in Sec. 27.4. All the hallucinogens have the capacity to induce disturbances of mood, thinking, and perception, sometimes even hallucinations.

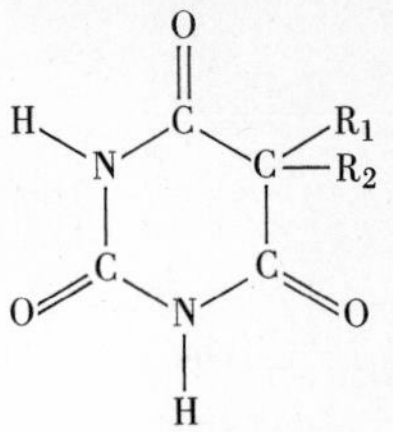

27.2   Drug action

A surprisingly large amount of understanding has developed with regard to the mechanism of drug action. To appreciate the discussion, however, some knowledge of neurobiology is necessary. In the following, we have, first, a highly simplified review of some general aspects of the nervous system and then a consideration of how drugs might affect the normal working of signal transmission. Primary emphasis is on the chemical side of the problem. Texts on biochemistry, physiology, and pharmacology should be consulted for amplification of specific details.

The nervous system is usually divided into two parts—the central

Fig. 27.2
Barbiturate.

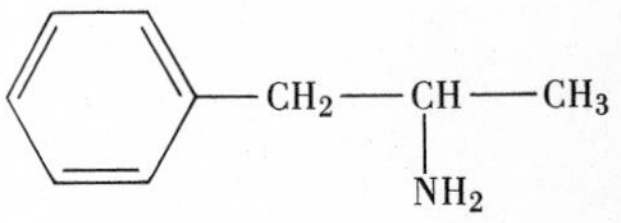

Amphetamine

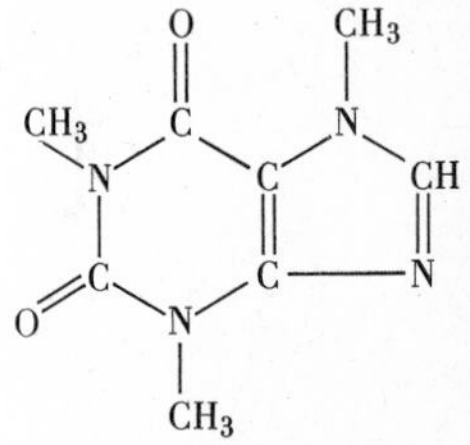

Caffeine

Fig. 27.3
Examples of stimulant drugs.

Fig. 27.4
Tranquilizers.

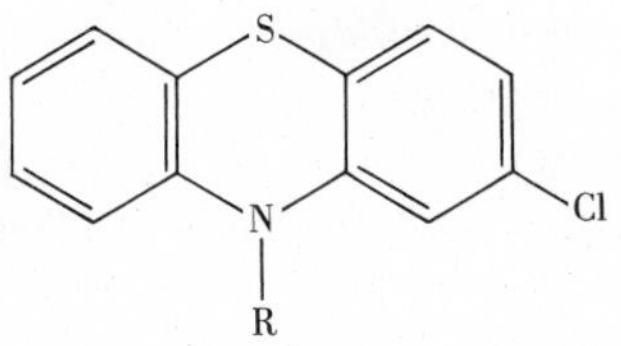

Phenothiazines

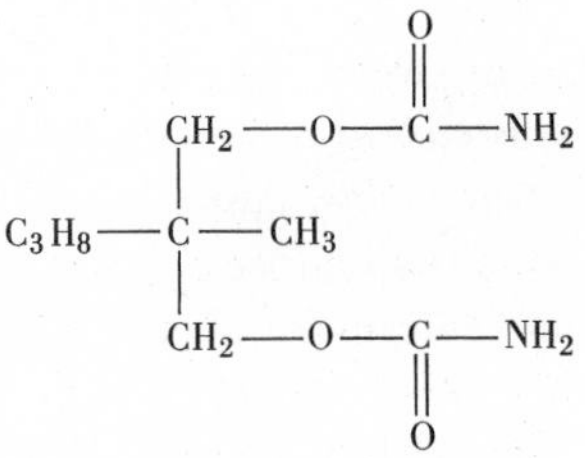

Meprobamate
(Miltown, Equanil)

nervous system (CNS) and the peripheral system. The central nervous system consists of the brain and the spinal column; the peripheral nervous system includes everything else, specifically the afferent nerves (which bring information to the CNS) and the efferent nerves (which take instructions to the muscles or glands). The best-understood part of the nervous system is the *neuron*, or nerve cell, which is shown schematically in Fig. 27.5. It consists of a cell body with nucleus to which are joined several short, branched processes, the dendrites, which carry impulses to the cell body, and a single, usually longer process,

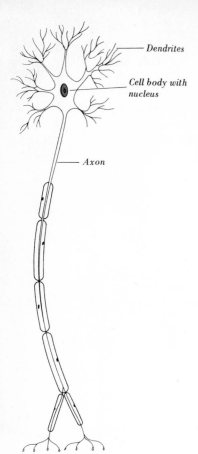

Fig. 27.5
Schematic representation of a neuron.

Dendrites

Cell body with nucleus

Axon

the axon, which carries pulses away from the cell body. There are two modes of nerve-pulse transmission—*axonic*, i.e., along the axon, where it is primarily electric, and *synaptic*, i.e., at the synapse, where it is primarily chemical. The *synapse*, also called the *synaptic gap*, is the junction between two neurons or between a neuron and muscle tissue. It contains most of the drug chemistry we are interested in.

In its normal resting state, the axon has a higher concentration of Na^+ outside itself than inside, whereas the K^+ concentration is higher inside. A sort of ion-exchange diffusion mechanism known as the *sodium pump* keeps these concentration gradients intact. The result is there is a certain steady potential difference between the interior of the axon and its surroundings, the so-called "membrane potential." When a nerve pulse passes along an axon, a transient electric depolarization wave moves along it, associated, it is believed, with a brief but reversible leakiness of the Na^+–K^+ barrier. Some Na^+ ions diffuse into the axon, resulting in a temporary elevation of the positive charge; K^+ ions move out, and the positive charge decreases. The spike of transient charge disturbance produces a current which stimulates neighboring regions of the axon membrane; so the impulse moves along to the end of the axon. The speed of the pulse may be as high as 100 m/sec.

When the axonic pulse reaches the synapse at the end of the axon, the electric disturbance is translated into release of a chemical transmitter agent into the synaptic gap. This chemical transmitter then diffuses across the gap from the end bulb of the axon where it was released on the so-called "presynaptic side" to some kind of receptor (either the dendrite of another neuron or muscle tissue) on the post-synaptic side. The geometry of a typical situation is shown schematically in Fig. 27.6. Storage vesicles in the end bulb of the axon contain the neurotransmitter until it is to be released into the gap.

Figure 27.7 shows two of the most likely neurotransmitters—*acetylcholine,* a sort of organic derivative of ammonium ion, and *norepinephrine,* also known as noradrenaline. Also shown is *serotonin,* which is believed to be a neurotransmitter in the brain. (A remarkable and interesting fact is that the structure of serotonin has major features in common with the structures of LSD, psilocybin, and other hallucinogenic drugs. As we shall see below, the ability of some molecules to have interesting physiological properties is often related to their ability to mimic vital compounds.) Any given synapse has its characteristic transmitter. If the transmitter is acetylcholine, the synapse is called a *cholinergic* gap; if norepinephrine, *adrenergic.* Practically all the terminal synapses of the sympathetic nerve system are adrenergic; those of the parasympathetic system, cholinergic. (Sympathetic and parasympathetic operate against each other. As an example, the pupil of the eye is dilated by stimulus of a sympathetic nerve and constricted by a parasympathetic one.) Intermediate synapses are cholinergic.

Once the chemical neurotransmitter has traveled across the synaptic gap (a typical distance traversed would be 50 nm) and interacted with the receptor so as to generate a pulse in another neuron or activate

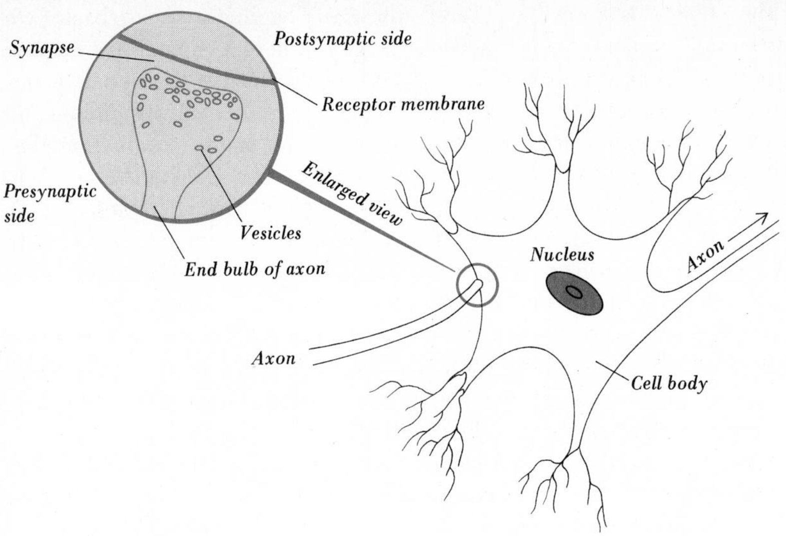

Fig. 27.6
Synapse, showing relative placing of storage vesicles from which chemical neurotransmitters are released for diffusion across the gap to the receptor membrane.

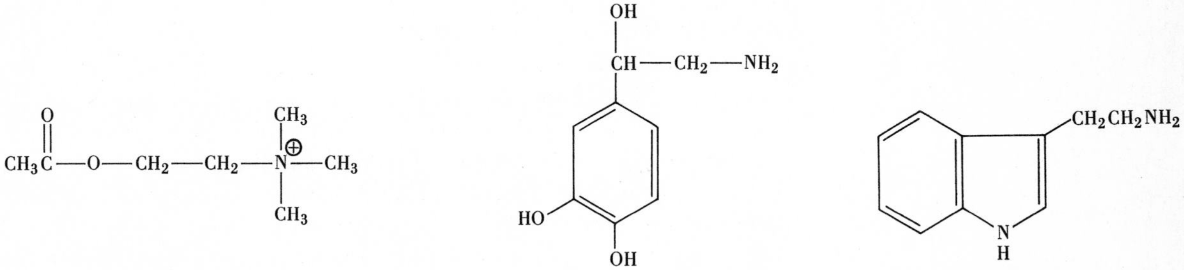

Acetylcholine *Norepinephrine* *Serotonin*

Fig. 27.7
Structures of some important neurotransmitters.

a muscle or gland, the neurotransmitter molecule has to be destroyed. Otherwise it would stay in the synaptic gap and continue to generate new pulses at the receptor surface. To eliminate the transmitter molecule and restore the synapse to its resting condition, chemical inactivation occurs. If the transmitter is acetylcholine, it reacts with an enzyme *cholinesterase* which helps hydrolyze it into acetic acid and choline.

$$\underset{\textit{Acetylcholine}}{CH_3\overset{O}{\overset{\|}{C}}-O-CH_2CH_2-\overset{CH_3}{\overset{|}{\underset{\underset{CH_3}{|}}{N^{\oplus}}}}-CH_3} \xrightarrow{\textit{cholinesterase}}$$

$$\underset{\textit{Acetic acid}}{CH_3\overset{O}{\overset{\|}{C}}-OH} + \underset{\textit{Choline}}{HOCH_2CH_2-\overset{CH_3}{\overset{|}{\underset{\underset{CH_3}{|}}{N^{\oplus}}}}-CH_3}$$

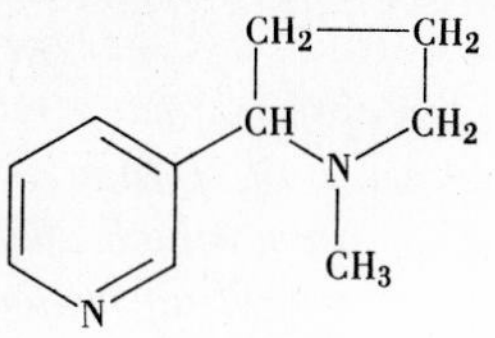

CH_2—CH_2
CH N CH_2
   CH_3

Nicotine

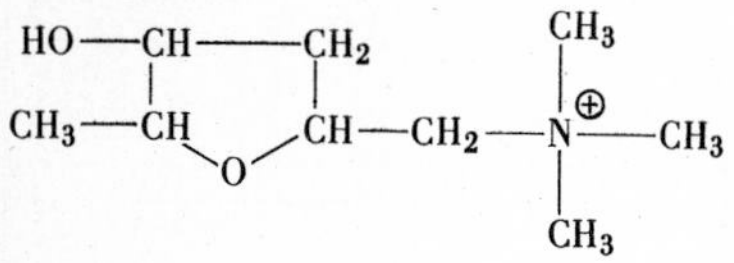

HO—CH——CH_2 CH_3
CH_3—CH CH—CH_2—$N^{\oplus}$—CH_3
      O CH_3

Muscarine

Fig. 27.8

Nicotine and muscarine.

The choline is then regenerated into acetylcholine and put back into a storage vesicle. In the adrenergic gap, where the neurotransmitter molecule is norepinephrine, the mechanism by which the signal carrier is eliminated is less clearly established. There is a corresponding enzyme, monoamine oxidase (MAO), that helps in the elimination, but its precise location is not known. Incidentally, LSD is believed to inhibit the action of MAO, which may explain part of its activity.

How do drugs affect signal transmission at the synaptic gap? There is no unique answer. It depends on the drug and its concentration, the nature of the particular synapse, and what other drugs or nondrugs may be present. As illustration of the synapse problem, not all cholinergic junctions respond the same way to drugs. One group is stimulated by nicotine, producing *nicotinic* symptoms (e.g., effects on voluntary muscles, paralysis, and twitching); another group is stimulated by muscarine, producing *muscarinic* symptoms (e.g., slowing of heart, constriction of pupils, urination, and salivation). Nicotine, as is well known, comes from tobacco; muscarine, from the red variety of *Amanita muscaria*, a mushroom. Their respective structures are shown in Fig. 27.8. The difference in the two kinds of action is believed to stem from the existence of two kinds of receptor sites. The neurotransmitter molecule, such as acetylcholine, may have a fairly floppy structure; so it might be able to adjust to either site. Drug molecules such as morphine, however, have a fairly rigid structure and may be able to bind to only one of the sites. As discussed below, many drugs are believed to act by competing with a neurotransmitter for binding to the receptor site. The drug molecule does not act by reacting with the neurotransmitter, but by being preferentially bound to the receptor site it may keep the neurotransmitter from accomplishing its regular function. Some drugs, however, such as ethyl alcohol, are structurally not very specific; they generally require a fairly high dose, and small changes in structure do not change the response very much. On the contrary, structurally specific drugs (these include morphine, methadone, tetrahydrocannabinol, LSD, psilocybin, and mescaline) act in trace amounts, are quite specific in their action, and may change drastically in action under relatively small changes in structure. Some of these points are illustrated further in the next three sections, where we look more specifically at morphine, marihuana, and LSD.

27.3   Morphine

Morphine is the classic example of a narcotic analgesic drug. The word "narcotic" comes from the Greek *narkotikos* meaning "deadening" or "numbing." A narcotic analgesic is a substance which combines three qualities: analgesic (pain-relieving), hypnotic (sleep-producing), and euphoriant (inducing a sense of well-being, loss of care). As a narcotic analgesic, morphine brings about elevation of the threshold of pain; the pain is still there, but one is indifferent to it. There is also a change in mood: anxiety disappears, feelings of inferiority vanish. Finally, on repeated ingestion morphine produces tolerance (increasing doses are needed to make the effects appear) and dependence (severe psychological and physical disturbance appears on withdrawal of the drug). Addiction is another word for dependence.

Morphine comes from opium, which is obtained from the opium poppy *Papaver somniferum*. The unripe capsule of the plant is scratched, and the ooze collected, dried, and pressed into bricks. This is raw opium. Opium itself has been known since antiquity. It is described as the "plant of joy" in Egyptian texts dating back to 3000 B.C. Relics of the Stone Age lake dwellers in Switzerland (2000 B.C.) include seeds and seed heads of the poppy. Homer (1000 B.C.) in the "Odyssey" tells of the visit of Telemachus to Menelaus in Sparta, where Helen is described as pouring the drug into wine "to bring forgetfulness of sorrow," so no one shed a tear the whole day long. As has been pointed out, they must have been used to the stuff since a single dose of opium does not produce a day-long emotional effect. However, it was probably reserved for the upper classes and the warriors. The Greeks spread the use of opium to Rome, but apparently the Romans did not pass it on. The big spread came from the Arabs, especially to China. In 1644, a Manchu emperor unwittingly helped by forbidding the smoking of tobacco; so, they smoked opium instead. Inasmuch as heat destroys some 80 percent of the active ingredients, addiction through smoking, however, was relatively mild. Still it thrived mightily as the British East India Company, particularly after the Opium Wars of 1839 to 1842, pushed the trade in opium as a substitute medium of exchange in place of gold and silver. The use of opium came to the United States *via* the West Coast from China.

Opium is a complex mixture of over 20 alkaloids, complex, nitrogen-containing organic bases, among which are morphine (about 10 percent), codeine (0.5 percent), and thebaine (0.2 percent). Figure 27.9 shows a two-dimensional projection formula of morphine. If the lower OH is replaced by OCH_3, we have codeine, the well-known cough suppressant. If both OH's are replaced by OCH_3, we have thebaine, also known as paramorphine. Treatment of morphine with acetyl chloride (CH_3COCl) replaces the OH groups by acetyl groups, giving diacetylmorphine, or heroin.

Morphine was first isolated as an active component of opium in 1805, but it was not until 1926 that its structure was unraveled. As already indicated, morphine is great for alleviating pain, but it does have adverse side effects (respiratory depression, dependence, tolerance, etc.). As a result, many attempts have been made to synthesize substitutes and find out what are the essential parts of the molecule that give it the narcotic analgesic properties. Some of the clinically useful substitutes are shown in Fig. 27.10. *Pethidine* is about one-eighth as potent as morphine. It also produces respiratory depression and leads to drug dependence. Curiously, it has an interesting psychic dependence in that patients using it generally resist a change to more effective analgesics. The *morphinans* are essentially stripped-down morphine, about one-fifth as potent. Putting back the OH groups restores most of the potency. The *benzomorphans* created a real flurry of interest

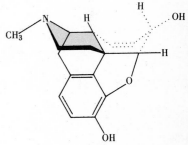

Fig. 27.9

Morphine. (Dotted lines are behind the plane of the diagram. Extra-heavy lines are in front.)

Fig. 27.10
Morphine replacements.

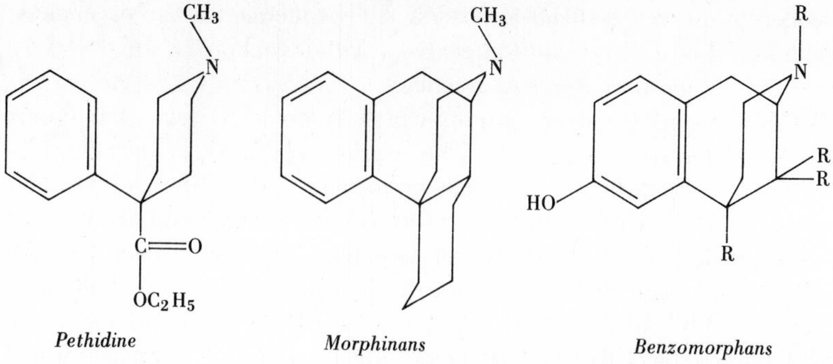

Pethidine   Morphinans   Benzomorphans

when in their first tests (i.e., with monkeys) they showed complete separation of analgesia from dependence liability. Unfortunately, the fine results did not carry over to humans. Still, there are lesser side effects with the benzomorphans than with the other narcotic analgesics.

What are the common characteristics of these molecules? Studies of the type mentioned above indicate that to be active the molecules need four characteristic features:

1   A quaternary carbon—i.e., a carbon that is attached to four other carbons.

2   An aromatic ring attached to the quaternary carbon.

3   A tertiary amine (i.e., a nitrogen that is attached to three carbons) which is two saturated CH_2 groups away from the quaternary carbon.

4   If the nitrogen is part of a ring, then there should be an OH group on the aromatic ring meta to the tertiary carbon.

Taking these minimal structural features, we assemble a molecule that looks like this:

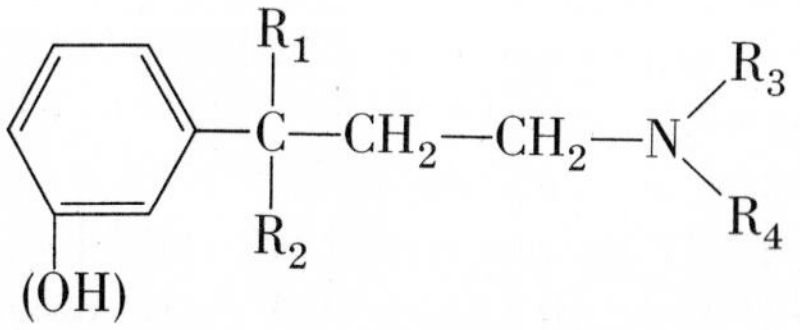

With but one exception, all strong narcotic analgesics fit such a pattern.

The rigid structural requirements have led to speculation as to how morphine acts. It is generally agreed that morphine does reduce the release of acetylcholine at the peripheral cholinergic junctions. It is suggested that this comes about because of a drug-receptor interaction that is very structure specific. In particular, three features appear to be essential in the receptor site:

1   A flat area that binds the aromatic ring, probably by van der Waals interaction

2 An anionic site that attracts the tertiary amine, which would be protonated and therefore positively charged under usual conditions

3 A suitably oriented trough between (1) and (2) that would be able to accommodate the —CH_2CH_2— bridge that sticks out in the front of Fig. 27.9

The suggested disposition of these features is shown in Fig. 27.11. It should be emphasized, however, that the above suggestions are only speculative and are still being vigorously debated.

Even after the local geometry of the receptor site has been established, it is by no means settled where the drug-receptor binding occurs. Some possibilities are shown in Fig. 27.12. The membrane on the post-synaptic side of the synapse (1) is but one possibility. Other possibilities include (2) the membrane that surrounds the storage vesicle and (3) the inner and (4) the outer surface of the membrane on the presynaptic side of the synapse. Other suggestions have included (5) the axon itself and (6)

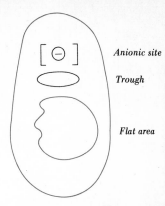

Fig. 27.11
Proposed geometry of receptor site needed to bind morphine molecule.

Fig. 27.12
Possible sites for drug-receptor interaction. (Numbers identify sites mentioned in text.)

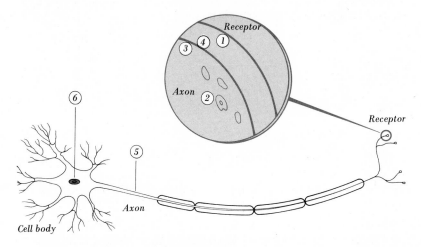

the nucleus of the cell body. Much interest has recently been expressed in sites (2) and (3) as possible explanations for withdrawal symptoms. If, for example, morphine adsorbed on the inner surface of the axon membrane prevents acetylcholine from getting out into the gap (*ergo* producing the narcotic effect), then unaccustomed deprivation of morphine could lead to massive release of acetylcholine, with consequent physical stress. When an addict gives up morphine, heroin, etc., he goes through a physiological hell which is largely chemically induced. Methadone, which mimics some of the effects of morphine, does not produce the same euphoria and does not entail the same severe withdrawal symptoms; so it finds use in treatment of drug addiction.

27.4 Marihuana

Marihuana comes from the hemp plant *Cannabis sativa*. This is a common weed that grows practically everywhere, although it prefers subtropical climate, and is used as a fiber to make rope. The name is believed to come from the Portuguese *mariguango*, meaning "intoxicant." It is also called weed, stuff, Indian hay, grass, pot, tea, maryjane, etc. In the United States, it is generally smoked; elsewhere it is often drunk or eaten.

There are three grades of marihuana, described by their Indian names as *bhang, ganja,* and *charas*. Bhang, which is the cheapest and least potent, is derived from the tops of uncultivated plants. It has a low resin content and in some countries is scorned as being fit only for the very poor; most United States marihuana is this grade. As indicated below, the active principle in marihuana is tetrahydrocannabinol (THC). The United States stuff, which mostly comes from Mexico, contains about 0.2% THC; Jamaica and southeast Asian bhang has 2 to 4% THC. Ganja, which is obtained from the flowering tops and leaves of carefully selected cultivated plants, generally has 4 to 8% THC. Charas, the highest grade, contains 5 to 12% THC. It is made from the sticky, yellow resin, carefully scraped from the tops of mature plants. It is the only grade that is properly called *hashish*.

The chemistry of marihuana is very complex and is not yet com-

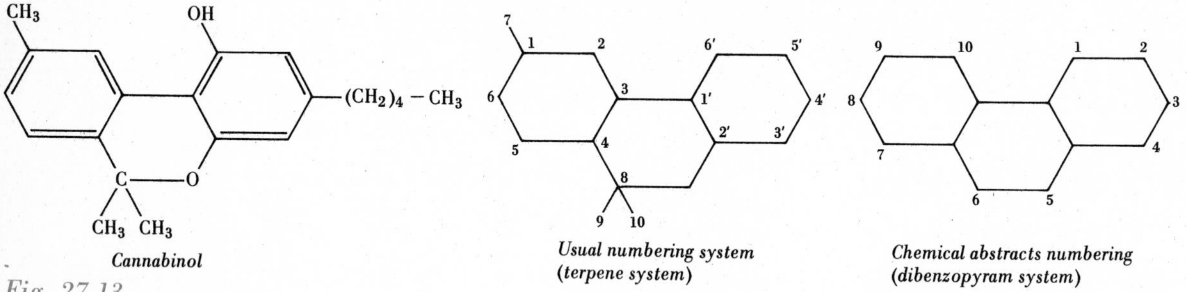

Fig. 27.13

Cannabinol and two numbering systems used to describe it.

Cannabinol

Usual numbering system
(terpene system)

Chemical abstracts numbering
(dibenzopyram system)

pletely understood. The plant contains a variety of organic chemicals including mono- and sesquiterpenes (compounds derived from the isoprene skeleton, $CH_2=C(CH_3)—CH=CH_2$, as found in rubber), plus carbohydrates, aromatics, and a variety of nitrogen-containing compounds. The resin gives what is known as the cannabinoid group, compounds which, unlike the other hallucinogens, contain no nitrogen. The parent compound is cannabinol, shown in Fig. 27.13. Tetrahydrocannabinol is derived from cannabinol by adding four hydrogens to the aromatic ring on the left in the molecule shown in Fig. 27.13. These four hydrogens are indicated by the prefix *tetrahydro-*. One double bond is left in the ring, and its position is indicated by a delta Δ with a superscript telling where the double bond is. Since there are two numbering systems used in describing the compounds, there is some risk of confusion. In the usual terpene numbering system, which is the one we use, Δ^1-THC means THC with the double bond starting at carbon 1 and going to carbon 2. (This same compound would be referred to as Δ^9-THC in the Chemical Abstracts numbering system.) If the double bond starts at carbon 1 and goes to carbon 6, we write

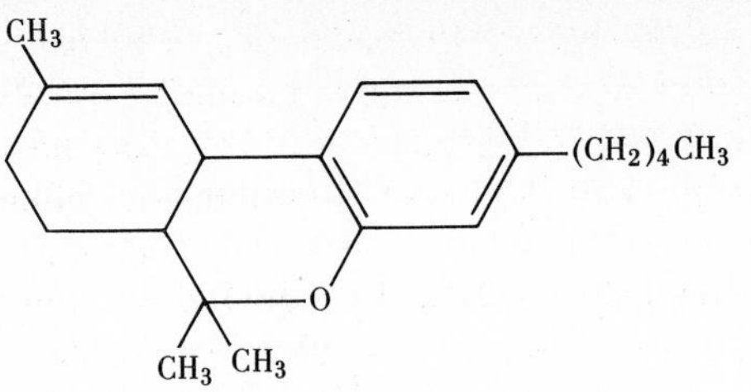

Fig. 27.14
Some components of marihuana and metabolic product.

Δ^1–Tetrahydrocannabinol
(active)

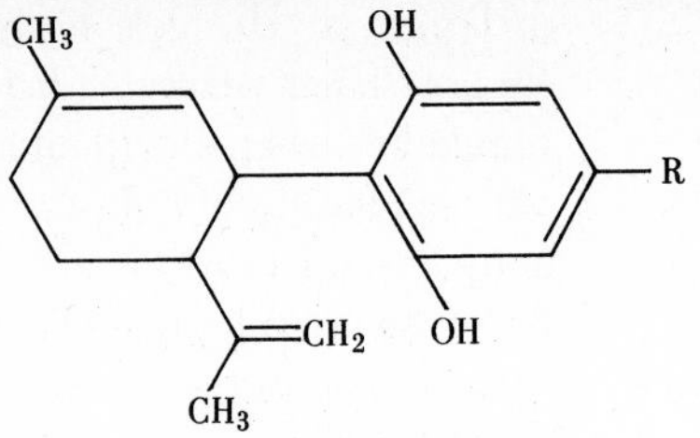

Cannabidiol (inactive)

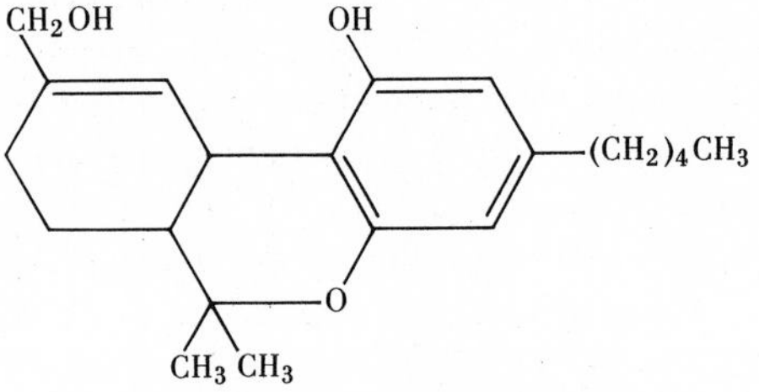

7-Hydroxy–tetrahydroxycannabinol
(metabolic product)

$\Delta^{1(6)}$-THC. The reason for making a fuss about the position of the double bond is that it apparently is critical for bringing about psychotomimetic properties. Only Δ^1-THC and $\Delta^{1(6)}$-THC have proved to be hallucinogenic. The other isomers either are not hallucinogenic or have not yet been prepared.

The structure of Δ^1-THC is shown in Fig. 27.14. Also shown is its principal companion, cannabidiol. The latter is not psychotomimetic but can generate THC under certain metabolic conditions. The principal metabolite of both the above is 7-hydroxy-THC, and there is currently a great deal of controversy as to whether perhaps this is the real active agent.

The main effect of THC appears to be on the central nervous system. In small amounts, it has a calming effect on the subject. In larger doses, it leads to vomiting, diarrhea, tremors, and failure of muscular coordination. The lethal dose is very large; for cats it is 3 g of charas, 8 g of ganja, or 10 g of bhang, all per kilogram of body weight. In man, no fatalities have been reported. Calculating from the lethal dose of 30 mg THC/kg for rats, one would expect the lethal dose in man to be about 2 g. This would correspond to about 40 cigarettes at 5% THC if all the THC were absorbed. However, it is estimated that 98% of the Δ^1-THC is lost on combustion.

The psychic effects of THC begin to appear at about 1×10^{-6} g/kg, which makes THC as potent as LSD, notorious for the trace amounts needed to produce a "high." Many interesting accounts of marihuana experience have appeared in the literature (e.g., Baudelaire, Gautier, and Dumas). Intoxication effects appear in 10 to 30 min. First there is an anxious period, which gives way to a definite euphoria. The subject becomes talkative, exhilarated, and begins to have an astounding feeling

Section 27.4
Marihuana

719

of lightness. He feels he has a deep understanding of the meaning of things. Time sense is distorted, and confusion appears on trying to remember what was thought. Sometimes the subject sees visual hallucinations, flashes of light or forms of vivid color which continually change and develop into geometric figures and pictures of great complexity. After about 6 h, the effects disappear, only to be followed in some cases by a mild sense of depression. There are no aftereffects, unlike the sometimes shattering recurrences that follow LSD ingestion. Curiously enough, novice users of marihuana frequently experience no effect at all, whereas experienced smokers need less and less to get a "high." THC is very high in fat solubility (it is insoluble in water and body fluids), and it is suspected that it accumulates in the fat cells, thus perhaps explaining why it shows reverse tolerance. The rate of excretion from the body is very slow—only 1 to 2 percent per day after intravenous injection.

Like LSD, there is very little information on the chemical way THC acts in the body. In some mysterious way, both are believed to affect the neurotransmitter serotonin (Fig. 27.7), which is particularly rich in the brain. However, there must be major differences in their mechanisms. Almost invariably THC produces a "happy high," whereas LSD "trips" are frequently beset with awe and fear.

27.5   LSD

LSD stands for *Lysergsäure Diethylamid,* which is the German name for the diethylamide derivative of lysergic acid. LSD is one of the most important of the hallucinogenic drugs, among which are to be counted also psilocybin, derived from the peyote cactus, and mescaline, from the "sacred" mushroom *Psilocybe mexicana.* Tetrahydrocannabinol is not normally included in the classification, but it rightfully belongs here.

Hallucinogens are substances that induce hallucinations, illusions, and distortions of perception and thinking. They are also called *psychotomimetics,* because in some cases their effects mimic psychosis, or *psychedelics,* because they sometimes are reputed to disclose unsuspected features of one's psyche.

Figure 27.15 shows the structures of three important hallucinogens—mescaline, psilocybin, and LSD; shown also is indole, which they all resemble, and serotonin, postulated to be a prime neurotransmitter in the brain. As can be seen, all five have in common the indole structure, except mescaline, which mimics it very well. Primitive societies have known the hallucinogens for ages, using them for contact with the gods, for divination, and for medical care. In the Western Hemisphere, the region extending from the southwestern part of the United States to the northwest basin of the Amazon is particularly noted for its rich folklore detailing such use of cactus and mushrooms. As a specific example, the Aztecs for their ceremonial rites used certain sacred mushrooms called *teonanacatyl,* which means "flesh of the

Fig. 27.15
*Some hallucinogenic drugs
and related structures.*

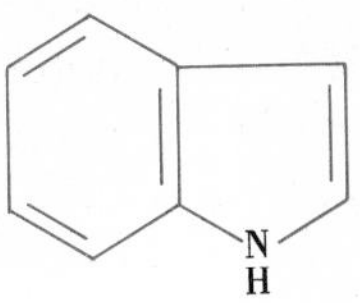

Indole 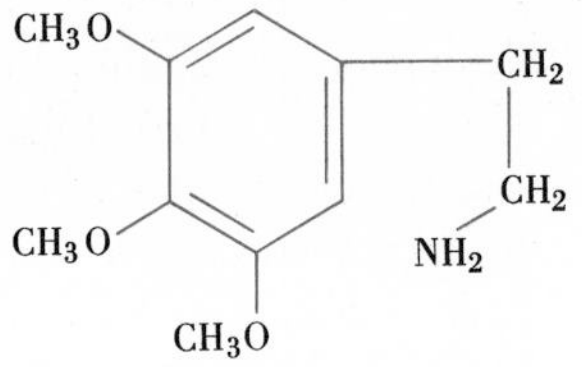   Mescaline 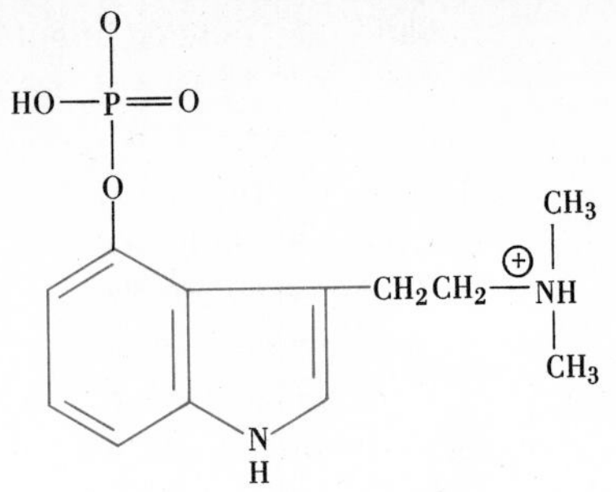   Psilocybin

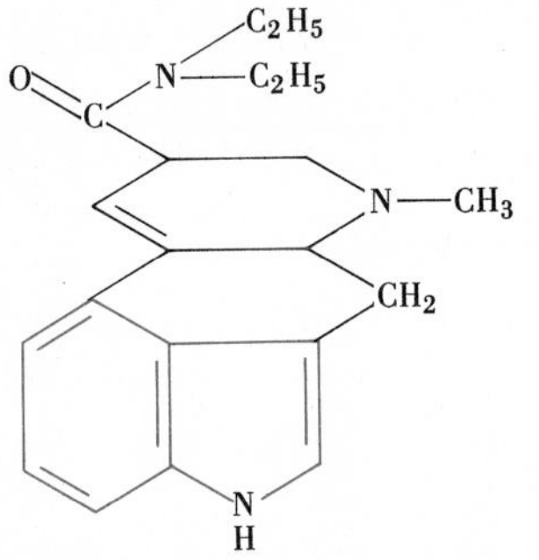

LSD 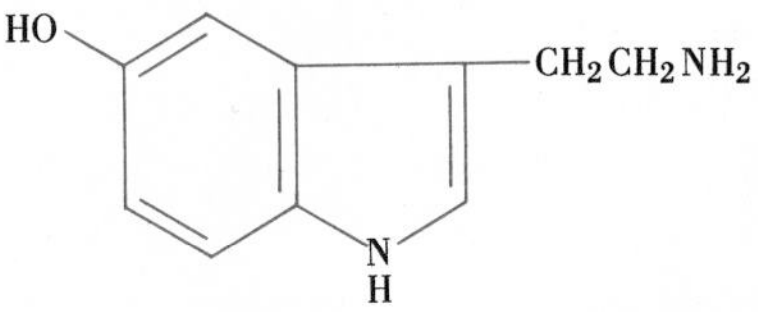   Serotonin (5–hydroxytryptamine)

gods." These are said to have been distributed at the coronation of Montezuma to make the ceremony more spectacular.

LSD was discovered in 1943 as a synthetic derivative of lysergic acid. It was prepared by Dr. Albert Hofmann of the Sandoz Chemical Works in Switzerland and is often designated as LSD-25 because that was the number of the derivative in the Sandoz study. Lysergic acid, which has the same structure as the LSD molecule shown in Fig. 27.15 except that the $N(C_2H_5)_2$ grouping is replaced by OH, is obtained from *ergot*, the product of a fungus (*Claviceps purpurea*) that grows on rye and other grains. Extract of ergot has been known for ages as a potent stimulant for contraction of the pregnant uterus. In the Middle Ages, periodic epidemics of a disorder called ergotism used to sweep Europe, apparently as a result of poisoning from eating rye bread infected with ergot. It was also called Saint Anthony's fire (after the patron saint whose intercession was supposed to give relief to the victims). Characteristic symptoms included itching and tingling of the skin, painful burning sensations in the extremities, and in severe cases gangrene of the feet, legs, hands, and arms. Severe mental disturbance was also likely to accompany the disease. It is no wonder that such terrible complications appeared since ergot is known to contain more than 30 compounds with poisonous properties. One puzzling aspect is that at present it is almost impossible to reproduce all the symptoms, and it is surmised that they may have been compounded by malnutrition; it was the poor and starving who were most likely to eat ergot-infested bread. Incidentally, ergotism is by no means a thing of the past. As recently as 1951, an outbreak appeared in France, though in a very mild form.

There is presently great interest in hallucinogens as a possible key to understanding natural psychoses. The evident chemical similarity between LSD and the neurotransmitter serotonin as well as the extraordinary potency of LSD has led to speculation that mental disorders such as schizophrenia may be caused by an imbalance in the metabolism of serotonin. It is not yet clear whether too much or too little serotonin produces the dramatic effects observed, but certainly there is much resemblance between some symptoms of mental disorder and those encountered in an LSD "trip." As little as 0.1 mg of LSD is enough to show the full chemical effects (the usual dose is 20 μg). These include changes in visual perception (not always pleasant), complex auditory hallucinations (crossover of senses so that, for example, one "sees" music and "hears" colors), a sense of timelessness, and a merging of personal identity with the nonself. This last can be a shattering experience for weak personalities and may explain the attempted suicides that occasionally happen on "bad trips." Incidentally, the experiencing of a "good trip" or a "bad trip" is not always predictable. It depends on the basic personality traits of the individual, his mood at the time of LSD ingestion, and the social context in which the drug is taken. Apparently, LSD is not addictive in the physiological sense; tolerance, however, does develop; so increased doses may be needed to produce the same effects. Because the possibility of adverse mental reaction is relatively high and ever present, LSD should be considered a *very dangerous drug;* it is so labeled in federal law.

Neither the mode nor site of the chemical action of LSD is definitely known. It is not even known whether the various symptoms are due to primary action of LSD or involve different mechanisms and different sites. It is known that LSD is a strong antagonist to serotonin; i.e., it prevents serotonin's normal action. For example, the clam heart, which is extraordinarily sensitive to serotonin and has been used to help map its distribution in the body, does not respond when LSD is present. One hypothesis suggests that LSD produces its psychotomimetic effects by blocking the serotonin receptor sites in the central nervous system. Other people object to this hypothesis because, as they point out, 2-Br-LSD, which contains Br instead of H at the 2 position of the indole ring, is psychotomimetically *inactive,* even though it is also a strong serotonin antagonist. The objection may not be crucial, however. One of the strange things observed about 2-Br-LSD is that, even though it itself is not hallucinogenic, it does prevent hallucinogenic effects from subsequently administered LSD. In other words, it may be that 2-Br-LSD blocks serotonin sites just as LSD does but cannot produce some needed secondary action that triggers hallucinogenic activity. The situation may be the same as the "lock-and-key" mechanism often invoked for enzyme reaction mechanisms. Not only must a key fit into a lock, but it must also turn. LSD and 2-Br-LSD may "fit into the same lock," but only LSD may "turn" to produce psychotomimetic action.

Another hypothesis of LSD action, based on the observation that most of the LSD has left the brain before the mental effects are seen, is that LSD triggers some other process. Alternatively, the LSD may have to be converted to some metabolite, which then goes on to produce the mental changes. Still another possibility is that LSD may act by its effect on various enzyme systems, as in its inhibition of the action of monoamine oxidase (MAO), which is important for chewing up norepinephrine and serotonin after they have done their neurotransmission.

Much work remains to be done to understand the working of the hallucinogenic drugs and how they may be used to unravel the chemical aspects of mental disorder. However, the work should not be done by amateurs. It is too dangerous.

27.6   Radiation hazards in the environment

The standard electromagnetic spectrum, which is shown in Fig. 27.16, extends from the cosmic rays at 0.0004 nm to the radio waves at 300 m. As can be seen, it can be divided into two segments, ionizing radiation

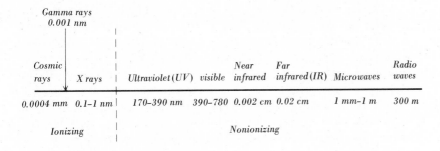

Fig. 27.16
Electromagnetic spectrum.

and nonionizing radiation. For the ionizing radiations, the radiation damage is linear with dose and so far as can be determined there is no no-effect level; for the nonionizing radiation the situation is not yet clear. As far as man is concerned, the main radiation hazard in the environment comes from the cosmic rays, which may produce genetic mutations in the reproductive cells. However, there is not much we can do about it; so we forget it. The next biggest hazard comes from X rays, which account for 95 percent of our radiation exposure other than cosmic rays. Most of this comes from medical diagnosis, but some of it stems from faulty color-television receivers. Distinctly lesser radiation hazards come from the UV (which causes sunburn and may cause skin cancer), the visible (where lasers are a new hazard for special circumstances), and the microwave (where radar cooking and small-craft radar may pose problems in special cases). Figure 27.17 shows the approximate relative distribution of the main sources of radiation exposure. As can be seen, to the electromagnetic spectrum we have to add

Source	Average per capita dose, millirem/yr
Natural background	130
Medical diagnostic X rays	90
Weapons-test fallout	5
Nuclear power plants	<0.01

Fig. 27.17
Radiation exposure in United States by source.

radiation exposure originating in nuclear disintegration, specifically from weapons-test fallout and nuclear power plants. Particle radiation should be classified with the ionizing type.

Radiation damage to individuals is generally classified as *somatic* or *genetic*. Somatic implies injury to the irradiated individual and may show up as skin rash, cataracts, cancer, etc. Genetic implies inheritable changes or mutations in the reproductive cells. The effects of acute exposure to ionizing radiation are fairly well characterized. Over the short term, they may include nausea, anemia, fatigue, blood and intestinal disorders, loss of hair, damage to the central nervous system, and even death; over the long term, cancer (e.g., leukemia) and cataracts. The effects of small-dose, chronic exposure are not yet understood, but it is widely accepted that there is no such thing as a no-effect level. For nonionizing radiation, particularly radar, there is wide disagreement where the no-effect level should be set. Microwaves shorter than 10 cm are usually reflected or absorbed by the skin and can be felt by heating of the surface tissue; waves between 10 and 30 cm can penetrate the skin and fat layer subject to individual variations; waves longer than 30 cm can penetrate deep tissue without subjective awareness of heating. Different organs have different sensibilities. The eyes and organs that cannot dissipate heat are most vulnerable. Until recently the hazard level for microwave power density, at least in United States installations, has been set at 0.01 W/cm^2 with a special limit of 0.001 W/cm^2 for eye exposure, especially around 10-cm wavelength. The Russians have consistently set lower exposure limits, claiming that effects such as headache, fatigue, dizziness, skin burns, eye injuries, and cataracts can appear below heat-injury levels. The radar hazard, especially in high-power installations, comes mainly from the heating effect. In aqueous systems this peaks around 10 cm, where there is a strong microwave absorption band by H_2O. In small-power installations, there may be quite other hazards, such as disrupting artificial pacemakers for the heart.

The units shown in Fig. 27.17 for describing exposure to ionizing radiation are but one of several kinds in use, depending on whether charge, energy, or biologic effect is of major interest. The units most widely used are the *roentgen, rep, rad,* and *rem.* The roentgen (designated as *r*) was the original unit, introduced to measure the ionization produced in air by X rays or γ rays. One roentgen (1 r) is the X or γ radiation that produces one electrostatic unit (esu) of plus-minus charge pairs in one cubic centimeter of air at STP. Since a single monopositive ion (or a single mononegative ion) has a charge of 4.80×10^{-10} esu (that is, 1.60×10^{-19} coulomb), 1 esu would correspond to $1/4.80 \times 10^{-10}$, or 2.08×10^9, ion pairs. Air at STP has a density of 0.001293 g/cm^3; so 1 r corresponds to $2.08 \times 10^9/1.29 \times 10^{-3}$, or 1.6×10^{12}, ion pairs per gram of air. Empirically, it is found that it takes 5.2×10^{-11} erg (that is, 5.2×10^{-18} J) to create an ion pair in air at STP. Consequently, 1 r = 83 ergs per gram of air.

The *rep*, which stands for "roentgen equivalent, physical," was introduced to take care of particulate radiation as well as X rays and γ rays. It measures the energy absorbed by soft tissue, or its equivalent, H_2O. One rep is approximately 93 ergs per gram of tissue. The precise value depends on tissue composition and the energy of the radiation. Soft tissue actually varies from 63 to 100 ergs/g; bone is about 1000 ergs/g. Thus, the rep is all right for describing the amount of radiation absorbed by soft tissue, but not by the whole body.

To get around the dependence of energy absorption on medium, the unit *rad* was introduced. This stands for "radiation absorbed dose" and is defined as 100 ergs/g. It describes absorption of energy of any type in a medium of any type.

The *rem,* standing for "roentgen equivalent, man," was designed to compensate for differences in ionization efficiency, energy transfer, etc., of various radiations. It is defined as the quantity of radiation of any type which produces the same biologic effect in man as that resulting from 1 r of X-ray or γ-ray irradiation. The biologic effect, however, depends on the part of the body irradiated and the type of radiation. As a specific example, the relative biological effectiveness, (RBE) for producing cataracts is 10 times as great for fast neutrons as for γ rays and 5 times as great for slow neutrons as for γ rays. To compute the total dose from mixed radiation, we multiply each dose by its RBE. Thus, for 0.4 rad of γ rays plus 0.3 rad of slow neutrons plus 0.2 rad of fast neutrons we have a total dose of 3.9 rem. Of this, 0.4 rem comes from the γ (0.4 rad × RBE of 1), 1.5 rem from the slow neutrons (0.3 rad × RBE of 5), and 2.0 rem from the fast neutrons (0.2 rad × RBE of 10).

How do the various particle radiations differ in their hazard? α particles dissipate their energy quickly. They do not penetrate very far and may even be stopped by a sheet of paper. It takes a 7.5-MeV α to penetrate the skin; so externally α emission is a negligible hazard. However, when α emitters are ingested, they can be very bad since for equal energies they produce more ion pairs than either β or γ. For example, in air a 1-MeV α produces about 100,000 ion pairs per centimeter; 1-MeV β produces only 100 ion pairs per centimeter; 1-MeV γ produces 10,000 ion pairs per centimeter.

β particles are generally more penetrating. Whereas it takes but a sheet of paper to stop an α, it takes a 1-mm thickness of aluminum to stop a β. β emitters external to the body are more hazardous than α emitters, since the β particles can penetrate from a few millimeters to a centimeter or so under the skin. Internally, β emitters are more hazardous than they are externally, but they are less so than are ingested α emitters, again because of the lesser specific ionization by β.

γ rays, like X rays, are the most dangerous. They have very high penetrating power (e.g., it takes about 5 cm of lead to stop a typical γ ray); they can destroy tissue and inflict serious burns quite rapidly. Both γ and X rays interact with matter in three ways: photoelectric (ejection of e^- by low $h\nu$ on high-atomic-weight atoms), Compton (moderate $h\nu$ on an atom of any atomic weight to produce e^- plus longer λ), and pair production (high $h\nu$ on high-atomic-weight atoms to produce e^- and e^+). The charged particles formed then go on to excite and ionize the biologic medium. Since γ rays and X rays can penetrate to extreme depths in tissue, they constitute a hazard for the entire body. Incidentally, γ emission frequently accompanies α or β emission.

The biologic effects of radiation come about because of alterations in the functions of the cells. Recognizing that biologic organisms are mostly H_2O, we can appreciate that radiation produces a whole host of species (such as H^+, H_2, H_2O^-, H_2O^+, e^-, e^+, HO_2, H_3O^-, and H_2O_2), some of which are highly reactive. These go on to react with the proteins and, in particular, appear to deactivate enzymes by breaking up the S—H···S hydrogen bond. With enzyme inhibition, cell growth may continue, but cell division may be stopped. Furthermore, proteins play an important role in the formation of cell membranes. Radiation damage may make cell membranes permeable, and non-normal interchange of material through an imperfect cell membrane can result in temporary or permanent injury. The sensitivity to radiation damage appears to be directly proportional to the cell's reproductive capacity and inversely proportional to its degree of differentiation.

27.7 Nuclear energy, reactors, and bombs

The nucleus is usually pictured as containing Z protons, where Z is the atomic number, and $A - Z$ neutrons, where A is the mass number, in a region which is about 10^{-13} cm in radius ($r \approx 10^{-13}\ A^{1/3}$ cm). The difficult thing to understand is how positive charges can be packed together into such a small space without flying apart as a result of coulomb repulsion. Neutrons must be at least partly responsible because, first, there is no nucleus consisting of more than one proton without simultaneous presence of one or more neutrons and, second, the more protons there are in a nucleus, the more neutrons per proton are needed for stability. This latter point is demonstrated by the *trough of stability* shown in Fig. 27.18, where a plot is made of all the known stable (nonradioactive) nuclei. Each point corresponds to a known nucleus containing the indicated number of protons and number of neutrons. Energy can be imagined as a third coordinate perpendicular to the plane of the diagram, and the points shown are minimum-energy nuclei. As one goes away from the plotted points, the energy rises so that one moves up the walls of an energy trough. The straight line represents the direction along which the minimum of the energy trough would lie if stability required an equal number of protons and neutrons. As can be seen, the lighter stable nuclei contain approximately equal numbers of neutrons and protons (for example, $^{14}_{7}N$, $^{16}_{8}O$, and $^{40}_{20}Ca$), but the heavier stable nuclei contain more neutrons than protons (for example, $^{138}_{56}Ba$, $^{202}_{80}Hg$, and $^{208}_{82}Pb$). Nuclei that do not fall in the trough of stability are radioactive—i.e., their neutron-proton ratios are either too high or too low, and nuclear rearrangement occurs so as to produce more stable nuclei.

How does a nucleus get up on the wall of the trough so that it is unstable? Two mechanisms are common: In one, target nuclei are

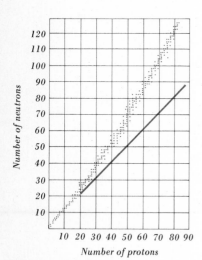

Fig. 27.18

Trough of stability for nuclei. (Energy is the coordinate perpendicular to the plane, and the dots correspond to nuclei in the bottom of the energy trough.)

bombarded by other nuclei that have been accelerated in a particle accelerator (e.g., cyclotron and synchrotron) so as to form a colliding-pair compound nucleus that has decay properties; in the other an unstable nucleus is formed as the result of radioactive decay of some other nucleus.

There are essentially three modes of radioactive decay:

1 If the n/p ratio is too high, a neutron may be ejected or a beta particle (negative electron) may be ejected. Simple neutron ejection is rarely observed because it usually occurs so rapidly it is hard to measure. For example, the decay

$$^5_2\text{He} \longrightarrow {}^4_2\text{He} + {}^1_0n$$

to produce an alpha particle (^4_2He) and a neutron (^1_0n) has been calculated to have a half-life (see Sec. 18.2) of 2×10^{-21} sec, much too short to be observed. There are, however, some neutron emissions from fission particles (see below) which are delayed long enough that they become observable. A specific example of this is

$$^{87}_{36}\text{Kr} \longrightarrow {}^{86}_{36}\text{Kr} + {}^1_0n$$

for which the half-life appears to be about 1 min.

Beta emission is much more common. It corrects a too-high neutron-proton ratio by emitting one unit of negative charge and thereby increasing the positive charge of the residual nucleus. Since the beta particle (generally designated $_{-1}^0e$) has essentially zero mass, its emission does not change the mass number of the emitting nucleus. A few examples of beta decay are

$$^{14}_6\text{C} \longrightarrow {}^{14}_7\text{N} + {}^0_{-1}e \qquad t_{1/2} = 5570 \text{ yr}$$

$$^{90}_{38}\text{Sr} \longrightarrow {}^{90}_{39}\text{Y} + {}^0_{-1}e \qquad t_{1/2} = 28 \text{ yr}$$

$$^{137}_{55}\text{Cs} \longrightarrow {}^{137}_{56}\text{Ba} + {}^0_{-1}e \qquad t_{1/2} = 30 \text{ yr}$$

The first of these examples is used in radiocarbon age dating; the second and third represent important fission products from uranium-bomb explosions.

2 If the n/p ratio is too low, the number of neutrons must be increased, protons decreased, or both. One device is to absorb into the nucleus one of the orbital electrons, usually an electron of the K shell. Such K capture (also designated EC, or electron capture) reduces the nuclear charge by one unit, leaving the mass number unchanged, as in the following example:

$$^{90}_{42}\text{Mo} \xrightarrow[K \text{ capture}]{} {}^{90}_{41}\text{Nb} \qquad t_{1/2} = 5.7 \text{ h}$$

This process frequently occurs in fission products. Invariably, when K capture occurs, an outer-shell electron drops into the K shell to fill the vacancy, thus liberating energy, usually as an X ray.

Another way to raise an n/p ratio that is too low is for the nucleus to emit a positron ($_1^0e$, a positive electron). This process, typified by

$$^{11}_6\text{C} \longrightarrow {}^{11}_5\text{B} + {}^0_1e \qquad t_{1/2} = 20.5 \text{ min}$$

decreases the nuclear charge by one unit and leaves the mass number unchanged.

3 Sometimes there are too many protons crammed into a single nucleus; so it is unstable no matter how many neutrons are present. This happens for all nuclei with 84 or more protons; they all lie beyond the trough of stability. No one of the above decay steps by itself can lead to stability. Instead, it is necessary to split off larger pieces, and even then a series of steps may be required. Most commonly, the piece split off is an α particle, and, in fact, most of the heavy nuclei are α emitters. With $^{212}_{84}\text{Po}$ a single step is enough to attain a stable nucleus:

$$^{212}_{84}\text{Po} \longrightarrow {}^{208}_{82}\text{Pb} + {}^{4}_{2}\text{He} \qquad t_{1/2} = 3 \times 10^{-7} \text{ sec}$$

With $^{234}_{92}\text{U}$ many steps involving a combination of α and β decays are required, as in the following:

$$^{234}_{92}\text{U} \xrightarrow{\alpha} {}^{230}_{90}\text{Th} \xrightarrow{\alpha} {}^{226}_{88}\text{Ra} \xrightarrow{\alpha} {}^{222}_{86}\text{Rn} \xrightarrow{\alpha} {}^{218}_{84}\text{Po} \xrightarrow{\alpha}$$

$$^{214}_{82}\text{Pb} \xrightarrow{\beta} {}^{214}_{83}\text{Bi} \xrightarrow{\alpha} {}^{210}_{81}\text{Tl} \xrightarrow{\beta} {}^{210}_{82}\text{Pb} \xrightarrow{\beta} {}^{210}_{83}\text{Bi} \xrightarrow{\beta}$$

$$^{210}_{84}\text{Po} \xrightarrow{\alpha} {}^{206}_{82}\text{Pb}$$

Other steps leading to the same stable nucleus, $^{206}_{82}\text{Pb}$, are also possible.

Besides ordinary radioactive decay, there is another kind of nuclear stability based on interconversion of mass and energy. The observed fact is that the mass of a nucleus is always less than the sum of the masses of the neutrons and protons believed to compose it. By the Einstein relation $E = mc^2$ (where E is energy in joules, m is mass in kilograms, and c is the speed of light, 2.9979×10^8 m/sec) a deficiency in mass is equivalent to a deficiency in energy. In other words, the assembled nucleus is lower in energy than the isolated component particles by an amount equal to the missing mass. Hence, the missing mass gives a measure of the binding energy of the nucleons (neutrons and protons) in the particular nucleus. How this works is illustrated below for the case of $^{56}_{26}\text{Fe}$, which can be regarded as made up from 26 protons and 30 neutrons. The mass of a proton is 1.00728 amu, and the mass of a neutron is 1.00866 amu.

26 protons = 26(1.00728) = 26.1893 amu

30 neutrons = 30(1.00866) = $\underline{30.2598}$ amu

56.4491 amu

The observed mass of an $^{56}_{26}\text{Fe}$ atom is 55.9349 amu. Subtracting 26 electrons at 0.0005486 amu gives us 55.9206 amu for the observed mass of an $^{56}_{26}\text{Fe}$ nucleus. The mass deficiency, therefore, is 0.5285 amu for the whole nucleus, or since it contains 56 nucleons, 0.009438 amu per nucleon. This is 1.567×10^{-26} g or, by Einstein equivalence, 1.408×10^{-12} J. Nuclear binding energies are usually expressed in million electron volts, where 1 MeV = 1.6×10^{-13} J; so for $^{56}_{26}\text{Fe}$

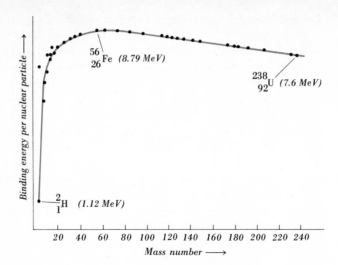

Fig. 27.19
Binding energy per nucleon
in the various nuclei.

$^{56}_{26}$Fe *(8.79 MeV)*

$^{238}_{92}$U *(7.6 MeV)*

$^{2}_{1}$H *(1.12 MeV)*

Binding energy per nuclear particle ⟶

20 40 60 80 100 120 140 160 180 200 220 240

Mass number ⟶

we have a nuclear binding energy of 8.79 MeV per nucleon. A similar calculation for $^2_1$H gives 1.12 MeV per nucleon; for $^{238}_{92}$U, 7.6 MeV per nucleon. Figure 27.19 shows how the binding energy compares in the various nuclei. As can be seen, intermediate elements of mass number about 60 have the highest binding energies and are the most stable. The other elements are unstable with respect to conversion to them. This means, for example, that if a heavy-element nucleus, such as uranium, is converted to iron, the difference in binding energy per nucleon should be liberated. Similarly, if a light-element nucleus, such as hydrogen, is converted to iron, energy should also be liberated. Such conversions are the bases for utilization of nuclear energy. Breakup of large nuclei to intermediate ones is called *nuclear fission*; merging of smaller nuclei to intermediate ones is called *nuclear fusion*.

A typical fission process is the following:

$$^1_0n + {}^{235}_{92}U \longrightarrow [{}^{236}_{92}U] \longrightarrow {}^{141}_{56}Ba + {}^{92}_{36}Kr + 3{}^1_0n + Q$$

A neutron impinging on a ^{235}U nucleus gets absorbed by it to produce temporarily a compound nucleus ^{236}U, which almost immediately breaks up into two approximately equal fragments, Ba and Kr, as well as three neutrons. Emitted at the same time is a large burst of energy Q, which is mostly kinetic and amounts to about 200 MeV per fission. The number of neutrons emitted per fission is variable. It is usually two or three but may go as high as six. The weighted average over all fissions, which was a closely guarded secret during World War II, is 2.43. The product nuclei resulting from a fission generally have high neutron-proton ratios, much higher than they should have for stable nuclei of that Z. Hence, they are β active. For the above case, we would have the following decays:

$$^{141}_{56}Ba \xrightarrow[18 \ min]{\beta} {}^{141}_{57}La \xrightarrow[3.9 \ h]{\beta} {}^{141}_{58}Ce \xrightarrow[33 \ days]{\beta} {}^{141}_{59}Pr(stable)$$

$$^{92}_{36}Kr \xrightarrow[3 \ sec]{\beta} {}^{92}_{37}Rb \xrightarrow[5 \ sec]{\beta} {}^{92}_{38}Sr \xrightarrow[2.7 \ h]{\beta} {}^{92}_{39}Y \xrightarrow[3.5 \ h]{\beta} {}^{92}_{40}Zr(stable)$$

*Section 27.7
Nuclear energy,
reactors, and bombs*

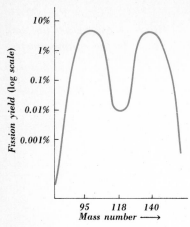

Fig. 27.20

Fission yields vs. mass number.

As can be seen, the half-lives of the fission products gradually get longer as the consecutive decay products approach stability. Actually, the fission shown is just one of the ways in which ^{235}U undergoes fission. Figure 27.20 shows what the statistical distribution of fission products looks like. Exact division of the starting nucleus into two equal fragments is not very probable; the process is asymmetric, with maximum yields observed at about 95 and 140.

About 90 percent of the energy release in fission occurs in the first 0.1 μsec. Of this immediate release, about 165 MeV shows up as kinetic energy of fission products, 5 MeV as kinetic energy of emitted neutrons, and 7 MeV as gamma rays. Delayed energy release comes from the fission products—6 MeV as gamma rays, 7 MeV as β particles, and 10 MeV in the form of delayed neutron emission. This 23 MeV of delayed energy release is what constitutes the *fallout* problem. When fissionable material such as ^{235}U is used in a nuclear explosion (see below), the fission products that have short half-lives disappear rather rapidly. Some of the decay products, however, such as ^{90}Sr and ^{137}Cs, have appreciable half-lives; so they persist—long enough to "fall out" over the environment, get into the food chain, be absorbed in the human system, and possibly create a radiation hazard.

An important feature of the fission process is that more neutrons are produced in the fission than are needed to initiate it. This means that the fission process can become self-sustaining as a chain reaction. When the chain propagates so that more neutrons are generated per unit time than are absorbed, even allowing for losses to the outside and to absorption by nonfissionable material, then the chain reaction can go on at an ever-increasing rate. Once the system passes what is called *criticality*, it tends to go into a runaway condition. In a bomb, geometric conditions are chosen so as to enhance the rapid exponential increase of the reaction rate. In fact, the problem there is to keep the system from blowing apart too soon. In a nuclear reactor, design requirements are quite the opposite. The idea is to keep the system just short of criticality so that the chain reaction continues but not in a runaway fashion. This is achieved by interspersing the uranium fuel material with cadmium control rods. Cadmium has a high neutron capture ability and acts as a control by keeping the neutron number down to some small, desirable level. Fission goes on, but at a slow, usable rate. In any case, because nuclear reactors are designed so differently from nuclear bombs, the oft-voiced fear of some people that nuclear power plants might blow up like a bomb is not justified. There may be a radiation hazard from an accident in a nuclear power plant, but not a nuclear-bomb hazard.

To get a better idea of what is involved in nuclear-fission energy, it is instructive to calculate how much uranium is needed for criticality. We can estimate this from the published information that the first atomic bomb, which was dropped on Hiroshima, was equivalent to 20,000 tons of TNT. One ton of TNT releases 4.18×10^9 J; so 20 kilotons would be 8.4×10^{13} J. If each nuclear fission liberates 200 MeV, we have per fission $(200 \text{ MeV})(1.6 \times 10^{-13} \text{ J/MeV})$, or

3.2 × 10⁻¹¹ J. One bomb then would be equal to 8.4×10^{13} J divided by 3.2×10^{-11} J per fission, or 2.6×10^{24} atoms. Dividing this by the Avogadro number gives 4.3 mol of ^{235}U, or 1.0 kg. Of course, the efficiency is less than 100 percent; so this figure is probably a lower limit. Given that the density of uranium is 19.05 g/cm³, we can calculate that the above critical mass would correspond to 52 cm³, or a sphere about 5 cm in diameter.

Criticality is not just a problem with metallic ^{235}U. It also applies to solutions of ^{235}U. However, in solutions the U atoms are farther apart; so there is more chance of neutron loss between fissions. The critical mass is therefore greater; it corresponds, for example, to 2.5 kg of ^{235}U in a 5.5-liter volume of solution. Criticality depends very much on geometry, and there have been a few processing accidents in which change of geometry has been a key factor. In one case, for instance, a solution went critical when it was siphoned from a flat, shallow container into a spherical flask; the resultant burst of radiation exceeded 10,000 rad.

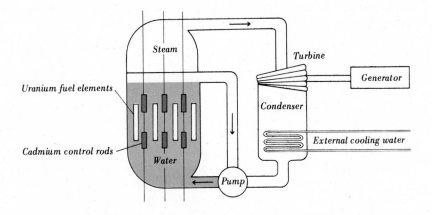

Fig. 27.21
Boiling-water reactor.

Nuclear-reactor safety is of ever-increasing concern, particularly as the number of nuclear power plants increases. At present, there are estimated to be about 100 such plants distributed in 40 countries (most are in the United States). It has taken approximately 17 yr to get the first 100 built; it is believed the next 100 will come in but 4 yr. Most reactor designs are very conservative, but this does not mean that a 100 percent no-risk situation has been achieved. The *boiling-water reactor*, for example, which is represented schematically in Fig. 27.21, has recently come under criticism for its emergency-cooling safeguards. The Atomic Energy Commission (AEC) has assumed that the worst credible accident would be failure of the pipe that carries water over the fuel elements. In such case the uranium would quickly heat up, even though the fail-safe mechanism on the cadmium control rods would shut down the reactor and stop the fission. Fission products already generated would continue to give off heat, which would eventually melt the uranium. Emergency sprays, flooding basins, isolation domes—all these are emergency backup devices. So far they have proved unnecessary, but as reactors get super large, one begins to wonder what would happen if they are called on to perform.

Actually the potential energy supply from uranium is currently rather limited because the amount of ^{235}U is limited. New reactor technology is being developed, however, to create *breeder* reactors. These operate with ^{235}U or ^{239}Pu as regular fuels, but in addition they are charged with nonfissionable materials such as ^{238}U and ^{232}Th. Under proper conditions of neutron irradiation, these can be converted, respectively, to ^{239}Pu and ^{233}U, both of which are fissionable. The idea, of course, is to have our cake and eat it. As we consume nuclear fuel, we generate more to replace it. Since ^{238}U and ^{232}Th are relatively abundant, their exploitation as nuclear fuels would put off the energy crisis by several hundred years.

A more dramatic possibility lies in the potential of exploiting nuclear *fusion* as a source of controllable energy. Nuclear fusion has been achieved in the hydrogen bomb, but its controlled application has persisted as an exercise in frustration. Nuclear fusion seeks to release energy by binding nucleons of very light elements into nuclei of heavier elements. Feasible reactions, as demonstrated in the hydrogen bomb, are as follows:

$$^{2}_{1}H + {}^{2}_{1}H \longrightarrow {}^{3}_{2}He + {}^{1}_{0}n + 3.2\ \text{MeV}$$

$$^{2}_{1}H + {}^{2}_{1}H \longrightarrow {}^{1}_{1}H + {}^{3}_{1}H + 4.0\ \text{MeV}$$

$$^{3}_{1}H + {}^{2}_{1}H \longrightarrow {}^{4}_{2}He + {}^{1}_{0}n + 17\ \text{MeV}$$

$$\overline{5{}^{2}_{1}H \longrightarrow {}^{3}_{2}He + {}^{1}_{1}H + {}^{4}_{2}He + 2{}^{1}_{0}n + 24\ \text{MeV}}$$

In the first reaction, two deuterium nuclei combine to give a helium-3 nucleus and a neutron; in the second, two deuterium nuclei combine to give a protium and a tritium; in the third reaction, tritium and deuterium combine to give ordinary helium 4 plus a neutron. All three reactions occur together, and the net result is the conversion of five deuterium nuclei to helium 3, helium 4, protium, and two neutrons. Energy emission is 24 MeV for 10 nucleons, or 2.4 MeV per nucleon. This is about three times as great as the 200 MeV for 235 nucleons, or 0.85 MeV per nucleon, obtainable from nuclear fission.

The deuterium-protium ratio in nature is $1:7000$. This may not seem to be very much, but there is such a fantastic amount of protium that the amount of deuterium is quite appreciable. It has been calculated that there is enough deuterium in a gallon of water, separable at a price of about 2 to 3 cents, to be equivalent in energy to 300 gal of gasoline. The problem with getting out that energy is that nuclear fusion occurs only at very high nuclear velocities and at very high nuclear densities. These are contradictory requirements in that the conditions giving high nuclear velocities are high temperatures, which mean high dispersal and low density of nuclei. Current efforts for peaceful use of nuclear fusion are concentrated on the use of plasmas (ionized gases) to attain the high temperatures needed to initiate the process; containment of the plasmas, a major problem, is being tried by strong magnetic fields. Nuclear fusion in bombs (i.e., thermonuclear weapons) is no problem. Particularly convenient is the reaction

$$\ce{^6_3Li} + \ce{^1_0}n \longrightarrow \ce{^4_2He} + \ce{^3_1H}$$

in lithium deuteride. It appears to be the middle phase of an especially clever three-phase bomb: fission-fusion-fission. An ordinary ^{235}U fission bomb sets off nuclear fusion in ^6LiD, which then sets off nuclear fission in a ^{238}U jacket on the warhead. ^{238}U normally does not undergo fission, but with fast neutrons it can be done. The diabolic part comes from the fact that ^{238}U costs only about \$25/kg whereas ^{235}U costs more than \$30,000/kg. Fusion bombs, incidentally, are described as being "cleaner" than fission bombs. This is because they do not produce fission products except from the fission starter. However, they can be made "dirty," i.e., characterized by lots of fallout with very long half-lives, by incorporating in the construction some nonfissionable material such as cobalt. The neutron irradiation of the cobalt would produce a very strong γ emitter.

Exercises

27.1 *Drug action.* Trace a possible sequence of chemical events illustrating drug interference with synaptic transmission.

27.2 *Marihuana.* Discuss the pros and cons for the legalization of marihuana possession.

27.3 *Marihuana.* Show how the structure of the active agent in marihuana is related to cannabinol. Show also how half of the cannabinol molecule is related to the isoprene skeleton.

27.4 *Hallucinogens.* What is a hallucinogen? What are the natural sources? How are the molecular structures related to each other?

27.5 *LSD.* Indicate the relation of LSD to serotonin. What might be the molecular mode of action of LSD? How would the mechanism account for the lack of psychotomimetic action in 2-Br-LSD?

27.6 *Radiation.* It is recommended that television viewers, particularly children, should remain at least 2 m from the screen. By referring to Thomson's old experiment (Fig. 1.5) suggest why this might be a good idea.

27.7 *Radiation.* The radiation hazard of a worker in a high-power radar installation is quite different from that of a dentist taking X-ray pictures of his patient's teeth. What is the essential difference? What protective measures would you recommend in each case?

27.8 *Radioactive decay.* Correlate the different kinds of radioactive disintegration observed with the neutron-proton ratio of the decaying nucleus. Write some typical decay reactions. Make a representative plot of number of nuclei vs. time, and show what half-life means on this plot.

** 27.9   Nucleon binding energy. For an atom of $^{16}_{8}O$, the mass is 15.9949 amu. Calculate the binding energy per nucleon, and compare with that in $^{40}_{20}Ca$ (mass 39.9626 amu).

** 27.10   Nuclear fusion. Thermonuclear weapons are generally described as equivalent to 100 kilotons or more of TNT. What mass of deuterium would you need for such a bomb?

*** 27.11   Drugs. What is a narcotic analgesic? How does it differ from a barbiturate? What chemical features characterize each? What do you suppose is their main difference in mode of molecular action?

*** 27.12   Morphine. Draw structure formulas of morphine, codeine, heroin, and methadone. Show how they are related. If you have access to an atom model kit, especially of the space-filling kind, build a model of the morphine molecule. Describe the features that are believed to be important for binding to the receptor site. What are some of the possible locations of the receptor sites?

*** 27.13   Radiation. In measuring radiation exposure the rad is usually more informative than the rep. Why might the rem be even more informative than the rad?

*** 27.14   Fission products. ^{90}Sr ($t_{1/2} = 28$ yr) is considerably more hazardous as a fallout product than is ^{137}Cs ($t_{1/2} = 30$ yr). Suggest a physiological reason why this is true.

In this chapter we examine material and energy demands on the environment, the problem of population control, and the prediction of limits to growth.

The world is finite; its material resources are limited. Yet man's population continues to skyrocket, and his demands on the environment increase exponentially. Obviously, we have a problem. There are arguments as to whether population increase or multiplying technological demand is the bigger sin, but the facts suggest that man's allowed time for coming up with a solution is rapidly running out.

In 1970 the United States population was about 200 million; for the year 2000 it is projected at 340 million. In 1970, the population of the world was estimated to be 3.5×10^9; for 2000 it is projected at 7×10^9. Each United States child is supposed to put about 50 times as great a resource demand on the environment as a child in a developing country. How long can we continue to ask so much from a limited resource base?

To understand the magnitude of the problem and to be able to evaluate intelligently the solutions being proposed, we need to acquaint ourselves with where we stand with respect to reserves of important materials. This is not easy since the term "reserve" has at best only a qualitative meaning. For minerals, reserves are defined as known stocks of economically exploitable mineral deposits. They are usually characterized as *proven reserves* (which are definitely known to exist), *probable reserves* (which are known with a fair degree of certainty), and *possible reserves* (which pertain to unexplored parts of a mineral deposit and are generally not even quantified but described as small or large). To these three, we could also add *potential reserves* to describe known occurrences that are not now exploitable for various reasons (e.g., too low grade, too far from consuming centers, too deep, and too difficult to treat chemically). Most interesting as an example of the reserve problem is that of oil shale. The proven oil reserves for the free world are 3.75×10^{11} barrels, of which about 0.71×10^{11} are offshore. At current production figures, the *reserve index*, which is defined as reserves-to-annual production, is 33 yr for the world and 11 yr for the United States. However, in the Western United States (mainly Colorado, Utah, and Wyoming), there is a vast area of shale containing kerogen, a complex organic material composed of 77 to 83% C, 5 to 10% H, 10 to 15% O, plus some N. It is the most common form of organic C and is a thousand times more abundant than coal. Probably formed by biochemical conversion of plant and animal material, it can be subjected to destructive distillation to produce hydrocarbons of the oil and gas type. At present, the extraction is not economically competitive, but a Swedish company, the Oil Shale Corporation, has recently developed a new process called TOSCO, which appears promising. The essence of the extraction is to crush the shale, add preheated alumina balls, and rotate in a horizontal furnace in the absence of air. The reserves of shale oil are apparently very large, estimated conservatively at 9.6×10^{11} barrels in the Green River formation alone, for example.

As far as metals are concerned, typical estimated figures for reserve-index values are given in Fig. 28.1. The actual values vary depending on the source and may change dramatically from year to year as new discoveries are made or new technology is applied to the problem of extraction. Another way of presenting reserve-index estimates is to divide reserves by annual consumption instead of by annual production. In some cases, this gives a significantly different picture. As an example, for iron the ratio of United States domestic reserves to United States domestic consumption is 120 instead of 180; the difference in the figures arises from the fact that the United States imports about a third of its steel consumption. One can argue that the United States cannot expect to live off the rest of the world's reserves for more than a few years at best. Still, very low reserve indexes are not necessarily cause for panic. In the case of aluminum, for example, the United States gets most of its bauxite mineral from friendly, nearby sources; so it is not even stockpiled. Furthermore, much of the technology for getting aluminum out of clay, for example, where abundant reserves exist, has been worked out, although the process as yet is still not competitive economically. However, in a real pinch it could be exploited.

More serious than the materials resource depletion, where history tells us there will be a rolling replacement of one material by another (e.g., steel by plastics), is the expected strain on food production. With world population doubling approximately every 35 yr, we will need to feed about 7×10^9 people by the year 2000, 14×10^9 by 2035, and 28×10^9 by 2070. It appears that increased use of fertilizers and improved agriculture practices can comfortably quadruple the present food production on land and perhaps double it from the sea. Beyond that, long-range prospects begin to look grim. Already, according to United Nations estimates the average food consumption over the world is less than that considered to be a desirable minimum (2200 nutritional Calories per day versus 2400; one nutritional Calorie = 10^3 cal = 4.18×10^3 J), even if we ignore local imbalances arising from poor distribution, social distortions, and natural catastrophes. It is believed that at maximum the world can feed 30×10^9 people at near-starvation levels. Much more desirable would be to have the population level off at about 10×10^9 people, which could be supported quite comfortably. We will return to this point in the final section of this chapter.

28.2 Energy crisis

The other great problem for man is energy. Where does it come from? Where does it go? How long before we run out of that commodity in the environment? Practically all our energy comes ultimately from the sun. It pours 3.8×10^{26} J into space per second, and at the earth's distance from the sun this solar flux amounts to 0.134 J/cm². Considering the earth as a disk of area πr^2, the total energy reaching us is 5.4×10^{24} J/yr. (Actually the earth is not a disk but a spinning sphere of area $4\pi r^2$, which means the average solar flux at the earth's surface would only be 0.033 J/cm².) What happens to our supply of $5.4 \times$

Metal	Reserve index (reserves/annual production)	
	World	U.S.
Iron	380	180
Molybdenum	100	200
Copper	40	26
Uranium	24	12
Manganese	146	6
Lead	18	4
Zinc	20	20
Tungsten	40	12
Aluminum	160	4
Chromium	532	6

Fig. 28.1
Reserve-index values in years for various metal minerals.

10^{24} J/yr? About 35 percent of this gets reflected back into space; so we really have only 3.5×10^{24} J/yr to play with.

Some gets converted into plants by the process of photosynthesis:

$$6CO_2 + 6H_2O \longrightarrow C_6H_{12}O_6 + 6O_2 \qquad \Delta H = +2800 \text{ kJ}$$

It is estimated that the net primary productivity averaged over the whole earth amounts to 320 g/m^2, in terms of number of dry grams of green plant matter produced per year. Multiplying by the surface of the earth we get a total of 1.6×10^{17} g/yr. In terms of the above reaction, this would correspond to about 2.5×10^{21} J/yr. If we allow another 25 percent for self-metabolism of the plants produced, then the energy consumption by plant growth is about 3.1×10^{21} J/yr. Compared with the total supply of 3.5×10^{24} J/yr, the plant requirement is only about 0.1 percent.

What fraction of this goes into food? If we take the average consumption as 2200 nutritional Calories per person per day, then we have $(2.2 \times 10^6$ cal$)(4.18$J/cal$)(3.5 \times 10^9$ people$)(365$ days/yr$)$, or 1.2×10^{19} J/yr. In other words, if all our food requirements were met through plants, we would be using for food only $1.2 \times 10^{19}/3.1 \times 10^{21}$, or one two-hundred-fiftieth, of the total energy that comes from the sun and goes into plant growth. (Actually, we like much of our food as meat, and the conversion from plant to animal is only about 10 percent efficient.) Of course, much of plant growth goes into wood. Some of this is used directly as fuel. In the past, much plant material was converted to the fossil fuels petroleum and coal, and presumably this process is still going on.

Where else does the energy from the sun go? Lest we think we might ultimately be able to use it all for food production, let us simply point out that approximately one-third of the total 3.5×10^{24} J/yr goes to power the hydrocycle. It is estimated there is about 5×10^{17} liters of rain per year. To evaporate this much water would take $(5 \times 10^{17}$ liters$)(55$ mol/liter$)(40.9$ kJ/mol$) = 1.1 \times 10^{24}$ J. Other weather cycles also have their energy requirements.

The above energy figures are a bit abstract, and they seem to show enough surplus that we should have little cause for worry. From where then comes all this noise about an "energy crisis"? Part of it comes from the fact that the rate of external energy consumption (i.e., other than food) is going up faster than the population. For instance, world energy consumption, which grew from 1.5×10^{19} J/yr in 1900 to 20×10^{19} J/yr in 1970 is increasing at an annual rate of 3.6 percent per year, the per capita energy consumption $(2.8 - 0.9 = 1.9$ growth) this means per capita consumption is going up at 1.6 percent per year. For the United States, the growth rate in energy consumption is smaller (i.e., 2.8 percent per year), but it starts from a higher base. Even so, because population growth in the United States is only 0.9 percent per year, the per capita energy consumption growth $(2.8 - 0.9 = 1.9)$ is greater than for the world average. It is a fact that, although the United States accounts for only $6\frac{1}{2}$ percent of the world's population, it accounts for 30 percent of its energy consumption.

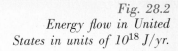

Fig. 28.2
*Energy flow in United
States in units of 10^{18} J/yr.*

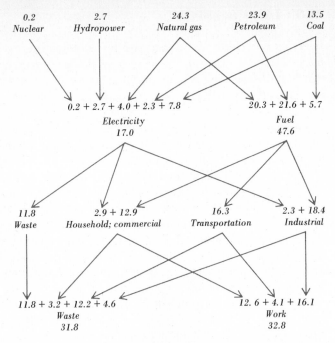

Figure 28.2 shows a schematic representation of the energy flow from source to use in the United States. As can be seen, most of the energy comes from the fossil fuels, natural gas, petroleum, and coal. About a quarter of it goes into electricity, and three-quarters into fuel. It may come as a shock that half of the energy is wasted and only half ends up as useful work, but the facts of life are that losses in the generation and transmission of power are inevitable.

More interesting than representative figures for energy consumption are the reserve-index figures. How long will our energy sources last? We can get an idea of this by assuming all the United States consumption is furnished by a single energy source. If it all came from coal, proven reserves would last 125 yr; potential reserves might bring this figure up to 1300 yr. If it all came from petroleum, proven United States reserves would last only 5 yr. Potential reserves would stretch this out to 280 yr. Exploitation of oil shale would last us another 2500 yr. Natural gas is worth only 5 yr but may go for 110 yr.

Nuclear reactors of the conventional fission type would carry us for only 2.3 yr, possibly for 15 yr. If breeder reactors were operational, then we could go for 115 yr, possibly 750 yr. The spectacular source is nuclear fusion. If we could make it work, it would last us 10^6 yr, maybe 10^9 yr. In other words, nuclear fusion offers the possibility of an essentially unlimited supply of energy, which would last us "forever." As we will see in the next section, unlimited energy is one of the parameters that has to be fed into the computer-simulated world model in order to prevent a serious decline in the quality of life within about 50 yr.

At present, practical energy sources are severely limited. Already, local crises of supply are beginning to occur. They are certain to become more frequent in the near future.

28.3 Limits to growth

In trying to predict where man and his environment are going, mere mortals are at a disadvantage because of the many *ifs, ands,* and *buts* that are involved in making the predictions. The cross influence of how fast a change in one parameter can bring about a change in another parameter that affects the first parameter is generally only dimly seen; so extrapolations other than linear ones are rarely made. With the advent of computers, much of the guesswork in the predicting process has been removed, but, as we shall see, there is a surprising amount of controversy left about interpretation of the results obtained.

In 1968, an Italian industrialist, Dr. Aurelio Peccei, brought together a group of 70 people from 25 countries, the so-called "Club of Rome," to study the present and future predicament of mankind. The first stage of the assignment was to take presently known facts about man and his environment and determine from straightforward computer analysis what would happen if present trends in resource depletion, population growth, etc., were allowed to continue. The problem was essentially a computer-simulation problem in which certain parameters were defined as to level and rate of change and various interconnections of cross influence were established. A so-called "world model" was set up, and five parameters chosen to describe it: population, pollution, resources, capital investment, and capital investment in agriculture. The idea was that certain mathematical equations could be set up relating these variables to each other, in particular, the rate at which a change in one would effect a change in each of the others. The problem is referred to as a problem in *world dynamics*.

Once the parameters had been chosen and the connecting equations established, it was then simply a question of turning over present values of the parameters plus how fast they were changing and allowing the computer to map out the future course of events. For instance, population levels, birth rates, death rates, etc., are fairly well known figures, which can be fed into the problem. Admittedly some of the variables such as pollution could be only roughly quantified; so there was an air of unreality about the whole exercise. However, the results were almost uniformly pessimistic. No matter what man does, in about 50 to 100 yr he faces a real crisis in that "quality of living," a vague term that is gradually losing favor, goes into a rather steep decline sometime around A.D. 2020.

Figure 28.3 shows what is known as the *world-model standard run*. It assumes no major change in the relations that have historically governed the world system. Food, industrial output, and population grow at an exponentially increasing rate until the inevitable steep decline in resources forces a slowdown in industrial growth followed at some delay time later by a drop in pollution and population.

Successive changes are then introduced into the model to see what happens. What if we assume that resources are not limited but technology finds a way, e.g., nuclear power makes feasible extensive recycling of resources, to keep resources from showing the sharp drop shown

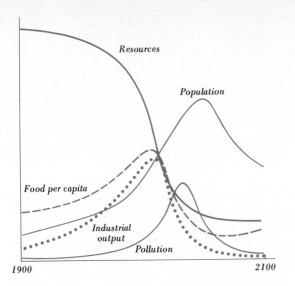

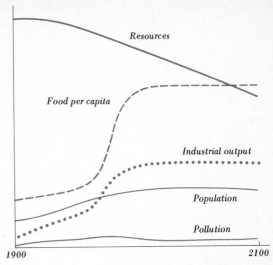

(on left) *Fig. 28.3*
*World-model standard-run
projections. (Adapted from
"Limits to Growth" by
Meadows, et al.)*

(above) *Fig. 28.4*
*Stabilized world-model
projections. (Adapted from
"Limits to Growth" by
Meadows, et al.)*

in Fig. 28.3? The result is pretty much the same except now it is pollution that goes out of control and brings about a declining population, food per capita, etc. Next, one puts in the assumption that unlimited nuclear energy when applied to pollution controls will eradicate pollution and remove that as a growth-limiting factor. Again, however, the world system goes to catastrophe, a bit later admittedly, but still rather clearly indicating a decline in the quality of living. This time the culprit is food shortage. How to resolve the dilemma? If population does not get you, pollution will; if pollution does not get you, eventually food shortage will. What is needed for a stabilized world model? Figure 28.4 shows the ideal solution. It would come about if population were made constant by equality of birth and death rates, if there were maximum resource recycling so that nature's stock was not depleted, and if pollution controls kept the environment from degrading beyond the unpleasant to the impossible.

Views of the world system at the extremes can be characterized as malthusian or technological-optimistic. In the malthusian view, population inevitably catches up with any improvement in quality of living so that overcrowding eventually degrades the quality of life. Equilibrium establishes at high death rate = high birth rate. Technology, in this view, can only temporarily alleviate the problem. On the other extreme, the technological optimist argues that resources never deplete as rolling adjustment occurs to substitute one material for another. Quality of life continually improves as increased technology increases productivity and standard of living. Equilibrium establishes at low death rate = low birth rate.

The Club of Rome world model has been characterized as malthusian. Misery, starvation, and low quality of life—these are the inevitable lot of man unless he takes draconian measures to limit population growth. It is certainly true that population cannot increase indefinitely. How to stop it? Pollution, food shortage, and resource depletion—these can all be effective. Is there, however, a better way? One of the criticisms of the Club of Rome world model is that it does not allow for social feedback. There is not enough scope for man to intervene when the system heads off in undesirable directions. As pointed out before, man is clever at getting into trouble but even more clever at getting out of it. It is to be hoped that in this case he will recognize the problem before it is too late.

World population increased rather slowly until about 1650. Estimated to be 0.25×10^9 in the year 1 and 0.50×10^9 in 1650, it went to 1.1×10^9 in 1850 and to 2.0×10^9 in 1930, and is estimated to reach 4.0×10^9 in 1975. The doubling period has shrunk from 1650 to 200 to 80 to 45 to 35 yr. There is some recent indication that it may be stabilizing at about 45 yr. In any event, predictions of future populations have to be made with great care since they have almost always been wrong, and that on the low side! Part of the problem arises from the fact that good predictions can be made only if we have good age profiles for the various countries, especially the developing countries, which, having generally younger populations, are likely to grow faster. For example, whereas advanced countries appear to be stabilizing, with an average annual population increase of 0.8 percent, the developing countries are growing at a rate of 2.8 percent per year. Because developing countries account for 69 percent of the current total, no quick deceleration in population growth is expected. The only encouraging fact is that advanced countries, when they become advanced, go through a demographic transition where death rates fall but birth rates fall even faster. So far, the later the time of the demographic transition, the faster the fall. The United Nations now makes its population projections separately for advanced and developing countries. Whereas the advanced countries from 1970 to 2000 are expected to go from 1.09×10^9 to 1.55×10^9 (an increase of 34 percent), the developing countries are predicted to go from 2.56×10^9 to 6.37×10^9 (an increase of 149 percent). The total expected for 2000 of 7.82×10^9 people is very much an upper limit; the lower estimation limit is 6.49×10^9.

How can one limit population? One suggestion has been to raise taxes, inflate the currency, make housing scarce, force wives to work to gain supplemental income, and starve the transit system. All these are best described as *déjà vu*, or at best "how to improve the quality of life by lowering it." More effective, apparently, is chemical contraception. This is shown by a comparison of the relative efficiency of various contraception methods as measured by the number of pregnancies per 100 woman-years of different exposure: douche, 31; rhythm, 24; jelly, 20; withdrawal, 18; condom, 14; diaphragm, 12; intrauterine device, 5; sequential steroids, 5; and combined steroids, 0.1. As can be seen, the combined steroids are practically 100 percent efficient in contraception.

The "Pill" really started in 1953, when it was discovered that there is ovulation inhibition by the chemical progesterone, shown in Fig. 28.5, and the estrogens, one of which is also shown in Fig. 28.5.

Both of these are steroid hormones. The steroids are complicated ring structures which have biologic activity, but they have not been elucidated as to mechanism of action. Synthetic derivatives having minor structure modifications have been prepared; some of them show considerably more biologic activity than the natural compounds. In sequential therapy, an estrogen is administered daily for 15 days and then estrogen plus progesterone or derivative for 5 days. In combined therapy, both hormones are administered daily for 20 or 21 days. In both cases dosage starts on the fifth day of the menstrual cycle.

How do these chemicals accomplish their action? They are chemicals normally produced in pregnancy; so they can be said to simulate chemically a pseudopregnancy. Their function is to prevent ovulation; so of course there is no conception. Still, we must admit that there is no adequate explanation for the mode of action. The problem basically is that the functioning of the natural hormones is not understood. Until it is, the "Pill" will be a mystery. Nevertheless, over 10×10^6 women in the United States have used the "Pill." Questions have been raised about the possibility of deleterious side effects (e.g., coagulability of the blood and liver disturbance), but the hazard is apparently minor, less than that of pregnancy itself.

It is manifestly unjust to assign the full responsibility for population control to womankind alone. Chemical contraception by the male, however, has not yet been developed. In any case, for both sexes there are complex questions of social, religious, and personal barriers that need to be overcome before effective birth control can become a world reality. The only sure thing at this point is that society must recognize that growth *per se* is not necessarily good and that something has to be saved for future generations.

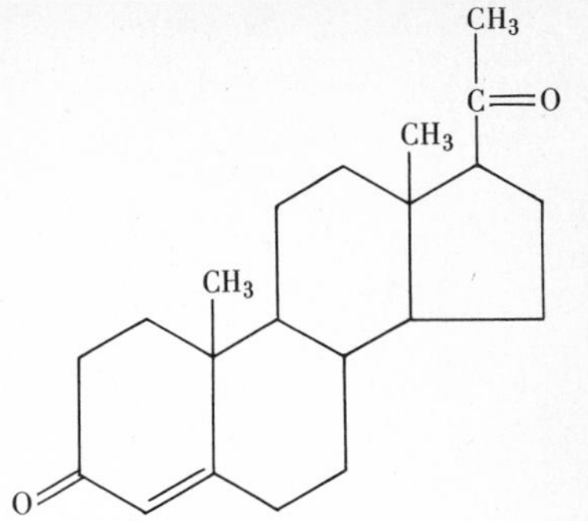

Progesterone

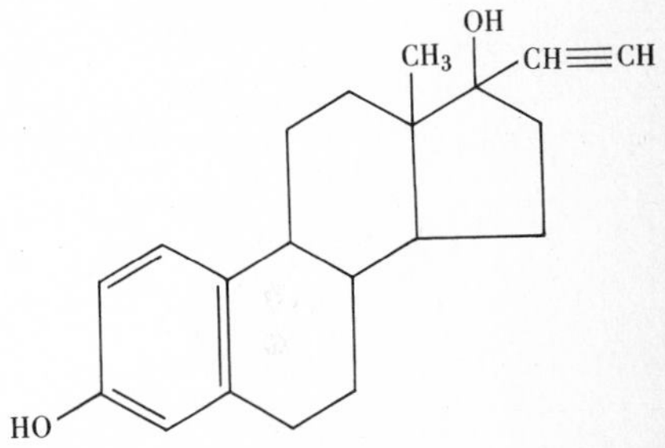

Fig. 28.5
Examples of steroid hormones.

Exercises

28.1 *Photosynthesis.* Given $\Delta H = 2800$ kJ for production of one mole of $C_6H_{12}O_6$ from CO_2 and H_2O, calculate what wavelength of light would be needed per molecule of CO_2 to achieve photosynthesis in a one-step process.

28.2 *Population.* Assuming that population growth is a first-order rate process $dP = kP\,dt$ and using world population figures of 2.4×10^9 for 1945 and 3.6×10^9 for 1970, show that the doubling time is 44 yr and annual growth is 1.56 percent.

28.3 *Limits to growth.* It has been pointed out that should other methods of population control fail it might become necessary to resort to such draconian measures as adding contraceptive chemicals to public water supplies, assuming such chemicals could be developed. Discuss the possibility in the context of the problem of overpopulation and what should be done to prevent it.

28.4 *Man.* What is your assessment of man's outlook for the future? What course of action should he follow to ensure a good quality of life for himself and his progeny?

Most of the units used in this text are SI units, as recommended by the International Committee on Weights and Measures. The International System of Units (usually designated SI, after Système International) is constructed from seven base units. These are the following:

Name of unit	*Physical quantity*	*Symbol for unit*
Meter	*Length*	m
Kilogram	*Mass*	kg
Second	*Time*	s (sec)
Ampere	*Electric current*	A (amp)
Kelvin	*Thermodynamic temperature*	K ($^\circ$K)
Candela	*Light intensity*	cd
Mole	*Amount of substance*	mol

The symbols given in parentheses, although not the officially recommended ones, are used in this text for the sake of clarity.

Decimal fractions or multiples of these units are indicated as follows:

Fraction	*Prefix*	*Symbol*	*Multiple*	*Prefix*	*Symbol*
10^{-1}	*Deci-*	d	10^{1}	*Deka-*	da
10^{-2}	*Centi-*	c	10^{2}	*Hecto-*	h
10^{-3}	*Milli-*	m	10^{3}	*Kilo-*	k
10^{-6}	*Micro-*	μ	10^{6}	*Mega-*	M
10^{-9}	*Nano-*	n	10^{9}	*Giga-*	G
10^{-12}	*Pico-*	p	10^{12}	*Tera-*	T
10^{-15}	*Femto-*	f			
10^{-18}	*Atto-*	a			

Derived from the base units are various specially named units. These include the following:

Name of unit	Physical quantity	Symbol for unit	Definition of unit
Newton	Force	N	$kg \cdot m/s^2$
Pascal	Pressure	Pa	$N/m^2 = kg\ m^{-1}\ s^{-2}$
Joule	Energy	J	$kg \cdot m^2/s^2$
Watt	Power	W	$J/s = kg \cdot m^2/s^3$
Coulomb	Electric charge	C	$A \cdot s$
Volt	Electric potential difference	V	$J\ A^{-1}\ s^{-1} = kg \cdot m^2\ s^{-3}\ A^{-1}$
Ohm	Electric resistance	Ω	$V/A = kg \cdot m^2\ s^{-3}\ A^{-2}$
Siemens	Electric conductance	S	$\Omega^{-1} = A/V = s^3 \cdot A^2\ kg^{-1}\ m^{-2}$
Farad	Electric capacitance	F	$A \cdot s/V = A^2 \cdot s^4\ kg^{-1}\ m^{-2}$
Hertz	Frequency	Hz	s^{-1} (cycle per second)

Use of the above units is recommended. There are also certain decimal fractions and multiples of SI units, having special names, which do not belong to the International System of Units; their use is to be *progressively discouraged*. Among such are the following:

Name of unit	Physical quantity	Symbol for unit	Definition of unit
Angstrom	Length	Å	$10^{-10}\ m = 10^{-8}\ cm$
Dyne	Force	dyn	$10^{-5}\ N$
Bar	Pressure	bar	$10^5\ N/m^2$
Erg	Energy	erg	$10^{-7}\ J$

There are, in addition, other units that are not simple fractions or multiples of SI units which can now be defined exactly in terms of SI units. They do not belong to the International System of Units and are recommended to be *abandoned*. These include the following:

Name of unit	Physical quantity	Symbol for unit	Definition of unit
Inch	Length	in	$2.54 \times 10^{-2}\ m$
Pound	Mass	lb	$0.453502\ kg$
Atmosphere*	Pressure	atm	$101,325\ N/m^2$
Torr	Pressure	Torr	$(101,325/760)\ N/m^2$
Millimeter of mercury	Pressure	mmHg	$13.5951 \times 980.665 \times 10^{-2}\ N/m^2$
Calorie	Energy	cal	$4.184\ J$

Appendix 1
SI units

* Use of this unit is sanctioned for a limited period of time.

Besides the above, there are certain natural units tied directly to the properties of microscopic constituents of matter. Use of these as natural units is acceptable. Important examples are the following:

Unit or physical quantity	Symbol	Conversion factor
Atomic mass unit	amu	1.6605×10^{-24} g
Avogadro number	N	6.0222×10^{23} molecules per mole
Boltzmann constant	k	1.3806×10^{-23} J/deg
Electron charge	e	1.6022×10^{-19} coulomb
Electron mass	m	9.1096×10^{-28} g
Electron volt	eV	1.6022×10^{-19} J
Faraday constant	$\mathcal{F}$	9.6487×10^4 coulombs/equiv
Gas constant	R	8.2057×10^{-2} l-atm mol^{-1} deg^{-1}
		8.3143 J mol^{-1} deg^{-1}
Planck's constant	h	6.6262×10^{-34} J-s
Speed of light	c	2.9979×10^8 m/s

2.1   *Inorganic compounds*

Compounds composed of but two elements have names derived directly from the elements. Usually the more electropositive element is named first, and the other element is given an *-ide* ending. Thus, we have sodium chloride, NaCl; calcium oxide, CaO; and aluminum nitride, AlN. If more than one atom of an element is involved, prefixes such as *di-* (for 2), *tri-* (3), *tetra-* (4), *penta-* (5), and *sesqui-* ($1\frac{1}{2}$) are used. For example, AlF_3 is aluminum trifluoride, Na_3P is trisodium phosphide, and N_2O_4 is dinitrogen tetroxide. When the same two elements form more than one compound, the compounds can be distinguished as in the following example:

$FeCl_2$	$FeCl_3$
(1) *Iron dichloride*	*Iron trichloride*
(2) *Ferrous chloride*	*Ferric chloride*
(3) *Iron(II) chloride*	*Iron(III) chloride*

In (1) distinction is made through use of prefixes; in (2) the endings *-ous* and *-ic* denote the lower and higher oxidation states, respectively, of iron; in (3), the Stock system, Roman numerals in parentheses indicate the oxidation states. In a given series of compounds, the suffixes *-ous* and *-ic* may not be sufficient for complete designation but may need to be supplemented by one of the other methods of nomenclature. For example, the oxides of nitrogen are usually named as follows: 747

N_2O	nitrous oxide
NO	nitric oxide
N_2O_3	dinitrogen trioxide, or nitrogen sesquioxide
NO_2	nitrogen dioxide
N_2O_4	dinitrogen tetroxide, or nitrogen tetroxide
N_2O_5	dinitrogen pentoxide, or nitrogen pentoxide

Compounds containing more than two elements are named differently depending on whether they are bases, acids, or salts. Since most bases contain hydroxide ion (OH^-), they are generally called hydroxides, e.g., sodium hydroxide (NaOH), calcium hydroxide [$Ca(OH)_2$], and arsenic trihydroxide [$As(OH)_3$]. The naming of acids and of salts derived from them is more complicated, as can be seen from the following series:

Acid	*Sodium salt*
HClO, *hypochlorous acid*	NaClO, *sodium hypochlorite*
$HClO_2$, *chlorous acid*	$NaClO_2$, *sodium chlorite*
$HClO_3$, *chloric acid*	$NaClO_3$, *sodium chlorate*
$HClO_4$, *perchloric acid*	$NaClO_4$, *sodium perchlorate*
H_2SO_3, *sulfurous acid*	Na_2SO_3, *sodium sulfite*
H_2SO_4, *sulfuric acid*	Na_2SO_4, *sodium sulfate*

When there are only two common oxyacids of a given element, the one corresponding to lower oxidation state is given the *-ous* ending, and the other the *-ic* ending. If there are more than two oxyacids of different oxidation states, the prefixes *hypo-* and *per-* may also be used. As indicated in the above example, the prefix *hypo-* indicates an oxidation state lower than that of an *-ous* acid, and the prefix *per-*, an oxidation state higher than that of an *-ic* acid. For salts derived from oxyacids the names are formed by replacing the ending *-ous* by *-ite* and *-ic* by *-ate*. Salts derived from polyprotic acids (for example, H_3PO_4) are best named so as to indicate the number of hydrogen atoms left unneutralized. For example, NaH_2PO_4 is monosodium dihydrogen phosphate, and Na_2HPO_4 is disodium monohydrogen phosphate. Frequently, the prefix *mono-* is left off. For monohydrogen salts of diprotic acids, such as $NaHSO_4$, the presence of hydrogen may also be indicated by the prefix *bi-*. Thus, $NaHSO_4$ is sometimes called sodium bisulfate, though the name "sodium hydrogen sulfate" is preferred.

Complex cations, such as $Cr(H_2O)_6^{3+}$, are named by giving the number and name of the groups attached to the central atom followed by the name of the central atom with its oxidation number indicated by Roman numerals in parentheses. Thus, $Cr(H_2O)_6^{3+}$ is hexaaquo-chromium(III). Complex anions, such as $PtCl_6^{2-}$, are named by giving the number and name of attached groups followed by the name of the element with an *-ate* ending and its oxidation number in parentheses. Thus, $PtCl_6^{2-}$ is hexachloroplatinate(IV). If the attached groups (*ligands*) are not all alike, it is customary to name the ligands in the same order in which they should be written in the formula—i.e., anion ligands generally precede neutral ligands. If more than one kind of anion ligand is present, the order is H^- (hydrido), O^{2-} (oxo), OH^- (hydroxo), other monatomic anions (in order of increasing electronegativity of the elements—for example, F^-, fluoro, last), polyatomic anions (in order of increasing number of atoms), and organic anions (in alphabetic order). If more than one kind of neutral ligand is present, the order is H_2O (aquo), NH_3 (ammine), other inorganic ligands (in order of increasing electronegativity of their central atom—for example, CO, carbonyl, precedes NO, nitrosyl), and organic ligands (in alphabetic order). To indicate the numbers of each kind of ligand, Greek prefixes are used: *mono-* (usually can be omitted), *di-*, *tri-*, *tetra-*, *penta-*, *hexa-*, *hepta-*, and *octa-*. Instead of these prefixes, *bis-* (twice), *tris-* (thrice), *tetrakis-* (four times), etc., may be used, especially when the name of the ligand itself contains a numerical designation (e.g., ethylenediamine, frequently abbreviated *en*). Some examples of the application of the above rules follow:

$CrCl_2(H_2O)_4{}^+$	dichlorotetraaquochromium(III)
$CrCl_4(H_2O)_2{}^-$	tetrachlorodiaquochromate(III)
$Cr(H_2O)(NH_3)_5{}^{3+}$	aquopentaamminechromium(III)
$Ga(OH)Cl_3{}^-$	hydroxotrichlorogallate(III)
cis-$PtBrCl(NO_2)_2{}^{2-}$	*cis*-bromochlorodinitroplatinate(II)
trans-$Co(OH)Clen_2{}^+$	*trans*-hydroxochlorobisethylenediaminecobalt(III)
$Mn(CO)_3(C_6H_6)^+$	tricarbonylbenzenemanganese(I)

In the case of complex-ion isomerism, the names *cis-* or *trans-* may precede the formula or the complex-ion name to indicate the spatial arrangement of the ligands. Cis means the ligands occupy adjacent coordination positions; trans means opposite positions.

2.2 Organic compounds

The key rules recommended by the International Union of Pure and Applied Chemistry (IUPAC) are summarized as follows:

1 Choose as the parent carbon skeleton the longest sequence of C atoms that contains the principal functional group.

2 Name the parent structure using the name of the alkane that contains the same number of C atoms as the chosen structure. Replace *-ane* by *-ene* for double bond or *-yne* for triple bond. If a functional group is present, drop the final *-e* and add suffixes as follows:

-ol for alcohol (OH)
-al for aldehyde (CHO)
-one for ketone (CO)
-oic acid for acid (COOH)

3 Use prefixes in alphabetic order to denote other substituents.

4 Locate substituents and points of unsaturation by numbering the C atoms of the parent skeleton with the following criteria used in decreasing order of priority:

(a) Assign the C atom of the principal functional group the number 1 if it is terminal.

(b) Assign numbers so that the location of the principal functional group is as low as possible if the group is non-terminal.

(c) Assign numbers so that substituents are located by lowest possible numbers. If there are two kinds of substituents, give low-number preference to the first named.

5 If an attached side chain bears substituents, it too must be numbered starting with the C atom which is attached to the parent carbon skeleton. Names of substituents on the side chain and numbers locating them are enclosed in parentheses with the name of the side chain:

$$\overset{\text{CH}_3}{\underset{|}{\text{CH}_3\text{CHCH}_2\text{CH}_3}}$$

2-Methylbutane

$$\overset{\text{CH}_3}{\underset{|}{\text{CH}_3\text{CH}_2\text{CHCH}=\text{CH}_2}}$$

3-Methyl-1-pentene

$$\overset{\text{CH}_3}{\underset{\underset{\text{CH}_3}{|}}{\underset{|}{\text{CH}_3\text{CH}_2\text{CHCHCH}_2\text{OH}}}}$$

2,3-Dimethyl-1-pentanol

$$\text{CH}_3\text{CH}_2\text{CH}=\overset{\overset{\text{O}}{||}}{\text{CHCCH}_3}$$

3-Hexene-2-one

Esters are named by replacing the suffix of the parent acid *-oic acid* by *-oate:*

$$\text{CH}_3\text{CH}_2\text{CH}_2\text{COOCH}_3$$

Methylbutanoate

$$\text{CH}_3\text{COOCH}_2\text{CH}=\overset{\overset{\text{CH}_3}{|}}{\text{CCH}_3}$$

3-Methyl-2-butenylethanoate

Cyclic aliphatic hydrocarbons are named by prefixing *cyclo-* to the name of the corresponding open-chain hydrocarbon having the same number of C atoms as the ring:

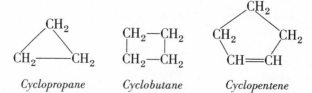

Cyclopropane   *Cyclobutane*   *Cyclopentene*

Rings containing atoms other than C (heterocycles) as well as aromatic rings are usually designated by trivial (nonsystematic) names:

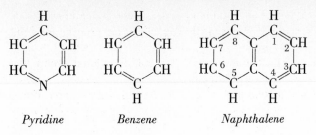

Pyridine   *Benzene*   *Naphthalene*

To locate substituents, rings are numbered clockwise around the periphery as shown for naphthalene.

Appendix 3 **Vapor pressure of water**

Temperature, °C	Pressure, atm	Pressure, mmHg	Temperature, °C	Pressure, atm	Pressure, mmHg
0	0.00603	4.58	23	0.0277	21.07
1	0.00648	4.93	24	0.0294	22.38
2	0.00697	5.29	25	0.0313	23.76
3	0.00748	5.69	26	0.0332	25.21
4	0.00803	6.10	27	0.0352	26.74
5	0.00861	6.54	28	0.0373	28.35
6	0.00923	7.01	29	0.0395	30.04
7	0.00989	7.51	30	0.0419	31.82
8	0.0106	8.04	35	0.0555	42.18
9	0.0113	8.61	40	0.0728	55.32
10	0.0121	9.21	45	0.0946	71.88
11	0.0130	9.84	50	0.1217	92.51
12	0.0138	10.52	55	0.1553	118.04
13	0.0148	11.23	60	0.1966	149.38
14	0.0158	11.99	65	0.2468	187.54
15	0.0168	12.79	70	0.3075	233.7
16	0.0179	13.63	75	0.3804	289.1
17	0.0191	14.53	80	0.4672	355.1
18	0.0204	15.48	85	0.5705	433.6
19	0.0217	16.48	90	0.6918	525.8
20	0.0231	17.54	95	0.8341	633.9
21	0.0245	18.65	100	1.0000	760.0
22	0.0261	19.83	105	1.1922	906.1

4.1   *Exponential numbers*

Multiplication by a positive power of 10 corresponds to moving the decimal point to the right; multiplication by a negative power of 10 corresponds to moving the decimal point to the left:

1.23×10^4 is 12,300.
1.23×10^{-4} is 0.000123.

Numbers expressed with powers of 10 can be added or subtracted directly only if the powers of 10 are the same:

$$1.23 \times 10^4 + 1.23 \times 10^5 = 1.23 \times 10^4 + 12.3 \times 10^4$$
$$= 13.5 \times 10^4$$

$$1.23 \times 10^{-4} - 1.23 \times 10^{-5} = 1.23 \times 10^{-4} - 0.123 \times 10^{-4}$$
$$= 1.11 \times 10^{-4}$$

When powers of 10 are multiplied, exponents are added; when divided, exponents are subtracted:

$$(1.23 \times 10^4)(1.23 \times 10^5) = (1.23 \times 1.23)(10^4 \times 10^5)$$
$$= 1.51 \times 10^9$$

$$\frac{1.23 \times 10^{-4}}{1.23 \times 10^{-5}} = \frac{1.23}{1.23} \times \frac{10^{-4}}{10^{-5}} = 1.00 \times 10$$

In taking square roots of powers of 10, the exponent is divided by   *753*

Logarithms

	0	1	2	3	4	5	6	7	8	9
10	0000	0043	0086	0128	0170	0212	0253	0294	0334	0374
11	0414	0453	0492	0531	0569	0607	0645	0682	0719	0755
12	0792	0828	0864	0899	0934	0969	1004	1038	1072	1106
13	1139	1173	1206	1239	1271	1303	1335	1367	1399	1430
14	1461	1492	1523	1553	1584	1614	1644	1673	1703	1732
15	1761	1790	1818	1847	1875	1903	1931	1959	1987	2014
16	2041	2068	2095	2122	2148	2175	2201	2227	2253	2279
17	2304	2330	2355	2380	2405	2430	2455	2480	2504	2529
18	2553	2577	2601	2625	2648	2672	2695	2718	2742	2765
19	2788	2810	2833	2856	2878	2900	2923	2945	2967	2989
20	3010	3032	3054	3075	3096	3118	3139	3160	3181	3201
21	3222	3243	3263	3284	3304	3324	3345	3365	3385	3404
22	3424	3444	3464	3483	3502	3522	3541	3560	3579	3598
23	3617	3636	3655	3674	3692	3711	3729	3747	3766	3784
24	3802	3820	3838	3856	3874	3892	3909	3927	3945	3962
25	3979	3997	4014	4031	4048	4065	4082	4099	4116	4133
26	4150	4166	4183	4200	4216	4232	4249	4265	4281	4298
27	4314	4330	4346	4362	4378	4393	4409	4425	4440	4456
28	4472	4487	4502	4518	4533	4548	4564	4579	4594	4609
29	4624	4639	4654	4669	4683	4698	4713	4728	4742	4757
30	4771	4786	4800	4814	4829	4843	4857	4871	4886	4900
31	4914	4928	4942	4955	4969	4983	4997	5011	5024	5038
32	5051	5065	5079	5092	5105	5119	5132	5145	5159	5172
33	5185	5198	5211	5224	5237	5250	5263	5276	5289	5302
34	5315	5328	5340	5353	5366	5378	5391	5403	5416	5428
35	5441	5453	5465	5478	5490	5502	5514	5527	5539	5551
36	5563	5575	5587	5599	5611	5623	5635	5647	5658	5670
37	5682	5694	5705	5717	5729	5740	5752	5763	5775	5786
38	5798	5809	5821	5832	5843	5855	5866	5877	5888	5899
39	5911	5922	5933	5944	5955	5966	5977	5988	5999	6010
40	6021	6031	6042	6053	6064	6580	6590	6599	6609	6618
41	6128	6138	6149	6160	6170	6675	6684	6693	6702	6712
42	6232	6243	6253	6263	6274	6767	6776	6785	6794	6803
43	6335	6345	6355	6365	6375	6857	6866	6875	6884	6893
44	6435	6444	6454	6464	6474	6946	6955	6964	6972	6981
45	6532	6542	6551	6561	6571	6075	6085	6096	6107	6117
46	6628	6637	6646	6656	6665	6180	6191	6201	6212	6222
47	6721	6730	6739	6749	6758	6284	6294	6304	6314	6325
48	6812	6821	6830	6839	6848	6385	6395	6405	6415	6425
49	6902	6911	6920	6928	6937	6484	6493	6503	6513	6522
50	6990	6998	7007	7016	7024	7033	7042	7050	7059	7067
51	7076	7084	7093	7101	7110	7118	7126	7135	7143	7152
52	7160	7168	7177	7185	7193	7202	7210	7218	7226	7235
53	7243	7251	7259	7267	7275	7284	7292	7300	7308	7316
54	7324	7332	7340	7348	7356	7364	7372	7380	7388	7396

	0	1	2	3	4	5	6	7	8	9
55	7404	7412	7419	7427	7435	7443	7451	7459	7466	7474
56	7482	7490	7497	7505	7513	7520	7528	7536	7543	7551
57	7559	7566	7574	7582	7589	7597	7604	7612	7619	7627
58	7634	7642	7649	7657	7664	7672	7679	7686	7694	7701
59	7709	7716	7723	7731	7738	7745	7752	7760	7767	7774
60	7782	7789	7796	7803	7810	7818	7825	7832	7839	7846
61	7853	7860	7868	7875	7882	7889	7896	7903	7910	7917
62	7924	7931	7938	7945	7952	7959	7966	7973	7980	7987
63	7993	8000	8007	8014	8021	8028	8035	8041	8048	8055
64	8062	8069	8075	8082	8089	8096	8102	8109	8116	8122
65	8129	8136	8142	8149	8156	8162	8169	8176	8182	8189
66	8195	8202	8209	8215	8222	8228	8235	8241	8248	8254
67	8261	8267	8274	8280	8287	8293	8299	8306	8312	8319
68	8325	8331	8338	8344	8351	8357	8363	8370	8376	8382
69	8388	8395	8401	8407	8414	8420	8426	8432	8439	8445
70	8451	8457	8463	8470	8476	8482	8488	8494	8500	8506
71	8513	8519	8525	8531	8537	8543	8549	8555	8561	8567
72	8573	8579	8585	8591	8597	8603	8609	8615	8621	8627
73	8633	8639	8645	8651	8657	8663	8669	8675	8681	8686
74	8692	8698	8704	8710	8716	8722	8727	8733	8739	8745
75	8751	8756	8762	8768	8774	8779	8785	8791	8797	8802
76	8808	8814	8820	8825	8831	8837	8842	8848	8854	8859
77	8865	8871	8876	8882	8887	8893	8899	8904	8910	8915
78	8921	8927	8932	8938	8943	8949	8954	8960	8965	8971
79	8976	8982	8987	8993	8998	9004	9009	9015	9020	9025
80	9031	9036	9042	9047	9053	9058	9063	9069	9074	9079
81	9085	9090	9096	9101	9106	9112	9117	9122	9128	9133
82	9138	9143	9149	9154	9159	9165	9170	9175	9180	9186
83	9191	9196	9201	9206	9212	9217	9222	9227	9232	9238
84	9243	9248	9253	9258	9263	9269	9274	9279	9284	9289
85	9294	9299	9304	9309	9315	9320	9325	9330	9335	9340
86	9345	9350	9355	9360	9365	9370	9375	9380	9385	9390
87	9395	9400	9405	9410	9415	9420	9425	9430	9435	9440
88	9445	9450	9455	9460	9465	9469	9474	9479	9484	9489
89	9494	9499	9504	9509	9513	9518	9523	9528	9533	9538
90	9542	9547	9552	9557	9562	9566	9571	9576	9581	9586
91	9590	9595	9600	9605	9609	9614	9619	9624	9628	9633
92	9638	9643	9647	9652	9657	9661	9666	9671	9675	9680
93	9685	9689	9694	9699	9703	9708	9713	9717	9722	9727
94	9731	9736	9741	9745	9750	9754	9759	9763	9768	9773
95	9777	9782	9786	9791	9795	9800	9805	9809	9814	9818
96	9823	9827	9832	9836	9841	9845	9850	9854	9859	9863
97	9868	9872	9877	9881	9886	9890	9894	9899	9903	9908
98	9912	9917	9921	9926	9930	9934	9939	9943	9948	9952
99	9956	9961	9965	9969	9974	9978	9983	9987	9991	9996

2; in taking cube roots, by 3:

> Square root of 9×10^4 is 3×10^2.
> Cube root of 8×10^{-12} is 2×10^{-4}.

4.2 Logarithms

A *logarithm* of a given number is the power to which a base number must be raised to equal the given number. There are in common usage two bases for logarithms: the base 10 and the base e ($e = 2.71828 \ldots$). These can be distinguished by writing "log" for the base-10 system and "ln" for the base-e system. The latter is derived from the name "natural logarithm" for reference to base e. The two systems are related by the equality

$2.303 \log x = \ln x$

For numerical calculations it is usually more convenient to use the base-10 logarithms because of the decimal nature of our number system. However, when dealing with equations from calculus natural logarithms arise because of relations such as $dx/x = d \ln x$. This relation between derivatives is true only for the case of natural logarithms. For numbers in the decimal system, tables of natural logarithms are quite extensive because each multiplication by a power of 10 does not simply add 1 to the natural logarithm. When needed, natural logarithms can be derived from a table of base-10 logarithms by use of the multiplier 2.303. The table on pages 754 to 755 gives the base-10 logarithms.

One principal use of logarithms in this text is in connection with pH, defined as the negative of the logarithm of the hydronium-ion concentration. For a hydronium-ion concentration of 0.00036 M the pH is found as follows:

$$
\begin{aligned}
\log 0.00036 &= \log (3.6 \times 10^{-4}) \\
&= \log 3.6 + \log 10^{-4} \\
&= 0.556 - 4 \\
&= -3.444
\end{aligned}
$$

$$\text{pH} = +3.444$$

Sometimes, the reverse procedure is required. For example, if a solution has a pH of 8.50, its hydronium-ion concentration can be found as follows:

$$\text{pH} = 8.50$$

$$\log [H_3O^+] = -8.50 = 0.50 - 9$$

$$[H_3O^+] = 3.2 \times 10^{-9}$$

The number 3.2 is the antilogarithm of 0.50 (the number whose logarithm is 0.50). Antilogarithms are obtained by using the table in reverse, i.e., by looking up the logarithm in the body of the table and then finding the number which corresponds to it.

4.3 Quadratic equations

A *quadratic equation* is an algebraic equation in which a variable is raised to the second power but no higher and which can be written in the form

$$ax^2 + bx + c = 0$$

The solution of such an equation is

$$x = \frac{-b \pm \sqrt{b^2 - 4ac}}{2a}$$

where the plus-or-minus sign indicates that there are two roots. Thus, the equation obtained in Example 2 of Sec. 12.2

$$1.8 \times 10^{-5} = \frac{(1.00 - y)(1.00 - y)}{y}$$

when rewritten gives

$$y^2 + (-2.00 - 1.8 \times 10^{-5})y + 1.00 = 0$$

for which the roots are

$$y = \frac{-(-2.000018) \pm \sqrt{(-2.000018)^2 - 4(1)(1.00)}}{2(1)}$$

$$= +1.004 \quad \text{or} \quad 0.996$$

The first root $y = 1.004$ is inadmissible from the nature of the problem (y cannot be greater than 1.00, which represents all the acid present). The second root $y = 0.996$ must be the correct one. It might be noted that the usual rules for carrying through significant figures do not apply when we operate with the quadratic formula.

4.4 Solving equations by successive approximations

Complicated algebraic equations can often be solved by the method of successive approximation. To use this method, the assumption is made that one or more terms can be neglected so as to give a simple approximate equation, which can quickly lead to a first approximate answer. This answer is then substituted into the terms that were neglected to give a better approximate equation, which is solved to give a second approximate answer (presumably better than the first). This second approximate answer is then fed back. The sequence continues until two successive trials give the same self-consistent value for the unknown.

Example
Try to solve $4x^3 - 0.800x^2 + 0.0500x - 0.00060 = 0$ by a method of successive approximation in which only the linear term is retained. (If you have a hunch that x is going to be larger than 1, you should throw away the low powers of x; if you think x is going to be smaller than 1, throw away the high powers.)

First approximation

Assume $x = 0$ in first two terms.

The equation becomes $0.0500x - 0.00060 = 0$, for which the solution is $x = 0.012$.

Second approximation

Assume $x = 0.012$ in first two terms.

The equation becomes $4(0.012)^3 - 0.800(0.012)^2 + 0.0500x - 0.00060 = 0$, which reduces to $0.0500x - 0.00071 = 0$, for which the solution is $x = 0.014$.

Third approximation

Assume $x = 0.014$ in first two terms.

The equation becomes $4(0.014)^3 - 0.800(0.014)^2 + 0.0500x - 0.00060 = 0$, which reduces to $0.0500x - 0.00075 = 0$, for which the solution is $x = 0.015$.

Fourth approximation

Assume $x = 0.015$ in first two terms.

The equation becomes $4(0.015)^3 - 0.800(0.015)^2 + 0.0500x - 0.00060 = 0$, which reduces to $0.0500x - 0.00077 = 0$, for which the solution is $x = 0.016$.

Fifth approximation

Assume $x = 0.016$ in first two terms.

The equation becomes $4(0.016)^3 - 0.800(0.016)^2 + 0.0500x - 0.00060 = 0$, which reduces to $0.0500x - 0.00078 = 0$, for which the solution is $x = 0.016$.

Since two successive trials lead to the same answer $x = 0.016$, we assume we have a self-consistent answer.

■　■　■

Normally, it takes no more than two or three successive trials to come up with a self-consistent answer, provided the first assumption is a reasonable one. If the first assumption is a bad one, then in general the succeeding steps will not converge on an answer, and the calculation using that assumption should be abandoned.

5.1   *Velocity and acceleration*

When an object changes its position, it is said to undergo a *displacement*. The rate at which displacement changes with time is called the *velocity* and has the dimensions of distance divided by time (e.g., centimeters per second). *Acceleration* is the rate at which velocity changes with time and has the dimensions of velocity divided by time (e.g., centimeters per second per second, or cm/sec^2).

5.2   *Force and mass*

Force can be thought of as a push or pull on an object which tends to change its motion, to speed it up or slow it down or to cause it to deviate from its path. Mass is a quantitative measure of the inertia of an object to having its motion changed. Thus, mass determines how difficult it is to accelerate an object. Quantitatively, force and mass are related by the equation

$F = ma$

where F is the force which produces acceleration a in mass m. If m is in kilograms and a is in meters per second per second, then F is in kilogram-meters per second per second, or newtons. (For reference, 1 newton is approximately the force exerted by an apple in the earth's gravity.) If m is in grams and a is in centimeters per second per second, then F is in gram-centimeters per second per second, or dynes. The   *759* recommended unit for force is the newton, which is equal to 10^5 dyn. Weight is an expression of force and arises because every object has mass and is being accelerated by gravity.

5.3   Momentum and impulse; angular momentum

In dealing with collision problems it is useful to have terms for describing the combined effect of mass and velocity and its change with time. Mass times velocity mv, called the *momentum,* determines the length of time required to bring a moving body to rest when decelerated by a constant force. Thus, for a particle of momentum mv to be stopped by a constant force F the time required t is mv/F.

The *impulse* is defined for the case of a constant force as Ft, where t is the time during which the force F acts. Thus, for the stopping of a particle originally of momentum mv by force F in time t the impulse is just

$$Ft = F\frac{mv}{F} = mv$$

This is true if the particle comes to a complete rest. If, however, the particle bounces back, as it would on collision with a rigid wall, the particle is reflected from the wall with momentum $-mv$ (the minus sign indicating that the velocity is now in the opposite direction). The total impulse, counting the time for deceleration to zero and acceleration to $-mv$, is twice what it was before, or $2mv$.

In considering the pressure exerted by a gas, impulse comes in as follows: The pressure, or force per unit area, is the rate of collision per unit area times the effect of each collision:

$$\text{Pressure} = \frac{\text{force}}{\text{area}} = \frac{\text{number of collisions}}{(\text{time})(\text{area})} \times \text{?}$$

$$\text{?} = \frac{(\text{force})(\text{time})}{\text{number of collisions}} = \text{impulse per collision}$$

In contrast to *linear momentum,* which measures mv along a straight line, there is also *angular momentum,* which describes an analogous quantity for spinning movement or movement along a curved path. For motion along a curved path, angular momentum is defined as mvr, where r is the radius of curvature of the path.

5.4   Work and energy

When a force F operates on (e.g., pushes) an object through a distance d, work W is done:

$$W = Fd$$

If force is expressed in newtons (kilogram-meters per second per second) and distance in meters, then work has the dimensions newton-meters ($\text{kg-m}^2/\text{sec}^2$), or joules. One joule is thus the work done in moving one kilogram through one meter so as to increase its velocity by one meter per second all in one second. If force is expressed in dynes (gram-centimeters per second per second) and distance in centimeters, then work has the dimensions dyne-centimeters ($g\text{-}cm^2/sec^2$), or ergs. One erg is thus the work done in moving one gram through one centimeter so as to increase its velocity by one centimeter per second all in one second. (For reference, 1 erg is approximately the work a fly does in one push-up.)

Energy is the ability to do work, and the dimensions of energy are the same as those of work. Kinetic energy is the energy a body possesses because of its motion and mass. It is equal to one-half the mass times the square of its velocity. Potential energy is the energy a body possesses because of its position or arrangement with respect to other bodies.

5.5 Electric charge and electric field

Electric charge is a property assigned to objects to account for certain observed attractions or repulsions which cannot be explained in terms of gravitational attraction between masses. Electric charge can be of two types, positive and negative. Objects which have the same type of electric charge repel each other; objects with opposite charges attract each other. Originally, a unit of charge was defined as the quantity of electric charge which at a distance of one centimeter from another identical charge produced a repulsive force of one dyne in a vacuum. This unit of charge was called the electrostatic unit (esu). An electron has a negative charge of 4.80×10^{-10} esu. The unit of electric charge is now defined as the coulomb, which is the amount of charge transferred by a current flow of one ampere (see below) for one second. In coulombs, the charge of an electron is 1.60×10^{-19} coulomb.

An electric field is said to exist at a point if a force of electric origin is exerted on any charged body placed at that point. The intensity of an electric field is defined as the magnitude of the electric force exerted on a unit charge. Any electrically charged body placed in an electric field moves unless otherwise constrained. The direction of a field is usually defined as the direction in which a positive charge would move.

5.6 Voltage and capacitance

An electric capacitor is a device for storing electric charge. In its simplest form a capacitor consists of two parallel, electrically conducting plates separated by some distance. The capacitor can be charged by making one plate positive and the other plate negative. In order to transfer a unit positive charge from the negative plate to the positive plate, work must be done against the electric field which exists between the charged plates. Therefore, the potential energy of the unit charge is increased in the process. In other words, there is a change in potential energy in going from one plate to the other. This difference in potential energy for a unit charge moved from one plate to the other is called the potential difference, or the voltage, of the capacitor. Voltage, or potential difference, is not restricted to capacitors but may exist between any two points so long as work must be done in transferring an electric charge from one point to the other. The potential difference between two points is said to be one volt if one joule (that is, 10^7 ergs) is required to move one coulomb of charge from one point to the other. To move an electron through a potential difference of one volt requires an amount of energy, called the electron volt, equal to 1.6×10^{-19} J.

Capacitance is the term used to describe quantitatively the amount of charge that can be stored on a capacitor. It is equal to the amount of charge that can be stored on the plates when the voltage difference between the plates is one volt. In general, the amount of charge a capacitor can hold is directly proportional to the voltage; the capacitance is simply the proportionality constant:

$$Q = CV$$

If Q, the charge, is one coulomb and if V, the voltage, is one volt, then C, the capacitance, is one farad. The capacitance of a capacitor depends on the capacitor design (e.g., area of the plates and distance between them) and on the nature of the material between the plates. For a parallel-plate capacitor the capacitance is given approximately by the following equation:

$$C = \frac{KA}{4\pi d}$$

where A is the area of the plates, d is the distance between the plates, and K is the dielectric constant of the material between the plates. For a vacuum the dielectric constant K is exactly equal to 1; for all other substances K is greater than 1. Some typical dielectric constants are 1.00059 for air at STP, 1.00026 for hydrogen gas at STP, 1.0046 for HCl gas at STP, 80 for liquid water at 20°C, 28.4 for ethyl alcohol at 0°C, 2 for petroleum, and 4 for solid sulfur.

5.7   Electric current

A collection of moving charges is called an electric current. The unit of current is the ampere, which is defined as the constant current which if maintained in two straight, parallel conductors of infinite length and negligible circular cross section that are placed one meter apart in a vacuum would produce between these conductors a force equal to 2×10^{-7} newton per meter of length. One ampere corresponds to a flow of one coulomb of charge past a point in one second. Since current specifies the rate at which charge is transferred, the current multiplied by time gives the total amount of charge transferred:

$$Q = It$$

If the current I is in amperes (coulombs per second) and the time t is in seconds, the charge Q is in coulombs.

The current that a wire carries is directly proportional to the voltage difference between the ends of the wire. The proportionality constant, called the conductance of the wire, is equal to the reciprocal of the resistance of the wire:

$$I = \frac{1}{R} V \qquad \text{or} \qquad V = IR$$

If V is the potential difference in volts and I is the current in amperes, R is the resistance in ohms.

There are two important kinds of current, direct and alternating. Direct current implies that the charge is constantly moving in the same direction along the wire. Alternating current implies that the current reverses its direction at regular intervals of time. The usual house current is 60-cycle alternating current; i.e., it goes through 60 complete back-and-forth oscillations per second.

Dissociation constants (*first step only*)

$CrOH^{2+}$	5×10^{-11}	$H_2AsO_4^-$	5.6×10^{-8}
$CuOH^+$	1×10^{-8}	$HAsO_4^{2-}$	3×10^{-13}
$ZnOH^+$	4×10^{-5}	H_2O	1.0×10^{-14}
H_3BO_3	6.0×10^{-10}	H_2S	1.1×10^{-7}
$CO_2 + H_2O$	4.2×10^{-7}	HS^-	1×10^{-14}
HCO_3^-	4.8×10^{-11}	H_2SO_3	1.3×10^{-2}
$HC_2H_3O_2$	1.8×10^{-5}	HSO_3^-	5.6×10^{-8}
HCN	4.0×10^{-10}	HSO_4^-	1.3×10^{-2}
$NH_3 + H_2O$	1.8×10^{-5}	H_2Se	1.9×10^{-4}
HNO_2	4.5×10^{-4}	H_2SeO_3	2.7×10^{-3}
H_3PO_3	1.6×10^{-2}	$HSeO_3^-$	2.5×10^{-7}
$H_2PO_3^-$	7×10^{-7}	H_2Te	2.3×10^{-3}
H_3PO_4	7.5×10^{-3}	HF	6.7×10^{-4}
$H_2PO_4^-$	6.2×10^{-8}	$HOCl$	3.2×10^{-8}
HPO_4^{2-}	10^{-12}	$HClO_2$	1.1×10^{-2}
H_3AsO_4	2.5×10^{-4}		

Solubility products

$Mg(OH)_2$	8.9×10^{-12}	NiS	3×10^{-21}
MgF_2	8×10^{-8}	PtS	8×10^{-73}
MgC_2O_4	8.6×10^{-5}	$Cu(OH)_2$	1.6×10^{-19}
$Ca(OH)_2$	1.3×10^{-6}	CuS	8×10^{-37}
CaF_2	1.7×10^{-10}	AgCl	1.7×10^{-10}
$CaCO_3$	4.7×10^{-9}	AgBr	5.0×10^{-13}
$CaSO_4$	2.4×10^{-5}	AgI	8.5×10^{-17}
CaC_2O_4	1.3×10^{-9}	AgCN	1.6×10^{-14}
$Sr(OH)_2$	3.2×10^{-4}	Ag_2S	5.5×10^{-51}
$SrSO_4$	7.6×10^{-7}	ZnS	1×10^{-22}
$SrCrO_4$	3.6×10^{-5}	CdS	1.0×10^{-28}
$Ba(OH)_2$	5.0×10^{-3}	Hg_2Cl_2	1.1×10^{-18}
$BaSO_4$	1.5×10^{-9}	Hg_2Br_2	1.3×10^{-22}
$BaCrO_4$	8.5×10^{-11}	Hg_2I_2	4.5×10^{-29}
$Cr(OH)_3$	6.7×10^{-31}	HgS	1.6×10^{-54}
$Mn(OH)_2$	2×10^{-13}	$Al(OH)_3$	5×10^{-33}
MnS	7×10^{-16}	SnS	1×10^{-26}
FeS	4×10^{-19}	$Pb(OH)_2$	4.2×10^{-15}
$Fe(OH)_3$	6×10^{-38}	$PbCl_2$	1.6×10^{-5}
CoS	5×10^{-22}	PbS	7×10^{-29}

Half-reaction	$E°$, V
$F_2(g) + 2H_3O^+ + 2e^- \longrightarrow 2HF + 2H_2O$	$+3.06$
$F_2(g) + 2e^- \longrightarrow 2F^-$	$+2.87$
$O_3(g) + 2H_3O^+ + 2e^- \longrightarrow O_2(g) + 3H_2O$	$+2.07$
$Ag^{2+} + e^- \longrightarrow Ag^+$	$+1.98$
$Co^{3+} + e^- \longrightarrow Co^{2+}$	$+1.82$
$H_2O_2 + 2H_3O^+ + 2e^- \longrightarrow 4H_2O$	$+1.77$
$MnO_4^- + 4H_3O^+ + 3e^- \longrightarrow MnO_2(s) + 6H_2O$	$+1.70$
$Au^+ + e^- \longrightarrow Au(s)$	$ca. +1.7$
$HClO_2 + 2H_3O^+ + 2e^- \longrightarrow HClO + 3H_2O$	$+1.64$
$HClO + H_3O^+ + e^- \longrightarrow \frac{1}{2}Cl_2(g) + 2H_2O$	$+1.63$
$Ce^{4+} + e^- \longrightarrow Ce^{3+}$	$+1.61$
$H_5IO_6 + H_3O^+ + 2e^- \longrightarrow IO_3^- + 4H_2O$	$+1.6$
$MnO_4^- + 8H_3O^+ + 5e^- \longrightarrow Mn^{2+} + 12H_2O$	$+1.51$
$Mn^{3+} + e^- \longrightarrow Mn^{2+}$	$+1.51$
$BrO_3^- + 6H_3O^+ + 5e^- \longrightarrow \frac{1}{2}Br_2 + 9H_2O$	$+1.50$
$Au^{3+} + 3e^- \longrightarrow Au(s)$	$+1.50$
$Cl_2(g) + 2e^- \longrightarrow 2Cl^-$	$+1.36$
$NH_3OH^+ + 2H_3O^+ + 2e^- \longrightarrow NH_4^+ + 3H_2O$	$+1.35$
$Cr_2O_7^{2-} + 14H_3O^+ + 6e^- \longrightarrow 2Cr^{3+} + 21H_2O$	$+1.33$
$2HNO_2 + 4H_3O^+ + 4e^- \longrightarrow N_2O(g) + 7H_2O$	$+1.29$
$Tl^{3+} + 2e^- \longrightarrow Tl^+$	$+1.25$
$MnO_2(s) + 4H_3O^+ + 2e^- \longrightarrow Mn^{2+} + 6H_2O$	$+1.23$
$O_2(g) + 4H_3O^+ + 4e^- \longrightarrow 6H_2O$	$+1.23$
$ClO_3^- + 3H_3O^+ + 2e^- \longrightarrow HClO_2 + 4H_2O$	$+1.21$
$IO_3^- + 6H_3O^+ + 5e^- \longrightarrow \frac{1}{2}I_2 + 9H_2O$	$+1.20$
$ClO_4^- + 2H_3O^+ + 2e^- \longrightarrow ClO_3^- + 3H_2O$	$+1.19$
$PuO_2^+ + 4H_3O^+ + e^- \longrightarrow Pu^{4+} + 6H_2O$	$+1.15$

Half-reaction	$E°$, V
$Br_2 + 2e^- \longrightarrow 2Br^-$	$+1.09$
$N_2O_4(g) + 2H_3O^+ + 2e^- \longrightarrow 2HNO_2 + 2H_2O$	$+1.07$
$Br_2(l) + 2e^- \longrightarrow 2Br^-$	$+1.07$
$PuO_2^{2+} + 4H_3O^+ + 2e^- \longrightarrow Pu^{4+} + 6H_2O$	$+1.04$
$N_2O_4(g) + 4H_3O^+ + 4e^- \longrightarrow 2NO(g) + 6H_2O$	$+1.03$
$V(OH)_4^+ + 2H_3O^+ + e^- \longrightarrow VO^{2+} + 5H_2O$	$+1.00$
$HNO_2 + H_3O^+ + e^- \longrightarrow NO(g) + 2H_2O$	$+1.00$
$Pu^{4+} + e^- \longrightarrow Pu^{3+}$	$+0.97$
$NO_3^- + 4H_3O^+ + 3e^- \longrightarrow NO(g) + 6H_2O$	$+0.96$
$2Hg^{2+} + 2e^- \longrightarrow Hg_2^{2+}$	$+0.92$
$2NO_3^- + 4H_3O^+ + 2e^- \longrightarrow N_2O_4(g) + 6H_2O$	$+0.80$
$Ag^+ + e^- \longrightarrow Ag(s)$	$+0.80$
$Hg_2^{2+} + 2e^- \longrightarrow 2Hg(l)$	$+0.79$
$Fe^{3+} + e^- \longrightarrow Fe^{2+}$	$+0.77$
$O_2(g) + 2H_3O^+ + 2e^- \longrightarrow H_2O_2 + 2H_2O$	$+0.68$
$UO_2^+ + 4H_3O^+ + e^- \longrightarrow U^{4+} + 6H_2O$	$+0.62$
$MnO_4^- + e^- \longrightarrow MnO_4^{2-}$	$+0.56$
$H_3AsO_4 + 2H_3O^+ + 2e^- \longrightarrow HAsO_2 + 4H_2O$	$+0.56$
$I_2 + 2e^- \longrightarrow 2I^-$	$+0.54$
$Cu^+ + e^- \longrightarrow Cu(s)$	$+0.52$
$VO^{2+} + 2H_3O^+ + e^- \longrightarrow V^{3+} + 3H_2O$	$+0.36$
$Fe(CN)_6^{3-} + e^- \longrightarrow Fe(CN)_6^{4-}$	$+0.36$
$Cu^{2+} + 2e^- \longrightarrow Cu(s)$	$+0.34$
$UO_2^{2+} + 4H_3O^+ + 2e^- \longrightarrow U^{4+} + 6H_2O$	$+0.33$
$Cu^{2+} + e^- \longrightarrow Cu^+$	$+0.15$
$Sn^{4+} + 2e^- \longrightarrow Sn^{2+}$	$+0.15$
$S(s) + 2H_3O^+ + 2e^- \longrightarrow H_2S(g) + 2H_2O$	$+0.14$
$HSO_4^- + 3H_3O^+ + 2e^- \longrightarrow SO_2 + 5H_2O$	$+0.11$
$P(s) + 3H_3O^+ + 3e^- \longrightarrow PH_3(g) + 3H_2O$	$+0.06$
$UO_2^{2+} + e^- \longrightarrow UO_2^+$	$+0.05$
$2H_3O^+ + 2e^- \longrightarrow H_2(g) + 2H_2O$	Zero
$Pb^{2+} + 2e^- \longrightarrow Pb(s)$	-0.13
$Sn^{2+} + 2e^- \longrightarrow Sn(s)$	-0.14
$Mo^{3+} + 3e^- \longrightarrow Mo(s)$	$ca. -0.2$
$Ni^{2+} + 2e^- \longrightarrow Ni(s)$	-0.25
$V^{3+} + e^- \longrightarrow V^{2+}$	-0.26
$H_3PO_4 + 2H_3O^+ + 2e^- \longrightarrow H_3PO_3 + 3H_2O$	-0.28
$Co^{2+} + 2e^- \longrightarrow Co(s)$	-0.28
$Tl^+ + e^- \longrightarrow Tl(s)$	-0.34
$In^{3+} + 3e^- \longrightarrow In(s)$	-0.34
$Cd^{2+} + 2e^- \longrightarrow Cd(s)$	-0.40
$Cr^{3+} + e^- \longrightarrow Cr^{2+}$	-0.41
$Eu^{3+} + e^- \longrightarrow Eu^{2+}$	-0.43
$Fe^{2+} + 2e^- \longrightarrow Fe(s)$	-0.44
$Ga^{3+} + 3e^- \longrightarrow Ga(s)$	-0.53
$U^{4+} + e^- \longrightarrow U^{3+}$	-0.61
$Cr^{3+} + 3e^- \longrightarrow Cr(s)$	-0.74

Appendix 7
Standard electrode
potentials

Half-reaction	$E°$, V
$Zn^{2+} + 2e^- \longrightarrow Zn(s)$	-0.76
$TiO^{2+} + 2H_3O^+ + 4e^- \longrightarrow Ti(s) + 3H_2O$	$ca. -0.9$
$V^{2+} + 2e^- \longrightarrow V(s)$	$ca. -1.2$
$Mn^{2+} + 2e^- \longrightarrow Mn(s)$	-1.18
$Zr^{4+} + 4e^- \longrightarrow Zr(s)$	-1.53
$Al^{3+} + 3e^- \longrightarrow Al(s)$	-1.66
$Hf^{4+} + 4e^- \longrightarrow Hf(s)$	-1.70
$U^{3+} + 3e^- \longrightarrow U(s)$	-1.80
$Be^{2+} + 2e^- \longrightarrow Be(s)$	-1.85
$Th^{4+} + 4e^- \longrightarrow Th(s)$	-1.90
$Pu^{3+} + 3e^- \longrightarrow Pu(s)$	-2.07
$Sc^{3+} + 3e^- \longrightarrow Sc(s)$	-2.08
$\frac{1}{2}H_2(g) + e^- \longrightarrow H^-$	-2.25
$Y^{3+} + 3e^- \longrightarrow Y(s)$	-2.37
$Mg^{2+} + 2e^- \longrightarrow Mg(s)$	-2.37
$Ce^{3+} + 3e^- \longrightarrow Ce(s)$	-2.48
$La^{3+} + 3e^- \longrightarrow La(s)$	-2.52
$Na^+ + e^- \longrightarrow Na(s)$	-2.71
$Ca^{2+} + 2e^- \longrightarrow Ca(s)$	-2.87
$Sr^{2+} + 2e^- \longrightarrow Sr(s)$	-2.89
$Ba^{2+} + 2e^- \longrightarrow Ba(s)$	-2.90
$Ra^{2+} + 2e^- \longrightarrow Ra(s)$	-2.92
$Cs^+ + e^- \longrightarrow Cs(s)$	-2.92
$Rb^+ + e^- \longrightarrow Rb(s)$	-2.93
$K^+ + e^- \longrightarrow K(s)$	-2.93
$Li^+ + e^- \longrightarrow Li(s)$	-3.05

Ac^{3+}	0.118	Br^-	0.196	Cu^+	0.096	Hf^0	0.144
Ag^0	0.134	Br^{5+}	0.047	Cu^{2+}	0.072	Hf^{4+}	0.078
Ag^+	0.126					Hg^0	0.144
Ag^{2+}	0.089	C^0	0.077	Dy^0	0.160	Hg^{2+}	0.110
Al^0	0.125	C^{4+}	0.016	Dy^{3+}	0.092	Ho^0	0.158
Al^{3+}	0.051	Ca^0	0.174			Ho^{3+}	0.091
Am^{3+}	0.107	Ca^{2+}	0.099				
Am^{4+}	0.092	Cd^0	0.141	Er^0	0.158		
As^0	0.121	Cd^{2+}	0.097	Er^{3+}	0.089	I^0	0.133
As^{3+}	0.058	Ce^0	0.165	Eu^0	0.185	I^-	0.220
As^{5+}	0.046	Ce^{3+}	0.107	Eu^{3+}	0.098	I^{5+}	0.062
Au^0	0.134	Ce^{4+}	0.094			I^{7+}	0.050
Au^+	0.137	Cl^0	0.099	F^0	0.064	In^0	0.150
Au^{3+}	0.085	Cl^-	0.181	F^-	0.133	In^{3+}	0.081
		Cl^{5+}	0.034	Fe^0	0.117	Ir^0	0.127
B^0	0.081	Cl^{7+}	0.027	Fe^{2+}	0.074	Ir^{4+}	0.068
B^{3+}	0.023	Co^0	0.116	Fe^{3+}	0.064		
Ba^0	0.198	Co^{2+}	0.072			K^0	0.203
Ba^{2+}	0.134	Co^{3+}	0.063			K^+	0.133
Be^0	0.089	Cr^0	0.118	Ga^0	0.125		
Be^{2+}	0.035	Cr^{3+}	0.063	Ga^{3+}	0.062		
Bi^0	0.15	Cr^{6+}	0.052	Gd^0	0.162	La^0	0.169
Bi^{3+}	0.096	Cs^0	0.235	Gd^{3+}	0.097	La^{3+}	0.114
Bi^{5+}	0.074	Cs^+	0.167	Ge^0	0.122	Li^0	0.123
Br^0	0.114	Cu^0	0.117	Ge^{2+}	0.073	Li^+	0.068
				Ge^{4+}	0.053	Lu^0	0.156

Lu^{3+}	0.085	P^{3+}	0.044	S^{2-}	0.184	Th^{0}	0.165
		P^{5+}	0.035	S^{4+}	0.037	Th^{4+}	0.102
Mg^{0}	0.136	Pa^{4+}	0.098	S^{6+}	0.030	Ti^{0}	0.132
Mg^{2+}	0.066	Pb^{0}	0.154	Sb^{0}	0.141	Ti^{3+}	0.076
Mn^{0}	0.117	Pb^{2+}	0.120	Sb^{3+}	0.076	Ti^{4+}	0.068
Mn^{2+}	0.080	Pb^{4+}	0.084	Sb^{5+}	0.062	Tl^{0}	0.155
Mn^{3+}	0.066	Pd^{0}	0.128	Sc^{0}	0.144	Tl^{+}	0.147
Mn^{4+}	0.060	Pd^{2+}	0.080	Sc^{3+}	0.081	Tl^{3+}	0.095
Mn^{7+}	0.046	Pd^{4+}	0.065	Se^{0}	0.117	Tm^{0}	0.158
Mo^{0}	0.130	Pm^{0}	0.163	Se^{2-}	0.198	Tm^{3+}	0.087
Mo^{4+}	0.070	Pm^{3+}	0.106	Se^{4+}	0.050		
Mo^{6+}	0.062	Po^{0}	0.153	Se^{6+}	0.042	U^{0}	0.142
		Pr^{0}	0.164	Si^{0}	0.117	U^{4+}	0.097
N^{0}	0.070	Pr^{3+}	0.106	Si^{4+}	0.042	U^{6+}	0.080
N^{3+}	0.016	Pr^{4+}	0.092	Sm^{0}	0.162		
N^{5+}	0.013	Pt^{0}	0.130	Sm^{3+}	0.100	V^{0}	0.122
Na^{0}	0.157	Pt^{2+}	0.080	Sn^{0}	0.140	V^{2+}	0.088
Na^{+}	0.097	Pt^{4+}	0.065	Sn^{2+}	0.093	V^{3+}	0.074
Nb^{0}	0.134	Pu^{3+}	0.108	Sn^{4+}	0.071	V^{4+}	0.063
Nb^{4+}	0.074	Pu^{4+}	0.093	Sr^{0}	0.191	V^{5+}	0.059
Nb^{5+}	0.069			Sr^{2+}	0.112		
Nd^{3+}	0.104	Ra^{2+}	0.143			W^{0}	0.130
Ni^{0}	0.115	Rb^{0}	0.216	Ta^{0}	0.134	W^{4+}	0.070
Ni^{2+}	0.069	Rb^{+}	0.147	Ta^{5+}	0.068	W^{6+}	0.062
Np^{3+}	0.110	Re^{0}	0.128	Tb^{0}	0.161		
Np^{4+}	0.095	Re^{4+}	0.072	Tb^{3+}	0.093	Y^{0}	0.162
		Re^{7+}	0.056	Tb^{4+}	0.081	Y^{3+}	0.092
O^{2-}	0.140	Rh^{0}	0.125	Tc^{0}	0.127		
O^{0}	0.066	Rh^{3+}	0.068	Tc^{7+}	0.056	Zn^{0}	0.125
Os^{0}	0.126	Ru^{0}	0.125	Te^{0}	0.137	Zn^{2+}	0.074
Os^{4+}	0.069	Ru^{4+}	0.067	Te^{2-}	0.221	Zr^{0}	0.145
				Te^{4+}	0.070	Zr^{4+}	0.079
P^{0}	0.110	S^{0}	0.104	Te^{6+}	0.056		

Appendix 8
Atomic and ionic
radii (in nanometers)

References of general utility covering many of the topics in this text are M. J. Sienko, R. A. Plane, and R. E. Hester, "Inorganic Chemistry: Principles and Elements," Benjamin; G. Barrow, "Physical Chemistry," McGraw-Hill; and F. A. Cotton and G. Wilkinson, "Advanced Inorganic Chemistry," Interscience.

Additional information and background material can be found in the following books, which are listed by the chapters to which they most apply:

Chapter 1
G. Herzberg, "Atomic Spectra and Atomic Structure," Dover; and J. C. Slater, "Modern Physics," McGraw-Hill.

Chapters 2, 3
O. K. Rice, "Electronic Structure and Chemical Binding," McGraw-Hill; L. Pauling, "Nature of the Chemical Bond," Cornell University Press; H. B. Gray, "Electrons and Chemical Bonding," Benjamin; A. Companion, "Chemical Bonding," McGraw-Hill; J. W. Linnett, "Electronic Structure of Molecules," Methuen; and C. A. Coulson, "Valence," Oxford.

Chapter 4
M. J. Sienko, "Chemistry Problems," Benjamin.

Chapters 5, 6
G. M. Barrow, "Physical Chemistry," McGraw-Hill; and J. H. Hildebrand, "An Introduction to Molecular Kinetic Theory," Reinhold.

Chapter 7

F. C. Brown, "The Physics of Solids," Benjamin.

Chapters 8, 9

J. P. Hunt, "Metal Ions in Solution," Benjamin; and R. A. Robinson and R. H. Stokes, "Electrolyte Solutions," Butterworth.

Chapter 10

E. L. King, "How Chemical Reactions Occur," Benjamin; and J. O. Edwards, "Inorganic Reaction Mechanisms," Benjamin.

Chapters 11, 12

M. J. Sienko, "Chemistry Problems," Benjamin.

Chapter 13

B. H. Mahan, "Elementary Chemical Thermodynamics," Benjamin; and K. S. Pitzer and L. Brewer, "Thermodynamics," 2d ed., McGraw-Hill.

Chapter 14

D. A. MacInnes, "The Principles of Electrochemistry," Reinhold; and W. M. Latimer, "Oxidation Potentials," Prentice-Hall.

Chapters 15–25

W. M. Latimer and J. H. Hildebrand, "Reference Book of Inorganic Chemistry," Macmillan; A. F. Wells, "Structural Inorganic Chemistry," Oxford; B. E. Douglas and D. H. McDaniel, "Concepts and Models of Inorganic Chemistry," Blaisdell; and G. Hägg, "General and Inorganic Chemistry," Wiley. For material on organic compounds, R. T. Morrison and R. N. Boyd, "Organic Chemistry," Allyn and Bacon, is highly recommended.

Chapters 26–28

H. S. Stoker and S. L. Seager, "Environmental Chemistry: Air and Water Pollution," Scott, Foresman; J. N. Pitts and R. L. Metcalf (eds.), "Advances in Environmental Science and Technology," vols. 1 and 2, Wiley; J. E. Zajic, "Water Pollution," vols. 1 and 2, Dekker; "Flavor Chemistry," American Chemical Society; O. S. Ray, "Drugs, Society and Human Behavior," Mosby; L. S. Goodman and A. Gilman (eds.), "Pharmacological Basis of Therapeutics," Macmillan; H. F. Henry, "Fundamentals of Radiation Protection," Wiley; National Academy of Sciences, "Resources and Man," Freeman; and D. H. Meadows, D. L. Meadows, J. Randers, and W. W. Behrens, "Limits to Growth," Potomac Associates.

Excellent references for qualitative analysis are E. J. King, "Qualitative Analysis and Electrolytic Solutions," Harcourt Brace Jovanovitch; and T. R. Hogness and W. C. Johnson, "Qualitative Analysis and Chemical Equilibrium," Holt.

Very useful also are the "Handbook of Chemistry and Physics," Chemical Rubber Co.; "The Merck Index of Chemicals and Drugs," Merck & Co.; and the "McGraw-Hill Encyclopedia of Science and Technology," McGraw-Hill.

Periodic Table of the Elements

H 1																	He 2
Li 3	Be 4											B 5	C 6	N 7	O 8	F 9	Ne 10
Na 11	Mg 12											Al 13	Si 14	P 15	S 16	Cl 17	Ar 18
K 19	Ca 20	Sc 21	Ti 22	V 23	Cr 24	Mn 25	Fe 26	Co 27	Ni 28	Cu 29	Zn 30	Ga 31	Ge 32	As 33	Se 34	Br 35	Kr 36
Rb 37	Sr 38	Y 39	Zr 40	Nb 41	Mo 42	Tc 43	Ru 44	Rh 45	Pd 46	Ag 47	Cd 48	In 49	Sn 50	Sb 51	Te 52	I 53	Xe 54
Cs 55	Ba 56	*	Hf 72	Ta 73	W 74	Re 75	Os 76	Ir 77	Pt 78	Au 79	Hg 80	Tl 81	Pb 82	Bi 83	Po 84	At 85	Rn 86
Fr 87	Ra 88	†	Ku 104	Ha 105													

* Lanthanides	La 57	Ce 58	Pr 59	Nd 60	Pm 61	Sm 62	Eu 63	Gd 64	Tb 65	Dy 66	Ho 67	Er 68	Tm 69	Yb 70	Lu 71
† Actinides	Ac 89	Th 90	Pa 91	U 92	Np 93	Pu 94	Am 95	Cm 96	Bk 97	Cf 98	Es 99	Fm 100	Md 101	No 102	Lr 103